Regenerative Medicine and Plastic Surgery

Dominik Duscher • Melvin A. Shiffman
Editors

Regenerative Medicine and Plastic Surgery

Skin and Soft Tissue, Bone, Cartilage, Muscle, Tendon and Nerves

Editors
Dominik Duscher
Department for Plastic Surgery and
Hand Surgery
Division of Experimental Plastic Surgery
Technical University of Munich
Munich
Germany

Melvin A. Shiffman
Private Practice
Tustin, CA
USA

ISBN 978-3-030-19964-7   ISBN 978-3-030-19962-3   (eBook)
https://doi.org/10.1007/978-3-030-19962-3

© Springer Nature Switzerland AG 2019
This work is subject to copyright. All rights are reserved by the Publisher, whether the whole or part of the material is concerned, specifically the rights of translation, reprinting, reuse of illustrations, recitation, broadcasting, reproduction on microfilms or in any other physical way, and transmission or information storage and retrieval, electronic adaptation, computer software, or by similar or dissimilar methodology now known or hereafter developed.
The use of general descriptive names, registered names, trademarks, service marks, etc. in this publication does not imply, even in the absence of a specific statement, that such names are exempt from the relevant protective laws and regulations and therefore free for general use.
The publisher, the authors, and the editors are safe to assume that the advice and information in this book are believed to be true and accurate at the date of publication. Neither the publisher nor the authors or the editors give a warranty, expressed or implied, with respect to the material contained herein or for any errors or omissions that may have been made. The publisher remains neutral with regard to jurisdictional claims in published maps and institutional affiliations.

This Springer imprint is published by the registered company Springer Nature Switzerland AG
The registered company address is: Gewerbestrasse 11, 6330 Cham, Switzerland

To Aura, the love of my life
Dominik Duscher

Foreword

Plastic surgery, by its very nature, takes care of nearly every part of the human body including the skin, muscles, bones, nerves, and blood vessels. As Joseph G. McCarthy, my former Chief at NYU, once said, plastic surgeons are problem-solvers. They are called upon by other practitioners to develop creative solutions to unsolvable problems in nearly every anatomic region. Thus, it is not surprising that plastic surgeons have flocked to the burgeoning field of regenerative medicine, which promises even more elegant solutions to clinical problems found everywhere in the human body. Regenerative medicine proposes using cellular and molecular processes to recreate the exact same tissues that plastic surgeons perform long operations to recreate.

It is logical that plastic surgery would be at the forefront of the field of regenerative medicine. And this is indeed the case at stem cell conferences and tissue engineering symposia, where plastic surgeons often outnumber all the other clinical attendees combined. However, up until now, there did not exist an authoritative reference documenting all the ways that plastic surgical practice and regenerative medicine science overlap or provide a road map for the future of both specialties. Drs. Duscher and Shiffman have provided a valuable service by gathering in one place the leading voices in these two fields in a clear and concise manner.

Reading through this work, one is impressed by both the breadth of plastic surgery practice and the enormous potential of regenerative medicine to cure a multitude of human diseases. One also sees the potential for regenerative medicine to be integrated into clinical plastic surgery. With beautiful clinical images and artwork, this book will be a central companion to both practicing plastic surgeons who wish to remain abreast of upcoming technological advances and regenerative medicine researchers who wish to understand the current state of the art of surgical reconstruction. The analogies between the two disciplines are clearly laid out, and the possibilities for advances in clinical care leap off the pages.

Ultimately, regenerative medicine may make many operations plastic surgeons perform obsolete. This is familiar territory for plastic surgeons who always need to look for the next clinical arena for innovation. I am confident that clinical plastic surgeons will remain at the forefront and become leaders in this emerging field. In your hands is a comprehensive encyclopedia of two rapidly converging fields. Drs. Duscher and Shiffman have done an outstanding job of highlighting the interdependent relationship between plastic surgery and regenerative medicine. Ultimately, this is to the benefit of both fields.

Stanford, CA   Geoffrey C. Gurtner

Preface

A surgical success will impact the patient in question, but a research success can impact a global population. This thought came to me at the end of medical school and still is the main reason for my fascination for science today. Driven by my desire to create knowledge and discover new things, I found the seemingly endless possibilities of the young field of regenerative medicine particularly enchanting.

Through channeling the power of stem cells to repair or replace damaged tissues, regenerative therapies are making their way into mainstay clinical routine. As both case-based stem cell therapy and global understanding evolve, we are entering an era in which we can design treatments for some of the world's most devastating diseases. Innovative therapeutic concepts born of the intersection of clinical medicine, engineering, and cell biology have the potential to change the way we practice medicine.

The international efforts put into this field have created an unyielding body of literature. Staying abreast of the genetic, epigenetic, cellular, stromal, hematopoietic, and pathologic research emerging each year is critical. A comprehensive, up-to-date reconnaissance of these parameters in the field of regenerative medicine is therefore a valuable tool. The expertise required to generate such a text far exceeded that of its editors, and the roots of this book are nourished in the soul of collaboration. I am indebted to the scores of renowned specialists who have contributed their expertise and ingenuity to this work. This book is intended for surgeons and scientists, for biologists and engineers, for students of medicine, biomedical engineering, cell biology, and biotechnology, and simply for everyone who is interested in the extraordinary potential that regenerative medicine has to offer. The first edition of this book represents an attempt to organize the current knowledge in the field. However, knowledge is of no value unless you put it into practice. Thus, we all need to strive for the clinical translation of the principles presented here to make the dream of tissue and organ regeneration a reality.

You can have results or excuses. Not both.
—Arnold Schwarzenegger

Munich, Germany   Dominik Duscher

Contents

Part I Skin and Soft Tissue Regeneration

1 Induction of the Fetal Scarless Phenotype in Adult Wounds: Impossible? 3
Michael S. Hu, Mimi R. Borrelli, Michael T. Longaker,
and H. Peter Lorenz

2 Scar Treatment and Prevention: Know Thine Enemy 19
Elizabeth A. Brett and Dominik Duscher

3 Challenges and Opportunities in Drug Delivery and Wound Healing 27
Matthias M. Aitzetmüller, Hans-Günther Machens,
and Dominik Duscher

4 Harvesting, Processing, and Injection of Lipoaspirate for Soft-Tissue Reconstruction: Details Make the Difference . . . 39
Matthias A. Sauter, Elizabeth A. Brett,
Matthias M. Aitzetmüller, and Dominik Duscher

5 Adipose Tissue Complex (ATC): Cellular and Biocellular Uses of Stem/Stromal Cells and Matrix in Cosmetic Plastic, Reconstructive Surgery and Regenerative Medicine 45
Robert W. Alexander

6 Preparation, Characterization, and Clinical Implications of Human Decellularized Adipose Tissue Extracellular Matrix . . 71
Derek A. Banyard, Christos Sarantopoulos, Jade Tassey,
Mary Ziegler, Evangelia Chnari, Gregory R. D. Evans,
and Alan D. Widgerow

7 Mesenchymal Cells that Support Human Skin Regeneration . . 91
Joanne K. Gardner, Zalitha Pieterse, and Pritinder Kaur

8 Stem Cells and Burn 109
Anesh Prasai, Amina El Ayadi, David N. Herndon,
and Celeste C. Finnerty

9 Skin Tissue Engineering in Severe Burns: A Review on Its Therapeutic Applications 117
Alvin Wen Choong Chua, Chairani Fitri Saphira, and Si Jack Chong

10 Skin Substitutes for Burn Wounds 137
Daniel Popp, Christian Tapking, and Ludwik K. Branski

11 Wnt Signaling During Cutaneous Wound Healing 147
Khosrow Siamak Houschyar, Dominik Duscher, Susanne Rein, Zeshaan N. Maan, Malcolm P. Chelliah, Jung Y. Cha, Kristian Weissenberg, and Frank Siemers

12 Drug Delivery Advances for the Regeneration of Aged Skin ... 157
Daniela Castillo Pérez, Matthias M. Aitzetmüller, Philipp Neßbach, and Dominik Duscher

Part II Bone Regeneration

13 Innovative Scaffold Solution for Bone Regeneration Made of Beta-Tricalcium Phosphate Granules, Autologous Fibrin Fold, and Peripheral Blood Stem Cells 167
Ciro Gargiulo Isacco, Kieu C. D. Nguyen, Andrea Ballini, Gregorio Paduanelli, Van H. Pham, Sergey K. Aityan, Melvin Schiffman, Toai C. Tran, Thao D. Huynh, Luis Filgueira, Vo Van Nhan, Gianna Dipalma, and Francesco Inchingolo

14 Ordinary and Activated Bone Substitutes 181
Ilya Y. Bozo, R. V. Deev, A. Y. Drobyshev, and A. A. Isaev

15 Absorbable Bone Substitute Materials Based on Calcium Sulfate as Triggers for Osteoinduction and Osteoconduction 211
Dominik Pförringer and Andreas Obermeier

16 Perivascular Progenitor Cells for Bone Regeneration 223
Carolyn Meyers, Paul Hindle, Winters R. Hardy, Jia Jia Xu, Noah Yan, Kristen Broderick, Greg Asatrian, Kang Ting, Chia Soo, Bruno Peault, and Aaron W. James

17 Bone Repair and Regeneration Are Regulated by the Wnt Signaling Pathway 231
Khosrow Siamak Houschyar, Dominik Duscher, Zeshaan N. Maan, Malcolm P. Chelliah, Mimi R Borrelli, Kamran Harati, Christoph Wallner, Susanne Rein, Christian Tapking, Georg Reumuth, Gerrit Grieb, Frank Siemers, Marcus Lehnhardt, and Björn Behr

Contents

Part III Cartilage Regeneration

18 **Cartilage Tissue Engineering: Role of Mesenchymal Stem Cells, Growth Factors, and Scaffolds** 249
Mudasir Bashir Gugjoo, Hari Prasad Aithal,
Prakash Kinjavdekar, and Amarpal

19 **Sox9 Potentiates BMP2-Induced Chondrogenic Differentiation and Inhibits BMP2-Induced Osteogenic Differentiation** 263
Junyi Liao, Ning Hu, Nian Zhou, Chen Zhao, Xi Liang,
Hong Chen, Wei Xu, Cheng Chen, Qiang Cheng,
and Wei Huang

20 **Stem Cells and Ear Regeneration** 281
Hamid Karimi, Seyed-Abolhassan Emami,
and Ali-MohammadKarimi

Part IV Muscle and Tendon Regeneration

21 **Muscle Fiber Regeneration in Long-Term Denervated Muscles: Basics and Clinical Perspectives** 301
Ugo Carraro, Helmut Kern, Sandra Zampieri, Paolo Gargiulo,
Amber Pond, Francesco Piccione, Stefano Masiero,
Franco Bassetto, and Vincenzo Vindigni

22 **Rejuvenating Stem Cells to Restore Muscle Regeneration in Aging** 311
Eyal Bengal and Maali Odeh

23 **Silk Fibroin-Decorin Engineered Biologics to Repair Musculofascial Defects** 325
Lina W. Dunne, Nadja Falk, Justin Hubenak,
Tejaswi S. Iyyanki, Vishal Gupta, Qixu Zhang,
Charles E. Butler, and Anshu B. Mathur

24 **Skeletal Muscle Restoration Following Volumetric Muscle Loss: The Therapeutic Effects of a Biologic Surgical Mesh** 347
Jenna L. Dziki, Jonas Eriksson, and Stephen F. Badylak

25 **Principles of Tendon Regeneration** 355
Jacinta Leyden, Yukitoshi Kaizawa, and James Chang

26 **Stem Cells and Tendon Regeneration** 369
Hamid Karimi, Kamal Seyed-Forootan,
and Ali-Mohammad Karimi

27 **Cell Therapies for Tendon: Treatments and Regenerative Medicine** 385
Anthony Grognuz, Pierre-Arnaud Aeberhard,
Murielle Michetti, Nathalie Hirt-Burri,
Corinne Scaletta, Anthony de Buys Roessingh,
Wassim Raffoul, and Lee Ann Laurent-Applegate

Part V Nerve Regeneration

28 Current Trends and Future Perspectives for Peripheral Nerve Regeneration 411
Georgios N. Panagopoulos, Panayiotis D. Megaloikonomos, and Andreas F. Mavrogenis

29 The Regeneration of Peripheral Nerves Depends on Repair Schwann Cells 425
Kristján R. Jessen and Rhona Mirsky

30 Adipose-Derived Stem Cells (ASCs) for Peripheral Nerve Regeneration 437
Mathias Tremp and Daniel D. Kalbermatten

31 Direct Reprogramming Somatic Cells into Functional Neurons: A New Approach to Engineering Neural Tissue In Vitro and In Vivo 447
Meghan Robinson, Oliver McKee-Reed, Keiran Letwin, and Stephanie Michelle Willerth

Index 463

Contributors

Pierre-Arnaud Aeberhard, MD Unit of Regenerative Therapy, Service of Plastic and Reconstructive Surgery, Department of Musculoskeletal Medicine, Lausanne University Hospital, Epalinges, Switzerland

Hari Prasad Aithal, MVSc, PhD Training and Education Centre, ICAR-Indian Veterinary Research Institute, College of Agriculture Campus, Pune, Maharashtra, India

Sergey K. Aityan, DSc, PhD Department of Business and Economics, Lincoln University, Oakland, CA, USA

Matthias M. Aitzetmüller, MD Department of Plastic and Hand Surgery, Klinikum rechts der Isar, Technical University of Munich, Munich, Germany

Section of Plastic and Reconstructive Surgery, Department of Trauma, Hand and Reconstructive Surgery, Westfaelische Wilhelms, University of Muenster, Muenster, Germany

Robert W. Alexander, MD, DMD Department of Surgery, Institute of Regenerative Medicine, University of Washington, Stevensville, MT, USA

Amarpal PhD Division of Surgery, ICAR-Indian Veterinary Research Institute, Izatnagar, Uttar Pradesh, India

Lee Ann Laurent-Applegate, PhD Unit of Regenerative Therapy, Service of Plastic and Reconstructive Surgery, Department of Musculoskeletal Medicine, Lausanne University Hospital, Epalinges, Switzerland

Greg Asatrian, DDS School of Dentistry, University of California, Los Angeles, CA, USA

Stephen F. Badylak, DVM, PhD, MD McGowan Institute for Regenerative Medicine, Pittsburgh, PA, USA

Department of Surgery, University of Pittsburgh School of Medicine, Pittsburgh, PA, USA

Department of Bioengineering, University of Pittsburgh, Pittsburgh, PA, USA

Andrea Ballini Department of Basic Medical Sciences, Neurosciences and Sense Organs, University of Bari Aldo Moro, Bari, Italy

Derek A. Banyard, MD, MBA Department of Plastic Surgery, Center for Tissue Engineering, University of California, Irvine, Orange, CA, USA

Franco Bassetto, MD Interdepartmental Research Centre of Myology, University of Padova, Padova, Italy

Björn Behr, MD Department of Plastic Surgery, BG University Hospital Bergmannsheil, Ruhr University Bochum, Bochum, Germany

Eyal Bengal, PhD Department of Biochemistry, The Ruth and Bruce Rappaport Faculty of Medicine, Technion-Israel Institute of Technology, Haifa, Israel

Mimi R. Borrelli, MBBS, MSc, BSc Stanford Institute for Stem Cell Biology and Regenerative Medicine, Stanford University School of Medicine, Stanford, CA, USA

Division of Plastic and Reconstructive Surgery, Department of Surgery, Stanford School of Medicine, Stanford, CA, USA

Hagey Laboratory for Pediatric Regenerative Medicine, Division of Plastic Surgery, Department of Surgery, Stanford University School of Medicine, Stanford, CA, USA

Ilya Y. Bozo, MD, PhD Department of Maxillofacial Surgery, A.I. Burnazyan Federal Medical Biophysical Center of FMBA of Russia, Moscow, Russia

"Histograft", LLC, Moscow, Russia

Ludwik K. Branski, MD Division of Plastic, Aesthetic and Reconstructive Surgery, Department of Surgery, Medical University of Graz, Graz, Austria

Elizabeth A. Brett, MSc Department of Plastic and Hand Surgery, Klinikum rechts der Isar, Technical University of Munich, Munich, Germany

Kristen Broderick, MD Department of Surgery, Johns Hopkins University, Baltimore, MD, USA

Charles E. Butler, MD Department of Plastic Surgery, Unit 602, The University of Texas MD Anderson Cancer Center, Houston, TX, USA

Ugo Carraro, MD, PhD IRCCS Fondazione Ospedale San Camillo, Venezia-Lido, Italy

Department of Neurorehabilitation, Foundation San Camillo Hospital, I.R.C.C.S., Venice, Italy

Daniela Castillo Pérez, MSc Biotechnology Research Center, Costa Rica Institute of Technology, Cartago, Costa Rica

Jung Y. Cha, MD, PhD Orthodontic Department, College of Dentistry, Yonsei University, Seoul, Republic of Korea

James Chang, MD Division of Plastic and Reconstructive Surgery, Stanford University Medical Center, Palo Alto, CA, USA

Malcolm P. Chelliah, MD, MBA Division of Plastic and Reconstructive Surgery, Department of Surgery, Stanford School of Medicine, Stanford, CA, USA

Cheng Chen, MD Department of Orthopaedic Surgery, The First Affiliated Hospital of Chongqing Medical University, Chongqing, China

Qiang Cheng, MD Department of Orthopaedic Surgery, The First Affiliated Hospital of Chongqing Medical University, Chongqing, China

Hong Chen, MD, PhD Department of Orthopaedic Surgery, The First Affiliated Hospital of Chongqing Medical University, Chongqing, China

Evangelia Chnari, PhD Musculoskeletal Transplant Foundation, Edison, NJ, USA

Si Jack Chong, MBBS, MRCS Department of Plastic, Reconstructive & Aesthetic Surgery, Singapore General Hospital, Singapore, Singapore, Singapore

Alvin Wen Choong Chua, PhD Department of Plastic, Reconstructive & Aesthetic Surgery, Singapore General Hospital, Singapore, Singapore

Roman V. Deev, MD, PhD "Histograft", LLC, Moscow, Russia

Department of General Pathology, I.P. Pavlov Ryazan State Medical University, Ryazan, Russia

PJSC "Human Stem Cells Institute", Moscow, Russia

Gianna Dipalma Department of Basic Medical Sciences, Neurosciences and Sense Organs, University of Bari Aldo Moro, Bari, Italy

Alexey Y. Drobyshev, MD, PhD, DSc Department of Maxillofacial and Plastic Surgery, A.I. Evdokimov Moscow State University of Medicine and Dentistry, Moscow, Russia

Lina W. Dunne, PhD Department of Plastic Surgery, Unit 602, The University of Texas MD Anderson Cancer Center, Houston, TX, USA

Dominik Duscher, MD Department for Plastic Surgery and Hand Surgery, Division of Experimental Plastic Surgery, Technical University of Munich, Munich, Germany

Jenna L. Dziki, PhD McGowan Institute for Regenerative Medicine, Pittsburgh, PA, USA

Department of Surgery, University of Pittsburgh School of Medicine, Pittsburgh, PA, USA

Amina El Ayadi, PhD Department of Surgery, University of Texas Medical Branch, Galveston, TX, USA

Seyed-Abolhassan Emami, MD Department of Plastic and Reconstructive Surgery, Iran University of Medical Sciences, Tehran, Iran

Jonas Eriksson, PhD McGowan Institute for Regenerative Medicine, Pittsburgh, PA, USA

Department of Surgery, University of Pittsburgh School of Medicine, Pittsburgh, PA, USA

Gregory R.D. Evans Department of Plastic Surgery, Center for Tissue Engineering, University of California, Irvine, Orange, CA, USA

Nadja Falk, MD Department of Plastic Surgery, Unit 602, The University of Texas MD Anderson Cancer Center, Houston, TX, USA

Luis Filgueira, PhD Department of Human Biology and Anatomy, University of Fribourg, Fribourg, Switzerland

Celeste C. Finnerty, PhD Department of Surgery, University of Texas Medical Branch, Galveston, TX, USA

Joanne K. Gardner, BSc, PhD Epithelial Stem Cell Biology Group, School of Biomedical Sciences, Curtin University, Perth, WA, Australia

Curtin Health and Innovation Research Institute, Curtin University, Perth, WA, Australia

Ciro Gargiulo Isacco, PhD, MSc, PG Department of Regenerative Medicine, Human Medicine International Clinic, HSC/Euro Stem Cells Research Center and International Clinic, Ho Chi Minh City, Vietnam

Paolo Gargiulo, PhD Clinical Engineering and Information Technology, Landspitali—University Hospital, Reykjavik, Iceland

Gerrit Grieb, MD Department of Plastic Surgery and Hand Surgery, Gemeinschaftskrankenhaus Havelhoehe, Teaching Hospital of the Charité Berlin, Berlin, Germany

Anthony Grognuz, PhD Unit of Regenerative Therapy, Service of Plastic and Reconstructive Surgery, Department of Musculoskeletal Medicine, Lausanne University Hospital, Epalinges, Switzerland

Mudasir Bashir Gugjoo, MVSc, PhD Faculty of Veterinary Sciences and Animal Husbandry, SKUAST-Kashmir, Jammu & Kashmir, India

Vishal Gupta, PhD Department of Plastic Surgery, Unit 602, The University of Texas MD Anderson Cancer Center, Houston, TX, USA

Kamran Harati, MD Department of Plastic Surgery, BG University Hospital Bergmannsheil, Ruhr University Bochum, Bochum, Germany

Winters R. Hardy, PhD Orthopedic Hospital Research Center, University of California, Los Angeles, CA, USA

David N. Herndon, MD Department of Surgery, University of Texas Medical Branch, Galveston, TX, USA

Paul Hindle, PhD Department of Trauma and Orthopaedic Surgery, The University of Edinburgh, Edinburgh, UK

Nathalie Hirt-Burri, PhD Unit of Regenerative Therapy, Service of Plastic and Reconstructive Surgery, Department of Musculoskeletal Medicine, Lausanne University Hospital, Epalinges, Switzerland

Khosrow Siamak Houschyar, MD Department of Plastic Surgery, BG University Hospital Bergmannsheil, Ruhr University Bochum, Bochum, Germany

Burn Unit, Department for Plastic and Hand Surgery, Trauma Center Bergmannstrost Halle, Halle (Saale), Germany

Wei Huang, MD, PhD Department of Orthopaedic Surgery, The First Affiliated Hospital of Chongqing Medical University, Chongqing, China

Justin Hubenak, BS Department of Plastic Surgery, Unit 602, The University of Texas MD Anderson Cancer Center, Houston, TX, USA

Michael S. Hu, MD, MSc, MPh Department of Plastic Surgery, University of Pittsburgh School of Medicine, Pittsburgh, PA, USA

Stanford Institute for Stem Cell Biology and Regenerative Medicine, Stanford University School of Medicine, Stanford, CA, USA

Division of Plastic and Reconstructive Surgery, Department of Surgery, Stanford School of Medicine, Stanford, CA, USA

Hagey Laboratory for Pediatric Regenerative Medicine, Division of Plastic Surgery, Department of Surgery, Stanford University School of Medicine, Stanford, CA, USA

Ning Hu, MD, PhD Department of Orthopaedic Surgery, The First Affiliated Hospital of Chongqing Medical University, Chongqing, China

Thao D. Huynh Department of Embryology, Genetics and Stem Cells, Pham Ngoc Thach University of Medicine, Ho Chi Minh City, Vietnam

Francesco Inchingolo Department of Basic Medical Sciences, Neurosciences and Sense Organs, University of Bari Aldo Moro, Bari, Italy

Artur A. Isaev, MD PJSC "Human Stem Cells Institute", Moscow, Russia

Tejaswi S. Iyyanki, MS Department of Plastic Surgery, Unit 602, The University of Texas MD Anderson Cancer Center, Houston, TX, USA

Aaron W. James, MD, PhD Department of Pathology, Johns Hopkins University, Baltimore, MD, USA

Kristján R. Jessen, MSc, PhD Department of Cell and Developmental Biology, University College London, London, UK

Yukitoshi Kaizawa, M.D., Ph.D. Division of Plastic and Reconstructive Surgery, Stanford University Medical Center, Palo Alto, CA, USA

Daniel F. Kalbermatten, MD, PhD Department of Plastic, Reconstructive, Aesthetic and Hand Surgery, University Basel, Basel, Switzerland

Ali-Mohammad Karimi, BS School of Medicine, Iran University of Medical Sciences, Tehran, Iran

Hamid Karimi, MD Department of Plastic and Reconstructive Surgery, Hazrat Fatemeh Hospital, Iran University of Medical Sciences, Tehran, Iran

Pritinder Kaur, PhD, MSc Epithelial Stem Cell Biology Group, School of Biomedical Sciences, Curtin University, Perth, WA, Australia

Curtin Health and Innovation Research Institute, Curtin University, Perth, WA, Australia

Helmut Kern, MD, PhD Physiko- und Rheumatherapie, St. Poelten, Austria

Prakash Kinjavdekar, BVSc, AHMVSc, PhD Division of Surgery, ICAR-Indian Veterinary Research Institute, Izatnagar, Uttar Pradesh, India

Marcus Lehnhardt, MD Department of Plastic Surgery, BG University Hospital Bergmannsheil, Ruhr University Bochum, Bochum, Germany

Keiran Letwin, MD Biomedical Engineering Program, University of Victoria, Victoria, BC, Canada

Jacinta Leyden, BS Stanford University School of Medicine, Stanford, CA, USA

Xi Liang, MD, PhD Department of Orthopaedic Surgery, The First Affiliated Hospital of Chongqing Medical University, Chongqing, China

Junyi Liao, MD, PhD Department of Orthopaedic Surgery, The First Affiliated Hospital of Chongqing Medical University, Chongqing, China

Michael T. Longaker, MD, MBA Stanford Institute for Stem Cell Biology and Regenerative Medicine, Stanford University School of Medicine, Stanford, CA, USA

Division of Plastic and Reconstructive Surgery, Department of Surgery, Stanford School of Medicine, Stanford, CA, USA

Hagey Laboratory for Pediatric Regenerative Medicine, Division of Plastic Surgery, Department of Surgery, Stanford University School of Medicine, Stanford, CA, USA

H. Peter Lorenz, MD Stanford Institute for Stem Cell Biology and Regenerative Medicine, Stanford University School of Medicine, Stanford, CA, USA

Hagey Laboratory for Pediatric Regenerative Medicine, Division of Plastic Surgery, Department of Surgery, Stanford University School of Medicine, Stanford, CA, USA

Zeshaan N. Maan, MD Division of Plastic and Reconstructive Surgery, Department of Surgery, Stanford School of Medicine, Stanford, CA, USA

Hans-Günther Machens, MD Department of Plastic and Hand Surgery, Klinikum rechts der Isar, Technical University of Munich, Munich, Germany

Stefano Masiero, MD, PhD Interdepartmental Research Centre of Myology, University of Padova, Padova, Italy

Anshu B. Mathur, PhD Department of Plastic Surgery, Unit 602, The University of Texas MD Anderson Cancer Center, Houston, TX, USA

Andreas F. Mavrogenis, MD First Department of Orthopaedics, National and Kapodistrian University of Athens, School of Medicine, Athens, Greece

Oliver Mckee-Reed, MD Department of Biochemistry, University of Victoria, Victoria, BC, Canada

Panayiotis D. Megaloikonomos, MD First Department of Orthopaedics, National and Kapodistrian University of Athens, School of Medicine, Athens, Greece

Carolyn Meyers, BS Department of Pathology, Johns Hopkins University, Baltimore, MD, USA

Murielle Michetti, BS Unit of Regenerative Therapy, Service of Plastic and Reconstructive Surgery, Department of Musculoskeletal Medicine, Lausanne University Hospital, Epalinges, Switzerland

Rhona Mirsky, PhD Department of Cell and Developmental Biology, University College London, London, UK

Philipp Neßbach, MSc Department of Plastic and Hand Surgery, Klinikum Rechts der Isar, Technical University of Munich, Munich, Germany

Kieu Cao Diem Nguyen, MD, MBA Department of Stem Cell Research, HSC International Clinic, Ho Chi Minh City, Vietnam

Andreas Obermeier, MSc Klinik und Poliklinik für Unfallchirurgie, Klinikum rechts der Isar, Technische Universität München, Munich, Germany

Maali Odeh, MD, PhD Department of Biochemistry, The Ruth and Bruce Rappaport Faculty of Medicine, Technion-Israel Institute of Technology, Haifa, Israel

Gregorio Paduanelli Department of Basic Medical Sciences, Neurosciences and Sense Organs, University of Bari Aldo Moro, Bari, Italy

Georgios N. Panagopoulos, MD First Department of Orthopaedics, National and Kapodistrian University of Athens, School of Medicine, Athens, Greece

Bruno Peault, PhD Orthopedic Hospital Research Center, University of California, Los Angeles, CA, USA

BHF Center for Vascular Regeneration and MRC Center for Regenerative Medicine, University of Edinburgh, Edinburgh, UK

Dominik Pförringer, MD, MBA Klinik und Poliklinik für Unfallchirurgie, Klinikum rechts der Isar, Technische Universität München, Munich, Germany

Van H. Pham Department of Microbiology, Nam Khoa-Bioteck Microbiology Laboratory and Research Center, Ho Chi Minh City, Vietnam

Department of Microbiology, Nam Khoa-Bioteck Microbiology Laboratory and Research Center, Ho Chi Minh City, VN, USA

Francesco Piccione, MD, PhD IRCCS Fondazione Ospedale San Camillo, Venezia-Lido, Italy

Zalitha Pieterse, MSc Epithelial Stem Cell Biology Group, School of Biomedical Sciences, Curtin University, Perth, WA, Australia

Curtin Health and Innovation Research Institute, Curtin University, Perth, WA, Australia

Amber Pond, PhD Anatomy Department, Southern Illinois University School of Medicine, Carbondale, IL, USA

Daniel Popp, MD Division of Plastic, Aesthetic and Reconstructive Surgery, Department of Surgery, Medical University of Graz, Graz, Austria

Anesh Prasai, PhD Department of Surgery, University of Texas Medical Branch, Galveston, TX, USA

Wassim Raffoul, MD Unit of Regenerative Therapy, Service of Plastic and Reconstructive Surgery, Department of Musculoskeletal Medicine, Lausanne University Hospital, Epalinges, Switzerland

Susanne Rein, MD Department of Plastic and Hand Surgery, Burn Center-Clinic St. Georg, Leipzig, Germany

Georg Reumuth, MD Department of Plastic and Hand Surgery, Burn Unit, Trauma Center Bergmannstrost Halle, Halle, Germany

Meghan Robinson, BEng Biomedical Engineering Program, University of Victoria, Victoria, BC, Canada

Anthony De Buys Roessingh, MD, PhD Unit of Regenerative Therapy, Service of Plastic and Reconstructive Surgery, Department of Musculoskeletal Medicine, Lausanne University Hospital, Epalinges, Switzerland

Chairani Fitri Saphira, MD Department of Plastic, Reconstructive & Aesthetic Surgery, Singapore General Hospital, Singapore, Singapore

Department of Surgery, Dr. Mohamad Soewandhie General Hospital, Surabaya, Indonesia

Christos Sarantopoulos, BS Department of Plastic Surgery, Center for Tissue Engineering, University of California, Irvine, Orange, CA, USA

Corinne Scaletta, BS Unit of Regenerative Therapy, Service of Plastic and Reconstructive Surgery, Department of Musculoskeletal Medicine, Lausanne University Hospital, Epalinges, Switzerland

Matthias A. Sauter, MD Department of Plastic and Hand Surgery, Klinikum rechts der Isar, Technical University of Munich, Munich, Germany

Kamal Seyed-Forootan, MD Department of Plastic and Reconstructive Surgery, Iran University of Medical Sciences, Tehran, Iran

Frank Siemers, MD Department of Plastic Surgery and Hand Surgery, Gemeinschaftskrankenhaus Havelhoehe, Teaching Hospital of the Charité Berlin, Berlin, Germany

Chia Soo, MD Orthopedic Hospital Research Center, University of California, Los Angeles, CA, USA

Department of Surgery, University of California, Los Angeles, Los Angeles, CA, USA

Christian Tapking, MD Department of Surgery, Shriners Hospital for Children-Galveston, University of Texas Medical Branch, Galveston, TX, USA

Department of Hand, Plastic and Reconstructive Surgery, Burn Trauma Center, BG Trauma Center Ludwigshafen, University of Heidelberg, Heidelberg, Germany

Jade Tassey, BS Department of Plastic Surgery, Center for Tissue Engineering, University of California, Irvine, Orange, CA, USA

Kang Ting, DMD, DMedSci School of Dentistry, University of California, Los Angeles, CA, USA

Toai C. Tran Stem Cells, Embryology and Immunity Department, Pham Ngoc Thach University of Medicine, Ho Chi Minh City, Vietnam

Mathias Tremp, MD Department of Plastic, Reconstructive, Aesthetic and Hand Surgery, University Basel, Basel, Switzerland

Nhan V. Van University of Pharmacy and Medicine, Ho Chi Minh City, Vietnam

Nha Khoa Nham Tam, Policlinic and International Dental Implant Center, Ho Chi Minh City, Vietnam

Vincenzo Vindigni, MD, PhD Unit of Plastic Surgery, Department of Neuroscience, University of Padova, Padova, Italy

Christoph Wallner, MD Department of Plastic Surgery, BG University Hospital Bergmannsheil, Ruhr University Bochum, Bochum, Germany

Kristian Weissenberg, MD Burn Unit, Department for Plastic and Hand Surgery, Trauma Center Bergmannstrost Halle, Halle (Saale), Germany

Alan D. Widgerow, MBBCh, MMed Department of Plastic Surgery, Center for Tissue Engineering, University of California, Irvine, Orange, CA, USA

Stephanie Michelle Willerth, PhD, PEng Department of Mechanical Engineering and Division of Medical Sciences, University of Victoria, Victoria, BC, Canada

Jia Jia Xu, PhD Department of Pathology, Johns Hopkins University, Baltimore, MD, USA

Wei Xu, MD Department of Orthopaedic Surgery, The First Affiliated Hospital of Chongqing Medical University, Chongqing, China

Noah Yan, BS Department of Pathology, Johns Hopkins University, Baltimore, MD, USA

Sandra Zampieri, PhD Interdepartmental Research Centre of Myology, University of Padova, Padova, Italy

Qixu Zhang, MD, PhD Department of Plastic Surgery, Unit 602, The University of Texas MD Anderson Cancer Center, Houston, TX, USA

Chen Zhao, MD, PhD Department of Orthopaedic Surgery, The First Affiliated Hospital of Chongqing Medical University, Chongqing, China

Nian Zhou, MD, PhD Department of Orthopaedic Surgery, The First Affiliated Hospital of Chongqing Medical University, Chongqing, China

Mary Ziegler, PhD Department of Plastic Surgery, Center for Tissue Engineering, University of California, Irvine, Orange, CA, USA

Part I

Skin and Soft Tissue Regeneration

Induction of the Fetal Scarless Phenotype in Adult Wounds: Impossible?

1

Michael S. Hu, Mimi R. Borrelli, Michael T. Longaker, and H. Peter Lorenz

M. S. Hu (✉)
Department of Plastic Surgery, University of Pittsburgh School of Medicine, Pittsburgh, PA, USA

Stanford Institute for Stem Cell Biology and Regenerative Medicine, Stanford University School of Medicine, Stanford, CA, USA

Division of Plastic and Reconstructive Surgery, Department of Surgery, Stanford School of Medicine, Stanford, CA, USA

Hagey Laboratory for Pediatric Regenerative Medicine, Division of Plastic Surgery, Department of Surgery, Stanford University School of Medicine, Stanford, CA, USA
e-mail: hums2@upmc.edu

M. R. Borrelli · M. T. Longaker
Stanford Institute for Stem Cell Biology and Regenerative Medicine, Stanford University School of Medicine, Stanford, CA, USA

Division of Plastic and Reconstructive Surgery, Department of Surgery, Stanford School of Medicine, Stanford, CA, USA

Hagey Laboratory for Pediatric Regenerative Medicine, Division of Plastic Surgery, Department of Surgery, Stanford University School of Medicine, Stanford, CA, USA
e-mail: longaker@stanford.edu

H. P. Lorenz
Division of Plastic and Reconstructive Surgery, Department of Surgery, Stanford School of Medicine, Stanford, CA, USA

Hagey Laboratory for Pediatric Regenerative Medicine, Division of Plastic Surgery, Department of Surgery, Stanford University School of Medicine, Stanford, CA, USA
e-mail: plorenz@stanford.edu

1.1 Introduction

Wound healing is essential to restore the barrier and protective functions of the skin. After hemostasis and formation of a clot, mammalian epidermal healing occurs via three predictable and overlapping phases: inflammation; proliferation and fibroplasia with production of granulation tissue; and maturation. In adult wounds, excess accumulation of extracellular matrix (ECM) results in a scar, which is defined as a macroscopic fibrous disturbance in the normal tissue architecture. The healing process is efficient and quickly reestablishes epithelial integrity; however, the newly formed skin is incompletely regenerated. Dermal appendages, such as hair follicles, sebaceous glands, and sweat glands, are missing. The epidermis is flattened, and epidermal rete ridges are absent. Newly synthesized collagen is arranged into dense and unorganized matrices [1], resulting in a fibrotic scar of reduced tensile strength [2, 3]. Scars can restrict growth, impair mobility across joints, and be cosmetically disfiguring with consequent detrimental psychological and social impact. Adult wounds of humans can also develop into pathological keloidor hypertrophic scars when healing involves the deposition of excess collagen. Wounds that occur early to mid-gestation in human fetuses, however, are able to heal without the formation of a scar [4]. The ability of embryonic epidermis to heal without scarring is a feature observed across numerous mammalian species, as well as ex vivo in fetal

© Springer Nature Switzerland AG 2019
D. Duscher, M. A. Shiffman (eds.), *Regenerative Medicine and Plastic Surgery*,
https://doi.org/10.1007/978-3-030-19962-3_1

skin [4–7]. Unlike adult wounds, fetal wounds rapidly heal and completely regenerate the dermis and epidermis, including the dermal appendages. Collagen is synthesized into matrices identical to those found in uninjured tissue [1]. Scarless cutaneous wound healing is a feature intrinsic to the fetal epidermal tissue, rather than the conditions in the intrauterine environment [8]. It is dependent upon gestational age, tissue size, and site. The transition from the fetal to adult epidermal healing phenotype occurs around 24 weeks of gestation in human embryos [9], and around embryonic day 18.5 (E19) in mice [10] and 17.5 (E18) in rats [10, 11]. Larger wounds undergo this phenotypic transition at an earlier gestational age [9]. Wounds in the oral mucosal heal at an accelerated rate, and rarely produce scars, including keloid and hypertrophic scars [12], even in mammalian adults (Fig. 1.1) [13–17].

The possibility of scarless wound healing gives rise to tremendous clinical potential. In-depth understanding of the principles and mechanisms underlying tissue regeneration in fetal epidermis is essential to be able to use this knowledge to

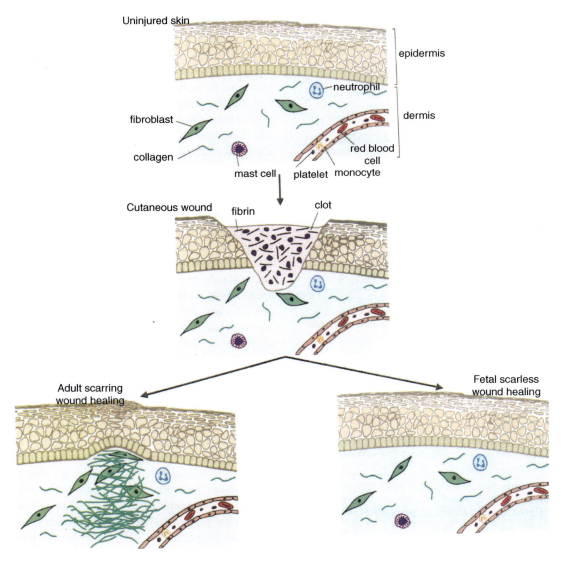

Fig. 1.1 The different healing phenotypes of adult and fetal skin. Adult cutaneous wounds heal via repair and scarring. Fetal cutaneous wounds heal with scarless skin regeneration

1.2 Inflammatory Response

1.2.1 Inflammatory Cells

promote scar-free healing in adult wounds. Although the exact cellular and molecular mechanisms of scarless healing are yet to be elucidated, there are clear differences in the inflammatory response, cellular medications, genes expressed, and function of stem cells. Current developments in tissue regeneration applications, such as cell-based therapies, are making progress towards reducing scarring in adult wounds. This chapter outlines the current understanding of the biological and biomechanical processes underlying fetal scarless healing, how this differs from responses in adult wound healing via scarring, and how this knowledge has been applied to promote scarless healing in adult wounds (Table 1.1) (Fig. 1.2).

Table 1.1 Table of differences in fetal and adult wound healing

Feature	Fetal wounds	Adult wounds
Inflammatory cells		
Neutrophils	Low	High
Macrophages	Low	High
Mast cells	Low	High
Inflammatory signaling molecules		
Pro-inflammatory: IL-6, IL-8	Low	High
Anti-inflammatory: IL-10	High	Low
VEGF	Variable	High
Extracellular matrix		
ECM rate of synthesis	High	Low
Collagen		
Type III:I ratio	High	Low
Synthesis rate	High	Low
Histological pattern of fibers	Fine, reticular, large	Dense, parallel, small
Cross-linking	Low	High
Glycosaminoglycans		
HA expression	High	Low
HA receptors	High	Low
HASA	High	Low
Chondroitin sulfate	High	Low
Proteoglycans		
Fibromodulin	Increased	Decreased
Decorin	Decreased	Increased
Myofibroblasts	Absent	Present
Adhesion proteins	Rapid increase	Diminished increase
MMP:TIMP ratio	High	Low

HA hyaluronic acid, *HASA* hyaluronic acid synthase, *MMP* matrix metalloproteinase, *TIMP* tissue inhibitors of metalloproteinase

The inflammatory component sets in within minutes of cutaneous tissue damage, and is much reduced in fetal, compared to adult, wounds. Aggregation and degranulation of platelets at the site of epidermal tissue damage are responsible both for the initial hemostasis and for attracting neutrophils, the first migrating inflammatory cells, to the lesion site [18]. Fetal platelets aggregate less when exposed to collagen than adult platelets [19], and produce less inflammatory signaling molecules upon degranulation, including less transforming growth factor β1 (TGFβ1), platelet-derived growth factor (PDGF) [20], tumor necrosis factor-α (TNF-α), and interleukin-1 (IL-1). Fewer chemoattractant molecules attract less circulating inflammatory cells. Low levels of TNF-α and IL-1 also lead to diminished upregulation of neutrophil adhesion molecules on the surface of fetal neutrophils [21], limiting neutrophil–endothelial cell interactions, and the consequent migration of neutrophils during fetal wound healing [22, 23].

Monocytes are the second type of inflammatory cells that migrate from blood vessels and differentiate into macrophages at the site of cutaneous tissue damage. They transform into macrophages around 48–96 h after injury onset [23]. Macrophages contribute to both the inflammatory and proliferative phases of wound healing. Macrophages secrete further interleukins and TNF which stimulate fibroblasts to make collagen, mediate angiogenesis, and produce nitric oxide [24]. The degree of macrophages remaining at wound sites directly correlates with the degree of scar formed [25]. In murine fetal and embryonic skin wounds, macrophages are almost absent prior to gestational day 14 (E14), except when the tissue damage is in excess [26]. TGFβ1 is a growth factor partly responsible for transitioning circulating monocytes into activated macrophages. Low TGFβ1 levels in fetal wounds likely contribute to recruitment of fewer macrophages [27].

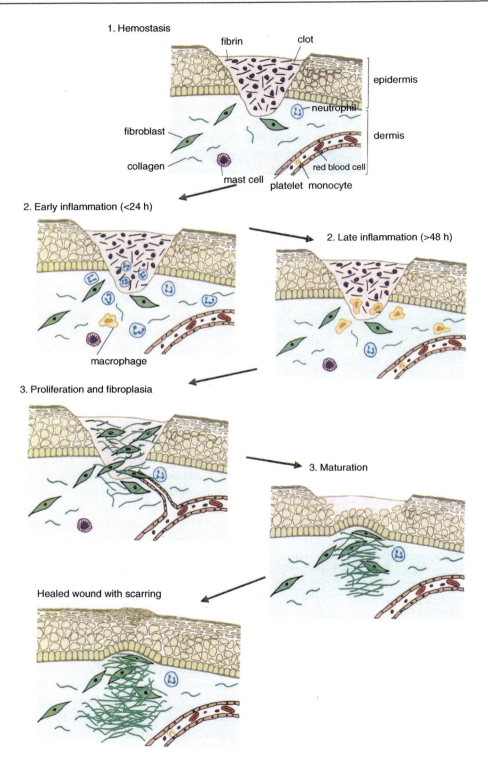

Fig. 1.2 The different phases of wound healing. Stage 1 is hemostasis where tissue injury initiates a coagulation cascade to stop bleeding. Stage 2 is the inflammatory phase. Early in inflammation neutrophils are increased at the wound site. Late in inflammation macrophages are recruited which stimulate angiogenesis and re-epithelialization. Stage 3 is the proliferative phase where granulation tissue is formed and a network of collagen, fibronectin, and hyaluronic acid. Collagen is further deposited as the wound site matures and cross-links, resulting in scar tissue

Mast cells are another inflammatory cell which are predominantly resident cells found in the vicinity of connective tissue of vessels, skin, and mucosa [28], but which migrate to wound site within 24 h. Mast cells can degranulate to release cytokines (IL-6 and IL-8), vascular endothelial growth factor (VEGF), and histamine, which can initiate a substantial inflammatory reaction. Serine proteases, such as chymase and tryptase, are released early in inflammation, which break down the ECM and prepare the site for subsequent repair. Tryptase also has an important role in the synthesis and deposition of collagen [29, 30]. Mast cells are not required for wound healing, however, and cutaneous wounds in mice are able to heal whether mast cells are present or not [31]. Compared to adult skin wounds, fewer mast cells are found in fetal wounds, and those mast cells present are less able to degranulate and release less histamine, TGFβ, and VEGF upon degranulation. This contributes to the decreased chemotaxis and extravasation of neutrophils [32].

1.2.2 Inflammatory Molecules

The balance of inflammatory signaling molecules is in favor of anti-inflammation in fetal wounds, but pro-inflammation in adult wounds. TGFβ is a growth factor influencing all phases of healing including inflammation, angiogenesis, fibroblast proliferation, collagen synthesis, deposition, and remodeling of the ECM [33, 34]. There are three isoforms of TGFβ in humans: TGFβ1, TGFβ2, and TGFβ3 [35]. TGFβ1 and TGFβ2 levels are decreased [36], and TGFβ3 levels are increased, in fetal compared to adult wounds, conducive to less scar formation [37–40]. TGFβ1 promotes protein deposition and collagen gene expression, and inhibits the degradation of the ECM by increasing the expression of TIMPs which inhibit the matrix metalloproteinases (MMPs). TGFβ1 attracts fibroblasts and macrophages to the wound site and enhances angiogenesis [41]. Fetal wounds treated with TGFβ1 scar [42], and adult wounds treated with anti-TGFβ1 or anti-TGFβ-2 antibodies, heal without scarring [43]. TGFβ1 autoregulates its own production via autocrine signaling, which can lead to TGFβ1 overproduction and scar formation [43]. It also prevents its degradation by releasing tissue inhibitors of metalloproteinases (TIMPs) and downregulating proteases. In fetal wounds this positive feedback loop is diminished [44]. TGFβ3 is a potent anti-scarring cytokine, and maintains cells in a relatively undifferentiated state. Levels of TGFβ3 peak in the fetal period of scarless wound healing [37, 38]. Expression of TGFβ3 is correlated with hypoxic inducible factor I (HIF1), which is released in hypoxic environments characteristic of the environment of the developing fetal epidermis [43, 45]. Addition of TGFβ3 to adult wounds decreases scar formation [45].

The pro-inflammatory interleukins, IL-6 and IL-8, are highly expressed in adult cutaneous wounds, and minimally expressed in fetal wounds [23, 46]. IL-10 is an anti-inflammatory interleukin which inhibits expression of IL1, IL6, IL8, TNF-α, and inflammatory cell migration, and permits normal deposition of collagen and reconstruction of a normal dermal architecture [47]. IL-10 is overexpressed in fetal wounds [47], and IL-10 knockout mice fetuses scar when they would otherwise not [48].

VEGF is an oxygen-dependent factor released in response to high HIF1 levels characteristic of the hypoxic fetal epidermis [43]. The role of VEGF in fetal scarless healing is not fully understood and its effects are likely multifactorial. VEGF both stimulates angiogenesis and has pro-inflammatory action, involved in attracting inflammatory cells to the wound site. VEGF expression has been observed greater in E16, the scar-free period, compared to the scarring E18 period in mice, but this did not translate to any histologic differences in neovascularization [49]. High VEGF levels, on the other hand, are associated with the development of keloid and hypertrophic scars in humans [50–54]. Neutralization of VEGF with antibodies reduced scar width [55]. It is likely that VEGF can both up- and downregulate scarring.

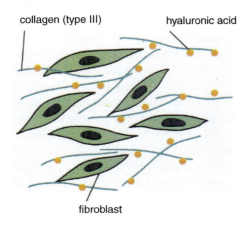 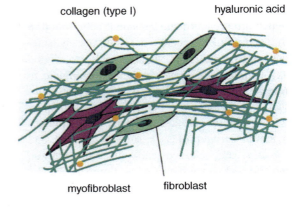

Fig. 1.3 The different phenotypes of the extracellular matrix (ECM) in adult and fetal cutaneous wounds. (Left) In the fetal wound there is an abundance of hyaluronic acid and fibroblasts which synthesize predominantly type III collagen. The collagen is arranged in a fine reticular pattern, with minimal cross-linking, in a pattern indistinguishable from uninjured skin. (Right) In the adult wound, there are fewer fibroblasts, and less hyaluronic acid. Type I collagen predominates, which is dense and extensively cross-linked. Some fibroblasts differentiate into myofibroblasts which contract to change the orientation of the newly synthesized collagen, and close the cutaneous wound, resulting in scarring

1.3 Extracellular Matrix (Fig. 1.3)

The ECM is rich in proteins including fibrous adhesion proteins, glycosaminoglycans (GAGs), proteoglycans, as well as resident fibroblasts responsible for synthesizing these components. The composition and architecture of the ECM differ in adult and fetal wounds, and its biological and biophysical properties have substantial influences on cell proliferation, differentiation, and adhesion, which likely impact wound healing.

1.3.1 Glycosaminoglycans

GAGs may play a role in scarless wound healing [56]. Compared to the adult wound, the fetal wound environment is rich in hyaluronic acid (HA), one of the main GAGs, responsible for accelerated cell proliferation, motility, and morphogenesis [56]. HA has a negative charge which attracts water molecules preventing the healing skin from becoming deformed and helping cellular migration [57]. The production of HA is accelerated and sustained in fetal compared to adult wounds [58, 59]. Fetal fibroblasts express more HA receptors [60, 61]. HA synthase is differentially regulated in fetal and adult fibroblasts via inflammatory cytokines; there are fewer proinflammatory cytokines, including IL-1 and TNF, which serve to downregulate HA expression [62]. Beyond the late fetal period, the HA content of ECM declines [6]. Chondroitin sulfate is another GAG which is also significantly produced in scarless fetal, but not fibrotic adult wounds [63, 64].

1.3.2 Adhesion Molecules

ECM adhesion molecules, including fibronectin and tenascin, are produced in greater abundance at earlier time points in fetal compared to adult wounds [63, 65, 66]. These molecules help organize the ECM, minimize scarring, and attract and bind fibroblasts and endothelial cells to the wound. Fibronectin helps anchor cells to the wound site and tenascin facilitates the movement of cells. Keratinocytes lining the wound edge bind ECM proteins, including fibronectin, tenascin collagen, and laminin, through integrin receptors. The fetal keratinocytes rapidly increase their expression of integrin receptors during healing, more so than the analogous keratinocytes in adult wounds [67]. The αvβ6 integrin receptor co-localizes with both TGFβ1 and TGFβ3 during wound healing and is thought to activate both growth factors. Prolonged TGFβ3 and αvβ6 integrin expression may protect from scar formation [39].

1.3.3 Proteoglycans

Proteoglycan matrix modulators are important in the production, organization, and degradation of collagen. Small leucine-rich proteoglycans (SLRPs) are polyanionic macromolecules of the ECM found covalently bound to linear sulfated GAG chains. SLRPs interact with collagen molecules to modulate fibrillogenesis and collagen turnover. SLRPs show different expressions in fetal and adult wound healing phenotypes. Decorin is upregulated during the transition to scar formation [68]. Chondroitin sulfate is present in scarless wounds but absent in wounds that scar. Fibromodulin, a SLRP that binds and inactivates TGFβ, is induced in wounds that are scar free, but is decreased in adult wounds which lead to scar formation [27].

1.3.4 Collagen

Collagen is a central component of the ECM. In adult skin type I collagen predominates, whereas type III collagen is the most abundant collagen isoform in fetal skin. Increasing amounts of collagen type I are made as gestational age increases, and this transition from the predominance of type I to type III correlates with the transition from scarring to scarless healing [61, 69, 70]. Consistent with these findings, the synthesis of procollagen 1α1 is increased, and procollagen 3 expression decreased, as gestational age progresses [71]. HA, abundant in fetal wounds, upregulates type III collagen deposition by fibroblasts [56, 61]. The type III collagen in fetal wounds is fine, synthesized into a reticular network that is undistinguishable from that found in uninjured skin [61]. The newly formed type I collagen in adult wounds and fetal wounds late in gestation, however, is laid down parallel to skin, in dense bundles, with more extensive cross-linking. This gives rise to a rigid and fibrotic ECM characteristic of scarring, and may impair cell migration and regeneration [1, 72]. Lysyl oxidase is expressed more in adult wounds, and is the enzyme important to enable cross-linking of collagen fibers [68].

1.3.5 Matrix Metalloproteinases

MMPs are responsible for remodeling the ECM during wound healing. They work antagonistically to the TIMPs. In scarless fetal wounds the ratio of MMP:TIMP is higher than in adult wounds, which favors an environment of matrix remodeling, as opposed to accumulation of collagen [73].

1.4 Mechanotransduction

Micromechanical forces present during wound healing modulate fibroblast activity, production, aggregation, and orientation collagen [74]. Tension predisposes to the formation of scars, and reducing wound tension produces less scarring [75]. Excessively taut collagen bundles are thought to contribute to the development of hypertrophic scars [76, 77]. Mechanical stress stimulates expression of TGFβ1,which causes fibroblasts to differentiate into myofibroblasts in the proliferative phases of wound healing [78]. Myofibroblasts are contractile and responsible for wound closure in adults; their activity is correlated with contraction and degree of scarring [79]. Myofibroblast contraction causes the ECM to undergo conformational change and leads to scar formation by aligning the collagen fiber architecture [80]. Histologically, collagen fibers align along the vector of myofibroblast contraction [81, 82]. Fetal fibroblasts, even in the presence of prostaglandin E2, are less contractile than adult fibroblasts, and myofibroblasts are almost absent in fetal wounds. Fetal fibroblasts do differentiate into myofibroblasts in response to TGFβ1, but this response happens more readily in postnatal cells [83]. Fetal wounds are thought to instead close via contraction of actin casts in a purse-string fashion [84–86]. Since tension itself increases fibroblast differentiation and myofibroblast activity, fetal wounds are thought to experience less tension than adult wounds. The laxity of fetal skin may relate to the organization and type of collagen fibers. The organization of collagen fibers creates vectors of mechanical tension of the skin which are thought to correlate with Langer's lines [87–89]. The focal adhesion kinase (FAK) pathway has been identified as the pathway linking mechanotransduction with fibrosis. Disruption of this pathway, using inhibitory components or knockout models in mice, can attenuate formation of scars [90]. This pathway is linked to human fibrotic disorders [91].

1.5 Extracellular Matrix Cells

Fibroblasts are the cells that play an integral role in wound healing and remodeling of the ECM. They are the principal cells responsible for synthesizing collagen. Fetal and adult fibroblasts exhibit numerous differences. Fetal fibroblasts are able to simultaneously proliferate and synthesize ECM, whereas adult fibroblasts must proliferate before collagen can be synthesized [92]. Fetal fibroblasts show enhanced migratory capabilities [45] and proliferate faster [93]. Low oxygen conditions stimulate the proliferation of fibroblasts [94], and the hypoxic environment may facilitate fetal fibroblast function. Fetal fibroblasts produce more total collagen and higher proportions of type III and IV collagen than adult fibroblasts. Consistent with this, higher amounts of prolyl hydroxylase, the rate-limiting enzyme in collagen synthesis, are found at earlier points in gestation in fetal wounds [35]. The discoid domain receptors (DDRs) are tyrosine kinase receptors, with an extracellular "discoidin" domain, found on the surface of fibroblasts which bind collagen fibers. The DDRs regulate cell proliferation, differentiation, and wound healing. DDR-1 receptors can lead to production of collagen akin to regeneration as opposed to scar formation. DDR-1 is activated by collagen types I, IV, and V, but DDR-2 is mainly activated by collagen type I. Prolonged activation of DDR-2 has been associated with increased MMP activity [95]. Fetal fibroblasts have DDR-1 early in gestation, but as gestational age increases DDR-2 expression is steadily increased [96]. HA receptors are also more plentiful in fetal, compared to adult, fibroblasts [60], which facilitate fibroblast migration in wounds and thereby accelerate healing time. TGFβ1 inhibits HA synthesis and thereby mitigates fibroblast migration [97].

Recently, advances in lineage tracing have given rise to a better understanding of fibroblast heterogeneity. Distinct fibroblast lineages have been shown to give rise to the upper dermis and lower dermis. In wounded skin of adult mice, dermal repair is mediated by lower dermal fibroblasts. Fibroblasts of the upper dermis are only recruited during re-epithelialization [98]. Another study showed that a subpopulation of dermal fibroblasts, derived from engrailed-1 (En1)-expressing progenitors, are responsible for depositing connective tissue late in embryonic development and during cutaneous wound healing [99]. Flow cytometry revealed that dipeptidyl peptidase-4 (DDP4) was a surface marker for this lineage. Disrupting DPP4 enzymatic activity with diprotin A reduced scarring. Additionally, a set of perivascular cells, found in muscle and dermis, have been identified that are activated in acute injury, via expression of a disintegrin and metalloproteinase 12 (ADAM12). When these cells were disrupted with ablation or knockout techniques, scarring and fibrosis were decreased [100].

1.6 Tissue Engineering and Regenerative Medicine

A greater understanding of the molecular mechanisms underlying scarless regeneration of fetal wounds has led to a number of potential therapeutic applications to improve scarring in adult wounds, ranging from stem cells and molecules to scaffolds and devices to reduce tension.

1.6.1 Stem Cells

Cell-based therapies have typically involved the transplantation of stem cells, such as epidermal stem cells or mesenchymal stromal cells (MSCs), into a wound bed in the hope that the cells promote regeneration [101]. Xenografted human bone marrow-derived mesenchymal stromal cells (BM-MSCs) are able to survive long term within fetal sheep wounds and can develop into multiple cell lineages including chondrocytes, epithelial cells, skeletal muscle, and cardiac muscle [102]. Implanted amniotic fluid stem cells in lamb fetal wounds were able to hone the wounds and differentiate to form cells that were indistinguishable from the surrounding chondrocytes [103]. Stem cells typically have poor survival in the hypoxic cytokine-rich environment of adult cutaneous wounds. Stem cells taken from bone marrow, umbilical cord blood, adipose tissue, and hair follicles have all been used to improve healing of human adult skin wounds. Stem cell-based therapies are in the preclinical stages [104, 105].

1.6.2 Molecules

Another tissue engineering technique has been to topically apply specific growth factors and cytokines, or molecular antagonists, to help create an environment akin to the scarless healing of fetal wounds. Fibrillosis has been successfully blocked in an in vitro model using antibodies to collagen telopeptides (anti-alpha2Ct) [106], receptors which interact with triple helices of neighboring collagen molecules to form cross-links. HA-treated cutaneous wounds in mice in vivo resulted in a more organized connective tissue matrix on histological examination, characteristic of the scarless healing pattern, compared to wounds not treated with HA [107].

A large focus has been on attempting to recreate the fetal ratio of TGFβ isoforms thought optimal for scarless wound healing, with increased TGFβ3 and decreased TGFβ1 and TGFβ2 [108]. Natural products have been investigated to manipulate TGFβ. Chen et al. [109] used astragaloside IV, which antagonizes TGFβ1 and regulates the collagen type I/III ratio in mice, and found that it reduced scarring. Other natural products including crocodile oil [110] and curcumin [111] suppress TGFβ1 and reduce scarring in an in vivo model. Bioactive protein enzymes and growth factors have also been used. Choi et al. [112] used an antisense RNA to decrease the expression of TGFβ1 in dermal wounds of mice and found that this resulted in the formation of less fibrotic scar tissue. Antibodies which neutralize TGFβ1 and TGFβ2, alone or simultaneously, improve scarring in adult rodent wounds [113–115]. Mannose-6-phosphate (marketed as Juvidex® by Renovo) competitively inhibits activation of TGFβ1 and TGFβ2 and has been reported to improve healing in knockout mice [17]. Addition of TGFβ3 to adult rodent wounds can decrease scarring [114]. Attempts have been made to create synthetic inhibitors of TGFβ1, including celecoxib, chitosan, and TGFβ1 antibodies. Currently, however, there are no approved synthetic TGFβ1 antagonists or other licensed therapeutics effective in ameliorating scarring that have withstood testing clinical trials beyond phase I or II [116]. The danger of pharmaceutically manipulating cytokines and growth factors involved in scar formation pertains to their involvement in numerous additional non-fibrosis-related biological functions. The physical properties of skin change with aging, specifically the stress of skin decreases with increasing age, occurring more in women than in men [117]. Aschroft et al. [118] observed that the skin of elderly human females is of reduced quality and has enhanced healing in terms of scarring. Histologically, these age-related differences were associated with reduced levels of TGFβ1 in the skin of elder women and hormone replacement therapy reversed the macroscopic and microscopic changes. The selective estrogen receptor modulator (SERM) tamoxifen is thought to have antagonistic action in the skin [119], and has been shown to inhibit the proliferation and contraction of fibroblasts [120]. Systemically manipulating estrogen levels, however, is unfeasible for obvious and undesirable fertility consequences, but there is potential for developing SERMs that have tropism for the dermis and epidermis.

Using a gel which blocks connexin 43 (Cx43), a gap junction channel mediating the cell signaling involved in inflammation, reduces inflammation and blocks the activation of leukocytes and subsequent scarring [121]. Interestingly, in the mouse buccal mucosa, an area which experiences privileged wound healing, Cx43s are rapidly downregulated within 6 h from mucosal injury and this downregulation is thought to contribute to rapid regenerative healing [122].

1.6.3 Scaffolds

Scaffolds are three-dimensional structures that can be implanted into a wound, and can be designed to act like the supportive extracellular environment, integral to scarless wound healing. The scaffolds can be made out of naturally occurring substances, including collagen, HA, fibrin, and chitosan. New bioprinting and electrospinning techniques have opened up the opportunity of creating degradable synthetic polymer scaffolds with physical properties that can be precisely manipulated [123]. Scaffolds can be impregnated with growth factors, cytokines, and stem cells to promote scarless healing [124].

1.6.4 Reducing Tension

Wounds under greater tension scar more. Strategies to manipulate biomechanical signaling have the potential to reduce tension on fibroblasts, decreasing their production of collagen and differentiation into myofibroblasts, and thus reduce scarring. Botulinum toxin A, an injectable neurotoxin that paralyzes muscles and may reduce the mechanical tension on fibroblasts, was able to improve cosmesis of facial wounds [125], but made no difference to the development of keloid scars [126]. Materials have been applied topically to relieve tension on healing wounds. Paper tapes applied across scars reduced scarring in a randomized controlled trial [127] and a large blinded study [128]. Silicone sheets may confer a more scarless fetal repair phenotype [129, 130]

and were thought to able to prevent keloid and hypertrophic scars [131, 132], but a recent Cochrane review concluded that the evidence on silicone sheets is of poor quality [133]. The Embrace device is a dressing made of silicone sheet-based polymer, designed to apply compressive forces to incisional wounds and off-load mechanical tension. Two randomized controlled clinical trials have shown that the Embrace device is able to significantly reduce the formation of scars [134, 135].

1.7 Conclusions

Significant advances have been made towards characterizing the differences in scarless fetal and scarring adult wound healing phenotypes. Scarless fetal wounds heal with relatively little inflammation. There are fewer inflammatory cells and less pro-inflammatory cytokines and growth factors released. The ECM in fetal wounds is optimized to promote cell adhesion, proliferation, and differentiation. Despite the increased understanding, the exact mechanisms underlying scarless wound healing remain to be elucidated fully. There are a number of therapeutic options but treatment of scarless healing is still in the preclinical phases. A greater understanding of the precise molecular processes responsible for scar formation may improve future clinical treatment of scars. Continued research promises the induction of the fetal scarless phenotype in adult wounds.

References

1. Beanes SR, Hu FY, Soo C, Dang CM, Urata M, Ting K, Atkinson JB, Benhaim P, Hedrick MH, Lorenz HP. Confocal microscopic analysis of scarless repair in the fetal rat: defining the transition. Plast Reconstr Surg. 2002;109(1):160–70.
2. Singer AJ, Clark RA. Cutaneous wound healing. N Engl J Med. 1999;341(10):738–46.
3. Gurtner GC, Werner S, Barrandon Y, Longaker MT. Wound repair and regeneration. Nature. 2008;453(7193):314–21.
4. Rowlatt U. Intrauterine wound healing in a 20 week human fetus. Virchows Arch. 1979;381(3):353–61.

5. Lorenz HP, Longaker MT, Perkocha LA, Jennings RW, Harrison MR, Adzick NS. Scarless wound repair: a human fetal skin model. Development. 1992;114(1):253–9.
6. Adzick NS, Longaker MT. Animal models for the study of fetal tissue repair. J Surg Res. 1991;51(3):216–22.
7. Adzick NS, Longaker MT. Scarless fetal healing. Therapeutic implications. Ann Surg. 1992;215(1):3.
8. Longaker MT, Whitby DJ, Ferguson MW, Lorenz HP, Harrison MR, Adzick NS. Adult skin wounds in the fetal environment heal with scar formation. Ann Surg. 1994;219(1):65–72.
9. Cass DL, Bullard KM, Sylvester KG, Yang EY, Longaker MT, Adzick NS. Wound size and gestational age modulate scar formation in fetal wound repair. J Pediatr Surg. 1997;32(3):411–5.
10. Colwell AS, Krummel TM, Longaker MT, Lorenz HP. An in vivo mouse excisional wound model of scarless healing. Plast Reconstr Surg. 2006;117(7):2292–6.
11. Ihara S, Motobayashi Y, Nagao E, Kistler A. Ontogenetic transition of wound healing pattern in rat skin occurring at the fetal stage. Development. 1990;110(3):671–80.
12. Wong JW, Gallant-Behm C, Wiebe C, Mak K, Hart DA, Larjava H, Häkkinen L. Wound healing in oral mucosa results in reduced scar formation as compared with skin: evidence from the red Duroc pig model and humans. Wound Repair Regen. 2009;17(5):717–29.
13. Szpaderska AM, Zuckerman JD, DiPietro LA. Differential injury responses in oral mucosal and cutaneous wounds. J Dent Res. 2003;82(8):621–6.
14. Sciubba JJ, Waterhouse JP, Meyer J. A fine structural comparison of the healing of incisional wounds of mucosa and skin. J Oral Pathol Med. 1978;7(4):214–27.
15. Häkkinen L, Uitto VJ, Larjava H. Cell biology of gingival wound healing. Periodontology. 2000;24:127.
16. Szpaderska AM, Walsh CG, Steinberg MJ, DiPietro LA. Distinct patterns of angiogenesis in oral and skin wounds. J Dent Res. 2005;84(4):309–14.
17. Ferguson MW, O'Kane S. Scar–free healing: from embryonic mechanisms to adult therapeutic intervention. Philos Trans R Soc Lond Ser B Biol Sci. 2004;359(1445):839–50.
18. Arnardottir HH, Freysdottir J, Hardardottir I. Two circulating neutrophil populations in acute inflammation in mice. Inflamm Res. 2012;61(9):931–9.
19. Olutoye OO, Alaish SM, Carr ME Jr, Paik M, Yager DR, Cohen IK, Diegelmann RF. Aggregatory characteristics and expression of the collagen adhesion receptor in fetal porcine platelets. J Pediatr Surg. 1995;30(12):1649–53.
20. Olutoye OO, Barone EJ, Yager DR, Cohen IK, Diegelmann RF. Collagen induces cytokine release by fetal platelets: implications in scarless healing. J Pediatr Surg. 1997;32(6):827–30.
21. Naik-Mathuria B, Gay AN, Zhu X, Yu L, Cass DL, Olutoye OO. Age-dependent recruitment of neutrophils by fetal endothelial cells: implications in scarless wound healing. J Pediatr Surg. 2007;42(1):166–71.

22. Olutoye OO, Zhu X, Cass DL, Smith CW. Neutrophil recruitment by fetal porcine endothelial cells: implications in scarless fetal wound healing. Pediatr Res. 2005;58(6):1290–4.
23. Satish L, Kathju S. Cellular and molecular characteristics of scarless versus fibrotic wound healing. Dermatol Res Pract. 2010;2010:790234.
24. Witte MB, Barbul A. Role of nitric oxide in wound repair. Am J Surg. 2002;183(4):406–12.
25. Robson MC, Barnett RA, Leitch IO, Hayward PG. Prevention and treatment of postburn scars and contracture. World J Surg. 1992;16(1):87–96.
26. Hopkinson-Woolley J, Hughes D, Gordon S, Martin P. Macrophage recruitment during limb development and wound healing in the embryonic and foetal mouse. J Cell Sci. 1994;107(5):1159–67.
27. Soo C, Hu FY, Zhang X, Wang Y, Beanes SR, Lorenz HP, Hedrick MH, Mackool RJ, Plaas A, Kim SJ, Longaker MT, Freymiller E, Ting K. Differential expression of fibromodulin, a transforming growth factor-β modulator, in fetal skin development and scarless repair. Am J Pathol. 2000;157(2):423–33.
28. Woolf N. Pathology: basic and systemic. London: WB Saunders; 1998.
29. Garbuzenko E, Nagler A, Pickholtz D, Gillery P, Reich R, Maquart FX, Levi-Schaffer F. Human mast cells stimulate fibroblast proliferation, collagen synthesis and lattice contraction: a direct role for mast cells in skin fibrosis. Clin Exp Allergy. 2002;32(2):237–46.
30. Abe M, Kurosawa M, Ishikawa O, Miyachi Y. Effect of mast cell-derived mediators and mast cell-related neutral proteases on human dermal fibroblast proliferation and type I collagen production. J Allergy Clin Immunol. 2000;106(1):S78–84.
31. Nauta AC, Grova M, Montoro DT, Zimmermann A, Tsai M, Gurtner GC, Galli SJ, Longaker MT. Evidence that mast cells are not required for healing of splinted cutaneous excisional wounds in mice. PLoS One. 2013;8(3):e59167.
32. Wulff BC, Parent AE, Meleski MA, DiPietro LA, Schrementi ME, Wilgus TA. Mast cells contribute to scar formation during fetal wound healing. J Invest Dermatol. 2012;132(2):458–65.
33. Roberts AB, Sporn MB, Assoian RK, Smith JM, Roche NS, Wakefield LM, Heine UI, Liotta LA, Falanga V, Kehrl JH, et al. Transforming growth factor type beta: rapid induction of fibrosis and angiogenesis in vivo and stimulation of collagen formation in vitro. Proc Natl Acad Sci. 1986;83(12):4167–71.
34. Nall AV, Brownlee RE, Colvin CP, Schultz G, Fein D, Cassisi NJ, Nguyen T, Kalra A. Transforming growth factor β1 improves wound healing and random flap survival in normal and irradiated rats. Arch Otolaryngol Head Neck Surg. 1996;122(2):171–7.
35. Bullard KM, Longaker MT, Lorenz HP. Fetal wound healing: current biology. World J Surg. 2003;27(1):54–61.
36. Soo C, Beanes SR, Hu FY, Zhang X, Dang C, Chang G, Wang Y, Nishimura I, Freymiller E, Longaker MT, Lorenz HP, Ting K. Ontogenetic transition in

fetal wound transforming growth factor-β regulation correlates with collagen organization. Am J Pathol. 2003;163(6):2459–76.

37. Caniggia I, Mostachfi H, Winter J, Gassmann M, Lye SJ, Kuliszewski M, Post M. Hypoxia-inducible factor-1 mediates the biological effects of oxygen on human trophoblast differentiation through TGFβ3. J Clin Invest. 2000;105(5):577–87.

38. Barrientos S, Stojadinovic O, Golinko MS, Brem H, Tomic-Canic M. Growth factors and cytokines in wound healing. Wound Repair Regen. 2008;16(5):585–601.

39. Eslami A, Gallant-Behm CL, Hart DA, Wiebe C, Honardoust D, Gardner H, Häkkinen L, Larjava HS. Expression of integrin αvβ6 and TGF-β in scarless vs. scar-forming wound healing. J Histochem Cytochem. 2009;57(6):543–57.

40. Hsu M, Peled ZM, Chin GS, Liu W, Longaker MT. Ontogeny of expression of transforming growth factor-beta 1 (TGF-beta 1), TGF-beta 3, and TGF-beta receptors I and II in fetal rat fibroblasts and skin. Plast Reconstr Surg. 2001;107(7):1787–94.

41. Sporn MB, Roberts AB. Transforming growth factor-beta: recent progress and new challenges. J Cell Biol. 1992;119(5):1017–21.

42. Krummel TM, Michna BA, Thomas BL, Sporn MB, Nelson JM, Salzberg AM, Cohen IK, Diegelmann RF. Transforming growth factor beta (TGF-β) induces fibrosis in a fetal wound model. J Pediatr Surg. 1988;23(7):647–52.

43. Shah M, Rorison P, Ferguson MW. The role of transforming growth factors beta in cutaneous scarring. In: Garg HG, Longaker MT, editors. Scarless wound healing. New York, NY: Marcel Dekker; 2000. p. 213–26.

44. Rolfe KJ, Irvine LM, Grobbelaar AO, Linge C. Differential gene expression in response to transforming growth factor-β1 by fetal and postnatal dermal fibroblasts. Wound Repair Regen. 2007;15(6):897–906.

45. Scheid A, Wenger RH, Schäffer L, Camenisch I, Distler O, Ferenc A, Cristina H, Ryan HE, Johnson RS, Wagner KF, Stauffer UG, Bauer C, Gassmann M, Meuli M. Physiologically low oxygen concentrations in fetal skin regulate hypoxia-inducible factor 1 and transforming growth factor-β3. FASEB J. 2002;16(3):411–3.

46. Liechty KW, Adzick NS, Crombleholme TM. Diminished interleukin 6 (IL-6) production during scarless human fetal wound repair. Cytokine. 2000;12(6):671–6.

47. Peranteau WH, Zhang L, Muvarak N, Badillo AT, Radu A, Zoltick PW, Liechty KW. IL-10 overexpression decreases inflammatory mediators and promotes regenerative healing in an adult model of scar formation. J Invest Dermatol. 2008;128(7):1852–60.

48. Liechty KW, Kim HB, Adzick NS, Crombleholme TM. Fetal wound repair results in scar formation in interleukin-10–deficient mice in a syngeneic murine model of scarless fetal wound repair. J Pediatr Surg. 2000;35(6):866–73.

49. Colwell AS, Beanes SR, Soo C, Dang C, Ting K, Longaker MT, Atkinson JB, Lorenz HP. Increased angiogenesis and expression of vascular endothelial growth factor during scarless repair. Plast Reconstr Surg. 2005;115(1):204–12.

50. Amadeu T, Braune A, Mandarim-de-Lacerda C, Porto LC, Desmoulière A, Costa A. Vascularization pattern in hypertrophic scars and keloids: a stereological analysis. Pathol Res Pract. 2003;199(7):469–73.

51. Mak K, Manji A, Gallant-Behm C, Wiebe C, Hart DA, Larjava H, Häkkinen L. Scarless healing of oral mucosa is characterized by faster resolution of inflammation and control of myofibroblast action compared to skin wounds in the red Duroc pig model. J Dermatol Sci. 2009;56(3):168–80.

52. Mogili NS, Krishnaswamy VR, Jayaraman M, Rajaram R, Venkatraman A, Korrapati PS. Altered angiogenic balance in keloids: a key to therapeutic intervention. Transl Res. 2012;159(3):182–9.

53. van der Veer WM, Niessen FB, Ferreira JA, Zwiers PJ, de Jong EH, Middelkoop E, Molema G. Time course of the angiogenic response during normotrophic and hypertrophic scar formation in humans. Wound Repair Regen. 2011;19(3):292–301.

54. Wilgus TA. Immune cells in the healing skin wound: influential players at each stage of repair. Pharmacol Res. 2008;58(2):112–6.

55. Wilgus TA, Ferreira AM, Oberyszyn TM, Bergdall VK, Dipietro LA. Regulation of scar formation by vascular endothelial growth factor. Lab Investig. 2008;88(6):579–90.

56. Mast BA, Diegelmann RF, Krummel TM, Cohen IK. Hyaluronic acid modulates proliferation, collagen and protein synthesis of cultured fetal fibroblasts. Matrix. 1993;13(6):441–6.

57. Clark R. Wound repair. In: Clark R, editor. The molecular and cellular biology of wound repair. New York: Springer; 1988. p. 3–50.

58. Longaker MT, Chiu ES, Adzick NS, Stern M, Harrison MR, Stern R. Studies in fetal wound healing. V. A prolonged presence of hyaluronic acid characterizes fetal wound fluid. Ann Surg. 1991;213(4):292.

59. Hu MS, Maan ZN, Wu JC, Rennert RC, Hong WX, Lai TS, Cheung AT, Walmsley GG, Chung MT, McArdle A, Longaker MT, Lorenz HP. Tissue engineering and regenerative repair in wound healing. Ann Biomed Eng. 2014;42(7):1494–507.

60. Alaish SM, Yager D, Diegelmann RF, Cohen IK. Biology of fetal wound healing: hyaluronate receptor expression in fetal fibroblasts. J Pediatr Surg. 1994;29(8):1040–3.

61. Longaker MT, Adzick NS, Hall JL, Stair SE, Crombleholme TM, Duncan BW, Bradley SM, Harrison MR, Stern R. Studies in fetal wound healing VI. Second and early third trimester fetal wounds demonstrate rapid collagen deposition without scar formation. J Pediatr Surg. 1990;25(1):63–9.

62. Kennedy CI, Diegelmann RF, Haynes JH, Yager DR. Proinflammatory cytokines differentially regulate

hyaluronan synthase isoforms in fetal and adult fibroblasts. J Pediatr Surg. 2000;35(6):874–9.
63. Whitby D, Ferguson M. The extracellular matrix of lip wounds in fetal, neonatal and adult mice. Development. 1991;112(2):651–68.
64. Coolen NA, Schouten KC, Middelkoop E, Ulrich MM. Comparison between human fetal and adult skin. Arch Dermatol Res. 2010;302(1):47–55.
65. Whitby DJ, Longaker MT, Harrison MR, Adzick NS, Ferguson MW. Rapid epithelialisation of fetal wounds is associated with the early deposition of tenascin. J Cell Sci. 1991;99(3):583–6.
66. Longaker MT, Whitby DJ, Jennings RW, Duncan BW, Ferguson MW, Harrison MR, Adzick NS. Fetal diaphragmatic wounds heal with scar formation. J Surg Res. 1991;50(4):375–85.
67. Cass DL, Bullard KM, Sylvester KG, Yang EY, Sheppard D, Herlyn M, Adzick NS. Epidermal integrin expression is upregulated rapidly in human fetal wound repair. J Pediatr Surg. 1998;33(2):312–6.
68. Beanes SR, Dang C, Soo C, Wang Y, Urata M, Ting K, Fonkalsrud EW, Benhaim P, Hedrick MH, Atkinson JB, Lorenz HP. Down-regulation of decorin, a transforming growth factor–beta modulator, is associated with scarless fetal wound healing. J Pediatr Surg. 2001;36(11):1666–71.
69. Burd DAR, Longaker MT, Adzick NS, Harrison MR, Ehrlich HP. Foetal wound healing in a large animal model: the deposition of collagen is confirmed. Br J Plast Surg. 1990;43(5):571–7.
70. Merkel JR, DiPaolo BR, Hallock GG, Rice DC. Type I and type III collagen content of healing wounds in fetal and adult rats. Proc Soc Exp Biol Med. 1988;187(4):493–7.
71. Carter R, Jain K, Sykes V, Lanning D. Differential expression of procollagen genes between mid- and late-gestational fetal fibroblasts. J Surg Res. 2009;156(1):90–4.
72. Lovvorn HN 3rd, Cheung DT, Nimni ME, Perelman N, Estes JM, Adzick NS. Relative distribution and cross-linking of collagen distinguish fetal from adult sheep wound repair. J Pediatr Surg. 1999;34(1):218–23.
73. Dang CM, Beanes SR, Lee H, Zhang X, Soo C, Ting K. Scarless fetal wounds are associated with an increased matrix metalloproteinase-to-tissue-derived inhibitor of metalloproteinase ratio. Plast Reconstr Surg. 2003;111(7):2273–85.
74. Huang C, Leavitt T, Bayer LR, Orgill DP. Effect of negative pressure wound therapy on wound healing. Curr Probl Surg. 2014;51(7):301–31.
75. Wong VW, Beasley B, Zepeda J, Dauskardt RH, Yock PG, Longaker MT, Gurtner GC. A mechanomodulatory device to minimize incisional scar formation. Adv Wound Care (New Rochelle). 2013;2(4):185–94.
76. Junker JP, Kratz C, Tollbäck A, Kratz G. Mechanical tension stimulates the transdifferentiation of fibroblasts into myofibroblasts in human burn scars. Burns. 2008;34(7):942–6.
77. Baur P, Larson D, Stacey T. The observation of myofibroblasts in hypertrophic scars. Surg Gynecol Obstet. 1975;141(1):22–6.

78. Desmoulière A, Geinoz A, Gabbiani F, Gabbiani G. Transforming growth factor-beta 1 induces alpha-smooth muscle actin expression in granulation tissue myofibroblasts and in quiescent and growing cultured fibroblasts. J Cell Biol. 1993;122(1):103–11.
79. Estes JM, Vande Berg JS, Adzick NS, MacGillivray TE, Desmoulière A, Gabbiani G. Phenotypic and functional features of myofibroblasts in sheep fetal wounds. Differentiation. 1994;56(3):173–81.
80. Wipff PJ, Rifkin DB, Meister JJ, Hinz B. Myofibroblast contraction activates latent TGF-β1 from the extracellular matrix. J Cell Biol. 2007;179(6):1311–23.
81. Ferdman AG, Yannas LV. Scattering of light from histologic sections: a new method for the analysis of connective tissue. J Invest Dermatol. 1993;100(5):710–6.
82. Yannas IV. Similarities and differences between induced organ regeneration in adults and early foetal regeneration. J R Soc Interface. 2005;2(5):403–17.
83. Rolfe KJ, Richardson J, Vigor C, Irvine LM, Grobbelaar AO, Linge C. A role for TGF-β1-induced cellular responses during wound healing of the non-scarring early human fetus? J Invest Dermatol. 2007;127(11):2656–67.
84. Martin P, Lewis J. Actin cables and epidermal movement in embryonic wound healing. Nature. 1992;360(6400):179–83.
85. Brock J, Midwinter K, Lewis J, Martin P. Healing of incisional wounds in the embryonic chick wing bud: characterization of the actin purse-string and demonstration of a requirement for Rho activation. J Cell Biol. 1996;135(4):1097–107.
86. Yates CC, Hebda P, Wells A. Skin wound healing and scarring: fetal wounds and regenerative restitution. Birth Defects Res Part C Embryo Today. 2012;96(4):325–33.
87. Ridge M, Wright V. The directional effects of skin: a bio-engineering study of skin with particular reference to Langer's lines. J Invest Dermatol. 1966;46(4):341–6.
88. Cox H. The cleavage lines of the skin. Br J Surg. 1941;29(114):234–40.
89. Piérard GE, Lapière CM. Microanatomy of the dermis in relation to relaxed skin tension lines and Langer's lines. Am J Dermatopathol. 1987;9(3):219–24.
90. Wong VW, Rustad KC, Akaishi S, Sorkin M, Glotzbach JP, Januszyk M, Nelson ER, Levi K, Paterno J, Vial IN, Kuang IA, Longaker MT, Gurtner GC. Focal adhesion kinase links mechanical force to skin fibrosis via inflammatory signaling. Nat Med. 2012;18(1):148–52.
91. Wynn T. Cellular and molecular mechanisms of fibrosis. J Pathol. 2008;214(2):199–210.
92. Larson BJ, Longaker MT, Lorenz HP. Scarless fetal wound healing: a basic science review. Plast Reconstr Surg. 2010;126(4):1172–80.
93. Nodder S, Martin P. Wound healing in embryos: a review. Anat Embryol. 1997;195(3):215–28.
94. Falanga V, Kirsner RS. Low oxygen stimulates proliferation of fibroblasts seeded as single cells. J Cell Physiol. 1993;154(3):506–10.

95. Vogel W, Gish GD, Alves F, Pawson T. The discoidin domain receptor tyrosine kinases are activated by collagen. Mol Cell. 1997;1(1):13–23.

96. Chin GS, Lee S, Hsu M, Liu W, Kim WJ, Levinson H, Longaker MT. Discoidin domain receptors and their ligand, collagen, are temporally regulated in fetal rat fibroblasts in vitro. Plast Reconstr Surg. 2001;107(3):769–76.

97. Ellis IR, Schor SL. Differential effects of TGF-β1 on hyaluronan synthesis by fetal and adult skin fibroblasts: implications for cell migration and wound healing. Exp Cell Res. 1996;228(2):326–33.

98. Driskell RR, Lichtenberger BM, Hoste E, Kretzschmar K, Simons BD, Charalambous M, Ferron SR, Herault Y, Pavlovic G, Ferguson-Smith AC, Watt FM. Distinct fibroblast lineages determine dermal architecture in skin development and repair. Nature. 2013;504(7479):277–81.

99. Rinkevich Y, Walmsley GG, Hu MS, Maan ZN, Newman AM, Drukker M, Januszyk M, Krampitz GW, Gurtner GC, Lorenz HP, Weissman IL, Longaker MT. Skin fibrosis: identification and isolation of a dermal lineage with intrinsic fibrogenic potential. Science. 2015;348(6232):aaa2151.

100. Dulauroy S, Di Carlo SE, Langa F, Eberl G, Peduto L. Lineage tracing and genetic ablation of ADAM12(+) perivascular cells identify a major source of profibrotic cells during acute tissue injury. Nat Med. 2012;18(8):1262–70.

101. Degen KE, Gourdie RG. Embryonic wound healing: a primer for engineering novel therapies for tissue repair. Birth Defects Res C Embryo Today. 2012;96(3):258–70.

102. Mackenzie TC, Flake AW. Human mesenchymal stem cells persist, demonstrate site-specific multipotential differentiation, and are present in sites of wound healing and tissue regeneration after transplantation into fetal sheep. Blood Cells Mol Dis. 2001;27(3):601–4.

103. Klein JD, Turner CG, Steigman SA, Ahmed A, Zurakowski D, Eriksson E, Fauza DO. Amniotic mesenchymal stem cells enhance normal fetal wound healing. Stem Cells Dev. 2010;20(6):969–76.

104. Wu SC, Marston W, Armstrong DG. Wound care: the role of advanced wound healing technologies. J Vasc Surg. 2010;52(3 Suppl):59S–66S.

105. Gauglitz GG, Jeschke MG. Combined gene and stem cell therapy for cutaneous wound healing. Mol Pharm. 2011;8(5):1471–9.

106. Chung HJ, Steplewski A, Chung KY, Uitto J, Fertala A. Collagen fibril formation: a new target to limit fibrosis. J Biol Chem. 2008;283(38):25879–86.

107. Iocono JA, Ehrlich HP, Keefer KA, Krummel TM. Hyaluronan induces scarless repair in mouse limb organ culture. J Pediatr Surg. 1998;33(4):564–7.

108. Frank S, Madlener M, Werner S. Transforming growth factors 1, 2, and 3 and their receptors are differentially regulated during normal and impaired wound healing. J Biol Chem. 1996;271(17):10188–93.

109. Chen X, Peng LH, Li N, Li QM, Li P, Fung KP, Leung PC, Gao JQ. The healing and anti-scar effects of astragaloside IV on the wound repair in vitro and in vivo. J Ethnopharmacol. 2012;139(3):721–7.

110. Li HL, Chen LP, Hu YH, Qin Y, Liang G, Xiong YX, Chen QX. Crocodile oil enhances cutaneous burn wound healing and reduces scar formation in rats. Acad Emerg Med. 2012;19(3):265–73.

111. Jia S, Xie P, Hong SJ, Galiano R, Singer A, Clark RA, Mustoe TA. Intravenous curcumin efficacy on healing and scar formation in rabbit ear wounds under nonischemic, ischemic, and ischemia–reperfusion conditions. Wound Repair Regen. 2014;22(6):730–9.

112. Choi BM, Kwak HJ, Jun CD, Park SD, Kim KY, Kim HR, Chung HT. Control of scarring in adult wounds using antisense transforming growth factor-β1 oligodeoxynucleotides. Immunol Cell Biol. 1996;74(2):144–50.

113. Shah M, Foreman DM, Ferguson MW. Control of scarring in adult wounds by neutralising antibody to transforming growth factor β. Lancet. 1992;339(8787):213–4.

114. Shah M, Foreman DM, Ferguson M. Neutralisation of TGF-beta 1 and TGF-beta 2 or exogenous addition of TGF-beta 3 to cutaneous rat wounds reduces scarring. J Cell Sci. 1995;108(3):985–1002.

115. Shah M, Foreman DM, Ferguson M. Neutralising antibody to TGF-beta 1, 2 reduces cutaneous scarring in adult rodents. J Cell Sci. 1994;107(5):1137–57.

116. Rhett JM, Ghatnekar GS, Palatinus JA, O'Quinn M, Yost MJ, Gourdie RG. Novel therapies for scar reduction and regenerative healing of skin wounds. Trends Biotechnol. 2008;26(4):173–80.

117. Diridollou S, Vabre V, Berson M, Vaillant L, Black D, Lagarde JM, Grégoire JM, Gall Y, Patat F. Skin ageing: changes of physical properties of human skin in vivo. Int J Cosmet Sci. 2001;23(6):353–62.

118. Ashcroft GS, Dodsworth J, van Boxtel E, Tarnuzzer RW, Horan MA, Schultz GS, Ferguson MW. Estrogen accelerates cutaneous wound healing associated with an increase in TGF-β1 levels. Nat Med. 1997;3(11):1209–15.

119. Nelson L. The role of oestrogen in skin. PhD Thesis. Bradford: School of Life Sciences, University of Bradford; 2006.

120. Hu D, Hughes MA, Cherry GW. Topical tamoxifen—a potential therapeutic regime in treating excessive dermal scarring? Br J Plast Surg. 1998;51(6):462–9.

121. Qiu C, Coutinho P, Frank S, Franke S, Law LY, Martin P, Green CR, Becker DL. Targeting connexin43 expression accelerates the rate of wound repair. Curr Biol. 2003;13(19):1697–703.

122. Davis NG, Phillips A, Becker DL. Connexin dynamics in the privileged wound healing of the buccal mucosa. Wound Repair Regen. 2013;21(4):571–8.

123. Debels H, Hamdi M, Abberton K, Morrison W. Dermal matrices and bioengineered skin substitutes: a critical review of current options. Plast Reconstr Surg Global Open. 2015;3(1):e284.
124. Hodgkinson T, Bayat A. Dermal substitute-assisted healing: enhancing stem cell therapy with novel biomaterial design. Arch Dermatol Res. 2011;303(5):301–15.
125. Ziade M, Domergue S, Batifol D, Jreige R, Sebbane M, Goudot P, Yachouh J. Use of botulinum toxin type A to improve treatment of facial wounds: a prospective randomised study. J Plast Reconstr Aesthet Surg. 2013;66(2):209–14.
126. Gauglitz G, Bureik D, Dombrowski Y, Pavicic T, Ruzicka T, Schauber J. Botulinum toxin A for the treatment of keloids. Skin Pharmacol Physiol. 2012;25(6):313–8.
127. Atkinson JA, McKenna KT, Barnett AG, McGrath DJ, Rudd M. A randomized, controlled trial to determine the efficacy of paper tape in preventing hypertrophic scar formation in surgical incisions that traverse Langer's skin tension lines. Plast Reconstr Surg. 2005;116(6):1648–56.
128. Rosengren H, Askew DA, Heal C, Buettner PG, Humphreys WO, Semmens LA. Does taping torso scars following dermatologic surgery improve scar appearance? Dermatol Pract Conceptual. 2013;3(2):75–83.
129. Lo DD, Zimmermann AS, Nauta A, Longaker MT, Lorenz HP. Scarless fetal skin wound heal-

ing update. Birth Defects Res C Embryo Today. 2012;96(3):237–47.
130. Aarabi S, Bhatt KA, Shi Y, Paterno J, Chang EI, Loh SA, Holmes JW, Longaker MT, Yee H, Gurtner GC. Mechanical load initiates hypertrophic scar formation through decreased cellular apoptosis. FASEB J. 2007;21(12):3250–61.
131. Puri N, Talwar A. The efficacy of silicone gel for the treatment of hypertrophic scars and keloids. J Cutan Aesthet Surg. 2009;2(2):104–6.
132. Fulton JE. Silicone gel sheeting for the prevention and management of evolving hypertrophic and keloid scars. Dermatol Surg. 1995;21(11):947–51.
133. O'Brien L, Jones DJ. Silicone gel sheeting for preventing and treating hypertrophic and keloid scars. Cochrane Database Syst Rev. 2013;(9):CD003826.
134. Lim AF, Weintraub J, Kaplan EN, Januszyk M, Cowley C, McLaughlin P, Beasley B, Gurtner GC, Longaker MT. The embrace device significantly decreases scarring following scar revision surgery in a randomized controlled trial. Plast Reconstr Surg. 2014;133(2):398–405.
135. Longaker MT, Rohrich RJ, Greenberg L, Furnas H, Wald R, Bansal V, Seify H, Tran A, Weston J, Korman JM, Chan R, Kaufman D, Dev VR, Mele JA, Januszyk M, Cowley C, McLaughlin P, Beasley B, Gurtner GC. A randomized controlled trial of the embrace advanced scar therapy device to reduce incisional scar formation. Plast Reconstr Surg. 2014;134(3):536–46.

Scar Treatment and Prevention: Know Thine Enemy

2

Elizabeth A. Brett and Dominik Duscher

2.1 Introduction

Adult mammalian skin damage is healed with a scar. Understanding our desired end point (healthy, unscarred skin) and the model we wish to mimic (fetal wound healing) is allowing better identification of the pitfalls of current treatments (Fig. 2.1). Decades of research have shown attempts at recreating fetal wound healing in vivo and in vitro. This chapter aims to outline scarless healing, examine new research, and outline the current options for scar treatment.

In the short term, scar tissue helps wound contraction, and will bring wound boundaries closer together. However, the long-term effects of scarring are significantly less advantageous. Cosmesis and function are two aspects affected greatly by scarring [1]. Injury-related and postoperative scars have an effect which is felt economically. In the USA alone, 2010 saw an estimated 38 million

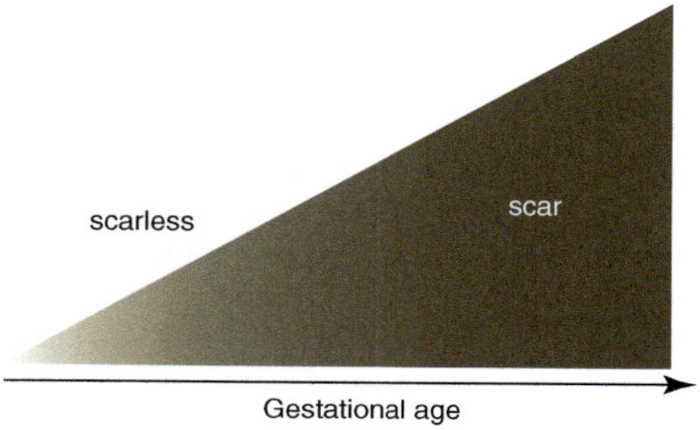

Fig. 2.1 Mammals lose their fetal scarless wound healing phenotype with a specific gestational age

patients coming for post-op wound care, funding a market which was set to reach \$15.3 bn by the same year [2]. The sharp need for functional wound healing therapy has put regenerative medicine into the light. Here, the fetal model is idealized among researchers. The mammalian fetus retains the ability to heal not by reparation, but by regeneration. Biomimetics of fetal wound healing is a complex and multifaceted area, meaning an equally complex series of experimental and research avenues.

The desire to create "scarless wounds" has complicated the required end product, not stratified fibroblasts and keratinocytes, but a complex, multifunctional layer. The overall dermis was described in a recent study using various principles of light scattering, to analyze the difference in the skin modalities [3].

E. A. Brett
Department of Plastic and Hand Surgery, Klinikum rechts der Isar, Technical University of Munich, Munich, Germany
e-mail: eliza.brett@tum.de

D. Duscher (✉)
Department for Plastic Surgery and Hand Surgery, Division of Experimental Plastic Surgery, Technical University of Munich, Munich, Germany
e-mail: Dominik.duscher@mri.tum.de

© Springer Nature Switzerland AG 2019
D. Duscher, M. A. Shiffman (eds.), *Regenerative Medicine and Plastic Surgery*,
https://doi.org/10.1007/978-3-030-19962-3_2

2.2 Wound Creation and Healing

The model of adult wound healing is one that has been well elucidated. The "adult" wound healing pertains to humans at the third trimester of development onwards, when inflammatory pathways and scarring become activated (thought to be the function of mast cell activation) [4–6]. What is of interest is the tissue that is created after the wound has closed and healed. This skin is not normal, native tissue, but fibrotic scar. Chiefly comprised of type 1 collagen, scarred dermis and overlying epidermis will not retain the strength or resilience of its healthy surrounding tissue [7]. Adding to the impaired dermal regeneration is the hindrance of cell migration into the wound, caused by the diminished upregulation of cell adhesion molecules throughout the matrix [8].

Abnormal scar development extends across two main phenotypes, keloid and hypertrophic. Hypertrophic scars will generally develop on high-tension anatomy, neck, shoulders, and knees for instance. Keloids are found to develop frequently on ears, cheeks, and upper arms, skin of low tension. Keloid scars also have the capacity to sprawl beyond the borders of the original injury, consuming healthy nearby skin [9]. The suboptimal adult dermal repair system is best seen in contrast with an ideal model. The shortcomings of adult wound healing are clear when observing fetal wound healing.

Since the 1970s, we have known that the mammalian fetus, in the first two trimesters of development, will heal its dermis without scarring [10]. There are several factors that distinguish fetal wound healing from adult wound healing: some obvious, some not so. First, we observe the sterile, thermostable, constant environment afforded to the fetus while in utero. The fetus is hypoxic, relative to the mother, having an arterial PO_2 of 20 mmHg [11]. While completely aseptic, the fetus does not have to deal with the rigors of bacterial, viral, or fungal infection as adult wounds do. This environment is a vital necessity, as the fetus will not have a developed immune system, or a functional analogue of adaptive immunity. Physical trauma faced by the fetus can vary. It is generally in the format of acute pressure, motor vehicle accidents, falls, and domestic violence, causing increased uterine pressure and possible abruption of the placenta [12].

Upon wounding, the fetal dermis produces a high level of matrix metalloproteases (MMPs) to clear away the damaged tissue, and to help make way for the regeneration of native, healthy skin. Fetal wounds also have increased GAGs, like chondroitin sulfate, and hyaluronic acid [13]. Fetal fibroblastic movement into the wound is much more rapid than in adults, owed to the fact that there is a high expression of cell adhesion molecules. The fibroblasts proceed to lay down a brand new procollagen matrix, which is highly ordered and non-excessive [14]. In scarring, there is a higher amount of collagen type 3 relative to type 1, resulting in thinner, frail fibers [15]. The role of TGF-β was explored in the mid-1990s. A highly expressed molecule in adult wounds, simple immunohistochemistry tests revealed no positive staining for TGF-β in fetal wounds. It was postulated to be bound by decorin [16]. The function of TGF-β in scarring was then characterized by an interventional study, where TGF-β1 was injected into fetal dermis and scar formation was observed [17]. Transplanted fetal skin was grafted to an adult wound, and was shown to heal without scarring. This showed that even out of context, fetal tissue will still heal by scarless regeneration [18]. Alongside neutrophil inactivation is a downregulation of mast cell production and degranulation in fetal skin. In turn, this means that vicious cytokine cascades, pro-inflammatory pathways, and damaging enzyme production will not occur. Fetal platelets do not aggregate or degranulate as much as adult platelets do [19]. Activity of immature B-cells is simulated in the fetus by the passage of maternal antibodies through the placenta, IgG, IgM, and IgA [20].

2.3 In the Laboratory

The inundation of the wound closure market with new products highlights the massive need for an ideal wound cover, and the absence of just such a therapy. Epidermal substitutes range among Epicel, Epidex, MySkin, and ReCell. Dermal substitutes are Alloderm, Dermagraft, Matriderm, and Integra. Dermoepidermal composite grafts include Orcel and Apligraf. Research has focused on the wound/graft interaction [21]. Evaluation of Integra has shown that it requires 3–4 weeks for sufficient vascularization to infiltrate the graft [22]. With this knowledge, Haifei et al. [23] have shown that paradoxically, a thinner graft of 0.5 mm will produce a more normal dermis in contrast to a thicker graft of 2 mm. We have also learned that a graft will not take if the grafted surface is not losing fluid via transudation/exudation, a situation seen in dry bone and tendon [24]. Pros and cons of each graft type dominate the literature on scar revision therapy and wound management. While a rapid fibrin clot is provided by fibrin sealant TISSEEL, it does not facilitate good neodermis formation. Dermalogen gives dermis strong elastin fibers, but has a propensity for squamous hyperplasia [25]. Meanwhile, in vitro and in vivo testing has become more elaborate, and we are embracing biomimetics of wounding and healing. The desired "single-step reconstruction" of full-thickness skin has been identified as a better alternative for closing full-thickness wounds, than the current standard of deep excision and split-thickness skin grafting. Replacement of nourished, healthy subcutaneous tissue can now be likened to the foundations of a building. Use of autologous fat in deep wound beds as a base for Matriderm sheets was described, and suggested an extremely viable way of ensuring graft adhesion by use of the patient's own preadipocytes [26].

Biomimetic wounds are becoming quick, cheap, and reproducible, by the concept of wounding the culture. Creating tissue trauma in vitro is becoming more advanced, recently seen with the addition of whole human blood to de-epidermized dermis. This was to allow thrombocyte lysates to exert their effect on keratinocyte migration. There was a near-normal dermal pattern observed under the blood scab [27]. The undulating nature of the DEJ, as described previously, has been identified as an issue in healed grafted skin. The oscillating DEJ pattern allows for stem cell migration patterns, normal keratinocyte behavior, and increased paracrine interaction between the two layers. It is for this reason that in vitro engineering of a DEJ was completed in a

model called "μDERM". Keratinocytes cultured above the junction showed a normal, native epidermal analogue, which shows the affinity of keratinocytes for an undulating topography [28]. This opens up a potential opportunity for composite grafts to be made, with a biomimetic dermoepidermal junction already formed.

New polymeric based treatments are currently being developed in the laboratory, and in clinical trial models. A study looked at burn injuries, and the potential uses for synthetic polymers. Cross-linked PEGDA and dextran in a hydrogel format have been shown to reconstitute the dermis, and its appendages, from a full-thickness burn. Epithelialization showed re-establishment of an undulating dermoepidermal junction, which means that the resultant healed skin does not resemble scar [29]. Clinical trials are also underway for hydrogels which ameliorate scars. An investigation carried out by Oculus encompassed a double-blind, multicenter randomized trial using their newly developed "Microcyn Scar Management Hydrogel." They have received FDA clearance as of the end of 2013, and aim to make the hydrogel available in the first half of 2014 [30, 31].

2.4 In the Clinic

2.4.1 Grading Scars

Clinically grading scars means that the severity can be documented, impact of therapy can be evaluated, and treatment can be augmented if necessary. Scars are assessed along the following parameters: pliability, firmness, color and visual appearance, thickness, vascular perfusion, and 3-D topography. Worldwide hospitals are not in agreement with one method of grading scars; there are at minimum five different scales used in the present day to classify scarring [32] (Table 2.1).

In 2013 the research of an Australian team called to attention the need for a more comprehensive scar scale for burn victims. This scale is designed to combine total burn surface area percentage with the grading of the scar. Its development could result in tailored, improved afterburn scar care for patients [33].

Table 2.1 An evaluation of scar scales in current use, and to be developed

Scale name	Unique property/benefit
Vancouver Scar Scale	Numerical (0–13). Heavily used in literature, and currently the best predictor for burn scars
Manchester Scar Scale	Numerical (5–18). Can be applied to a wide range of scar types
Patient and Observer Scar Assessment Scale	Numerical (5–50). Takes into account the patient's view in tandem with clinician's view on severity of scar
Visual Analogue Scale with Scar Ranking	Numerical and descriptive (0–100, "excellent" to "poor"). Simple scale, easy to conduct
The Stony Brook Scar Evaluation Scale	Numerical (0–5). Developed for short-term scar assessment from healed lacerations
The Modified Vancouver Scar Scale linked with TBSA (to be developed)	Scar scale for burns which elucidates variation in scar outcome across the burn surface area

2.4.2 Prevention of Scars

In the immediate treatment of wounds, there are methods of minimizing scar formation before the dermis has healed over. In any wound, it is vital that re-epithelialization occurs quickly, to help provide a barrier for the underlying damaged dermis, and to give support to the tissue while it is repairing [9]. New techniques of surgical closure are under development, which result in smaller, more manageable scars. In that regard, the "short scar" method of suturing was characterized, and was seen as a better alternative to methods which give long vertical scars and T scars [34]. Tissue sealants have opened up the opportunity for suture-free healing. "Tissucol" (Baxter) was adopted by clinicians and was in use by the start of the 1990s [35]. "Dermabond" (Ethicon) is used to close surgical lacerations nowadays, and can give very impressive results. However, it is still suboptimal. A report by Yeilding [36] drew an association between Dermabond use and necrotizing fasciitis and in a separate case secondary contact dermatitis [37]. Dermabond is also still cost ineffective when compared to Monocryl sutures [38]. Topical negative pressure wound therapy is used to bring the boundaries of an unhealed wound together, reducing the domain of scarring on the body surface. Regarding scar tissue, its use has been described for large surgically resected hypertrophic scars [39].

2.4.3 Treatment of Scars

A large body of research focuses on how to treat the scars after they have formed. This reflects the current treatment patterns seen worldwide. Indeed, the process of eliminating intractable scars can be compared to that of tattoos. Ridding the dermis of stubborn tattoo ink and fibrotic dermis is achieved by dermabrasion, salabrasion, thermal and chemical destruction, and lasers [40]. Indeed, surgical resection can be just as impactful, providing healthy skin boundaries. This corresponds to a new, fresh wound, which can be managed against scar development from the very start [41]. Other promising ways for scar mitigation are as follows:

2.4.3.1 Vitamin E
Topical solutions which contain vitamin E are frequently used to ameliorate scar appearance. The mechanism of action is carried out by a family of tocotrienol and tocol derivatives, which are antioxidants found in the skin; they serve to reduce the abundance of reactive oxygen species in the scar bed. In turn, the natural scar reduction pathways are thought to be unhindered [42].

2.4.3.2 Mesenchymal Stem Cell
The injection of mesenchymal stem cells is a relatively new therapy; many patients who receive this treatment are parts of clinical trials. MSCs have been seen to have extraordinary effects in the scar environment, seeming to "understand" what's wrong and what has to be changed. Along with producing antifibrotic factors, they have been seen to neutralize reactive oxygen species, enhance dermal fibroblastic function, and stimulate angiogenesis and stability

of resulting vessels [43]. MSC use in scars faces barriers preventing seamless clinical translation. The problems of teratoma/malignancy formation and cost efficiency are issues which the scientific world is attempting to overcome.

2.4.3.3 Retinoids

The use of retinoids to treat scars resultant from acne has increased proportionally to the knowledge of their benefits. A study by Haider and Shaw [44] in 2005 showed an extremely significant increase in the amelioration of scars created from acne vulgaris, a condition which 90% of adolescents are expected to suffer from at some stage in their development. Even earlier than this was the idea of combining dermabrasion and retinoids. Using these two anti-scar measures in tandem with one another has long been done, and leads to faster healing of the postoperative lesion [45].

2.4.3.4 Laser-Assisted Scar Healing (LASH)

The use of lasers on scars is clinically attractive. Light-induced healing is commonly used for hypertrophic scarring. Recently, the application of 595 nm waves of light is used on scars immediately once the sutures have been removed [46]. This process has evolved to the use of 810 nm diode laser treatment on skin immediately after surgery, and has shown promising results [47]. The light serves to disrupt the dermis and its scarred architecture, without ablating the overlying epidermis. It allows remodelling of the dermis into a similar state to its healthy, peripheral dermis. This has been further developed to include the beneficial effects of steroids on the disaggregated dermis. A study done by Waibel et al. [48] in 2013 took advantage of the dermis in its plastic state, and saw an opportunity to inject steroid, retinoid, drugs, and substances like triamcinolone acetonide, to better the scar appearance. An emerging new laser treatment of keloid scars was also published in 2013, delivering CO_2 in high-energy pulses to keloids. It has been shown to recontour raised keloids; however frequency of its use is directly linked to pronounced erythema of the area. Safety and efficacy of this method have been established and are undergoing further tests [49].

2.4.3.5 Silicone Treatment

Silicone is frequently used in topical treatments for scars. This is generally in a gel form, used topically, and is self-drying. The nanostructure is long chains of silicon polymers, called polysiloxanes, cross-linked with SiO_2. Products like HealGel act via growth factor modulation, and reduction in itching and discomfort of the scar itself [50]. Gel sheets are designed to be put on intact skin; fully formed keloid and hypertrophic scars [51] as placement of silicone gel sheets on non-epithelialized skin will be ineffective. The exact mechanisms of silicone action are unknown, even since its use for pediatric hypertrophic scarring after burn injuries in 1981 [52]. Theories have been constructed and corroborated; silicone gel is hypothesized to have an occluding, hydrating effect on damaged skin. The phenomenon of transepithelial water loss (TEWL) is one which skin will suffer from after any full-thickness injury. Silicone gel is shown to restore and maintain hydration to areas of heavy scarring [53].

2.4.3.6 Negative and Positive Pressure Treatment

Pressure garments and bandages are still widely used as a method of controlling scar formation. VSS evaluation of scars caused by burns has been shown to reduce in thickness after positive pressure therapy. This corresponds with the behavior of fibroblasts in pressurized culture, having lower migratory and mitotic rates [54]. Taking this a step further, the medical innovation company "i2r Medical" published a patent in August 2013 [55], detailing a "biocompatible dressing" through which positive pressure is delivered to the scar. The application of compressive force to the formed scar is claimed to reduce the scar. This is in contrast to Smith & Nephew's "Pico," a single-use negative pressure wound therapy device. This is used on fresh wounds, as well as fully formed scars. A preliminary report in 2012 showed a reduction in developed keloid scars after 1 month of negative pressure treatment using Pico [56].

2.4.3.7 Alternative Topical Treatments

Any attempt at downregulating collagen synthesis will ultimately help in the reduction of keloid scars. Since 1990, the intradermal injection of interferon gamma was done with the purpose of decreasing collagen bulk in the scar [57]. This has been translated into a topical cream, which is used in modern day. Imiquimod (Medicis Pharmaceutical Corporation) is applied onto the surface of the scar, and increases interferon alpha production within the scar [58]. This immune modulator has recently been shown to carry some extreme risk, causing massive disfiguring scars through a mechanism yet unknown [59]. Bleomycin, an antineoplastic agent, offers an alternative method of topical scar treatment, which has been elucidated through both keloid and hypertrophic scars. In a study, the effects of bleomycin were shown to be amplified with electroporation, giving significantly positive results in a noninvasive manner [60].

2.4.3.8 Cryotherapy

The use of cryosurgery as a monotherapy for keloid and hypertrophic scars was pioneered in 1982. Its procedure involved use of a hollow needle, inside which liquid nitrogen is forced around at a pressure of approximately 5 ppi. The needle is inserted along the long axis of the scar tissue, and its tip is pushed through, protruding out of the tissue at the other end. The scar is then completely frozen. Results showed promising effects, reducing scar volume, and pigmentation between 1 and 20 sessions [61]. However, only certain morphologies of scars may be treated in this manner. Such is the efficacy of freezing on human tissues that its use is indicated in other dermal pathologies. Cryosurgery has recently been successful in a human model of Paget's disease of the breast [62].

2.5   The Ideal Graft

Research focus must turn a corner where the aim is to eliminate the need for scar revision therapy. The design of a graft using a combination of the current knowledge and ideas has high potential for improved wound healing and decreased scarring. Specifically, the graft must take to the tissue successfully, must guide the regeneration of new dermis, and must do so in a way that leaves functional skin, not scar. We can now envision a dermal substitute that we could not 5 years ago by encompassing the concepts of new research, a graft with a ready-made DEJ [28] (Clement et al.), of optimal thickness [23], potentially strengthened with a degradable polymer [29], perhaps containing gland cells from patient tissue obtained in theatre [21], all set on a foundation of patient adipocytes [26]. The battle against scarring could be won by using autologous tissue in new ways.

2.6   Conclusions

The market for wound closure devices, strips, sealants, and hemostats, which leads to decreased scar formation, is expected to rise to a global value of $4.7 bn by 2019 (via a compound annual growth rate of 5%) [63]. Energy put into closure devices must be directed at halting scar development from the moment the device is introduced into the skin. This field is highly unique in the amount of potential developments, obstacles, and key market opportunities.

References

1. Shin D, Minn KW. The effect of myofibroblast on contracture of hypertrophic scar. Plast Reconstr Surg. 2004;113:633–40.
2. Sen CK, Gordillo GM, Roy S, Kirsner R, Lambert L, Hunt TK, Gottrup F, Gurtner GC, Longaker MT. Human skin wounds: a major and snowballing threat to public health and the economy. Wound Repair Regen. 2009;17:763–71.
3. Heuke S, Vogler N, Meyer T, Akimov D, Kluschke F, Rowert-Huber HJ, Lademann J, Dietzek B, Popp J. Multimodal mapping of human skin. Br J Dermatol. 2013;169:794–803.
4. Shimizu T, Wolfe LS. Arachidonic acid cascade and signal transduction. J Neurochem. 1990;55:1–15.
5. Wulff BC, Parent AE, Meleski MA, Dipietro LA, Schrementi ME, Wilgus TA. Mast cells contribute to scar formation during fetal wound healing. J Invest Dermatol. 2012;132:458–65.

6. Velnar T, Bailey T, Smrkolj V. The wound healing process: an overview of the cellular and molecular mechanisms. J Int Med Res. 2009;37:1528–42.
7. Simpson DG. Dermal templates and the wound-healing paradigm: the promise of tissue regeneration. Expert Rev Med Devices. 2006;3:471–84.
8. Larson BJ, Longaker MT, Lorenz HP. Scarless fetal wound healing: a basic science review. Plast Reconstr Surg. 2010;126:1172–80.
9. Gauglitz GG, Korting HC, Pavicic T, Ruzicka T, Jeschke MG. Hypertrophic scarring and keloids: pathomechanisms and current and emerging treatment strategies. Mol Med. 2011;17:113–25.
10. Wilgus TA, Bergdall VK, Tober KL, Hill KJ, Mitra S, Flavahan NA, Oberyszyn TM. The impact of cyclo-oxygenase-2 mediated inflammation on scarless fetal wound healing. Am J Pathol. 2004;165:753–61.
11. Adzick NS, Longaker MT. Scarless fetal healing—therapeutic implications. Ann Surg. 1992;215:3–7.
12. Mendez-Figueroa H, Dahlke JD, Vrees RA, Rouse DJ. Trauma in pregnancy: an updated systematic review. Am J Obstet Gynecol. 2013;209:1–10.
13. Rolfe KJ, Grobbelaar AO. A review of fetal scarless healing. ISRN Dermatol. 2012;2012:698034.
14. Mast BA, Diegelmann RF, Krummel TM, Cohen IK. Scarless wound healing in the mammalian fetus. Surg Gynecol Obstet. 1992;174:441–51.
15. Dale PD, Sherratt JA, Maini PK. A mathematical model for collagen fibre formation during foetal and adult dermal wound healing. Proc Biol Sci. 1996;263:653–60.
16. Sullivan KM, Lorenz HP, Meuli M, Lin RY, Adzick NS. A model of scarless human fetal wound repair is deficient in transforming growth factor beta. J Pediatr Surg. 1995;30:198–202.
17. Lin RY, Adzick NS. The role of the fetal fibroblast and transforming growth factor-beta in a model of human fetal wound repair. Semin Pediatr Surg. 1996;5:165–74.
18. Lorenz HP, Longaker MT, Perkocha LA, Jennings RW, Harrison MR, Adzick NS. Scarless wound repair: a human fetal skin model. Development. 1992;114:253–9.
19. Schwartzfarb E, Kirsner RS. Understanding scarring: scarless fetal wound healing as a model. J Invest Dermatol. 2012;132:260.
20. Ben-Hur H, Gurevich P, Elhayany A, Avinoach I, Schneider DF, Zusman I. Transport of maternal immunoglobulins through the human placental barrier in normal pregnancy and during inflammation. Int J Mol Med. 2005;16:401–7.
21. Biedermann T, Boettcher-Haberzeth S, Reichmann E. Tissue engineering of skin for wound coverage. Eur J Pediatr Surg. 2013;23:375–82.
22. Moiemen NS, Vlachou E, Staiano JJ, Thawy Y, Frame JD. Reconstructive surgery with Integra dermal regeneration template: histologic study, clinical evaluation, and current practice. Plast Reconstr Surg. 2006;117:160s–74s.
23. Haifei S, Xingang W, Shoucheng W, Zhengwei M, Chuangang Y, Chunmao H. The effect of collagen-chitosan porous scaffold thickness on dermal regeneration in a one-stage grafting procedure. J Mech Behav Biomed Mater. 2014;29:114–25.
24. Foong DP, Evriviades D, Jeffery SL. Integra permits early durable coverage of improvised explosive device (IED) amputation stumps. J Plast Reconstr Aesthet Surg. 2013;66:1717–24.
25. Truong ATN, Kowal-Vern A, Latenser BA, Wiley DE, Walter RJ. Comparison of dermal substitutes in wound healing utilizing a nude mouse model. J Burns Wounds. 2005;4:e4.
26. Keck M, Selig HF, Kober J, Lumenta DB, Schachner H, Gugerell A, Kamolz LP. Erratum to: first experiences with a new surgical approach in adult full-thickness burns: single step reconstruction of epidermal, dermal and subcutaneous defects by use of split-thickness skin grafting, a dermal collagen matrix and autologous fat-transfer. Eur Surg. 2013;45:282.
27. Van Kilsdonk JW, Van Den Bogaard EH, Jansen PA, Bos C, Bergers M, Schalkwijk J. An in vitro wound healing model for evaluation of dermal substitutes. Wound Repair Regen. 2013;21(6):890–6.
28. Clement AL, Moutinho TJ Jr, Pins GD. Micropatterned dermal-epidermal regeneration matrices create functional niches that enhance epidermal morphogenesis. Acta Biomater. 2013;9:9474–84.
29. Sun G, Zhang X, Shen YI, Sebastian R, Dickinson LE, Fox-Talbot K, Reinblatt M, Steenbergen C, Harmon JW, Gerecht S. Dextran hydrogel scaffolds enhance angiogenic responses and promote complete skin regeneration during burn wound healing. Proc Natl Acad Sci U S A. 2011;108(52): 20976–81.
30. Landsman A, Blume PA, Jordan DA Jr, Vayser D, Gutierrez A. An open-label, three-arm pilot study of the safety and efficacy of topical Microcyn Rx wound care versus oral levofloxacin versus combined therapy for mild diabetic foot infections. J Am Podiatr Med Assoc. 2011;101:484–96.
31. Oculus Innovative Sciences, Inc. Microcyn Scar Management Hydrogel. https://www.accessdata.fda.gov/cdrh_docs/pdf13/K131672.pdf.
32. Fearmonti R, Bond J, Erdmann D, Levinson H. A review of scar scales and scar measuring devices. Eplasty. 2010;10:e43.
33. Gankande TU, Wood FM, Edgar DW, Duke JM, Dejong HM, Henderson AE, Wallace HJ. A modified Vancouver Scar Scale linked with TBSA (mVSS-TBSA): inter-rater reliability of an innovative burn scar assessment method. Burns. 2013;39:1142–9.
34. Shalom A, Friedman T, Schein O, Hadad E. A novel short-scar breast reduction technique in large breasts. Aesthet Plast Surg. 2013;37:336–40.
35. Romanos GE, Strub JR. Effect of Tissucol on connective tissue matrix during wound healing: an immuno-

histochemical study in rat skin. J Biomed Mater Res. 1998;39:462–8.

36. Yeilding RH, O'Day DM, Li C, Alexander PT, Mawn LA. Periorbital infections after Dermabond closure of traumatic lacerations in three children. J AAPOS. 2012;16:168–72.

37. Howard BK, Downey SE. Contact dermatitis from Dermabond. Plast Reconstr Surg. 2010;125:252e–3e.

38. Wong EM, Rainer TH, Ng YC, Chan MS, Lopez V. Cost-effectiveness of Dermabond versus sutures for lacerated wound closure: a randomised controlled trial. Hong Kong Med J. 2011;17(Suppl 6):4–8.

39. Aldunate JLCB, Vana LPM, Fontana C, Ferreira MC. Uso de matriz dérmica associado ao curativo por pressão negativa na abordagem da contratura em pacientes queimados. Rev Bras Cirurg Plást. 2012;27:369–73.

40. Kirby W, Chen CL, Desai A, Desai T. Causes and recommendations for unanticipated ink retention following tattoo removal treatment. J Clin Aesthet Dermatol. 2013;6:27–31.

41. Gauglitz GG. Management of keloids and hypertrophic scars: current and emerging options. Clin Cosmet Investig Dermatol. 2013;6:103–14.

42. Liu A, Moy RL, Ozog DM. Current methods employed in the prevention and minimization of surgical scars. Dermatol Surg. 2011;37:1740–6.

43. Jackson WM, Nesti LJ, Tuan RS. Mesenchymal stem cell therapy for attenuation of scar formation during wound healing. Stem Cell Res Ther. 2012;3:20.

44. Haider A, Shaw JC. Treatment of acne vulgaris. J Am Med Assoc. 2004;292:726–35.

45. Mandy SH. Tretinoin in the preoperative and postoperative management of dermabrasion. J Am Acad Dermatol. 1986;15:878–9, 888–9.

46. Capon A, Iarmarcovai G, Mordon S. Laser-assisted skin healing (LASH) in hypertrophic scar revision. J Cosmet Laser Ther. 2009;11:220–3.

47. Capon A, Iarmarcovai G, Gonnelli D, Degardin N, Magalon G, Mordon S. Scar prevention using Laser-Assisted Skin Healing (LASH) in plastic surgery. Aesthet Plast Surg. 2010;34:438–46.

48. Waibel JS, Wulkan AJ, Shumaker PR. Treatment of hypertrophic scars using laser and laser assisted corticosteroid delivery. Lasers Surg Med. 2013;45:135–40.

49. Nicoletti G, De Francesco F, Mele CM, Cataldo C, Grella R, Brongo S, Accardo M, Ferraro GA, D'Andrea F. Clinical and histologic effects from CO_2 laser treatment of keloids. Lasers Med Sci. 2013;28:957–64.

50. Puri N, Talwar A. The efficacy of silicone gel for the treatment of hypertrophic scars and keloids. J Cutan Aesthet Surg. 2009;2:104–6.

51. O'Brien L, Pandit A. Silicon gel sheeting for preventing and treating hypertrophic and keloid scars. Cochrane Database Syst Rev. 2006;(1):Cd003826.

52. McCarty M. An evaluation of evidence regarding application of silicone gel sheeting for the management of hypertrophic scars and keloids. J Clin Aesthet Dermatol. 2010;3:39–43.

53. Mustoe TA. Evolution of silicone therapy and mechanism of action in scar management. Aesthet Plast Surg. 2008;32:82–92.

54. Li-Tsang CW, Feng BB, Li KC. Pressure therapy of hypertrophic scar after burns and related research. Zhonghua Shao Shang Za Zhi. 2010;26:411–5.

55. Banwell P, Heaton KP, Hardman IJ. Scar reduction apparatus. US 20150032035 A1. https://www.google.com/patents/US20150032035.

56. Fraccalvieri M, Sarno A, Gasperini S, Zingarelli E, Fava R, Salomone M, Bruschi S. Can single use negative pressure wound therapy be an alternative method to manage keloid scarring? A preliminary report of a clinical and ultrasound/colour-power-Doppler study. Int Wound J. 2013;10:340–4.

57. Larrabee WF Jr, East CA, Jaffe HS, Stephenson C, Peterson KE. Intralesional interferon gamma treatment for keloids and hypertrophic scars. Arch Otolaryngol Head Neck Surg. 1990;116:1159–62.

58. Ud-Din S, Bayat A. Strategic management of keloid disease in ethnic skin: a structured approach supported by the emerging literature. Br J Dermatol. 2013;16(Suppl 3):71–81.

59. Qiu Y, Ma G, Lin X, Jin Y, Chen H, Hu X. Treating protruding infantile hemangiomas with topical imiquimod 5% cream caused severe local reactions and disfiguring scars. Pediatr Dermatol. 2013;30:342–7.

60. Manca G, Pandolfi P, Gregorelli C, Cadossi M, De Terlizzi F. Treatment of keloids and hypertrophic scars with bleomycin and electroporation. Plast Reconstr Surg. 2013;132:621e–30e.

61. Har-Shai Y, Amar M, Sabo E. Intralesional cryotherapy for enhancing the involution of hypertrophic scars and keloids. Plast Reconstr Surg. 2003;111:1841–52.

62. Rzaca M, Tarkowski R. Paget's disease of the nipple treated successfully with cryosurgery: a series of cases report. Cryobiology. 2013;67:30–3.

63. GBI Research Report Guidance. Wound closure devices market to 2019—new product launches and favorable clinical outcomes for tissue sealants and hemostats drive physician adoption. https://www.marketresearch.com/product/sample-7721928.pdf.

Challenges and Opportunities in Drug Delivery and Wound Healing

3

Matthias M. Aitzetmüller, Hans-Günther Machens, and Dominik Duscher

3.1 Introduction

There is an ongoing shift in the distribution of the world's population towards old age and we recently experience an increase in comorbidities like diabetes or cardiovascular insufficiency. This results in a rising number of chronic wounds which have become not only an individual medical but also a significant economic burden, consuming 2–4% of healthcare budgets worldwide [1].

Wound healing is a complex systems biology interplay depending on the timed coordination of several cell types, intra- and extracellular mechanisms, proteins, and pathways, but also on several external factors like infections or mechanical irritation. Either defect, loss, or dominance of one factor of this convoluted interaction can cause a breakdown of the whole system resulting in chronic wounds and a loss of quality of life. A famous example for the fragility of the cellular mechanism for tissue homeostasis and repair is the connection between vitamin C deficiency and scurvy resulting in nonhealing wounds and spontaneous bleeding known since the sixteenth century [2, 3]. Mentioned first in journey books of Christopher Columbus as a result of monotone diet, the pathomechanism remained unclear until the twentieth century when it could be demonstrated that vitamin C represents a main cofactor for collagen cross-linking and an important factor to reduce oxidative stress [4]. This example of how impactful just minimal perturbations in the underlying processes can be illustrates that a highest possible understanding of all molecular and cellular players involved in wound healing is pivotal for developing treatment strategies and effective drugs.

However, no drug can be effective when its sustained and targeted delivery to the wound site cannot be assured. Specific drug delivery systems (DDS) are key for achieving this goal. An efficacious DDS addresses the obstacles presented by the harsh wound environment and prevents the wound from mechanical, oxidative, and enzymatic stress and from bacterial contamination and provides enough oxygen while maximizing localized and sustained drug delivery to the target tissue (Fig. 3.1). In this chapter, we

M. M. Aitzetmüller (✉)
Department of Plastic and Hand Surgery, Klinikum rechts der Isar, Technical University of Munich, Munich, Germany

Section of Plastic and Reconstructive Surgery, Department of Trauma, Hand and Reconstructive Surgery, Westfaelische Wilhelms, University of Muenster, Muenster, Germany
e-mail: matthias.aitzetmueller@tum.de

H.-G. Machens
Department of Plastic and Hand Surgery, Klinikum rechts der Isar, Technical University of Munich, Munich, Germany
e-mail: Hans-Guenther.Machens@mri.tum.de

D. Duscher
Department for Plastic Surgery and Hand Surgery, Division of Experimental Plastic Surgery, Technical University of Munich, Munich, Germany

© Springer Nature Switzerland AG 2019
D. Duscher, M. A. Shiffman (eds.), *Regenerative Medicine and Plastic Surgery*,
https://doi.org/10.1007/978-3-030-19962-3_3

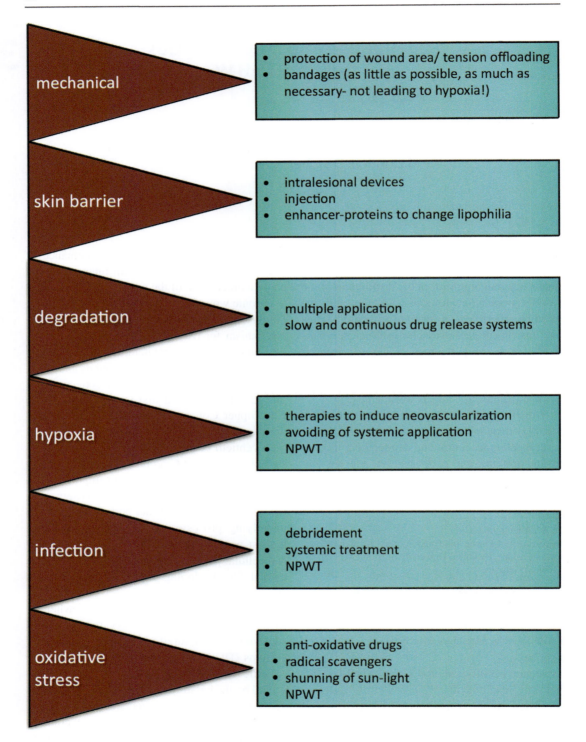

Fig. 3.1 Problems and preventive/solving strategies for effective drug delivery to wounds

summarize the most promising recent advances in wound healing therapeutics with the corresponding delivery challenges and shed light on possible solutions for effective application.

3.2 Drug Delivery Routes for Wound Healing Applications

The importance of drug delivery and the challenge for translational medicine to develop effective DDS for wound healing applications is represented by the large amount of recent studies regarding this topic (176 ongoing clinical trials registered) [5]. Different routes of drug application warrant strategies to face the obstacles represented by the systemic circulation and the harsh wound environment.

3.2.1 Systemic Application

Systemic drug administration is the most general form of application and common clinical practice for various substances. "Dosis facit venenum" (Solely the dose determines that a thing is not a poison) represents a rule postulated by Paracelsus already in the fifteenth century, which is still true today [6]. The limitation of dosing for orally or intravenously delivered drugs meant to act locally in a wound is essential to prevent adverse effects outweighing benefits. An additional major problem of systemic application in chronic wounds is the reduced blood circulation caused by common underlying comorbidities like diabetes, vessel diseases, tobacco abuse, or aging. Consequentially only few studies showed the benefit of systemically administrated drugs for wound healing applications. For example, pyrvinium, a small molecule, has been shown to modulate the Wnt pathway resulting in enhanced tissue repair [7, 8]. These findings are limited by the cross-reactivity of pyrvinium and observed gastrointestinal side effects [9, 10]. Another molecule that has shown promising results as a systemically administered wound enhancer is deferoxamine (DFO), an iron-chelating agent also acting as an antioxidant. While DFO has been in clinical use for short-term systemic treatment of hemochromatosis for decades, the long-term high-dose application necessary to efficiently stimulate wound healing results in inside effects as well as acute and chronic toxicity. Together with its short plasma half-life these significant downsides limit the applicability of DFO as a systemic wound therapeutic in the clinical setting [9, 11].

Recombinant erythropoietin (rEPO), a drug stimulating erythropoiesis, is emerging as an additive in the treatment of various chronic diseases and was shown to significantly improve outcomes of critically ill patients [12, 13]. Anemia is evident early in the courses of critical illnesses, and hemoglobin concentrations fall throughout stays in the ICU [14, 15]. A feature of anemia of critical illness is a lack of appropriate elevation of circulating erythropoietin concentrations in response to physiological stimuli [16]. Considering these mechanisms of critical illness and anemia, it is not surprising that EPO treatment for patients with severe burn wounds has shown quite promising effects on main organ function, including kidney, liver, heart, lungs, and central nerve system (preliminary unpublished data from our group). This novel application of a known substance might lead to a new standard in the treatment of burn wounds [17].

The need for high dosages and considerable side effects lead to the fact that systemic drug application currently is not part of common clinical practice, although some substances are promising to become a valuable additive for wound treatment. Systemic supplementation to optimize the nutritional state with protein and vitamins, however, increases wound healing abilities and already represents a crucial supporting therapy.

3.2.2 Local Application

Considering the risks for systemic toxicity and less predictable drug delivery to the target tissue, there has been a significant shift in clinical focus towards localized delivery of drugs for wound healing. Localized drug delivery permits convenient self-administration for patients while avoiding issues with gastrointestinal tract absorption and hepatic first-pass metabolism, thereby improving bioavailability and maintenance of drug concentration within the therapeutic window [18]. Furthermore, local delivery enables transmission of the largest fraction of drug molecules to the target area, maximizing therapeutic potential and reducing systemic drug toxicity [19]. Despite the advantages of localized delivery, many challenges still remain, including penetration of the stratum corneum in skin at risk for ulceration, maintaining cell survival after delivery, and development of effective mechanisms for sustained delivery.

Growth factor and progenitor cell-based therapy research has recently centered on identifying new delivery mechanisms to overcome biological degradation and poor cell survival in the harsh wound environment. Nanoparticles, for example, have successfully been used to increase the half-life of therapeutic growth factors delivered to wounds in diabetic rats [20]. Recent advances in negative pressure wound therapy have allowed for intermittent fluid instillation, which may enable an alternative delivery method for aqueous wound therapies with the added benefit of providing local debridement during instillation [21]. Cell delivery methods such as fibrin sealant sprays or hydrogel scaffolds have been shown to improve cell retention and functional capacity at the application site both in vitro and in vivo [22, 23]. Experience with these various methods shows that an ideal delivery system is nontoxic, facilitates access of therapies to the wound site, and protects these therapies from premature degradation.

3.3 Recent Advances in Wound Therapeutics and Delivery Challenges

3.3.1 Biological Therapies

Utilized for their ability to treat both complex acute and chronic wounds, biological therapies functionally aim to restore the body's natural repair capabilities. Creating microenvironments that encourage proliferation of both matrix-depositing stromal cells and endothelial cells at the site of injury, biological therapies may facilitate the formation of a vascular network in newly forming tissue. Biologic approaches include bioactive scaffolds, growth factor-based therapies, and stem cell-based therapies.

3.3.2 Bioactive Scaffolds

Acellular scaffolds function by providing coverage to the wound site, establishing a matrix for resident cell infiltration, and fostering granulation tissue formation. Two clinically established acellular matrices (biologic skin equivalents, BSEs) derived from human dermis that have attempted to answer this challenge are AlloDerm (LifeCell Corp., Branchburg, NJ) and GraftJacket (Wright Medical Technology, Inc., Arlington, TX).

AlloDerm is a nonliving dermal replacement composed of human cadaveric skin which has been formed by salt processing. Though initially developed for the treatment of full-thickness burns, AlloDerm was quickly adapted for use in soft-tissue injuries and has been shown to undergo a host-cell infiltration and neovascularization following application to the wound site. It can be used to immediately cover large wounds and reduce the need for skin grafts. However, data suggests that the take rate of split-thickness skin grafts (STSGs) applied over AlloDerm is decreased when compared to STSG application alone [24].

Formed from cadaveric skin, GraftJacket is an allogeneic human tissue which uses specialized technology to maintain its basic matrix and biochemical structure after the removal of the epidermis and cellular components during processing [25]. GraftJacket has demonstrated efficacy for the treatment of diabetic chronic lower extremity ulcers in two midsized randomized, controlled trials (86 and 28 patients), where results revealed accelerated healing and clinically significant reductions in wound depth and volume when compared to standard wet-to-dry dressings [26]. In chronic wounds, the ECM is often dysfunctional due to the inflammatory and proteolytic environment. GraftJacket provides an intact, acellular dermal matrix that retains natural biological components, is repopulated by the patient's cells, and allows the body to initiate a normal tissue regeneration process.

Nonhuman-derived acellular biological matrices have also been used for wound care. OASIS Wound Matrix (Cook Biotech, Inc., West Lafayette, IN), derived from the submucosal layers of porcine jejunum, received FDA clearance in 2006. In separate, midsized randomized, controlled trials (120 and 50 patients), a significantly higher percentage of OASIS-treated chronic lower extremity wounds healed when compared to compression or moist dressing therapies [27, 28].

Despite these promising findings, a general uncertainty regarding the optimal source of tissue and processing technique forces surgeons to choose products based on familiarity, cost, and availability—rather than efficacy. Lack of randomized controlled, head-to-head trials between products, and studies often sponsored by the commercial manufacturers themselves, increases the risk of potential bias. Additionally, a constant fear surrounding biological scaffolds is the risk of disease transmission and donor rejection of the graft. Because wounds vary in vascularity, presence of infection, and amount of debris, it is essential for surgeons to prepare the wound site in order to allow these scaffolds to become successful [29]. Lastly, though BSEs hold promise, a more robust clinical comparison of host tissue must be conducted before general acceptance among surgeons can be achieved.

3.3.3 Growth Factor-Based Therapies

Growth factor-based therapies are vested on an understanding that specific regulatory pathways govern the host response to wound healing and are used to stimulate wound angiogenesis, matrix deposition, and re-epithelialization [30]. Regranex (Ortho-McNeil, Raritan, NJ), a recombinant platelet-derived growth factor (PDGF)-based therapy, is currently the only growth factor-based biological therapy with FDA approval. Regranex accelerates the regenerative process by promoting fibroblast migration and wound re-epithelialization. Randomized trials have shown that Regranex application signifi-

cantly increased both the probability and time course for complete healing of leg and foot ulcers when compared to placebo gel [31]. In addition, Regranex application to pressure ulcers results in higher incidences of complete healing, as well as a significant reduction in ulcer volume compared to the application of placebo gel [32]. However, despite these promising findings, the value of Regranex for wound healing is unclear, as a recent randomized control trial found that the application of Regranex gel was not superior to a simple hydrogel dressing for the healing of hypertensive leg ulcers [33]. Moreover, an FDA review concluded that Regranex has the potential to increase the risk of cancer death in diabetic patients resulting in an FDA black box warning for this product [34].

Evidence is accumulating that a mono-factor therapy like Regranex is less efficient in promoting wound closure than approaches enhancing local concentration of all growth factors involved in healing. Negative pressure wound therapies (NWPT = vacuum-assisted closure = VAC therapy) provide an elegant way of increasing effectors of wound healing and neovascularization locally. NWPT temporarily creates relative hypoxia in the wound region resulting in significant higher levels of main regenerative factors such as VEGF, TGFβ, FGF, angiopoietin 1, and BMP 2 and its application shows benefits regarding bacterial contamination [35–39]. Additionally, several studies suggest that the micro-deformation of the wound surface leads to accelerated cell migration and matrix production. By using silver-coated foams the NWPT can be even more effective in preventing or treating bacterial contamination [40]. Rowan et al. [41] demonstrate further that NWPT therapy not only influences bacterial contamination by removal of microorganisms, but also enhances local concentration of systemically applied antibiotics making it an ideal therapeutic for contaminated wounds.

Another promising approach for multifactor therapy is to deliver a cocktail of growth factors by an injectable scaffold. Hadjipanayi et al. [42] used fibroblast-loaded collagen scaffolds and treated them under hypoxic conditions. The hypoxic stimulus led to an increased production of angiogenic growth factors which could be trapped in the surrounding matrix. Utilizing this matrix as a cell-free growth factor carrier system provided a minimally invasive method for localized delivery of growth factor mixtures, as a tool for physiological induction of spatiotemporally controlled angiogenesis. Further developing this approach peripheral blood cells (PBCs) were identified as the ideal factor-providing candidates due to their autologous nature, ease of harvest, and ample supply [42]. Engineered PBC-derived factor mixtures could be harvested within cell-free gel and microsphere carriers. The angiogenic effectiveness of factor-loaded carriers could be demonstrated by the ability to induce endothelial cell tubule formation and directional migration in in vitro Matrigel assays, and microvessel sprouting in the aortic ring assay. This approach could facilitate the controlled release of these factors both at the bed side, as an angiogenic therapy in wounds and peripheral ischemic tissue, and pre-, intra-, and postoperatively as angiogenic support for central ischemic tissue, grafts, flaps, and tissue-engineered implants.

3.3.4 Stem Cell Therapies

Stem cells are characterized by their capacity for self-renewal and ability to differentiate into various tissue types via asymmetric replication. The trophic activity of these cells has led to the development of cell-based approaches for the treatment of chronic wounds. Growth factors released from stem cells stimulate local cell proliferation and migration, increased angiogenesis, organized ECM production, and antimicrobial activity [43]. Stem cells are thought to be immuno-privileged as they are able to modulate the immune response, at least partially through the recruitment of regulatory T-cells [44]. The pro-regenerative cytokine release and unique differentiation capacity of stem cells are believed to underlie their capacity to promote healing [45]. Stem cells implanted in wounds also draw in endogenous circulating progenitor cells, further emphasizing their role as initiators of wound repair [46, 47].

Stem cells harvested from various sources such as the bone marrow, adipose tissue, epidermis, and circulating adult blood have been uti-lized for wound therapy [48]. However, there are still various concerns related to the clinical application of stem cell therapeutics. To date, no study has confirmed the optimal cell source and potency (autologous vs. allogeneic; multipotent vs. pluripotent). Moreover, there is a need to promote cell survival and activity within the harsh wound environment. Biomimetic hydrogel dressings can be used to promote the survival of stem cells within a wound [45] while administering either topical or systemic EPO provides a promising approach for enhancing stem cell functionality leading to improved healing in scald wounds [49, 50]. While clearly still in early phase development, stem cell therapies, used either independently or in conjunction with skin graft substitutes such as a decellularized matrix, appear to be another promising step for the treatment of nonhealing wounds [48].

Looking to the future, biological products will continue to be an intriguing area of potential clinical therapy as spatially controlled drug-release systems aim to minimize the quantities of drug being delivered, reduce migratory effects on surrounding tissues, and reduce overall cost [18]. Furthermore, their eventual clinical acceptance will be contingent on the development of evidence-based wound care guidelines for these biologic therapies.

3.3.5 Small-Molecule Therapies

Though the advances in growth factor- and progenitor-based therapies hold promise for the treatment of acute and chronic wounds, they are still limited by the high costs involved, potential antigenicity, and legal and ethical issues surrounding stem cell research [51]. From a translational perspective, the application of small molecules in lieu of cells and proteins has significant advantages in terms of sterility, shelf life, and regulatory hurdles [52]. Emerging small molecule-based therapies for wound healing enhancement focus on the modulation of key signaling pathways involved in tissue repair such as the Wnt and hypoxia inducible factor-1 (HIF-1) pathways.

Wnt proteins are highly conserved signaling molecules, which regulate embryonic development and cell fate [53]. They bind to cell surface receptors of the frizzled (Fz) and lipoprotein receptor-related protein (LRP) family. Wnt signaling is transduced by beta-catenin, which enters the nucleus and forms a complex with the T-cell factor (TCF) transcription factor family to activate transcription of Wnt target genes [54, 55]. Wnt signaling not only is essential in development, but has also been linked to mammalian cutaneous wound repair as a potential therapeutic target [56]. Thorne et al. recently identified the FDA-approved small molecule pyrvinium as a potent Wnt inhibitor [57]. Applying this molecule in an in vivo model of wound healing demonstrated that temporary inhibition of Wnt leads to increased cell proliferation, granulation tissue formation, and vascularity [8]. Furthermore, pyrvinium positively affected the engraftment and regenerative capacity of mesenchymal progenitor cells, which are promising therapeutic modalities for wound healing [7].

Another prominent pathway that can be manipulated by small molecules to enhance wound healing involves the response of tissue to hypoxia, largely regulated by the transcription factor hypoxia inducible factor-1 (HIF-1) [58]. HIF-1 includes an α-subunit that is degraded in the presence of oxygen and iron (Fe^{2+}) by the enzyme prolyl hydroxylases (PHD) [59, 60]. Hypoxia impairs HIF-1α degradation, resulting in the expression of a number of pro-regenerative proteins, including vascular endothelial growth factor (VEGF) and stromal cell-derived factor-1 (SDF-1) [61, 62]. Numerous animal studies have demonstrated improvements in wound healing due to enhancement of the HIF-1 pathway [63, 64] through positive effects of VEGF on neovascularization, and SDF-1 on progenitor cell homing [65, 66]. A promising approach for the therapeutic modulation of HIF-1 signaling is the application of deferoxamine (DFO), an FDA-approved iron chelator that has been in clinical use for decades. DFO augments neovascularization and consequentially wound healing by the inhibition of HIF-1α degradation and the decrease of oxidative stress in the wound environment [67–69]. These effects synergistically promote wound healing by decreasing tissue necrosis [70, 71].

Novel small-molecule treatment strategies offer tremendous opportunities in the often frustrating management of problematic wounds. However, translation into clinical practice remains challenging. Both pyrvinium and DFO are associated with significant toxic side effects when delivered systemically, limiting dosing and duration of possible application [8, 11]. To maximize efficacy while minimizing potential side effects, localized targeted delivery directly to wound sites is essential. Controlled local drug delivery would also improve the bioavailability and enhance the uptake of small molecules by maintaining drug concentration within the therapeutic window [18]. Packaging an existing FDA-approved drug into a controlled-release formulation may not only improve its performance but also extend its commercial patent life. The average cost and time required to develop a new drug delivery system (DDS) (approximately $20–50 million and 3–4 years) is significantly lower than that for a new drug (approximately $500 million and over 10 years) [72]. This has led to significant growth in the US market for advanced DDSs from $75 million in 2001 to $121 billion in 2010, with the worldwide market for polymer-based controlled-release system alone being estimated at $60 billion in 2010 [73]. Acknowledging the benefits of sustained local small-molecule delivery, a transdermal delivery system containing DFO was recently designed [74]. This transdermal polymer patch overcomes the challenge of delivering the hydrophilic DFO molecules through the normally impermeable stratum corneum. This demonstrates that the use of a modern polymeric dressing for the delivery of an active substance to the wound site offers a promising therapeutic approach of the future.

3.3.6 RNAi Therapies

Unlike traditional pharmaceutical approaches, the silencing of gene function through RNAi offers selective targeting of molecules that have been difficult to regulate using growth factor and small molecule-based therapies. RNAi is a powerful gene silencing mechanism with enormous potential for therapeutic application [75]. Inhibiting gene expression at the posttranscriptional level, RNAi (either endogenous miRNAs (micro-RNAs) or synthetical siRNAs (small interfering RNAs)) targets specific mRNA molecules for destruction, and offers an exciting therapeutic approach to wound healing (Fig. 3.2). A recent study illustrated the potential of this modality, demonstrating that the use of RNAi to silence Smad2 enhances wound regeneration and improves the overall wound quality [76]. Despite broad therapeutic potential, the effective delivery of RNAi to target cells in vivo remains a significant challenge due to the high rates of degradation by ubiquitous RNases, targeting of specific tissues, and maintenance of long-term silencing [77–79]. Developing a controlled DDS for RNAi is crucial to realize the full potential of these next-generation therapeutics. Allowing for low-dose application, minimal systemic side effects, site-specific delivery, and lower costs, local delivery of RNAi offers several advantages over systemic delivery with regard to the potential for clinical translation [79]. Current approaches for the delivery of siRNA in vivo include the direct injection of siRNA in saline, incorporation into liposomes, and delivery in the form of nanoparticles. However, these approaches have not demonstrated sustained RNAi activity and offer low rates of cell uptake [80]. Recent data has shown that hydrogel scaffolds offer the ability to retain siRNA locally and release it directly at the sight of interest. Using a collagen-based hydrogel, the delivery of siRNA to a specific location in vivo was recorded for up to 6 days with a low fraction of siRNA being released to the surrounding host tissue. These hydrogel scaffolds were applied topically and offer an exciting new platform for siRNA delivery in vivo [80].

Currently the only RNAi-based drug in clinical trials for wound healing (phase II) is RXi-109 (RXi Pharmaceuticals, Marlborough, MA) which aims to prevent the development of scarring by targeting connective tissue growth factor (CTGF) [81]. RXi-909 employs a collagen/silicone membrane bilayer BSE combined with trimethyl chitosan (TMC) and siRNA complexes to induce suppression of the transforming growth factor-$\beta1$ (TGF-$\beta1$) pathway, resulting in a functionalized matrix for scar reduc-

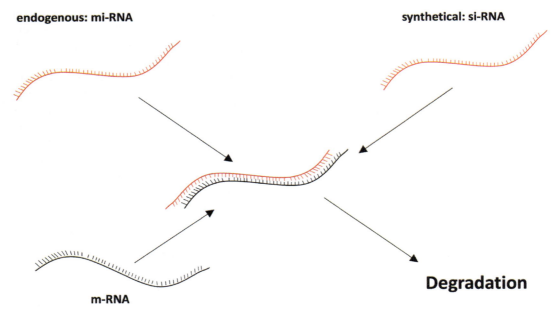

Fig. 3.2 Principle of RNA interference: body-own RNA (mi-RNA) or synthetic RNA (si-RNA) binds to m-RNA and leads to its degradation

tion. The RNAi-BDE scaffold demonstrated high viability and suppressed TGF-β1 expression for 2 weeks. Additionally, the expressions of collagen type I, collagen type III, and α-smooth muscle actin (α-SMA) were shown to be downregulated in large animals [82]. The use of RNAi-BSE parallels the structural development typically seen in normal skin repair and offers a unique delivery scaffold for the future of local RNAi delivery.

Technical difficulties restrict the development of RNAi, including stability, off-target effects, immune stimulation, and delivery problems. Researchers have attempted to surmount these barriers and improve the bioavailability and safety of RNAi-based therapeutics. However, with minimal clinical trials currently being conducted for RNAi therapeutics in wound healing, their clinical translation remains elusive. Looking ahead, as delivery methods are fine-tuned, RNAi therapeutics may develop into a drug class with the potential to exert a transformational effect on modern regenerative medicine.

3.4 Conclusions and Future Perspectives

The significant disability and cost to society associated with chronic wounds highlight the inadequacy of current therapeutic approaches. The current armamentarium of therapeutic options does not fully address the impaired cellular and molecular mechanisms underlying nonhealing wounds. Emerging therapeutic options have embraced the need to correct the deficits in critical signaling pathways, cellular dysfunction, and impaired neovascularization associated with chronic wounds but are largely experimental or in early phases of development. Bioactive dressings and scaffolds, growth factor- and cell-based therapies, small-molecule delivery, and RNAi therapeutics all appear to be promising, but do not fully replicate the precise spatiotemporal gradients of molecules and factors during wound healing. Imperfect processing techniques and risk of rejection continue to impair biological scaffolds, though their ability to provide volume replacement, wound cover,

and a matrix for cell engraftment is encouraging. Growth factor therapies are exciting but studies are inconclusive over their clinical benefits, potentially due to the complexity and dynamic nature of growth factor expression during the response to injury. Stem cell-based therapies are limited by the capacity of the cells to survive and function in a harsh wound environment, while small molecule-based treatments rely on efficient, targeted, and sustainable delivery systems. Finally, RNAi potentially enables the modification of downstream targets, limiting potential side effects, but is limited by its propensity for degradation by ubiquitous RNases. A recently merging area of research is the rapid progress occurring in nanomedicine. The use of a microfluidic platform as DDS as well as controlled-release applications for wound scaffolds represents a novel application for drug delivery which may hold the key to unlocking clinical wound healing therapy in the future.

References

1. Lindholm C, Searle R. Wound management for the 21st century: combining effectiveness and efficiency. Int Wound J. 2016;13(Suppl 2):5–15.
2. Lanman TH, Ingalls TH. Vitamin C deficiency and wound healing: an experimental and clinical study. Ann Surg. 1937;105(4):616–25.
3. Tiesler V, Coppa A, Zabala P, Cucina A. Scurvy-related morbidity and death among Christopher Columbus' Crew at La Isabela, the First European Town in the New World (1494–1498): An Assessment of the Skeletal and Historical Information. Int J Osteoarchaeol. 2014;26(2):191–202.
4. Grinnell F, Fukamizu H, Pawelek P, Nakagawa S. Collagen processing, crosslinking, and fibril bundle assembly in matrix produced by fibroblasts in long-term cultures supplemented with ascorbic acid. Exp Cell Res. 1989;181(2):483–91.
5. Clinical trials regarding wound healing and drug delivery system. https://clinicaltrials.gov/. Accessed 2 May 2017.
6. Deichmann W, Henschler D, Holmstedt B, Keil G. What is there that is not poison? A study of the Third Defense by Paracelsus. Arch Toxicol. 1986;58(4):207–13.
7. Saraswati S, Deskins DL, Holt GE, Young PP. Pyrvinium, a potent small molecule Wnt inhibitor, increases engraftment and inhibits lineage commitment of mesenchymal stem cells (MSCs). Wound Repair Regen. 2012;20(2):185–93.

8. Saraswati S, Alfaro MP, Thorne CA, Atkinson J, Lee E, Young PP. Pyrvinium, a potent small molecule Wnt inhibitor, promotes wound repair and post-MI cardiac remodeling. PLoS One. 2010;5(11):e15521.

9. Lim M, Otto-Duessel M, He M, Su L, Nguyen D, Chin E, et al. Ligand-independent and tissue-selective androgen receptor inhibition by pyrvinium. ACS Chem Biol. 2014;9(3):692–702.

10. Antonijevic B, Stojiljkovic MP. Unequal efficacy of pyridinium oximes in acute organophosphate poisoning. Clin Med Res. 2007;5(1):71–82.

11. Hom DB, Goding GS Jr, Price JA, Pernell KJ, Maisel RH. The effects of conjugated deferoxamine in porcine skin flaps. Head Neck. 2000;22(6):579–84.

12. Zarychanski R, Turgeon AF, McIntyre L, Fergusson DA. Erythropoietin-receptor agonists in critically ill patients: a meta-analysis of randomized controlled trials. Canad Med Assoc J. 2007;177(7):725–34.

13. John MJ, Jaison V, Jain K, Kakkar N, Jacob JJ. Erythropoietin use and abuse. Indian J Endocrinol Metab. 2012;16(2):220–7.

14. Corwin HL, Parsonnet KC, Gettinger A. RBC transfusion in the ICU. Is there a reason? Chest. 1995;108(3):767–71.

15. Corwin HL, Gettinger A, Pearl RG, Fink MP, Levy MM, Abraham E, MacIntyre NR, Shabot MM, Duh MS, Shapiro MJ. The CRIT study: Anemia and blood transfusion in the critically ill—current clinical practice in the United States. Crit Care Med. 2004;32(1):39–52.

16. Corwin HL, Gettinger A, Fabian TC, May A, Pearl RG, Heard S, An R, Bowers PJ, Burton P, Klausner MA, Corwin MJ, EPO Critical Care Trials Group. Efficacy and safety of epoetin alfa in critically ill patients. N Engl J Med. 2007;357(10):965–76.

17. Gunter CI, Bader A, Dornseifer U, Egert S, Dunda S, Grieb G, Wolter T, Pallua N, von Wild T, Siemers F, Mailänder P, Thamm O, Ernert C, Steen M, et al. A multi-center study on the regenerative effects of erythropoietin in burn and scalding injuries: study protocol for a randomized controlled trial. Trials. 2013;14:124.

18. Zhang Y, Chan HF, Leong KW. Advanced materials and processing for drug delivery: the past and the future. Adv Drug Deliv Rev. 2013;65(1):104–20.

19. Weiser JR, Saltzman WM. Controlled release for local delivery of drugs: barriers and models. J Controlled Rel. 2014;190:664–73.

20. Chu Y, Yu D, Wang P, Xu J, Li D, Ding M. Nanotechnology promotes the full-thickness diabetic wound healing effect of recombinant human epidermal growth factor in diabetic rats. Wound Repair Regen. 2010;18(5):499–505.

21. Jerome D. Advances in negative pressure wound therapy: the VAC instill. J Wound Ostomy Continence Nurs. 2007;34(2):191–4.

22. Wong VW, Rustad KC, Glotzbach JP, Sorkin M, Inayathullah M, Major MR, Longaker MT, Rajadas J, Gurtner GC. Pullulan hydrogels improve mesenchymal stem cell delivery into high-oxidative-stress wounds. Macromol Biosci. 2011;11(11):1458–66.

23. Zimmerlin L, Rubin JP, Pfeifer ME, Moore LR, Donnenberg VS, Donnenberg AD. Human adipose stromal vascular cell delivery in a fibrin spray. Cytotherapy. 2013;15(1):102–8.

24. Wainwright D, Madden M, Luterman A, Hunt J, Monafo W, Heimbach D, et al. Clinical evaluation of an acellular allograft dermal matrix in full-thickness burns. J Burn Care Rehab. 1996;17(2):124–36.

25. Kim PJ, Heilala M, Steinberg JS, Weinraub GM. Bioengineered alternative tissues and hyperbaric oxygen in lower extremity wound healing. Clin Podiatr Med Surg. 2007;24(3):529–46.

26. Reyzelman A, Crews RT, Moore JC, Moore L, Mukker JS, Offutt S, Tallis A, Turner WB, Vayser D, Winters C, Armstrong DG. Clinical effectiveness of an acellular dermal regenerative tissue matrix compared to standard wound management in healing diabetic foot ulcers: a prospective, randomised, multicentre study. Int Wound J. 2009;6(3):196–208.

27. Mostow EN, Haraway GD, Dalsing M, Hodde JP, King D, Group OVUS. Effectiveness of an extracellular matrix graft (OASIS Wound Matrix) in the treatment of chronic leg ulcers: a randomized clinical trial. J Vasc Surg. 2005;41(5):837–43.

28. Romanelli M, Dini V, Bertone MS. Randomized comparison of OASIS wound matrix versus moist wound dressing in the treatment of difficult-to-heal wounds of mixed arterial/venous etiology. Adv Skin Wound Care. 2010;23(1):34–8.

29. Scimeca CL, Bharara M, Fisher TK, Kimbriel H, Mills JL, Armstrong DG. An update on pharmacological interventions for diabetic foot ulcers. Foot Ankle Spec. 2010;3(5):285–302.

30. Gurtner GC, Werner S, Barrandon Y, Longaker MT. Wound repair and regeneration. Nature. 2008;453(7193):314–21.

31. Smiell JM, Wieman TJ, Steed DL, Perry BH, Sampson AR, Schwab BH. Efficacy and safety of becaplermin (recombinant human platelet-derived growth factor-BB) in patients with nonhealing, lower extremity diabetic ulcers: a combined analysis of four randomized studies. Wound Repair Regen. 1999;7(5):335–46.

32. Rees RS, Robson MC, Smiell JM, Perry BH. Becaplermin gel in the treatment of pressure ulcers: a phase II randomized, double-blind, placebo-controlled study. Wound Repair Regen. 1999;7(3):141–7.

33. Senet P, Vicaut E, Beneton N, Debure C, Lok C, Chosidow O. Topical treatment of hypertensive leg ulcers with platelet-derived growth factor-BB: a randomized controlled trial. Arch Dermatol. 2011;147(8):926–30.

34. Nair DG, Miller KG, Lourenssen SR, Blennerhassett MG. Inflammatory cytokines promote growth of intestinal smooth muscle cells by induced expression of PDGF-Rbeta. J Cell Molec Med. 2014;18(3):444–54.

35. Ingber DE. Mechanical control of tissue growth: function follows form. Proc Natl Acad Sci U S A. 2005;102(33):11571–2.

36. Ingber DE. The mechanochemical basis of cell and tissue regulation. Mech Chem Biosyst. 2004;1(1):53–68.
37. Nie B, Yue B. Biological effects and clinical application of negative pressure wound therapy: a review. J Wound Care. 2016;25(11):617–26.
38. Zhang YG, Wang X, Yang Z, Zhang H, Liu M, Qiu Y, Guo X. The therapeutic effect of negative pressure in treating femoral head necrosis in rabbits. PLoS One. 2013;8(1):e55745.
39. Li X, Liu J, Liu Y, Hu X, Dong M, Wang H, Hu D. Negative pressure wound therapy accelerates rats diabetic wound by promoting agenesis. Int J Clin Exp Med. 2015;8(3):3506–13.
40. Valente PM, Deva A, Ngo Q, Vickery K. The increased killing of biofilms in vitro by combining topical silver dressings with topical negative pressure in chronic wounds. Int Wound J. 2016;13(1):130–6.
41. Rowan MP, Niece KL, Rizzo JA, Akers KS. Wound penetration of cefazolin, ciprofloxacin, piperacillin, tazobactam, and vancomycin during negative pressure wound therapy. Adv Wound Care. 2017;6(2):55–62.
42. Hadjipanayi E, Bauer AT, Moog P, Salgin B, Kuekrek H, Fersch B, Hopfner U, Meissner T, Schlüter A, Ninkovic M, Machens HG, Schilling AF. Cell-free carrier system for localized delivery of peripheral blood cell-derived engineered factor signaling: towards development of a one-step device for autologous angiogenic therapy. J Control Release. 2013;169(1–2):91–102.
43. Caplan AI, Correa D. The MSC: an injury drugstore. Cell Stem Cell. 2011;9(1):11–5.
44. Dazzi F, Lopes L, Weng L. Mesenchymal stromal cells: a key player in 'innate tolerance'? Immunology. 2012;137(3):206–13.
45. Rustad KC, Wong VW, Sorkin M, Glotzbach JP, Major MR, Rajadas J, Longaker MT, Gurtner GC. Enhancement of mesenchymal stem cell angiogenic capacity and stemness by a biomimetic hydrogel scaffold. Biomaterials. 2012;33(1):80–90.
46. Shin L, Peterson DA. Human mesenchymal stem cell grafts enhance normal and impaired wound healing by recruiting existing endogenous tissue stem/progenitor cells. Stem Cells Transl Med. 2013;2(1):33–42.
47. Kosaraju R, Rennert RC, Maan ZN, Duscher D, Barrera J, Whittam AJ, Januszyk M, Rajadas J, Rodrigues M, Gurtner GC. Adipose-derived stem cell-seeded hydrogels increase endogenous progenitor cell recruitment and neovascularization in wounds. Tissue Eng Part A. 2016;22(3–4):295–305.
48. Rennert RC, Rodrigues M, Wong VW, Duscher D, Hu M, Maan Z, Sorkin M, Gurtner GC, Longaker MT. Biological therapies for the treatment of cutaneous wounds: phase III and launched therapies. Expert Opin Biol Ther. 2013;13(11):1523–41.
49. Giri P, Ebert S, Braumann UD, Kremer M, Giri S, Machens HG, Bader A. Skin regeneration in deep second-degree scald injuries either by infusion pumping or topical application of recombinant human erythropoietin gel. Drug Des Devel Ther. 2015;9:2565–79.
50. Bader A, Ebert S, Giri S, Kremer M, Liu S, Nerlich A, Günter CI, Smith DU, Machens HG. Skin regeneration with conical and hair follicle structure of deep second-degree scalding injuries via combined expression of the EPO receptor and beta common receptor by local subcutaneous injection of nanosized rhEPO. Int J Nanomedicine. 2012;7:1227–37.
51. Boateng JS, Matthews KH, Stevens HN, Eccleston GM. Wound healing dressings and drug delivery systems: a review. J Pharm Sci. 2008;97(8):2892–923.
52. Ekenseair AK, Kasper FK, Mikos AG. Perspectives on the interface of drug delivery and tissue engineering. Adv Drug Deliv Rev. 2013;65(1):89–92.
53. Nusse R. Cell biology: relays at the membrane. Nature. 2005;438(7069):747–9.
54. Nusse R, Varmus HE. Wnt genes. Cell. 1992;69(7):1073–87.
55. Tamai K, Semenov M, Kato Y, Spokony R, Liu C, Katsuyama Y, Hess F, Saint-Jeannet JP, He X. LDL-receptor-related proteins in Wnt signal transduction. Nature. 2000;407(6803):530–5.
56. Bielefeld KA, Amini-Nik S, Alman BA. Cutaneous wound healing: recruiting developmental pathways for regeneration. Cell Mol Life Sci. 2013;70(12):2059–81.
57. Thorne CA, Hanson AJ, Schneider J, Tahinci E, Orton D, Cselenyi CS, Jernigan KK, Meyers KC, Hang BI, Waterson AG, Kim K, Melancon B, Ghidu VP, Sulikowski GA, LaFleur B, Salic A, Lee LA, Miller DM 3rd, Lee E. Small-molecule inhibition of Wnt signaling through activation of casein kinase 1alpha. Nat Chem Biol. 2010;6(11):829–36.
58. Semenza GL. Hypoxia-inducible factor 1: master regulator of O_2 homeostasis. Curr Opin Genet Dev. 1998;8(5):588–94.
59. Jaakkola P, Mole DR, Tian YM, Wilson MI, Gielbert J, Gaskell SJ, von Kriegsheim A, Hebestreit HF, Mukherji M, Schofield CJ, Maxwell PH, Pugh CW, Ratcliffe PJ. Targeting of HIF-alpha to the von Hippel-Lindau ubiquitylation complex by O_2-regulated prolyl hydroxylation. Science. 2001;292(5516):468–72.
60. Ivan M, Kondo K, Yang H, Kim W, Valiando J, Ohh M, Salic A, Asara JM, Lane WS, Kaelin WG Jr. HIFalpha targeted for VHL-mediated destruction by proline hydroxylation: implications for O_2 sensing. Science. 2001;292(5516):464–8.
61. Kimura H, Weisz A, Ogura T, Hitomi Y, Kurashima Y, Hashimoto K, D'Acquisto F, Makuuchi M, Esumi H. Identification of hypoxia-inducible factor 1 ancillary sequence and its function in vascular endothelial growth factor gene induction by hypoxia and nitric oxide. J Biol Chem. 2001;276(3):2292–8.
62. Ceradini DJ, Kulkarni AR, Callaghan MJ, Tepper OM, Bastidas N, Kleinman ME, Capla JM, Galiano RD, Levine JP, Gurtner GC. Progenitor cell trafficking is regulated by hypoxic gradients through HIF-1 induction of SDF-1. Nat Med. 2004;10(8):858–64.
63. Kajiwara H, Luo Z, Belanger AJ, Urabe A, Vincent KA, Akita GY, Cheng SH, Mochizuki S, Gregory RJ, Jiang C. A hypoxic inducible factor-1 alpha

hybrid enhances collateral development and reduces vascular leakage in diabetic rats. J Gene Med. 2009;11(5):390–400.

64. Sarkar K, Fox-Talbot K, Steenbergen C, Bosch-Marce M, Semenza GL. Adenoviral transfer of HIF-1alpha enhances vascular responses to critical limb ischemia in diabetic mice. Proc Natl Acad Sci U S A. 2009;106(44):18769–74.

65. Wetterau M, George F, Weinstein A, Nguyen PD, Tutela JP, Knobel D, Cohen Ba O, Warren SM, Saadeh PB. Topical prolyl hydroxylase domain-2 silencing improves diabetic murine wound closure. Wound Repair Regen. 2011;19(4):481–6.

66. Gallagher KA, Liu ZJ, Xiao M, Chen H, Goldstein LJ, Buerk DG, Nedeau A, Thom SR, Velazquez OC. Diabetic impairments in NO-mediated endothelial progenitor cell mobilization and homing are reversed by hyperoxia and SDF-1 alpha. J Clin Invest. 2007;117(5):1249–59.

67. Thangarajah H, Yao D, Chang EI, Shi Y, Jazayeri L, Vial IN, Galiano RD, Du XL, Grogan R, Galvez MG, Januszyk M, Brownlee M, Gurtner GC. The molecular basis for impaired hypoxia-induced VEGF expression in diabetic tissues. Proc Natl Acad Sci U S A. 2009;106(32):13505–10.

68. Bergeron RJ, Wiegand J, McManis JS, Bussenius J, Smith RE, Weimar WR. Methoxylation of desazadesferrithiocin analogues: enhanced iron clearing efficiency. J Med Chem. 2003;46(8):1470–7.

69. Andrews NC. Disorders of iron metabolism. N Engl J Med. 1999;341(26):1986–95.

70. Botusan IR, Sunkari VG, Savu O, Catrina AI, Grunler J, Lindberg S, Pereira T, Ylä-Herttuala S, Poellinger L, Brismar K, Catrina SB. Stabilization of HIF-1alpha is critical to improve wound healing in diabetic mice. Proc Natl Acad Sci U S A. 2008;105(49):19426–31.

71. Sundin BM, Hussein MA, Glasofer S, El-Falaky MH, Abdel-Aleem SM, Sachse RE, Klitzman B. The role of allopurinol and deferoxamine in preventing pressure ulcers in pigs. Plast Reconstr Surg. 2000;105(4):1408–21.

72. Verma RK, Garg S. Drug delivery technologies and future directions. Pharmaceut Technol On-Line. 2001;25(2):1–14.

73. Zhang L, Pornpattananangku D, Hu CM, Huang CM. Development of nanoparticles for antimicrobial drug delivery. Curr Med Chem. 2010;17(6):585–94.

74. Duscher D, Neofytou E, Wong VW, Maan ZN, Rennert RC, Inayathullah M, Rodrigues M, Malkovskiy AV, Whitmore AJ, Walmsley GG, Galvez MG, Whittam AJ, Brownlee M, Rajadas J, Gurtner GC. Transdermal deferoxamine prevents pressure-induced diabetic ulcers. Proc Natl Acad Sci U S A. 2015;112(1):94–9.

75. Gavrilov K, Saltzman WM. Therapeutic siRNA: principles, challenges, and strategies. Yale J Biol Med. 2012;85(2):187–200.

76. Gao Z, Wang Z, Shi Y, Lin Z, Jiang H, Hou T, Wang Q, Yuan X, Zhao Y, Wu H, Jin Y. Modulation of collagen synthesis in keloid fibroblasts by silencing Smad2 with siRNA. Plast Reconstr Surg. 2006;118(6):1328–37.

77. Gary DJ, Puri N, Won YY. Polymer-based siRNA delivery: perspectives on the fundamental and phenomenological distinctions from polymer-based DNA delivery. J Control Release. 2007;121(1–2):64–73.

78. Song E, Zhu P, Lee SK, Chowdhury D, Kussman S, Dykxhoorn DM, Feng Y, Palliser D, Weiner DB, Shankar P, Marasco WA, Lieberman J. Antibody mediated in vivo delivery of small interfering RNAs via cell-surface receptors. Nat Biotechnol. 2005;23(6):709–17.

79. Dykxhoorn DM, Palliser D, Lieberman J. The silent treatment: siRNAs as small molecule drugs. Gene Ther. 2006;13(6):541–52.

80. Krebs MD, Jeon O, Alsberg E. Localized and sustained delivery of silencing RNA from macroscopic biopolymer hydrogels. J Am Chem Soc. 2009;131(26):9204–6.

81. Kanasty R, Dorkin JR, Vegas A, Anderson D. Delivery materials for siRNA therapeutics. Nat Mater. 2013;12(11):967–77.

82. Liu X, Ma L, Liang J, Zhang B, Teng J, Gao C. RNAi functionalized collagen-chitosan/silicone membrane bilayer dermal equivalent for full-thickness skin regeneration with inhibited scarring. Biomaterials. 2013;34(8):2038–48.

Harvesting, Processing, and Injection of Lipoaspirate for Soft-Tissue Reconstruction: Details Make the Difference

4

Matthias A. Sauter, Elizabeth A. Brett, Matthias M. Aitzetmüller, and Dominik Duscher

4.1 Introduction

Adipose tissue transfer is a well-established process, which has evolved massively since its inception. The first reported treatment using autologous fat was performed in 1889 by Van der Meulen [1], who grafted a free omental flap between the liver and the diaphragm to treat a hernia. Four years later the first free lipotransfer was done by Gustav Neuber [2] harvesting upper arm fat to correct adherent, depressed scar sequelae of osteomyelitis. Surgeons soon came to understand the qualities of fat tissue equate to perfect filling material. Fat is readily available and simple to harvest, and exhibits ideal plastic properties for remodelling soft-tissue defects. While at its baseline autologous fat is innocuous, studies show that grafted fat tissue stands to benefit the surrounding tissue in a regenerative manner. The fat grafting surgery itself shows a low donor-site morbidity, and is inexpensive and repeatable. However, with the rapidly expanding repertoire of techniques, the optimal fat grafting methodology is ill defined. Classically fat grafting has three main stages: harvesting, processing, and injection [3]. This chapter seeks to describe these steps in detail and examine the factors which govern graft outcome.

M. A. Sauter (✉)
Department of Plastic and Hand Surgery,
Klinikum rechts der Isar, Technical University
of Munich, Munich, Germany
e-mail: matthias.sauter@tum.de

E. A. Brett
Department of Plastic and Hand Surgery, Klinikum
rechts der Isar, Technical University of Munich,
Munich, Germany
e-mail: eliza.brett@tum.de

M. M. Aitzetmüller
Department of Plastic and Hand Surgery, Klinikum
rechts der Isar, Technical University of Munich,
Munich, Germany

Section of Plastic and Reconstructive Surgery,
Department of Trauma, Hand and Reconstructive
Surgery, Westfaelische Wilhelms, University of
Muenster, Muenster, Germany
e-mail: matthias.aitzetmueller@tum.de

D. Duscher
Department for Plastic Surgery and Hand Surgery,
Division of Experimental Plastic Surgery,
Technical University of Munich, Munich, Germany

4.2 Harvesting

The first step in lipofilling is the harvest of the transplantation tissue [4]. The largest consideration of the harvesting procedure is obtaining healthy tissue which translates to a healthy graft. The loss of fat at the harvest site itself can provide an aesthetic benefit for the patient; a "two-in-one" process is resultant from fat harvest from the inguinal, thigh, or abdominal area. Before the suction, the chosen site is usually pretreated with a tumescent solution often containing an anes-

© Springer Nature Switzerland AG 2019
D. Duscher, M. A. Shiffman (eds.), *Regenerative Medicine and Plastic Surgery*,
https://doi.org/10.1007/978-3-030-19962-3_4

thetic (lidocaine) to loosen the tissue and facilitate suction. Though useful in achieving postoperative pain relief, the anesthetics alter the metabolism and growth of mature adipocytes, and therefore initial differentiation from preadipocytes [5, 6]. Other studies focusing on the viability and total number of adipose-derived stem cells (ASCs)as part of the stromal vascular fraction (SVF) of cells in the graft harvested using tumescent fluid with lidocaine reported significantly decreased numbers, compared to control groups pretreated with tumescent solution without lidocaine [7]. Given that the functionality of these cells is key to survival of the graft [8], a preinjection step removing lidocaine with a saline wash has shown increased graft survival, decreased lipid reservoirs, and decreased levels of fibrosis [9]. Another additive to a standard solution could be epinephrine or a derivate. This vasoconstrictive drug shuts the vessels in the harvesting area and therefore decreases the bleeding. Conversely, a solution of lidocaine/epinephrine as tumescent fluid in a study of male Wistar rats showed no difference in fat graft survival compared to prilocaine or saline control [10]. The dichotomy in the literature on this one aspect of fat grafting serves to illustrate the nebulous stance on the optimal harvesting technique.

There is a plethora of suction methods from which the surgeons may choose. The standard techniques are the manual (Coleman technique) or suction-assisted liposuction (SAL). With the Coleman technique and SAL negative pressure is applied in combination with gentle back-and-forth movement of the cannula, causing physical disruption, thereby allowing adipose tissue harvest [11]. These methods represent the current gold standard, and reports show a lack of ASC damage and preservation of their regenerative potential [12]. There are multiple modifications of this method, including power-assisted liposuction (PAL), water jet-assisted liposuction (WAL), laser-assisted liposuction (LAL), and ultrasound-assisted liposuction (UAL). They all are developed to further ease the process of suctioning the tissue, with minimal trauma to the donor site and maximal outcome in the requested aesthetic result. However, their design is generally not optimized to enhance trans-

plantable graft tissue quality. The water jet-assisted liposuction uses a fan-shaped jet of water which decomposes the fat tissue into pieces. Meanwhile these fragments can easily be sucked in through the opening of the cannula [13]. When compared with the manual liposuction as a standard, no significant difference in the condition of the harvested ASCs can be found, which makes this harvest method applicable for fat grafting [14, 15]. In PAL the cannula is oscillating, which manually disrupts the tissue before aspiration [16]. When the number and viability and the ASCs harvested by PAL are compared to SAL, there is no significant difference [17]. Another approach is the utilization of cavitational ultrasound waves to dissociate adipose tissue. Currently available in its third generation, UAL also provides a stable outcome of viable ASCs, comparable to SAL [18, 19]. The basic principle of LAL is the utilization of a laser beam of a certain wavelength. The laser is used to create thermal energy which physically breaks down the fat tissue, which may be then easily extracted. Here however, scientists found impairments in the overall quality of the harvested ASCs. Viability, proliferation, and regenerative potential are reduced compared to ASCs harvested by SAL [12]. In conclusion, apart from the LAL, all harvesting methods seem to be sufficient in providing functional grafts, at least in regard to ASC quality.

Another aspect of fat harvest to consider is the size of the cannula used. For effective fat grafting, it is pivotal to reduce the mechanical and shear forces on the harvested fat. Several findings support the rheological impact on the viability of the fat graft. When comparing diameters of 6, 4, and 2 mm, Erdim et al. [20] found that cellular viability was increased with the use of each bigger cannula. Kirkham et al. [21] saw that 5 mm cannulas are superior to 3 mm cannulas in leading to fat graft retention. Large-bore cannulas decrease the mechanical sheer stress on the harvested cells, and subsequently increase the total number and health of the aspirated cells. This is per the fluid dynamic equation of $\tau = F/A$ (tau (shear stress) is equal to force divided by area with area being the cross section of the cannula). Corroborating these results a study using ex vivo panniculectomies showed that differences in applied negative pressure did not impact fat viability; however the inner diameter of the utilized cannulas did [22]. This is a very accurate tableau showing the impact of cannula size as the chief regulator of harvested fat health.

4.3 Processing

The harvested fat usually undergoes processing before injection, usually meaning the separation of the actual fat tissue from unnecessary other components of the aspirate such as an oily and a liquid portion containing blood and cell debris. The optimal way of graft preparation is currently discussed; this also includes further effectors of graft outcome, the amount of viable fat cells injected, inflammation induced by intraoperative contaminants, or addition of cells from the SVF [23]. Similarly, the physical processing of the fat yields significant differences in graft take. Girard et al. compared simple decantation and single centrifugation with multiple repeated centrifugations, combined with saline washes of the graft material. Implanted fat grafts in mice showed that engraftment suffered if the fat was just decanted, and graft quality was better after multiple centrifugations combined with several washing steps [9]. Different studies looked at another method called cotton (Telfa™) gauze rolling. Here, the graft material is rolled on a sterile gauze, which serves to absorb the oil and liquid portions of the aspirate. It is a method known to result in increased graft survival, and functional adipocytes when compared to centrifugation [24]. Alternately, a recent study compared centrifuged to phase-separated fat for breast augmentation in post-mastectomy patients. They performed volumetric analysis of grafts via MRI, and determined no significant difference in clinical outcome [25]. These disparate findings illustrate the ongoing lack of clarity on adipose tissue processing, and lack of a comprehensive protocol. Two extensive literature reviews in 2015 also came to this conclusion. They compared studies where fat was decanted, cotton gauze rolled, centrifuged, filtered, or washed. The inconclusive findings presented are assigned to the paucity of quality, clinical data, and lack of profound research [23, 26].

4.4 Injection

With the actual grafting of the prepared tissue, the artistry of the plastic surgeons is required to achieve a satisfactory aesthetic outcome. First a small incision is placed at the recipient site, in which a delivery cannula is inserted. Through the cannula the fat is injected into the transplantation zone. There is some discord about the optimal way of executing the injection. Given the research on cannula gauge and air pressure while fat harvesting, it is unsurprising that these two aspects overlap into the grafting phase. A recent study in 2017 used human cases to exhibit the differences between a manually controlled directional injection (akin to syringe) and a device which constantly puts pressure on the grafting material, creating only the need for directing the flow of material. There was an appreciable decrease in unevenness, ecchymosis, and nodules on the side of the face treated with the device, compared to that of the manual injection [27]. While this pertains to technique of grafting fat, there have been questions about the main delivery effectors of graft viability. Lee et al. [22] compared cannula pressure levels and injection speeds. It was observed that higher injection speeds/flow rate (and therefore higher shear stress) of 3–5 mL/s negatively affected the graft viability to low injection speed of only 0.5–1 mL/s. No detectable viability difference resulted from using different delivery pressures. An interesting evolving topic is preparation of the recipient site. The typical method of site preparation for breast reconstruction is use of tissue expanders; short-term physical devices implanted subcutaneous/submuscular to create space [28]; a newer idea is the external volume expansion (EVE) prior to grafting [30, 31]. This has found its home in larger fat grafting applications, as in breast reconstruction/augmentation. It involves adhering a suction device to the outer chest wall, and applying negative pressure. The purpose of this is twofold; to primarily stretch the skin and loosen the underlying stroma/breast parenchyma, and to induce various mechanisms like ischemia, edema, and inflammation. These are known processes for stimulating cell proliferation and angiogenesis [29]. Studies researching this theoretical benefit widely confirm it in practice. The accepted proposition is that the initial

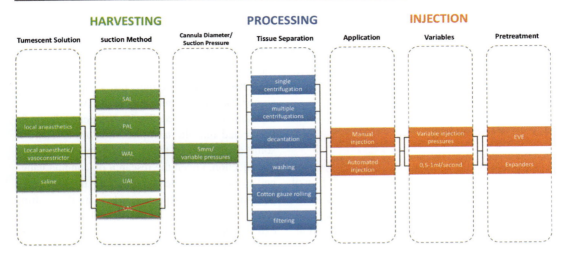

Fig. 4.1 Summary of the different options available to execute the different grafting steps. Red crosses mark procedures that are not recommended

preemptive cavitation of the recipient site decreases the pressure on the grafted fat, allowing facile oxygen and nutrient diffusion through the new graft [30, 31].

4.5 Conclusions

The ranging options in each step of the fat grafting process make it difficult to pin down what exactly is the right combination for the healthiest fat graft (Fig. 4.1). Given the recent literature, it would appear that disparate factors like anesthesia, injection speed and consistency, use of Telfa™, phase separation of the grafting material, and physical preparation of recipient site all play roles in graft quality over time. With the current and emerging techniques, it is powerful and important to grasp fully these methods. Only then can legitimate protocols be established through informed reading of the literature.

References

1. Bellini E, Grieco MP, Raposio E. The science behind autologous fat grafting. Ann Med Surg (Lond). 2017;24:65–73.
2. Mazzola RF, Mazzola IC. The fascinating history of fat grafting. J Craniofac Surg. 2013;24(4):1069–71.
3. Coleman SR. Structural fat grafts: the ideal filler? Clin Plast Surg. 2001;28(1):111–9.
4. Shiffman MA, Mirrafati S. Fat transfer techniques: the effect of harvest and transfer methods on adipocyte viability and review of the literature. Dermatol Surg. 2001;27(9):819–26.
5. Keck M, Zeyda M, Gollinger K, Burjak S, Kamolz LP, Frey M, Stulnig TM. Local anesthetics have a major impact on viability of preadipocytes and their differentiation into adipocytes. Plast Reconstr Surg. 2010;126(5):1500–5.
6. Moore JH Jr, Kolaczynski JW, Morales LM, Considine RV, Pietrzkowski Z, Noto PF, Caro JF. Viability of fat obtained by syringe suction lipectomy: effects of local anesthesia with lidocaine. Aesthet Plast Surg. 1995;19(4):335–9.
7. Goldman JJ, Wang WZ, Fang XH, Williams SJ, Baynosa RC. Tumescent liposuction without lidocaine. Plast Reconstr Surg Global Open. 2016;4(8):e829.
8. Piccinno MS, Veronesi E, Loschi P, Pignatti M, Murgia A, Grisendi G, Castelli I, Bernabei D, Candini O, Conte P, Paolucci P, Horwitz EM, De Santis G, Iughetti L, Dominici M. Adipose stromal/stem cells assist fat transplantation reducing necrosis and increasing graft performance. Apoptosis. 2013;18(10):1274–89.
9. Girard AC, Mirbeau S, Gence L, Hivernaud V, Delarue P, Hulard O, Festy F, Roche R. Effect of washes and centrifugation on the efficacy of lipofilling with or without local anesthetic. Plast Reconstr Surg Global Open. 2015;3(8):e496.
10. Livaoglu M, Buruk CK, Uraloglu M, Ersoz S, Livaogglu A, Sozen E, Agdoğan Ö. Effects of lidocaine plus epinephrine and prilocaine on autologous fat graft survival. J Sraniofac Surg. 2012;23(4):1015–8.
11. Venkataram J. Tumescent liposuction: a review. J Cutan Aesthet Surg. 2008;1(2):49–57.
12. Duscher D, Luan A, Rennert RC, Atashroo D, Maan ZN, Brett EA, Whittam AJ, Ho N, Lin M, Hu MS,

Walmsley GG, Wenny R, Schmidt M, Schilling AF, Machens HG, Huemer GM, Wan DC, Longaker MT, Gurtner GC. Suction assisted liposuction does not impair the regenerative potential of adipose derived stem cells. J Transl Med. 2016;14(1):126.

13. Taufig AZ. Water-jet-assisted liposuction. In: Shiffman MA, editor. Liposuction: principles and practice. Berlin: Springer; 2006. p. 326–30.

14. Bony C, Cren M, Domergue S, Toupet K, Jorgensen C, Noel D. Adipose mesenchymal stem cells isolated after manual or water-jet-assisted liposuction display similar properties. Front Immunol. 2015;6:655.

15. Meyer J, Salamon A, Herzmann N, Adam S, Kleine HD, Matthiesen I, Ueberreiter K, Peters K. Isolation and differentiation potential of human mesenchymal stem cells from adipose tissue harvested by water jet-assisted liposuction. Aesthet Surg J. 2015;35(8):1030–9.

16. Coleman WP 3rd. Powered liposuction. Dermatol Surg. 2000;26(4):315–8.

17. Keck M, Kober J, Riedl O, Kitzinger HB, Wolf S, Stulnig TM, Zeyda M, Gugerell A. Power assisted liposuction to obtain adipose-derived stem cells: impact on viability and differentiation to adipocytes in comparison to manual aspiration. J Plast Reconstr Aesthet Surg. 2014;67(1):e1–8.

18. Duscher D, Maan ZN, Luan A, Aitzetmuller MM, Brett EA, Atashroo D, Whittam AJ, Hu MS, Walmsley GG, Houschyar KS, Schilling AF, Machens HG, Gurtner GC, Longaker MT, Wan DC. Ultrasound-assisted liposuction provides a source for functional adipose-derived stromal cells. Cytotherapy. 2017;19(12):1491–500.

19. Duscher D, Atashroo D, Maan ZN, Luan A, Brett EA, Barrera J, Khong SM, Zielins ER, Whittam AJ, Hu MS, Walmsley GG, Pollhammer MS, Schmidt M, Schilling AF, Machens HG, Huemer GM, Wan DC, Longaker MT, Gurtner GC. Ultrasound-assisted liposuction does not compromise the regenerative potential of adipose-derived stem cells. Stem Cells Translat Med. 2016;5(2):248–57.

20. Erdim M, Tezel E, Numanoglu A, Sav A. The effects of the size of liposuction cannula on adipocyte survival and the optimum temperature for fat graft storage: an experimental study. J Plast Reconstr Aesthet Surg. 2009;62(9):1210–4.

21. Kirkham JC, Lee JH, Medina MA 3rd, McCormack MC, Randolph MA, Austen WG Jr. The impact of liposuction cannula size on adipocyte viability. Ann Plast Surg. 2012;69(4):479–81.

22. Lee JH, Kirkham JC, McCormack MC, Nicholls AM, Randolph MA, Austen WG Jr. The effect of pressure and shear on autologous fat grafting. Plast Reconstr Surg. 2013;131(5):1125–36.

23. Cleveland EC, Albano NJ, Hazen A. Roll, spin, wash, or filter? Processing of lipoaspirate for autologous fat grafting: an updated, evidence-based review of the literature. Plast Reconstr Surg. 2015;136(4):706–13.

24. Canizares O Jr, Thomson JE, Allen RJ Jr, Davidson EH, Tutela JP, Saadeh PB, Warren SM, Hazen A. The effect of processing technique on fat graft survival. Plast Reconstr Surg. 2017;140(5):933–43.

25. Sarfati I, van la Parra RFD, Terem-Rapoport CA, Benyahi D, Nos C, Clough KB. A prospective randomized study comparing centrifugation and sedimentation for fat grafting in breast reconstruction. J Plast Reconstr Aesthet Surg. 2017;70(9):1218–28.

26. Gupta R, Brace M, Taylor SM, Bezuhly M, Hong P. In search of the optimal processing technique for fat grafting. J Craniofac Surg. 2015;26(1):94–9.

27. Song M, Liu Y, Liu P, Zhang X. A promising tool for surgical lipotransfer: a constant pressure and quantity injection device in facial fat grafting. Burns Trauma. 2017;5:17.

28. Bellini E, Pesce M, Santi P, Raposio E. Two-stage tissue-expander breast reconstruction: a focus on the surgical technique. Biomed Res Int. 2017;2017:1791546.

29. Lancerotto L, Chin MS, Freniere B, Lujan-Hernandez JR, Li Q, Valderrama Vasquez A, Bassetto F, Del Vecchio DA, Lalikos JF, Orgill DP. Mechanisms of action of external volume expansion devices. Plast Reconstr Surg. 2013;132(3):569–78.

30. Reddy R, Iyer S, Sharma M, Vijayaraghavan S, Kishore P, Mathew J, Unni AK, Reshmi P, Sharma R, Prasad C. Effect of external volume expansion on the survival of fat grafts. Indian J Plast Surg. 2016;49(2):151–8.

31. Khouri RK, Khouri RER, Lujan-Hernandez JR, Khouri KR, Lancerotto L, Orgill DP. Diffusion and perfusion: the keys to fat grafting. Plast Reconstr Surg Global Open. 2014;2(9):e220.

Adipose Tissue Complex (ATC): Cellular and Biocellular Uses of Stem/Stromal Cells and Matrix in Cosmetic Plastic, Reconstructive Surgery and Regenerative Medicine

5

Robert W. Alexander

5.1 Evolution of Regenerative Medicine in Plastic Surgery

For more than three decades, aesthetic, plastic, and reconstructive surgeons have devoted themselves to understand the intricate management of both acute and chronic wounds, including the fundamentals of healing and repair. Through those years, the importance of examination of the homeostatic and replacement mechanisms have afforded us the opportunity to explore how the body successfully maintains and repairs defects from aging, damage or degeneration. Throughout, the value of understanding how various tissues accomplished such task has evolved to one of doing everything possible to encourage vascularization of sites and activations of reparative cells (both remote and locally encouraged via signal proteins and growth factors) which are needed in essentially all groups of patients. In the early 1990s, appreciation of the value of using the most available and rich source of such molecules could be isolated and concentrated, known as platelet-rich plasma (PRP), with platelets serving as a

R. W. Alexander (✉)
Department of Surgery, University of Washington, Seattle, WA, USA

Institute of Regenerative Medicine, Stevensville, MT, USA
e-mail: rwamd@cybernet1.com

very rich source for a multitude of growth factors, cytokines, and signal proteins from within their alpha granules stored in their cytoplasm. First reports were examined in the area of maxillofacial and craniofacial bony reconstruction in a variety of situations in the early 1990s. Wound healing was felt to significantly improve in the presence of such elements. At the time, users were required to utilize bulky and costly "cell-saving" technology, meaning that major OR uses were a limiting factor, in that most required presence of technicians and cardiovascular ORs to be able to accomplish. In the most complicated surgeries, this was felt by many to be a "value-added" contribution to the intraoperative surgical care and follow-up care requirements.

Early in the 2000s, these bulky and somewhat inefficient machines were replaced by a number of medical device companies creating smaller, efficient, and more simple centrifugation systems which provided consistent and high concentration abilities. At the same time, this expanded delivery capabilities in more cost-effective and consistent protocols, for use in both the operating room and the outpatient surgical centers. With the advent of small footprint and affordable systems, the uses and advantages of utilizing these autologous elements expanded exponentially. The area of chronic wound healing, devascularizing injuries, and a vast range of soft-tissue repair became ubiquitous. As of the writing of this contribution, many

© Springer Nature Switzerland AG 2019
D. Duscher, M. A. Shiffman (eds.), *Regenerative Medicine and Plastic Surgery*,
https://doi.org/10.1007/978-3-030-19962-3_5

researchers and clinical scientists are developing more sophisticated understanding of optimization of effects, including which product or content element is contributing in critical situations of healing or regeneration. As the applications are tested for safety and efficacy, these values are expected to gradually evolve as are the clinical uses.

Poorly understood is the fact that MANY products have rushed to the market, each claiming capabilities and concentrations that are simply incorrect. During the evolution, many surgeons (especially orthopedic surgery and related areas) were disappointed at the outcome improvement. A significant issue evolved with lesser understanding that simply isolating some platelets did NOT yield high enough, or consistent enough, products which could deliver the most optimal outcomes. It has become very clear that those which provide a low concentration (single-spin centrifugation yielding 1.5–2.5 times measured baselines) often can be of value in facial and dermal uses, whereas much higher (>4–6 times baseline) concentrates are now consistently and easily obtained in point of care within outpatient or operating room settings.

Since it is now well established that there is a linear increase in available critical growth factors and signal proteins, most surgical and regenerative applications are favored with the higher concentrations. In addition, the level of understanding is being evolved where decisions of choosing a high hematocrit version or a low hematocrit version have a clear indication tied to the specific locations where they may be utilized. In aesthetic plastic surgery, it is considered that the use of the highest concentrations that can be currently isolated is of significant clinical advantage over the lower concentrations.

Over the past 15+ years, many indications and uses have evolved. One of the most important related to the addition of these platelet concentrates to uses in autologous fat grafting (in small and large volumes) became more clear. The statistically significant reduction in lipid cysts and microcalcifications, coupled with more stable volume retention in use of autologous fat graft breast augmentation, has enhanced the use of platelet products in such cases. In addition, observational findings of important skin vascularity and texture changes in the areas

of the face/neck and the body areas have come to the forefront.

These concepts have made a major contribution that has spread well beyond the wound healing and aesthetic applications. For the past decade and one-half, uses in regenerative medical and surgical fields have become known as highly effective and safe alternatives to use of steroids and non-autologous products, and changed some indications for invasive surgical interventions. This contribution will expand on this background and evolving protocols aimed at skin, tissue supplementation, and extensive applications using targeted guidance in chronic pain, musculoskeletal applications, and systemic cellular therapeutic options.

5.2 Evolution of Cell-Based Therapies

Over the past decade, great strides have been made in the understanding and potentials of systemic and targeted cell-based therapies. Starting decades ago, use of an irritant solution to stimulate inflammatory reactions has been replaced in the past few years with transition to injecting various platelet-rich plasma (PRP) concentrates for supporting an effective inflammatory reaction at damaged or degenerative sites. Use of the contained growth factors and signal proteins became recognized as offering a significant improvement in tissue-healing responses, but seemed to be limited by incomplete repair while requiring a series (often 4–6) to achieve long-term clinical improvement. Current biocellular (orthobiologic) approaches combine the trophic growth factors and important signal proteins which are added with cellular and bioactive matrix elements synergistically to enhance the reparative and regenerative abilities in areas of need. The importance of signaling mechanisms and paracrine responses have been expanding rapidly since 2000. Aesthetic and reconstructive applications led the way, as constant challenges of injury, loss of circulatory capabilities, degenerative changes, repair, etc. demanded an optimal approach to structural augmentation and regenerative needs. In-depth examination of how our body maintains itself revealed that undesignated cells were integrally important to replace aging cells (such as skin, hair, bowel lining, etc.). Early on, fat was not thought of as undergoing such homeostatic mechanisms, since typical mitotic activities were not observed. We now recognize that, rather than a static number of cells varying only in size, mature adipocytes do actually undergo total replacement at a rate of 10–20% per year, but do so in a different form of cell division, known as "asymmetric cell division." It is very important to understand that many of the biocellular therapies also rely on the presence and reactivity of pre-existing "resident" cell populations which also contribute to local niche (microenvironment) responses. The asymmetric cellular replacement is the reason that one of the cells may be differentiated into a specific cell type or contribute molecular component needs (paracrine effects) locally, while the other one retains its "undesignated" cell capability and remains available for future responses. This is the basis for abilities of tissues & cells to respond to demands and retain the "self-renewal" capabilities ongoing and future homeostasis demands [1].

With the advent of FDA-approved tabletop programmable, centrifugal devices for custom high-density platelet concentrations via closed system, use of a simple blood draw yielded more than four to six times patient's own circulating baselines levels. It has been very well shown that the higher the achieved concentrations, the proportionally higher delivery of important factors intrinsically involved in all wound healing and repair (Fig. 5.1).

It has become clear that certain tissue characteristics are most favorable for use in cell-based therapies including easy and safe access coupled with plentiful autologous stores of a group of cells possessing multipotent potential. Multipotency is important in that such cells have the capability of responding to local signals and possess the ability to transform or replenish signals needed at damaged or diseased site for repair or regeneration. Researchers are gradually understanding the complexities of contributions of undesignated cellular elements in combination with secretory signaling (paracrine effects) in chemotactic and critical "up or down" regulation of regenerative processes within the environment.

Research has confirmed that the vast majority of such undesignated cells are associated and stored in proximity to the microvascular capillary system and adventitia (Fig. 5.2). Essentially all tissues (with blood supply) have some of these

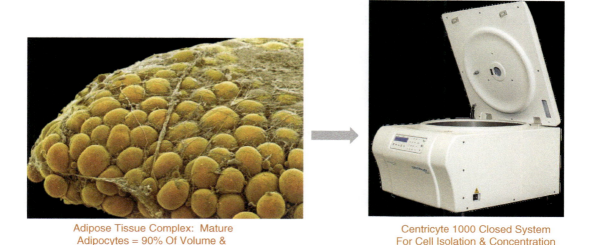

Adipose Tissue Complex: Mature Adipocytes = 90% Of Volume & Only 10% Of Nuclei In tSVF

[Turnover Rate: 10-20% Per Year]

Centricyte 1000 Closed System For Cell Isolation & Concentration

Fig. 5.1 Left: Adipose tissue complex (mature adipocytes provide approximately 90% by volume; 10% of nuclei in tSVF). Right: Closed system for centrifugation, cell isolation/concentration at point of care (CC1000 Healeon Medical, Newbury Park, CA, USA)

multipotent cells available to deal with local and isolated demands. The body retains the ability to chemotactically attract and mobilize cells from local and remote storage points in response to chemical and physical signaling in the body. Approximately 15 years ago, an important scientific advance was made by researchers in finding that adipose tissue (fat) contained high numbers of such cells [2, 3]. This is not totally surprising considering that fat also represents the largest microvascular organ in the body.

Enhancement of cellular and biologic therapies comes directly with the ability for providers to be able to identify, target, and guide the cellular vascular fraction (SVF) or cellular-biologic combination to areas of injury or degeneration. Management of protocols has led to improved structural grafting successes in addition to the contribution in orthopedic and sports medicine application using the same cell components. In that regard, ultrasonography has clearly become a MAJOR feature to clinical responses and success. As an example, in medium and deep targets, or those difficult to access, guided MSK ultrasound capabilities offer the optimal integral part of successful responses. Over the past decade, thousands of treatments using Biocellular Regenerative Medicine© techniques have proven safe and remarkably effective. The information provided in this chapter is intended as an introduction to important concepts and describes the current logic believed to be involved. Major steps have been taken, moving from the laboratory to the bedside. Today, musculoskeletal (MSK) and aesthetic-plastic surgical patients are routinely treated with this combination of cellular or biocellular elements [4–6].

5.3 What Is Biocellular Medicine?

The term "biocellular" refers to the combination of important *biological chemicals* (such as growth factors, signal proteins, and chemicals important to wound healing) with *undesignated cells* (often referred to as adult stem/stromal cells, pericyte/endothelial, periadventitial, or mesenchymal cells) found widely spread within our body, and which participate in tissue maintenance, repair, and regeneration. Since 2010, science and medicine have advanced which is termed "translational phase," where proven laboratory science has demonstrated important contributions along with the clinical application of science in human applications in the last decade. Since then there has been controversy concerning the use of the term "stem cells" in the current practice of medicine. Unfortunately, these arguments typically occur with the misuse or overuse of the term "stem cell," as being interpreted as uses of pure "embryonic" or fetal stem cells, implying destruction of embryo or fetal tissues. In the past decade, the recognition of the safety and efficacy of using a person's own (autologous) adult stem/stromal cells has advanced to the point that it is now widely documented and published (Fig. 5.3).

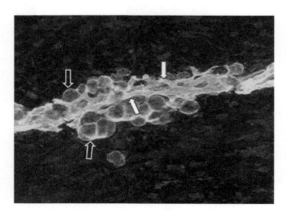

Fig. 5.2 Micrograph demonstrating the relationship of intimate relationship of microcapillaries and attached adipocyte during development (photo: fetal pig during extremity formation)

- Return To Full Form or Function
- Eliminate Or Markedly Decrease Pain
- Resist Recurrence Of Injury Or Damage
- Reverse, Stabilize, Resist Degeneration
- Avoid Immune Reactions
- Prefer *Autologous* Cells/Tissues For Repair
- Accelerate Healing Processes
- Restore Tissues With Minimal Scarring
- Minimal Tissue Distortion & Morbidity
- Prefer Minimal Manipulation Requirements
- Predictable & Reproducible Outcomes

Fig. 5.3 General goals in regenerative medical applications

5.3.1 Cellular Components

Biocellular Regenerative Medicine® within the United States currently refers to the use of autologous, ADULT (non-embryonic) multipotent cells capable of participating in maintaining our tissues (homeostasis), healing, and regeneration. Since 2006, the scientific studies demonstrating the values of the highly variable stromal cell populations have exploded, now to the point that active reports and studies of component cells of adipose origin now exceed the study of non-hematopoietic stromal cells in bone marrow in MSK and aesthetic-plastic surgical applications. The importance of such studies and understanding that adipose tissue deposits have gained such recognition due to the much greater numbers of stem/stromal cells (other than blood forming element) in the body is coupled with the important overlap of potential cellular functions. Essentially every tissue in our bodies that contains microvascular supply maintains a reservoir of such cells within the stroma and matrix. That said, we now recognize that adipose tissue complex (ATC) possesses the greatest microvascular organ in the body and serves as an excellent option to access for a variety of uses. Many important peer-reviewed scientific reports suggest that adipose-derived stem/stromal cells of mesodermal origin provide between 1000 and 2000 times the actual numbers found in bone marrow [7]. With the relatively easy collection of adipose tissue, less penetration, widely heterogeneous cellular populations, and important immune-privileged properties, subdermal fat deposits effectively and safely serve as a primary source for gathering stem/stromal cells (Figs. 5.4 and 5.5).

Efforts at using mechanical separation (ultrasound, nutational, emulsification efforts) have not proven to be able to separate the very complex chemical binding of stem-capable group due to the cell-to-cell or cell-to-matrix (ECM and periadventitial) attachments. The complex multifaceted connectivity necessary for cells to act on area signaling that exists with the ATC simply will not yield optimal separation and cellular purity needed to create a cSVF pellet. At this point in time, use of incubation, agitation, and digestive processes (blends of collagenase and

- Easy Harvest Access
- High Quantity Of Viable Stem/Stromal Cells
- Minimum Morbidity Of Donor Site
- Safety After Implantation (Autologous Best)
- Multipotent & Proliferative Cell Groups
- May Be Isolated & Concentrated If Desired
- Prefer Inclusion Stromal Elements (Bioactive)
- Paracrine Functions Encouraged
- Secrete Immunomodulatory Factors
- Secrete Pro-Inflammatory Factors (Benign)
- Immunopriviledged Cell Groups Preferred

Fig. 5.4 Adipose tissue complex: optimal cell source

neutral proteases) remains the optimal way to isolate and concentrate the desired cellular groups [8] (Figs. 5.6 and 5.7).

The small nucleated cells found closely associated within the vascular tissues are now recognized as serving important roles in maintaining normal tissue content (homeostasis). PLUS have the ability to respond to injury or disease processes in a constant effort to maintain, heal, or repair damaged cells (as in aging, arthritis, musculoskeletal tissues, neurological disorders, etc.) in the body. The remarkable design of the human body uses these reservoirs of available, non-differentiated, multipotent cells as the tissue "first responders" in the situations of major or microtrauma and aging. By secretion and signaling of certain chemicals from a degenerating or injured site, these multipotent cells (i.e., can become various types of cells) can be called upon to participate in the repairs needed to restore tissues and functions. At first, it was felt that the cellular elements were the critical, and perhaps the only important, ones; experience now shows that the **paracrine** secretory capabilities play critical roles (exosomes, microvesicles). There are many peer-reviewed publications which provide examples of how the cells involved in this process can be enhanced by combined provision of the cellular, native scaffolding, and biologically active components.

5.3.2 Biologic Components

The "biological" components in this context refer specifically to the availability of a diverse and

KEY Multipotent Cells Found In AD-tSVF

❏ Mesenchymal Stem/Stromal Cells

❏ Pericyte-Endothelial Precursor Cells**

❏ Adipocyte Precursor Cells

Mature Adipocytes (Not Multipotent Temporary But Contribute In Signaling)

Tissue Resident Cell Populations & Bioactive NATIVE Structural Matrix Contribute to the Regenerative Process

** May Be The "Origin" Of All MSCs

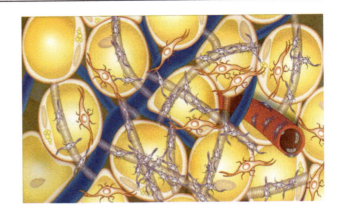

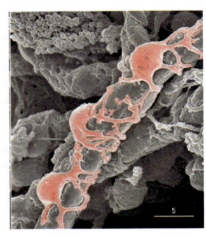

Fig. 5.5 Multipotent cells found in tissue stromal vascular fraction (tSVF)

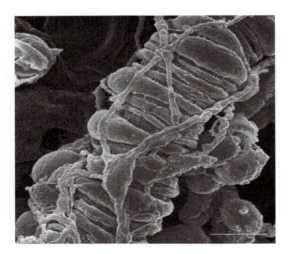

Fig. 5.6 SEM image of pericytes' relationship to microcapillary. Note: See many complex cell-to-cell and cell-to-matrix attachments

important variety of growth factors and signal proteins which interact with the cells of the degenerative or damaged sites to help recruit needed reparative cells and materials to repair the area. There are two major "biological" components in common use at this time. First is found within recognized contents of platelets, which store and release a wide variety of needed growth factors and proteins to act on available cells to begin the wound-healing processes [9]. For many years, we thought that the only important roles of platelets were to become "sticky," adhere to each other, and participate in clotting mechanisms. We now realize that this may be their LEAST important contribution to wounds and wound healing (with exception of providing a fibrin clot to permit gradual release of platelet contents). Platelets represent a storehouse of small gran-

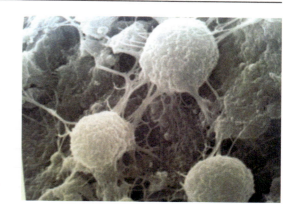

MSCs Cell-To-Matrix
& Cell-To-Cell Contacts

Native Adipose Extracellular Matrix
(Bioactive – Storage GF - Secretive)

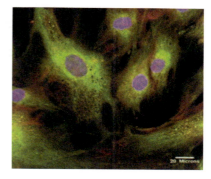

Cell-To-Cell Contacts

Fig. 5.7 Left: Native scaffolding adipose tissue complex after cellular removal. Right top: SEM mesenchymal stem cells' intimate relationship with extracellular matrix and cell-to-cell elements. Right bottom: Mesenchymal stem cell micrograph with NAPI nuclear stain

ules, each containing very important growth factors and signal proteins that serve to "quarterback" to the healing cascade, and do this for a significant time during the healing phases. For example, an important chemical available from these granules is essential for blood vessel replacement and repair in order to improve the circulation ability critical to healing of all wounds. Without adequate blood flow, needed oxygen cannot reach the area of damage, nor permit migration of a variety of cells from nearby or distant cell sites (Figs. 5.8 and 5.9).

The second source of biological contributors is found in bone marrow aspirates. Bone marrow has been used for many decades, and it is commonly used in blood-related disorders. However bone marrow does demonstrate microvasculature and, therefore, does have some undesignated reparative cells (stem/stromal cells). The vast

HD Platelet Concentrates

• Higher Growth Factors & Signal Proteins

• Directly Impact *Proliferation & Migration*

• Platelets "*Quarterback*" Healing Cascade

• Contributes Tissue "Autoamplification" Of Critical *Signaling* Within Repairing Sites

Robert W. Alexander, MD, FiCS

Fig. 5.8 Why use of high-density platelet concentrates (HD-PRP)?

majority of stem cells located in marrow belong to the "hematopoietic" stem cell group, and are not considered extremely valuable in the case of regeneration or repair cellular group. However, the desired stem/stromal cells are found in rela-

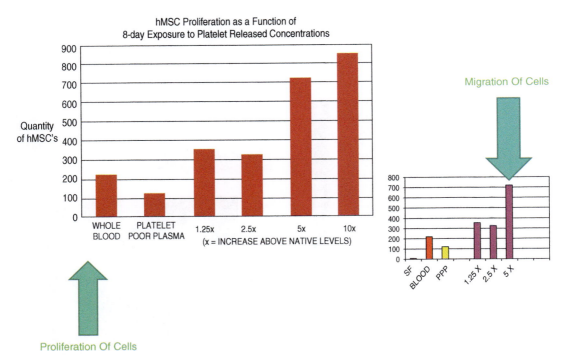

Fig. 5.9 Importance of HD-PRP on proliferation of cells and migration of cells [note: linear response as concentration increases] (Hayneworth, S.E, et al. Mitogenic Stimulation of Human MSCs by Platelet Concentrates. Orth Res Soc. (2002))

tively very low numbers, compared to adipose tissues and its microvasculature. Therefore, many now consider bone marrow as primarily a valuable biologic and platelet source. In order to become a valuable cell contributor for such reparative group, it is required that bone marrow aspirates be isolated, concentrated, and culture-expanded to achieve meaningful numbers needed in regenerative and healing applications. This source is technically a bit more invasive to obtain, poses higher complication-sequelae rates, and is significantly more expensive to the patients. In addition, the actual number of "reparative" cells (including mesenchymal, pericyte and periadventicial cells) available with the marrow are markedly lower than found in adipose microcapillary tissues. The bone marrow undesignated group is heavily weighted to the hematopoietic stem cell groups. Many now consider that the bone marrow should be considered as a "super" PRP source and included as a valuable biological growth factor (as a platelet concentrate) rather than a primary reparative cell source. At this time there is very little evidence of significant contribution to the regeneration process of MSK tissues derived from the HSC group.

The primary importance and value of concentrates are the ability to provide important growth factors and cytokines/chemokines to optimize earlier healing conditions and abilities. Of even more importance in the cellular therapeutic based effects is their important paracrine secretory influences, rather than contributions of individual cellular components and physical engraftment. Further, it is well established that the mesenchymal group (MSC) of multipotent cells may actually originate from the pericyte-endothelial cell groups [10]. These offer a great amount of overlapping capabilities, in vitro, suggesting that all tissues having some microvasculature have resident stem/stromal elements capable of providing "first responders" to sites of damage or degenerative effects. It is suggested that the multi-tissue MSC-like groups overlap at greater than 95% in their differential capabilities, at least in in vitro conditions. Host-site interaction with these stem/stromal cells, growth factors, and signal proteins seems to create a complex, heterogeneous precursor population that is considered "site specific" in many of their responses [11].

5.4 How Did Biologic and Cellular Therapeutic Concepts Evolve?

Aesthetic and plastic surgeons have traditionally dealt with wound healing and scarring issues for many years. During that time, careful study of the processes of homeostasis, remodeling, and repair led to a better understanding of how the body tissues managed to maintain themselves. For many years, the importance of biologics as a derivative part of platelets was appreciated, not only for its clotting functions, but also for the gradual release of critical chemical components essential to the healing processes in individual sites. These biocellular concentrates are felt to immediately begin to participate in secretions capable of site-specific repair and regeneration, while local cells begin to actively contribute. In addition to these elements, appreciation of the importance of the bioactive, native adipose (3-D) scaffolding (matrix) in provision of essential contact points which serve to encourage microenvironment changes (including cellular proliferation and chemotactic migration) has come to the forefront [12–14]. It has become more clear that site specificity greatly influences cellular changes within the non-designated, heterogeneous, multipotent populations found in essentially ALL tissues which have microvasculature. In addition, appreciation of the importance of cellular secretions (paracrine and autocrine) within these undifferentiated cell groups has been reported to be of as great, or greater in some instances, as the multipotent cellular differentiation effects [15].

Once the complex processes of repair and regeneration were examined closely, it became apparent that specific interactions of any single cell or chemical are not able to be determined at this time. At this time, the ability to create an "in vitro" environment to duplicate the "in vivo" niche remains elusive. This makes the ability to develop "optimal" components or interpret activities which can be translated to the in vivo applications impossible. It is well known that the process of cellular isolation and culture expansion likely introduces variables which cannot yet be interpreted accurately.

Key adult multipotent cells are found in essentially every tissue and organ in the body. The determination that some of the highest concentrations of these adult stem/stromal cell populations were found within adipose tissue complex (ATC) has led to a major trend shift to more closely evaluate the activities of such tissues, and how they can be easily and safely acquired, and concentrated, for uses in wound healing and repair. Early on, since adipocytes within the ATC were determined not to undergo mitosis, it was assumed that these were relatively static in number, and only changed in size according to lipid storage droplets. It is now clear that adipocytes do have a life cycle, replacing all mature adipocytes every 5–10 years [16]. Examination of how they accomplished this replacement, via a process of "asymmetrical cell division," was found to be the mechanism, thereby preserving one stem cell during the process. This process results in activation of a precursor cell population capable of reacting to secretions from a senescing adipocyte, allowing the process of the precursor replacement cell to occur, while preserving the precursor cell. This replication by the asymmetric cell division results in a replacement immature adipocyte, while the other portion retaining its precursor form and abilities. This is logical, in that, if otherwise, we would accumulate a massive number of precursor cells in our tissues (Fig. 5.10).

- Safe & Easy Harvest Via Closed Syringe System
- *Prefer Use Of <u>Own Cells</u>* (Autologous)
- Optimal To Include Native Matrix For Structural Volume Enhancements
- Transplant In *Same Surgical Session, Same Day*
- Predictable & Reproducible Outcomes
- Induce Minimal Recipient Site Inflammation
- Result In Healing With Minimal Scarring
- Optional Ability To Utilize Parenteral Uses For Systemic Disorders
- Not Require Manipulation (But Do Offer Future Culture/Expansion)
- Transplantation Of Intact Micro-Environments

Fig. 5.10 Adipose tissue complex: advantages for use

It was Zuk et al. who identified the multipotent capabilities of adipose-derived *cellular* stromal vascular fraction (AD-cSVF), with capabilities of differentiation to a variety of tissues, including bone, cartilage, tendon-ligament, muscle, fat, nerve, etc. (Figs. 5.11 and 5.12). Once this capability was identified, efforts to isolate specific cell types started. This has proven somewhat difficult to interpret, in that it is currently impossible to imitate the in vivo microenvironment in the laboratory. Over the past decade, exploration of concentrating the identified ATC undifferentiated cell population coupled with high-density platelet concentrates (HD-PRP) has received a great deal of attention. Identifying stem/stromal cells which can participate in the processes has revealed what an "ideal" cell-based therapy may represent (Fig. 5.13).

In the early 1990s, a method of closed-syringe microcannula lipoaspiration was patented, permitting less traumatic and efficient means of acquiring ATC for use as a small-particle structural graft [17]. This has since evolved to a disposable, microcannula option which permits safe and efficacious low-pressure acquisition of AD-tSVF. In the past 15 years, clinicians and laboratory researchers have identified several important cell types which interact to provide remarkable contributions in tissue repair and regeneration. These have been identified as a

Partial List of stem/stromal cells in tSVF
- Pericytes/Endothelial Cells & Adventitial Cells (Key Group)
- Mesenchymal Stem Cells
- Pre-Adipocytes (Progenitors)
- Fibroblasts
- Macrophages (Type I & II)
- Vascular Smooth Muscle Cells
- T Lymphocytes (TREG)
- Miscellaneous Native Blood Derived Cells
- [NOTE: Reminder - These React With Local Site Cells]

Fig. 5.12 Adipose tissue complex main cellular elements

IMPORTANT

Mesenchymal Stem-Stromal Cells Capabilities Overlap >98+% Regardless Of Tissue Origin

Almost Every Tissue In The Body Contains Pericyte/EPC & MSCs (Vasculary Related)

Adipose & Bone Marrow MSCs Are Virtually *Interchangeable* In Capabilities *In Vitro*

Adipose Provides >1-2000 *TIMES* The Actual MSC Numbers Compared To Bone Marrow (per cc)

Adipose Does *NOT* Require Isolation, Culture-Expansion To Achieve Therapeutic Numbers

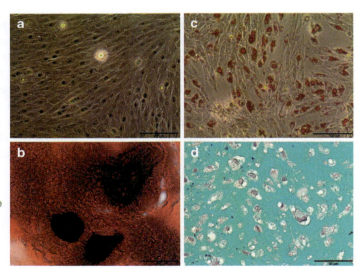

a). Control MSC; b) Bone (Alazarin Red); c) Adipose (Oil Red O); Cartilage (Hematoxylin Mayer's & Alcian Blue)

Fig. 5.11 Mesenchymal stem cell differentiation (MSC) potentials (basic): (**a**) Top left, micrograph control MSC in vitro. (**b**) Bottom left, bone. (**c**) Top right, adipose. (**d**) Bottom right, cartilage [note: overlapping differentiation capabilities of all MSCs is extensive]

very complex and heterogeneous population, closely related to cellular, adventitial areas, and extracellular matrix contacts. At first, mesenchymal cell group (MSC or ADSC) was thought to be the most important multipotent "stem" cell. Further examination, however, now suggests that these may serve a "sentry" capacity, and that the actual cell group is known as pericyte/endothelial stem/stromal cells [10] (Figs. 5.14 and 5.15).

There is much confusion in interpretation of the scientific and clinical published materials caused by a lack of explanation of the difference between *tissue* stromal vascular fraction (tSVF) and *cellular* stromal vascular fraction (cSVF) (Fig. 5.16). For clarification, cSVF is the isolated cellular elements in the ATC created via use of certain collagenase-enzyme blends to separate the complex and multiple attachments comprising the cell-to-cell or cell-to-matrix connections (Fig. 5.7). The utilization of such cSVF is currently the subject of multiple clinical trial applications (see www.clinicaltrial.gov), and is heavily utilized in cell isolation, culture expansion, and cell characterization studies. This creates an "information gap" between clinical applications and those strictly of research value. If clinicians

"Ideal" Cell-Based Therapy

- *Use Patient's Own Cells* (Autologous)
- Safe & Easy Harvest Via Closed System
- Optimal To Include Native Matrix
- Transplant *Same* Surgical Session & Day
- Predictable & Reproducible Outcomes
- Ability To Isolate/Concentrate For cSVF Parenteral Uses For Systemic Disorders
- Not Require Manipulation (But Do Offer Optional Culture/Expansion)

Robert W. Alexander, MD, FICS

Fig. 5.13 Ideal cell-based therapy advantages

Disposable Microcannula System

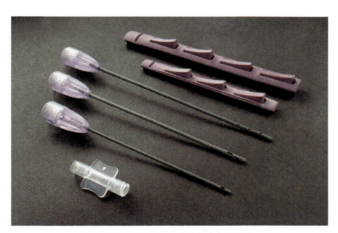

GEMS Disposable Closed Syringe System & Closed Transfer

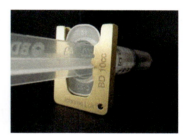

External Lock

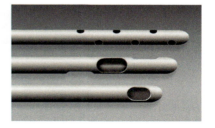

Microcannula Tips

Fig. 5.14 Closed, sterile disposable, microcannula system (GEMS) (Tulip Medical, San Diego, CA, USA). Left top: Internal lock sample (purple), microcannulas (2.11 mm OD) showing multiport infiltrator, spiral cannula (2.11 mm), and single port injector (1.65 mm), sterile clear luer-to-luer transfer. Top right: External universal non-disposable lock. Bottom right: Microcannula tips

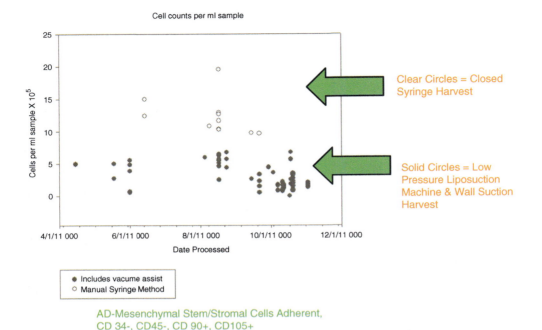

Fig. 5.15 Comparative cellular recovery: closed-syringe microcannula harvest versus machine vacuum on mesenchymal cells in ATC

TERMS: tSVF & cSVF

Tissue Stromal Vascular Fraction *(tSVF)*
- Includes ALL Cellular Components Of Tissue
- Includes ALL Biologic Components
- Includes Native Bioactive Matrix (Secretive)
- Requires NO Manipulation

Cellular Stromal Vascular Fraction *(cSVF)*
- *Requires Digestion, Incubation, Isolation*
- *Common Uses Reported In Research Settings*
- *Does Not Have Native Matrix Component*
- *Often Being Use As "Cell-Enrichment" Protocols In Tissue Augmentation & In Degenerative Disorders*

Robert W. Alexander, MD, FICS

Fig. 5.16 Important understanding of terms: tissue stromal vascular fraction (tSVF) and cellular stromal vascular fraction (cSVF)

read only the peer-reviewed clinical journals, they will miss more than 85% of the pertinent information and data evolving on almost a daily basis, as the important advances appear with basic scientific and engineering publications. It is reported that important information publications are doubling the existing knowledge base every 4-6 months, making the ability to remain current in the scientific and clinical applications most challenging.

In clinical applications, use of AD-tSVF has taken the primary role in aesthetic and regenerative uses, as it is a product that provides the full complement of structural (stroma, ECM) elements plus the resident cellular population of the AD-cSVF. The existing native stroma of ATC is now recognized as of great importance, not only due to the available attachment sites, but also due to the actual secretory bioactivity of the tissue. This dual role is considered to be of great importance, making use of existing native scaffolding of ATC, considering the "mini-microenvironmental attachments" felt to positively interact and contribute to the local recipient sites in need (Fig. 5.17).

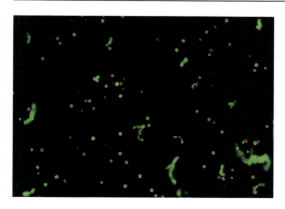

Fig. 5.17 Flow cytometry (live/dead) stain AO-acridine orange and PI-propidium iodide. (NOTE: The green "strings" represent actual viable cells attached to the extracellular matrix, making the grafting a minute, living microenvironment state (tSVF) thought to enhance recipient site responses)

In biological aspects, it is important to clearly recognize that not all platelet-rich plasma preparations and concentrates are the same. It is clearly shown that the amount of growth factors, signal proteins, and important chemical agents has a DIRECT, *linear* relationship to the concentration of platelets actually achieved. It is confusing to follow the variety of processes used in creating what is being called PRP, particularly since most practitioners do not have the capability of confirming actual patient measured baselines to compare with achieved concentrations. To qualify as a true "high-density" PRP, we utilize the minimal concentrations to be four to six times an actual measured baseline, not a calculated extrapolation. This is very important based on the correlation of such concentration to cellular proliferation and migration capabilities. Refinements in HD-PRP options are now recognized as many subdivisions, such as low- and high-hematocrit solutions which have definite clinical implications to tissue tolerance and reactivity [18].

Use of centrifugation has increased in biocellular applications, as it creates a very effective "gravity density separation," which is important to avoid cellular debris, unwanted fluids and local anesthetics, and isolation of the unwanted free lipid layer from the upper portions of the lipoaspirate. In addition, it permits decrease of the interstitial fluid load, a factor requiring "overcorrection" of grafts or small joint placements. This unneeded load is felt to potentially impact site perfusion, as a factor of importance in many plastic surgical, reconstructive poor perfusion wounds, and musculoskeletal (MSK) applications [19].

The final area of importance in MSK applications relates to the ability of optimal targeting of areas of damage, degeneration, or inflammation. Without the use of high-definition ultrasonography, it is virtually impossible to assure accurate placement of the biocellular therapeutic modality. With the use of ultrasonography, coupled with compressed and thoroughly mixed biocellular components, patients respond more rapidly, show metrics of responses, and achieve earlier final outcome than when placed via palpation only [20].

Within the past 2 years, an interesting option of removing the unwanted mature adipocytes from the AD-tSVF has become available. It is well documented that the large, mature adipocytes do not contribute a significant value to an injection site (including when performing structural fat grafting in aesthetic surgery) as they are gradually lost and removed following their anoxic exposure. It is likewise clear that the stem/stromal cells in the ATC are not as susceptible to those conditions, and in fact may be stimulated in low-oxygen-tension environments [21]. Recent publication of viability and numbers of stem/stromal cells remaining after emulsification process confirm that the relative numbers of such cells remain statistically the same as those not submitted for emulsification. One of the advantages of this process is that not only the AD-tSVF retains valuable stromal tissue, but also the entire specimen (mixed with HD PRP) can be easily injected through small-bore needles (25–30 gauge). This facilitates uses in scars, radiated damage skin, and hair loss and permits more patient comfort in MSK injections (including small joint targets) [22].

In regenerative medicine, the main goals are relatively well established (Fig. 5.3). Likewise, description of "optimal" features of cellular based therapy in both aesthetic and regenerative applications is becoming standardized. It is important to recognize that the combination of platelet concentrates and AD-tSVF appears to be more effective than either of the entities by themselves.

5.5 Understanding the "Workers and Bricks" Analogy

Considering chronic wound and musculoskeletal (MSK) applications, a simple analogy is often helpful in understanding the importance of both the biological and the cellular elements to achieve more rapid and complete healing and repair (Fig. 5.18).

If you have a brick wall that is beginning to break down, some of the mortar holding the bricks together is lost or crumbling. What is needed to repair the wall would be hiring *"WORKERS"* to come in, clean up the site, and repair and replace the damaged mortar. Once completed, the wall is repaired and functions as originally intended. These workers are found in great quantities in platelet concentrates, and comprise the "biological" contribution of the biocellular regenerative treatments.

In the event, however, that your wall is losing mortar holding the bricks in place, imagine if you have lost or broken many of the bricks in the wall. This would require not only the "workers," but also *"BRICKS"* to replace the lost and damaged ones. The "bricks" in this analogy come from the cellular source (cSVF). Combining biologics + cell sources has proved to be more successful than use of either of the agents by themselves.

It is well established that there are many more of these undifferentiated cells located in the largest microvascular organ of your body, within the adipose (fat) matrix. Therefore, the readily available and safely accessible "cellular" contributor of choice has become adipose tissue retrieved from subdermal fat deposits in the abdomen and thigh areas. These are gently removed via closed-syringe lipoaspiration, compressed by centrifu-

gation, and mixed by the platelet concentrates (>4–6 times patient circulating platelets) to form the therapeutic mixture known as "biocellular regenerative matrix."

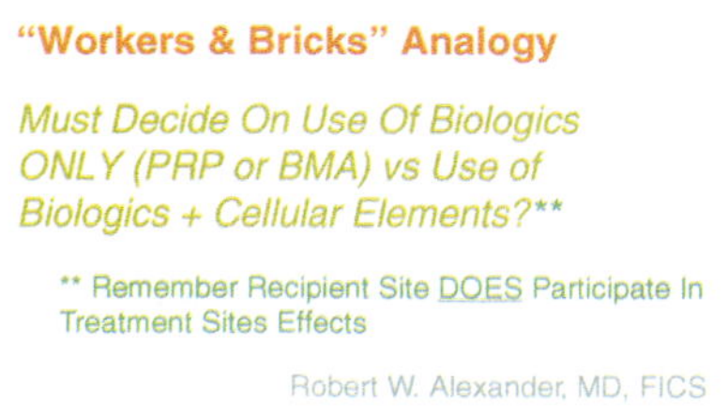

This mixture is in current use in aesthetic (plastic), reconstructive and wound healing, sports and pain medicine, orthopedic medicine and surgery, neurological disorders, musculoskeletal and arthritic applications, and a wide area of overlapping disorders.

5.6 What Are Adipose-Derived "Adult Stem/Stromal Cells"?

These are a diverse group of "non-designated" cells found throughout essentially all the tissues of our bodies. They serve as a reservoir of replacement and repair cells, which react to injury, aging, or disease. "ADULT" cells in this category are often referred to as "stem/stromal cells" or "stromal" cells, and should be clearly separated from embryonic cells. They are also called by confusing names, such as "progenitor" or "precursor" cells, which means that they have the capability to differentiate into different types of cells, via responses to growth factors and signal proteins within the microenvironment where they are located. For example, if you have a muscle or ligament tear, local tissue components (native to site) plus these non-differentiated cells are felt to participate in healing or repairing the damage providing replacement muscle or ligament tissues, rather than resulting in scarified tissue. Scar tissue is not as functional or tolerant of future stresses, and is NOT the ideal goal in wound healing. The terms differentiating "benign" versus "toxic" inflammation is becoming important. A highly reactive toxic inflammatory response (such as seen in many animals and young humans) as a protective mechanism, typically result in scarring as the primary result. The term, benign, on the other hand refers to a needed inflammatory reaction, but one modulated by various factors and resulting in little or no scar build up. A core goal of use of concentrated biocellular elements is to encourage the healing while minimizing residual scarring. In example, this is very important in ten-

"Workers & Bricks" Analogy

*Must Decide On Use Of Biologics ONLY (PRP or BMA) vs Use of Biologics + Cellular Elements?***

** Remember Recipient Site DOES Participate In Treatment Sites Effects

Robert W. Alexander, MD, FICS

Fig. 5.18 "Workers and bricks" analogy

don repair, where restoration of myotendinous elasticity is critical to avoid re-injury when scar would be exposed to rapid loading of tension (as represented in the Achilles Tendon images above) [23–25] (Figs. 5.19 and 5.20).

There are many experiences in such cases over the past 15 years in musculoskeletal area, and for more than 30 years in aesthetic surgical practice. These are often reported on small case series or case reports of treatment and outcomes, and are now being further studied in many clinical trials [26]. Evolving clinical trials include both guided placement of stem/stromal elements and biological agents in orthopedic medicine and surgery, but also intravenous and central nervous system placement in a variety of complex disorders which do minimally or not respond to conventional therapy (such as diabetes, multiple sclerosis, Alzheimer's disease, Parkinson's disease, severe limb ischemia, traumatic brain injuries) [27, 28]. Early reports of improvement in chronic conditions, including pain, arthritis, damaged tendons-ligaments, low back and facet degeneration, etc., are driving many to select this option to improve surgical outcomes or avoid surgical interventions and shorten the demands for physical therapy.

Many remained confused about the potentials or best source of stem/stromal cells, often believing that this only refers to the use of embryonic tissues or nonautologous sources such as placental amniotic or umbilical cord-derived cells. In the past 15 plus years, much evidence has led us to understand that our own fat may be a much more plentiful and optimal cell source, avoiding the need to destroy fetal or embryonic tissues, or undergo more invasive marrow access in order to acquire cells and culture-expand them to achieve optimal potential. Further, the stem cells found in bone are heavily weighted to the blood-forming side, rather than the reparative group of cells.

Considering ready availability of fat and minimally invasive access (using closed-syringe liposuction for example), adipose now has become an

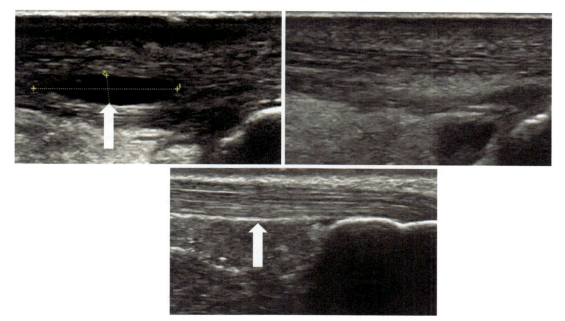

Fig. 5.19 Ultrasound-guided biocellular therapy in torn Achilles tendon. Top left: Tendon tear pretreatment [note: rest of tendon showing tendinosis]. Top right: Posttreatment tendon (at 6–8 weeks); posttreatment 1 year [note: resolution of tear without scarring, rest of tendon returning to more normal fibrillar tendon echotexture]

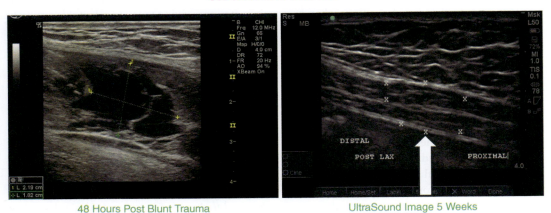

Fig. 5.20 Use of tSVF + high-density platelet concentrate (biocellular therapy) in torn rectus abdominus trauma. Left: Pretreatment image. Right: Area of trauma at 5 weeks post-trauma [note: return of muscle echotexture without scarring]

optimal source for these cells with a high safety profile for patients. As previously described, ATC is the largest microvascular organ in the body, and as such has become well recognized as the largest depository of undifferentiated stem/stromal cells in the entire body. The ease of gathering fat tissues on an outpatient basis and local anesthetic has led to the evolution of "biocellular therapy" and "cell-based therapy" for a wide variety of disorders and conditions. With the advent of closed cell isolation/concentration systems (incubation, agitative digestion of cellular contact points) at *point-of-care* (POC) availability, the opportunity of cell enrichment in grafting procedures plus the availability of parenteral intravascular deployments can be achieved. It is most common for these procedures to be performed in outpatient ambulatory surgical centers (ASC) or dedicated clinic procedural facilities.

Specific "key" cells that have been credited for promotion of healing and repair reside in tissue microenvironments, where they comprise parts of tissues, and organ system identification of pathways, however, remains somewhat elusive. The complex components within the AD-tSVF may be considered to offer "smorgasbord" of elements which can become available to any site or tissue. Analyses of growth factors and signal chemicals would suggest that the intact AD-tSVF may offer contributions over and above those as isolated elements [9]. The cell groups participating in the healing or repair are subject to important contributions of native cell components in vascularized tissues, and by introduction of concentrates of cells and biologics appear to "auto-enhance" the site controls and effects. These native site cell groups are also called "niches," and are the locations where injury or disease must be addressed to permit the body to repair or regenerate itself. It is believed that when that process is underway, addition of needed cell types and biological elements specifically targeted (via ultrasound guidance for example) can effectively utilize your own tissues to heal themselves in a more efficient and effective manner.

5.7 What Is Involved in Providing Cellular or Biocellular Regenerative Therapy?

Cellular therapy begins with the exact same harvesting and processing of ATC via microcannulas, including centrifugation to remove the remaining suspensory fluid derived from the infusion of very diluted local/epinephrine solution in normal saline. Once removed, the tSVF derivative is exposed to digestion (typically using a blend of collagenase and neutral proteases), incubation, and concentration of the separation of cSVF portions. This cSVF is then neutralized and rinsed to remove the collagenase and protease additives. A final centrifugation results in a concentration "pellet" of cSVF (Fig. 5.21).

At this point, cellular isolates/concentrates may be used in multiple applications, such as cell enrichment (where these isolates are added back to a tSVF portion) to provide higher stem/stromal cell numbers for potential graft and injection applications. The other option is to resuspend this cSVF group in 500 cc normal saline for delivery parenterally (IV, IA, intrathecal, intraperitoneal, or mucosal spray) depending on desired locations and uses. If IV or IA, typical very-fine-in-line tubing filters (150 μm) ensure removal of any fibrin or other components in a manner common during blood product transfusions (which utilize a standard 170 μm filter).

In the case of parenteral deployment, most clinical researchers are using this within an FDA/NIH-approved Clinical Trial situation, where safety and efficacy data is developed and reported (*see www.ClinicalTrials.gov*).

The biocellular therapies refer to the use of tSVF (harvested via microcannula syringe systems) plus the mixed additive of HD PRP (or lower concentrations when used on radiation damage and hair and for dermal uses). The tSVF may be used after a centrifugation (typically 800–1200 g-Force for 3–4 min), mixed with an appropriate concentration (by volume percentage) of PRP, and then guided or intradermally placed to targets. Plastic surgery literature has

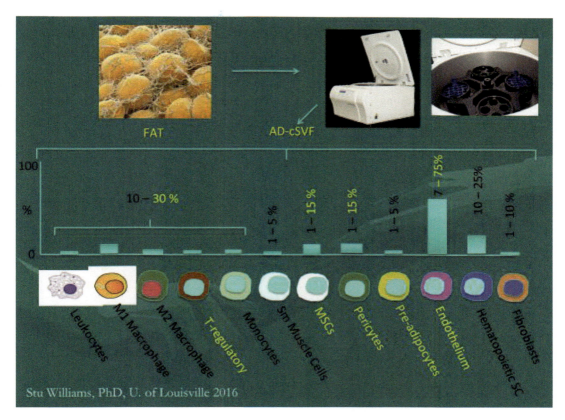

Fig. 5.21 Main component cells in cellular stromal vascular fraction [note: approximately one-third have multipotent capabilities, balance supportive cells of importance] (courtesy of Dr. Stu Williams, University of Louisville)

many examples of improved structural grafting accomplished using cell-enriched tSVF as previously discussed (Fig. 5.22).

As the platelet concentrates are understood to be a key contributor for provision of critical healing growth factors and signal proteins, we recommend striving for greater concentrations achieved directly with linear increases of those elements. Acting as a central component in the inflammatory and healing cascade, they help begin and maintain the healing processes in conjunction with the local site stroma and cells. This effect is recognized as an "auto-amplification" effect, wherein the site-specific needs are boosted in response during the most important regenerative or healing processes. Such factors as vascular endothelial growth factor (VEGF) and many others contribute to this process with encouragement of microvessel formation and improved perfusion. Thousands of patients have undergone treatments using these concentrates with quality results in many inflammatory or aging conditions.

An autologous tSVF sample is easily harvested from subdermal fat deposits under sterile protocols, using the patented closed-syringe system. This is often referred to as microcannula lipoaspiration or lipoharvesting [17]. This adipose tissue complex (ATC) may be cleaned and compressed (centrifuged) and unwanted liquid layers separated by centrifugation. This process not only helps with removal of unwanted liquids, but also compresses the adipose cellular components to provide a more effective cell and bioactive matrix with less intercellular fluid load (Fig. 5.23). By effectively reducing the volume of injection materials, earlier recovery of comfort and ambulation is common. Biologicals such as high-density platelet concentrates (HD-PRP) can then be added via closed, sterile luer-to-luer transfer to create a mixture of cells and the important growth factors/signal proteins provided from within the platelet alpha-granules. There is a direct correlation between concentration achieved and delivery to targets [29].

Cellular and biocellular therapies are commonly performed in many areas of within and outside of the United States; FDA-suggested *guidelines* being discussed currently confuse these issues employing digestive chemicals. Many sites are actively providing these services in the United States, often acting within controlled Institutional Research Based (IRB) trials or study groups within specialties. The author currently stresses the importance of information documentation and reporting on safety and efficacy. Until these existing trials are concluded and reported, it is common to perform these newer options within an approved IRB channel. Following multiple

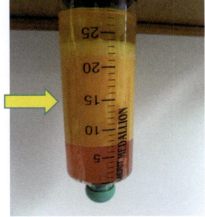

Fig. 5.22 Value of centrifugation for layer separation compared to gravity decantation. Note: Density gradient separation at 800–1000 g-Force/3 min provides effective excess liquid/debris (infranatant), compressed graft (tSVF), and well-separated unwanted free lipid (supranatant)

Graft Separation-Density Gradient

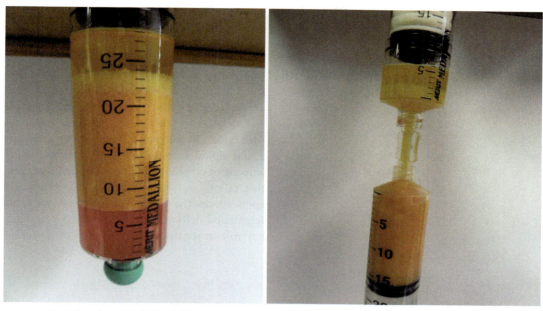

Centrifugation 800-1000g 3 Minutes   Aerobic Transfer Loading Graft ONLY

Fig. 5.23 Left: Showing centrifuged specimen. Right: Loading of compressed graft via clear luer-to-luer transfer, leaving supranatant layer (not desired for injection)

institutional and organizational IRB guidelines using specific trial studies by providing the approved trial protocols, both within the United States and internationally, is important.

The cellular isolates are currently being utilized for a wide variety of human clinical applications on a global basis. Following a myriad of basic research studies, animal models were tested and reported in the bioscientific literature, and are gradually being reported in translational clinical journals in the medical literature. As the cellular therapy group is well recognized as favoring immune privilege and pro-"benign" inflammation, those local, systemic, and autoimmune issues are included in many clinical studies (Figs. 5.24 and 5.25).

How the cSVF actually works is under intense study at this time. At first, the thought that the cellular components were the most important (like the incorrect belief that autologous fat grafting relied on the presence of mature adipocytes) has proven not to be correct. In fact, it is now believed that the actual graft successes are a result of the stem/stromal cells and bioactive matrix (extracellular matrix and peri-adventitial cells) and their secretory abilities to impact healing and graft sites. This secretory action is considered of greater importance than the undesignated cell populations within the tSVF [30, 31].

Current Biocellular Uses
- Aesthetic-Reconstructive Surgery
- Neurodegenerative Diseases
- Autoimmune Disorders
- Ischemic and Devascularized Wounds
- Cardiovascular - COPD
- Crohn's and Ulcerative Diseases
- Skin & Anti-Aging Applications
- Musculoskeletal Applications
- Chronic Wounds and Pain

Fig. 5.24 Sample of current cellular and biocellular uses

The importance of this understanding has led to appreciation of the paracrine capabilities via exosomes and microvesicles, and their roles in transferring signals via mitochondrial RNA (miRNA) and messenger RNA (mRNA). It is now believed that the exchange of these elements is the means of communication and stimulation of nuclear change leading to the proliferative and trophic effects to guide the differentiation for cell activities or secretions in specific sites. There are currently important studies utilizing both autologous and non-autologous exosomes and microvesicles as an important contribution to the activation and guidance of cells to respond [32] (Figs. 5.26 and 5.27).

In the discussions on "nonhomologous" uses when using tSVF or cSVF, it is most misunderstood that the actual cells that are of importance are NOT the adipocytes, but rather the supportive and undesignated cell group found in association with the ECM and peri-adventitial elements that actual stimulate precursor replacement cells of adipose or other cell groups in applied regenerative efforts in wound healing and orthopedic applications. The concept of these terminally differentiated cells (mature adipocytes) playing a key role in repair and regeneration of musculo-

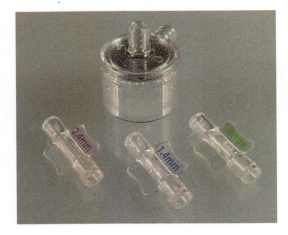

Fig. 5.25 Disposable system for emulsification (micronization, "Nanofat"). Top: screen (offset 600/400 μm screen). Bottom: shows "partial-emulsification" luer-to-luer series of progressively smaller internal sizes (2.4, 1.4, and 1.2, respectively) (photo provided by Tulip Medical, San Diego, CA, USA)

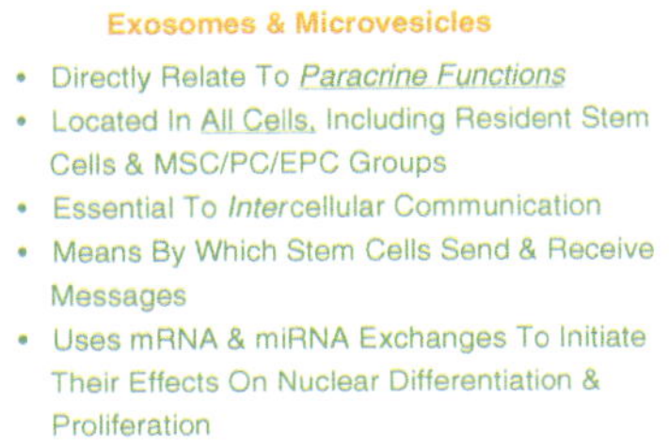

Exosomes & Microvesicles
- Directly Relate To *Paracrine Functions*
- Located In *All Cells*, Including Resident Stem Cells & MSC/PC/EPC Groups
- Essential To *Inter*cellular Communication
- Means By Which Stem Cells Send & Receive Messages
- Uses mRNA & miRNA Exchanges To Initiate Their Effects On Nuclear Differentiation & Proliferation

Fig. 5.26 Paracrine secretions and intercellular communications: exosomes and microvesicles as means of signaling for stem cell activation and proliferation

Fig. 5.27 Diagrammatic representation of exosomes and microvesicles

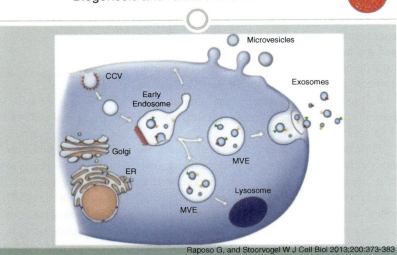

skeletal components has been mostly discarded. The FDA concept that adipose structural grafting should not be allowed in large-volume breast augmentation is based on the statement that it is a nonhomologous use based on the fact that fat cells "cannot secrete milk" as the only function or tissue comprising the breast is ludicrous.

5.8 A Recent Advance in Use of Biocellular Uses: Micronized and "Nanofat™" (Emulsified AD-tSVF)

Over the past few years, several alternative advances in processing the lipoaspirated tSVF via mechanical emulsification have evolved [32]. However many applications still favor the same biocellular product creation (including use of additive advantages offered by addition of HD-PRP concentrates) while retaining small fragment tSVF capable of injection via small-bore needles (down to 30 gauge). Recent published evidence has shown that creation of the mechanically emulsified "nanofat" does not have a detrimental impact on stem/stromal cellular numbers or viabilities while markedly reducing the volume of ATC provided by mature adipocytes (Fig. 5.25).

Recent advances now offer the opportunity to have a disposable system to achieve maximal lysis of adipocytes while preserving the much smaller cSVF component (Figs. 5.28 and 5.29). It is important to note that this emulsified (micronized) tSVF product is NOT ABLE to produce true cSVF and, therefore, CANNOT be safely used in *any* intravascular uses. Although not believed to be as effective as using larger fragment tSVF for structural grafting, it can still be combined with PRP concentrates to provide improved dermal vascularity and configuration.

With the ability to inject through small-bore needles, patient comfort in dermal and hair injections is enhanced, and it offers a range of radia-

Emulsified AD-tSVF

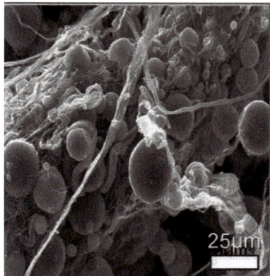

AD-tSVF **Emulsified AD-tSVF**

Fig. 5.28 Emulsified adipose tissue complex (tSVF). Left: Shows SEM of AD-tSVF prior to emulsification. Right: SEM of AD-tSVF after emulsification [note: essential removal of all intact mature adipocytes, leaving perivascular and extracellular matrix (ECM) intact] (Feng, J, Doi, K. et al. Micronized Cellular Adipose Matrix as a Therapeutic Injectable in Diabetic Ulcer. Regen Med. 2015; 10(6))

Fig. 5.29 Emulsification of adipose tissue complex (tSVF). Top left: Shows AD-tSVF prior to emulsification (stain showing vascularity on left, and intact adipocytes (green) on right of micrograph). Bottom right: Shows AD-tSVF after emulsification of mature adipocytes [note: multiple blue-stained nuclei remaining intact (small fragment tSVF, not a cSVF)]

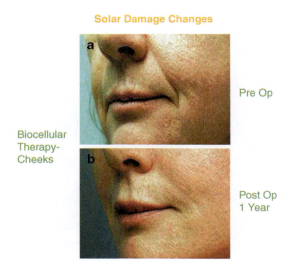

Fig. 5.30 Sample of post-radiation skin aging/damage (sun). Top: Pretreatment biocellular mix. Bottom: Posttreatment biocellular grafting (1 year) of cheeks, lips, and nasolabial folds [note: to show true texture change as a result of biocellular mix, placed subdermally]

tion/solar-damaged skin and small joint targeted applications in orthopedic medicine. The abilities of biocellular modalities to promote wound healing and regenerative capabilities via intradermal placement have opened opportunities to permit improved skin circulation and texture; improve skin aging and radiation damage and hair regeneration; and participate in chronic wound applications, as well as many small joint and superficial targets in musculoskeletal applications (Fig. 5.30).

5.9 What Is the Future in Stem/Stromal Cellular and Biocellular Treatments?

There are now safe, reproducible capabilities of sterile, closed isolation of the large numbers of stem and stromal cells from the adipose tissues. Within such semiautomated and automated closed systems, this ability is becoming practical even in outpatient procedural rooms and carefully prepared within sterile protocols. Once this was exclusively possible only in very costly laboratory settings, requiring extensive equipment and technician costs. Today such isolation is being done in the United States under Institutional Review Board settings to insure reporting of patient safety and determining effectiveness. Clinical trials are gradually being released at this time, many requiring several years to acquire data, compile, and report. The vast majority of such reports are providing clear clinical evidence of patient safety and effective clinical treatment outcomes.

With the changes in legislation and within the FDA, new and exciting possibilities of IND/RMAT designations are being considered and suggested guidelines are evolving. Timing of these changes is optimal, as much of the safety and early efficacy opportunities will already be in place by the time such designations are better understood. The twenty-first-century Cures Act and the "Right To Try" 2018 Legislation are examples of some of these changes.

Isolation of these cells permits creation of what is termed "cell-enriched" biocellular grafts. These grafts, higher in numbers of the heterogeneous undifferentiated cells, are believed to provide an even more potent guided injectable therapy. For example, there are many peer-reviewed clinical articles providing strong evidence of enhanced outcomes within the aesthetic-plastic surgical literature. Over the past decade, there are estimated numbers of use of biocellular therapies in musculoskeletal application exceeding 150,000 human clinical uses, with a remarkable efficacy and safety profile. These are reported in case series or reports, and should not be discarded out of hand, simply because they are not participating in specific trial settings. It remains of a pivotal value to insure very accurate diagnostics and guided placement to defined targets. Ultrasonography, with its dynamic abilities during examination, will remain as a needed core competency for those taking care of musculoskeletal and chronic wound cases.

In the future, it is very likely that such isolated cells will provide parenteral (intravenous, intra-arterial, intra-thecal, intra-peritoneal, etc.) pathways, and become effective for a very expansive treatment for such disorders as neurodegenerative diseases (MS, Alzheimer's, ALS, Parkinson's, brain injuries/stroke, etc.), diabetes, chronic lung disease, heart disease and damage, chronic wound healing, fibromyalgia/causalgia, ulcerative bowel disease, Crohn's, colitis, and so forth.

5.10 Who Provides Biocellular Treatment?

Initially the realm of cosmetic plastic surgeons advanced in uses for sun damage, and deep structural and superficial dermal injections, and a variety of indications have been adopted by a very wide variety of providers. Patients are now cared for by a diverse group of providing doctors (e.g., primary care healthcare providers, internists/neurologists, aestheticians, hair regenerative specialists, aesthetic-plastic surgeons, general surgeons, orthopedic surgeons, emergency/sports medical specialists, pain management specialists, wound care centers, etc.). Those nonsurgical trained practitioners are now deciding whether candidates have a condition which has reasonable potential for improvement through the use of combinations of biologic and stem/stromal cellular treatment. Thorough physical and pretreatment evaluations, proper training, and informed consent delivery to patients are essential in diagnostic, treatment planning, and care delivery. Circulatory, neurological, and indicated systemic conditions should be documented and thoroughly evaluated regardless of the background of the medical provider. In the case of orthopedic applications, use of metrics such as range of motion, indicated MRI studies, and high-quality ultrasonographic imaging combined can determine the specific locations of problems and guide proper placement. In the vast majority of orthopedic cases, use of high-quality MSK ultrasonography is considered a KEY part of such treatment. Proper diagnostic imaging and ultrasound evaluation are considered a key to the most success, particularly considering that this modality plays a central role for providers to effectively "hit" the desired targets. Palpation may provide fairly accurate placement in experienced providers; targeted and tracked therapy consistently correlates with earlier and improved clinical outcomes. Use of metrics that are more objective, successful monitoring of many patients including range of motion, remodeling of tissues in repeated-interval ultrasound studies, and return of strength, range of motion, and comfort during function provide very valuable informative standards. For many years, simple dextrose prolotherapy major benchmarks were limited to patient-reported pain levels, activity levels, and perceived improvement as their primary metrics.

Most times these procedures are completed on outpatient, ambulatory basis using local anesthesia, nitrous oxide, or occasionally light sedation depending on patient needs and desires. These cases are designed and planned to be completed within the same day. Providers handle tissues using standard aseptic protocols. Since the advent of a variety of mechanical emulsification systems have evolved, it has become easier to permit guided (ultrasound) targeting of damaged or degenerative tissues via very small bore needles (e.g. 23-30 gauge). In addition, this mechanical reduction in particle size of the tSVF does not statistically reduce the numbers, viability or clinical outcomes from it use. The ability to provide improved patient comfort and access to skin, hair, and small joints is rapidly becoming an appealing option.

5.11 Conclusion

"USING YOUR OWN TISSUES TO HEAL" represents a major healthcare paradigm change, and is one of the most exciting minimally invasive options currently available. Both cellular and biocellular regenerative therapies are rapidly improving in documentation and cellular analyses, and gaining very good safety and efficacy profiles. Once considered purely experimental, it has entered into an accepted, translational period to clinical providers, backed by improving science supporting the basic hypotheses. It is a well-recognized and reported alternative to many traditional medical/surgical interventions. Clinical trials will prove to be of great value going forward in providing evidence of the specific area of optimal use.

There are many evolving clinical trials (NIH/FDA oversight) recognizing the very remarkable abilities of cellular components of adipose tissue complex (ATC). The number of active and evolving trials using ATC is surpassing the uses of bone marrow due mostly to the much higher numbers of reparative cells within the marrow. Marrow remains the gold standard for various blood-related diseases, but it is not clear that it possesses very few of the non-hematologic (reparative) stem cells, and it is thought by many to be only useful as a biologic contributor to the wound healing and regenerative processes discussed in this chapter.

To date, researchers and bioengineers are not able to effectively reproduce or provide true three-dimensional scaffolding at this time. This inability limits the development of understanding how to mimic a true in vivo culture/expansion capability, and continues to delay the understanding of how these cells act and are controlled at the local tissue level. Translation from tissue culture and the inherent changes potentially introduced in the process leave us with the need to cautiously advance therapeutic modalities into clinical practices. Classic tissue culture/expansion remains problematic, in that it is known to introduce variables which are not able to be translated into how the body uses these undesignated cell populations.

Acknowledgements The author wishes to thank our clinical and laboratory staff, Ms. Nancy L. Smith and Ms. Susan Riley, for their endless hours and efforts in order to facilitate patients, sample gathering, and testing. Without such devotees, clinical papers cannot be reported.

There are no conflicts of interest reported in this chapter or its content.

References

1. Alexander RW. Understanding adipose-derived stromal vascular fraction (AD-SVF) cell biology and use on the basis of cellular, chemical, structural and paracrine components: a concise review. J Prolother. 2012;4(1):e855–69.
2. Zuk PA, Zhu M, Mizuno H, et al. Multi-lineage cells from human adipose tissue: implications for cell-based therapies. Tissue Eng. 2001;7(2):211–38.
3. Zuk P. Adipose-derived stem cells in tissue regeneration: a review. ISRN Stem Cells 2013;2013:35 pages. Article ID 713959.
4. Alexander RW. Understanding adipose-derived stromal vascular fraction (SVF) cell biology in recon-

structive and regenerative applications on the basis of mononucleated cell components. J Prolother. 2013;10:15–29.

5. Alderman D, Alexander RW, Harris G. Stem cell prolotherapy in regenerative medicine: background, research, and protocols. J Prolother. 2011;3(3):689–708.

6. Sadati KS, Corrado AC, Alexander RW. Platelet-rich plasma (PRP) utilized to promote greater graft volume retention in autologous fat grafting. Am J Cosmet Surg. 2006;23(4):627–31.

7. Alexander RW. Use of PRP in autologous fat grafting. In: Shiffman M, editor. Autologous fat grafting, vol. 14. Berlin: Springer; 2010. p. 87–112.

8. Alexander RW. Biocellular regenerative medicine: use of adipose-derived stem/stromal cells and it's bioactive matrix. Phys Med Rehabil Clin North America. 2016;27:871–91.

9. Blaber S, Webster R, Cameron J, et al. Analysis of in vitro secretion profiles from adipose-derived cell populations. J Transl Med. 2012;10:172.

10. Crisan M, Yap S, Casteilla L, et al. A perivascular origin for mesenchymal stem cells in multiple human organs. Cell Stem Cell. 2008;3:301–13.

11. Guilak F, Cohen DM, Estes BT, Gimble JM, Liedtke W, Chen CS. Control of stem cell fate by physical interactions with the extracellular matrix. Cell Stem Cell. 2009;5(1):17–26.

12. Brizzi M, Tarone G, Defilippi P. Extracellular matrix, integrins, and growth factors as tailors of stem cell niche. Curr Opin Cell Biol. 2012;24(5):645–51.

13. Soo-Yyun K, Turnbull J, Guimond S. Extracellular matrix and cell Signaling: the dynamic cooperation of integrin, proteoglycan, and growth factor receptor. J Endocrinol. 2011;209:139–51.

14. Alexander RW. Fat transfer with platelet-rich plasma for breast augmentation. In: Shiffman M, editor. Breast augmentation: principles and practice. Berlin: Springer; 2009. p. 451–70.

15. Kato H. Short- and long-term cellular events in adipose tissue remodeling after non-vascularized grafting. Paper presented at the International Federation for Adipose Therapeutics and Science Miami, 9th Annual Symposium on Adipose Stem Cells and Clinical Applications of Adipose Tissue; 2011 Nov 4–6; Miami, Florida, USA.

16. Alexander RW. Liposculpture in the superficial plane: closed syringe system for improvement in fat removal and free fat transfer. Am J Cosm Surg. 1994;11(2):127–34.

17. Alexander RW, Harrell DB. Autologous fat grafting: use of closed syringe microcannula system for enhanced autologous structural grafting. Clin Cosmet Investig Dermatol. 2013;6:91–102.

18. Alexander R. Unpublished data. 2019.

19. Kurita M, Matsumoto D, Shigeura T, et al. Influences of centrifugation on cells and tissues in liposuction aspirates: optimized centrifugation for lipotransfer and cell isolation. Plast Reconstr Surg. 2008;121(3):1033–41.

20. Alexander RW. Introduction to biocellular medicine. In: Moore R, editor. Sonography of the extremities: techniques & protocols. 4th ed. Cincinnati: General Musculoskeletal Imaging Inc.; 2015. p. 97–104.

21. Eto H, Suga H, Inoue K, et al. Adipose injury-associated factors mitigate hypoxia in ischemic tissues through activation of adipose-derived stem/progenitor/stromal cells and induction of angiogenesis. Am J Pathol. 2011;178(5):2322–32.

22. Alexander RW. Understanding mechanical emulsification (nanofat) versus enzymatic isolation of tissue stromal vascular fraction (tSVF) cells from adipose tissue: potential uses in biocellular regenerative medicine. J Prolother. 2016;8:e947–60.

23. Oliver K, Alexander RW. Combination of autologous adipose-derived tissue stromal vascular fraction plus high-density platelet-rich plasma or bone marrow concentrates in Achilles tendon tears. J Prolother. 2013;5:e895–912.

24. Albano J, Alexander RW. Autologous fat grafting as a mesenchymal stem cell source and living bioscaffold in a patellar tendon tear. Am J Sports Med. 2011;21(4):359–61.

25. Alderman D, Alexander RW. Advances in regenerative medicine: high-density platelet-rich plasma and stem cell prolotherapy. J Pract Pain Management. 2011;10:49–90.

26. Alexander RW. Autologous fat grafts as mesenchymal stromal stem cell source for use in prolotherapy: a simple technique to acquire lipoaspirants. J Prolother. 2011;3(3):680–8.

27. Jackson M, Morrison D, Doherty D, et al. Mitochondrial transfer via tunneling nanotubes is important mechanism by which mesenchymal stem cells enhance macrophage phagocytosis in the in vitro and in vivo models or ARDS. Stem Cells. 2016;34:2210–23.

28. Alexander RW. Use of software analytics of brain MRI (with & without contrast) as objective metric in neurological disorders and degenerative diseases. Int Phys Med Rehab J. 2017;2(2):41–4. https://doi.org/10.15406/ipmrj.2017.02.00046.

29. Tobita M, Tajima S, Mizuno H. Adipose tissue-derived mesenchymal stem cells and platelet-rich plasma: stem cell transplantation methods that enhance stemness. Stem Cell Res Ther. 2015;6:215–22.

30. Alexander RW. Use of PIXYL software analysis of brain magnetic resonance imaging (with & without contrast) as valuable metric in clinical trial tracking in study of multiple sclerosis (MS)and related neurodegenerative processes. Clin Trial Degen Dis. 2017;2(1):1–7.

31. Lamichhane T, Sokic S, Schardt J, et al. Emerging roles for extracellular vesicles in tissue engineering and regenerative medicine. Tiss Eng B Rev. 2015;21(1):45–54.

32. Tonnard P, et al. Nanofat grafting: basic research and clinical applications. Plast Reconstr Surg. 2013;152:1017–26.

Preparation, Characterization, and Clinical Implications of Human Decellularized Adipose Tissue Extracellular Matrix

6

Derek A. Banyard, Christos Sarantopoulos, Jade Tassey, Mary Ziegler, Evangelia Chnari, Gregory R. D. Evans, and Alan D. Widgerow

6.1 Introduction

Plastic and reconstructive surgeons are continually seeking alternative strategies to improve the clinical results attained by soft-tissue filling. Autologous fat grafting is routinely implemented to treat a variety of soft-tissue defects and contour abnormalities. However, unpredictable resorption rates, ranging anywhere from 25% to 80%, often result in sub-optimal volumizing effects that necessitate repeat harvesting and grafting procedures [1, 2]. One such alternative strategy is cell-assisted lipotransfer (CAL). In this setting, regenerative components, such as adipose-derived stem cells (ADSCs), are isolated from adipose tissue, enriched, and then added back to the lipoaspirate prior to lipofilling [3]. Various studies demonstrate the regenerative benefits of ADSC enrichment [4], while others show that the retention of adipose tissue after

D. A. Banyard · C. Sarantopoulos · J. Tassey
M. Ziegler · G. R. D. Evans · A. D. Widgerow (✉)
Department of Plastic Surgery, Center for Tissue Engineering, University of California, Irvine, Orange, CA, USA
e-mail: dbanyard@uci.edu; csaranto@uci.edu; jtassey@uci.edu; zieglerm@uci.edu; gevans@uci.edu; awidgero@uci.edu

E. Chnari
Musculoskeletal Transplant Foundation, Edison, NJ, USA
e-mail: Evangelia_Chnari@mtf.org

CAL is superior to fat grafting alone [1, 3, 5]. However, these techniques are limited by their need for a separate harvesting procedure, extended cell-culturing times, and to date are generally not approved for use in the United States by the Food and Drug Administration (FDA) [6].

In general, there are two accepted theories that describe the pathophysiology of fat grafting results. The first is that once grafted, there is at least partial survival of the adipocytes to achieve a lipofilling effect [7]. The second theory is that the grafted adipocytes undergo necrosis, and it is the remaining stromal components and stem cells that recruit host tissue to fill the void through adipogenesis, neovascularization, and fibrosis [2]. A critical review of the literature, however, supports the notion that both of the above factors contribute to the volume effects observed with fat grafting. Debate will continue on the subject, but a recent study demonstrates that the majority of neovascularization and adipogenesis observed in fat grafting occur through recipient tissue recruitment [8]. Moreover, ADSCs act more substantially in a paracrine nature, rather than via direct engraftment or differentiation [9].

Frequently, substantial amounts of adipose tissue are discarded as medical waste during abdominoplasty, liposuction, body contouring, and breast reduction procedures. This has led to burgeoning research that has focused on the optimization of fat harvesting and processing, or on

© Springer Nature Switzerland AG 2019
D. Duscher, M. A. Shiffman (eds.), *Regenerative Medicine and Plastic Surgery*,
https://doi.org/10.1007/978-3-030-19962-3_6

its cellular components, such as the stromal vascular fraction (SVF). However, the noncellular component, or the human decellularized adipose extracellular matrix (hDAM), may be as important. Several groups have started to examine the use of hDAM as a scaffold for tissue engineering, which shows great promise as a vehicle for adipose stem cell delivery as well as a construct that promotes soft-tissue regeneration [10–28]. The term hDAM is synonymous with decellularized adipose tissue (DAT), adipose-derived matrix (ADM), and allograft adipose matrix (AAM). In addition, there is the possibility of using cadaver tissue as the source of hDAM, which, thus, creates an ideal off-the-shelf soft-tissue filler.

Young and Christman [29] examined the use of commercially available soft-tissue fillers and synthetic and natural polymers, with and without extracellular matrix (ECM)-based materials for adipose tissue engineering and found that ECM-based products have the greatest potential to promote de novo adipogenesis and, hence, promote long-term retention. They also identified several qualities that would enhance the adipo-inductive properties of the grafted material, including minimal in vivo immunogenicity, the ability to induce angiogenesis in vivo, and a composition similar to native adipose ECM. With these findings in mind, decellularized adipose matrices are thought to provide the ideal biochemical and biomechanical microenvironment for ADSCs to proliferate and recruit native tissue [30]. The focus of this chapter is hDAM, which may serve as an ideal soft-tissue filler both alone and as an adjunct to fat grafting.

Several studies were published early on that examined hDAM as a scaffold for ADSC supplementation to enhance adipogenesis [10, 11, 15, 25, 27, 31–37]. Additionally, evidence suggests that the addition of cross-linking agents to hDAM may modify the matrix properties to achieve the superior preclinical results already observed with this soft-tissue filler [31, 36]. Here, we examine how different research groups isolate, characterize, and use hDAM for tissue-engineering applications. We then discuss the potential for hDAM as an off-the-shelf lyophilized adjunct in alloge-

neic and autologous fat grafting to facilitate both adipogenesis and angiogenesis, and to improve volume retention. In addition, we discuss the potential of hDAM as a stand-alone material for soft-tissue reconstruction without the need for autologous fat or ADSCs due to the inductive properties of the ECM to recruit host cells and facilitate de novo adipogenesis.

6.2 Decellularization of Adipose Tissue

The optimal method of adipose tissue matrix preparation includes the removal of all cellular components, and hence immunogenicity, while maintaining an ideal 3D configuration of the ECM and key components, such as type IV collagen and laminin (Table 6.1) [11, 14, 25, 38]. It is widely accepted that the method of hDAM isolation impacts the ECM makeup and configuration, which carry functional and pathophysiological implications with regard to its regenerative capacity [11, 39]. Additionally, Gilbert et al. [40] demonstrated that not all cellular components are removed in various decellularization processes used in commercially available products. Measurable amounts of DNA (0.01–0.1%) were found in many of these tissues using histological staining and gel electrophoresis, which could potentially elicit an immune response. In order to compare and validate the results of different studies assessing the efficacy of hDAM as a scaffold for tissue engineering, a standardized method of decellularization and characterization is desired to limit variation in the composition, purity, and configuration of the hDAM being evaluated.

Most published work on the decellularization of adipose tissue consists of physical, chemical, and biological treatment stages [15]. Flynn was the first to describe comprehensive methods for the decellularization of adipose tissue [10]. This 5-day protocol starts with the mechanical disruption of the tissue achieved through multiple freeze-thaw cycles in hypotonic buffer. An overnight enzymatic digestion is followed by a 2-day polar solvent extraction in isopropanol to

Table 6.1 The optimal method of adipose tissue matrix preparation includes the removal of all cellular components

Technique	Types	Detects/confirms
Histological stains	(1) Hematoxylin and eosin (2) Oil red O (3) Masson's trichrome (4) Alkaline phosphatase (ALP) (5) Alizarin red S (ARS) (6) Von Kossa staining	(1) Cell and vascular architecture (2) Lipid and adipocyte content (3) Collagen structure (4) and (5) Osteoblastic differentiation (6) Matrix mineralization (calcium deposits)
Immunohistochemistry	(1) Hoechst (2) DAPI (3) AO/PI (4) Receptors/CD markers	(1), (2), and (3) Nucleic acid and retrained cellular components (4) Presence and location of markers (e.g., CD31, Col IV, laminin)
Biochemical assays	(1) DNA extraction kits (2) RT-PCR (3) Western blot	(1) and (2) Quantification and identification of specific component (e.g., DNA, genes, GAGs, elastin) (3) Protein expression
Electron microscopy	Scanning transmission	3D ultrastructure of extracellular matrix
Mechano-stress testing	(1) Young modulus (2) Storage and loss modulus (3) Small-angle oscillatory shear rheology (SAOS)	(1) Measure of elasticity (2) Mechanical integrity (3) Rheological properties/mechanical stiffness
Mass spectrometry	Secondary ion mass spectrometry	Identity of molecules
Atomic force microscopy	Scanning probe microscopy	Surface topography (e.g., surface area roughness)
Second harmonic generation microscopy	Laser scanning microscopy	Fiber orientation and density

remove the lipid content. Next, the tissue is exposed to a number of wash and enzymatic digestion steps, including DNase. Finally, the tissue is subjected to one final polar solvent extraction before it is ready for use [10]. Flynn was able to achieve complete decellularization of adipose tissue samples up to 25 g in mass that, after hydration, typically represented 30–45% of the original tissue mass. Hematoxylin and eosin (H&E) staining and scanning electron microscopy (SEM) confirmed the absence of cells and cellular debris [10].

Brown et al. [11] compared three different methods for the decellularization of adipose tissue. The first protocol, originally developed for the decellularization of liver tissue, included "mechanical massaging" prior to exposure to the strong detergent, Triton X-100. An n-propanol incubation step was eventually added due to its cost-effective ability to remove residual lipid components. The final two methods involved the use of various enzymes or detergents and acids,

but both methods produced wet materials that were high in lipid content as well as DNA of high base pair length. Decellularization was assessed by H&E and 4′,6-diamidino-2-phenylindole (DAPI) staining as well as DNA quantification. Despite the finding that the first method was superior, in terms of the preservation of the native collagen structure, the secretion of basic fibroblastic growth factor, and an observed increase in the retention of glycosaminoglycans (GAGs), all three methods resulted in the complete removal of laminin, a key component of native adipose extracellular matrix.

Another method described by Choi et al. [14] employed a tissue homogenizer for mechanical disruption prior to treating the tissue with a hyperosmolar salt solution. The lipoaspirate was then treated with the detergent sodium dodecyl sulfate (SDS), followed by the enzymes DNase and RNase. A near-complete elimination of DNA content was quantified via gel electrophoresis, and decellularization was confirmed with acridine orange (AO) and propidium iodide (PI) staining followed by SEM. The presence of laminin, however, was not analyzed in this study.

In addition, Wang et al. [25] prepared hDAM using a modified protocol they previously developed to decellularize musculofascial ECM. Their protocol utilized very similar components to the Flynn methods, including a polar solvent extraction. However, they swapped the enzymatic digestion step for a Triton X-100 incubation. H&E and DAPI staining confirmed the absence of cellular content that was later confirmed via SEM. DNA was also extracted and quantified to reveal a low 2.1 ng·mg^{-1} of processed tissue. While SEM demonstrated that this extraction technique was very successful at decellularization while maintaining the native 3D architecture, laminin was not detected in the final tissue.

The most comprehensive comparison of hDAM isolation methods was published by Sano et al. in 2014 [24]. Particularly, they compared purely mechanical, acid-based, and detergent-based methods to the original Flynn protocol and applied variations to each technique in an effort to define the optimal implementation. The mechanical based method A was the least successful; even after 18 freeze-thaw cycles, all of the cells and cellular components still remained, yet the matrix structures were destroyed. The acid-based (method B) and enzyme-based (method C) protocols were successful at maintaining the matrix architecture. Cells and cellular fragments remained even after extending the incubation periods eightfold. The only method that showed completely successful decellularization was the Flynn protocol. However, Sano et al. [24] only achieved complete decellularization after extending the enzymatic digestion steps. Additionally, they found that 0.8 g of adipose tissue was the upper limit of en bloc tissue that could be processed at any given time. This method did result in the preservation of collagen IV and laminin.

Recently, Wang et al. [41] employed the concept of using supercritical carbon dioxide (SC-CO$_2$) to decellularize adipose ECM. They washed the lipoaspirate twice in ethanol and loaded the samples into a reaction vessel with ethanol as a modifier. The liquid CO$_2$ was then compressed to 1.8×10^4 kPa and was passed through the reaction vessel at 37 °C for 3 h before collection and storage at 4 °C. Oil Red O staining demonstrated delipidization, while H&E staining with PicoGreen assay showed acceptable levels of DNA (43.52 ± 6.17 ng/mg), according to the threshold level for clinical applications. Immunostaining demonstrated the preservation of collagen, fibronectin, elastin, and laminin, which are essential to the ECM. The conservation of growth factors, such as vascular endothelial growth factor (VEGF) and fibroblast growth factor (bFGF), is reassuring for their potential contribution to angiogenesis and adipogenesis of tissue. Due to the evaporation rates of SC-CO$_2$, the treatment resulted in a solvent-free hDAM that did not require further steps, such as lyophilization, thus making it a prime candidate for an ideal decellularization protocol due to its low cytotoxicity and immunogenicity interactions [41].

Optimally, the decellularization process strikes a balance between the efficiency of cell removal and the maintenance of the native matrix structure. Flynn avoided detergents, like SDS, because their retention can be cytotoxic to cells, and they cause significant alterations, swelling, or irreversible macroscopic degradation of the matrix components [10]. Poon et al. [22] also avoided detergents because of the increased risk of matrix protein denaturation and degradation. Similarly, Wang et al. [41] used SC-CO$_2$ to avoid abrasive detergents, retaining key proteins and growth factors native to ECM. Wu et al. [36] used peracetic acid in combination with Triton X-100 to create a form of hDAM. The low concentration of the acid (0.1%) resulted in samples with a high DNA content and incomplete lipid removal, while a high concentration (5%) resulted in a significant degradation of collagen. The group determined that a 3% peracetic acid step yielded the optimal effect. However, they did not assay for the presence of laminin.

Over the past decade, proteomic capabilities have dramatically increased due to technological improvements that include better sample preparation protocols and enhanced capabilities in mass spectrometry (MS), database searching, and bioinformatics analysis. The use of proteomics to examine ECM has been extensively done for certain types of ECM, specifically cardiac ECM [42]. However, for hDAM, to the best of our knowledge, these types of studies have not been performed. We propose that the utilization of a proteomic approach to examine hDAM will contribute to our understanding of what actually constitutes this matrix. Furthermore, we hypothesize that, during the collection of adipose tissue, the fascia component, which is normally ignored, might actually contain key ECM proteins that are essential to adipogenesis and angiogenesis [43]. Future studies, some of which are being conducted in our lab, should examine these prospects.

6.3 Characterization of hDAM

Currently, there is a need for a standard set of characterization assays that would not only allow for a more accurate comparison of the various forms of hDAM, but also predict consistent results upon hDAM use in the clinical setting (Table 6.2). Among all studies reviewed, there is a great deal of variability among the methods used to characterize the structure of hDAM as well as the resulting microenvironment upon its implantation. So far, each study analyzing hDAM consistently utilizes at least one form of simple histological staining or scanning electron microscopy (SEM) to assess for the presence of cells and cellular components as well as to visually characterize the matrix. In order to decipher the components of hDAM as well as the efficacy of the decellularization process, some studies

Table 6.2 Methods used to characterize the structure of hDAM as well as the resulting microenvironment upon its implantation

Group	Classification	Key steps/ components	Processing time	Advantages	Disadvantages
Flynn [10]	Nondetergent based	(1) 3 freeze/thaw cycles (2) Trypsin digestion (3) Polar solvent extraction (4) Specialized rinsing buffer (5) Specialized enzymatic digestion (6) Specialized rinsing buffer (7) Polar solvent extraction	~120–156 h [8, 10]	• Retention of native hDAM structures including LN, Col IV • Complete decellularization consistently observed	• Time consuming • Maximum fat parcel size that can be processed limited to 800 mg [10]
Choi et al. [12]	Detergent based	(1) Tissue homogenization (2) Salt-solution treatment (3) SDS incubation (4) DNase/RNase treatment	~54 h	• DNA and RNA almost completely undetectable • Grafts showed no signs of inflammatory response in vivo	• Processing led to 24% decrease in acid/ pepsin-soluble collagen and 21% decrease in soluble elastin • LN and Col IV not analyzed
Young et al. [14, 27]	Detergent based	(1) Tissue thawed (2) SDS incubation (3) Specialized enzymatic digestion (4) Lyophilization	~88–124 h	• Complete decellularization based on IHC staining • Good preservation of Col I–IV with some preservation of LN	• High-retained DNA content compared with other studies • Demonstrated good host integration in in vivo model

(continued)

Table 6.2 (continued)

Group	Classification	Key steps/components	Processing time	Advantages	Disadvantages
Brown et al. [11]	Detergent based	(1) Tissue thawed (2) Mechanical massaging (3) Trypsin digestion (4) Mechanical massaging (5) Triton X-100 incubation (6) Ethanol/peracetic acid treatment (7) N-propanol polar solvent extraction (8) Lyophilization	~24 h	• Relatively quick assay • Results in retention of collagen (mainly Col III) and GAGs	• LN completely removed by decellularization process • Some lipid droplets remained in matrix
Wang et al. [27]	Detergent based	(1) 3 freeze/thaw cycles (2) Extensive wash step (3) Salt-solution treatment (4) Overnight wash (5) Salt-solution treatment (6) Trypsin digestion (7) Polar solvent extraction (8) Triton X-100 incubation	~214 h	• Complete decellularization with extremely low DNA content observed • Maintenance of collagen, VEGF, and GAGs and removal of MHC-I	• Long preparation time required • LN completely removed during processing
Wang et al. [40]	Nondetergent based	(1) Lipoaspirate washed in DI water (2) Rinsed twice in ethanol (3) Loaded into reaction vessel with ethanol as modifier (4) SC-CO$_2$ (1.8×10^4 kPa) passed through	~3 h	• Low levels of DNA content • Preservation of collagen, laminin, fibronectin, and elastin • Fast processing time • Retention of GAGs, bFGF, and VEGF	• Some residual DNA content • Limited sample size due to size of reaction vessel

employ techniques such as DNA isolation/quantification, immunohistochemistry (IHC), biochemical assays, mechanostress testing, and particle size separation [20]. This section summarizes these methods and discusses their implications.

Simple histological staining is the quickest and most direct way to assess for decellulariza-

tion. Additional methods commonly used include Masson's trichrome staining, which, in addition to staining cell fragments, allows for the detailed detection of collagen [10, 28, 36], and oil red O staining, which allows for the identification of adipocytes and lipid content [25, 27, 28, 36]. The majority of studies reviewed also utilize H&E staining, which allows for the direct visualization of cell membranes, proteins, and nuclei. H&E staining is also conducted to observe cellular infiltration, angiogenesis, collagen organization, adipogenesis, and structural integration of hDAM into the host tissue. Studies that do not use H&E staining, however, typically employ some form of nucleic acid staining such as Hoechst, DAPI, or AO/PI for the detection of retained cellular components [11, 14, 23, 25, 27, 33].

Additionally, biochemical assays are powerful tools for the assessment of hDAM constituents. One of the more commonly studied aspects includes the extraction, characterization, and quantification of DNA [11, 14, 23, 25, 27, 33], a surrogate marker of decellularization. It is suggested that a DNA content <50 ng·mg^{-1} of tissue is optimal for proper decellularization [11, 23]. Other assays in this category allow for the assessment of components like GAGs [27], elastin [14], and specific collagen content [36]. Also, gel electrophoresis can be used to quantify specific matrix components such as fibronectin and laminin [22].

Immunohistochemical staining (IHC) is a common method used to characterize hDAM. Immunohistochemistry enables scientists to detect specific cellular and structural markers, such as CD31 [23, 36] and VEGF [25], as well as important extracellular components, such as collagen, laminin, fibronectin, and vitronectin. A large number of studies implement IHC to quantify the presence of type IV collagen and laminin [10, 11, 24, 25, 27], the primary constituents of adipose extracellular matrix basement membrane [10, 11, 29]. Recent studies also increasingly utilize IHC to assess for the presence of collagen IV, a key basement membrane protein, to further ensure that their decellularization procedure is successful in retaining this component of hDAM [20]. Another study implemented IHC to screen for collagen VI, which is positively correlated with obesity-associated fibrosis [44]. Additionally, IHC is used to detect the presence or absence of major histocompatibility complex class I (MHC-I), the absence of which indicates the removal of alloantigenicity.

In addition to matrix proteins, IHC is used to screen for bioactive materials related to adipogenesis, angiogenesis, and macrophage response. One group implemented IHC to screen for CD68 (pan-macrophage marker) and CD163 (M2 macrophage marker) following hDAM implantation in a murine model [30]. M2 macrophages are well known for their pro-angiogenic capabilities and are linked to the regenerative properties observed with hDAM implantation [45]. Another protein of interest, perilipin, is associated with intracellular lipid droplets and is used to observe increases in lipid-containing macrophages and adipocytes as a result of an hDAM injection [30]. Two other adipogenic markers that are screened for include peroxisome proliferative activated receptor gamma (PPARγ) and anti-inflammatory adipokine adiponectin. PPARγ localizes in the nucleus upon initiation of the adipogenesis pathway [30]. Also, adiponectin is secreted by mature adipocytes as well as lipid-containing M2 macrophages and is reported to increase after hDAM implantation [30].

hDAM is comprised of several structural proteins, such as collagen, chondroitin sulfate, elastin, and non-proteoglycans, as well as several cell adhesive proteins, such as fibronectin, vitronectin, laminin, and tenascins [46]. These cell adhesive properties make it a favorable compound to be used in conjunction with synthetic scaffolds. For example, hDAM coating around the synthetic scaffold poly(sebacoyl diglyceride) (PSeD) allows for enhanced bone marrow mesenchymal stem cell (BMSC) adhesion because of the hydrophilic properties as well as the porous nature of hDAM [46]. This study indicates that the ECM allows for growth factors to bind, become stored, and then be released when necessary following ECM protein degradation [46]. Wei et al. [46] performed ELISA and demonstrated that human adipose-derived matrix secreted the growth factors bone morphogenetic protein-2 (BMP-2) and insulin-like growth factor-1 (IGF-1), which facilitate the formation of new bone tissue. Using a western blot analysis, this group also showed that bone-related gene markers, such as Ocn, OPN, Runx2, and Osx, were present in BMSCs cultured on hDAM. The group also compared the protein expression of collagen I, fibronectin, and laminin from adipose-derived matrix and bone marrow-derived matrix. These techniques permit scientists to determine whether hDAM is the optimal matrix to be used for regeneration of a desired tissue type.

Another highly specific method for analyzing the protein composition of hDAM components is mass spectroscopy. However, interestingly, this has not been done for hDAM, but has been done for a variety of other ECM tissues. The characterization of the ECM, as a fundamental component of multicellular organisms, has been underway for some time. In 1984, Martin et al. [47] coined the term "matrisome" to define the ECM components in terms of the basement membrane. Then, in 2012 the term was expanded to include all the genes that encode functional and structural ECM proteins and also ECM interacting and remodeling proteins [48]. To capitalize on the development of ECM proteomics, the same group, from the Massachusetts Institute of Technology, described new bioinformatic tools and experimental strategies for ECM research, and they introduced a Web platform called "the matrisome project" and a database called MatrisomeDB, which compiles in silico and in vivo matrisome data. This essentially provides an ECM atlas based on proteomics data from the ECM components of numerous tissue and tumor types [49]. We utilized mass spectrometry to identify proteins in hDAM and in ECM derived from adipose fascia from four cadaveric donors. After identifying all of the proteins, we selected only those associated with the matrisome using the database and Web platform described above. The proteins were then compared and identified in each type of ECM and found that there were 115 and 98 proteins in the hDAM and fascia ECM, respectively, and these were identified in at least 2 of the 4 donors. There were 86 proteins in common and 29 and 12 were unique in the hDAM and fascia ECM, respectively (Fig. 6.1). The unique proteins were then annotated using Gene Ontology (GO). The GO analysis revealed that the fascia proteins were more related to angiogenesis than the hDAM proteins. These data suggest that a combination of hDAM and fascia ECM component might enhance angiogenesis to improve graft retention. These are the areas of continued research in our lab.

Another emerging method of hDAM characterization is the examination of particle size [46]. Following a milling procedure, hDAM was separated and grouped into three size ranges utilizing stainless steel sieves of varying mesh sizes, including small (<45 μm sieve), medium (100 and 150 μm sieves), and large (250 and 300 μm). Using a laser particle size analyzer, the size distribution was evaluated by using the refractive index of interstitial collagen (1.547). Isolating varying particle sizes from hDAM allows researchers to observe the optimal particle size necessary for regeneration and adipogenesis to occur. For example, smaller hDAM particles seem to promote adipogenic differentiation by increasing surface area for cell-to-cell interaction [46]. Thus, understanding the particle size-to-tissue induction relationship is paramount creating hDAMs for various regenerative applications.

Analysis of collagen structure is another useful technique to characterize changes in extracellular matrix. One study conducted

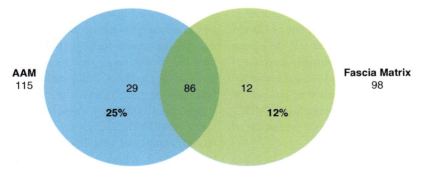

Fig. 6.1 Gene ontology analysis following mass spectometry to identify proteins in the human decellularized adipose tissue extracellular matrix (AAM) and the adipose fascia from 4 cadaveric donors. Proteins only associated with the matrisome as described in the text were analyzed. The two tissue types shared 86 proteins, while the AAM contained 29 unique proteins; the fascia matrix with 12 unique proteins

multiphoton autofluorescence second harmonic generation (SHG) imaging of obese and healthy hDAM collagen structures and analyzed the images through a custom-built autocorrelation analysis algorithm [50]. SHG imaging is utilized as a microscopic imaging contrast mechanism to determine the structures and functions of cells and tissues. Interestingly, this study utilized SHG imaging and found that obese hDAM is structurally thicker in comparison to healthy hDAM through collagen observation. Also, obesity-derived hDAM is associated with functional defects and an increase in malignant behavior [46]. This type of imaging provides a tool to characterize the structure of hDAM components in healthy and diseased tissue states and is therefore a crucial tool used to characterize healthy and non-healthy hDAM. Virtually every study employs SEM or transmission electron microscopy in order to visualize the microstructure of hDAM [10, 11, 14, 24, 25, 27, 28, 33, 36]. In addition to enabling the confirmation of decellularization, this high-level visualization also allows one to assess the ultrastructural characteristics of the ECM [24].

Examining the cellular deposition and remodeling of fibronectin are also equally important when characterizing hDAM. The same study utilized a Förster resonance energy transfer (FRET) analysis to characterize the conformation of fibronectin in hDAM from obese patients [46]. The study confirmed previous results that obesity-associated hDAM exhibits increased fibronectin unfolding, contributing to a more dysfunctional matrix. This may be crucial to ensuring that hDAM will be functionally effective in a regenerative capacity if it is derived from obese patients. Screening for these proteins allows for a greater understanding of the changes in the structural makeup of hDAM under a chronic disease state, such as obesity.

Additionally, the mechanical properties of hDAM must be taken into consideration when defining the structure of hDAM. Mechanostress testing is commonly overlooked in hDAM characterization, but it is critical to its evaluation. A way to measure matrix stiffness is by calculating the compressive modulus using the linear region of the stress–strain curve. To do this, printed hDAM constructs are tested using an electromechanical universal testing machine. One group performing this testing determined that their printed hDAM constructs have a compressive modulus of 122.56 ± 20.23 kPa, which is stiffer than the compressive modulus of native adipose tissue (19 ± 7 kPa) [20]. Kochhar et al. [51] recently reported that their detergent-based hDAM demonstrated viscoelastic properties similar to standard lipoaspirate and claimed that their hDAM, therefore, is an ideal candidate to substitute for standard lipofilling. Mechanical properties can also be communicated as rheological or elasticity data in the form of the storage and loss modulus [31, 51] or Young's modulus [14, 37, 52], respectively. Obtaining the ideal matrix stiffness is important because hDAM scaffolds that are too soft may result in structural collapse, whereas those that are too rigid may lead to irritation and scar-tissue formation [52].

Finally, matrix porosity is a component thought to influence cellular autocrine and paracrine function [53]. Reporting on this structural element is achieved using SEM [24], ethanol displacement [54], and capillary flow porometry [16]. One group engineered a dome-shaped hDAM construct and was able to control for porosity and reported long-term in vitro ADSC culture capabilities and an adequate transfer of oxygen and nutrients, and obtained a high cell viability [20]. Adequate hDAM porosity allows for a suitable environment for homogenous cellular distribution and tissue infiltration, and supports tissue remodeling at the implantation site [20]. Obtaining the ideal matrix porosity is challenging. Therefore, constructing a 3D printed hDAM hydrogel or using an electrospray protocol may be the two most efficient methods to ensure adequate matrix porosity.

6.4 Emerging Trends in Basic Science and Translational Applications

Several groups are currently testing hDAM in vitro and in vivo for clinical-translatability and tissue-engineering applications because natural ECMs possess an inherent property of biological recognition, which includes ligand presentation for cellular interactions, susceptibility to cell-triggered proteolytic enzymes, and subsequent remodeling required for morphogenesis [55]. The properties of hDAM that lead to optimal tissue formation are currently an area of investigation. For example, a preclinical study suggests that the presence of collagen IV and laminin, key constituents of the basement membrane, is associated with increased adipogenesis [38]. Another study indicates that the optimal particle size of the hDAM components that results in optimal adipogenesis is <45 μm. Although older studies examining hDAM used it in combination with progenitor cells, several scientists have recently examined hDAM and show that tissue regeneration is possible without the need for cell supplementation.

6.5 hDAM Supplementation

The use of biomaterial scaffolds in combination with cells expressing regenerative potential rather than utilizing just one of the two alone is gaining popularity in the field of regenerative medicine [30]. Currently the mechanisms of ADSC-mediated adipose tissue regeneration are not very well understood. However, a recent study indicates that ADSC-mediated adipogenesis is enhanced upon co-implantation with exogenous hDAM in a female rat model [30]. Co-culturing ADSCs and hDAM leads to enhanced blood vessel formation in ADSC-seeded hDAM scaffolds and significantly greater levels of VEGF compared to unseeded hDAM at 4 weeks and, also, shows an increased blood vessel diameter. However, there are no significant differences in levels of CD31+ and VEGF+ cells at later time points [30]. hDAM also plays a surprising role in immune cell recruitment. Following an ADSC-seeded hDAM scaffold implantation in a rat model, CD68+ macrophage accumulation was induced by hDAM rather than by ADSCs. Furthermore, CD163+ macrophages are a subtype of the macrophages responsible for matrix deposition and tissue remodeling. The authors of this study hypothesize that following hDAM implantation, M2b macrophages may be induced to secrete IL10, leading to activation of CD163+ macrophages [30]. Increased CD163+ macrophages are more present in ADSC-seeded hDAM compared to non-seeded implants at 8 and 12 weeks, and these macrophages are reported to have adipogenic potential. Also, an increase in the M2 macrophage subtype can contribute to the sprouting of new blood vessels and stimulate pericyte recruitment and vascular remodeling [30, 45]. The effects of hDAM are likely a result of changes in the microenvironment involving the secretion of beneficial paracrine factors that promote localized adipose tissue regeneration [30]. There are additional studies that demonstrate the capacity of hDAM to fill soft-tissue defects when supplemented with ADSCs. One group used ultrasonic homogenization followed by pancreatic digestion to create an hDAM as well as an ECM from porcine small intestine mucosa. The hDAM was superior for promoting ADSC viability and proliferation when co-cultured in vitro, inducing the upregulation of adipose-related genes, such as PPARγ and leptin, and also demonstrating superior adipogenesis in an in vivo model [28]. Ultimately, hDAM seeded with ADSCs promotes the recruitment of beneficial host cell populations that can aid in adipogenesis, and this proves to be a promising method in tissue regeneration. For this reason, many scientists are investigating the benefits of using hDAM in combination with ADSCs.

Cells are extremely sensitive to the environment in which they are cultured and introducing hDAM to the microenvironment improves cell culture outcomes. In a recent study by Marinkovic et al. [56], bone marrow-derived MSCs (BM-MSCs), ADSCs, as wells as cervical cancer cell line, HeLa, and breast cancer cell lines, MDA-MB-231 and MCF-7, were maintained in classical 2D culture (TCP), in BM-ECM, and in human adipose-derived ECM (AD-ECM). The authors discovered that stem cell proliferation for both the BM-MSCs and ADSCs on the two ECMs was significantly greater compared to the TCP. They also determined that the highest cell number was achieved when the cells were cultured in a tissue-specific manner with their native

ECMs. For example, BM-MSCs proliferated significantly more on BM-ECM compared to AD-ECM, while ADSCs proliferated significantly more on AD-ECM compared to BM-ECM. The authors also examined cancer cell proliferation on the two ECMs and TCP. The authors determined that HeLa cell proliferation is significantly reduced when the cells are cultured on BM-ECM and AD-ECM compared to TCP, and interestingly AD-ECM provides the most significant reduction compared to both BM-ECM and TCP. MCF-7 cells experienced a similar trend in reduction. However, this was only when the MCF-7 cells were cultured on AD-ECM, and for MDA-MB-213 cells proliferation was only significantly inhibited when they were cultured on BM-ECM. Further studies are needed to examine the in vivo effects of hDAM on cancer cell inhibition.

Another promising future application of hDAM is enrichment with catalase to reduce the negative effects of a hypoxic environment. Hypoxia leads to the formation of reactive oxygen species (ROS), such as H_2O_2, which in excess amounts leads to ischemic conditions, cell death, formation of necrotic tissue, and graft failure. A failure to provide oxygen to cells is one of the major reasons for unsuccessful tissue regeneration [57]. One group recently developed a catalase-enriched hDAM 3D scaffold with the ability to detoxify H_2O_2 through its conversion to oxygen and water. The group implanted catalase-enriched and catalase-free scaffolds in rats and observed greater tissue regeneration and new blood vessel formation in catalase-enriched groups compared to catalase-free scaffolds. Through the presence of oxygen, angiogenesis is supported and catalase protects cells and tissues from oxidative stress. Rijal et al. [57] discovered that scaffolds without catalase were unable to regenerate tissue in the center of large-sized scaffolds in a rat model, indicating that preventing oxidative tissue damage is imperative for large-tissue regeneration. The ability to utilize hDAM as a delivery mechanism for antioxidant biomaterials and for promotion of angiogenesis is just beginning to be explored and proves to be a promising application of hDAM for the enhancement of tissue grafting and regeneration.

6.6 Decellularization

A revolutionary method for the delipidization and decellularization of lipoaspirate-derived ECM is the use of supercritical carbon dioxide (SC-CO_2). SC-CO_2-treated ECM shows promise for future applications in wound healing [41]. Immortalized human keratinocyte cells (HaCAT cells) are commonly used to study wound healing, and keratinocyte migration plays an important role in full-thickness and epidermal wound healing [41]. HaCAT cells cultured on SC-CO_2-treated ECM-coated tissue culture plastic (TCP) show significantly improved migration and proliferation. SC-CO_2-treated ECM-coated TCP also significantly improves the proliferation of ADSCs and human umbilical vein endothelial cells (HUVECs) compared to TCP controls. SC-CO_2-treated ECM also prevents an unfavorable inflammatory immune response. THP-1-derived macrophages were cultured on SC-CO_2-ECM-coated TCP, and the inflammatory response was measured by TNF-α expression. THP-1-derived macrophages secrete significantly less TNF-α when they are cultured on ECM-coated TCP compared to TCP alone. SC-CO_2-ECM shows promise as a material that exhibits low cytotoxicity and low immunogenicity. This raises the question of whether SC-CO_2-ECM can be used as an off-the-shelf product to enhance patient outcomes. However, the research is still in its early stages, and further in vivo studies should be performed.

Wang et al. [25] created a detergent-based hDAM that they extensively characterized in terms of complete decellularization, low DNA count, porosity, and maintenance of collagen. This hDAM, which was devoid of laminin, showed comparable results to native tissue fat grafting after supplementation with 4×10^5 cells/mL of ADSCs in an in vivo model. The hDAM alone, similar to the previously mentioned detergent-based adipose ECMs, did not promote adipogenesis. This reinforces the notion that hDAM must be processed in a manner that preserves the key components of the basement membrane in order to be utilized without cellular supplementation; otherwise, an external cross-linking agent or the addition of stem cells is needed for hDAM to serve as a soft-tissue filler.

6.7 hDAM 3D Architecture

A moving trend in the utilization of hDAM is to manipulate it to obtain characteristics that are most favorable for tissue regeneration. One method of manipulation of hDAM is electro-spraying, which allows for the generation of spherical hDAM microcarriers of a desirable diameter. One group manipulated hDAM to synthesize porous microcarriers to an average diameter of 428 ± 41 μm via electrospraying [58]. hDAM microcarriers are porous, spherical, soft, compliant, and stable in long-term culture without chemical cross-linking. hDAM microcarriers are generated using a mild digestion protocol with the glycolytic enzyme α-amylase. Then, the collagen fibrils from hDAM are dispersed in solution and electrosprayed into liquid nitrogen to form spherical particles. Finally, a lyophilization step is required to generate porous hDAM microcarriers. Comparing hDAM microcarriers to commercially available Cultispher-S microcarriers, ADSC proliferation is enhanced in culture with hDAM microcarriers. When assessing the potential of ADSCs to differentiate to adipocytes, osteocytes, and chondrocytes following in vitro culture in either hDAM microcarriers or Cultispher-S microcarriers, there is a significant upregulation of the adipogenic markers PPARγ on hDAM microcarriers and LPL for both ADSCs expanded on hDAM microcarriers and Cultispher-S [58]. ADSCs expanded on DAT microcarriers formed more uniform intracellular lipids as well as enhanced matrix mineralization and alkaline phosphatase (ALP) staining, demonstrating increased osteoblastic differentiation compared to Cultispher-S microcarriers and TCP. The group also assessed for chondrogenic differentiation by collagen II staining and discovered that ADSCs expanded on DAT microcarriers exhibited enhanced differentiation compared to Cultispher-S and TCP-expanded ADSCs. Quantitative real-time-PCR was also performed to measure chondrogenic gene expression, and

ADSCs expanded on DAT microcarriers had a significantly higher expression of COLL2, COMP, and AGG compared to Cultispher-S or TCP. There is a need for scaffolds with a design that can mimic the complex microenvironment that occurs in the stem cell niche to promote differentiation and proliferation. The porous nature of DAT microcarriers makes them ideal for cellular infiltration and remodeling. Future studies utilizing DAT microcarriers should seek to identify the endogenous growth factors present in the microcarriers. The use of electrosprayed microcarriers is promising, since it yields positive outcomes without the need for chemical cross-linking. However, even though the potential for hDAM to facilitate different tissue formation is great, researchers are still investigating the details about the ideal hDAM characteristics that are necessary for each tissue type, such as the stiffness, chemical cross-linking, composition, mechanical properties, and surface topography.

Cross-linking appears throughout the literature in various forms as a method to augment the beneficial properties of decellularized adipose tissue matrix. Lu et al. [59] obtained murine decellularized adipose tissue extracellular matrix (mDAM) and cross-linked heparin to the matrix to create a delivery vehicle for basic fibroblast growth factor (bFGF). Their enhanced mDAM resulted in the formation of highly vascularized adipose tissue after 6 weeks compared with mDAM alone in a mouse model. Additionally, the explanted mDAM exhibited higher levels of the key components of adipogenesis, including CEBPα, adiponectin, and glucose transporter-4. The efficacy of an injectable hDAM, which was cross-linked with hexamethylene diisocyanate and 1-ethyl-3-[3-dimethylaminopropyl]carbodiimide (EDC) to improve volume retention, was also investigated [36]. The injectable hDAM, which was derived using acids and detergent, supports the growth and differentiation of ADSCs in vitro. Additionally, in a rat model cross-linking increases the hDAM's resistance to enzymatic degradation, promoting host-cell migration with adipose-tissue development and vascularization without the need for ADSC supplementation. Long-term in vivo studies demonstrate that acellular injectables develop into newly formed vascularized adipose tissue [36]. Young et al. [31] also compared a detergent-based hDAM gel for soft-tissue filling in a mouse model that was supplemented with either ADSCs or transglutaminase (TG), which is a cross-linking agent. They found that both were superior to hDAM alone and that the addition of TG was as effective as ADSCs at improving neovascularization and the soft-tissue-filling effect. This study contributes to the argument that progenitor cell supplementation of hDAM is not a necessity in achieving positive outcomes.

The efficacy of a purely mechanically based hDAM powder for soft-tissue-filling applications is currently under investigation [15]. Lipoaspirate was subjected to multiple washes with distilled water, homogenization, and centrifugation, forming a gel-like tissue suspension that was freeze-dried and ground to a fine powder. After sterilization, the gel was cultured and injected into nude mice. The hDAM powder seeded with ADSCs displayed superior proliferation, viability, and distribution in vitro and demonstrated improved neovascularization and adipogenesis in vivo when compared with the hDAM powder alone. In a subsequent study, Choi et al. [16] fabricated hDAM scaffolds into shapes, such as round discs, hollow tubes, and beads. In vitro testing revealed that the frequency of ADSC attachment was directly correlated to the scaffold pore size, a parameter for which the magnitude is directly proportional to the freezing temperature used for the scaffold preparation. More recently, this group investigated the use of various forms of hDAM for tissue-specific applications [13, 33, 60].

Initially, Flynn [10] demonstrated that hDAM seeded with ADSCs exhibited an increased expression of PPARγ and human CCAAT-enhanced binding protein alpha (CEBPα), which are both key regulators of adipogenesis, when compared with control adipogenic differentiated ADSCs. Glyceraldehyde 3-phosphate dehydrogenase (GAPDH) activity (a marker of adipose tissue maturation) was inversely proportional to the body mass index of the ADSC donors. Turner and Flynn [35] advanced these studies by designing composite hydrogel hDAM scaffolds that

serve as microcarriers for ADSCs. They implanted this hDAM subcutaneously in rats with or without ADSC supplementation. Macroscopically, there was no difference between the retained grafts, which were both superior to the gelatin-carrier control group. When analyzed under a microscope, both the seeded and unseeded hDAM grafts exhibited good integration with the host tissue, whereas the gelatin microcarriers were encapsulated by a dense fibrous tissue [61]. Because the seeded hDAM microcarriers exhibited the highest degree of cellularity and angiogenesis when compared with the unseeded graft, the researchers conducted further experiments on this model and found that the addition of the cross-linking agent methacrylated chondroitin sulfate increased its overall performance [32].

6.8 Clinical Implications

The utilization of hDAM as a therapeutic is currently being intensely studied in the field of regenerative medicine. A large quantity of adipose tissue is discarded as medical waste during surgical procedures including liposuction, reduction mammoplasty, abdominoplasty, and various other body contouring procedures. Rather than being discarded, this adipose tissue could serve as a reservoir of hDAM. As soon as it is harvested, the adipose tissue can be frozen and sent to a central processing facility where decellularization could proceed. This is practical because the first stage in many decellularization protocols consists of multiple freeze-thaw cycles. In addition, cadaveric sources of adipose tissue and subsequent preparation and refinement offer excellent opportunities for mass production.

Preliminary studies show a great potential for the use of hDAM in a variety of tissue-engineering approaches with obvious clinical ramifications, and these applications are not limited to soft-tissue filling. For instance, one group demonstrated the ability to significantly improve the healing of full-thickness wounds using sheets of detergent-based hDAM impregnated with ADSCs [62]. Another group combined hDAM with mDAM to construct an adipose tissue-derived acellular matrix thread that, when seeded with ADSCs, improved erectile function after 3 months in a rat model for cavernous nerve injury [34].

6.8.1 Breast Reconstruction

One study investigated using hDAM without cell supplementation from various adipose depots to assess the potential for breast reconstruction in patients who have undergone either a mastectomy or a lumpectomy [63]. The objective of this study was to investigate whether hDAM can be used as a material for breast reconstruction and identify the adipose depot ideally suited as a source of hDAM for this procedure. This study demonstrated, through a finite element model, that following its application, no hDAM sample had detectable contour irregularities following changes in body position to supine or to upright, similar to that of the normal breast. They identified that the hDAM, which can be isolated from pericardial, thymic, and omental adipose depots, causes higher breast deformation compared to subcutaneous and breast adipose depot following post-mastectomy reconstruction. As a result, the most promising depots to isolate hDAM for use in post-mastectomy breast reconstruction are the subcutaneous abdominal and breast depots, since they both lead to a deformation that is similar to a normal breast. In the post-lumpectomy scenario, the impact of hDAM is minimal since the majority of breast volume is occupied by native breast tissue. Therefore, in a lumpectomy, there is no restraint from which the adipose depot hDAM can be derived from because the deformation differences are not significantly different among hDAM derived from the different adipose depots. Therefore, subcutaneous adipose tissue can be potentially used as a universal source of adipose tissue from which to isolate hDAM for treatment in volume grafting.

There is a growing awareness of the connection between the adipose ECM and the diseased state. The altered properties of obese ECM may be responsible for obese patients' increased predisposition and worse clinical prognosis observed in the setting of breast cancer. Evidence shows a connection between obesity, ECM remodeling, inflammation, and mechanotransduction [50]. The extracellular matrix of obese patients is much stiffer in comparison to the stiffness of hDAM from lean patients. The mechanisms by which obesity causes these changes within hDAM are currently unknown [63]. It is hypothesized that obesity-associated chronic inflammation or obesity-associated hypoxia may play a role in matrix stiffness and increased tumorigenesis in these patients. Obesity also presents a challenge in early clinical screening with mammography for breast cancer, since women with highly dense breasts have a higher risk of breast cancer. Since adipose tissue is radiolucent, the breast tissue density of obese women appears lower. Thus, alternative imaging is needed to monitor tumorigenic regions [63]. The changes in obese hDAM may possibly contribute to the increased mechanosignaling in obese tumors and contribute to a poor clinical prognosis for obese cancer patients. Obese ECM from mice increases the malignant behavior of human breast cancer cells. However, the obesity-associated mechanisms causing changes in the physicochemical properties of hDAM are still unknown and are a future area of investigation.

6.8.2 Wound Healing

Wound healing is an area of regenerative medicine, in which hDAM is providing promising solutions. An ideal wound dressing exhibits a functionality and structure that is similar to endogenous skin tissue. hDAM is an ideal candidate to be used in wound healing because of its abundance, easy accessibility, and plethora of adipokines it contains, such as bFGF, VEGF, insulin-like growth factor-1 (IGF-1), and transforming growth factor-β1 (TGF-β1) [64]. The high porosity, surface area, physical property manipulability, and morphology of hDAM enable its application in deep-penetrating wound therapy. Recently, a group developed a novel bilayer wound dressing with a top layer comprised of an antibacterial titanium dioxide (TiO_2)-incorporated chitosan membrane and a sublayer of human adipose-derived extracellular matrix (ECM). The group utilized their wound dressing for treatment of full-thickness wounds in rats [64]. The bilayer-treated group expressed many CD31 endothelial cells, and the microvessel density was greater in comparison to the no bilayer treatment wound group. In addition, there was also reduced scar formation and an expedited regeneration of granulation tissue and epidermis in the bilayer composites. This work provides an excellent example of one of the future potential uses of hDAM in wound healing through its combination with an antibacterial material that can serve as a wound dressing without the need for progenitor cell supplementation.

6.8.3 Diabetes Mellitus

A recent study observing the effects of nondiabetic and diabetic hDAM on adipocyte metabolic function suggests that hDAM is a potential target for manipulating adipose tissue metabolism [65]. Diabetes mellitus (DM) is associated with adipocyte metabolic dysfunction as well as abnormal changes in the adipose extracellular matrix, since hDAM regulates adipocyte metabolism. Visceral hDAM from DM and non-DM patients was compared to assess hDAM-regulated adipocyte metabolism. Co-culturing diabetic hDAM and nondiabetic adipocytes inhibits insulin-stimulated glucose uptake in non-DM adipocytes. Also, nondiabetic hDAM partially rescues the effect of isoproterenol-stimulated lipolysis in DM preadipocytes. Nondiabetic hDAM also rescues glucose uptake and basal lipolysis in DM adipocytes, while DM hDAM impairs glucose uptake in non-DM adipocytes. hDAM, as a means to manipulate adipocyte metabolism, presents novel avenues for diabetes therapeutics, which should be further explored. This therapy is still in its early stages, and many in vivo studies are needed to determine its translatability.

6.8.4 Bone Generation

The use of ADSC-derived matrix (ADM) is gaining attention for its efficacy in bone tissue gen-eration. Utilizing ECM from ADSCs is a promising alternative over the use of ECM from BMSCs in repairing defective bone tissue for several reasons [46]. First, ADSCs are in abundance in the adipose tissue and it is much easier to isolate them in comparison to the painful procedure of obtaining BMSCs. Second, the ethical concerns surrounding the use of ADSCs are virtually nonexistent, and finally ADSCs demonstrate a greater proliferation capacity making the isolation of their ECM more readily available. A method of manipulating ADM is coating it on PSeD for bone tissue regeneration. PSeD is a novel polymer functional material that contains the necessary hardness and stiffness for bone repair. Human-ADM was isolated from human-ADSCs and coated onto PSeD to examine bone defect repair. In situ bone formation was examined by seeding a PSeD-ADM scaffold with rat BMSCs, PSeD with rBMSCs, and a defect-only group. New bone formation was observed to a greater extent in the PSeD/ADM + rBMSC group than the PSeD + rBMSC or the defect-only group. A morphometric analysis indicated that the total bone volume/total volume ratio in the PSeD/ADM group was significantly higher compared to the other two groups, indicating that PSeD/ADM promotes more bone tissue regeneration than PSeD and rBMSCs alone. Aside from promoting BMSC proliferation, ADM also demonstrates BMP-2 and IGF-1 growth factor release over a period of 30 days; exhibits strong osteogenic stimulant effects, with increases in the osteogenic genes Ocn, BSP, ALPL, and OPN; and also demonstrates a greater expression of collagen 1 and fibronectin than BMSC-ECM. The growth factors that are intrinsically stored in ADM are released and do have an effect on osteoblastic differentiation. Blocking BMP-2 in the ADM suppresses osteogenic gene expression (Ocn and OPN) and decreases ALP activity, and alizarin red S (ARS) staining demonstrates that the intrinsic growth factors in hDAM actively contribute to tissue regeneration. It would have been interesting if the same study had been performed with ECM derived from whole adipose tissue instead of only ADM, because the composition of the ECM from the different components of adipose tissue and the role they play in tissue regeneration are not completely understood and are currently an area of investigation.

6.8.5 Soft-Tissue Replacement

The use of hDAM as a stand-alone injectable filler to substitute for traditional fat grafting in situations requiring volumetric replacement provides a major advantage to current processes as it obviates the need for same-setting fat harvesting but still uses natural human adipose-derived scaffold for soft-tissue replacement. Though the majority of studies reviewed focus on hDAM as a framework for ADSC delivery, a further emphasis is needed on the regenerative benefits of hDAM alone, a product that could be lyophilized, stored, and used in an off-the-shelf manner. For instance, Turner and Flynn [35] created a nondetergent-based, unsupplemented hDAM that exhibited good host-tissue integration after implantation in rats when compared with the control. Similarly, Poon et al. created a detergent-free porcine decellularized adipose tissue extracellular matrix (pDAM) hydrogel that was capable of promoting adipogenesis in rats [18, 22]. Not only does hDAM lack immunogenicity, but also its properties would align it under section 361 of the Public Health Service Act by the US FDA as being a "minimally manipulated" tissue, much like the many brands of acellular dermal matrix used today. Turner et al. [61] did not detect any macroscopic or microscopic differences between seeded and unseeded hDAM grafts in a rat model, which is supported by recent research. Unfortunately, removing the unpredictability in fat grafting using seeded hDAM results in a process that requires an initial harvest procedure, extensive time in cell culture, and approval as an investigational new drug in the United States. The recent revolutionary preparation methods of hDAM along with enzyme and biomaterial supplementation as well as cross-linking may lead to the development of an hDAM that does not require cellular supplementation. Recent preparation protocols are allowing for intrinsic growth factors to remain present within the matrix leading to their continuous release to promote tissue regeneration [64].

6.9 Conclusions

It is well established that hDAM has the potential to promote the attachment and proliferation of progenitor cells. The functionality of hDAM in vivo is not as clearly understood. A majority of researchers explore the utilization of hDAM as a method to deliver progenitor cells and few studies seek to examine the effects of utilizing hDAM alone without progenitor cell enrichment to observe native cell infiltration and tissue enhancement. The supplementation of hDAM with cells in culture to be provided for patients may pose regulatory challenges; therefore, more studies should seek to explore the potential of hDAM without exogenous cellular supplementation in tissue regeneration. The low immunogenicity and low cytotoxicity of hDAM along with its ability to enhance progenitor cell proliferation make it a strong candidate for use in soft-tissue regeneration and may result in the off-the-shelf use of this product. Furthermore, there are several methods to derive hDAM; yet, no research group, thus far, has compared the most promising methods of hDAM isolation against one another to examine the results of specific tissue regeneration. hDAM is promising in many areas of tissue regeneration, such as bone formation, adipose tissue formation, cartilage formation, and wound healing. Also, the role hDAM plays in cancer cell suppression as well as metabolic regulation is promising for the development of novel therapeutics. Creating the optimal microenvironment for tissue regeneration may prove to be tissue specific, and recent research on hDAM manipulation and supplementation with synthetic scaffolds is paving the way for novel therapies in the field of regenerative medicine.

References

1. Kolle SF, Fischer-Nielsen A, Mathiasen AB, Elberg JJ, Oliveri RS, Glovinski PV, Kastrup J, Kirchhoff M, Rasmussen BS, Talman ML, Thomsen C, Dickmeiss E, Drzewiecki KT. Enrichment of autologous fat grafts with ex-vivo expanded adipose tissue-derived stem cells for graft survival: a randomised placebo-controlled trial. Lancet. 2013;382:1113–20.
2. Harrison BL, Malafa M, Davis K, Rohrich RJ. The discordant histology of grafted fat: a systematic review of the literature. Plast Reconstr Surg. 2015;135:542e–55e.
3. Yoshimura K, Sato K, Aoi N, Kurita M, Hirohi T, Harii K. Cell-assisted lipotransfer for cosmetic breast augmentation: supportive use of adipose-derived stem/stromal cells. Aesthet Plast Surg. 2008;32:48–55.
4. Banyard DA, Salibian AA, Widgerow AD, Evans GR. Implications for human adipose-derived stem cells in plastic surgery. J Cell Mol Med. 2015;19:21–30.
5. Paik KJ, Zielins ER, Atashroo DA, Maan ZN, Duscher D, Luan A, Walmsley GG, Momeni A, Vistnes S, Gurtner GC, Longaker MT, Wan DC. Studies in fat grafting: Part V. Cell-assisted lipotransfer to enhance fat graft retention is dose dependent. Plast Reconstr Surg. 2015;136:67–75.
6. Vogt PM, Reimers K. Cell-assisted autologous fat grafting. Dtsch Arztebl Int. 2015;112:253–4.
7. Khouri RK Jr, Khouri RE, Lujan-Hernandez JR, Khouri KR, Lancerotto L, Orgill DP. Diffusion and perfusion: the keys to fat grafting. Plast Reconstr Surg Glob Open. 2014;2:e220.
8. Dong Z, Peng Z, Chang Q, Zhan W, Zeng Z, Zhang S, Lu F. The angiogenic and adipogenic modes of adipose tissue after free fat grafting. Plast Reconstr Surg. 2015;135:556e–67e.
9. Lo Sicco C, Reverberi D, Balbi C, Ulivi V, Principi E, Pascucci L, Becherini P, Bosco MC, Varesio L, Franzin C, Pozzobon M, Cancedda R, Tasso R. Mesenchymal stem cell-derived extracellular vesicles as mediators of anti-inflammatory effects: endorsement of macrophage polarization. Stem Cells Transl Med. 2017;6:1018–28.
10. Flynn LE. The use of decellularized adipose tissue to provide an inductive microenvironment for the adipogenic differentiation of human adipose-derived stem cells. Biomaterials. 2010;31:4715–24.
11. Brown BN, Freund JM, Han L, Rubin JP, Reing JE, Jeffries EM, Wolf MT, Tottey S, Barnes CA, Ratner BD, Badylak SF. Comparison of three methods for the derivation of a biologic scaffold composed of adipose tissue extracellular matrix. Tissue Eng Part C Methods. 2011;17:411–21.
12. Choi JS, Kim BS, Kim JD, Choi YC, Lee EK, Park K, Lee HY, Cho YW. In vitro expansion of human adipose-derived stem cells in a spinner culture system using human extracellular matrix powders. Cell Tissue Res. 2011;345:415–23.
13. Choi JS, Kim BS, Kim JD, Choi YC, Lee HY, Cho YW. In vitro cartilage tissue engineering using adipose-derived extracellular matrix scaffolds seeded with adipose-derived stem cells. Tissue Eng Part A. 2012;18:80–92.
14. Choi JS, Kim BS, Kim JY, Kim JD, Choi YC, Yang HJ, Park K, Lee HY, Cho YW. Decellularized extracellular matrix derived from human adipose tissue as a potential scaffold for allograft tissue engineering. J Biomed Mater Res Part A. 2011;97:292–9.
15. Choi JS, Yang HJ, Kim BS, Kim JD, Kim JY, Yoo B, Park K, Lee HY, Cho YW. Human extracellular matrix (ECM) powders for injectable cell delivery and adipose tissue engineering. J Control Release. 2009;139:2–7.
16. Choi JS, Yang HJ, Kim BS, Kim JD, Lee SH, Lee EK, Park K, Cho YW, Lee HY. Fabrication of porous extracellular matrix scaffolds from human adipose tissue. Tissue Eng Part C Methods. 2010;16:387–96.
17. Choi YC, Choi JS, Kim BS, Kim JD, Yoon HI, Cho YW. Decellularized extracellular matrix derived from porcine adipose tissue as a xenogeneic biomaterial for tissue engineering. Tissue Eng Part C Methods. 2012;18:866–76.
18. Debels H, Gerrand YW, Poon CJ, Abberton KM, Morrison WA, Mitchell GM. An adipogenic gel for surgical reconstruction of the subcutaneous fat layer in a rat model. J Tissue Eng Regen Med. 2017;11(4):1230–41.
19. Francis MP, Sachs PC, Madurantakam PA, Sell SA, Elmore LW, Bowlin GL, Holt SE. Electrospinning adipose tissue-derived extracellular matrix for adipose stem cell culture. J Biomed Mater Res A. 2012;100:1716–24.
20. Pati F, Ha DH, Jang J, Han HH, Rhie JW, Cho DW. Biomimetic 3D tissue printing for soft tissue regeneration. Biomaterials. 2015;62:164–75.
21. Pati F, Jang J, Ha DH, Won Kim S, Rhie JW, Shim JH, Kim DH, Cho DW. Printing three-dimensional tissue analogues with decellularized extracellular matrix bioink. Nat Commun. 2014;5:3935.
22. Poon CJ, Pereira ECMV, Sinha S, Palmer JA, Woods AA, Morrison WA, Abberton KM. Preparation of an adipogenic hydrogel from subcutaneous adipose tissue. Acta Biomater. 2013;9:5609–20.
23. Porzionato A, Sfriso MM, Macchi V, Rambaldo A, Lago G, Lancerotto L, Vindigni V, De Caro R. Decellularized omentum as novel biologic scaffold for reconstructive surgery and regenerative medicine. Eur J Histochem. 2013;57:e4.
24. Sano H, Orbay H, Terashi H, Hyakusoku H, Ogawa R. Acellular adipose matrix as a natural scaffold for tissue engineering. J Plast Reconstr Aesthet Surg. 2014;67:99–106.
25. Wang L, Johnson JA, Zhang Q, Beahm EK. Combining decellularized human adipose tissue extracellular matrix and adipose-derived stem cells for adipose tissue engineering. Acta Biomater. 2013;9:8921–31.
26. Young DA, Choi YS, Engler AJ, Christman KL. Stimulation of adipogenesis of adult adipose-

26. derived stem cells using substrates that mimic the stiffness of adipose tissue. Biomaterials. 2013;34:8581–8.

27. Young DA, Ibrahim DO, Hu D, Christman KL. Injectable hydrogel scaffold from decellularized human lipoaspirate. Acta Biomater. 2011;7:1040–9.

28. Wang JQ, Fan J, Gao JH, Zhang C, Bai SL. Comparison of in vivo adipogenic capabilities of two different extracellular matrix microparticle scaffolds. Plast Reconstr Surg. 2013;131:174e–87e.

29. Young DA, Christman KL. Injectable biomaterials for adipose tissue engineering. Biomed Mater. 2012;7:024104.

30. Han TT, Toutounji S, Amsden BG, Flynn LE. Adipose-derived stromal cells mediate in vivo adipogenesis, angiogenesis and inflammation in decellularized adipose tissue bioscaffolds. Biomaterials. 2015;72:125–37.

31. Adam Young D, Bajaj V, Christman KL. Award winner for outstanding research in the PhD category, 2014 Society for Biomaterials annual meeting and exposition, Denver, Colorado, April 16–19, 2014: Decellularized adipose matrix hydrogels stimulate in vivo neovascularization and adipose formation. J Biomed Mater Res Part A. 2014;102:1641–51.

32. Cheung HK, Han TT, Marecak DM, Watkins JF, Amsden BG, Flynn LE. Composite hydrogel scaffolds incorporating decellularized adipose tissue for soft tissue engineering with adipose-derived stem cells. Biomaterials. 2014;35:1914–23.

33. Kim BS, Choi JS, Kim JD, Choi YC, Cho YW. Recellularization of decellularized human adipose-tissue-derived extracellular matrix sheets with other human cell types. Cell Tissue Res. 2012;348:559–67.

34. Lin G, Albersen M, Harraz AM, Fandel TM, Garcia M, McGrath MH, Konety BR, Lue TF, Lin CS. Cavernous nerve repair with allogenic adipose matrix and autologous adipose-derived stem cells. Urology. 2011;77:1509 e1–8.

35. Turner AE, Flynn LE. Design and characterization of tissue-specific extracellular matrix-derived microcarriers. Tissue Eng Part C Methods. 2012;18:186–97.

36. Wu I, Nahas Z, Kimmerling KA, Rosson GD, Elisseeff JH. An injectable adipose matrix for soft-tissue reconstruction. Plast Reconstr Surg. 2012;129:1247–57.

37. Yu C, Bianco J, Brown C, Fuetterer L, Watkins JF, Samani A, Flynn LE. Porous decellularized adipose tissue foams for soft tissue regeneration. Biomaterials. 2013;34:3290–302.

38. Mori S, Kiuchi S, Ouchi A, Hase T, Murase T. Characteristic expression of extracellular matrix in subcutaneous adipose tissue development and adipogenesis; comparison with visceral adipose tissue. Int J Biol Sci. 2014;10:825–33.

39. Pellegrinelli V, Heuvingh J, du Roure O, Rouault C, Devulder A, Klein C, Lacasa M, Clément E, Lacasa D, Clément K. Human adipocyte function is impacted by mechanical cues. J Pathol. 2014;233:183–95.

40. Gilbert TW, Freund JM, Badylak SF. Quantification of DNA in biologic scaffold materials. J Surg Res. 2009;152:135–9.

41. Wang JK, Luo B, Guneta V, Li L, Foo SEM, Dai Y, Tan TTY, Tan NS, Choong C, Wong MTC. Supercritical carbon dioxide extracted extracellular matrix material from adipose tissue. Mater Sci Eng C Mater Biol Appl. 2017;75:349–58.

42. Lindsey ML, Hall ME, Harmancey R, Ma Y. Adapting extracellular matrix proteomics for clinical studies on cardiac remodeling post-myocardial infarction. Clin Proteomics. 2016;13:19.

43. Su X, Lyu Y, Wang W, Li D, Wei S, Du C, Geng B, Sztalryd C, Xu G. Fascia origin of adipose cells. Stem Cells. 2016;34:1407–19.

44. Park J, Scherer PE. Adipocyte-derived endotrophin promotes malignant tumor progression. J Clin Invest. 2012;122:4243–56.

45. Zajac E, Schweighofer B, Kupriyanova TA, Juncker-Jensen A, Minder P, Quigley JP, Deryugina EI. Angiogenic capacity of M1- and M2-polarized macrophages is determined by the levels of TIMP-1 complexed with their secreted proMMP-9. Blood. 2013;122:4054–67.

46. Wei W, Li J, Chen S, et al. In vitro osteogenic induction of bone marrow mesenchymal stem cells with a decellularized matrix derived from human adipose stem cells and in vivo implantation for bone regeneration. J Mater Chem B. 2017;5:2468–82.

47. Martin GR, Kleinman HK, Terranova VP, Ledbetter S, Hassell JR. The regulation of basement membrane formation and cell-matrix interactions by defined supramolecular complexes. Ciba Found Symp. 1984;108:197–212.

48. Naba A, Hoersch S, Hynes RO. Towards definition of an ECM parts list: an advance on GO categories. Matrix Biol. 2012;31:371–2.

49. Naba A, Clauser KR, Ding H, Whittaker CA, Carr SA, Hynes RO. The extracellular matrix: tools and insights for the "omics" era. Matrix Biol. 2016;49:10–24.

50. Seo BR, Bhardwaj P, Choi S, Gonzalez J, Andresen Eguiluz RC, Wang K, Mohanan S, Morris PG, Du B, Zhou XK, Vahdat LT, Verma A, Elemento O, Hudis CA, Williams RM, Gourdon D, Dannenberg AJ, Fischbach C. Obesity-dependent changes in interstitial ECM mechanics promote breast tumorigenesis. Sci Transl Med. 2015;7:301ra130.

51. Kochhar A, Wu I, Mohan R, Condé-Green A, Hillel AT, Byrne PJ, Elisseeff JH. A comparison of the rheologic properties of an adipose-derived extracellular matrix biomaterial, lipoaspirate, calcium hydroxylapatite, and cross-linked hyaluronic acid. JAMA Facial Plast Surg. 2014;16:405–9.

52. Omidi E, Fuetterer L, Reza Mousavi S, Armstrong RC, Flynn LE, Samani A. Characterization and assessment of hyperelastic and elastic properties of decellularized human adipose tissues. J Biomech. 2014;47:3657–63.

53. Banyard DA, Bourgeois JM, Widgerow AD, Evans GR. Regenerative biomaterials: a review. Plast Reconstr Surg. 2015;135:1740–8.
54. Dunne LW, Huang Z, Meng W, Fan X, Zhang N, Zhang Q, An Z. Human decellularized adipose tissue scaffold as a model for breast cancer cell growth and drug treatments. Biomaterials. 2014;35:4940–9.
55. Sharma NS, Nagrath D, Yarmush ML. Adipocyte-derived basement membrane extract with biological activity: applications in hepatocyte functional augmentation in vitro. FASEB J. 2010;24:2364–74.
56. Marinkovic M, Block TJ, Rakian R, Li Q, Wang E, Reilly MA, Dean DD, Chen XD. One size does not fit all: developing a cell-specific niche for in vitro study of cell behavior. Matrix Biol. 2016;52-54:426–41.
57. Rijal G, Kim BS, Pati F, Ha DH, Kim SW, Cho DW. Robust tissue growth and angiogenesis in large-sized scaffold by reducing H_2O_2-mediated oxidative stress. Biofabrication. 2017;9:015013.
58. Yu C, Kornmuller A, Brown C, Hoare T, Flynn LE. Decellularized adipose tissue microcarriers as a dynamic culture platform for human adipose-derived stem/stromal cell expansion. Biomaterials. 2017;120:66–80.
59. Lu Q, Li M, Zou Y, Cao T. Delivery of basic fibroblast growth factors from heparinized decellularized adipose tissue stimulates potent de novo adipogenesis. J Control Release. 2014;174:43–50.
60. Choi YC, Choi JS, Woo CH, Cho YW. Stem cell delivery systems inspired by tissue-specific niches. J Control Release. 2014;193:42–50.
61. Turner AE, Yu C, Bianco J, Watkins JF, Flynn LE. The performance of decellularized adipose tissue micro-carriers as an inductive substrate for human adipose-derived stem cells. Biomaterials. 2012;33:4490–9.
62. Lin YC, Grahovac T, Oh SJ, Ieraci M, Rubin JP, Marra KG. Evaluation of a multi-layer adipose-derived stem cell sheet in a full-thickness wound healing model. Acta Biomater. 2013;9:5243–50.
63. Haddad SM, Omidi E, Flynn LE, Samani A. Comparative biomechanical study of using decellularized human adipose tissues for post-mastectomy and post-lumpectomy breast reconstruction. J Mech Behav Biomed Mater. 2016;57:235–45.
64. Woo CH, Choi YC, Choi JS, Lee HY, Cho YW. A bilayer composite composed of TiO_2-incorporated electrospun chitosan membrane and human extracellular matrix sheet as a wound dressing. J Biomater Sci Polym Ed. 2015;26:841–54.
65. Baker NA, Muir LA, Washabaugh AR, Neeley CK, Chen SY, Flesher CG, Vorwald J, Finks JF, Ghaferi AA, Mulholland MW, Varban OA, Lumeng CN, O'Rourke RW. Diabetes-specific regulation of adipocyte metabolism by the adipose tissue extracellular matrix. J Clin Endocrinol Metab. 2017;102:1032–43.

Mesenchymal Cells that Support Human Skin Regeneration

7

Joanne K. Gardner, Zalitha Pieterse, and Pritinder Kaur

7.1 Introduction

Skin is a large, complex organ that covers the surface of the body and undergoes continuous renewal throughout life. The most important function of skin is forming a protective barrier between the body and external environment, thereby defending the host against external agents, such as pathogens, mechanical and chemical insults, heat, and radiation. Skin also prevents water and electrolyte loss, provides insulation, mediates thermoregulation and sensory function, and is a site of immune surveillance and synthesis of biological mediators, such as vitamin D [1–3]. The skin consists of two major layers:

1. The epidermis, an upper layer of stratified squamous epithelium
2. The dermis, a lower layer of connective tissue (Fig. 7.1) [1, 3]

J. K. Gardner · Z. Pieterse · P. Kaur (✉)
Epithelial Stem Cell Biology Group, School of
Biomedical Sciences, Curtin University,
Perth, WA, Australia

Curtin Health and Innovation Research Institute,
Curtin University, Perth, WA, Australia
e-mail: Joanne.gardner@postgrad.curtin.edu.au;
zalitha.pieterse@curtin.edu.au;
pritinder.kaur@curtin.edu.au

The epidermis and dermis are separated by a basement membrane [1, 3], and in human skin the epidermal-dermal interface displays a characteristic undulating structure of rete ridges and dermal papillae (Fig. 7.1) [4]. Beneath the dermis, there is a layer of subcutaneous adipose tissue, called the hypodermis, which provides additional shock protection and insulation, as well as participating in energy metabolism and storage [1].

The protective skin barrier can be compromised by injuries and pathologies, such as burns, ulcers, and cancer [2, 5, 6]. A major therapeutic approach for the treatment of skin injury or loss is culturing skin cells *in vitro* to generate epidermal sheets or skin equivalents for transplant to the patient's affected skin area [5, 7, 8]. The development of *in vitro* epithelial sheets or skin equivalents for therapeutic purposes requires an understanding of the process of wound healing and epidermal regeneration, especially which cell types and mechanisms are involved. Cross talk between cells of the epidermis and dermis, mediated by soluble factors, is critical for successful epidermal regeneration [8]. In this chapter, we provide an overview of the epidermal and dermal cell types known to be involved in skin regeneration, followed by a discussion of experimental approaches used to date involving these cell types to grow skin equivalents *in vitro*, and the potential underlying mechanisms that mediate cross talk of epidermal and dermal cells that may contribute to skin regeneration.

© Springer Nature Switzerland AG 2019
D. Duscher, M. A. Shiffman (eds.), *Regenerative Medicine and Plastic Surgery*,
https://doi.org/10.1007/978-3-030-19962-3_7

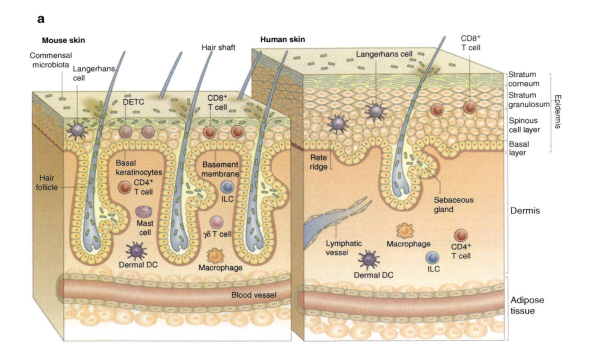

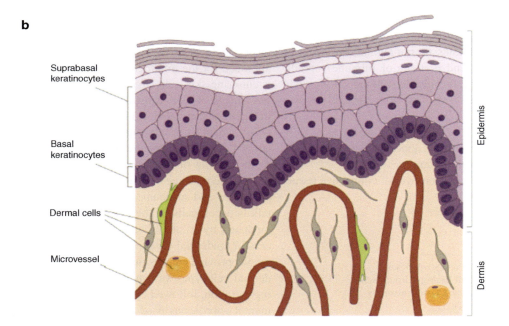

Fig. 7.1 Human skin structure: (**a**) Skin is comprised of (1) the epidermis, a stratified squamous epithelium; (2) the dermis, a connective tissue layer which is subdivided into papillary (superficial) and reticular (deep) layers, and contains skin-associated structures such as hair follicles, sweat glands, and sebaceous glands; and (3) a layer of subcutaneous adipose tissue, termed the hypodermis. The epidermal-dermal interface of human skin displays a characteristic undulating structure of epidermal/rete ridges and dermal papillae. (**b**) A close-up view of the epidermal-dermal junction shows that each dermal papilla is supplied by a dermal microvessel, and depicts the location of basal keratinocytes (which contain stem cell and transit-amplifying populations), relative to dermal cells such as fibroblasts (grey), pericytes lining dermal capillaries (green), and immune cells (yellow) (reproduced with permission)

7.2 Epidermis

The epidermis consists of a multilayered epithelium of keratinocytes, and associated structures, such as hair follicles (Fig. 7.1); the areas of the epidermis between hair follicles are referred to as the interfollicular epidermis [3]. Keratinocytes are the predominant cell type in the epidermis, and are organized into distinct, stratified layers based on their proliferation/differentiation status. The basal layer, located just above the dermis, contains keratinocytes with high proliferative capacity. Above the basal layer are the suprabasal spinous and granular layers, and the outermost layer, the stratum corneum. Keratinocytes lose their proliferative capacity and become progressively more differentiated as they move upwards through the suprabasal layers and to the stratum corneum [1]. Keratinocyte differentiation is associated with profound changes in their morphology, in which they become progressively flatter, and anucleated when they reach the stratum corneum [1]. The different epidermal layers also display distinct expression patterns of keratins, structural proteins synthesized by keratinocytes. Keratin-5, -14, and -15 are expressed in keratinocytes of the basal layer, whilst keratin-1 and -10 are expressed by differentiated keratinocytes in the spinous layer [4, 9, 10].

7.3 Basal Keratinocytes/Stem Cells

The continuous renewal of the epidermis throughout life, and skin regeneration following injury, is mediated by epidermal stem cells [1, 2, 10]. These cells have high proliferative and regenerative potential, and are located in the basal layer of the interfollicular epidermis (Fig. 7.1) [2, 10]. A population of epidermal stem cells can also be found in hair follicles, and whilst their main role is to contribute to hair regeneration, they can also participate in epidermal wound healing [10]; however, we will focus our discussion on keratinocyte stem cells in the interfollicular epidermis. In human skin, the exact location of stem cell niches in the basal keratinocyte layer along the

rete ridges remains uncertain [4]. Depending on the markers used to identify epidermal stem cells, and the type of skin examined (different body sites, adult versus neonatal), stem cell niches have been described at various locations along the rete ridges [4]. Expression of high levels of $\beta1$ integrin has been associated with keratinocytes with stem cell properties, specifically high proliferative potential and slow cycling [10, 11]. In human skin in vivo, high levels of $\beta1$ integrin expression have been observed at the top of the dermal papillae where the dermis is closest to the outer surface (Fig. 7.1) in adult breast and scalp skin and neonatal foreskin, suggesting that these are the regions at which stem cells are localized [12, 13]. Use of the markers melanoma chondroitin sulfate proteoglycan [14] and Lrig-1 [15] to identify epidermal stem cells also places them at the top of the dermal papillae. In contrast, palm and foot skin display high $\beta1$ integrin levels along the deep rete ridge, suggesting that stem cells identified using this marker vary in location depending on body site [12, 13]. Studies by our group have identified that interfollicular epidermal stem cells exhibit high expression of $\alpha6$ integrin, and low expression of CD71 (transferrin receptor), and also express keratin-15 in adult skin, although this keratin is also expressed in the dividing progeny of stem cells termed transit-amplifying (TA) cells [16–18]. Based on keratin-15 distribution in vivo, epidermal stem cells and their progeny appear to be located at the deep rete ridge in adult skin [19].

Regardless of their location in the basal layer, epidermal stem cells are responsible for skin renewal and generation of a stratified epidermis. There are different models to explain how epidermal renewal and stratification may be achieved, with the orientation of basal keratinocyte division being an important factor [9, 10]. In the symmetric division model, the majority of basal keratinocyte divisions occur in a parallel orientation relative to the basement membrane, generating two basal daughter cells. Slow-cycling basal keratinocyte stem cells give rise to TA intermediate cells, which are highly proliferative and undergo four to five symmetric divisions, replenishing the basal cell population. Following this,

TA cells delaminate and differentiate, migrating upwards through the suprabasal skin layers, and eventually to the stratum corneum [9]. In the asymmetric division model, which was first demonstrated in embryonic murine skin [20], the majority of basal keratinocytes reorient their mitotic spindle from being parallel to perpendicular, relative to the basement membrane. During cell division, the basal daughter receives proliferation-associated factors, whilst differentiation-associated factors are distributed towards the suprabasal daughter, thereby enabling differentiated, stratified skin layers to form [20]. *In vitro*, when given the appropriate support and conditions (for example, growth factors, signals, and nutrients), human primary basal keratinocytes (which are a mixture of slow-cycling stem cells and highly proliferative TA cells) can proliferate and differentiate into a stratified epidermis, which includes a basement membrane, and resembles *in vivo* skin [21].

7.4   Basement Membrane

The skin basement membrane, which separates the epidermis and dermis, is an extracellular matrix composed predominantly of collagen type IV, laminins including the LAMA3 and LAMA5 isoforms, nidogen, and perlecan [3, 22]. A key function of the basement membrane is to provide anchoring and support to basal keratinocytes, thereby helping to maintain the structural organization and integrity of the epidermis [22]. Importantly, the basement membrane is a dynamic interface that permits diffusion of soluble factors/mediators, enabling cross talk between cells of the dermal and epidermal compartments [22], which plays a critical role in directing epidermal regeneration [8].

7.5   Dermis

The dermis is a layer of connective tissue composed of an extracellular matrix containing collagen and elastic fibers, and it provides mechanical support and nutrition to the epidermis [1]. It is organized into the papillary dermis (also called the superficial dermis), which is the layer closest to the epidermis, and the reticular dermis (also called the deep dermis), which is the layer beneath the papillary dermis and above the hypodermis (Fig. 7.1) [3]. The dermis contains skin-associated appendages, such as hair follicles, sweat glands and sebaceous glands, as well as networks of blood vessels, capillaries and lymphatic vessels (Fig. 7.1) [1]. Fibroblasts are the most abundant cell type in the dermis, and are located in the areas of connective tissue matrix in between hair follicles (Fig. 7.1), as well as specific fibroblast subtypes associated with hair follicles [8]. Other cell types found in the dermis include endothelial cells and pericytes of dermal capillaries, immune cells (Fig. 7.1), as well as adipocytes and adipose-derived mesenchymal stem cells (MSCs) of the hypodermis [1, 4].

7.6   Fibroblasts

Fibroblasts are the most abundant cell type in the dermis, where their main role is production of collagen, elastic fibers, and matrix proteins of the extracellular matrix [23, 24]. Additionally, fibroblasts are important regulators of skin physiology due to their ability to communicate with each other, and other skin cells (including keratinocytes) via direct cell-to-cell contact and/or soluble factors [23]. Fibroblasts are a morphologically and functionally heterogeneous population, and can be categorized into distinct subtypes [23]. In skin, fibroblasts located in the areas of the dermis in between hair follicles are referred to as interfollicular dermal fibroblasts, and are further classified as:

1. Papillary dermal fibroblasts, located in the papillary (superficial) dermis
2. Reticular dermal fibroblasts, located in the reticular (deep) dermis [8, 23]

Another group of fibroblasts found in skin is hair follicle-associated fibroblasts, which include hair follicle dermal sheath fibroblasts, located along the hair follicle shaft, and hair follicle dermal papilla fibroblasts, located at the base of the hair follicle [8, 23]. Whilst the main role of hair follicle-associated fibroblasts is regulation of hair follicle morphogenesis and hair cycle, they can also participate in epidermal regeneration and wound healing [24]. Skin fibroblast subtypes display different characteristics, including behavior in *in vitro* cultures, rates of division, and secretion profiles of extracellular matrix components and soluble factors, such as growth hormones and cytokines [8, 23]. Differences in the characteristics of skin fibroblast subtypes, particularly growth factor and cytokine secretion profiles, confer differing abilities to support keratinocytes [8].

7.7 Other Dermal Cells: Endothelial Cells, Pericytes, Mesenchymal Stem Cells, and Immune Cells

In human skin, dermal capillaries are located in the dermal papillae, in between the rete ridges (Fig. 7.1), where they supply the epidermis with nutrients and oxygen [25]. Dermal capillaries consist of endothelial cells, which form the vessel wall and are in direct contact with a layer of pericytes, which form a protective sheath around the vessel wall and provide an interface between the endothelial cells and surrounding dermal extracellular matrix [25, 26]. Pericytes are cells of mesenchymal origin, and in addition to their main roles in stabilizing blood vessels and regulating vascular development, they exhibit a range of other functions, including involvement in blood coagulation, modulating immune function, and phagocytosis [26], and can act as MSCs [26, 27]. The MSC activity of pericytes means that they have the capacity to differentiate into different tissue cell lineages, enabling them to participate in tissue regeneration, including skin [27]. Given the close proximity of pericytes and endothelial cells in the capillary loops of dermal papillae to the basal keratinocyte layer (Fig. 7.1), it is possible that they are able to influence the process of epidermal regeneration [4].

In addition to the MSC-like pericyte population, other MSC subtypes can be found in skin, such as adipose-derived MSCs in the hypodermis [28]. Similarly to pericytes, adipose-derived MSCs may contribute to skin regeneration via their multilineage potential [28]. Despite their distance from the basal keratinocyte layer, they may also contribute to epidermal regeneration indirectly via secretion of factors that modulate the dermal microenvironment and/or dermal cells [28].

Immune cells represent another important cell population in skin, due to their role in immunosurveillance and defense against pathogens [29, 30]. The skin contains cells of both the innate (antigen-presenting dendritic cells, phagocytic macrophages, and neutrophils) and adaptive immune systems (lymphocytes). Although their role in epidermal regeneration has not yet been directly characterized, immune cells secrete a variety of cytokines and inflammatory mediators during an immune response, and it is possible that these factors may modulate keratinocytes. Additionally, macrophages are known to play a role in wound healing in a variety of tissues, including skin [31], providing further support for the notion that immune cells contribute to epidermal regeneration.

7.8 Approaches to Grow Keratinocytes In Vitro for Regenerative Purposes

The development and advancement of skin regeneration techniques require an understanding of which cell types support and promote keratinocyte growth and renewal, as well as the mechanisms by which support cells perform this role. The earliest method for clonal human keratinocyte expansion *in vitro*, developed by Rheinwald and Green [32], used a two-dimensional (2D) monolayer co-culture system in which human primary keratinocytes were grown on an irradiated or mitomycin C-treated feeder layer of the murine Swiss 3T3 J2 strain fibroblastic cell line. This resulted in generation of keratinocyte colonies that displayed epidermal stratification, and features of *in vivo* skin, such as desmosomes [32]. Furthermore, epidermal sheets generated using this technique have been successfully used to treat patients with burn injuries [7], and this method remains the current gold standard for expanding human keratinocytes in vitro. Rheinwald and Green's [32] approach has formed the basis for further development and improvement of *in vitro* skin equivalent cultures and bioengineered skin models for therapeutic purposes [5, 7, 33]. Subsequent research has focused on identifying optimal feeder cell types to support keratinocyte expansion. In particular, replacement of animal (rodent)-derived feeder cells/components with human-derived feeder cells is important from a therapeutic viewpoint, and for safe clinical application of *in vitro*-derived skin therapies. Another important improvement in the field of skin regeneration is the shift from 2D monolayer cultures to three-dimensional (3D) organotypic models. Organotypic cultures are produced by culturing support cells alongside keratinocytes at an air-liquid interface, enabling establishment of the stratified, differentiating layers of the epidermis, and thus more accurately reflect the 3D microenvironment of skin *in vivo* [21, 34]. Additionally, the type of matrix/structural support provided to keratinocytes as well as the type of media and supplementation with specific growth factors are other important aspects that require optimization for successful keratinocyte culture and generation of skin equivalents *in vitro*. Of all these factors, a critical component for the success of *in vitro* keratinocyte expansion is the inclusion of feeder cells [8]. There is strong evidence that dermal/mesenchymal cell-derived factors and signals are necessary for keratinocyte growth and regeneration, based on observations from Rheinwald and Green's initial work [32], and subsequent studies performed using a variety of 2D and 3D *in vitro* skin models, as keratinocytes expand poorly *in vitro* in the absence of support cells [8]. Thus, we will discuss the dermal/mesenchymal cell types that have been used to support *in vitro* keratinocyte growth to date, potential mechanisms by which these feeder cells may promote keratinocyte growth and expansion, as well as future directions and important considerations in the field of epithelial regeneration.

7.8.1 The Role of Fibroblasts as Keratinocyte Support Cells

To date, the most widely studied and used feeder cell type to support keratinocyte expansion *in vitro* is fibroblasts. This began with the pioneering studies of Rheinwald and Green [32], who used the murine Swiss 3T3 J2 fibroblast cell line as a feeder cell for human primary keratinocytes in 2D monolayer cultures. Murine Swiss 3T3 fibroblasts have subsequently been used as the benchmark for comparing different approaches to culture human keratinocytes *in vitro* [8], as summarized in Table 7.1.

The replacement of rodent-derived feeder cells with those of human origin is important from a clinical perspective. Several studies have demonstrated that human primary dermal fibroblasts from adult skin and neonatal foreskin can support keratinocyte growth and expansion in 2D and 3D *in vitro* co-cultures, under a variety of growth media and matrix conditions [8], as summarized in Table 7.1. Additionally, with certain media and growth matrices, human dermal fibroblasts have been shown to be as efficient as murine Swiss 3T3 J2 fibroblasts in promoting keratinocyte expansion/proliferation, epidermal stratification, and expression profiles of phenotypic/differentiation markers (Table 7.1) [38–42, 48–52]. Furthermore, *in vitro* skin equivalents generated using human dermal fibroblasts have been transplanted onto burn patients to produce stable skin regeneration [46, 47].

The aforementioned studies involving human dermal fibroblasts used the entire fibroblast population, but it is now recognized that dermal fibroblasts are a heterogeneous population comprised of several different subtypes, which have differing abilities to promote keratinocyte growth and regeneration. Interfollicular dermal fibroblasts can be further categorized into papillary and reticular subpopulations, and there is a consensus that human papillary dermal fibroblasts

7 Mesenchymal Cells that Support Human Skin Regeneration

Table 7.1 Summary of studies using different mesenchymal feeder cells to support 2D and 3D human (rat and rabbit) keratinocyte cultures

Source and type of feeder cell	Source of keratinocytes	Co-culture model and measurement of keratinocyte growth/regeneration	References
Current gold standard			
Murine Swiss 3T3 J2 strain fibroblasts (γ-irradiated or mitomycin C treated)	Human adult skin or neonatal foreskin	2D monolayer; fetal calf serum (FCS) in media • Stratified, keratinized colonies with desmosomes formed	[7, 32, 35–37]
Unfractionated fibroblast population			
Human adult dermal fibroblasts (nonirradiated)	Human adult skin	2D monolayer; FCS in media • Keratinocyte proliferation rate and K14 and K10 expression comparable to murine Swiss 3T3 feeder layers	[38]
Human adult dermal fibroblasts (irradiated or nonirradiated)	Human adult skin	2D monolayer; plasma copolymer surface, serum-free conditions • Keratinocytes proliferate and form colonies • Keratinocyte viability comparable to Swiss 3T3 feeder layers	[39–41]
Human adult dermal fibroblasts (γ-irradiated)	Human adult skin	3D skin equivalent; serum-free media with growth factors • Keratinocytes formed stratified epithelial layers with appropriate K1, K10, and K14 expression comparable to Swiss 3T3 feeder layers	[42]
Human adult dermal fibroblasts	Human adult skin	3D organotypic culture; FCS in media • Keratinocytes proliferate, form stratified epidermis, expressing differentiation marker loricrin and β1-integrin and K15 • Display basement membrane and ultrastructural features resembling in vivo skin	[43–45]
Human adult dermal fibroblasts	Human adult skin	3D organotypic culture; clotted plasma scaffold, FCS in media • Keratinocytes formed stratified layers, expressed differentiation markers (keratin-5 and -10, loricrin), complete basement membrane present • *In vitro*-generated skin engrafted onto burn patients and produced stable skin regeneration	[46, 47]
Human neonatal foreskin and adult dermal fibroblasts (γ-irradiated)	Human neonatal foreskin and adult skin	2D monolayer; FCS in media • Keratinocyte growth rate, clonogenic potential, size, and morphology comparable to Swiss 3T3 feeder layers	[48, 49]
Human neonatal foreskin and adult dermal fibroblasts (nonirradiated)	Human neonatal foreskin and adult skin	3D organotypic culture; fibrin gel support, FCS in media • Keratinocyte colony number, doubling rate, stratification, expression of proliferation (PCNA, p63), differentiation (involucrin) and dermo-epidermal junction markers (integrin β4, collagen IV, laminin) comparable to Swiss 3T3 feeder layers	[50–52]
Human neonatal foreskin dermal fibroblasts	Human neonatal foreskin	3D skin equivalent; collagen matrix, serum-free conditions • Keratinocytes formed stratified, differentiated layers, expressed K1, K10, and K14, basement membrane and desmosomes present	[53, 54]

(continued)

Table 7.1 (continued)

Source and type of feeder cell	Source of keratinocytes	Co-culture model and measurement of keratinocyte growth/regeneration	References
Fibroblast subpopulations			
Human adult dermal fibroblasts separated into papillary and reticular subtypes	Human adult skin	3D organotypic culture; collagen matrix, FCS in media Papillary fibroblasts were superior to reticular fibroblasts with respect to: • Promoting epidermal stratification with differentiated keratinocyte layers, polarized basal cells, and cornification of upper layers • Promoting basal keratinocyte proliferation • Basement membrane formation and expression of associated proteins (laminin-5, nidogen, collagens IV and VII) • Uniform expression of keratinocyte differentiation markers (filaggrin, K5, loricrin, small proline-rich protein-2)	[55–61]
Human adult hair follicle dermal papilla and dermal sheath fibroblasts	Human adult skin	2D monolayer • Hair follicle-derived fibroblasts promoted increased keratinocyte proliferation and colony numbers compared to human interfollicular dermal fibroblasts and murine Swiss 3T3 fibroblasts	[62]
Human adult hair follicle dermal papilla and dermal sheath fibroblasts	Human neonatal foreskin	3D organotypic culture • Keratinocyte proliferation and differentiation were comparable to hair interfollicular dermal fibroblasts • Hair follicle dermal sheath fibroblasts had a superior ability to promote basement membrane formation, compared to hair follicle dermal papilla and interfollicular dermal fibroblasts	[63]
Human adult hair follicle dermal papilla and dermal sheath fibroblasts	Human adult skin	3D in vitro skin equivalent transplanted onto dorsal wound of BALB/c nude mice • Skin grafts with hair follicle dermal papilla fibroblast feeder layers promoted superior basement membrane formation and collagen synthesis, compared to grafts with hair follicle dermal sheath or interfollicular dermal fibroblast feeder layers	[64]
Human adult hair follicle dermal sheath fibroblasts	Human neonatal foreskin	3D skin equivalent; FCS in media Hair follicle dermal sheath fibroblasts were superior to neonatal foreskin interfollicular dermal fibroblasts with respect to: • Inducing formation of thicker epidermal layers • Promoting expression of keratinocyte differentiation markers (filaggrin and K1) and basement membrane components (integrin α6) • Promoting desmosomal ultrastructure development	[65]
Rat (PVG/C strain) hair follicle dermal papilla fibroblasts	Rat (PVG/C strain) skin	2D monolayer • Keratinocyte proliferation, aggregation, and formation of organotypic structures and basement membrane were comparable for hair follicle dermal papilla fibroblast and interfollicular dermal fibroblast feeder layers	[66]
Other dermal/mesenchymal cells			
Human neonatal foreskin dermal pericytes	Human neonatal foreskin	3D organotypic culture Pericytes were superior to dermal fibroblasts with respect to: • Restoring regenerative capacity of differentiating keratinocytes • Promoting laminin-511/521 deposition at dermal-epidermal junction	[27]

7 Mesenchymal Cells that Support Human Skin Regeneration

Table 7.1 (continued)

Source and type of feeder cell	Source of keratinocytes	Co-culture model and measurement of keratinocyte growth/regeneration	References
Human adult adipose tissue MSCs (X-ray irradiated)	Human adult skin	2D monolayer; FCS in media Human adipose MSCs were comparable to or better than murine Swiss 3T3 fibroblasts with respect to: • Promoting keratinocyte proliferation and forming a multilayered epidermis • Maintaining keratinocytes in a proliferative rather than differentiated state	[67]
Human adult adipose tissue MSCs (X-ray irradiated)	Human adult skin	2D monolayer and 3D transplantable epithelial sheet • Comparable keratinocyte colony size induced by human adipose MSC and murine Swiss 3T3 fibroblast feeder layers	[68]
Human adult adipose tissue MSCs	Human neonatal foreskin	3D skin equivalent Compared to neonatal foreskin fibroblasts, adipose MSCs promoted: • More complete epidermal differentiation and filaggrin expression • Similar levels of keratinocyte proliferation (assessed by p63 expression)	[69]
Human adult adipose tissue MSCs	Human adult skin	3D organotypic culture; collagen matrix, FCS in media • Adipose MSCs promoted increased primary keratinocyte proliferation, and similar epidermal morphology and thickness, comparable K1, K10, and K14 expression, increased K19 expression, and abnormal K5 distribution, compared to adult human dermal fibroblast feeder cells	[70]
Wistar rat MSCs from heart, lung, spleen, liver, and kidney	Wistar rat skin	3D organotypic culture; collagen matrix • Non-skin-derived MSCs promoted formation of stratified epidermal layers, expression of the proliferative marker p63 in the basal keratinocyte layer, and patterns of K10, K14, and involucrin resembling in vivo rat skin	[71]
Wistar rat bone marrow MSCs and subcutaneous pre-adipocytes	Wistar rat skin	3D organotypic culture; collagen matrix Bone marrow MSCs and subcutaneous pre-adipocytes were comparable feeder cells to dermal fibroblast FR cell line with respect to: • Promoting epidermal stratification and ridgelike structures • Inducing K14 in basal keratinocytes • Inducing K10 expression pattern similar to *in vivo* rat skin • Promoting keratinocyte proliferation • Inhibiting keratinocyte apoptosis	[72]
Human adult bone marrow MSCs	Human adult skin	3D organotypic culture Bone marrow MSCs were comparable feeder cells to human adult dermal fibroblasts with respect to: • Promoting formation of stratified epidermal layers • Inducing K6 and K10 expression patterns similar to *in vivo* skin • Inducing basal keratinocyte proliferation • Expression of basement membrane marker collagen IV	[73]

(continued)

Table 7.1 (continued)

Source and type of feeder cell	Source of keratinocytes	Co-culture model and measurement of keratinocyte growth/ regeneration	References
Wistar rat adipocytes	Wistar rat skin	3D organotypic culture; collagen matrix • Keratinocytes displayed greater proliferation, formation of thick, stratified layers, and expression of keratins more closely resembling *in vivo* rat skin, with adipocyte feeder layers, compared to dermal fibroblast feeder layers	[74]
Murine corneal endothelial cells	Rabbit corneal epithelium	3D organotypic culture; collagen matrix Compared to organotypic cultures with only keratinocytes and fibroblasts, inclusion of endothelial cells improved: • Basement membrane formation and morphology • Epithelial differentiation	[75]

are superior compared to their reticular counterparts with respect to promoting keratinocyte proliferation, morphogenesis, stratification, expression of differentiation markers, and basement membrane formation in 3D organotypic cocultures (Table 7.1) [55–61]. In addition to the fibroblast subpopulations contained in the interfollicular dermis, more recent studies have started to examine the supportive potential of hair follicle fibroblast subtypes, namely, hair follicle dermal papilla and dermal sheath fibroblasts (summarized in Table 7.1). These studies have shown that rat [66] and human [62–65] hair follicle-associated fibroblasts are comparable or superior to human interfollicular dermal fibroblasts and/or murine Swiss 3T3 J2 fibroblasts in supporting keratinocyte proliferation, differentiation, and basement membrane formation *in vitro* (summarized in Table 7.1). Thus, hair follicle-associated fibroblasts represent a promising alternative feeder cell type that warrants further investigation.

A major mechanism by which fibroblasts support keratinocyte growth and expansion is via secretion of paracrine factors, in particular, cytokines and growth factors. A key concept in fibroblast-keratinocyte cross talk is reciprocal paracrine signaling, whereby each cell type secretes factors that act on the other cell type, which has an additive effect to further promote epidermal proliferation, differentiation, and growth [8, 23, 76, 77]. Fibroblasts can secrete many different cytokines and growth factors, and some of the key factors they produce that are central to promoting keratinocyte growth, proliferation, and migration are keratinocyte growth factor (KGF, also known as fibroblast growth factor-7; FGF-7), FGF-10, hepatocyte growth factor (HGF), granulocyte-macrophage colony-stimulating factor (GM-CSF), interleukin-6 (IL-6), and IL-8 [8, 23, 78]. In addition, fibroblast-derived transforming growth factor (TGF)-β1 is an important inducer of keratinocyte differentiation [8, 79], and is likely to promote the development of epidermal stratification in organotypic cultures *in vitro*. Keratinocytes also secrete a variety of cytokines and growth factors, and IL-1α has been identified as a primary factor produced by keratinocytes that induces fibroblasts to produce keratinocyte growth-promoting factors, such as KGF, GM-CSF, pleiotrophin, and stromal cell-derived factor-1 [23, 76, 77, 80]. As well as cytokine and growth factor-mediated fibroblast-keratinocyte cross talk, it is important to consider the potential role of other soluble factors, such as hormones and vitamins, as binding of these ligands to neurohormonal and/or nuclear receptors is known to influence epidermal proliferation and differentiation [8].

Another important outcome of reciprocal fibroblast-keratinocyte paracrine signaling is the production of basement membrane components and correct basement membrane assembly [81–83]. In organotypic cultures, the presence of fibroblasts is required for production of basement membrane components by keratinocytes, such as collagens IV and VII; laminin-1, -5, and -10/11; nidogens; and bullous pemphigoid antigens

[8, 23, 78, 82–84]. TGF-β2, KGF, GM-CSF, and epidermal growth factor (EGF) secreted by fibroblasts are some of the main factors which induce keratinocytes to produce basement membrane components [8, 23, 78, 82, 83, 85–88]. Additionally, fibroblasts themselves can produce basement membrane molecules, including collagens IV and VII, laminin-5, and nidogen [78, 85], and promote the ultrastructural assembly of these components, thereby contributing to basement membrane formation [78]. Thus, factors secreted by fibroblasts, and subsequent paracrine cross talk between fibroblasts and keratinocytes, represent a critical underlying mechanism by which fibroblasts support epidermal growth and regeneration in *in vitro* skin cultures.

Different fibroblast subtypes have varying abilities to promote keratinocyte growth and regeneration, and this is likely related to differences in their cytokine and growth factor secretion profiles. The superior ability of papillary fibroblasts, compared to reticular fibroblasts, to promote keratinocyte growth could be attributed to the observation that papillary dermal fibroblasts produce a higher ratio of GM-CSF to KGF than their reticular counterparts [55], as differences in the ratio of these two factors impact upon keratinocyte growth status [76]. Additionally, papillary fibroblasts secrete more IL-6, compared to reticular fibroblasts [55], which could suggest that the former subtype is more likely to induce keratinocyte proliferation. Conversely, reticular fibroblasts have been shown to express higher levels of TGF-β1 than papillary fibroblasts [89, 90], which may suggest that reticular fibroblasts are more likely to promote keratinocyte differentiation, rather than proliferation. Furthermore, papillary and reticular fibroblasts demonstrate differences in the profiles of extracellular matrix molecules they secrete, such as collagens, decorin, and versican [23, 59], which may impact matrix and basement membrane formation in organotypic cultures *in vitro*. For example, papillary fibroblasts have higher expression of collagen type VI α2, which may help promote formation of a more intact basement membrane *in vitro*, compared to reticular fibroblasts [59].

Of the few studies to date examining the role of hair follicle-associated fibroblasts as keratinocyte feeder cells (Table 7.1), only one has directly examined potential support mechanisms [62]. Hill et al. [62] observed that human hair follicle dermal fibroblasts expressed higher levels of secreted protein acidic and rich in cysteine (SPARC, a protein involved in tissue repair), compared to interfollicular dermal fibroblasts; however, this was not involved in the improved keratinocyte support offered by hair follicle-associated fibroblasts, suggesting that this effect is likely mediated by other factors. Hair follicle dermal papilla and dermal sheath fibroblasts are known to secrete factors that promote keratinocyte growth, such as IL-6 and insulin-like growth factor (IGF)-1 [91]. Differences in hair follicle-associated fibroblast and dermal interfollicular fibroblast growth factor and cytokine secretion profiles have been hypothesized to contribute to the differential supportive effects of these fibroblast subtypes [65, 91]. For example, hair follicle dermal papilla and dermal sheath fibroblasts produce higher levels of TGF-β2, compared to dermal interfollicular fibroblasts [91, 92]. Fibroblast-derived TGF-β2 promotes keratinocytes to produce basement membrane components, and the higher production of TGF-β2 by hair follicle-associated fibroblasts could contribute to improved basement membrane formation in organotypic cultures with hair follicle dermal cells [64, 65]. Additionally, hair follicle-associated fibroblasts contribute to wound healing in skin, potentially via deposition of matrix components [93, 94], which represents another mechanism by which they may contribute to improved basement membrane assembly in organotypic cultures.

7.8.2 Other Dermal/Mesenchymal Cells that Can Support Keratinocytes

7.8.2.1 Pericytes
Although fibroblasts are the major dermal cell population, and the most widely used keratinocyte feeder cell, it is important to consider other

cell types present in the dermis, as there is growing evidence that they can also play a supportive role in keratinocyte growth. A study by our group has shown that dermal pericytes derived from human neonatal foreskins promote human keratinocyte growth in 3D organotypic co-cultures in an angiogenesis-independent manner [27]. The support offered by pericytes was superior to that from dermal fibroblasts, as pericytes, but not fibroblasts, enhanced the low regenerative capacity of differentiating keratinocytes, as well as enhancing LAMA5 deposition at the dermal-epidermal junction in organotypic cultures (Table 7.1) [27].

These observations are consistent with the role of pericytes in promoting repair and regeneration in other tissues, such as muscle [95]. Although the mechanisms by which pericytes promote tissue regeneration have not yet been fully characterized, paracrine effects of pericyte-derived growth factors and/or cytokines are thought to play a central role [95–97]. In the context of keratinocyte/epidermal regeneration, pericytes secrete several of the aforementioned factors that promote keratinocyte proliferation. Specifically, human embryonic and fetal muscle pericytes secrete KGF [97], human CD146+ MSCs which are classified as pericytes produce GM-CSF [98], and murine brain pericytes secrete GM-CSF and IL-6 [99]. Thus, secretion of these factors may contribute to the ability of pericytes to induce keratinocyte proliferation [27]. Additionally, bovine retinal pericytes have been shown to produce several extracellular matrix components, such as fibronectin, laminin, thrombospondin, and collagens, which suggests that pericytes may contribute to matrix formation during wound healing [96], and potentially to matrix and/or basement membrane formation in organotypic skin cultures.

7.8.2.2 Mesenchymal Stem Cells

MSCs have also been examined as potential support cells for keratinocytes. Human MSCs derived from the subcutaneous adipose layer of skin have been shown to be comparable to, or better than, murine Swiss 3T3 fibroblasts and human dermal fibroblasts with respect to promoting keratinocyte proliferation, colony formation and differentiation (Table 7.1) [67–69]. One study has also observed that human bone marrow-derived MSCs are comparable feeder cells to dermal fibroblasts and induce keratinocytes to form an epidermal structure and phenotype similar to that of *in vivo* skin [73], and similar findings have been reported for rat bone marrow MSCs [72]. An additional study using a rat model has shown that MSCs from other tissues (heart, lung, spleen, liver, and kidney) are capable of promoting epidermal stratification and keratin expression profiles in 3D organotypic co-cultures *in vitro* that are comparable to *in vivo* rat skin (Table 7.1) [71]. However, whether findings from the study by Aoki et al. [71] are also applicable to human MSCs derived from tissues other than skin and bone marrow needs to be determined.

MSCs play a significant role in promoting tissue regeneration and wound healing, including that occurring in skin [100–103]. MSCs secrete many wound-healing mediators, and several of these are known to influence keratinocyte growth, proliferation, and/or migration [100–103], and could represent potential mechanisms by which MSCs support keratinocyte growth and regeneration in *in vitro* organotypic cultures. For example, Alexaki et al. [70] showed that human adipose-derived MSCs promoted increased keratinocyte migration and proliferation, compared to dermal fibroblasts, and these effects were mediated by adipose MSC-derived KGF-1 and platelet-derived growth factor-BB (PDGF-BB), as blocking these two factors using neutralizing antibodies abrogated the effects on keratinocytes. Other wound-healing mediators secreted by bone marrow- and adipose-derived MSCs that may provide support to keratinocytes in organotypic cultures include IL-6, TGF-β1, EGF, IGF-1, and collagen type I [100–103].

7.8.2.3 Endothelial Cells, Adipocytes, and Immune Cells

Other cell types found in the dermis, such as endothelial cells of dermal microvessels, and immune cells, as well as hypodermal cells, such as adipocytes, may also potentially support keratinocyte growth and expansion, although the role of these cell types in this context has not yet been well studied. One study has shown that co-culturing rat keratinocytes with a feeder layer of rat adipocytes leads to greater keratinocyte proliferation and generation of skin equivalents more closely resembling *in vivo* skin, compared to cultures that include fibroblasts in the feeder layer (Table 7.1) [74]. Another study has demonstrated that inclusion of endothelial cells in 3D organotypic co-cultures with fibroblasts and keratinocytes led to improved basement membrane organization and epithelial differentiation, compared to organotypic cultures with only fibroblasts and keratinocytes (Table 7.1) [75]. However, these two studies used rat cells [74] or murine and rabbit cells [75], and whether human adipocytes and endothelial cells can induce the same effects remains to be determined. Two other studies have shown that tri-cultures of human endothelial cells, keratinocytes, and fibroblasts can form vascularized skin equivalents [104, 105], and another study has demonstrated successful incorporation of human primary keratinocytes, fibroblasts, and monocyte-derived dendritic cells into a 3D organotypic model [106]. Although these three studies [104–106] did not directly examine whether incorporation of endothelial cells or dendritic cells affected keratinocyte growth/regeneration, endothelial cells and dendritic cells secrete factors that are known to stimulate keratinocytes, such as IL-6, IL-8, GM-CSF, and TGF-β [107, 108]; therefore it is possible that they may influence keratinocyte growth and regeneration. In regard to organotypic cultures containing more than two cell types, it is important to consider not only interactions of each feeder cell type with keratinocytes, but also how cross talk between different feeder cell types with each other may then influence keratinocytes. For example, fibroblasts and endothelial cells have been demonstrated to interact with each other [109], and this cross talk may result in production of factors that modulate keratinocytes.

7.9 Conclusions

The inclusion of support cells is an essential component for successful keratinocyte growth and formation of epidermal equivalents *in vitro*. Whilst the role of fibroblasts as feeder cells is well established, recent studies have started to recognize that different fibroblast subtypes have differential capacities to support keratinocytes, suggesting that development of future organotypic skin models should consider fibroblast subtypes separately, rather than using the fibroblast population as a whole. Additionally, there is growing evidence that other dermal cell types, such as pericytes, endothelial cells, MSCs, and immune cells, can support keratinocyte growth and regeneration and may even offer superior alternatives to the current gold standard, and therefore warrant further investigation. Another important area for future studies is characterizing the mechanisms by which feeder cells promote keratinocyte growth and regeneration, in particular the identification of soluble factors and signaling pathways that mediate paracrine keratinocyte-feeder cell cross talk. Elucidation of the factors involved and the support cells that secrete them will assist in identification of optimal cell types to be used as feeder layers, and may also provide useful insights for developing optimal media compositions and growth factor supplementation. Capturing the complexity of the *in vivo* skin microenvironment under *in vitro* conditions is difficult, although this can be assisted by approaches such as 3D organotypic models, as well as the simultaneous incorporation of several different feeder cell types into these cultures. The continued and more in-depth investigation of support cells and the underlying mechanisms by which they promote keratinocyte growth and regeneration will help to inform the development of improved *in vitro* models that more accurately reflect *in vivo* skin. This will provide an informative basis for studying skin regeneration in the context of different disease states, and enable the design of improved skin equivalents for therapeutic purposes.

References

1. Baroni A, Buommino E, De Gregorio V, Ruocco E, Ruocco V, Wolf R. Structure and function of the epidermis related to barrier properties. Clin Dermatol. 2012;30(3):257–62.
2. Ojeh N, Pastar I, Tomic-Canic M, Stojadinovic O. Stem cells in skin regeneration, wound healing, and their clinical applications. Int J Mol Sci. 2015;16(10):25476–501.
3. Watt FM. Mammalian skin cell biology: at the interface between laboratory and clinic. Science. 2014;346(6212):937–40.
4. Lawlor KT, Kaur P. Dermal contributions to human interfollicular epidermal architecture and self-renewal. Int J Mol Sci. 2015;16(12):28098–107.
5. Rowan MP, Cancio LC, Elster EA, Burmeister DM, Rose LF, Natesan S, Chan RK, Christy RJ, Chung KK. Burn wound healing and treatment: review and advancements. Crit Care. 2015;19:243.
6. Simões MC, Sousa JJ, Pais AA. Skin cancer and new treatment perspectives: a review. Cancer Lett. 2015;357(1):8–42.
7. Green H. The birth of therapy with cultured cells. BioEssays. 2008;30(9):897–903.
8. Sriram G, Bigliardi PL, Bigliardi-Qi M. Fibroblast heterogeneity and its implications for engineering organotypic skin models in vitro. Eur J Cell Biol. 2015;94(11):483–512.
9. Fuchs E. Skin stem cells: rising to the surface. J Cell Biol. 2008;180(2):273–84.
10. Blanpain C, Fuchs E. Epidermal stem cells of the skin. Annu Rev Cell Dev Biol. 2006;22:339–73.
11. Jones PH, Watt FM. Separation of human epidermal stem cells from transit amplifying cells on the basis of differences in integrin function and expression. Cell. 1993;73(4):713–24.
12. Jones PH, Harper S, Watt FM. Stem cell patterning and fate in human epidermis. Cell. 1995;80(1):83–93.
13. Jensen UB, Lowell S, Watt FM. The spatial relationship between stem cells and their progeny in the basal layer of human epidermis: a new view based on whole-mount labelling and lineage analysis. Development. 1999;126(11):2409–18.
14. Legg J, Jensen UB, Broad S, Leigh I, Watt FM. Role of melanoma chondroitin sulphate proteoglycan in patterning stem cells in human interfollicular epidermis. Development. 2003;130(24):6049–63.
15. Jensen KB, Collins CA, Nascimento E, Tan DW, Frye M, Itami S, Watt FM. Lrig1 expression defines a distinct multipotent stem cell population in mammalian epidermis. Cell Stem Cell. 2009;4(5):427–39.
16. Li A, Simmons PJ, Kaur P. Identification and isolation of candidate human keratinocyte stem cells based on cell surface phenotype. Proc Natl Acad Sci U S A. 1998;95(7):3902–7.
17. Li A, Pouliot N, Redvers R, Kaur P. Extensive tissue-regenerative capacity of neonatal human keratinocyte stem cells and their progeny. J Clin Invest. 2004;113(3):390–400.
18. Schluter H, Paquet-Fifield S, Gangatirkar P, Li J, Kaur P. Functional characterization of quiescent keratinocyte stem cells and their progeny reveals a hierarchical organization in human skin epidermis. Stem Cells. 2011;29(8):1256–68.
19. Webb A, Li A, Kaur P. Location and phenotype of human adult keratinocyte stem cells of the skin. Differentiation. 2004;72(8):387–95.
20. Lechler T, Fuchs E. Asymmetric cell divisions promote stratification and differentiation of mammalian skin. Nature. 2005;437(7056):275–80.
21. Gangatirkar P, Paquet-Fifield S, Li A, Rossi R, Kaur P. Establishment of 3D organotypic cultures using human neonatal epidermal cells. Nat Protoc. 2007;2(1):178–86.
22. Breitkreutz D, Koxholt I, Thiemann K, Nischt R. Skin basement membrane: the foundation of epidermal integrity—BM functions and diverse roles of bridging molecules nidogen and perlecan. Biomed Res Int. 2013;2013:179784.
23. Sorrell JM, Caplan AI. Fibroblast heterogeneity: more than skin deep. J Cell Sci. 2004;117(Pt 5):667–75.
24. Driskell RR, Watt FM. Understanding fibroblast heterogeneity in the skin. Trends Cell Biol. 2015;25(2):92–9.
25. Braverman IM. The cutaneous microcirculation. J Investig Dermatol Symp Proc. 2000;5(1):3–9.
26. Birbrair A, Zhang T, Wang Z-M, Messi ML, Mintz A, Delbono O. Pericytes at the intersection between tissue regeneration and pathology. Clin Sci. 2015;128(2):81–93.
27. Paquet-Fifield S, Schluter H, Li A, Aitken T, Gangatirkar P, Blashki D, Koelmeyer R, Pouliot N, Palatsides M, Ellis S, Brouard N, Zannettino A, Saunders N, Thompson N, Li J, Kaur P. A role for pericytes as microenvironmental regulators of human skin tissue regeneration. J Clin Invest. 2009;119(9):2795–806.
28. Gaur M, Dobke M, Lunyak VV. Mesenchymal stem cells from adipose tissue in clinical applications for dermatological indications and skin aging. Int J Mol Sci. 2017;18(1):E208.
29. Nestle FO, Di Meglio P, Qin J-Z, Nickoloff BJ. Skin immune sentinels in health and disease. Nat Rev Immunol. 2009;9(10):679–91.
30. Mann ER, Smith KM, Bernardo D, Al-Hassi HO, Knight SC, Hart HL. Review: skin and the immune system. J Clin Exp Dermatol Res. 2012;S2:003.
31. Minutti CM, Knipper JA, Allen JE, Zaiss DMW. Tissue-specific contribution of macrophages to wound healing. Semin Cell Dev Biol. 2017;61(Suppl C):3–11.
32. Rheinwald JG, Green H. Serial cultivation of strains of human epidermal keratinocytes: the formation of keratinizing colonies from single cells. Cell. 1975;6(3):331–43.
33. Green H, Rheinwald JG, Sun TT. Properties of an epithelial cell type in culture: the epidermal keratinocyte and its dependence on products of the fibroblast. Prog Clin Biol Res. 1977;17:493–500.

34. Oh JW, Hsi TC, Guerrero-Juarez CF, Ramos R, Plikus MV. Organotypic skin culture. J Invest Dermatol. 2013;133(11):1–4.
35. Nanba D, Matsushita N, Toki F, Higashiyama S. Efficient expansion of human keratinocyte stem/progenitor cells carrying a transgene with lentiviral vector. Stem Cell Res Ther. 2013;4(5):127.
36. Pellegrini G, Ranno R, Stracuzzi G, Bondanza S, Guerra L, Zambruno G, Micali G, De Luca M. The control of epidermal stem cells (holoclones) in the treatment of massive full-thickness burns with autologous keratinocytes cultured on fibrin. Transplantation. 1999;68(6):868–79.
37. Ronfard V, Rives JM, Neveux Y, Carsin H, Barrandon Y. Long-term regeneration of human epidermis on third degree burns transplanted with autologous cultured epithelium grown on a fibrin matrix. Transplantation. 2000;70(11):1588–98.
38. Jubin K, Martin Y, Lawrence-Watt DJ, Sharpe JR. A fully autologous co-culture system utilising non-irradiated autologous fibroblasts to support the expansion of human keratinocytes for clinical use. Cytotechnology. 2011;63(6):655–62.
39. Higham MC, Dawson R, Szabo M, Short R, Haddow DB, MacNeil S. Development of a stable chemically defined surface for the culture of human keratinocytes under serum-free conditions for clinical use. Tissue Eng. 2003;9(5):919–30.
40. Sun T, Higham M, Layton C, Haycock J, Short R, MacNeil S. Developments in xenobiotic-free culture of human keratinocytes for clinical use. Wound Repair Regen. 2004;12(6):626–34.
41. Bullock AJ, Higham MC, MacNeil S. Use of human fibroblasts in the development of a xenobiotic-free culture and delivery system for human keratinocytes. Tissue Eng. 2006;12(2):245–55.
42. Mujaj S, Manton K, Upton Z, Richards S. Serum-free primary human fibroblast and keratinocyte coculture. Tissue Eng Part A. 2010;16(4):1407–20.
43. El-Ghalbzouri A, Gibbs S, Lamme E, Van Blitterswijk CA, Ponec M. Effect of fibroblasts on epidermal regeneration. Br J Dermatol. 2002;147(2):230–43.
44. El Ghalbzouri A, Lamme E, Ponec M. Crucial role of fibroblasts in regulating epidermal morphogenesis. Cell Tissue Res. 2002;310(2):189–99.
45. Boehnke K, Mirancea N, Pavesio A, Fusenig NE, Boukamp P, Stark HJ. Effects of fibroblasts and microenvironment on epidermal regeneration and tissue function in long-term skin equivalents. Eur J Cell Biol. 2007;86(11–12):731–46.
46. Llames S, Garcia E, Garcia V, del Rio M, Larcher F, Jorcano JL, López E, Holguín P, Miralles F, Otero J, Meana A. Clinical results of an autologous engineered skin. Cell Tissue Bank. 2006;7(1):47–53.
47. Llames SG, Del Rio M, Larcher F, Garcia E, Garcia M, Escamez MJ, Jorcano JL, Holguín P, Meana A. Human plasma as a dermal scaffold for the generation of a completely autologous bioengineered skin. Transplantation. 2004;77(3):350–5.
48. Bisson F, Rochefort É, Lavoie A, Larouche D, Zaniolo K, Simard-Bisson C, Damour O, Auger FA, Guérin SL, Germain L. Irradiated human dermal fibroblasts are as efficient as mouse fibroblasts as a feeder layer to improve human epidermal cell culture lifespan. Int J Mol Sci. 2013;14(3):4684–704.
49. Auxenfans C, Thepot A, Justin V, Hautefeuille A, Shahabeddin L, Damour O, Hainaut P. Characterisation of human fibroblasts as keratinocyte feeder layer using p63 isoforms status. Biomed Mater Eng. 2009;19(4–5):365–72.
50. Panacchia L, Dellambra E, Bondanza S, Paterna P, Maurelli R, Paionni E, Guerra L. Nonirradiated human fibroblasts and irradiated 3T3-J2 murine fibroblasts as a feeder layer for keratinocyte growth and differentiation in vitro on a fibrin substrate. Cells Tissues Organs. 2010;191(1):21–35.
51. Meana A, Iglesias J, Del Rio M, Larcher F, Madrigal B, Fresno MF, Martin C, San Roman F, Tevar F. Large surface of cultured human epithelium obtained on a dermal matrix based on live fibroblast-containing fibrin gels. Burns. 1998;24(7):621–30.
52. Kamolz LP, Luegmair M, Wick N, Eisenbock B, Burjak S, Koller R, Meissl G, Frey M. The Viennese culture method: cultured human epithelium obtained on a dermal matrix based on fibroblast containing fibrin glue gels. Burns. 2005;31(1):25–9.
53. Ng W, Ikeda S. Standardized, defined serum-free culture of a human skin equivalent on fibroblast-populated collagen scaffold. Acta Derm Venereol. 2011;91(4):387–91.
54. Chen C-SJ, Lavker RM, Rodeck U, Risse B, Jensen PJ. Use of a serum-free epidermal culture model to show deleterious effects of epidermal growth factor on morphogenesis and differentiation. J Invest Dermatol. 1995;104(1):107–12.
55. Sorrell JM, Baber MA, Caplan AI. Site-matched papillary and reticular human dermal fibroblasts differ in their release of specific growth factors/cytokines and in their interaction with keratinocytes. J Cell Physiol. 2004;200(1):134–45.
56. Varkey M, Ding J, Tredget EE. Fibrotic remodeling of tissue-engineered skin with deep dermal fibroblasts is reduced by keratinocytes. Tissue Eng Part A. 2014;20(3–4):716–27.
57. Varkey M, Ding J, Tredget EE. Superficial dermal fibroblasts enhance basement membrane and epidermal barrier formation in tissue-engineered skin: implications for treatment of skin basement membrane disorders. Tissue Eng Part A. 2014;20(3–4):540–52.
58. Janson D, Saintigny G, Mahé C, Ghalbzouri AE. Papillary fibroblasts differentiate into reticular fibroblasts after prolonged in vitro culture. Exp Dermatol. 2013;22(1):48–53.
59. Janson D, Rietveld M, Mahe C, Saintigny G, El Ghalbzouri A. Differential effect of extracellular matrix derived from papillary and reticular fibroblasts on epidermal development in vitro. Eur J Dermatol. 2017;27(3):237–46.

60. Mine S, Fortunel NO, Pageon H, Asselineau D. Aging alters functionally human dermal papillary fibroblasts but not reticular fibroblasts: a new view of skin morphogenesis and aging. PLoS One. 2009;3(12):e4066.

61. Pageon H, Zucchi H, Asselineau D. Distinct and complementary roles of papillary and reticular fibroblasts in skin morphogenesis and homeostasis. Eur J Dermatol. 2012;22(3):324–32.

62. Hill RP, Gardner A, Crawford HC, Richer R, Dodds A, Owens WA, Lawrence C, Rao S, Kara B, James SE, Jahoda CA. Human hair follicle dermal sheath and papilla cells support keratinocyte growth in monolayer coculture. Exp Dermatol. 2013;22(3):236–8.

63. Higgins CA, Roger MF, Hill RP, Ali-Khan AS, Garlick JA, Christiano AM, Jahoda CAB. Multifaceted role of hair follicle dermal cells in bioengineered skins. Br J Dermatol. 2017;176(5):1259–69.

64. Shin YH, Seo YK, Yoon HH, Yoo BY, Song KY, Park JK. Comparison of hair dermal cells and skin fibroblasts in a collagen sponge for use in wound repair. Biotechnol Bioprocess Eng. 2011;16(4):793.

65. Cho HJ, Bae IH, Chung HJ, Kim DS, Kwon SB, Cho YJ, Youn SW, Park KC. Effects of hair follicle dermal sheath cells in the reconstruction of skin equivalents. J Dermatol Sci. 2004;35(1):74–7.

66. Reynolds AJ, Jahoda CA. Hair follicle stem cells? A distinct germinative epidermal cell population is activated in vitro by the presence of hair dermal papilla cells. J Cell Sci. 1991;99(2):373–85.

67. Tosca MC, Chlapanidas T, Galuzzi M, Antonioli B, Perteghella S, Vigani B, Mantelli M, Ingo D, Avanzini MA, Vigo D, Faustini M, Torre ML, Marazzi M. Human adipose-derived stromal cells as a feeder layer to improve keratinocyte expansion for clinical applications. Tissue Eng Regen Med. 2015;12(4):249–58.

68. Sugiyama H, Maeda K, Yamato M, Hayashi R, Soma T, Hayashida Y, Yang J, Shirakabe M, Matsuyama A, Kikuchi A, Sawa Y, Okano T, Tano Y, Nishida K. Human adipose tissue-derived mesenchymal stem cells as a novel feeder layer for epithelial cells. J Tissue Eng Regen Med. 2008;2(7):445–9.

69. Huh CH, Kim SY, Cho HJ, Kim DS, Lee WH, Kwon SB, Na JI, Park KC. Effects of mesenchymal stem cells in the reconstruction of skin equivalents. J Dermatol Sci. 2007;46(3):217–20.

70. Alexaki VI, Simantiraki D, Panayiotopoulou M, Rasouli O, Venihaki M, Castana O, Alexakis D, Kampa M, Stathopoulos EN, Castanas E. Adipose tissue-derived mesenchymal cells support skin re-epithelialization through secretion of KGF-1 and PDGF-BB: comparison with dermal fibroblasts. Cell Transplant. 2012;21(11):2441–54.

71. Aoki S, Takezawa T, Uchihashi K, Sugihara H, Toda S. Non-skin mesenchymal cell types support epidermal regeneration in a mesenchymal stem cell or myofibroblast phenotype-independent manner. Pathol Int. 2009;59(6):368–75.

72. Aoki S, Toda S, Ando T, Sugihara H. Bone marrow stromal cells, preadipocytes, and dermal fibroblasts promote epidermal regeneration in their distinctive fashions. Mol Biol Cell. 2004;15(10):4647–57.

73. Ojeh NO, Navsaria HA. An in vitro skin model to study the effect of mesenchymal stem cells in wound healing and epidermal regeneration. J Biomed Mater Res A. 2014;102(8):2785–92.

74. Sugihara H, Toda S, Yonemitsu N, Watanabe K. Effects of fat cells on keratinocytes and fibroblasts in a reconstructed rat skin model using collagen gel matrix culture. Br J Dermatol. 2001;144(2):244–53.

75. Zieske JD, Mason VS, Wasson ME, Meunier SF, Nolte CJ, Fukai N, Olsen BR, Parenteau NL. Basement membrane assembly and differentiation of cultured corneal cells: importance of culture environment and endothelial cell interaction. Exp Cell Res. 1994;214(2):621–33.

76. Maas-Szabowski N, Szabowski A, Andrecht S, Kolbus A, Schorpp-Kistner M, Angel P, Fusenig NE. Organotypic cocultures with genetically modified mouse fibroblasts as a tool to dissect molecular mechanisms regulating keratinocyte growth and differentiation. J Invest Dermatol. 2001;116(5):816–20.

77. Maas-Szabowski N, Shimotoyodome A, Fusenig NE. Keratinocyte growth regulation in fibroblast cocultures via a double paracrine mechanism. J Cell Sci. 1999;112(Pt 12):1843–53.

78. Nolte SV, Xu W, Rennekampff HO, Rodemann HP. Diversity of fibroblasts—a review on implications for skin tissue engineering. Cells Tissues Organs. 2008;187(3):165–76.

79. Choi Y, Fuchs E. TGF-beta and retinoic acid: regulators of growth and modifiers of differentiation in human epidermal cells. Cell Regul. 1990;1(11):791–809.

80. Maas-Szabowski N, Stark HJ, Fusenig NE. Keratinocyte growth regulation in defined organotypic cultures through IL-1-induced keratinocyte growth factor expression in resting fibroblasts. J Invest Dermatol. 2000;114(6):1075–84.

81. El Ghalbzouri A, Ponec M. Diffusible factors released by fibroblasts support epidermal morphogenesis and deposition of basement membrane components. Wound Repair Regen. 2004;12(3):359–67.

82. Smola H, Stark HJ, Thiekotter G, Mirancea N, Krieg T, Fusenig NE. Dynamics of basement membrane formation by keratinocyte-fibroblast interactions in organotypic skin culture. Exp Cell Res. 1998;239(2):399–410.

83. El Ghalbzouri A, Jonkman MF, Dijkman R, Ponec M. Basement membrane reconstruction in human skin equivalents is regulated by fibroblasts and/or exogenously activated keratinocytes. J Invest Dermatol. 2005;124(1):79–86.

84. Lee DY, Cho KH. The effects of epidermal keratinocytes and dermal fibroblasts on the formation of cutaneous basement membrane in three-dimensional culture systems. Arch Dermatol Res. 2005;296(7):296–302.

85. Wong T, McGrath JA, Navsaria H. The role of fibroblasts in tissue engineering and regeneration. Br J Dermatol. 2007;156(6):1149–55.

86. Smola H, Thiekotter G, Baur M, Stark HJ, Breitkreutz D, Fusenig NE. Organotypic and epidermal-dermal co-cultures of normal human keratinocytes and dermal cells: regulation of transforming growth factor alpha, beta1 and beta2 mRNA levels. Toxicol In Vitro. 1994;8(4):641–50.

87. Konig A, Bruckner-Tuderman L. Transforming growth factor-beta stimulates collagen VII expression by cutaneous cells in vitro. J Cell Biol. 1992;117(3):679–85.

88. Konig A, Bruckner-Tuderman L. Transforming growth factor-beta promotes deposition of collagen VII in a modified organotypic skin model. Lab Investig. 1994;70(2):203–9.

89. Varkey M, Ding J, Tredget EE. Differential collagen-glycosaminoglycan matrix remodeling by superficial and deep dermal fibroblasts: potential therapeutic targets for hypertrophic scar. Biomaterials. 2011;32(30):7581–91.

90. Wang J, Dodd C, Shankowsky HA, Scott PG, Tredget EE. Deep dermal fibroblasts contribute to hypertrophic scarring. Lab Investig. 2008;88(12):1278–90.

91. Chiu HC, Chang CH, Chen JS, Jee SH. Human hair follicle dermal papilla cell, dermal sheath cell and interstitial dermal fibroblast characteristics. J Formos Med Assoc. 1996;95(9):667–74.

92. Inoue K, Aoi N, Yamauchi Y, Sato T, Suga H, Eto H, Kato H, Tabata Y, Yoshimura K. TGF-beta is specifically expressed in human dermal papilla cells and modulates hair folliculogenesis. J Cell Mol Med. 2009;13(11–12):4643–56.

93. Jimenez F, Poblet E, Izeta A. Reflections on how wound healing-promoting effects of the hair follicle can be translated into clinical practice. Exp Dermatol. 2015;24(2):91–4.

94. Jahoda CA, Reynolds AJ. Hair follicle dermal sheath cells: unsung participants in wound healing. Lancet. 2001;358(9291):1445–8.

95. Crisan M, Corselli M, Chen WC, Peault B. Perivascular cells for regenerative medicine. J Cell Mol Med. 2012;16(12):2851–60.

96. Mills SJ, Cowin AJ, Kaur P. Pericytes, mesenchymal stem cells and the wound healing process. Cell. 2013;2(3):621–34.

97. Chen CW, Montelatici E, Crisan M, Corselli M, Huard J, Lazzari L, Péault B. Perivascular multi-lineage progenitor cells in human organs: regenerative units, cytokine sources or both? Cytokine Growth Factor Rev. 2009;20(5–6):429–34.

98. Sorrentino A, Ferracin M, Castelli G, Biffoni M, Tomaselli G, Baiocchi M, Fatica A, Negrini M, Peschle C, Valtieri M. Isolation and characterization of CD146+ multipotent mesenchymal stromal cells. Exp Hematol. 2008;36(8):1035–46.

99. Dohgu S, Banks WA. Brain pericytes increase the lipopolysaccharide-enhanced transcytosis of HIV-1 free virus across the in vitro blood-brain barrier: evidence for cytokine-mediated pericyte-endothelial cell crosstalk. Fluids Barriers CNS. 2013;10(1):23.

100. Walter MN, Wright KT, Fuller HR, MacNeil S, Johnson WE. Mesenchymal stem cell-conditioned medium accelerates skin wound healing: an in vitro study of fibroblast and keratinocyte scratch assays. Exp Cell Res. 2010;316(7):1271–81.

101. Isakson M, de Blacam C, Whelan D, McArdle A, Clover AJP. Mesenchymal stem cells and cutaneous wound healing: current evidence and future potential. Stem Cells Int. 2015;2015:12.

102. Ghieh F, Jurjus R, Ibrahim A, Geagea AG, Daouk H, El Baba B, Chams S, Matar M, Zein W, Jurjus A. The use of stem cells in burn wound healing: a review. Biomed Res Int. 2015;2015:9.

103. Wu Y, Zhao RC, Tredget EE. Concise review: bone marrow-derived stem/progenitor cells in cutaneous repair and regeneration. Stem Cells. 2010;28(5):905–15.

104. Black AF, Berthod F, L'Heureux N, Germain L, Auger FA. In vitro reconstruction of a human capillary-like network in a tissue-engineered skin equivalent. FASEB J. 1998;12(13):1331–40.

105. Marino D, Luginbuhl J, Scola S, Meuli M, Reichmann E. Bioengineering dermo-epidermal skin grafts with blood and lymphatic capillaries. Sci Transl Med. 2014;6(221):221ra14.

106. Chau DY, Johnson C, MacNeil S, Haycock JW, Ghaemmaghami AM. The development of a 3D immunocompetent model of human skin. Biofabrication. 2013;5(3):035011.

107. Mai J, Virtue A, Shen J, Wang H, Yang XF. An evolving new paradigm: endothelial cells—conditional innate immune cells. J Hematol Oncol. 2013;6:61.

108. Banchereau J, Briere F, Caux C, Davoust J, Lebecque S, Liu YJ, Pulendran B, Palucka K. Immunobiology of dendritic cells. Ann Rev Immunol. 2000;18(1):767–811.

109. Sorrell JM, Baber MA, Caplan AI. Human dermal fibroblast subpopulations; differential interactions with vascular endothelial cells in coculture: nonsoluble factors in the extracellular matrix influence interactions. Wound Repair Regen. 2008;16(2):300–9.

Stem Cells and Burn

8

Anesh Prasai, Amina El Ayadi, David N. Herndon, and Celeste C. Finnerty

8.1 Introduction

Tremendous clinical advancement has been made in the last four decades in terms of treating burn victims [1]. Thus, currently, mortality is no longer an uprising issue following a severe burn injury. However, improving the quality of life following a severe burn injury is an accomplished challenge [2]. Researchers are actively investigating different synthetic substitutes [3], biological mediators [4, 5], and pharmacological compounds [6] to increase the rate of wound healing and decrease scarring and finally improve the quality of life following burn injury. Among the biological mediators, the use of stem cells, especially the adult stem cells to accelerate the rate of wound healing, decreases the inflammation and improves scarring and pain outcome has been intensely studied in recent years [5]. Although stem cells in conjunction with tissue engineering were hoped to synergistically change the field of burn and tissue injury, almost four decades back, when the first epidermal keratinocytes forming colonies were expanded, however, the changes have been slow. But, with the

advancement in technology, characterization techniques, and increase in public interest in the use of adult stem cells and tissue engineering to address the comorbidities associated with the wound healing in general, the field has again garnered a momentum in the last two decades. This book chapter examines the current application, outcomes, and therapeutic effects and challenges from a safety, efficacy, and ethical point of view following the application of different types of stem cells in burn wound healing models in a clinical and laboratory setting.

There has been a steady rise in the use of stem cells to address the comorbidities associated with the burn [5, 7]. The applied stem cells on the burn patients can be sourced from either (1) allogeneic tissue (obtained from the same species) or (2) autogenic tissue (obtained from own body parts) and there are pros and cons with the usage of each source of stem cells (Fig. 8.1).

An autologous stem cell refers to stem cells isolated from the patient's own body tissue from either bone marrow, fat, or skin, while allogeneic refers to stem cells isolated from a donor's tissue. Limited number of early animal [8] and human [9] studies had shown stem cells in general to be immune privileged. However, recently, a number of studies have reported immune rejection and other clinical complications including tumor formation following the transplantation of allogeneic stem cells [10–12]. Thus, one of the main advantages of using autologous stem cells over

A. Prasai · A. El Ayadi · D. N. Herndon
C. C. Finnerty (✉)
Department of Surgery, University of Texas Medical Branch, Galveston, TX, USA
e-mail: anprasai@utmb.edu; amelayad@utmb.edu; dherndon@utmb.edu; ccfinner@utmb.edu

© Springer Nature Switzerland AG 2019
D. Duscher, M. A. Shiffman (eds.), *Regenerative Medicine and Plastic Surgery*,
https://doi.org/10.1007/978-3-030-19962-3_8

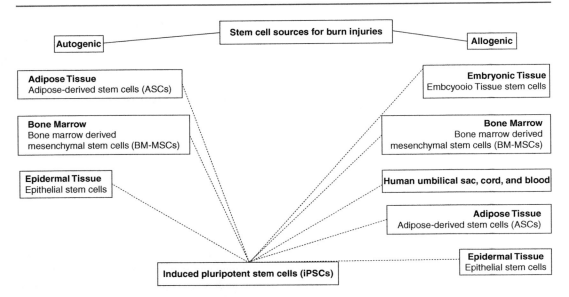

Fig. 8.1 Current sources of stem cells to treat burn-related injuries

allogeneic is to circumvent the immune rejection that may occur with the allogenic stem cell source. However, due to scarcity of tissue following severe injury or other clinical limitations allogeneic sources may be used to obtain the stem cells.

Both allogeneic and autogenic stem cells isolated from various sources are currently used to treat burn injury. A new major breakthrough—the development and usage of induced pluripotent stem cells (iPSCs) from mature differentiated somatic cells, mostly the skin fibroblasts, as a source to treat the burn-related wounds—is also on the rise [13].

8.2 Human Umbilical Cord-Derived Stem Cells

The use of fetal membrane to treat wound can be traced back almost a century when Sabella [14] first described its usage; however the use of isolated and enriched human umbilical cord-derived mesenchymal stem cell (HU-MSC) to treat burn wound has taken a pace in the last two decades [15, 16]. Unlike bone marrow-derived mesenchymal stem cells and embryonic stem cells, HU-MSC are free of ethical concerns, are immune privileged, and are non-tumorigenic. Researchers have isolated both mesenchymal and epithelial stem cells from the human umbilical cord and are actively studying its usage in treating wounds in general [17].

A recent study by Moenadjat et al. [18] has reported that the topical application of HU-MSC along with silver sulfadiazine (SSD) cream on the comparable deep partial-thickness wound size on 20 patients has showed faster wound closure, reduction in the pain, and no immune rejection or infection. Additionally, the authors also noted the increase in formation of dermal-epidermal junction and rete ridges examined via histological examination in HU-MSC-treated patients. Another clinical study done by Eskandarlou et al. [15] on 32 patients observed rapid epithelization, wound healing, increase in mobility, and decrease in pain. A study done by our group also observed significant decrease in the rate of dressing change, following the application of amnion; however no significant differences were observed in terms of wound healing, length of hospital stay, and development of hypertrophic scarring [16].

The use of HU-MSC to treat burn-related injuries is on the rise. Currently as of 2017, according to clinicaltrials.gov two clinical trials have been registered with the application of HU-MSC to treat different comorbidities associated with the burn. In the first study HU-MSC has been used to repair sweat glands and body features following the burn injury, while the other study focuses on the use of HU-MSC to treat the ocular corneal burn damage [7].

The actual mechanism of how HU-MSC aid in burn wound healing has not been clearly elucidated yet; however researchers have used different in vivo and in vitro animal burn wound model to decipher various possible molecular mechanisms [19]. Although the actual mechanism was not examined in the study, Liu et al. [20] noted accelerated wound healing and decrease in infiltration of inflammatory cells and inflammatory mediators in the wounds treated with HU-MSC. The researchers also observed significant increase in neovascularization and increase in the ratio of collagen type I-to-type III ratio in the treated group. Researchers have also stated that the therapeutic effect of the HU-MSC is derived from its exosomes. Li et al. [21] observed that application of HU-MSC exosomes overexpressing microRNA miR-181c favorably modulated the macrophage response by downregulating the TLR-4 signaling pathway, in a murine burn inflammation model.

In summary, HU-MSC are one of the ideal sources of stem cells for clinical application, on account of their ease of collection, less ethical issues, high number of stem cells [22], low immunogenicity [23], and therapeutic properties in wound repair [24].

8.3 Bone Marrow-Derived Mesenchymal Stem Cells (BM-MSCs)

Although human umbilical cord and placenta sac tissues were first used to treat burn injuries, with the discovery of bone marrow-derived mesenchymal stem cells (BM-MSCs) it became the benchmark for the treatment of chronic and acute wound injuries. The bone marrow contains two distinct types of precursor stem cells, namely hematopoietic cells and mesenchymal cells, that

have stem cell-like properties. Hematopoietic cells have been rarely utilized to treat wound injuries, but mesenchymal stem cells from bone marrow have been regularly sourced to treat wound injuries. Published studies report that the BM-MSCs contribute to wound healing either via paracrine signaling or via its high differentiation potential abilities [25–29]. Chimeric murine studies have shown that during wound healing, bone marrow-derived cells can differentiate into CD45+antigen-presenting fibroblasts known as the fibrocyte [26, 30] and CD34+endothelial progenitor cells differentiate into endothelial progenitor cells which aid in angiogenesis [31].

Shumakov et al. [32] first described the usage of huBM-MSCs to treat deep burn wounds on a murine model. This study was followed by another clinical study by Rasulov's group, where huBM-MSCs were used to treat skin burn on a female patient. The author reported that application of BM-MSCs resulted in higher wound resolution, angiogenesis, and accelerated rehabilitation of the burn patient [33]. Falanga et al. [34] also reported the use of fibrin and autologous bone marrow-derived mesenchymal stem cells in murine and human cutaneous wounds. The study concluded that the BM-MSCs can be safely and effectively applied to wounds for better wound resolution. This study was followed by outburst of several other studies using different in vitro and in vivo models to better understand and verify the role of BM-MSCs in burn-related wound healing and other disease models [28, 33, 35–40].

One of the main problems in skin following burn injury is the loss of sweat glands. BM-MSCs are currently also being explored for their regenerative properties to restore the perspiration function of the skin via regeneration of sweat glands [41]. A study done by Ohyama and Okano reported that when BM-MSCs were transplanted into wounds from deep burn injury, the stem cell-transplanted areas showed positive iodine-starch perspiration test, which is a confirmation for the presence of sweat glands. In mice chimeric study, other researchers also reported the incorporation of GFP+ BM-MSCs in the sweat glands [41, 42].

8.4 Induced Pluripotent Stem Cells (iPSCs)

Induced pluripotent stem cell (iPSC) is an artificial technique of reprogramming the somatic cells into pluripotent state via nuclear transplantation of four specific stemness-related genes. iPSCs exhibit similar differentiation abilities, molecular markers, and functional and differentiation properties to those of embryonic stem cells. These cells can be prepared from easily accessible somatic cells like skin fibroblast, thus obtaining an autologous pool of iPSCs for regenerative medicine or cellular therapy will not be an issue. Additionally, in contrast to allogeneic stem cells, iPSCs are mostly made from autologous sources; thus there is a negligible risk of immune rejection [43–45]. Geron Corporation was one of the first institutions to initiate first stem cell clinical trial for spinal cord injuries in 2010 [46]; thereafter many other institutions followed the study in other disease models like heart failure [47], type 1 diabetes [48], macular degeneration [49], and burns [5].

The use of iPSCs as a potential source of stem cells is in early stage of research and development; thus instead of iPSCs some researchers have used embryonic stem cells as a surrogate for tissue engineering purposes [50, 51]. In recent years, researchers have successfully induced skin fibroblast into functional keratinocytes that can be incorporated into tissue engineering purposes [13, 52]. In terms of burn care, the primary focus of iPSCs would be to generate primary epidermal lineage stem cells that cannot be generated by the induction of other types of stem cells. iPSCs can also be utilized to generate the cells that can abrogate the hypermetabolic and inflammatory state by overexpressing anti-inflammatory mediators and growth factors that can accelerate wound healing.

Initially iPSCs were generated via retroviral transfection methods due to which there were concerns of epigenetic aberrations, unwanted mutations, and carcinogenesis. However, recently nonviral gene transfer methods have been developed to address these issues [52, 53]. In summary, iPSCs hold great promises in treating burn injuries; however, the method of isolation, optimum source of tissue to isolate the primary cells, amount of cells to be applied to the wound sites, and other clinical and administrative hurdles need to be addressed before iPSCs can be transferred from bench to the bedside.

8.5 Adipose-Derived Stem Cells (ASCs)

Following the discovery of adipose-derived stem cells (ASCs) in adipose tissue [54], there has been a surge in clinical trials using ASCs to treat different wound healing models including burns. Compared to other types of stem cells, ASCs offer better option for wound healing purposes, as they are readily available; each gram of fat can yield up to 5 K cells [55–57] and like other types of stem cells they have been reported for angiogenesis [58], wound healing [59], and immunomodulation [60] in different disease and animal models.

Autologous ASCs are especially favorable for burn wound injury. Following burn injury, as a standard of care, burned tissue along with some healthy periphery tissues was excised and discarded. In the discarded tissue lies a layer of subcutaneous fat where ASCs reside. Since burn may affect the resident ASCs that are residing in the fat tissue, study done by our group in an established rat burn model has shown that both stromal vascular fraction and ASCs isolated 3 days following burn injury have similar levels of differentiation potential, proliferation rate, cytokine production, and expression of stemness-related cell surface markers [4] indicating that the discarded tissue can be used as a source to obtain the autologous stem cells.

ASCs has been actively studied to address the hypertrophic scarring following burn injury. Yun and colleagues showed that local application of ASCs reduces normal scar size and pliability in the Yorkshire pig. At the molecular level, the authors observed a decrease in mast cell activity and an increase in the expression of TGF-β pathway components [61]. In another full-thickness burn wound model, authors reported that application of ASCs in the combination with a poly-3-hydroxybutyrate-co-hydroxyvalerate matrix improved wound healing, increased expression of VEGF and basic fibroblast growth factor, downregulated expression of TGF-β1 and alpha smooth muscle actin, and increased TGF-β3 mRNA expression, all of which resulted in an anti-fibrotic phenotype. Similar findings were reported following the injection of ASCs under grafted skin, with increased levels of VEGF and TGF-β3 and importantly increased ratio of collagen I to collagen III [62, 63]. Given these findings, ASCs will most likely have a beneficial role in modulating hypertrophic scar formation by accelerating the rate of angiogenesis and wound healing, attenuating inflammation, and most importantly modulating ECM matrix production following the burn injury.

Yun et al. [61] injected ASCs into the wound 50 days post-burn and after the scar started forming. ASCs were injected three times at 10-day intervals and the scars examined every 10 days until 100 days after surgery. Scars treated with ASCs had improved color and pliability, decreased area, and a reduction in the total number of mast cells. ASCs promoted more mature arrangement of collagen in this model, probably via the increase of TGF-β3 and MMP1 in the early phase and the decrease in alpha-smooth muscle actin and tissue inhibitor of metalloprotease-1 in the late phase of scar remodeling [61]. These data suggest that ASCs accelerate wound healing and reduce hypertrophic scarring through autocrine or paracrine regulation of different phases of wound healing or via suppression of fibrosis-related pathways including the p38/MAPK signaling pathway [64].

8.6 Conclusions

Application of stem cells to address burn wound healing comorbidities has a promising future. Nonetheless, still many unaddressed questions including the optimum source of the stem cells, method of isolation, characterization, and especially follow-up on the molecular mechanism following the application of stem cells on the burn wounds need to be further studied in detail.

References

1. Herndon DN. Total burn care. Philadelphia: Saunders Elsevier; 2007.
2. Finnerty CC, Jeschke MG, Branski LK, Barret JP, Dziewulski P, Herndon DN. Hypertrophic scarring: the greatest unmet challenge after burn injury. Lancet. 2016;388(10052):1427–36.
3. van Zuijlen P, Gardien K, Jaspers M, Bos EJ, Baas DC, van Trier A, Middelkoop E. Tissue engineering in burn scar reconstruction. Burns Trauma. 2015;3:18.
4. Prasai A, El Ayadi A, Mifflin RC, Wetzel MD, Andersen CR, Redl H, Herndon DN, Finnerty CC. Characterization of adipose-derived stem cells following burn injury. Stem Cell Rev. 2017;13(6):781–92.
5. Butler KL, Goverman J, Ma H, Fischman A, Yu YM, Bilodeau M, Rad AM, Bonab AA, Tompkins RG, Fagan SP. Stem cells and burns: review and therapeutic implications. J Burn Care Res. 2010;31(6):874–81.
6. Bhatia A, O'Brien K, Chen M, Wong A, Garner W, Woodley DT, Li W. Dual therapeutic functions of F-5 fragment in burn wounds: preventing wound progression and promoting wound healing in pigs. Mol Ther Methods Clin Dev. 2016;3:16041.
7. Clinicaltrails.gov. Umbilical cord mesenchymal stem cells and burns. 2017.
8. Bartholomew A, Sturgeon C, Siatskas M, Ferrer K, McIntosh K, Patil S, Hardy W, Devine S, Ucker D, Deans R, Moseley A, Hoffman R. Mesenchymal stem cells suppress lymphocyte proliferation in vitro and prolong skin graft survival in vivo. Exp Hematol. 2002;30(1):42–8.
9. Le Blanc K, Rasmusson I, Sundberg B, Gotherstrom C, Hassan M, Uzunel M, Ringdén O. Treatment of severe acute graft-versus-host disease with third party haploidentical mesenchymal stem cells. Lancet. 2004;363(9419):1439–41.
10. Ankrum JA, Ong JF, Karp JM. Mesenchymal stem cells: immune evasive, not immune privileged. Nat Biotechnol. 2014;32(3):252–60.
11. Campeau PM, Rafei M, Francois M, Birman E, Forner KA, Galipeau J. Mesenchymal stromal cells engineered to express erythropoietin induce anti-erythropoietin antibodies and anemia in allorecipients. Mol Ther. 2009;17(2):369–72.
12. Beggs KJ, Lyubimov A, Borneman JN, Bartholomew A, Moseley A, Dodds R, Archambault MP, Smith AK, McIntosh KR. Immunologic consequences of multiple, high-dose administration of allogeneic mesenchymal stem cells to baboons. Cell Transplant. 2006;15(8–9):711–21.
13. Itoh M, Umegaki-Arao N, Guo Z, Liu L, Higgins CA, Christiano AM. Generation of 3D skin equivalents fully reconstituted from human induced pluripotent stem cells (iPSCs). PLoS One. 2013;8(10):e77673.
14. Sabella N. Use of fetal membranes in skin grafting. Med Rec NY. 1913;83:478.

15. Eskandarlou M, Azimi M, Rabiee S, Seif Rabiee MA. The healing effect of amniotic membrane in burn patients. World J Plast Surg. 2016;5(1):39–44.
16. Branski LK, Herndon DN, Celis MM, Norbury WB, Masters OE, Jeschke MG. Amnion in the treatment of pediatric partial-thickness facial burns. Burns. 2008;34(3):393–9.
17. Reza HM, Ng BY, Phan TT, Tan DT, Beuerman RW, Ang LP. Characterization of a novel umbilical cord lining cell with CD227 positivity and unique pattern of P63 expression and function. Stem Cell Rev. 2011;7(3):624–38.
18. Moenadjat Y, Merlina M, Surjadi CF, Sardjono CT, Kusnadi Y, Sandra F. The application of human umbilical cord blood mononuclear cells in the management of deep partial thickness burn. Med J Indones. 2013;22(2):92–7.
19. Arno AI, Amini-Nik S, Blit PH, Al-Shehab M, Belo C, Herer E, Tien CH, Jeschke MG. Human Wharton's jelly mesenchymal stem cells promote skin wound healing through paracrine signaling. Stem Cell Res Ther. 2014;5(1):28.
20. Liu L, Yu Y, Hou Y, Chai J, Duan H, Chu W, Zhang H, Hu Q, Du J. Human umbilical cord mesenchymal stem cells transplantation promotes cutaneous wound healing of severe burned rats. PLoS One. 2014;9(2):e88348.
21. Li X, Liu L, Yang J, Yu Y, Chai J, Wang L, Ma L, Yin H. Exosome derived from human umbilical cord mesenchymal stem cell mediates mir-181c attenuating burn-induced excessive inflammation. EBioMedicine. 2016;8:72–82.
22. Panepucci RA, Siufi JL, Silva WA Jr, Proto-Siquiera R, Neder L, Orellana M, Rocha V, Covas DT, Zago MA. Comparison of gene expression of umbilical cord vein and bone marrow-derived mesenchymal stem cells. Stem Cells. 2004;22(7):1263–78.
23. Lee M, Jeong SY, Ha J, Kim M, Jin HJ, Kwon SJ, Chang JW, Choi SJ, Oh W, Yang YS, Kim JS, Jeon HB. Low immunogenicity of allogeneic human umbilical cord blood-derived mesenchymal stem cells in vitro and in vivo. Biochem Biophys Res Commun. 2014;446(4):983–9.
24. El Omar R, Beroud J, Stoltz JF, Menu P, Velot E, Decot V. Umbilical cord mesenchymal stem cells: the new gold standard for mesenchymal stem cell-based therapies? Tissue Eng Part B Rev. 2014;20(5):523–44.
25. Krause DS, Theise ND, Collector MI, Henegariu O, Hwang S, Gardner R, Neutzel S, Sharkis SJ. Multi-organ, multi-lineage engraftment by a single bone marrow-derived stem cell. Cell. 2001;105(3):369–77.
26. Badiavas EV, Abedi M, Butmarc J, Falanga V, Quesenberry P. Participation of bone marrow derived cells in cutaneous wound healing. J Cell Physiol. 2003;196(2):245–50.
27. Bara JJ, Richards RG, Alini M, Stoddart MJ. Concise review: bone marrow-derived mesenchymal stem cells change phenotype following in vitro culture: implications for basic research and the clinic. Stem Cells. 2014;32(7):1713–23.
28. Fu X, Fang L, Li X, Cheng B, Sheng Z. Enhanced wound-healing quality with bone marrow mesenchymal stem cells autografting after skin injury. Wound Rep Regen. 2006;14(3):325–35.
29. Pastides PS, Welck MJ, Khan WS. Use of bone marrow derived stem cells in trauma and orthopaedics: a review of current concepts. World J Orthop. 2015;6(6):462–8.
30. Bucala R, Spiegel LA, Chesney J, Hogan M, Cerami A. Circulating fibrocytes define a new leukocyte subpopulation that mediates tissue repair. Mol Med. 1994;1(1):71–81.
31. Asahara T, Murohara T, Sullivan A, Silver M, van der Zee R, Li T, Witzenbichler B, Schatteman G, Isner JM. Isolation of putative progenitor endothelial cells for angiogenesis. Science. 1997;275(5302):964–7.
32. Shumakov VI, Onishchenko NA, Rasulov MF, Krasheninnikov ME, Zaidenov VA. Mesenchymal bone marrow stem cells more effectively stimulate regeneration of deep burn wounds than embryonic fibroblasts. Bull Exp Biol Med. 2003;136(2):192–5.
33. Rasulov MF, Vasilchenkov AV, Onishchenko NA, Krasheninnikov ME, Kravchenko VI, Gorshenin TL, Pidtsan RE, Potapov IV. First experience of the use bone marrow mesenchymal stem cells for the treatment of a patient with deep skin burns. Bull Exp Biol Med. 2005;139(1):141–4.
34. Falanga V, Iwamoto S, Chartier M, Yufit T, Butmarc J, Kouttab N, Shrayer D, Carson P. Autologous bone marrow-derived cultured mesenchymal stem cells delivered in a fibrin spray accelerate healing in murine and human cutaneous wounds. Tissue Eng. 2007;13(6):1299–312.
35. Chen L, Tredget EE, Liu C, Wu Y. Analysis of allogenicity of mesenchymal stem cells in engraftment and wound healing in mice. PLoS One. 2009;4(9):e7119.
36. Chen L, Tredget EE, Wu PY, Wu Y. Paracrine factors of mesenchymal stem cells recruit macrophages and endothelial lineage cells and enhance wound healing. PLoS One. 2008;3(4):e1886.
37. Borena BM, Pawde AM, Amarpal AHP, Kinjavdekar P, Singh R, Kumar D. Evaluation of autologous bone marrow-derived nucleated cells for healing of full-thickness skin wounds in rabbits. Int Wound J. 2010;7(4):249–60.
38. Huang S, Lu G, Wu Y, Jirigala E, Xu Y, Ma K, Fu X. Mesenchymal stem cells delivered in a microsphere-based engineered skin contribute to cutaneous wound healing and sweat gland repair. J Dermatol Sci. 2012;66(1):29–36.
39. Kwon DS, Gao X, Liu YB, Dulchavsky DS, Danyluk AL, Bansal M, Chopp M, McIntosh K, Arbab AS, Dulchavsky SA, Gautam SC. Treatment with bone marrow-derived stromal cells accelerates wound healing in diabetic rats. Int Wound J. 2008;5(3):453–63.
40. Li H, Fu X, Ouyang Y, Cai C, Wang J, Sun T. Adult bone-marrow-derived mesenchymal stem cells contribute to wound healing of skin appendages. Cell Tissue Res. 2006;326(3):725–36.

41. Sheng Z, Fu X, Cai S, Lei Y, Sun T, Bai X, Chen M. Regeneration of functional sweat gland-like structures by transplanted differentiated bone marrow mesenchymal stem cells. Wound Repair Regen. 2009;17(3):427–35.
42. Fathke C, Wilson L, Hutter J, Kapoor V, Smith A, Hocking A, Isik F. Contribution of bone marrow-derived cells to skin: collagen deposition and wound repair. Stem Cells. 2004;22(5):812–22.
43. Kimbrel EA, Lu SJ. Potential clinical applications for human pluripotent stem cell-derived blood components. Stem Cells Int. 2011;2011:273076.
44. Rami F, Beni SN, Kahnamooi MM, Rahimmanesh I, Salehi AR, Salehi R. Recent advances in therapeutic applications of induced pluripotent stem cells. Cell Reprogram. 2017;19(2):65–74.
45. Daley GQ, Lensch MW, Jaenisch R, Meissner A, Plath K, Yamanaka S. Broader implications of defining standards for the pluripotency of iPSCs. Cell Stem Cell. 2009;4(3):200–1.
46. Scott CT, Magnus D. Wrongful termination: lessons from the Geron clinical trial. Stem Cells Transl Med. 2014;3(12):1398–401.
47. Menasche P, Vanneaux V, Fabreguettes JR, Bel A, Tosca L, Garcia S, Bellamy V, Farouz Y, Pouly J, Damour O, Périer MC, Desnos M, Hagège A, Agbulut O, Bruneval P, Tachdjian G, Trouvin JH, Larghero J. Towards a clinical use of human embryonic stem cell-derived cardiac progenitors: a translational experience. Eur Heart J. 2015;36(12):743–50.
48. Soejitno A, Prayudi PK. The prospect of induced pluripotent stem cells for diabetes mellitus treatment. Ther Adv Endocrinol Metab. 2011;2(5):197–210.
49. Chang YC, Chang WC, Hung KH, Yang DM, Cheng YH, Liao YW, Woung LC, Tsai CY, Hsu CC, Lin TC, Liu JH, Chiou SH, Peng CH, Chen SJ. The generation of induced pluripotent stem cells for macular degeneration as a drug screening platform: identification of curcumin as a protective agent for retinal pigment epithelial cells against oxidative stress. Front Aging Neurosci. 2014;6:191.
50. Itoh M, Kiuru M, Cairo MS, Christiano AM. Generation of keratinocytes from normal and recessive dystrophic epidermolysis bullosa-induced pluripotent stem cells. Proc Natl Acad Sci U S A. 2011;108(21):8797–802.
51. Guenou H, Nissan X, Larcher F, Feteira J, Lemaitre G, Saidani M, Del Rio M, Barrault CC, Bernard FX, Peschanski M, Baldeschi C, Waksman G. Human embryonic stem-cell derivatives for full reconstruction of the pluristratified epidermis: a preclinical study. Lancet. 2009;374(9703):1745–53.
52. Kim D, Kim CH, Moon JI, Chung YG, Chang MY, Han BS, Ko S, Yang E, Cha KY, Lanza R, Kim KS. Generation of human induced pluripotent stem cells by direct delivery of reprogramming proteins. Cell Stem Cell. 2009;4(6):472–6.
53. Miyazaki S, Yamamoto H, Miyoshi N, Takahashi H, Suzuki Y, Haraguchi N, Ishii H, Doki Y, Mori M. Emerging methods for preparing iPS cells. Jap J Clin Oncol. 2012;42(9):773–9.
54. Zuk PA, Zhu M, Ashjian P, De Ugarte DA, Huang JI, Mizuno H, Alfonso ZC, Fraser JK, Benhaim P, Hedrick MH. Human adipose tissue is a source of multipotent stem cells. Mol Biol Cell. 2002;13(12):4279–95.
55. Gimble JM, Katz AJ, Bunnell BA. Adipose-derived stem cells for regenerative medicine. Circ Res. 2007;100(9):1249–60.
56. Lindroos B, Suuronen R, Miettinen S. The potential of adipose stem cells in regenerative medicine. Stem Cell Rev. 2011;7(2):269–91.
57. Beeson W, Woods E, Agha R. Tissue engineering, regenerative medicine, and rejuvenation in 2010: the role of adipose-derived stem cells. Facial Plast Surg. 2011;27(4):378–87.
58. Sumi M, Sata M, Toya N, Yanaga K, Ohki T, Nagai R. Transplantation of adipose stromal cells, but not mature adipocytes, augments ischemia-induced angiogenesis. Life Sci. 2007;80(6):559–65.
59. Bliley JM, Argenta A, Satish L, McLaughlin MM, Dees A, Tompkins-Rhoades C, Marra KG, Rubin JP. Administration of adipose-derived stem cells enhances vascularity, induces collagen deposition, and dermal adipogenesis in burn wounds. Burns. 2016;42(6):1212–22.
60. Cho KS, Park HK, Park HY, Jung JS, Jeon SG, Kim YK, Roh HJ. IFATS collection: immunomodulatory effects of adipose tissue-derived stem cells in an allergic rhinitis mouse model. Stem Cells. 2009;27(1):259–65.
61. Yun IS, Jeon YR, Lee WJ, Lee JW, Rah DK, Tark KC, Lew DH. Effect of human adipose derived stem cells on scar formation and remodeling in a pig model: a pilot study. Dermatol Surg. 2012;38(10):1678–88.
62. Wang J, Hao H, Huang H, Chen D, Han Y, Han W. The effect of adipose-derived stem cells on full-thickness skin grafts. Biomed Res Int. 2016;2016:1464725.
63. Domergue S, Bony C, Maumus M, Toupet K, Frouin E, Rigau V, Vozenin MC, Magalon G, Jorgensen C, Noël D. Comparison between stromal vascular fraction and adipose mesenchymal stem cells in remodeling hypertrophic scars. PLoS One. 2016;11(5):e0156161.
64. Li Y, Zhang W, Gao J, Liu J, Wang H, Li J, Yang X, He T, Guan H, Zheng Z, Han S, Dong M, Han J, Shi J, Hu D. Adipose tissue-derived stem cells suppress hypertrophic scar fibrosis via the p38/MAPK signaling pathway. Stem Cell Res Ther. 2016;7(1):102.

Skin Tissue Engineering in Severe Burns: A Review on Its Therapeutic Applications

9

Alvin Wen Choong Chua, Chairani Fitri Saphira, and Si Jack Chong

9.1 Background

Despite the recent question on whether skin is the largest organ in the human body [1], no one can dispute its protective, perceptive, regulatory, and cosmetic functions. The top layer of the skin, the epidermis which comprised mainly of keratinocytes, is critical for survival as it provides the barrier against exogenous substances, chemicals, and pathogens and prevents dehydration through the regulation of fluid loss. Other cells within the epidermis include melanocytes which give pigmentation and Langerhans cells which provide immune surveillance. Beneath the epidermis, the dermis is a thicker layer of connective tissues that consists mainly of extracellular matrix (ECM) or structural components (predominantly collagen and elastin) which give mechanical strength, elasticity, and a vascular plexus for skin nourish-

ment. Cells interspersed within the ECM include fibroblasts, endothelial cells, smooth muscle cells, and mast cells [2]. These two morphologically distinct layers—the epidermis and the dermis—are in constant communication across various levels (for example at the molecular or cellular level, growth factor exchange, paracrine effects) to establish, maintain, or restore tissue homeostasis. Between the epidermis and dermis is the basement membrane (BM), a highly specialized ECM structure (composed of a set of distinct glycoproteins and proteoglycans) that physically separates the two layers rendering primarily a stabilizing though still dynamic interface and a diffusion barrier [3]. In general, the BM contains at least one member of the four protein families or subtypes of laminin, type IV collagen, nidogen, and perlecan, a heparan sulfate proteoglycan [4].

Populating the epidermal and dermal layers are the various skin appendages such as the hair follicles, sweat glands, sebaceous glands, blood vessels, and nerves. Extreme loss of skin function and structure due to injury and illness will result in substantial physiological imbalance and may ultimately lead to major disability or even death. As much as it is claimed that tissue-engineered skin is now a reality to treat severe and extensive burns, the fact remains that current skin substitutes available are still fraught with limitations for clinical use. It is clearly evident amongst burns or wound-care physicians that there is cur-

---

A. W. C. Chua · S. J. Chong (✉)
Department of Plastic, Reconstructive & Aesthetic Surgery, Singapore General Hospital,
Singapore, Singapore

C. F. Saphira
Department of Plastic, Reconstructive & Aesthetic Surgery, Singapore General Hospital,
Singapore, Singapore

Department of Surgery, Dr. Mohamad Soewandhie General Hospital, Surabaya, Indonesia
e-mail: chairani-fs-fk06@unair.ac.id

© Springer Nature Switzerland AG 2019
D. Duscher, M. A. Shiffman (eds.), *Regenerative Medicine and Plastic Surgery*,
https://doi.org/10.1007/978-3-030-19962-3_9

rently no single tissue-engineered substitute which can fully replicate the split-thickness skin autografts for permanent coverage of deep dermal or full-thickness wounds in a one-step procedure. Indeed, clinical practice for severe burn treatment has since evolved (Table 9.1) to incorporate some of these tissue-engineered skin substitutes (Table 9.2), usually as an adjunct to speed up epithelization for wound closure and/or to improve quality of life by improving functional and cosmetic results long-term.

9.2 Birth of Skin Tissue Engineering

The year 1975 seems to be a special year for skin tissue engineering, even before the term "tissue engineering" was officially adopted more than a decade later by the Washington National Science Foundation bioengineering panel meeting in 1987 [5] and later its definition was elucidated further by Langer and Vacanti [6] in 1993. The beginnings of skin tissue engineering can be attributed to the pioneering work of two groups in the United States 40 years ago. First, Rheinwald and Green [7] reported the successful serial cultivation of human epidermal keratinocytes in vitro and later made possible the expansion of these cells into multiple epithelia suitable for grafting [8] from a small skin biopsy.

In today's term, the work is termed "tissue engineering of the skin epidermis." Concurrently, Yannas et al. [9] reported their maiden work on the in vitro and in vivo characterization of collagen degradation rate which we believe paved the way for the design of artificial biological dermal substitute [10], resulting in the "tissue engineering of the skin dermis."

In 1981, both groups independently reported the clinical use of their respective tissue-engineered substitutes for the treatment of severe and extensive burns, albeit in different approaches. O'Connor et al. [11] and Greene [12] reported the world's first grafting of extensive burns with sheets of cultured epithelium (expanded from autologous epidermal cells) on two adult patients with success at the Peter Bent Brigham Hospital. These autologous cultured sheets (Fig. 9.1) termed cultured epidermal autografts (CEA) were also subsequently demonstrated to provide permanent coverage of extensive full-thickness burns

Table 9.1 Timeline of cultured epithelial autograft (CEA) supported on a fibrin mat [38] used at the Singapore General Hospital (SGH) Burns Centre to treat major burn

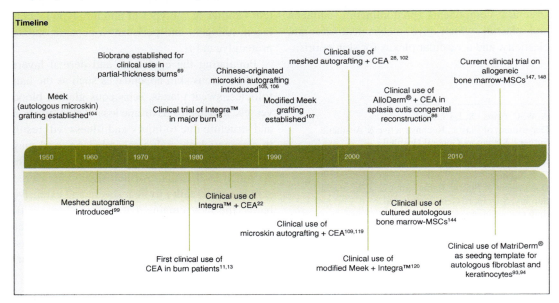

Table 9.2 Tissue-engineered skin substitutes and current surgical techniques

	Skin substitute/surgical technique	Structure	Advantage	Disadvantage	References
EPIDERMAL	Cultured epithelial autograft (CEA)	Confluent autologous keratinocytes	In vitro expansion for large burn area, permanent	Fragility, infection, high cost, variable take rate	[7, 8, 11–13, 19–22, 37]
	Cuono's method (CEA with split-thickness allograft)		Extensive burns	Two-stage procedure, precise grafting time coordination	[23, 24, 26, 27, 33]
	CEA with meshed split-thickness skin autograft		Expansion 1:4, no rejection	Beyond 1:4 expansion: poor cosmetic and functional results, delayed re-epithelialization	[28, 102, 103]
	CEA with microskin autograft		Expansion 1:9–15, no rejection, high take rate, shorter epithelization time	Time-consuming, labor-intensive, hypertrophic scarring	[101, 109, 119]
DERMAL	Integra™	Cross-linked bovine tendon collagen-based dermal matrix linked with glycosaminoglycan (GAG)	Good long-term aesthetic and functional outcome	Two-stage procedure, infection, hematoma, seroma	[59]
	Integra™ with CEA			High cost, poor adhesion	[82]
	Integra™ with Meek				[120, 121]
Artificial biological materials	MatriDerm®	Bovine non-cross-linked lyophilized dermis, coated with alpha-elastin hydrolysate	One-stage procedure, promotes vascularization, improves stability and elasticity of regenerating tissue	Need more scientific evidence to verify efficacy of one-step procedure	[44, 56–59, 66]
	Composite skin substitute	Matriderm as a template, seeded with expanded autologous skin fibroblast and keratinocytes	Full wound closure		[92–94]
	Biobrane®	Silicone membrane and nylon mesh impregnated with porcine dermal collagen	One-stage procedure, coverage of partial-thickness burns	Intolerant to contaminated wound bed	[68–70]

(continued)

Table 9.2 (continued)

	Skin substitute/surgical technique	Structure	Advantage	Disadvantage	References
Natural biological materials	AlloDerm®	Human acellular lyophilized dermis	Acellular, immunologically inert, provide natural dermal porosities for regeneration and vascularization on the wound bed	High cost, risk of transmitting disease, two-stage procedure	[62–65]
	AlloDerm® with CEA			Multiple applications	[85, 86]
	Permacol™	Porcine acellular lyophilized dermis	Good aesthetic and functional outcome	Infection, hematoma, seroma	
Synthetic materials	Transcyte®	Porcine collagen-coated nylon mesh seeded with allogeneic neonatal human foreskin fibroblasts	Immediate availability, ease of storage	Temporary	[72]
	Dermagraft®	Bioabsorbable polyglactin mesh scaffold seeded with cryopreserved allogeneic neonatal human foreskin fibroblasts	Ease of handling, no rejection, chronic wounds—diabetic ulcers	Poor ECM structure, infection, cellulitis	[44, 67]
	Suprathel®	Absorbable dressing based on synthetic copolymer of DL-lactide, trimethylene carbonate, e-caprolactone lactide	Similar elasticity, permeability to water and impermeability to bacteria as human skin	Applied only on acute burns wound (within 24 h of injury), high cost	[75, 80, 81]
Dermo-epidermal	PermaDerm™	Collagen-glycosaminoglycan substrates containing autologous fibroblasts and keratinocytes	Permanent replacement of both dermal and epidermal layers, one-step procedure	No clinical trial reported yet	[87–91]
	DenovoSkin	Plastically compressed collagen type I hydrogels engineered with human keratinocytes and fibroblasts	Near-normal skin architecture with phase 1 clinical trial reported	Long culture time	[95, 96]

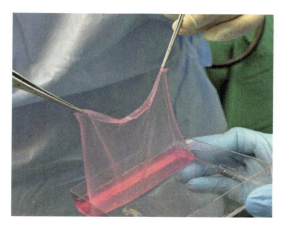

Fig. 9.1 Cultured epithelial autograft (CEA) supported on a fibrin mat [38] used at the Singapore General Hospital (SGH) Burns Centre to treat major burn

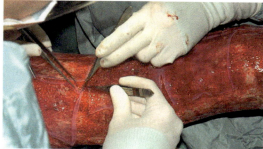

Fig. 9.2 Grafting of CEA on alloderms at SGH Burns Centre based on CUONO's two-stage method

in another two pediatric patients [13]. Meanwhile, Burke et al. [14], a few months after O'Connor et al.'s [11] article, reported the successful use of a physiologically acceptable artificial dermis in the treatment of extensive burn injuries with full-thickness component on ten patients [14]. This was followed by a randomized clinical trial for major burns led by Heimbach et al. [15] on the use of this artificial dermis, now known as Integra™ Dermal Regeneration Template. This successful multicenter study involving 11 centers and many other studies [16, 17] might have inevitably given this dermal substitute a "gold standard" status for full-thickness burn treatment [18]. While groundbreaking, the work of the above two groups is still far from reaching the ultimate goal of replacing skin autografts for permanent coverage of deep dermal or full-thickness wounds in extensive burns.

9.3 CEA Sheets

9.3.1 Importance of Cuono's Method

One of the main disadvantages of the CEA technology was apparently the lack of consistency in engraftment, with poor "take" reported mainly on wounds devoid of dermal elements, even with properly cultured keratinocytes [19–22]. It was later demonstrated in the mid-1980s by Cuono et al. [23, 24] the importance of having the dermal component present when they reported good graft take of the CEA laid on healthy vascularized allogeneic dermis in a full-thickness wound bed. For the Cuono's method to be effective, a two-stage procedure is required. First, human skin allografts must be available ready to be grafted on excised full-thickness wound. This is followed by a wait of about 2–3 weeks which would provide the patient with necessary protection and coverage as the underlying cadaver dermis vascularizes while the autologous epithelial sheets from the harvested small skin biopsy can be prepared simultaneously by culture. When the cultures are ready, the highly immunogenic cadaver epidermis placed on the patient earlier will have to be removed by dermabrasion to make way for the CEA to be grafted (Fig. 9.2). This two-stage composite allodermis/cultured autograft technique has been adopted by several centers with fairly reproducible success since the 1990s [25–27]. One relatively recent success story came from the Indiana University experience that reported a final graft take of 72.7% with a 91% overall survival rate on 88 severe burn patients. This result as the authors mentioned "gives much optimism for continuing to use CEA in critically burned patient" [28].

9.3.2 The Detractors

However, there are still detractors to this Cuono's method for a number of reasons. Firstly, there might not be readily available skin allografts, especially in the East Asian region where organ and tissue donation is still not prevalent [29, 30]. In addition, skin allografts carry some risks of infection and antigen exposure [31]. Secondly, the timing of the CEA placement could be a tricky balancing act. It was mentioned that if cadaver skin or epithelium is rejected or sloughed off prior to the availability of cultured epidermal grafts for the burn patients, the opportunity to use the cadaver dermis as vascularized dermal support (based on Cuono's method) might be lost [32]. The coordination of CEA use with the timing of surgery is therefore a concern. In another scenario, the wound bed might be ready for CEA grafting but yet the cultured keratinocytes were not ready or sufficient for grafting. On the other hand, there were situations where the CEA cultures were ready for grafting but the wound bed was not or the patient was too sick to undergo surgery. It is known that once the keratinocytes form a sheet in culture, the sheets need to be used within the shortest time as possible to maintain efficacy especially for treatment of full-thickness burns [28, 33]. Otherwise, the keratinocyte stem cell population in the cultures would be compromised and these critical cells for regeneration would move towards an irreversible unidirectional process from holoclones (stem cells) to paraclones (highly differentiated cells) [34–36]. In such a case, the efficacy of the CEA would drop drastically, rendering poor engraftment and sub-optimal wound healing [37]. Even though there was a recommendation to use colony-forming efficiency assay of keratinocytes (Fig. 9.3) as an indirect and simple quality check for the "regenerative property" of CEA cultures [36, 38], there were not too many adopters.

CEA sheets are fragile in nature and extreme care must be taken to avoid tangential and shearing forces while moving the patient's limb or repositioning the patient to prevent any loss of the cell layers. Therefore not surprising, it was reported that CEAs placed on anterior sites were amendable to improved take rates [28]. However with the need to keep the grafted site completely immobile [39] and given the limited sites for grafting of CEAs (recommended to be placed on "nonpressure sites" to prevent shearing off of these friable grafts), these led to some form of resistance to CEA use by certain burn surgeons. In addition, the higher vulnerability of CEA to bacterial contamination on the wound site which could result in almost complete loss of the grafts compared to meshed autograft [22, 40] also exacerbates the reluctance of CEA use in the clinical setting.

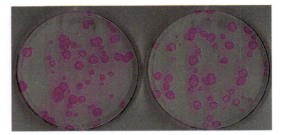

Fig. 9.3 Colony-forming efficiency assay: a simple way of measuring the clonogenic ability of keratinocytes and estimating the growth capacity of these cells

9.3.3 Issue of Cost

Finally, the high cost of production of CEA has often been quoted as one of the major hindrance for its widespread use in many review papers [37, 39, 41]. This cost is going to escalate further as there is a trend of directing cellular therapeutic products with "substantial manipulation" (this would include keratinocyte expansion) to be produced in a good manufacturing practice (GMP) setting for administrative demands like quality, safety controls, and regulations [42]. GMP is a pharmaceutical quality system which ensures that products are consistently produced in a tightly controlled cleanroom environment according to stringent quality standards. Typically, adoption of this practice especially for autologous human cellular therapeutic products would entail much higher cost in terms of overheads such as manpower and facility resources as there is no economy of scale for such tailored cellular products unlike the manufacturing of allogeneic cells [43].

9.4 Dermal Substitutes

9.4.1 Two-Stage Procedure

Based on the knowledge that there are now many dermal substitute products available commercially and with many of such products widely reviewed and tested in both preclinical and clinical settings [2, 18, 32, 41, 43–46], it is self-evident that the challenge for their therapeutic use (especially for acellular ones) is less than CEA (cellular-autologous products) insofar as their respective functional requirements (dermal versus epidermal) are totally different. If epidermis is "life," providing the protection crucial for our survival, then dermis is the "quality of life." Most current biocompatible dermal substitutes are to a certain extent able to mimic the basic properties of the ECM in the human skin by providing some form of structural integrity, elasticity, and a vascular bed. However, the fact remains that these products lack an epithelial layer and in most cases the use of such products will need to be followed up with grafting of split-thickness skin autograft for permanent coverage, usually in a two-stage procedure. While there are advantages of harvesting thinner split-thickness skin autografts and that donor sites heal faster [15], there is still harvest-site morbidity with a possibility of insufficient donor sites in extensive burns.

9.4.2 Integra™

Being the most widely accepted artificial biological dermal substitute [47], the use of Integra™ which is made up of bovine collagen and chondroitin 6-sulfate has been reported to give good aesthetic and functional outcomes when compared to using split-thickness skin autograft alone [48]. However, it is known that infection still remains the most commonly reported complication of Integra™ [49–51]. Meticulous wound bed preparation before the use of this template (or similar type of artificial biological materials) has been reported to be critical to ensure good take. Otherwise with the collection of hematomas and seromas beneath the material, the product is susceptible to infection resulting in a costly loss of an expensive tissue-engineered product and manpower time while increasing the length of hospital stay for the patient.

But with much progress in the development of newer wound care products, the use of advanced antimicrobial silver dressing such as Acticoat dressing as an overlay to Integra™ [44] as well as the use of topical negative pressure or vacuum-assisted closure (VAC) in combination with Integra™ [52–54] have been reported to mitigate the rates of infection with positive results. In one study, it was reported that the application of topical negative pressure dressings to dermal templates can reduce shearing forces, restrict seroma and hematoma formation, simplify wound care, and improve patient tolerance, even as it was reported that the negative pressure did not accelerate vascularization of the Integra™ dermal template based on histological assessment [55].

9.4.3 MatriDerm®

Another newer generation of artificial biological dermal substitute that is gaining wider acceptance for use in the clinics recently is MatriDerm®. Made up of bovine collagen and an elastin hydrolysate, this product is touted for use in a single-stage procedure. MatriDerm® was shown to be able to accommodate split-thickness skin autograft safely in one step with no compromise in take on burn injuries [56, 57]; and it seemed to be feasible for use in critically ill patients [58]. It was suggested that unlike Integra™ which has antigenic properties due to the presence of chondroitin-6-sulfate, the combination of collagen and elastin in MatriDerm® can promote vascularization quicker through the support of ingrowth cells and vessels while improving stability and elasticity of regenerating tissue [44]. Furthermore, higher rate of degradation and difference in neodermal thickness of MatriDerm® compared to Integra™ [59] might give the former an extra edge, even though there is still relatively weak scientific evidence on their comparison in the current literature [58].

9.4.4 Other Dermal Substitutes

There are also other categories of dermal substitutes available commercially. On top of substitutes made from "artificial biological materials" described above for Integra™ and MatriDerm®, the other two commonly recognized classifications are "natural biological materials" and "synthetic materials" [43, 44]. Decellularized human skin allografts (such as AlloDerm®) and decellularized porcine xenografts (such as Permacol™) are dermal products derived from "natural biological materials" as typically these products are "de-epidermalized" and processed to remove the antigenic cellular components while retaining the structure of the native dermis. Known as acellular dermal matrix (ADM), the advantage of using this class of product is that the templates derived from decellularized tissues provide natural dermal porosities for regeneration and vascularization on the wound bed in vivo. In vitro studies have shown that such products support adhesion, growth, and function of several cell types [60, 61]. In addition, there is partial conservation of BM which might aid in epidermal cell attachment [62]. Nevertheless these products are known for their high cost with the risk of transmitting infectious diseases and they are usually used in two surgical procedures [63]. But with advancement in processing of human skin allografts and also with the use of negative pressure therapy, studies using a one-stage procedure of co-grafting with human ADM (CG derm) and autologous split-thickness skin grafts have been reported with some success [64, 65].

Finally, dermal substitutes using synthetic materials seem to be less widely used since their inception in the 1990s for burn treatment. Such products include Transcyte®, a porcine collagen-coated nylon mesh seeded with allogeneic neonatal human foreskin fibroblasts bonded to a silicon membrane, and Dermagraft®, a bioabsorbable polyglactin mesh scaffold seeded with cryopreserved allogeneic neonatal human foreskin fibroblasts. It was reported that both of these products are currently off the market but their technologies have been licensed to Advanced BioHealing for further production and marketing to improve the product [44].

This brings to the issue about cost of dermal substitutes. In general, dermal substitutes are deemed to be costly for clinical usage as mentioned in a report comparing the clinical outcome of MatriDerm® and Integra™ [66]. Based on a tabulated comparison of cost per cm^2 between different dermal substitutes in 2007, it was noted that Dermagraft® was about twice the cost of Integra™ [67], and that might explain why Dermagraft® is presently off-market.

9.4.5 Biobrane*

As opposed to Transcyte®, Biobrane® is still widely used as a synthetic skin substitute as it is known for its success in the definitive management of partial-thickness burns (Fig. 9.4) in many centers [68–70]. Biobrane® is the exact product of Transcyte® less the neonatal human fibroblasts and is also used as a dressing to hold meshed autografts and cultured keratinocyte suspension [69, 71]. On top of the versatility in usage, the popularity of Biobrane® is likely due to its lower cost and yet it is as efficacious in treating partial-thickness burns as Transcyte® [72]. In a recent comparison of Biobrane® and cadaveric allograft for temporizing the acute burn wound, Austin et al. [73] concluded that Biobrane® is superior in terms of lower procedural time and associated cost largely due to the relative ease of application of this product. Indeed, Greenwood et al. [69] in a sharing of their experience using Biobrane® on 703 patients concluded that Biobrane® is relatively inexpensive; easy to store, apply, and fix; and reliable when used according to guidelines.

Currently, there is also an increasing trend to use Biobrane® as an alternative to cadaver allografts as temporizing dressings after excision of major burn injuries [68, 69, 73]. However, the caveat of using this technique is that the wound bed must be meticulously prepared to prevent any infection and there is still the lack of existing literature and published clinical protocols [68] to prove that it can be a worthy replacement of the human skin allografts, especially in the treatment of full-thickness burn wounds.

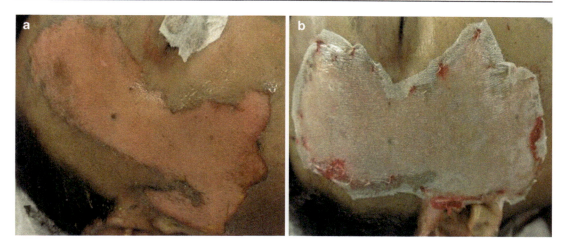

Fig. 9.4 Application of Biobrane. (**a**) Before application. (**b**) After application

9.4.6 Suprathel®

Suprathel®, a permanent and degradable skin substitute for the treatment of dermal wounds, is an absorbable new-generation dressing based on a copolymer of three compounds. Suprathel® is a polylactic-co-glycolic acid scaffold, similar with other polylactic-co-glycolic acid scaffolds such as TransCyte® and Dermagraft® [74]. It is produced from a synthetic copolymer mainly based on DL-lactide (>70%); the other components are trimethylene carbonate and e-caprolactone lactide [75]. It was developed analogous to the human skin and thus shares the same properties such as elasticity, permeability to water vapor, and impermeability to bacteria [75].

Suprathel® is used in the healing of mainly partial-thickness and also superficial burns [76]. Elasticity makes it possible to apply the dressing to the wound at the sites which are difficult to treat, and also ensures that the dressing will closely adhere to all the curves and depressions on the damaged skin surface [77, 78]. Water permeability prevents accumulation of exudate in the wound while at the same time creating the optimum humid environment contributing to wound epithelization [77, 78]. The inherent viscosity and the molecular weight, the maximum strength, and the elongation at break of Suprathel® decrease during the degradation phase, facilitating the detachment process on the completely epithelialized wound.

In clinical applications at the Marienhospital in Stuttgart it was observed that Suprathel® becomes transparent on the wound and therefore an evaluation of the wound ground without moving Suprathel® is feasible [79]. The drawbacks of Suprathel® as burn dressings are that it should be applied on acute burn wound (within 24 h of injury) [80] and the relatively high cost [81].

Towards a composite skin substitute for permanent replacement

9.5 Combining CEA and Integra™

The first hint of CEA and Integra™ combined use was in 1984 when Gallico et al. [13] reported the permanent coverage of large burn wounds with autologous culture. In the study, it was mentioned that Patient 1 with flame burns of 97% total body surface area had received excision to the level of muscle fascia on certain parts of the body and was covered temporarily by human cadaver skin allograft or a collagen-glycosaminoglycan silastic sheet (later known as Integra™). This was followed by grafting with CEA even though it was not mentioned whether the Integra™ was replaced with the cultured epithelium. It was only in 1998 that the use of cultured autologous keratinocytes with Integra™ in resurfacing of acute burns was presented in a case report by Pandya et al. [82]. Used as a two-step procedure, the authors resurfaced the neodermis (vascularized Integra™) by the third week with ultrathin meshed autografts and CEA on the anterior torso of the patient in two mirror-image halves. It was found that the CEA performed as well as the side covered with split-thickness autograft in terms of appearance, durability, and speed of healing. This positive result was not surprising as a month earlier in the same journal, another group [31] reported that vascularized collagen-glycosaminoglycan matrices produced a favorable substrate for cultured epithelial autografts in a porcine model.

There were practically no subsequent bigger clinical series which describe the two-stage use of Integra™ followed by the grafting of CEA. One of the reasons as alluded by Pandya et al. [82] was that of cost when they mentioned that the combination of Integra™ and autologous cultured keratinocytes was very expensive. The other reason quoted was that direct application of cultured keratinocytes to an Integra™ wound bed was found to be problematic due to the poor adhesion of the cells to the template [43]. This might be attributed to the lack of fibroblasts migrated into the Integra™ which delayed the maturation of the BM between the epithelial grafts and the neodermis. In a bilayered skin equivalent tested in vitro, the presence of fibroblasts with keratinocytes was reported to be important for the formation of high levels of collagen type IV and laminin, some of the key elements of the BM [32, 83]. In fact it was further validated later in another skin equivalent model that only in the presence of fibroblasts or of various growth factors, laminin 5 and laminin 10/11, nidogen, uncein, and type IV and type VII collagen (all of which are components of the BM) were decorating the dermal/epidermal junction [84].

9.6 Combining CEA and Other Skin Substitutes

Similarly, it was also observed that there were scanty clinical reports on the two-stage use of AlloDerm® (a decellularized human ADM product that was first approved by the FDA to treat burns in 1992 [85]) and CEA. One notable case report in 2009 was the successful treatment of aplasia cutis

congenita using the combination of first applying on the defect with AlloDerm® followed by CEA grafting 2 weeks later. It was reported that during a 2-year follow-up period, there were no complications such as motion limits resulting from hypertrophic scarring or scar contracture. Coincidentally, there was also an earlier attempt in 2000 to use allogeneic dermis and CEA as a one-stage procedure to reconstruct aplasia cutis congenita of the trunk in a newborn infant [86]. While the results were reported to be promising, it was noted that three additional applications of CEAs were required for 90% of the wound to be healed.

9.7 Autologous Dermo-Epidermal Composite Skin Substitutes

One of the most promising autologous dermo-epidermal (composite) skin substitutes reported is the cultured skin substitutes (CSS) developed in Cincinnati in the United States. This substitute is composed of collagen glycosaminoglycan substrates which contain autologous fibroblasts and keratinocytes. Reported to be able to provide permanent replacement of both dermal and epidermal layers in a single grafting procedure [2, 87–91], this product was later commercialized as PermaDerm™ [43]. PermaDerm™ can currently be engineered within 30 days. It is indicated for the treatment of large full-thickness skin defects; however it has not obtained Food and Drug Administration (FDA) approval and clinical trials on its efficacy remain to be seen. More recently, a German group reported the development of an engraftable tissue-cultured composite skin autograft using MatriDerm® as a template for the seeding of expanded autologous skin fibroblasts and keratinocytes [92]. They reported that this developed skin composite has strong homology to healthy human skin based on the characterization of the epidermal strata, comparison of the differentiation and proliferation markers, and presence of a functional basal lamina. This skin substitute was subsequently used clinically on two patients with full-thickness wounds. While the wounds are relatively small in size (the largest being 9 × 6 cm), there was positive outcome with full wound closure for all the defects treated [93, 94].

There are many promising autologous cellular bilayered skin substitutes proposed out there such as DenovoSkin developed at Tissue Biology Research Unit, University Children's Hospital, Zurich, Switzerland. This product is based on plastically compressed collagen type I hydrogels engineered with human keratinocytes and fibroblasts from a small skin biopsy [95, 96]. The same group has further reported for the first time a more advanced bioengineered human dermoepidermal skin graft containing functional dermal blood and lymphatic vessels using human keratinocytes, fibroblasts, and microvascular endothelial cells [97, 98]. However, the challenge for the utilization of such products remains; that is, how soon can sufficient autologous cells are cultured, how soon can they be impregnated into the scaffold, and how soon can the substitute be made ready for grafting. Time is of essence especially for a massive burn case with little donor site and options.

Adapting the use of skin tissue engineering products to current practice in the clinics

9.8 Combining CEA and Widely Meshed Autografting

One of the solutions adopted in the clinical setting autografting to quickly treat extensive full-thickness burn wounds is to use widely meshed split-thickness skin grafts to cover the large injured surfaces after the technique of meshing was introduced by Tanner et al. in 1964 [99]. However, at expansion rate greater than 1:4, such meshed grafts have been reported to be difficult to handle. Worse still, re-epithelialization might be delayed or even absent when a meshed piece of skin was expanded beyond a ratio of 1:6 [100], and with substantial areas left uncovered in the interstices, there would be cosmetically unsatisfactory "string vest" appearance [101]. To address these disadvantages, use of CEA in combination with widely meshed autografts (Fig. 9.5) has been reported with success in a clinical series of

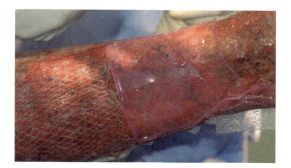

Fig. 9.5 Combining CEA and widely meshed autografting

12 children with major burns. As the authors in the study mentioned, this synergistic combination of autografts and autologous cultured epidermis sheets appeared more effective than one of these techniques applied alone [102]. Based on the Indiana University experience of 88 patients who received CEA (an earlier mentioned study deemed to be one of the success stories in CEA usage), the authors also reported that if an insufficient amount of cadaver dermis remains after allografting (Cuono's method), 1:6 meshed split-thickness autografts (if available) would be placed onto recipient wound bed under the CEA sheets. This was to minimize shear forces and hasten graft take in areas with inadequate allodermis [28]. Other variant techniques involving the use of sprayed cultured autologous keratinocytes in combination with meshed autografts to accelerate wound closure in difficult-to-heal burn patients were also reported [103].

9.9 Resurgence of Microskin Autografting

Based on the current literature, there seems to be a resurgent towards the use of autologous microskin grafting (Fig. 9.6) even though the concept of using small skin bits for autografting was described by Meek in 1958 [104], before the use of meshed grafts. Chinese-originated microskin autografting was described in the 1980s for the treatment of extensive burns [105, 106]. Later in 1993, Kreis et al. [107] improved on Meek's original tech-

Fig. 9.6 Microskin autografting on an extensive-burn patient at the SGH Burns Centre. (**a**) Split-thickness skin autografts were cut into small pieces and laid in close proximity with one another on cadaveric allografts. (**b**) Sheets of autologous microskin-allografts were grafted onto recipient wound bed

nique and popularized the so-called modified Meek method which was found to be superior to widely meshed autografts when higher expansion rates (up to 1:9) were used in adult patients with major burns [108]. While the modified Meek method or the Chinese-originated microskin grafting method (expansion rate of up to 1:15) is still time consuming and laborious with the need for more staff in the operating theatre [109], these problems do not seem to serve as a deterrent because this procedure which can be performed almost immediately is seen as lifesaving [110]. Outcome is generally positive with reliable take rate even on difficult wound bed [111], shorter epithelization time [109, 112, 113], less proneness to loss due to infection [99, 107], as well as satisfactory functional and aesthetic results [114–116]. Moreover, if the Meek graft fails, it is restricted to a partial area without affecting the neighboring skin islands [111] formed from the epithelial migration from the borders of each of the skin bits. More recently, the use of micrograft transplantation with immediate 100-fold expansion for epidermal regeneration on both healthy and diabetic wounds in porcine models was reported [117]. In the same report, it was mentioned that early clinical results confirmed the utility of this technique in a case report of a civilian patient with 54% total body surface area burn admitted to a US Army military hospital in Iraq and successfully treated with the described micrografting technique [118].

9.9.1 Combining CEA and Microskin Autografting

However, scar contracture and hypertrophic scar formation (as would be seen in cases using widely meshed autografts) are problems frequently associated with microskin autografting, especially where high expansion ratios are used for the treatment of extensive burns with high percentage of deep dermal or full-thickness component [100, 101]. Therefore, as what was described earlier for widely meshed skin autografts, CEA was also reported to be used in combination with microskin autografting to accelerate wound closure [101, 109, 119]. Results reported have been positive with one of the earliest studies by Raff et al. [109] describing that the combination of widely expanded postage stamp split-thickness grafts and CEA provided an excellent take rate and durable wound closure within a short time while avoiding the problems associated with engraftment of CEA on fascia. Menon et al. [101] also reported that with the use of sprayed CEA and modified Meek technique, they observed no cases of blistering or scar contracture in those treated sites but unfortunately the problem of hypertrophic scar remained.

9.10 Modified Meek Technique and Integra™

The modified Meek technique in combination with Integra™ dermal template in a two-stage procedure has been reported in extensive burns with some success in a case report involving three patients [120]. As well, radical resection and reconstruction of a giant congenital melanocytic nevus with Meek-graft-covered Integra™ were also reported [121]. However, there are very few reports that utilized the above-described technique subsequently. On top of cost and issue of infection, it can be speculated that the lack of popularity of this two-stage procedure is that it would incur a delay in utilizing the microskin for epithelization which is the main strength of the micrografting technique.

9.11 Stem Cells

Advances in research of adult stem cells and embryonic stem cells offer hope for the therapeutic deficiencies in severe burn treatment using existing skin tissue-engineered products. The therapeutic power of stem cells resides in their clonogenicity and potency [122] and these can be delivered in conjunction with skin composites or by various other methods, including direct application [123]. More recently, there is a burgeoning interest in human induced pluripotent stem cells (hiPSCs) as this Nobel-winning technology pioneered by Yamanaka and his team [124, 125] enables the reprogramming of adult somatic cells to embryonic stage cells. hiPSC technology therefore allows for patient- and disease-specific stem cells to be used for the development of therapeutics, including more advanced products for skin grafting and treatment of cutaneous wounds [123]. However, the recent suspension of the world's first clinical trial involving hiPSCs to treat age-related macular degeneration continues to raise questions about the safety of this new technology. hiPSCs often acquire mutations with epigenetic and chromosomal changes in culture [126]. Hence, human epidermal and mesenchymal stem cells remain the more promising options for clinical use to treat severe burns, at least in the near term.

9.11.1 Enriching for Epidermal Stem Cells

Poor engraftment of CEA even on a properly prepared vascularized wound bed with dermal ele-

ment is thought to be due to epidermal stem cell depletion during graft preparation. A solution for this would be to start with a pure population or higher percentage of these stem cells as suggested by Charruyer and Ghadially [127]. Epidermal stem cells can be enriched from the patient's own skin and a recent study demonstrated that ABCG2, a member of the ATP-binding cassette (ABC) transporter family, was a robust stem cell indicator in the human interfollicular keratinocytes that could potentially be used to quickly enrich for keratinocyte stem cells [128]. Mavilio et al. showed that sheets of epithelium grown from autologous holoclones or keratinocyte stem cells (modified genetically) could be used to treat a patient with junctional epidermolysis bullosa [129], demonstrating the power of this graft refinement. The use of enriched population epidermal stem cells for the preparation of cultured grafts for patients offers hope of overcoming several limitations of current skin substitutes as in a suitable microenvironment; keratinocyte stem cells can also form appendages such as hair, epidermis, and sebaceous glands [130, 131]. However finding or creating that elusive microenvironment (in vivo or in vitro)—to provide the necessary molecular or cellular signals for the stem cells to regenerate a fully functional skin with all its appendages—remains a challenge.

9.11.2 Harnessing Allogeneic Mesenchymal Stem Cells

During the past decade, adult tissue-derived MSCs have rapidly moved from in vitro and animal studies into human trials as a therapeutic modality for a diverse range of clinical applications. MSCs raise great expectations in regenerative medicine, not only because of their multipotent differentiation characteristics and trophic and immunomodulatory effects but also for their extensive sources and biostability when cultured and expanded in vitro [132]. Apart from bone marrow and adipose tissues, human MSCs can also be isolated from a variety of other tissues such as the amniotic membrane [133], umbilical cord [134, 135], cord blood [136], as well as hair follicle dermal papilla [137] and sheath [138, 139]. MSCs have demonstrated a number of properties in vitro that can promote tissue repair, including the production of multiple growth factors, cytokines, collagens, and matrix metalloproteinases [140, 141], in addition to the ability to promote migration of other skin cells such as keratinocytes [142]. MSCs have also been reported to enhance wound healing through differentiation and angiogenesis [143]. In the current literature, several clinical cases on the use of cultured autologous bone marrow MSCs for localized and topical treatment of chronic wounds have been reported. Yoshikawa et al. treated 20 patients with various nonhealing wounds (i.e., burns, lower extremity ulcers, and decubitus ulcers) using autologous bone marrow-derived mesenchymal stem cells expanded in culture and a dermal replacement with or without autologous skin graft [144]. The authors reported that 18 of the 20 wounds appeared healed completely with the cell-composite graft transfer, and the addition of mesenchymal stem cells facilitated regeneration of the native tissue by histologic examination. For allogeneic MSC usage, Hanson et al. [145] reported the use of allogeneic bone marrow- or adipose-derived MSCs to treat partial-thickness wounds of Göttingen minipigs and demonstrated the safety, feasibility, and potential efficacy of these MSCs for treatment of wounds.

The immunomodulatory effect of MSCs could be key to the immediate utilization of these cells for rapid treatment of severe burns. It is now clear that MSCs modulate both innate and adaptive responses and evidence is now emerging that the local microenvironment is important for the activation or licensing of MSCs to become immunosuppressive [146]. The regenerative and pro-angiogenic effects of the MSCs can be harnessed owing to this property.

The first worldwide clinical trial which uses allogeneic bone marrow MSCs to treat ten patients with large severe deep burns is in progress in Argentina. This is done by treating the wound with the application of MSCs through a fibrin-based polymer spray over an acellular dermal biological matrix [147]. The same group, Mansilla et al. [148], has just reported their preliminary experience treating a patient with 60% total body surface burned with positive results. A search using "allogeneic mesenchymal stem cells for burns" in ClinicalTrials.gov (as at Nov 2015) also revealed that two of such trials have been filed [149] which further reinforce the hypothesis that allogeneic MSCs might have a role in major burn treatment.

9.12 Conclusions

Similar to what was mentioned that no single treatment can be recommended in the management of diabetic foot ulcers based on the current and emerging therapies [150], there is no particular approach that is definitely superior for the treatment of severe burns. But based on existing technologies and products available for rapid coverage of extensive burn wounds, the use of Biobrane or similar products to cover the partial-thickness component whilst covering the deep dermal or full-thickness component with skin allografts after excision, followed by a definite closure with autografts (meshed, microskin, CEA or in combination), seems to be one of the efficacious and cost-effective management approaches. If the quality of life of the patients is to be considered such as to reduce scarring and contractures, tissue-engineered dermal templates can be used but they typically come at a cost. Therefore, before technology can catch up in terms of producing a truly functional substitute that comes at a reasonable cost, the need for skin allograft tissue banks, whether local or regional, to serve healthcare centers that treat severe burns cannot be overstated. This is especially true in the event of mass casualty [151]. Having a facility that can double up as both a skin allograft bank and an autologous epithelial cell sheet culture laboratory would be a bonus as we seek to train and build up a critical mass of skin tissue engineers, scientists, as well as administrators specializing in finance, quality assurance, and regulatory affairs. Only by working closely with clinicians to fully appreciate the requirements for the patients can this specialized pool of personnel innovate, harness emerging technologies, manage cost, and navigate through the regulatory minefields for a realistic advancement of this exciting field of skin-based regenerative medicine.

References

1. Sontheimer RD. Skin is not the largest organ. J Invest Dermatol. 2014;134(2):581–2.
2. Supp DM, Boyce ST. Engineered skin substitutes: practices and potentials. Clin Dermatol. 2005;23(4):403–12.
3. Breitkreutz D, Mirancea N, Nischt R. Basement membranes in skin: unique matrix structures with diverse functions? Histochem Cell Biol. 2009;132(1):1–10.
4. Breitkreutz D, Koxholt I, Thiemann K, Nischt R. Skin basement membrane: the foundation of epidermal integrity–BM functions and diverse roles of bridging molecules nidogen and perlecan. Biomed Res Int. 2013;2013:179784.
5. Nerem RM. Tissue engineering in the USA. Med Biol Eng Comput. 1992;30(4):CE8–12.
6. Langer R, Vacanti JP. Tissue engineering. Science. 1993;260(5110):920–6.
7. Rheinwald JG, Green H. Serial cultivation of strains of human epidermal keratinocytes: the formation of keratinizing colonies from single cells. Cell. 1975;6(3):331–43.
8. Green H, Kehinde O, Thomas J. Growth of cultured human epidermal cells into multiple epithelia suitable for grafting. Proc Natl Acad Sci U S A. 1979;76(11):5665–8.
9. Yannas IV, Burke JF, Huang C, Gordon PL. Correlation of in vivo collagen degradation rate with in vitro measurements. J Biomed Mater Res. 1975;9(6):623–8.
10. Yannas IV, Burke JF. Design of an artificial skin. I. Basic design principles. J Biomed Mater Res. 1980;14(1):65–81.
11. O'Conner NE, Mulliken JB, Banks-Schlegel S, Kehinde O, Green H. Grafting of burns with cultured epithelium prepared from autologous epidermal cells. Lancet. 1981;1(8211):75–8.
12. Green H. The birth of therapy with cultured cells. BioEssays. 2008;30(9):897–903.
13. Gallico GG 3rd, O'Connor NE, Compton CC, Kehinde O, Green H. Permanent coverage of large burn wounds with autologous cultured human epithelium. N Engl J Med. 1984;311(7):448–51.
14. Burke JF, Yannas IV, Quinby WC Jr, Bondoc CC, Jung WK. Successful use of a physiologically acceptable artificial skin in the treatment of extensive burn injury. Ann Surg. 1981;194(4):413–28.
15. Heimbach D, Luterman A, Burke J, Cram A, Herndon D, Hunt J, Jordan M, McManus W, Solem L, Warden G, et al. Artificial dermis for major burns. A multi-center randomized clinical trial. Ann Surg. 1988;208(3):313–20.
16. Heimbach DM, Warden GD, Luterman A, Jordan MH, Ozobia N, Ryan CM, Voigt DW, Hickerson WL, Saffle JR, DeClement FA, Sheridan RL, Dimick AR. Multicenter postapproval clinical trial of Integra dermal regeneration template for burn treatment. J Burn Care Rehabil. 2003;24(1):42–8.
17. Heitland A, Piatkowski A, Noah EM, Pallua N. Update on the use of collagen/glycosaminoglycate skin substitute-six years of experiences with artificial skin in 15 German burn centers. Burns. 2004;30(5):471–5.
18. Shevchenko RV, James SL, James SE. A review of tissue-engineered skin bioconstructs available for skin reconstruction. J R Soc Interface. 2010;7(43):229–58.
19. Eldad A, Burt A, Clarke JA, Gusterson B. Cultured epithelium as a skin substitute. Burns Incl Therm Inj. 1987;13(3):173–80.
20. De Luca M, Albanese E, Bondanza S, Megna M, Ugozzoli L, Molina F, Cancedda R, Santi PL, Bormioli M, Stella M, et al. Multicentre experience in the treatment of burns with autologous and allogenic cultured epithelium, fresh or preserved in a frozen state. Burns. 1989;15(5):303–9.
21. Herzog SR, Meyer A, Woodley D, Peterson HD. Wound coverage with cultured autologous keratinocytes: use after burn wound excision, including biopsy follow-up. J Trauma. 1988;28(2):195–8.
22. Munster AM. Whither [corrected] skin replacement? Burns. 1997;23(1):v.
23. Cuono C, Langdon R, McGuire J. Use of cultured epidermal autografts and dermal allografts as skin replacement after burn injury. Lancet. 1986;1(8490):1123–4.
24. Cuono CB, Langdon R, Birchall N, Barttelbort S, McGuire J. Composite autologous-allogeneic skin replacement: development and clinical application. Plast Reconstr Surg. 1987;80(4):626–37.
25. Nave M. Wound bed preparation: approaches to replacement of dermis. J Burn Care Rehabil. 1992;13(1):147–53.
26. Compton CC, Hickerson W, Nadire K, Press W. Acceleration of skin regeneration from cultured epithelial autografts by transplantation to homograft dermis. J Burn Care Rehabil. 1993;14(6):653–62.
27. Hickerson WL, Compton C, Fletchall S, Smith LR. Cultured epidermal autografts and allodermis combination for permanent burn wound coverage. Burns. 1994;20(Suppl 1):S52–5. discussion S5–6
28. Sood R, Roggy D, Zieger M, Balledux J, Chaudhari S, Koumanis DJ, Mir HS, Cohen A, Knipe C, Gabehart K, Coleman JJ. Cultured epithelial autografts for coverage of large burn wounds in eighty-eight patients: the Indiana University experience. J Burn Care Res. 2010;31(4):559–68.
29. Nivatvongs S, Dhitavat V, Jungsangasom A, Attajarusit Y, Sroyson S, Prabjabok S, Pinmongkol C. Thirteen years of the Thai red cross organ donation centre. Transplant Proc. 2008;40(7):2091–4.
30. Oniscu GC, Forsythe JL. An overview of transplantation in culturally diverse regions. Ann Acad Med Singap. 2009;38(4):365.

31. Orgill DP, Butler C, Regan JF, Barlow MS, Yannas IV, Compton CC. Vascularized collagen-glycosaminoglycan matrix provides a dermal substrate and improves take of cultured epithelial autografts. Plast Reconstr Surg. 1998;102(2):423–9.

32. Hansbrough JF, Franco ES. Skin replacements. Clin Plast Surg. 1998;25(3):407–23.

33. Siwy BK, Compton CC. Cultured epidermis: Indiana University Medical Center's experience. J Burn Care Rehabil. 1992;13(1):130–7.

34. Barrandon Y, Green H. Three clonal types of keratinocyte with different capacities for multiplication. Proc Natl Acad Sci U S A. 1987;84(8):2302–6.

35. Pellegrini G, Ranno R, Stracuzzi G, Bondanza S, Guerra L, Zambruno G, Micali G, De Luca M. The control of epidermal stem cells (holoclones) in the treatment of massive full-thickness burns with autologous keratinocytes cultured on fibrin. Transplantation. 1999;68(6):868–79.

36. Ronfard V, Rives JM, Neveux Y, Carsin H, Barrandon Y. Long-term regeneration of human epidermis on third degree burns transplanted with autologous cultured epithelium grown on a fibrin matrix. Transplantation. 2000;70(11):1588–98.

37. Pellegrini G, Bondanza S, Guerra L, De Luca M. Cultivation of human keratinocyte stem cells: current and future clinical applications. Med Biol Eng Comput. 1998;36(6):778–90.

38. Chua AW, Ma DR, Song IC, Phan TT, Lee ST, Song C. In vitro evaluation of fibrin mat and Tegaderm wound dressing for the delivery of keratinocytes—implications of their use to treat burns. Burns. 2008;34(2):175–80.

39. Atiyeh BS, Costagliola M. Cultured epithelial autograft (CEA) in burn treatment: three decades later. Burns. 2007;33(4):405–13.

40. De Luca M, Bondanza S, Cancedda R, Tamisani AM, Di Noto C, Muller L, Dioguardi D, Brienza E, Calvario A, Zermani R, et al. Permanent coverage of full skin thickness burns with autologous cultured epidermis and re-epithelialization of partial skin thickness lesions induced by allogeneic cultured epidermis: a multicentre study in the treatment of children. Burns. 1992;18(Suppl 1):S16–9.

41. Clark RA, Ghosh K, Tonnesen MG. Tissue engineering for cutaneous wounds. J Invest Dermatol. 2007;127(5):1018–29.

42. Bottcher-Haberzeth S, Biedermann T, Reichmann E. Tissue engineering of skin. Burns. 2010;36(4):450–60.

43. MacNeil S. Progress and opportunities for tissue-engineered skin. Nature. 2007;445(7130):874–80.

44. Shahrokhi S, Arno A, Jeschke MG. The use of dermal substitutes in burn surgery: acute phase. Wound Repair Regen. 2014;22(1):14–22.

45. van der Veen VC, Boekema BK, Ulrich MM, Middelkoop E. New dermal substitutes. Wound Repair Regen. 2011;19(Suppl 1):s59–65.

46. Philandrianos C, Andrac-Meyer L, Mordon S, Feuerstein JM, Sabatier F, Veran J, Magalon G, Casanova D. Comparison of five dermal substitutes in full-thickness skin wound healing in a porcine model. Burns. 2012;38(6):820–9.

47. Jones I, Currie L, Martin R. A guide to biological skin substitutes. Br J Plast Surg. 2002;55(3):185–93.

48. Nguyen DQ, Potokar TS, Price P. An objective long-term evaluation of Integra (a dermal skin substitute) and split thickness skin grafts, in acute burns and reconstructive surgery. Burns. 2010;36(1):23–8.

49. Bargues L, Boyer S, Leclerc T, Duhamel P, Bey E. Incidence and microbiology of infectious complications with the use of artificial skin Integra in burns. Ann Chir Plast Esthet. 2009;54(6):533–9.

50. Lohana P, Hassan S, Watson SB. Integra in burns reconstruction: our experience and report of an unusual immunological reaction. Ann Burns Fire Disasters. 2014;27(1):17–21.

51. Dantzer E, Braye FM. Reconstructive surgery using an artificial dermis (Integra): results with 39 grafts. Br J Plast Surg. 2001;54(8):659–64.

52. Pollard RL, Kennedy PJ, Maitz PK. The use of artificial dermis (Integra) and topical negative pressure to achieve limb salvage following soft-tissue loss caused by meningococcal septicaemia. J Plast Reconstr Aesthet Surg. 2008;61(3):319–22.

53. Leffler M, Horch RE, Dragu A, Bach AD. The use of the artificial dermis (Integra) in combination with vacuum assisted closure for reconstruction of an extensive burn scar—a case report. J Plast Reconstr Aesthet Surg. 2010;63(1):e32–5.

54. Sinna R, Qassemyar Q, Boloorchi A, Benhaim T, Carton S, Perignon D, Robbe M. Role of the association artificial dermis and negative pressure therapy: about two cases. Ann Chir Plast Esthet. 2009;54(6):582–7.

55. Moiemen NS, Yarrow J, Kamel D, Kearns D, Mendonca D. Topical negative pressure therapy: does it accelerate neovascularisation within the dermal regeneration template, Integra? A prospective histological in vivo study. Burns. 2010;36(6):764–8.

56. Kolokythas P, Aust MC, Vogt PM, Paulsen F. Dermal substitute with the collagen-elastin matrix Matriderm in burn injuries: a comprehensive review. Handchir Mikrochir Plast Chir. 2008;40(6):367–71.

57. van Zuijlen PP, van Trier AJ, Vloemans JF, Groenevelt F, Kreis RW, Middelkoop E. Graft survival and effectiveness of dermal substitution in burns and reconstructive surgery in a one-stage grafting model. Plast Reconstr Surg. 2000;106(3):615–23.

58. Haslik W, Kamolz LP, Manna F, Hladik M, Rath T, Frey M. Management of full-thickness skin defects in the hand and wrist region: first long-term experiences with the dermal matrix Matriderm. J Plast Reconstr Aesthet Surg. 2010;63(2):360–4.

59. Bottcher-Haberzeth S, Biedermann T, Schiestl C, Hartmann-Fritsch F, Schneider J, Reichmann E, Meuli M. Matriderm® 1 mm versus Integra® Single Layer 1.3 mm for one-step closure of full thickness skin defects: a comparative experimental study in rats. Pediatr Surg Int. 2012;28(2):171–7.

60. Conconi MT, De Coppi P, Di Liddo R, Vigolo S, Zanon GF, Parnigotto PP, Nussdorfer GG. Tracheal matrices, obtained by a detergent-enzymatic method, support in vitro the adhesion of chondrocytes and tracheal epithelial cells. Transpl Int. 2005;18(6):727–34.

61. Burra P, Tomat S, Conconi MT, Macchi C, Russo FP, Parnigotto PP, Naccarato R, Nussdorfer GG. Acellular liver matrix improves the survival and functions of isolated rat hepatocytes cultured in vitro. Int J Mol Med. 2004;14(4):511–5.

62. van der Veen VC, van der Wal MB, van Leeuwen MC, Ulrich MM, Middelkoop E. Biological background of dermal substitutes. Burns. 2010;36(3):305–21.

63. Wainwright DJ. Use of an acellular allograft dermal matrix (AlloDerm) in the management of full-thickness burns. Burns. 1995;21(4):243–8.

64. Kim EK, Hong JP. Efficacy of negative pressure therapy to enhance take of 1-stage allodermis and a split-thickness graft. Ann Plast Surg. 2007;58(5):536–40.

65. Yi JW, Kim JK. Prospective randomized comparison of scar appearances between cograft of acellular dermal matrix with autologous split-thickness skin and autologous split-thickness skin graft alone for full-thickness skin defects of the extremities. Plast Reconstr Surg. 2015;135(3):609e–16e.

66. Greenwood JE, Mackie IP. Neck contracture release with matriderm collagen/elastin dermal matrix. Eplasty. 2011;11:e16.

67. Yildirimer L, Thanh NT, Seifalian AM. Skin regeneration scaffolds: a multimodal bottom-up approach. Trends Biotechnol. 2012;30(12):638–48.

68. Tan H, Wasiak J, Paul E, Cleland H. Effective use of Biobrane as a temporary wound dressing prior to definitive split-skin graft in the treatment of severe burn: a retrospective analysis. Burns. 2015;41(5):969–76.

69. Greenwood JE, Clausen J, Kavanagh S. Experience with biobrane: uses and caveats for success. Eplasty. 2009;9:e25.

70. Cheah AKW, Chong SJ, Tan BK. Early experience with Biobrane in Singapore in the management of partial thickness burns. Proc Singapore Healthcare. 2014;23(3):196–200.

71. Farroha A, Frew Q, El-Muttardi N, Philp B, Dziewulski P. The use of Biobrane(R) to dress split-thickness skin graft in paediatric burns. Ann Burns Fire Disasters. 2013;26(2):94–7.

72. Pham C, Greenwood J, Cleland H, Woodruff P, Maddern G. Bioengineered skin substitutes for the management of burns: a systematic review. Burns. 2007;33(8):946–57.

73. Austin RE, Merchant N, Shahrokhi S, Jeschke MG. A comparison of Biobrane and cadaveric allograft for temporizing the acute burn wound: cost and procedural time. Burns. 2015;41(4):749–53.

74. Debels H, Hamdi M, Abberton K, Morrison W. Dermal matrices and bioengineered skin substitutes: a critical review of current options. Plast Reconstr Surg Global Open. 2015;3(1):e284.

75. Uhlig C, Rapp M, Hartmann B, Hierlemann H, Planck H, Dittel KK. Suprathel-an innovative, resorbable skin substitute for the treatment of burn victims. Burns. 2007;33:221–9.

76. Kamolz LP, Lumenta DB, Kitzinger HB, Frey M. Tissue engineering for cutaneous wounds: an overview of current standards and possibilities. Eur Surg. 2008;40(1):19–26.

77. Highton L, Wallace C, Shah M. Use of Suprathel® for partial thickness burns in children. Burns. 2013;39(1):136–41.

78. Rashaan ZM, Krijnen P, Allema JH, Vloemans AF, Schipper IB, Breederveld RS. Usability and effectiveness of Suprathel® in partial thickness burns in children. Eur J Trauma Emerg Surg. 2017;43(4):549–56.

79. Schwarze H, Küntscher M, Uhlig C, Hierlemann H, Prantl L, Noack N, Hartmann B. Suprathel, a new skin substitute, in the management of donor sites of split-thickness skin grafts: results of a clinical study. Burns. 2007;33:850–4.

80. Madry R, Struzyna J, Stachura-Kulach A, Drozdz Ł, Bugaj M. Effectiveness of Suprathel® application in partial thickness burns, frostbites and Lyell syndrome treatment. Pol Przegl Chir. 2011;83:541–8.

81. Fischer S, Kremer T, Horter J, Schaefer A, Ziegler B, Kneser U, Hirche C. Suprathel® for severe burns in the elderly: case report and review of the literature. Burns. 2016;42(5):e86–92.

82. Pandya AN, Woodward B, Parkhouse N. The use of cultured autologous keratinocytes with integra in the resurfacing of acute burns. Plast Reconstr Surg. 1998;102(3):825–8.

83. Cooper ML, Andree C, Hansbrough JF, Zapata-Sirvent RL, Spielvogel RL. Direct comparison of a cultured composite skin substitute containing human keratinocytes and fibroblasts to an epidermal sheet graft containing human keratinocytes on athymic mice. J Invest Dermatol. 1993;101(6):811–9.

84. El Ghalbzouri A, Jonkman MF, Dijkman R, Ponec M. Basement membrane reconstruction in human skin equivalents is regulated by fibroblasts and/or exogenously activated keratinocytes. J Invest Dermatol. 2005;124(1):79–86.

85. Eweida AM, Marei MK. Naturally occurring extracellular matrix scaffolds for dermal regeneration: do they really need cells? Biomed Res Int. 2015;2015:839694.

86. Simman R, Priebe CJ Jr, Simon M. Reconstruction of aplasia cutis congenita of the trunk in a newborn infant using acellular allogenic dermal graft and cultured epithelial autografts. Ann Plast Surg. 2000;44(4):451–4.

87. Boyce ST, Goretsky MJ, Greenhalgh DG, Kagan RJ, Rieman MT, Warden GD. Comparative assessment of cultured skin substitutes and native skin autograft for treatment of full-thickness burns. Ann Surg. 1995;222(6):743–52.

88. Boyce ST, Kagan RJ, Meyer NA, Yakuboff KP, Warden GD. The 1999 clinical research award. Cultured skin substitutes combined with Integra Artificial Skin to replace native skin autograft and allograft for the closure of excised full-thickness burns. J Burn Care Rehabil. 1999;20(6):453–61.

89. Boyce ST, Kagan RJ, Yakuboff KP, Meyer NA, Rieman MT, Greenhalgh DG, Warden GD. Cultured skin substitutes reduce donor skin harvesting for closure of excised, full-thickness burns. Ann Surg. 2002;235(2):269–79.

90. Boyce ST, Kagan RJ, Greenhalgh DG, Warner P, Yakuboff KP, Palmieri T, Warden GD. Cultured skin substitutes reduce requirements for harvesting of skin autograft for closure of excised, full-thickness burns. J Trauma. 2006;60(4):821–9.

91. Hansbrough JF, Boyce ST, Cooper ML, Foreman TJ. Burn wound closure with cultured autologous keratinocytes and fibroblasts attached to a collagen-glycosaminoglycan substrate. JAMA. 1989;262(15):2125–30.

92. Golinski PA, Zoller N, Kippenberger S, Menke H, Bereiter-Hahn J, Bernd A. Development of an engraftable skin equivalent based on matriderm with human keratinocytes and fibroblasts. Handchir Mikrochir Plast Chir. 2009;41(6):327–32.

93. Golinski P, Menke H, Hofmann M, Valesky E, Butting M, Kippenberger S, Bereiter-Hahn J, Bernd A, Kaufmann R, Zoeller NN. Development and characterization of an engraftable tissue-cultured skin autograft: alternative treatment for severe electrical injuries. Cells Tissues Organs. 2014;200(3–4):227–39.

94. Zoller N, Valesky E, Butting M, Hofmann M, Kippenberger S, Bereiter-Hahn J, Bernd A, Kaufmann R. Clinical application of a tissue-cultured skin autograft: an alternative for the treatment of non-healing or slowly healing wounds? Dermatology. 2014;229(3):190–8.

95. Pontiggia L, Klar A, Bottcher-Haberzeth S, Biedermann T, Meuli M, Reichmann E. Optimizing in vitro culture conditions leads to a significantly shorter production time of human dermo-epidermal skin substitutes. Pediatr Surg Int. 2013;29(3):249–56.

96. Hartmann-Fritsch F, Biedermann T, Braziulis E, Luginbuhl J, Pontiggia L, Bottcher-Haberzeth S, van Kuppevelt TH, Faraj KA, Schiestl C, Meuli M, Reichmann E. Collagen hydrogels strengthened by biodegradable meshes are a basis for dermo-epidermal skin grafts intended to reconstitute human skin in a one-step surgical intervention. J Tissue Eng Regen Med. 2016;10:81–91.

97. Marino D, Luginbuhl J, Scola S, Meuli M, Reichmann E. Bioengineering dermo-epidermal skin grafts with blood and lymphatic capillaries. Sci Transl Med. 2014;6(221):221ra14.

98. Marino D, Reichmann E, Meuli M. Skingineering. Eur J Pediatr Surg. 2014;24(3):205–13.

99. Tanner JC Jr, Vandeput J, Olley JF. The mesh skin graft. Plast Reconstr Surg. 1964;34:287–92.

100. Hsieh CS, Schuong JY, Huang WS, Huang TT. Five years' experience of the modified Meek technique in the management of extensive burns. Burns. 2008;34(3):350–4.

101. Menon S, Li Z, Harvey JG, Holland AJ. The use of the Meek technique in conjunction with cultured epithelial autograft in the management of major paediatric burns. Burns. 2013;39(4):674–9.

102. Braye F, Oddou L, Bertin-Maghit M, Belgacem S, Damour O, Spitalier P, Guillot M, Bouchard C, Gueugniaud PY, Goudeau M, Petit P, Tissot E. Widely meshed autograft associated with cultured autologous epithelium for the treatment of major burns in children: report of 12 cases. Eur J Pediatr Surg. 2000;10(1):35–40.

103. James SE, Booth S, Dheansa B, Mann DJ, Reid MJ, Shevchenko RV, Gilbert PM. Sprayed cultured autologous keratinocytes used alone or in combination with meshed autografts to accelerate wound closure in difficult-to-heal burns patients. Burns. 2010;36(3):e10–20.

104. Meek CP. Successful microdermagrafting using the Meek-Wall microdermatome. Am J Surg. 1958;96(4):557–8.

105. Zhang ML, Wang CY, Chang ZD, Cao DX, Han X. Microskin grafting. II. Clinical report. Burns Incl Therm Inj. 1986;12(8):544–8.

106. Zhang ML, Chang ZD, Wang CY, Fang CH. Microskin grafting in the treatment of extensive burns: a preliminary report. J Trauma. 1988;28(6):804–7.

107. Kreis RW, Mackie DP, Vloemans AW, Hermans RP, Hoekstra MJ. Widely expanded postage stamp skin grafts using a modified Meek technique in combination with an allograft overlay. Burns. 1993;19(2):142–5.

108. Kreis RW, Mackie DP, Hermans RR, Vloemans AR. Expansion techniques for skin grafts: comparison between mesh and Meek island (sandwich-) grafts. Burns. 1994;20(Suppl 1):S39–42.

109. Raff T, Hartmann B, Wagner H, Germann G. Experience with the modified Meek technique. Acta Chir Plast. 1996;38(4):142–6.

110. McHeik JN, Barrault C, Levard G, Morel F, Bernard FX, Lecron JC. Epidermal healing in burns: autologous keratinocyte transplantation as a standard procedure: update and perspective. Plast Reconstr Surg Glob Open. 2014;2(9):e218.

111. Lumenta DB, Kamolz LP, Frey M. Adult burn patients with more than 60% TBSA involved-Meek and other techniques to overcome restricted skin harvest availability—the Viennese Concept. J Burn Care Res. 2009;30(2):231–42.

112. Zermani RG, Zarabini A, Trivisonno A. Micrografting in the treatment of severely burned patients. Burns. 1997;23(7–8):604–7.

113. Lari AR, Gang RK. Expansion technique for skin grafts (Meek technique) in the treatment of severely burned patients. Burns. 2001;27(1):61–6.

114. Lee SS, Tsai CC, Lai CS, Lin SD. An easy method for preparation of postage stamp autografts. Burns. 2000;26(8):741–9.

115. Lee SS, Lin TM, Chen YH, Lin SD, Lai CS. "Flypaper technique" a modified expansion method for preparation of postage stamp autografts. Burns. 2005;31(6):753–7.

116. Lee SS, Chen YH, Sun IF, Chen MC, Lin SD, Lai CS. "Shift to right flypaper technique" a refined method for postage stamp autografting preparation. Burns. 2007;33(6):764–9.

117. Hackl F, Bergmann J, Granter SR, Koyama T, Kiwanuka E, Zuhaili B, Pomahac B, Caterson EJ, Junker JP, Eriksson E. Epidermal regeneration by micrograft transplantation with immediate 100fold expansion. Plast Reconstr Surg. 2012;129(3):443e–52.

118. Danks RR, Lairet K. Innovations in caring for a large burn in the Iraq war zone. J Burn Care Res. 2010;31(4):665–9.

119. Dorai AA, Lim CK, Fareha AC, Halim AS. Cultured epidermal autografts in combination with MEEK Micrografting technique in the treatment of major burn injuries. Med J Malaysia. 2008;63(Suppl A):44.

120. Papp A, Harma M. A collagen based dermal substitute and the modified Meek technique in extensive burns. Report of three cases. Burns. 2003;29(2):167–71.

121. Kopp J, Magnus Noah E, Rubben A, Merk HF, Pallua N. Radical resection of giant congenital melanocytic nevus and reconstruction with meek-graft covered integra dermal template. Dermatol Surg. 2003;29(6):653–7.

122. Butler KL, Goverman J, Ma H, Fischman A, Yu YM, Bilodeau M, Rad AM, Bonab AA, Tompkins RG, Fagan SP. Stem cells and burns: review and therapeutic implications. J Burn Care Res. 2010;31(6):874–81.

123. Sun BK, Siprashvili Z, Khavari PA. Advances in skin grafting and treatment of cutaneous wounds. Science. 2014;346(6212):941–5.

124. Takahashi K, Tanabe K, Ohnuki M, Narita M, Ichisaka T, Tomoda K, Yamanaka S. Induction of pluripotent stem cells from adult human fibroblasts by defined factors. Cell. 2007;131(5):861–72.

125. Yamanaka S. The winding road to pluripotency (Nobel Lecture). Angew Chem. 2013;52(52):13900–9.

126. Garber K. RIKEN suspends first clinical trial involving induced pluripotent stem cells. Nat Biotechnol. 2015;33(9):890–1.

127. Charruyer A, Ghadially R. Stem cells and tissue-engineered skin. Skin Pharmacol Physiol. 2009;22(2):55–62.

128. Ma D, Chua AW, Yang E, Teo P, Ting Y, Song C, Lane EB, Lee ST. Breast cancer resistance protein identifies clonogenic keratinocytes in human interfollicular epidermis. Stem Cell Res Ther. 2015;6:43.

129. Mavilio F, Pellegrini G, Ferrari S, Di Nunzio F, Di Iorio E, Recchia A, Maruggi G, Ferrari G, Provasi E, Bonini C, Capurro S, Conti A, Magnoni C, Giannetti A, De Luca M. Correction of junctional epidermolysis bullosa by transplantation of genetically modified epidermal stem cells. Nat Med. 2006;12(12):1397–402.

130. Oshima H, Rochat A, Kedzia C, Kobayashi K, Barrandon Y. Morphogenesis and renewal of hair follicles from adult multipotent stem cells. Cell. 2001;104(2):233–45.

131. Claudinot S, Nicolas M, Oshima H, Rochat A, Barrandon Y. Long-term renewal of hair follicles from clonogenic multipotent stem cells. Proc Natl Acad Sci U S A. 2005;102(41):14677–82.

132. Charbord P. Bone marrow mesenchymal stem cells: historical overview and concepts. Hum Gene Ther. 2010;21(9):1045–56.

133. Miki T, Mitamura K, Ross MA, Stolz DB, Strom SC. Identification of stem cell marker-positive cells by immunofluorescence in term human amnion. J Reprod Immunol. 2007;75(2):91–6.

134. Kita K, Gauglitz GG, Phan TT, Herndon DN, Jeschke MG. Isolation and characterization of mesenchymal stem cells from the sub-amniotic human umbilical cord lining membrane. Stem Cells Dev. 2010;19(4):491–502.

135. Baksh D, Yao R, Tuan RS. Comparison of proliferative and multilineage differentiation potential of human mesenchymal stem cells derived from umbilical cord and bone marrow. Stem Cells. 2007;25(6):1384–92.

136. Zhang X, Hirai M, Cantero S, Ciubotariu R, Dobrila L, Hirsh A, Igura K, Satoh H, Yokomi I, Nishimura T, Yamaguchi S, Yoshimura K, Rubinstein P, Takahashi TA. Isolation and characterization of mesenchymal stem cells from human umbilical cord blood: reevaluation of critical factors for successful isolation and high ability to proliferate and differentiate to chondrocytes as compared to mesenchymal stem cells from bone marrow and adipose tissue. J Cell Biochem. 2011;112(4):1206–18.

137. Driskell RR, Clavel C, Rendl M, Watt FM. Hair follicle dermal papilla cells at a glance. J Cell Sci. 2011;124(Pt 8):1179–82.

138. Richardson GD, Arnott EC, Whitehouse CJ, Lawrence CM, Hole N, Jahoda CA. Cultured cells from the adult human hair follicle dermis can be directed toward adipogenic and osteogenic differentiation. J Invest Dermatol. 2005;124(5):1090–1.

139. Ma D, Kua JE, Lim WK, Lee ST, Chua AW. In vitro characterization of human hair follicle dermal sheath mesenchymal stromal cells and their potential in enhancing diabetic wound healing. Cytotherapy. 2015;17(8):1036–51.

140. Fathke C, Wilson L, Hutter J, Kapoor V, Smith A, Hocking A, Isik F. Contribution of bone marrow-derived cells to skin: collagen deposition and wound repair. Stem Cells. 2004;22(5):812–22.

141. Kim DH, Yoo KH, Choi KS, Choi J, Choi SY, Yang SE, Yang YS, Im HJ, Kim KH, Jung HL, Sung KW, Koo HH. Gene expression profile of cytokine and growth factor during differentiation of bone marrow-derived mesenchymal stem cell. Cytokine. 2005;31(2):119–26.

142. Akino K, Mineda T, Akita S. Early cellular changes of human mesenchymal stem cells and their inter-

143. Wu Y, Chen L, Scott PG, Tredget EE. Mesenchymal stem cells enhance wound healing through differentiation and angiogenesis. Stem Cells. 2007;25(10):2648–59.

144. Yoshikawa T, Mitsuno H, Nonaka I, Sen Y, Kawanishi K, Inada Y, Takakura Y, Okuchi K, Nonomura A. Wound therapy by marrow mesenchymal cell transplantation. Plast Reconstr Surg. 2008;121(3):860–77.

145. Hanson SE, Kleinbeck KR, Cantu D, Kim J, Bentz ML, Faucher LD, Kao WJ, Hematti P. Local delivery of allogeneic bone marrow and adipose tissue-derived mesenchymal stromal cells for cutaneous wound healing in a porcine model. J Tissue Eng Regen Med. 2016;10(2):E90–E100.

146. English K. Mechanisms of mesenchymal stromal cell immunomodulation. Immunol Cell Biol. 2013;91(1):19–26.

147. Mansilla E, Aquino VD, Roque G, Tau JM, Maceira A. Time and regeneration in burns treatment: heading into the first worldwide clinical trial with cadaveric mesenchymal stem cells. Burns. 2012;38(3):450–2.

148. Mansilla E, Marin G, Berges M, Scafatti S, Rivas J, Nunez A, Menvielle M, Lamonega R, Gardiner C, Drago H, Sturla F, Portas M, Bossi S, et al. Cadaveric bone marrow mesenchymal stem cells: first experience treating a patient with large severe burns. Burns Trauma. 2015;3:17.

149. U.S. National Library of Medicine. Clinical Trials. gov. https://clinicaltrials.gov/ct2/home. Accessed 10 Nov 2015.

150. Karri VV, Kuppusamy G, Talluri SV, Yamjala K, Mannemala SS, Malayandi R. Current and emerging therapies in the management of diabetic foot ulcers. Curr Med Res Opin. 2015;32(3):519–42.

151. Chua A, Song C, Chai A, Chan L, Tan KC. The impact of skin banking and the use of its cadaveric skin allografts for severe burn victims in Singapore. Burns. 2004;30(7):696–700.

Skin Substitutes for Burn Wounds

10

Daniel Popp, Christian Tapking, and Ludwik K. Branski

10.1 Introduction

Progress in the care of severely burned patients has been achieved over the past decade and led to significantly decreased morbidity and mortality [1]. Mortality decrease could be seen especially in the severely burned children population [2]. The main fields of improvement in burn care have been (1) fluid resuscitation and early patient management, (2) control of infection, (3) modulation of the hypermetabolic response, and (4) surgery and wound care [3].

10.2 Burn Wound Care

Extensive burn injuries are characterized by a local and systemic inflammatory as well as by a hyper-

---

D. Popp · L. K. Branski (✉)
Division of Plastic, Aesthetic and Reconstructive Surgery, Department of Surgery, Medical University of Graz, Graz, Austria
e-mail: danpopp@utmb.edu; lubransk@utmb.edu

C. Tapking
Department of Surgery, Shriners Hospital for Children-Galveston, University of Texas Medical Branch, Galveston, TX, USA

Department of Hand, Plastic and Reconstructive Surgery, Burn Trauma Center, BG Trauma Center Ludwigshafen, University of Heidelberg, Heidelberg, Germany
e-mail: chtapkin@utmb.edu

metabolic response. Inflammatory mediators, produced and released by infiltrating immune cells, and toxins of wound-colonizing microorganisms would lead to burn sepsis if burn eschar is not removed. Early excision and early wound coverage are generally accepted as standard of care since the early 1970s [4]. Debridement of the burn wound and eschar should be done as soon as possible after the patient has been resuscitated and stabilized. That would be usually within the first 48–72 h postburn. It has been shown that early excision significantly reduces blood loss, amount of circulating endotoxin levels, hypermetabolic response, wound infection, overall hospital stay, and ultimately postburn morbidity and mortality [5–8].

As an effective alternative to surgical and nonsurgical standard of care in partial- and full-thickness burns, Debriding Gel Dressing (DGD) (NexoBrid®, formerly also known as Debrase®), a bromelain-based enzymatic agent, can be used safely and is nowadays widely used in burn centers [9, 10].

For partial-thickness burn wounds, silver sulfadiazine (SSD) has been the standard of care for many years. In the last three to four decades, a huge variety of new dressing came onto the market [11]. Nowadays, they are standard of care due to the below-mentioned superiority compared over SSD. In partial-thickness burn wounds, the wound coverage should provide an occlusive, moist environment conducive to wound healing and preventive to infection. The ultimate goal is to decrease treatment time, pain, and discomfort.

© Springer Nature Switzerland AG 2019
D. Duscher, M. A. Shiffman (eds.), *Regenerative Medicine and Plastic Surgery*,
https://doi.org/10.1007/978-3-030-19962-3_10

10.3 Skin Substitutes

Skin is a highly complex organ. The two highly specialized layers of the skin contribute to their function as a whole as follows: The epidermis serves as a barrier against vaporization and bacteria. The dermis provides mechanical strength and elasticity. Loss of that barrier function leads to a loss of fluid and protein. The loss of the epidermis makes the tissue prone for inflammation, bacterial colonization, infection, and sepsis. Prolonged wound healing leads to higher rates of scarring [12].

Despite the known benefits of early autografting [4], in many cases it is not safe, as in unstable patients or if not enough donor sites are available due to extensive burns, or simply not possible, as on the battlefield, in mass casualties, or due to limited operating room resources. In these circumstances the burn surgeon needs to resort to alternatives, either for temporary covering or as a definitive dermal replacement.

The ideal skin substitute is constantly available off the shelf; is durable, flexible, and easy to handle; can be applied in one single operation; provides an effective barrier layer to prevent water and heat loss as well as to bacterial invasion; does not become hypertrophic; is nonantigenic; grows with children; is cost efficient; and provides permanent wound coverage but unfortunately to date also does not exist [12].

Generally, skin substitutes can be classified based on their usage (temporary vs. permanent) and on their origin (biologic vs. synthetic). Many of them can be applied in the treatment of partial-thickness as well as full-thickness burns.

In this chapter, we elucidate a selection of currently available skin substitutes for temporary and (semi-) permanent coverage. We describe the origin of the material (biologic, synthetic, combination) and indications of its application (either for partial-thickness or for full-thickness burn wounds). Furthermore, we outline the current study situation and illustrate product-related characteristics and limitations.

10.3.1 Temporary Skin Substitutes: Clinical Use, Advantages, Limitations, Prospects

10.3.1.1 Biological Tissues

Human Allograft (Cadaver Skin)

Fresh allograft skin possesses many of the ideal features of a biologic dressing, wherefore it is the "gold standard" for temporary coverage of extensive full-thickness burn wounds when not enough autologous tissue is available. It basically replaces the lost physiologic barrier and reduces water, electrolyte, and protein loss; prevents wound desiccation; suppresses microbial proliferation; is non-immunogenic; and prepares the wound bed for definitive wound coverage and can serve as an indicator as to if the wound bed is ready for autografting. This can be crucial, as in large burns successful autografting can be essential for survival [13]. It also reduces pain, which makes occupational and physiotherapy easier for the patient.

Human allograft skin can be used as viable tissue up to 14 days when kept refrigerated at 4 °C and the nutrient solution is changed frequently [14]. Cryopreserved skin can be used up to 5 years [15]. It can also be used in a nonviable state after lyophilization [16].

Viable allograft fulfills its role as a biologic cover usually for 3–4 weeks until it gets rejected. Furthermore, meshed allograft is used as an overlay for widely meshed autograft (overlay technique) [17].

Glycerolized allograft is useful as permanent coverage for partial-thickness burns until re-epithelialization occurs. It is particularly useful in scald burns of children, as it makes dressing changes easy and less painful [18]. Following FDA and AATB regulations [19], the use of human cadaveric skin is generally considered safe. Nevertheless, there is still a risk for transmitting viral diseases, especially CMV. But with regard to the benefits, these risks are clinically negligible [20].

Human Amnion

Human amniotic membrane has been used for centuries as a biological wound dressing. After the first report of its usage in skin transplantations by Davis in 1910 [21], Sabella [22] described the use of amnion in burn patients.

Beneficial effects as faster wound epithelialization, lower rate of burn wound infections, pain relief, fluid loss, and scar reduction as well as shortening of the hospital stay have been proven [23, 24]. Furthermore it is easy to handle for the surgeon and adheres well to the wound bed [25].

Amnion is usually used as viable tissue. Since amnion is gained from living donors, consent has to be taken prior to caesarean section. Apart from that, it has to undergo a very similar process as allograft skin and is screened for any viral or bacterial diseases prior to grafting. Furthermore, the donor is screened to prevent transmission of diseases [17] and finally it is sterilized. Those standardized procedures provide safe usage and make amnion broadly available for specialized burn care providers [26]. When preserved in glycerol, it is a long-time storable nonviable biologic dressing that is enormously valuable in developing countries due to its cost-effectiveness [27].

Recently, there has been a method developed to preserve amnion/chorion that can be stored for up to 5 years under ambient conditions and though keep its biologic activity [28–30]. The nonviable and sterilized product still contains growth factors, chemokines, and other regulatory proteins that are important for wound healing, in much higher concentrations compared to other processing methods. The two-layer composition seems to contribute to that, especially chorion [31]. Dehydrated human amnion/chorion membrane (dHACM) is commercially available (EpiFix; AmnioFix; EpiBurn; MiMedx Group, Inc., Marietta, GA) and has been used to treat partial-thickness burns as well as full-thickness burns as a temporary treatment and also as overlay [32, 33]. Moreover, it is an ideal scaffold for stem cells in tissue engineering [34].

At our institution, until present, we use amnion mainly for second-degree facial burns because of its advantageously good plasticity. In a previous study, notably less frequent dressing changings

and related patient comfort at no higher infection rate with comparable cosmetic outcome were seen when compared to topical antimicrobials [35].

Xenograft

Among the different animal skins being studied in the past, only pig skin turned out to be useful due to its histologic structure close to human skin [36, 37]. It shows very little immunologic properties and gets more "ejected" by epithelialization underneath than rejected and should rather be classified as dressing [38]. It provides similar beneficial effects as allograft, but does not show vascularization or capillary ingrowth [39] and therefore xenografts cannot be used to prove readiness of the wound for autografting [40]. In some populations they might also not be used due to ethnic or religious reasons [41] and there is a theoretic risk of zoonoses. Porcine xenograft can be used as a temporary cover for partial-thickness as well as for full-thickness burns or for coverage of donor sites. It is processed and stored similar to allograft [42, 43].

10.3.1.2 Synthetic and Biosynthetic Materials

Up to date there exists a huge variety of synthetic wound dressings. The below mentioned are a selection of dressings routinely used at our institution.

Biobrane®

Biobrane® (Bertek Pharmaceuticals Inc., Morgantown, WV, USA) is a bilayer biosynthetic composite wound dressing, consisting of porcine-derived collagen chemically bound to a nylon mesh that is partially embedded into an ultrathin porous silicone. The silicone film serves as a semipermeable epidermal substitute that allows wound water vapor but still maintains a moist healing environment and serves as a bacterial barrier. Its translucent properties allow for wound judgment without removing the product, and its flexibility enables its usage over joints. Sera and blood clot within this matrix and firmly adhere the fabric to the wound bed until epithelialization occurs and Biobrane® can be easily removed [3, 13, 44, 45]. It accelerates wound healing and lowers pain overall and during dressing changes in partial-thickness burns [45]. It is a safe alternative to allograft as a temporary coverage in third-degree burn wounds when applied to thoroughly debrided, noninfected wound beds. A further advantage is that early mobilization can be performed, while after allograft transplantation the patient or at least the burned area has to be immobilized for a few days. That has clear benefits especially in hand and extremity burns. Overall costs seem not to differ significantly, even though if applied faster than, e.g., allograft, OR time can be saved [13, 46, 47]. It has to be changed usually after 10 days. For some reason, it is currently off the market. There are already existing products that claim to be its successor, but there is still a lack of clinical data [48].

TransCyte

TransCyte (Advanced Tissue Sciences, La Jolla, CA, USA) is also a bilayer biosynthetic composite wound dressing with similar properties as Biobrane with additional neonatal in vitro-cultured human fibroblasts integrated into the nylon mesh. Those fibroblasts secrete human dermal collagen, matrix proteins, and growth factors [49, 50]. It can also be used for treatment of partial-thickness burn wounds as well as a temporary substitute for full-thickness wounds [51–54]. There is evidence that it leads to faster re-epithelialization and fewer dressing changes when compared to Biobrane [50], but it is currently also off the market, probably due to high costs.

Suprathel®

Suprathel® (PolyMedics Innovations GmbH, Denkendorf, Germany) is a synthetic copolymer membrane that serves as a temporary replacement of the epithelium and imitates the same. It contains mainly DL-lactide (>70%); the other components are trimethylene carbonate and ε-caprolactone. The membrane features a porosity of 80% that enables exudate to drain and it can be elongated up to 2.5 times of its size, which gives the product a very good plasticity. Furthermore it supports wound healing and re-epithelialization [55].

Once applied after meticulous debridement of the wound, it attaches nicely to the moist wound bed. At our institution, we cover it with at least one layer of paraffin gauze under normal gauze to absorb the wound fluid. During healing, it becomes—at least partly—transparent, which allows the physician to judge the wound without removing the membrane. It will consecutively detach from the areas that already show epithelialization and should be trimmed in a circular manner until the whole wound has healed and it can be peeled off painless.

The major advantages of this product are its potent pain-reducing potential and its excellent handling. However, it is quite expensive compared to allogenic material or other products, used for second-degree burns [55, 56].

Especially in patients with extensive burns, STSGs can be saved for coverage of third-degree burns when Suprathel® is applied to the second-degree burn wounds [57]. It may be used not only for superficial partial-thickness burns, but also for mixed-depth partial-thickness burns [58]. Furthermore, it can be used also in an outpatient setting for adults as well as for children [56].

Mepilex® Ag

Mepilex Ag® (Mölnlycke Healthcare, Göteborg, Sweden) is an absorbable, silver-coated foam pad. Its innermost silicone layer Safetec® prevents adhesion to the wound bed and therefore reduces pain during dressing changes while the silicone foam absorbs exudate, yet keeps the wound in a moist condition. The broad-spectrum antimicrobial effect of Mepilex® Ag is due to therein comprised silver-sulfate ions and activated carbon [59]. Dressing changes need to take place usually every 3–7 days but are quite easy to handle and relatively pain free compared to dressings without a silicone layer. It furthermore may increase healing time and is more cost efficient than for example Suprathel® ($0.8/cm^2 vs. $0.56/cm^2) [56, 60]. At our institution it is the standard of care for superficial burn wounds.

Aquacel®

The first Aquacel® (ConvaTec Inc., Greensboro, North Carolina, USA) contains a core hydrofiber layer with carboxymethylcellulose and carboxymethylation [61]. An update was Aquacel® Ag, which includes ionic silver. The controlled release of ionic silver absorbs fluids to form a cohesive gel [62]. It provides an antimicrobial protection and protects the wound for up to 14 days [63]. A dressing, which is slightly larger than the wound, is placed on the wound and covered with a sterile secondary layer. Aquacel® is used for burn injuries as well as for chronic wounds and was shown to be safe and effective in partial-thickness burns [61, 64]. Especially in chronic wounds, which tend to develop infections, Aquacel® Ag was proven to decrease wound size and rate of infections [65]. Aquacel® was shown to be more cost effective than other dressings, because it normally does not require a lot of dressing changes [62]. Furthermore, Aquacel® seems to increase the comfort for patients and nurses [66].

10.3.2 Dermal Replacements/ Analogues

10.3.2.1 Biologic Materials

Split-Thickness Skin Graft (STSG)
Split-thickness skin grafts (STSG) are typically indicated for temporary or permanent coverage of cutaneous defects [67]. It consists of epidermis and parts of the dermis, depending on the graft thickness (0.2–0.7 mm). STSG are harvested with a dermatome (constant pressure at a 45° angle to the skin) from thigh or back and other areas, if necessary [68]. Some of the dermal skin appendages remain at the donor site. After harvesting, the graft may be meshed or Meek technique is used and then placed on the clean wound. The STSG can be kept in moist gauze and hydrated until ready to be applied [69]. STSG are well known and accepted for soft-tissue coverage, especially in burns and plastic surgery reconstruction, but also in ulcers [70]. STSG are usually fixed via staples or (sometimes) with sutures. For large sheet grafts, to leave the graft uncovered to allow rolling of fluids is an option [71]. Grafts initially survive via diffusion until a subsequent revasculariza-

tion occurs. A major limitation of STSG is its often unsatisfying functional and cosmetic results, which affects the patients' quality of life especially when used in exposed or joint areas. Hypertrophic scarring and poor elasticity and scar contractures are common problems [72]. In order to increase cosmetic and functional results, dermal matrices such as described below have been developed [73, 74]. Nevertheless, STSG remain the gold standard by now.

1. Indications
   Immediate coverage of clean soft-tissue defects and accelerated wound healing
   Prevention of scar contracture and enhanced cosmetic in superficial wounds
   Immediate coverage of burn defects and reduced fluid loss from the wounds
2. Contraindications
   Infected wounds or necrotic tissue
   Exposure of tendons or bones
   Exposure of blood vessels or nerves
3. Donor-site morbidity
   The donor site, which is often a large surface of the body, heals by epithelialization and is expected to heal like any abrasion [75]. It needs to be kept in mind that skin grafting produces a wound at the donor site which enlarges the unprotected wound area [75, 76]. It has been shown that scarring in donor site is proportional to the thickness of the graft and to the occurrence of infections [77]. Intensive itching may occur due to exposed nerve endings. It has been shown that returning harvested skin, which is not needed, to the donor site may decrease healing time and wound morbidity [78].

Mesh
The technique of meshing was introduced by Tanner et al. in 1963 [79]. To increase the coverable surface, the STSG can be enlarged up to a 1:4 ratio. Larger ratios can be difficult to handle, because the skin tends to curl on itself. Meshing can be performed by hand or the STSG is placed on a plastic sheet and rolled through a machine which cuts the skin sheet on several points, so that a net with preset interstices is pro-

duced [80]. The interstices prevent an accumulation of the fluid, which leads to better and safer healing [81]. The location and size of the wound as well as possible donor site determine the meshing ratios [80].

Meek first described this technique in 1958 [82] and it was later modified by other authors [79, 83, 84]. The expansion is efficient and effective. STSG are placed on a cork plate, which is then cut vertically and horizontally into 1×1 to 3×3 mm squares. The grafts are then transferred to a carrier with aluminum foil backing, the cork plate is removed, and the graft is sprayed with an adhesive spray. After waiting for 5–8 min, the aluminum foil is expanded and the graft can be placed on the wound. An expansion ratio up to 1:9 may be reached. This technique allows to cover larger wounds and if there is a lack of donor sites. That is why severely burned patients can often benefit from this technique [83, 85].

Acellular Dermal Matrix

Alloderm® (Life Cell Corporation, Branchburg, NJ, USA) is an acellular human matrix, which is processed from cadaveric dermis and does not contain epithelial elements [86]. The substitute is freeze-dried, which allows the graft to adapt to the dermal structure, and screened for potentially transmissible pathogens [72, 87]. Comparable to Integra®, Alloderm® is placed over the wound after full excision of nonviable tissue. The dermal matrix incorporated with the patients' own tissue and a thin layer of split-thickness skin graft is placed on top of the Alloderm® graft. Since the cells have been removed, Allograft® is not rejected by the immune system [88]. The outcome is similar to other dermal replacements with favorable results [89]. Recent studies have shown that Alloderm®, aside from burns [88], is also suitable for breast reconstructions, head and neck reconstructions, and abdominal wall/hernia surgery [90, 91]. Since Alloderm® contains elastin and collagen, there is less tension and increased elasticity compared to other dermal substitutes, which results in a less contractions [73].

10.3.3 Biosynthetic Materials

10.3.3.1 Integra

Integra® (Integra Life Sciences Corporation, Plainsboro, NJ, USA) consists of two layers: one bovine tendon collagen matrix and one silicone layer. The silicone layer, which prevents water loss and protects the dermis, is peeled away during wound healing and the bovine layer integrates with the human skin [92]. It is used as a dermal skin substitute and placed over the wound after full excision of nonviable tissue. After initial healing of approximately 3 weeks, a thin autograft is placed onto the neo-dermis [92]. In several studies, Integra® seemed to have a better outcome regarding wound healing time compared to autograft, allograft, or xenograft, but had a higher rate of infections than other substitutes such as Biobrane® [77, 93, 94]. Long-term use and outcomes and outcomes in terms of length of hospital stay, cosmetic results, and functional outcome are mentioned to be favorable [95]. In very large burns, it can be used under widely meshed autografts (4:1–10:1) with an overlay (e.g., allograft or Biobrane).

10.3.3.2 Matriderm

Matriderm® (MedSkin solutions Dr. Suwelack AG, Billerbeck, Germany) is a highly porous membrane composed of three-dimensionally coupled collagen and elastin. The collagen is gained from a bovine dermis and the elastin from a bovine nuchal ligament by hydrolysis [72]. After being sterilized and freeze-dried, Matriderm® can be stored at room temperature [72]. Matriderm® can be engrafted in a one-step procedure with a thin skin graft after full excision of the nonviable tissue [96]. Due to its good dermal wound bed preparation with extensive formation of rete ridges and capillary loops, the skin barrier and elasticity are close to the normal human skin, which is surrounding the wound [72, 97, 98]. It is reported to have minimal complications and good clinical outcomes and was proved valuable in restoring skin elasticity and skin barrier [72]. Survival rate is reported similar to other dermal matrices [99, 100].

10.4 Partial- Versus Full-Thickness Burns: Using the Right Substitute

Given the huge number of different skin substitutes available, the selection of product to use for a certain patient is always an individual decision based on the experience and personal preference of the surgeon. The clinician has to take the advantages and disadvantages of the product into account and ultimately, in the era of cost pressure on our healthcare system, cost-effectiveness. Given the fact that procedure time makes up around 40% of operating room time in a burn OR [101], not only material costs but also applicability in a timely manner have to be considered.

All abovementioned temporary substitutes are used at our institution. In partial-thickness facial burns—especially in children—amnion is a good option, as well as for hand burns. Here, also Suprathel is a very good alternative. If infection is present, biologic products or products containing silver may be preferred in combination with frequent dressing changes and/or debridements. For full-thickness burns, our standard of care is either allograft or xenograft until enough donor sites are available, even though the above mentioned are used if needed. In the end it may vary between institutions and every clinician has his or her preferred products.

References

1. Pereira C, Murphy K, Herndon D. Outcome measures in burn care: is mortality dead? Burns. 2004;30(8):761–71.
2. Jeschke MG, Pinto R, Kraft R, Nathens AB, Finnerty CC, Gamelli RL, Gibran NS, Klein MB, Arnoldo BD, Tompkins RG, Herndon DN, Inflammation and the Host Response to Injury Collaborative Research Program. Morbidity and survival probability in burn patients in modern burn care. Crit Care Med. 2015;43(4):808.
3. Jeschke MG, Shahrokhi S, Finnerty CC, Branski LK, Dibildox M. Wound coverage technologies in burn care: established techniques. J Burn Care Res 2018; 39:313–8.
4. Janžekovic Z. A new concept in the early excision and immediate grafting of burns. J Trauma Acute Care Surg. 1970;10(12):1103–8.
5. Merrell SW, Saffle JR, Larson CM, Sullivan JJ. The declining incidence of fatal sepsis following thermal injury. J Trauma Acute Care Surg. 1989;29(10):1362–6.
6. Herndon DN, Barrow RE, Rutan RL, Rutan TC, Desai MH, Abston S. A comparison of conservative versus early excision. Therapies in severely burned patients. Ann Surg. 1989;209(5):547.
7. Dobke MK, Simoni J, Ninnemann JL, Garrett J, Harnar TJ. Endotoxemia after burn injury: effect of early excision on circulating endotoxin levels. J Burn Care Res. 1989;10(2):107–11.
8. Barret J, Herndon D. Modulation of inflammatory and catabolic responses in severely burned children by early burn wound excision in the first 24 hours. Arch Surg. 2003;138(2):127–32.
9. Rosenberg L, Shoham Y, Krieger Y, Rubin G, Sander F, Koller J, David K, Egosi D, Ahuja R, Singer AJ. Minimally invasive burn care: a review of seven clinical studies of rapid and selective debridement using a bromelain-based debriding enzyme (NexoBrid®). Ann Burns Fire Disasters. 2015;28(4):264–74.
10. Ziegler B, Hirche C, Horter J, Kiefer J, Grützner PA, Kremer T, Kneser U, Münzberg M. In view of standardization part 2: management of challenges in the initial treatment of burn patients in Burn Centers in Germany, Austria and Switzerland. Burns. 2017;43(2):318–25.
11. Heyneman A, Hoeksema H, Vandekerckhove D, Pirayesh A, Monstrey S. The role of silver sulphadiazine in the conservative treatment of partial thickness burn wounds: a systematic review. Burns. 2016;42(7):1377–86.
12. Sheridan RL, Tompkins RG. Skin substitutes in burns. Burns. 1999;25(2):97–103.
13. Austin RE, Merchant N, Shahrokhi S, Jeschke MG. A comparison of Biobrane™ and cadaveric allograft for temporizing the acute burn wound: cost and procedural time. Burns. 2015;41(4):749–53.
14. Robb EC, Bechmann N, Plessinger RT, Boyce ST, Warden GD, Kagan RJ. Storage media and temperature maintain normal anatomy of cadaveric human skin for transplantation to full-thickness skin wounds. J Burn Care Res. 2001;22(6):393–6.
15. Ben-Bassat H, Chaouat M, Segal N, Zumai E, Wexler M, Eldad A. How long can cryopreserved skin be stored to maintain adequate graft performance? Burns. 2001;27(5):425–31.
16. Mackie DP. The Euro Skin Bank: development and application of glycerol-preserved allografts. J Burn Care Res. 1997;18(1):s7–9.
17. Kagan RJ, Robb EC, Plessinger RT. Human skin banking. Clin Lab Med. 2005;25(3):587–605.
18. Khoo T, Halim A, Saad AM, Dorai A. The application of glycerol-preserved skin allograft in the treatment of burn injuries: an analysis based on indications. Burns. 2010;36(6):897–904.

19. American Association of Tissue Banks. AATB Standards for tissue banking. 14th ed. 2016 archive. constantcontact.com/fs146/1102056357439/archive/1125303416087.html. Accessed 17 Mar 2018.

20. Herndon DN, Rose JK. Cadaver skin allograft and the transmission of human cytomegalovirus in burn patients: benefits clearly outweigh risks. J Am Coll Surg. 1996;182(3):263–4.

21. Davis JS. Skin transplantation. Johns Hopkins Hosp Rep. 1910;15:307–96.

22. Sabella N. Use of the fetal membranes in skin grafting. Med Rec NY. 1913;83:478–80.

23. Kesting MR, Wolff K-D, Hohlweg-Majert B, Steinstraesser L. The role of allogenic amniotic membrane in burn treatment. J Burn Care Res. 2008;29(6):907–16.

24. Mostaque AK, Rahman KBA. Comparisons of the effects of biological membrane (amnion) and silver sulfadiazine in the management of burn wounds in children. J Burn Care Res. 2011;32(2):200–9.

25. Gajiwala K, Gajiwala AL. Evaluation of lyophilised, gamma-irradiated amnion as a biological dressing. Cell Tissue Bank. 2004;5(2):73–80.

26. Herndon DN, Branski LK. Contemporary methods allowing for safe and convenient use of amniotic membrane as a biologic wound dressing for burns. Ann Plast Surg. 2017;78(2):S9–S10.

27. Maral T, Borman H, Arslan H, Demirhan B, Akinbingol G, Haberal M. Effectiveness of human amnion preserved long-term in glycerol as a temporary biological dressing. Burns. 1999;25(7):625–35.

28. Koob TJ, Rennert R, Zabek N, Massee M, Lim JJ, Temenoff JS, Li WW, Gurtner G. Biological properties of dehydrated human amnion/chorion composite graft: implications for chronic wound healing. Int Wound J. 2013;10(5):493–500.

29. Koob TJ, Lim JJ, Massee M, Zabek N, Denoziere G. Properties of dehydrated human amnion/chorion composite grafts: implications for wound repair and soft tissue regeneration. J Biomed Mater Res B Appl Biomater. 2014;102(6):1353–62.

30. Koob TJ, Lim JJ, Massee M, Zabek N, Rennert R, Gurtner G, Li WW. Angiogenic properties of dehydrated human amnion/chorion allografts: therapeutic potential for soft tissue repair and regeneration. Vasc Cell. 2014;6(1):10.

31. Koob TJ, Lim JJ, Zabek N, Massee M. Cytokines in single layer amnion allografts compared to multilayer amnion/chorion allografts for wound healing. J Bioml Mater Res B Appl Biomater. 2015;103(5):1133–40.

32. Tenenhaus M. The use of dehydrated human amnion/chorion membranes in the treatment of burns and complex wounds: current and future applications. Ann Plast Surg. 2017;78(2):S11–S3.

33. Reilly DA, Hickey S, Glat P, Lineaweaver WC, Goverman J. Clinical experience: using dehydrated human amnion/chorion membrane allografts for acute and reconstructive burn care. Ann Plast Surg. 2017;78(2):S19–26.

34. Niknejad H, Peirovi H, Jorjani M, Ahmadiani A, Ghanavi J, Seifalian AM. Properties of the amniotic membrane for potential use in tissue engineering. Eur Cells Mater. 2008;15:88–99.

35. Branski LK, Herndon DN, Celis MM, Norbury WB, Masters OE, Jeschke MG. Amnion in the treatment of pediatric partial-thickness facial burns. Burns. 2008;34(3):393–9.

36. Bromberg BE, Song IC, Mohn MP. The use of pig skin as a temporary biological dressing. Plast Reconstr Surg. 1965;36(1):80–90.

37. Ersek RA, Hachen H. Porcine xenografts in the treatment of pressure ulcers. Ann Plast Surg. 1980;5(6):464–70.

38. Burd A, Lam P, Lau H. Allogenic skin: transplant or dressing? Burns. 2002;28(4):358–66.

39. Song IC, Bromberg BE, Mohn MP, Koehnlein E. Heterografts as biological dressings for large skin wounds. Surgery. 1966;59(4):576–83.

40. Artz CP, Rittenbury MS, Yarbrough D 3rd. An appraisal of allografts and xenografts as biological dressings for wounds and burns. Ann Surg. 1972;175(6):934–8.

41. Chiu T, Burd A. "Xenograft" dressing in the treatment of burns. Clin Dermatol. 2005;23(4):419–23.

42. Weiss RA. Xenografts and retroviruses. Science. 1999;285(5431):1221–2.

43. Hermans MH. Porcine xenografts vs. (cryopreserved) allografts in the management of partial thickness burns: is there a clinical difference? Burns. 2014;40(3):408–15.

44. Greenwood JE, Clausen J, Kavanagh S. Experience with biobrane: uses and caveats for success. Eplasty. 2009;9:e25.

45. Lal S, Barrow RE, Wolf SE, Chinkes DL, Hart DW, Heggers JP, Herndon DN. Biobrane® improves wound healing in burned children without increased risk of infection. Shock. 2000;14(3):314–8.

46. Gonce S, Miskell P, Waymack J. A comparison of Biobrane vs. homograft for coverage of contaminated burn wounds. Burns. 1988;14(5):409–12.

47. Busche M, Herold C, Schedler A, Knobloch K, Vogt P, Rennekampff H. The Biobrane glove in burn wounds of the hand. Evaluation of the functional and aesthetic outcome and comparison of costs with those of conventional wound management. Handchir Mikrochir Plast Chir. 2009;41(6):348–54.

48. Woodroof EA, Phipps RR. Skin substitute and wound dressing with added anti-scar compound. https://patents.google.com/patent/US9439808B2/en. Accessed 17 Mar 2018.

49. Noordenbos J, Doré C, Hansbrough JF. Safety and efficacy of TransCyte for the treatment of partial-thickness burns. J Burn Care Res. 1999;20(4):275–81.

50. Kumar RJ, Kimble RM, Boots R, Pegg SP. Treatment of partial-thickness burns: a prospective, randomized trial using Transcyte™. ANZ J Surg. 2004;74(8):622–6.

51. Hansbrough J. Dermagraft-TC for partial-thickness burns: a clinical evaluation. J Burn Care Res. 1997;18(1):s25–s8.

52. Hansbrough JF, Mozingo DW, Kealey GP, Davis M, Gidner A, Gentzkow GD. Clinical trials of a biosynthetic temporary skin replacement, Dermagraft-Transitional Covering, compared with cryopreserved human cadaver skin for temporary coverage of excised burn wounds. J Burn Care Res. 1997;18(1):43–51.

53. Purdue GF, Hunt JL, Still JM Jr, , Law EJ, Herndon DN, Goldfarb IW, Schiller WR, Hansbrough JF, Hickerson WL, Himel HN, Kealey GP, Twomey J, Missavage AE, Solem LD, Davis M, Totoritis M, Gentzkow GD. A multicenter clinical trial of a biosynthetic skin replacement, Dermagraft-TC, compared with cryopreserved human cadaver skin for temporary coverage of excised burn wounds. J Burn Care Res 1997;18(1):52–57.

54. Demling RH, DeSanti L. Management of partial thickness facial burns (comparison of topical antibiotics and bio-engineered skin substitutes). Burns. 1999;25(3):256–61.

55. Schwarze H, Küntscher M, Uhlig C, Hierlemann H, Prantl L, Ottomann C, Hartmann B. Suprathel, a new skin substitute, in the management of partial-thickness burn wounds: results of a clinical study. Ann Plast Surg. 2008;60(2):181–5.

56. Hundeshagen G, Collins VN, Wurzer P, Sherman W, Voigt CD, Cambiaso-Daniel J, Nunez Lopez O, Sheaffer J, Herndon DN, Finnerty CC, Branski LK. A prospective, randomized, controlled trial comparing the outpatient treatment of pediatric and adult partial-thickness burns with Suprathel or Mepilex ag. J Burn Care Res. 2018;39(2):261–7.

57. Keck M, Selig H, Lumenta D, Kamolz L, Mittlböck M, Frey M. The use of Suprathel® in deep dermal burns: first results of a prospective study. Burns. 2012;38(3):388–95.

58. Highton L, Wallace C, Shah M. Use of Suprathel® for partial thickness burns in children. Burns. 2013;39(1):136–41.

59. Barrett S. Mepilex® Ag: an antimicrobial, absorbent foam dressing with Safetac® technology. Br J Nurs. 2009;18(20):S28. S30–6

60. Kee EG, Kimble R, Cuttle L, Khan A, Stockton K. Randomized controlled trial of three burns dressings for partial thickness burns in children. Burns. 2015;41(5):946–55.

61. Vloemans AF, Soesman AM, Kreis RW, Middelkoop E. A newly developed hydrofibre dressing, in the treatment of partial-thickness burns. Burns. 2001;27(2):167–73.

62. Kuo FC, Chen B, Lee MS, Yen SH, Wang JW. AQUACEL® Ag surgical dressing reduces surgical site infection and improves patient satisfaction in minimally invasive total knee arthroplasty: a prospective, randomized, controlled study. Biomed Res Int. 2017;2017:1262108.

63. Bowler PG, Jones SA, Walker M, Parsons D. Microbicidal properties of a silver-containing Hydrofiber® dressing against a variety of burn wound pathogens. J Burn Care. 2004;12(3):288–94.

64. Caruso DM, Foster KN, Blome-Eberwein SA, Twomey JA, Herndon DN, Luterman A, Silverstein P, Antimarino JR, Bauer GJ. Randomized clinical study of Hydrofiber dressing with silver or silver sulfadiazine in the management of partial-thickness burns. J Burn Care Res. 2006;27(3):298–309.

65. Coutts P, Sibbald RG. The effect of a silver-containing Hydrofiber dressing on superficial wound bed and bacterial balance of chronic wounds. Int Wound J. 2005;2(4):348–56.

66. Verbelen J, JHoeksema H, Heyneman A, Pirayesh A, Monstrey S. Aquacel® Ag dressing versus Acticoat™ dressing in partial thickness burns: a prospective, randomized, controlled study in 100 patients. Part 1: burn wound healing. Burns. 2014;40(3):416–27.

67. Snyder RJ, Doyle H, Delbridge T. Applying split-thickness skin grafts: a step-by-step clinical guide and nursing implications. Ostomy Wound Manage. 2001;47(11):20–6.

68. Høgsberg T, Bjarnsholt T, Thomsen JS, Kirketerp-Møller K. Success rate of split-thickness skin grafting of chronic venous leg ulcers depends on the presence of Pseudomonas aeruginosa: a retrospective study. PLoS One. 2011;6(5):e20492.

69. Boaheme K, Richmon J, Byrne P, Ishii L. Hinged forearm split-thickness skin graft for radial artery fasciocutaneous flap donor site repair. Arch Facial Plast Surg. 2011;13(6):392–4.

70. Roukis TS, Zgonis T. Skin grafting techniques for soft-tissue coverage of diabetic foot and ankle wounds. J Wound Care. 2005;14:173–6.

71. Yenidünya MO, Özdengill E, Emsen I. Split-thickness skin graft fixation with surgical drape. Plast Reconstr Surg. 2000;106(6):1429–30.

72. Min JH, Yun IS, Lew DH, Roh TS, Lee WJ. The use of matriderm and autologous skin graft in the treatment of full thickness skin defects. Arch Plast Surg. 2014;41(4):330–6.

73. Haslik W, Kamolz LP, Nathschläger G, Andel H, Meissl G, Frey M. First experiences with the collagen-elastin matrix Matriderm as a dermal substitute in severe burn injuries of the hand. Burns. 2007;33(3):364–8.

74. Ryssel H, Gazyakan E, Germann G, Ohlbauer M. The use of MatriDerm in early excision and simultaneous autologous skin grafting in burns: a pilot study. Burns. 2008;34(1):93–7.

75. Otene CI, Olaitan PB, Ogbonnaya IS, Nnabuko RE. Donor site morbidity following harvest of split-thickness skin grafts in South Eastern Nigeria. J West Afr Coll Surg. 2011;1(2):86–96.

76. Weingart D, Stoll P. The epithelialization of split skin graft donor sites—a test model for the efficacy of topical wound therapeutic agents. Eur J Plast Surg. 1993;16(1):22–5.

77. Heimbach D, Luterman A, Burke J, Cram A, Herndon D, Hunt J, Jordan M, McManus W, Solem L, Warden G, et al. Artificial dermis for major burns. A multicenter randomized clinical trial. Ann Surg. 1988;208(3):313–20.

78. Henderson J, Arya R, Gillespie P. Skin graft meshing, over-meshing and cross-meshing. Int J Surg. 2012;10(9):547–50.

79. Tanner JC, Vandeput J, Olley JF. The mesh skin graft. Plast Reconstr Surg. 1964;34:287–92.
80. Pripotnev S, Papp A. Split thickness skin graft meshing ratio indications and common practices. Burns. 2017;43(8):1775–81.
81. Simman R, Phavixay L. Split-thickness skin grafts remain the gold standard for the closure of large acute and chronic wounds. J Am Col Certif Wound Spec. 2011;3(3):55–9.
82. Meek CP. Successful microdermagrafting using the Meek-Wall microdermatome. Am J Surg. 1958;96(4):557–8.
83. Kreis RW, Mackie DP, Hermans RP, Vloemans AR. Expansion technique for skin grafts: comparison between mesh and Meek island (sandwiched-) grafts. Burns. 1994;20:39–42.
84. Nystrom G. Sowing of small skin graft particles as a method for epithelization especially of extensive wound surfaces. Plast Reconstr Surg Transplant Bull. 1959;23(3):226–39.
85. Hsieh CS, Schuong JY, Huang WS, Huang TT. Five years' experience of the modified Meek technique in the management of extensive burns. Burns. 2008;34(4):350–4.
86. Jeschke MG, Shahrokhi S, Finnerty CC, Branski LK, Dibildox M. Wound coverage technologies in burn care: established techniques. J Burn Care Res. 2014;10:e3182920d29.
87. Verkey M, Ding J, Tredget EE. Advances in skin substitutes—potential of tissue engineered skin for facilitating anti-fibrotic healing. J Funct Biomater. 2015;6(3):547–63.
88. Metcalfe AD, Ferguson MW. Tissue engineering of replacement skin: the crossroads of biomaterials, wound healing, embryonic development, stem cells and regeneration. J R Soc Interface. 2007;4(14):413–37.
89. Sheridan RL, Choucair RJ. Acellular allogenic dermis does not hinder initial engraftment in burn wound resurfacing and reconstruction. J Burn Care Rehabil. 1997;18:496–9.
90. Sobti N, Liao EC. Surgeon-controlled study and meta-analysis comparing flexHD and alloderm in immediate breast reconstruction outcomes. Plast Reconstr Surg. 2016;138(5):959–67.
91. Huntington CR, Cox TC, Blair LJ, Schell S, Randolph D, Prasad T, Lincourt A, Heniford BT, Augenstein VA. Biologic mesh in ventral hernia repair: outcomes, recurrence, and charge analysis. Surgery. 2016;160(6):1517–27.
92. Hansen SL, Voigt DW, Wiebelhaus P, Paul CN. Using skin replacement products to treat burns and wounds. Adv Skin Wound Care. 2001;14(1):37–44.
93. Peck MD, Kessler M, Meyer AA, Bonham Morris PA. A trial of the effectiveness of artificial dermis in the treatment of patients with burns greater than 45% total body surface area. J Trauma. 2002;52(5):971–8.
94. Heimbach DM, Warden GD, Luterman A, Jordan MH, Ozobia N, Ryan CM, Voigt DW, Hickerson WL, Saffle JR, DeClement FA, Sheridan RL, Dimick AR. Multicenter postapproval clinical trial of Integra dermal regeneration template for burn treatment. J Burn Care Rehabil. 2003;24(1):42–8.
95. Branski LK, Herndon DN, Pereira C, Mlcak RP, Celis MM, Lee JO, Sanford AP, Norbury WB, Zhang XJ, Jeschke MG. Longitudinal assessment of Integra in primary burn management: a randomized pediatric clinical trial. Crit Care Med. 2007;35(11):2615–23.
96. De Vries HJ, Mekkes JR, Middelkoop E, Hinrichs WL, Wildevuur CR, Westerhof W. Dermal substitutes for full-thickness wounds in a one-stage grafting model. Wound Repair Regen. 1993;1(4):244–52.
97. Jeon H, Kim J, Yeo H, Jeong H, Son D, Han K. Treatment of diabetic foot ulcer using matriderm in comparison with a skin graft. Arch Plast Surg. 2013;40(4):403–8.
98. Hamuy R, Kinoshita N, Yoshimoto H, Hayashida K, Houbara S, Nakashima M, Suzuki K, Mitsutake N, Mussazhanova Z, Kashiyama K, Hirano A, Akita S. One-stage, simultaneous skin grafting with artificial dermis and basic fibroblast growth factor successfully improves elasticity with maturation of scar formation. Wound Repair Regen. 2013;21(1):141–54.
99. Cervelli V, Brinci L, Spallone D, Tati E, Palla L, Lucarini L, De Angelis B. The use of MatriDerm® and skin grafting in post-traumatic wounds. Int Wound J. 2011;8(4):400–5.
100. Philandrianos C, Andrac-Meyer L, Mordon S, Feuerstein JM, Sabatier F, Veran J, Magalon G, Casanova D. Comparison of five dermal substitutes in full-thickness skin wound healing in a porcine model. Burns. 2012;38(6):820–9.
101. Madni TD, Imran JB, Clark A, Arnoldo BA, Phelan HA III, Wolf SE. Analysis of operating room efficiency in a burn center. J Burn Care Res. 2018;39:89–93.

Wnt Signaling During Cutaneous Wound Healing

11

Khosrow Siamak Houschyar, Dominik Duscher,
Susanne Rein, Zeshaan N. Maan,
Malcolm P. Chelliah, Jung Y. Cha,
Kristian Weissenberg, and Frank Siemers

11.1 Introduction

The skin is an intricate structure composed of the epidermis, dermis, and a dermal adipocyte layer [1]. It is the largest organ in the human body acting as a barrier against external microorganisms and dehydration [2]. The skin further contributes to homeostasis by participating in thermal regulation by sensing and responding to disturbances.

Regeneration is a process of restoration, renewal, and growth crucial to the ability of cells and organs to be resilient to damage [3]. It is important to distinguish between repair, healing via formation of scar tissue, and regeneration, which is restoration to the pre-injury state [4]. Full-thickness skin loss in adult mammals typically results in a reparative rather than regenerative response, leading to the formation of scar tissue [3]. Deposition of a collagen-rich matrix in the neo-dermis makes it prone to contracture, decreased elasticity, and tensile strength, and promotes hypertrophic scar formation [5]. Epithelialization without epidermal appendage development over a large surface area leads to alopecia and thermal imbalance [5]. This repair depends on the differentiation and proliferation of involved cells, including epidermal stem cells (ESCs), keratinocytes, and fibroblasts, together with the assistance of various biological signals [6].

Healing of skin wounds parallels embryonic skin development in many ways. Both processes involve the differentiation, migration, proliferation,

K. S. Houschyar (✉)
Department of Plastic Surgery, BG University,
Hospital Bergmannsheil, Ruhr University Bochum,
Bochum, Germany

Burn Unit, Department for Plastic and Hand Surgery,
Trauma Center Bergmannstrost Halle,
Halle (Saale), Germany
e-mail: Khosrow-Houschyar@gmx.de

D. Duscher
Department for Plastic Surgery and Hand Surgery,
Division of Experimental Plastic Surgery,
Technical University of Munich, Munich, Germany

S. Rein
Department of Plastic and Hand Surgery,
Burn Center-Clinic St. Georg, Leipzig,
Germany

Z. N. Maan · M. P. Chelliah
Division of Plastic and Reconstructive Surgery,
Department of Surgery, Stanford School of Medicine,
Stanford, CA, USA
e-mail: zmaan@stanford.edu; mchelliah@stanford.edu

J. Y. Cha
Orthodontic Department, College of Dentistry, Yonsei
University, Seoul, Republic of Korea

K. Weissenberg
Burn Unit, Department for Plastic and Hand Surgery,
Trauma Center Bergmannstrost Halle,
Halle (Saale), Germany

F. Siemers
Department of Plastic Surgery and Hand Surgery,
Gemeinschaftskrankenhaus Havelhoehe, Teaching
Hospital of the Charité Berlin, Berlin, Germany

© Springer Nature Switzerland AG 2019
D. Duscher, M. A. Shiffman (eds.), *Regenerative Medicine and Plastic Surgery*,
https://doi.org/10.1007/978-3-030-19962-3_11

and apoptosis of various cell types to create the multilayered tissue that constitutes the skin [3]. While skin wounds in early mammalian embryos regenerate without scar tissue formation and complete restitution of the normal skin architecture [7], this is not the case with adult wounds [8]. However, many of the same key signaling pathways that are activated during embryonic skin development are also activated during postnatal wound healing, e.g., Wnt/β-catenin, Notch, and Hedgehog pathways [9], creating interest in better understanding the role of these pathways.

Maintenance of epidermal homeostasis is achieved by separate populations of stem cells in the skin: stem cells that come from the bulb region of the hair follicles, interfollicular epidermis, as well as sebaceous gland [10]. While both epidermal and bulb stem cells have demonstrated the potential to regenerate epidermis, an effective cell-based approach utilizing these populations to promote "scarless" wound healing remains elusive [3]. Interestingly, recent data demonstrate that the epidermis of wounded adult mice can regenerate hair follicles under the influence of Wnt-responsive interfollicular stem cells [11]. Here, we present a summation of data, which provide strong evidence for an alternative approach for enhancing cutaneous regeneration after injury: augmenting the endogenous Wnt pathway to activate tissue-resident stem cells.

organogenesis, and stem cell renewal [13]. As the signaling pathways that play crucial role during embryogenesis are tightly regulated, the expressions of the Wnt proteins and Wnt antagonists are exquisitely restricted both temporally and spatially during development [14].

Intracellular Wnt signaling diversifies into three main branches: (1) the β-catenin pathway (canonical Wnt pathway), which activates target genes in the nucleus; (2) the planar cell polarity (PCP) pathway, which involves jun N-terminal kinase (JNK) and cytoskeletal rearrangements; and (3) the Wnt/Ca^{2+} pathway [3]. In humans, there are currently 19 different known Wnt proteins and ten different frizzled (Fzd) receptors [15]. Frizzled genes encode integral membrane proteins that function in multiple signal transduction pathways. They have been identified in diverse animals, from sponges to humans. The family is defined by conserved structural features, including seven hydrophobic domains and a cysteine-rich ligand-binding domain. Frizzled proteins are receptors for secreted Wnt proteins, as well as other ligands, and also play a critical role in the regulation of cell polarity. Frizzled genes are essential for embryonic development, tissue and cell polarity, formation of neural synapses, and regulation of proliferation, and many other processes in developing and adult organisms. Here we focus on canonical/ß-catenin-dependent Wnt signaling, which has been implicated in tissue regeneration and repair.

11.2 Three Wnt Signaling Pathways

The Wnt signaling pathway is an evolutionarily conserved pathway that regulates crucial aspects of cell fate determination, cell polarity, cell migration, neural patterning, and organogenesis during embryonic development [12]. The name Wnt is resultant from a fusion of the name of the Drosophila segment polarity gene wingless and the name of the vertebrate homolog, integrated or int-1 [13]. Wnt proteins regulate a dizzying array of cellular processes including cell fate determination, motility, polarity, primary axis formation,

11.3 Canonical Wnt Signaling

The hallmark of canonical Wnt signaling is the accumulation and translocation of the adherens junction-associated protein, β-catenin, into the nucleus [16]. β-Catenin has been shown to perform two apparently unrelated functions: cell-cell adhesion in addition to a signaling role as a component of the Wnt/wg pathway. Wnt/wg signaling results in β-catenin accumulation and transcriptional activation of specific target genes during development. Dysregulation of β-catenin signaling plays a role in the genesis of a number of malignancies, suggesting an important role in the control of cellular proliferation or cell death. Without Wnt signaling, cytoplasmic β-catenin is degraded by a β-catenin destruction complex [17]. Phosphorylation of β-catenin within this complex by casein kinase and GSK3 targets it for ubiquitination and subsequent proteolytic destruction by the proteosomal complex [18]. Binding of Wnt to its receptor complex composed of the Fz (frizzled) and the LRP5/6 triggers a series of events that disrupt the APC/Axin/GSK3 complex that is required for the targeted destruction of β-catenin [19, 20], allowing consequent stabilization and accumulation in the cytoplasm [3]. Stabilized β-catenin translocates into the nucleus, exerting its effect on gene transcription by functioning as a transcriptional coactivator [13]. A large number of binding partners for β-catenin in the nucleus have been uncovered and perhaps the best characterized are the members of the LEF/TCF DNA-binding transcription factors (Fig. 11.1) [21].

11.4 Wnt Signaling in Tissue Regeneration and Repair

The importance of Wnt in tissue regeneration has been highlighted by studies demonstrating impaired regenerative capacity in animals when Wnt signaling is reduced [22, 23]. Additionally, the Wnt pathway regulates cell proliferation in the adult epidermis, which directly impacts the rate and extent of skin wound healing [24]. Wnt proteins also serve as niche signals for at least two types of skin stem cells that contribute to skin wound healing: those in the bulge region of the hair follicle, and those in the basal layer of the interfollicular epidermis [25].

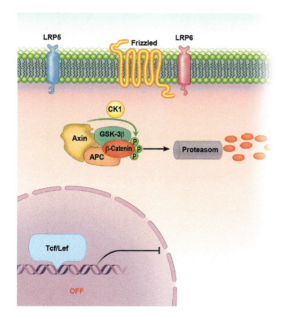
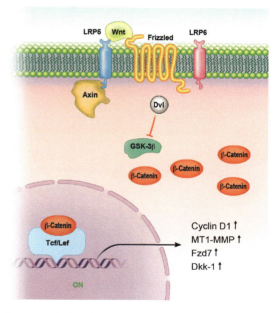

Fig. 11.1 Canonical Wnt signaling pathway. In the absence of signal, action of the destruction complex (CKIα, GSK-3β, APC, Axin) creates a hyperphosphorylated β-catenin, which is a target for ubiquitination and degradation by the proteosome. Binding of Wnt ligand to a frizzled/LRP-5/6 receptor complex leads to stabilization of hypophosphorylated β-catenin, which interacts with TCF/LEF proteins in the nucleus to activate transcription. In a canonical pathway, CKIα, GSK-3β, APC, and Axin act as negative regulators and all other components act positively. *APC* adenomatous polyposis coli, *CK* casein kinase, *GSK* glycogen synthase kinase, *Fzd* frizzled receptor, *LRP* low-density lipoprotein receptor-related protein, *Tcf/Lef* T-cell-specific transcription factor/lymphoid enhancer-binding factor

Recent studies have shown that fibroblast growth factor (FGF)-9 modulates hair follicle regeneration after skin injury in adult mice and that FGF-9 triggers Wnt activation in wound fibroblasts [26]. Through a unique feedback mechanism, activated fibroblasts then express FGF-9, thus amplifying Wnt activity throughout the wound dermis during a crucial phase of skin regeneration (Fig. 11.2). Skin wounds express various Wnt proteins during the early phases of healing, with transcripts of Wnts 1, 3, 4, 5a, and 10b being present in murine full-thickness cutaneous wounds up to 7 days after injury [9]. In the epithelium, Wnt 10b protein can be detected in migrating epithelial cells up to 3 days after wounding, while Wnt 4, 5a, and 10b localize to hair follicles [9]. Wnt 2a and 4 are expressed in the dermis, although reports vary with respect to the time course of their expression (range: 30 h to 7 days after wounding) [3]. It appears that Wnt signaling, through its ability to activate stem cells with induction of their self-renewal and proliferation, serves as a positive stimulus for wound repair [27]. Collectively, these data demonstrate that endogenous Wnt signaling is a prerequisite for tissue repair, but there are obvious caveats [22]. Most experimental methods used to study Wnt signaling in tissue healing rely on techniques that, in general, produce unrestrained Wnt pathway activation [22].

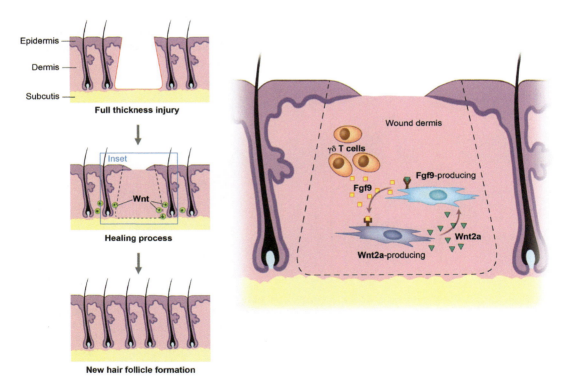

Fig. 11.2 Wnt signaling maintains the hair-inducing activity in skin repair. Fibroblast growth factor (Fgf) 9 is a secreted signaling molecule that is expressed in epithelium. Mesenchymal Fgf signaling interacts with β-catenin-mediated Wnt signaling in a feed-forward loop that functions to sustain mesenchymal Fgf responsiveness and mesenchymal Wnt/β-catenin signaling. Wnt2a is a canonical Wnt ligand that activates mesenchymal Wnt/β-catenin signaling, whereas Fgf9 is the only known ligand that signals to mesenchymal Fgf receptors (FGFRs). Mesothelial Fgf9 and mesenchymal Wnt2a are principally responsible for maintaining mesenchymal Fgf-Wnt/β-catenin signaling, whereas epithelial Fgf9 primarily affects epithelial branching. In summary, Fgf signaling is primarily responsible for regulating mesenchymal proliferation, whereas β-catenin signaling is a required permissive factor for mesenchymal Fgf signaling. *Fgf* fibroblast growth factor

11.5 Wnt Signaling and Stages of Wound Healing

11.5.1 Hemostasis and Inflammation

Wound healing is classically described as a process involving three overlapping phases. The first stage of physiological or acute wound healing is dedicated to hemostasis and formation of a provisional wound matrix, which occurs immediately after injury and is completed after some hours [28]. This matrix is comprised of activated platelets and fibrin molecules, fibronectin, vitronectin, and thrombospondins, forming a scaffold structure for the migration of leukocytes, keratinocytes, fibroblasts, and endothelial cells, while functioning as a reservoir of growth factors [29]. Recent data suggests that Wnt signaling is essential for development of megakaryocytes and for stimulating proplatelet function in vitro [30]. Interestingly, canonical Wnt has been shown to inhibit platelet aggregation whereas noncanonical Wnt-5A stimulates platelet aggregation [3].

Hemostasis triggers the inflammatory phase, which is characterized by the presence of erythema (rubor), warmth (calor), edema (tumor), and pain (dolor) [3]. At a cellular level, inflammation involves blood vessel dilation, increased vascular permeability, and leukocyte recruitment to the site of injury. Two leukocyte populations sequentially dominate the inflammatory events of wound healing: neutrophils and macrophages [31]. Both provide the critical function of wound debridement, whereas the latter population is critical in orchestrating the subsequent steps in wound healing. Wnt signaling has been shown to be involved in the regulation of inflammatory processes: Wnt5a is induced in human macrophages in response to mycobacteria and conserved bacterial structures, contributing to the regulation of pro-inflammatory cytokines via its receptor frizzled (Fzd) 5 [32]. Wnt5a is also induced in other infectious and inflammatory diseases such as tuberculosis, sepsis, psoriasis, rheumatoid arthritis, and atherosclerosis [33]. β-Catenin-dependent Wnt signaling enhances the inflammatory response [34].

11.5.2 Proliferation

During the proliferative phase of healing, approx. 3–10 days after injury, the body seeks to cover the wound surface through the formation of granulation tissue to restore the vascular network [35] and re-epithelialization. Under the control of regulating cytokines like IFN-γ and TGF-β, the synthesis of collagen, fibronectin, and other proteins by fibroblasts forms the basis for the new matrix of connective tissue and the restoration of mechanical strength to injured tissue [36]. Subsequently, the synthesis of collagen increases throughout the wound, while the proliferation of fibroblasts declines successively [33].

β-Catenin is an important regulator of fibroblast behavior during the proliferative phase of dermal wound repair [4]. β-Catenin protein levels and transcriptional activity are elevated in dermal fibroblasts during the proliferative phase of healing in murine cutaneous wounds and return to baseline during the remodeling phase [37]. Human wounds similarly show increased expression of β-catenin and its target genes, such as fibronectin and MMP7, during the proliferative phase [7]. While increased β-catenin activity during the proliferative phase is crucial for successful wound repair, prolonged or aberrant β-catenin activity beyond the normal parameters of healing contributes to excessive fibrosis and scar formation [38]. Indeed, human hypertrophic scars and keloids exhibit elevated β-catenin levels [39].

While Wnt ligands may participate in stimulating dermal β-catenin during wound repair, Wnt signaling is not crucial for maintaining elevated β-catenin levels during the proliferative phase of cutaneous healing [5]. This has been demonstrated in mice treated with an adenovirus expressing the Wnt signaling inhibitor Dickkopf (DKK1, which binds LRP6/Arrow), without a significant decline in β-catenin protein levels during the proliferative phase of skin wound healing, in contrast to the situation in bone repair [40]. This suggests that other factors play a role in regulating β-catenin levels during the proliferative phase of healing. Indeed, β-catenin levels in fibroblasts can be stimulated by growth factors, such as TGF-β1, which are released during the early stages of wound repair [41]. Furthermore, β-catenin activity in dermal fibroblasts is regulated by extracellular matrix (ECM) components, such as fibronectin, which activates β-catenin through a GSK3ß-dependent, β1 integrin-mediated pathway [42]. Hypertrophic scars and keloids represent a dysregulated response to cutaneous wounding, resulting in an excessive deposition of ECM, especially collagen [36]. TGF-β is believed to be responsible for excessive ECM deposition in hypertrophic scars, keloids, and other fibrotic conditions [39]. Since β-catenin is known to accumulate during fibroproliferation, it is speculated that it could play a role in the mechanisms that lead to hypertrophic/keloid scarring [43]. β-Catenin and Wnt signaling are intrinsically involved in the formation of the dermis and of epidermal structures, both during wound repair and during skin development [5]. It will be interesting to elucidate whether non-Wnt activators of β-catenin, such as ECM proteins and growth factors, modulate β-catenin during skin development as they do during the response to injury [22].

11.5.3 Remodeling

Remodeling is the final phase of wound healing and occurs from day 21 to up to 1 year after injury [44]. In the skin, remodeling consists of deposition of matrix and subsequent changes in its organization and composition over time [45]. During the maturation of the wound the components of the ECM undergo certain changes. Fibrin clot formed in the early inflammatory phase is replaced by granulation tissue that is rich in type III collagen and blood vessels during the proliferative phase and subsequently replaced by a collagenous scar predominantly of type I collagen [36]. This type of collagen is oriented in small parallel bundles and is, therefore, different from the basket-weave collagen in healthy dermis. Wnt is responsible for the differentiation of myofibroblasts causing wound contracture, decreasing the surface of the developing scar [46]. Wnt has also been shown to be critical in the process of angiogenesis and endogenous enhancement of Wnt can correct vascular defects [47]. As angiogenic processes diminish, wound blood flow declines, and acute wound metabolic activity slows, eventually stopping.

11.6 Fetal and Adult Wound Healing

In the fetus, at least through the second trimester, skin and bone wounds heal in a regenerative manner [46]. Cutaneous wound healing in the early gestation fetus is remarkably different from that in the adult [48]. The most striking features of the fetal wound response are the speed and the absence of obvious scarring [49]. Investigators have now begun applying more comprehensive transcriptomic techniques to the study of scarless wound healing. In particular, there has been a focus on the time during fetal gestation when regenerative healing changes to adult wound healing with scar formation in order to understand the phenomena occurring immediately before and after this transition [8]. In rats, wounds made on the 16th day of gestation (gestation period: 21 days) histologically regenerate, but wounds made on the 18th day of gestation are associated with scarring [50]. The major objective of skin wounding research is restoration of the extracellular matrix architecture, and a subsequent return of strength and function to the injured skin, and therefore must overcome the fibrotic nature of postnatal wound healing. It is clear from studies conducted in mammals that normal skin development absolutely depends on a tight regulation of the activities of secreted signaling molecules that display potent organizing properties in the embryo [51]. These signaling molecules include members of the Hedgehog (Hh), transforming growth factor-beta (TGF-beta), and Wnt families of secreted factors (Table 11.1).

Table 11.1 Summary of developmental signaling pathways in mammalian skin development and repair

Signaling pathway	Skin section	Skin development	Skin repair
TGF-β	Epidermis	No significant role in hair follicle development	Inhibitory role in re-epithelialization
	Dermis	Role previously unknown Expressed in developing dermis	Reconstitution of the dermis: fibroblast proliferation and behavior, myofibroblast formation, matrix production, wound contraction
Wnt	Wnt	Development and morphogenesis of hair follicles	Regeneration of hair follicles in large wounds
	Dermis	Development of the dermis	Reconstitution of the dermis: fibroblast numbers and behavior, matrix production
Sonic hedgehog	Epidermis	Development and morphogenesis of hair follicles	Present in regenerated hair follicles
	Dermis	Role previously unknown	Involved in dermal reconstitution: effects on matrix, cellularity and vascularity
Notch	Epidermis	Epidermal differentiation	Role previously unknown
	Dermis	Role previously unknown	Involved in dermal reconstitution: effects on macrophage behavior, angiogenesis

11.7   Conclusions

Converting tissue repair to tissue regeneration remains a lofty goal, but exciting new techniques and methods of investigation, including high-throughput transcriptomic analysis, make it an increasingly realistic objective. Over the past decade, considerable insights into the molecular pathways driving the animal healing response and impairment have suggested new therapeutic targets and provided scientific rationale for future clinical trials. The wound epithelium in adult mammals is capable of responding to morphogenic signals from the dermis, as it does in the embryo during hair placode formation. Adult stem cells have tremendous therapeutic potential, and the skin epithelium represents an enormous source of accessible stem cells that might be a starting point for generating cells to replace diseased tissue. Skin stem cells have already been used to replace skin lost to burns; whether it will be possible to use skin stem cell plasticity to engineer treatments for other disorders remains to be determined.

Although increasing evidence supports a role for Wnt signaling in skin epithelial stem cell maintenance and/or determination, deregulated Wnt signaling activation has long been implicated in human cancers. Wnt signaling is essential at multiple steps during the complex organogenesis of the skin and its appendages. It is required to induce the formation of the dorsal dermis and regulate the size of the different skin appendage tracts. Later, Wnt signaling is required for the very early stages of skin appendage formation. Skin appendage distribution and pigmentation are regulated, in part by Wnt signaling. Disruption of the pathway can lead to the formation of skin appendage tumors. Any strategy that attempts to target the Wnt pathway to augment tissue regeneration will have to take into consideration the need to selectively and locally activate signaling in the tissue or area of interest while simultaneously restricting Wnt signaling in other parts of the body.

The intricate and dynamic nature of the wound environment suggests that successful therapies for treating wound healing disorders will not rely upon a single all-encompassing agent, but will likely require a multitude of factors for a finely tuned attenuation of endogenous Wnt signaling during the wound-healing process. Recognition of the complexity of the wound-healing process and its diseases as well as acceptance of the seriousness and mortality associated with repair pathologies will be critical steps in these future efforts. Consequently, the combination of current knowledge in basic biology, identification of the limits of past clinical trials as well as translational research that includes development of improved animal models, harnessing of new technologies for more accurate imaging, and biomarker-based diagnostics will provide a strong basis to advance viable clinical approaches for treating patients with wound-healing pathologies. Identifying the relationships between developmental signaling pathways in adult wound repair and fetal skin development and/or regeneration will certainly propel the research community closer to this goal, and is a fruitful area of future investigation.

References

1. Takeo M, Lee W, Ito M. Wound healing and skin regeneration. Cold Spring Harb Perspect Med. 2015;5(1):a023267.
2. Dong L, Hao H, Liu J, Ti D, Tong C, Hou Q, Li M, Zheng J, Liu G, Fu X, Han W. A conditioned medium of umbilical cord mesenchymal stem cells overexpressing wnt7a promotes wound repair and regeneration of hair follicles in mice. Stem Cells Int. 2017;2017:3738071.
3. Houschyar KS, Momeni A, Pyles MN, Maan ZN, Whittam AJ, Siemers F. Wnt signaling induces epithelial differentiation during cutaneous wound healing. Organogenesis. 2015;11(3):95–104.
4. Darby IA, Laverdet B, Bonte F, Desmouliere A. Fibroblasts and myofibroblasts in wound healing. Clin Cosmet Investig Dermatol. 2014;7:301–11.
5. Fathke C, Wilson L, Shah K, Kim B, Hocking A, Moon R, Isik F. Wnt signaling induces epithelial differentiation during cutaneous wound healing. BMC Cell Biol. 2006;7:4.
6. Shi Y, Shu B, Yang R, Xu Y, Xing B, Liu J, Chen L, Qi S, Liu X, Wang P, Tang J, Xie J. Wnt and Notch signaling pathway involved in wound healing by targeting c-Myc and Hes1 separately. Stem Cell Res Ther. 2015;6:120.
7. Beare AH, Metcalfe AD, Ferguson MW. Location of injury influences the mechanisms of both regeneration and repair within the MRL/MpJ mouse. J Anat. 2006;209(4):547–59.
8. Hu MS, Rennert RC, McArdle A, Chung MT, Walmsley GG, Longaker MT, Lorenz HP. The role of stem cells during scarless skin wound healing. Adv Wound Care (New Rochelle). 2014;3(4):304–14.
9. Bielefeld KA, Amini-Nik S, Alman BA. Cutaneous wound healing: recruiting developmental pathways for regeneration. Cell Mol Life Sci. 2013;70(12):2059–81.

10. Blanpain C, Fuchs E. Epidermal homeostasis: a balancing act of stem cells in the skin. Nat Rev Mol Cell Biol. 2009;10(3):207–17.
11. Chueh SC, Lin SJ, Chen CC, Lei M, Wang LM, Widelitz R, Hughes MW, Jiang TX, Chuong CM. Therapeutic strategy for hair regeneration: hair cycle activation, niche environment modulation, wound-induced follicle neogenesis, and stem cell engineering. Expert Opin Biol Ther. 2013;13(3):377–91.
12. Gao B, Song H, Bishop K, Elliot G, Garrett L, English MA, Andre P, Robinson J, Sood R, Minami Y, Economides AN, Yang Y. Wnt signaling gradients establish planar cell polarity by inducing Vangl2 phosphorylation through Ror2. Dev Cell. 2011;20(2):163–76.
13. Komiya Y, Habas R. Wnt signal transduction pathways. Organogenesis. 2008;4(2):68–75.
14. Hiyama A, Yokoyama K, Nukaga T, Sakai D, Mochida J. A complex interaction between Wnt signaling and TNF-alpha in nucleus pulposus cells. Arthritis Res Ther. 2013;15(6):R189.
15. Willert K, Nusse R. Wnt proteins. Cold Spring Harb Perspect Biol. 2012;4(9):a007864.
16. Yang K, Wang X, Zhang H, Wang Z, Nan G, Li Y, Zhang F, Mohammed MK, Haydon RC, Luu HH, Bi Y, He TC. The evolving roles of canonical WNT signaling in stem cells and tumorigenesis: implications in targeted cancer therapies. Lab Investig. 2016;96(2):116–36.
17. Stamos JL, Weis WI. The beta-catenin destruction complex. Cold Spring Harb Perspect Biol. 2013;5(1):a007898.
18. Gao C, Xiao G, Hu J. Regulation of Wnt/beta-catenin signaling by posttranslational modifications. Cell Biosci. 2014;4(1):13.
19. Kim W, Kim M, Jho EH. Wnt/beta-catenin signalling: from plasma membrane to nucleus. Biochem J. 2013;450(1):9–21.
20. MacDonald BT, He X. Frizzled and LRP5/6 receptors for Wnt/beta-catenin signaling. Cold Spring Harb Perspect Biol. 2012;4(12):a007880.
21. Cadigan KM, Waterman ML. TCF/LEFs and Wnt signaling in the nucleus. Cold Spring Harb Perspect Biol. 2012;4(11):a007906.
22. Whyte JL, Smith AA, Helms JA. Wnt signaling and injury repair. Cold Spring Harb Perspect Biol. 2012;4(8):a008078.
23. Tanaka EM, Reddien PW. The cellular basis for animal regeneration. Dev Cell. 2011;21(1):172–85.
24. Whyte JL, Smith AA, Liu B, Manzano WR, Evans ND, Dhamdhere GR, Fang MY, Chang HY, Oro AE, Helms JA. Augmenting endogenous Wnt signaling improves skin wound healing. PLoS One. 2013;8(10):e76883.
25. Hsu YC, Li L, Fuchs E. Emerging interactions between skin stem cells and their niches. Nat Med. 2014;20(8):847–56.
26. Gay D, Kwon O, Zhang Z, Spata M, Plikus MV, Holler PD, Ito M, Yang Z, Treffeisen E, Kim CD, Nace A,

Zhang X, Baratono S, Wang F, Ornitz DM, Millar SE, Cotsarelis G. Fgf9 from dermal gammadelta T cells induces hair follicle neogenesis after wounding. Nat Med. 2013;19(7):916–23.

27. Fuchs E, Chen T. A matter of life and death: self-renewal in stem cells. EMBO Rep. 2013;14(1):39–48.

28. Reinke JM, Sorg H. Wound repair and regeneration. Eur Surg Res. 2012;49(1):35–43.

29. Olczyk P, Mencner L, Komosinska-Vassev K. The role of the extracellular matrix components in cutaneous wound healing. Biomed Res Int. 2014;2014:747584.

30. Macaulay IC, Thon JN, Tijssen MR, Steele BM, MacDonald BT, Meade G, Burns P, Rendon A, Salunkhe V, Murphy RP, Bennett C, Watkins NA, He X, Fitzgerald DJ, Italiano JE Jr, Maguire PB. Canonical Wnt signaling in megakaryocytes regulates proplatelet formation. Blood. 2013;121(1):188–96.

31. Greaves NS, Ashcroft KJ, Baguneid M, Bayat A. Current understanding of molecular and cellular mechanisms in fibroplasia and angiogenesis during acute wound healing. J Dermatol Sci. 2013;72(3):206–17.

32. Schaale K, Neumann J, Schneider D, Ehlers S, Reiling N. Wnt signaling in macrophages: augmenting and inhibiting mycobacteria-induced inflammatory responses. Eur J Cell Biol. 2011;90(6–7):553–9.

33. Papakonstantinou E, Roth M, Karakiulakis G. Hyaluronic acid: a key molecule in skin aging. Dermatoendocrinology. 2012;4(3):253–8.

34. Staal FJ, Luis TC, Tiemessen MM. WNT signalling in the immune system: WNT is spreading its wings. Nat Rev Immunol. 2008;8(8):581–93.

35. Landen NX, Li D, Stahle M. Transition from inflammation to proliferation: a critical step during wound healing. Cell Mol Life Sci. 2016;73(20):3861–85.

36. Xue M, Jackson CJ. Extracellular matrix reorganization during wound healing and its impact on abnormal scarring. Adv Wound Care (New Rochelle). 2015;4(3):119–36.

37. Poon R, Nik SA, Ahn J, Slade L, Alman BA. Beta-catenin and transforming growth factor beta have distinct roles regulating fibroblast cell motility and the induction of collagen lattice contraction. BMC Cell Biol. 2009;10:38.

38. Eming SA, Martin P, Tomic-Canic M. Wound repair and regeneration: mechanisms, signaling, and translation. Sci Transl Med. 2014;6:265):265–6.

39. Gauglitz GG, Korting HC, Pavicic T, Ruzicka T, Jeschke MG. Hypertrophic scarring and keloids: pathomechanisms and current and emerging treatment strategies. Mol Med. 2011;17(1–2):113–25.

40. Bielefeld KA, Amini-Nik S, Whetstone H, Poon R, Youn A, Wang J, Alman BA. Fibronectin and beta-catenin act in a regulatory loop in dermal fibroblasts to modulate cutaneous healing. J Biol Chem. 2011;286(31):27687–97.

41. Belacortu Y, Paricio N. Drosophila as a model of wound healing and tissue regeneration in vertebrates. Dev Dyn. 2011;240(11):2379–404.

42. Pataki CA, Couchman JR, Brabek J. Wnt Signaling cascades and the roles of Syndecan proteoglycans. J Histochem Cytochem. 2015;63(7):465–80.

43. Hahn JM, McFarland KL, Combs KA, Supp DM. Partial epithelial-mesenchymal transition in keloid scars: regulation of keloid keratinocyte gene expression by transforming growth factor-beta1. Burns Trauma. 2016;4(1):30.

44. Aller MA, Arias JI, Arraez-Aybar LA, Gilsanz C, Arias J. Wound healing reaction: a switch from gestation to senescence. World J Exp Med. 2014;4(2):16–26.

45. Pang C, Ibrahim A, Bulstrode NW, Ferretti P. An overview of the therapeutic potential of regenerative medicine in cutaneous wound healing. Int Wound J. 2017;14(3):450–9.

46. Yates CC, Hebda P, Wells A. Skin wound healing and scarring: fetal wounds and regenerative restitution. Birth Defects Res C Embryo Today. 2012;96(4):325–33.

47. Li F, Chong ZZ, Maiese K. Winding through the WNT pathway during cellular development and demise. Histol Histopathol. 2006;21(1):103–24.

48. Larson BJ, Longaker MT, Lorenz HP. Scarless fetal wound healing: a basic science review. Plast Reconstr Surg. 2010;126(4):1172–80.

49. Rowlatt U. Intrauterine wound healing in a 20 week human fetus. Virchows Arch A Pathol Anat Histol. 1979;381(3):353–61.

50. Kishi K, Okabe K, Shimizu R, Kubota Y. Fetal skin possesses the ability to regenerate completely: complete regeneration of skin. Keio J Med. 2012;61(4):101–8.

51. Veraitch O, Kobayashi T, Imaizumi Y, Akamatsu W, Sasaki T, Yamanaka S, Amagai M, Okano H, Ohyama M. Human induced pluripotent stem cell-derived ectodermal precursor cells contribute to hair follicle morphogenesis in vivo. J Invest Dermatol. 2013;133(6):1479–88.

Drug Delivery Advances for the Regeneration of Aged Skin

12

Daniela Castillo Pérez, Matthias M. Aitzetmüller, Philipp Neßbach, and Dominik Duscher

12.1 Introduction

The human skin is the main protective barrier between the body and the environment. It is composed of three different layers: the epidermis on the outside; the dermis, containing a complex network of nerves, vessels, and glands; and the hypodermis, composed of fat and connective tissue. Altogether, these layers, all equally important, constitute a physical and chemical barrier function. Additionally, the skin represents an organ, essential for biosynthetic, immunologic, homeostatic, and sensitive functions [1, 2].

Due to its external disposition in the human body, the skin is prone to traumas, potentially leading to chronic wounds or pathological scars. Additionally, exposure to external stimuli like UV radiation, chemical agents, cigarette smoke, and pollution has a great impact on skin integrity and leads to effects commonly summarized as extrinsic aging [3]. In parallel, skin gets also altered by intrinsic aging, which is time dependent and reflects the individual genetic background [4]. Intrinsic skin aging is also associated with increased formation of reactive oxygen species (ROS). ROS are products of cellular metabolism, which induce damage to essential cellular components like membranes, enzymes, and deoxyribonucleic acid (DNA) [5].

Extrinsic as well as intrinsic triggers of aging converge over time and cause broadly noticeable skin changes, such as loss of elasticity (so-called skin elastosis), irregular pigmentation, coarse wrinkles, or crow's feet [4, 6]. However, not only aging does lead to apparent changes, but also regenerative abilities have shown to be altered and lead to diminished wound healing and skin regeneration [7]. Skin damage repair requires a dynamic and highly regulated repair process, the coordination of different cell types and biochemical pathways [8]. Aging goes along with diminished cellular function and subsequent impairment of these pathways.

D. C. Pérez
Biotechnology Research Center, Costa Rica Institute of Technology, Cartago, Costa Rica

M. M. Aitzetmüller (✉)
Department of Plastic and Hand Surgery, Klinikum Rechts der Isar, Technical University of Munich, Munich, Germany

Section of Plastic and Reconstructive Surgery, Department of Trauma, Hand and Reconstructive Surgery, Westfaelische Wilhelms, University of Muenster, Muenster, Germany
e-mail: Matthias.aitzetmueller@tum.de

P. Neßbach
Department of Plastic and Hand Surgery, Klinikum Rechts der Isar, Technical University of Munich, Munich, Germany
e-mail: philipp.nessbach@tum.de

D. Duscher
Department for Plastic Surgery and Hand Surgery, Division of Experimental Plastic Surgery, Technical University of Munich, Munich, Germany

© Springer Nature Switzerland AG 2019
D. Duscher, M. A. Shiffman (eds.), *Regenerative Medicine and Plastic Surgery*,
https://doi.org/10.1007/978-3-030-19962-3_12

Sufficient skin repair also depends on several patient-dependent factors such as comorbidities, metabolic state, nutrition, and age [9]. The ongoing change in world's population towards older age is related to a higher prevalence of patients suffering from chronic diseases [10]. Disease states like diabetes make patients more vulnerable to chronical wounds. This vicious circle is putting increasing pressure on scientists and physicians to work on the advancement of skin regeneration [11].

For this purpose, the development of novel therapies through drug delivery system (DDS) technology is gaining importance. The use of intelligent delivery vehicles for skin regeneration has enabled the possibilities to exactly define administration routes (systemic or local), to limit the duration of effect, or to change pharmaceutical form and posology of drugs and bioactive compounds [12]. In this chapter, we provide a brief overview of recent strategies with the potential to improve or induce aging skin regeneration by using drug delivery systems

12.2 Drug Delivery Routes for Skin Regeneration Applications

The latest drug delivery systems (DDS) are working in a spatiotemporal manner and provide controlled release of a bioactive agent in a previously defined target tissue [13]. This technology has been progressing for more than 60 years with the ultimate goal to develop clinically useful formulations and to provide sustained drug release systems [14]. With ongoing development of drugs and bioactive compounds that promote wound healing and skin regeneration, also developing the optimal DDS for every new compound became more and more important. Additionally, DDSs can be used to enhance effect or change administration pathways of drugs that are already in clinical use [15].

Drug delivery systems are more than simple drug carriers. Their development has to be based and specifically adapted on the physicochemical characteristics of the drug and on the molecular drug-to-DDS interactions. In this manner, the appropriate coordination of pharmacokinetic and pharmacodynamic properties, efficacy, and safety can lay the foundation for the successful development of novel DDSs [12].

Skin regeneration still represents a major challenge for drug delivery. Different drug application forms have shown to hold different advantages and disadvantages. On the one hand, systemic administration faces the obstacles of systemic circulation such as drug liberation, absorption, first-step metabolism, drug protein binding, and excretion. On the other hand, local administration has to penetrate through the skin barrier, most importantly the stratum corneum, and address the sustained delivery needed for cutaneous, intradermal, and subcutaneous application [16].

12.2.1 Systemic Application

It is well understood that pharmacokinetic parameters (e.g., liberation, absorption, distribution, metabolism, and excretion of the intact active principle) must be taken into account for the development of a new systemically administered drug [17]. An advanced DDS potentially provides the solution for all of these problems. Certain pharmacokinetic properties can be modified through the application of a DDS; simultaneously, the drug's pharmacodynamics can be changed by using a particular DDS [17].

There are DDSs used for systemic administration that can modulate the drug release in response to physiological and biochemical stimuli. These stimuli can be unique for certain environments and thereby specifically activate DDS. For example, acid-sensitive polymers have been described to release their load at a specific and controlled manner by changing their architecture due to external stimuli [18]. External stimuli include redox conditions, pH, light, and enzymes. These factors could induce these polymeric carriers and thereby promote the sustained release of the drug [18, 19], whereas it can bind to the target structure [20, 21]. However, unlike systemic delivery, a localized delivery approaches this goal more effectively and is potentially associated with less adverse effects. To achieve a local drug concentration high enough to stimulate regeneration efficiently, the systemic concentration has to be increased with a subsequent increased risk of side effects.

12.2.2 Local Application

The local application of bioactive compounds or drugs for skin regeneration offers the advantage to avoid systemic circulation with all its problems and simultaneously enhances local efficacy [22]. Nevertheless, there remain several difficulties in preclinical and clinical studies. The efficient diffusion through the stratum corneum without risking ulceration, targeting the local cells, stimulating them to produce collagen for regeneration of the ECM, and chemotaxis of circulating regenerative cells represent the main challenges.

Topical drug administration pathway can follow two different absorption routes. Uptake can be either transepidermal through transcellular and/or intercellular transportation or via the cutaneous annexes, such as pilous follicles, sudoriferous glands, and sebaceous glands. These structures reach all the way to the subcutaneous tissue [19, 22]. DDSs should consider either one or both of these routes to guarantee efficient drug delivery and appropriate absorption [23–25].

Given that human skin has a lightly acid hydrolipidic film (pH 4.5–5.5), which facilitates regeneration and offers protection towards external agents, the delivery system must also overcome this barrier while protecting the drug and keeping it in a stable condition [22]. For this purpose, nanoparticles have been formulated to increase percutaneous absorption of bioactive compounds without damaging the skin barrier function. Nanoparticle constituents may be biodegradable polymers/copolymer (PLA, PLGA, PCL, chitosan, etc.) or lipids (Fig. 12.1) [26–28]. They can be classified into two types of particles: polymeric nanoparticles and lipid nanoparticles. Both enable drugs to penetrate deep into the skin. Their direct interaction helps to improve the efficacy of traditional topical drugs and thereby can also reduce the systemic dose and side effects [29].

Different nanoencapsulation techniques have been used to improve efficacy of local application [30–32]. For some drugs nanoencapsulation even represents the key mechanism of effect. For example, Raza et al. [33] developed ointments including

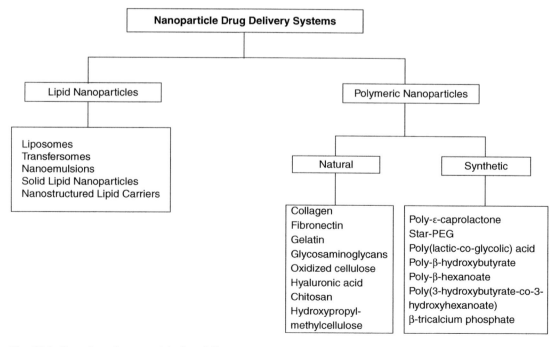

Fig. 12.1 Overview of nanoparticle drug delivery systems

different nanocarriers, to improve its skin delivery, biocompatibility, and pharmacodynamic efficacy of tretinoin (TRE). Further investigations showed that not TRE, but the nanocarriers represent the main effectors of the ointments.

Despite advantages of local application, some studies suggest possibilities with both administration routes for drug delivery. Such is the case of GHK (glycyl-L-histidyl-L-lysine)-Cu peptide, a human sequence common in proteins of the extracellular matrix (ECM). GHK-Cu has been shown to trigger positive effects on skin regeneration. GHK-Cu peptide is capable of up/downregulating about 4000 human genes, basically resetting DNA back to a healthier state and leading to a better regulation of certain cellular pathways [34]. Although the mechanism of action of this peptide is not completely elucidated, recent findings associate its ability to improve tissue regeneration with restored activity of genes involved in the TGF-beta pathway. Since GHK is present as an amino acid sequence in ECM proteins and is released after injury, it appears to serve as a natural modulator of dermal repair [35]. By stimulating skin remodeling, including cell migration, proliferation, and differentiation of skin cells, this compound could serve as an adhesion molecule between cells and ECM [34].GHK could be incorporated into different DDSs as additional supporting compound for dermal repair. However, further research is required to determine the best way of application to guarantee efficient stimulation of skin renewal [36]. Currently, many bioactive compounds and drugs share this scenario where their potential is restricted to find the best administration route and delivery system to control their release.

12.3 Drug Delivery Systems for Skin Regeneration

Skin regeneration involves three main steps: inflammation, cell proliferation, and dermal matrix remodeling. Every stage relies on a coordinated molecular interaction between different cell types and ECM components [18].

While healthy and young skin offers an optimal microenvironment and a fully functional growth factor (GF), peptide, and bioactive molecule pathway for cell attachment, proliferation, and differentiation, these functions are diminished in aged or pathological conditions [37]. Accordingly, recent strategies include incorporation of bioactive agents (small molecules, GF, nucleic acids, vitamins, antioxidants, stem cells, etc.) in novel drug delivery systems (liposomes, transfersomes, nanoemulsions, lipid nanoparticles, scaffolds, etc.). This section provides a brief overview of nanoparticles for DDSs and summarizes their use with bioactive agents (Fig. 12.1).

12.3.1 Liposomes

Liposomes encircle an aqueous core by self-assembled phospholipid bilayers [38]. These DDSs offer several advantages, such as the ability to deliver both hydrophilic and lipophilic drugs. Hydrophilic drugs are imbibed in the core and lipophilic ones are dispersed in the bilayer [39].

The use of liposomes can guarantee targeted delivery by fusion with other bilayers. However, untargeted delivery can also occur by fusing the bilayer with other cells (e.g., immune cells) and awake an immunity response [40]. Another difficulty with the use of liposomes is their instability. They are prone to lysis or fusion with other liposomes, which could eventually lead to loss or alteration of their content [38].

Yu and Liao [40] formulated a DDS for triamcinolone, a glucocorticoid locally applied for psoriatic plaque and keloid scar skin regeneration treatment. By using liposomes, they were able to enhance flux and permeability of the drug. Sinico et al. [41] reported the use of liposomes for dermal delivery of tretinoin, a derivative of vitamin A, that regulates cell proliferation and differentiation. The use of liposomes led to better drug retention and skin hydration results. Also pivotal factors for wound healing, such as the epidermal growth factor (EGF), can be encapsulated into liposomes and remain stable after a topical administration on lesions [42]. To sum up, drug permeability can be enhanced by utilizing a liposome bilayer structure. However, such features may be an inconvenience if absorption occurs at an undesirable site.

12.3.2 Transfersomes

Transfersomes are used as elastic vesicular carriers that have an aqueous interior surrounded by phospholipid edge activators—a special surfactant that causes destabilization of lipid bilayers and increases elasticity and deformability. The surfactant imparts flexibility to the transfersome membrane allowing penetrating pores smaller than its own size. Additionally, the structure of a transfersome also prevents aggregation and vesicular fusion [38].

The advantages offered by transfersomes were firstly observed with vitamin E and its derivatives. α-Tocopherol and tocopheryl acetate were encapsulated and delivered in them. Results showed higher encapsulation efficiency, improved skin deposition (four times higher than with liposomes), and enhanced photostability [43]. Fesq et al. [44] used transfersomes loaded with triamcinolone and thereby significantly increased drug activity (the same activity with tenfold lower dose compared to conventional formulations).

To summarize, due to their augmented vesicular flexibility transfersomes have the potential to be exploited for targeted topical delivery through cutaneous annexes.

12.3.3 Nanoemulsions

Nanoemulsions are dispersed systems of encapsulated hydrophilic/lipophilic species. The surface tension between both counterparts is stabilized by an emulsifier or a surfactant. Based on the proportion of hydrophilic/lipophilic agents nanoemulsions acquire a different encapsulation structure. They could be water-in-oil (w/o) systems, or oil-in-water systems (o/w) [38].

Nanoemulsions have been used for delivering cosmeceuticals to the skin to moisturize and improve its texture, but also for topical delivery of medications. For instance, hyaluronic acid formulations based on nanoemulsion have been used in topical application. This was offered as an alternative for the injectable application of the same therapeutic agent [38].

Another drug whose topical delivery could be enhanced via nanoemulsion technique was ubiquinone. Ubiquinone is an antioxidant found in both hydrophilic and hydrophobic compartments of skin cells. Its delivery as a nanoemulsion resulted in an increase of its natural antioxidant potential and a decrease in collagenase expression in fibroblasts. These results shed light on more possibilities to promote regeneration by using chemically similar compounds [45].

12.3.4 Lipid Nanoparticles

Lipid nanoparticles (LN) were developed to overcome certain difficulties presented by liposomes. LN include solid lipid nanoparticles (SLN) and nanostructured lipid carries (NLC). Both offer good tolerability, safe material content, and capability to encapsulate hydrophilic/lipophilic compounds and cosmetics for topical application [38, 46].

Lipid nanoparticles have been reported to release their load either in burst or in a controlled manner. A burst release enhances skin penetration, whereas sustained release is helpful, when the active compound is toxic at high concentrations [47, 48]. Bioactive agent distribution inside different LN types is variable and for some therapeutic agents or cosmetics NLC can load higher amounts than SLN. However, a lower potential for release of the active compound has been reported for NLC [49].

Another advantage described for both (SLN and NLC) is the augmented resistance against oxidation and hydrolysis [50]. SLN as well as NLC have been used successfully in creams and gels for topical application. In vitro experiments using glucocorticoid prednicarbate showed increased intake and targeted distribution when being loaded in SNL [51].

Gainza et al. [52] demonstrate that further SLN and NLCs loaded with recombinant human epidermal growth factor (rhEGF) significantly improved in vitro keratinocyte and fibroblast proliferation. Therefore, lipid nanoparticle encapsulation can guarantee sustained release of EGF and thereby improve EGF receptor binding capacity.

12.3.5 Scaffolds

Materials used as scaffolds are purposely designed to assure physical protection of the bioactive agent and to promote sustained release in the targeted tissue. Overall, these materials should be nonallergenic, nonmutagenic, nonhemolytic, and nonpyrogenic [51–53]. Scaffolds developed for skin regeneration have the main goal to mimic the extracellular matrix (ECM) and provide an appropriate environment for cell growth and tissue differentiation [54, 55]. Collagen, the most abundant protein of the ECM, has been used as a scaffold to deliver epidermal growth factor and TGF-β1 to promote wound closure [38, 56].

Other constituents such as fibronectin, gelatin, glycosaminoglycans, oxidized cellulose, hydroxypropyl methylcellulose, and alginates are alternatives to collagen. Nevertheless, their use is often limited by the possibility of immunological and inflammatory response after application, as well as difficulties to reproduce, characterize, and verify their properties [57].

These disadvantages have been solved for certain synthetic polymeric scaffolds. Star-PEG incorporating heparin (38), poly-ε-caprolactone [56], and poly(lactic-co-glycolic) acid [57] are commonly used for tissue regeneration purposes. However, unlike natural materials, synthetic scaffolds do not participate actively in intercellular communication and could hinder skin regeneration pathways [58].

Furthermore, scaffolds can be used as DDSs for delivering stem cells and serve as a protection layer to minimize cell loss. Moreover, the ECM-like environment offered by the scaffold itself represents an appropriate system for stem cell proliferation, migration, and differentiation to potentially regenerate the targeted tissue [59]. Huang et al. [60] reported the use of incorporated EGF nanosponges and microspheres as scaffolds to deliver stem cells. This approach led to direct cell differentiation and proliferation, slower degradation rate, and sustained release of EGF in contrast to injected EGF application. Undoubtedly, scaffolds are promising DDSs and their optimization has the potential to efficiently support skin regeneration.

12.4 Conclusions and Future Perspectives

The identification of the underlying mechanisms and key players of skin aging has become a major goal of regenerative medicine. Continuous research for innovative strategies to achieve tissue regeneration has involved holistic insight from different disciplines and integration of their knowledge. However, effective drug delivery strategies for skin regeneration still represent a major challenge for research and clinical translation. The delivery of bioactive agents into target tissues through DDS technology emerged as a therapeutic alternative to promote skin regeneration. New approaches aim to enhance and optimize DDS properties (biocompatibility, bioavailability, and safety) for specific bioactive agents to assure their sustained delivery, at the right concentration, for the appropriate period of time in the site of interest. Recent strategies incorporate bioactive agents in liposomes, transfersomes, nanoemulsions, lipid nanoparticles, and scaffolds to stimulate skin regeneration. However, further research is needed to compare different DDSs, as well as to define effective dosages and ways of administration (systemically or locally) that fit best for each compound. In the meantime, the development of promising synergistic strategies that combine more than one bioactive agent is a further challenge while we continue to shed light on new opportunities and more holistic approaches to achieve skin regeneration.

References

1. Reddy VJ, Radhakrishnan S, Ravichandran R, Mukherjee S, Balamurugan R, Sundarrajan S, Ramakrishna S. Nanofibrous structured biomimetic strategies for skin tissue regeneration. Wound Repair Regen. 2013;21(1):1–16.
2. Korrapati PS, Karthikeyan K, Satish A, Krishnaswamy VR, Venugopal JR, Ramakrishna S. Recent advancements in nanotechnological strategies in selection, design and delivery of biomolecules for skin regeneration. Mat Sci Eng C Mater Biol Appl. 2016;67:747–65.
3. Tobin DJ. Introduction to skin aging. J Tissue Viability. 2017;26(1):37–46.
4. Gragnani A, Mac Cornick S, Chominski V, de Noronha SMR, de Noronha SAAC, Ferreira LM. Review of major theories of skin aging. Adv Aging Res. 2014;3(04):265.
5. Farage MA, Miller KW, Elsner P, Maibach HI. Characteristics of the aging skin. Adv Wound Care. 2013;2(1):5–10.
6. Krutmann J. Skin aging. In: Krutmann J, Humbert P, editors. Nutrition for healthy skin. Berlin: Springer; 2010. p. 15–24.
7. Duscher D, Barrera J, Wong VW, Maan ZN, Whittam AJ, Januszyk M, Gurtner GC. Stem cells in wound healing: the future of regenerative medicine? A mini-review. Gerontology. 2016;62(2):216–25.
8. Kamolz LP, Griffith M, Finnerty C, Kasper C. Skin regeneration, repair, and reconstruction. Biomed Res Int. 2015;2015:892031.
9. Diegelmann RF, Evans MC. Wound healing: an overview of acute, fibrotic and delayed healing. Front Biosci. 2004;9(1):283–9.
10. National Institute on Aging. Global health and aging. 2011. https://www.nia.nih.gov/research/dbsr/global-aging. Accessed 3 Jun 2018.
11. Mustoe T. Understanding chronic wounds: a unifying hypothesis on their pathogenesis and implications for therapy. Am J Surg. 2004;187(5):S65–70.
12. Sheikhpour M, Barani L, Kasaeian A. Biomimetics in drug delivery systems: a critical review. J Control Release. 2017;253:97–109.
13. Devi VK, Jain N, Valli KS. Importance of novel drug delivery systems in herbal medicines. Pharmacogn Rev. 2010;4(7):27.
14. Park K. Controlled drug delivery systems: past forward and future back. J Control Release. 2014;190:3–8.
15. Yun Y, Lee BK, Park K. Controlled drug delivery systems: the next 30 years. Front Chem Sci Eng. 2014;8(3):276–9.
16. Bertens C, Gijs M, Nuijts R, van den Biggelaar F. Topical drug delivery devices: a review. Exp Eye Res. 2018;168:149–60.
17. Shargel L, Yu A, Wu-Pong S. Introduction to biopharmaceutics and pharmacokinetics. In: Shargel L, Yu A, Wu-Pong S, editors. Applied biopharmaceutics and pharmacokinetics. 6th ed. New York: McGraw-Hill; 2012. p. 1–17.
18. Ramasamy T, Ruttala HB, Gupta B, Poudel BK, Choi H-G, Yong CS, Kim JO. Smart chemistry-based nanosized drug delivery systems for systemic applications: a comprehensive review. J Control Release. 2017;258:226–53.
19. Tong R, Christian DA, Tang L, Cabral H, Baker JR, Kataoka K, Discher DE, Cheng J. Nanopolymeric therapeutics. MRS Bull. 2009;34(6):422–31.
20. Ramasamy T, Kim JO, Yong CS, Umadevi K, Rana D, Jiménez C, Campos J, Haidar ZS. Novel core–shell nanocapsules for the tunable delivery of bioactive rhEGF: formulation, characterization and cytocompatibility studies. J Biomater Tiss Eng. 2015;5(9):730–43.
21. Ramasamy T, Kim JH, Choi JY, Tran TH, Choi H-G, Yong CS, Kim JO. pH sensitive polyelectrolyte complex micelles for highly effective combination chemotherapy. J Mater Chem B. 2014;2(37):6324–33.
22. Tran TN. Cutaneous drug delivery: an update. J Investig Dermatol Symp Proc. 2013;16:S67–9.
23. Alvarez-Román R, Naik A, Kalia Y, Guy RH, Fessi H. Skin penetration and distribution of polymeric nanoparticles. J Control Release. 2004;99(1):53–62.
24. Blume-Peytavi U, Vogt A. Human hair follicle: reservoir function and selective targeting. Br J Dermatol. 2011;165(s2):13–7.
25. Chourasia R, Jain SK. Drug targeting through pilosebaceous route. Curr Drug Targets. 2009;10(10):950–67.
26. Baena-Aristizábal CM, Mora-Huertas CE. Micro, nano and molecular novel delivery systems as carriers for herbal materials. J Colloid Sci Biotechnol. 2013;2(4):263–97.
27. Laouini A, Jaafar-Maalej C, Limayem-Blouza I, Sfar S, Charcosset C, Fessi H. Preparation, characterization and applications of liposomes: state of the art. J Colloid Sci Biotechnol. 2012;1(2):147–68.
28. Jada A. A special issue on inorganic colloidal particles, synthesis, surface properties and applications. J Colloid Sci Biotechnol. 2014;3(1):1–2.
29. Sala M, Elaissari A, Fessi H. Advances in psoriasis physiopathology and treatments: up to date of mechanistic insights and perspectives of novel therapies based on innovative skin drug delivery systems (ISDDS). J Control Release. 2016;239:182–202.
30. Lira AAM, Cordo PL, Nogueira EC, Almeida EDP, Junior RAL, Nunes RS, Rogéria S, Bentley MVLB, Marchetti JM. Optimization of topical all-trans retinoic acid penetration using poly-DL-lactide and poly-DL-lactide-co-glycolide microparticles. J Colloid Sci Biotechnol. 2013;2(2):123–9.
31. Rosset V, Ahmed N, Zaanoun I, Stella B, Fessi H, Elaissari A. Elaboration of argan oil nanocapsules containing naproxen for cosmetic and transdermal local application. J Colloid Sci Biotechnol. 2012;1(2):218–24.
32. Zhao Y, Brown MB, Jones SA. Pharmaceutical foams: are they the answer to the dilemma of topical nanoparticles? Nanomedicine. 2010;6(2):227–36.

33. Raza K, Singh B, Lohan S, Sharma G, Negi P, Yachha Y, Katare OP. Nano-lipoidal carriers of tretinoin with enhanced percutaneous absorption, photostability, biocompatibility and anti-psoriatic activity. Int J Pharm. 2013;456(1):65–72.

34. Pickart L, Vasquez-Soltero JM, Margolina A. GHK and DNA: resetting the human genome to health. Biomed Res Int. 2014;2014:151479.

35. Maquart FX, Pickart L, Laurent M, Gillery P, Monboisse JC, Borel JP. Stimulation of collagen synthesis in fibroblast cultures by the tripeptide-copper complex glycyl-L-histidyl-L-lysine-Cu2+. FEBS Lett. 1988;238(2):343–6.

36. Pickart L, Vasquez-Soltero JM, Margolina A. GHK peptide as a natural modulator of multiple cellular pathways in skin regeneration. Biomed Res Int. 2015;2015:648108.

37. Pereira RF, Barrias CC, Granja PL, Bartolo PJ. Advanced biofabrication strategies for skin regeneration and repair. Nanomedicine (Lond). 2013;8(4):603–21.

38. Leonida MD, Kumar I. Bionanomaterials for skin regeneration. Cham: Springer Nature; 2016.

39. Gregoriadis G, Florence AT. Liposomes in drug delivery. Drugs. 1993;45(1):15–28.

40. Yu HY, Liao HM. Triamcinolone permeation from different liposome formulations through rat skin in vitro. Int J Pharm. 1996;127(1):1–7.

41. Sinico C, Manconi M, Peppi M, Lai F, Valenti D, Fadda AM. Liposomes as carriers for dermal delivery of tretinoin: in vitro evaluation of drug permeation and vesicle–skin interaction. J Control Release. 2005;103(1):123–36.

42. Brown GL, Curtsinger LJ, White M, Mitchell RO, Pietsch J, Nordquist R, von Fraunhofer A, Schultz GS. Acceleration of tensile strength of incisions treated with EGF and TGF-beta. Ann Surg. 1988;208(6):788.

43. Gallarate M, Chirio D, Trotta M, Eugenia Carlotti M. Deformable liposomes as topical formulations containing α-tocopherol. J Dispers Sci Technol. 2006;27(5):703–13.

44. Fesq H, Lehmann J, Kontny A, Erdmann I, Theiling K, Rother M, Ring J, Cevc G, Abeck D. Improved risk–benefit ratio for topical triamcinolone acetonide in Transfersome® in comparison with equipotent cream and ointment: a randomized controlled trial. Br J Dermatol. 2003;149(3):611–9.

45. Draelos ZD. Enhancement of topical delivery with nanocarriers. In: Nasir A, Friedman A, Wang S, editors. Nanotechnology in dermatology. New York: Springer; 2013. p. 87–93.

46. Müller RH, Radtke M, Wissing SA. Solid lipid nanoparticles (SLN) and nanostructured lipid carriers (NLC) in cosmetic and dermatological preparations. Adv Drug Deliv Rev. 2002;54:S131–S55.

47. Jenning V, Schäfer-Korting M, Gohla S. Vitamin A-loaded solid lipid nanoparticles for topical use: drug release properties. J Control Release. 2000;66(2–3):115–26.

48. Barry BW. Breaching the skin's barrier to drugs. Nat Biotechnol. 2004;22(2):165.

49. Mehnert W, Mäder K. Solid lipid nanoparticles: production, characterization and applications. Adv Drug Deliv Rev. 2012;64:83–101.

50. Schäfer-Korting M, Mehnert W, Korting H-C. Lipid nanoparticles for improved topical application of drugs for skin diseases. Adv Drug Deliv Rev. 2007;59(6):427–43.

51. Santos Maia C, Mehnert W, Schaller M, Korting H, Gysler A, Haberland A, Schäfer-Korting M. Drug targeting by solid lipid nanoparticles for dermal use. J Drug Target. 2002;10(6):489–95.

52. Gainza G, Pastor M, Aguirre JJ, Villullas S, Pedraz JL, Hernandez RM, Igartua M. A novel strategy for the treatment of chronic wounds based on the topical administration of rhEGF-loaded lipid nanoparticles: in vitro bioactivity and in vivo effectiveness in healing-impaired db/db mice. J Control Release. 2014;185:51–61.

53. Ueda H, Tabata Y. Polyhydroxyalkanonate derivatives in current clinical applications and trials. Adv Drug Deliv Rev. 2003;55(4):501–18.

54. Tateshita T, Ono I, Kaneko F. Effects of collagen matrix containing transforming growth factor (TGF)-β1 on wound contraction. J Dermatol Sci. 2001;27(2):104–13.

55. Inoue M, Ono I, Tateshita T, Kuroyanagi Y, Shioya N. Effect of a collagen matrix containing epidermal growth factor on wound contraction. Wound Repair Regen. 1998;6(3):213–22.

56. Augustine R, Dominic EA, Reju I, Kaimal B, Kalarikkal N, Thomas S. Electrospun polycaprolactone membranes incorporated with ZnO nanoparticles as skin substitutes with enhanced fibroblast proliferation and wound healing. RSC Adv. 2014;4(47):24777–85.

57. Shirazi RN, Aldabbagh F, Erxleben A, Rochev Y, McHugh P. Nanomechanical properties of poly (lactic-co-glycolic) acid film during degradation. Acta Biomater. 2014;10(11):4695–703.

58. Mohamed A, Xing MM. Nanomaterials and nanotechnology for skin tissue engineering. Int J Burns Trauma. 2012;2(1):29.

59. Wong VW, Levi B, Rajadas J, Longaker MT, Gurtner GC. Stem cell niches for skin regeneration. Int J Biomater. 2012;2012:926059.

60. Huang S, Lu G, Wu Y, Jirigala E, Xu Y, Ma K, Fu X. Mesenchymal stem cells delivered in a microsphere-based engineered skin contribute to cutaneous wound healing and sweat gland repair. J Dermatol Sci. 2012;66(1):29–36.

Part II

Bone Regeneration

Innovative Scaffold Solution for Bone Regeneration Made of Beta-Tricalcium Phosphate Granules, Autologous Fibrin Fold, and Peripheral Blood Stem Cells

13

Ciro Gargiulo Isacco, Kieu C. D. Nguyen,
Andrea Ballini, Gregorio Paduanelli, Van H. Pham,
Sergey K. Aityan, Melvin Schiffman, Toai C. Tran,
Thao D. Huynh, Luis Filgueira, Vo Van Nhan,
Gianna Dipalma, and Francesco Inchingolo

13.1 Introduction

Diseases, trauma, and surgical procedures can be the cause of bone paucity and defects. Due to the complexity of bone anatomy and physiology bone tissue degeneration and diseases can pose a big threat to doctors and physicians. However, modern bone tissue biomedical engineering has been considered as a valid substitute solution for these conditions [1]. Procedures applied to repair defects or

C. G. Isacco (✉) · G. Paduanelli · G. Dipalma
F. Inchingolo
Department of Interdisciplinary Medicine (DIM),
School of Medicine, University of Bari Aldo Moro,
Bari, Italy
e-mail: francesco.inchingolo@uniba.it

K. C. D. Nguyen
Department of Stem Cell Research, HSC
International Clinic, Ho Chi Minh City, Vietnam

A. Ballini
Department of Basic Medical Sciences,
Neurosciences and Sense Organs, University of
Bari Aldo Moro, Bari, Italy

V. H. Pham
Department of Microbiology, Nam Khoa-Bioteck
Microbiology Laboratory and Research Center,
Ho Chi Minh City, Vietnam

Department of Microbiology, Nam Khoa-Bioteck
Microbiology Laboratory and Research Center,
Ho Chi Minh City, VN, USA

S. K. Aityan
Multidisciplinary Research Center, Lincoln
University, Oakland, CA, USA
e-mail: aityan@lincolnuca.edu

M. Schiffman
Tustin, CA, USA

T. C. Tran
Stem Cells, Embryology and Immunity Department,
Pham Ngoc Thach University of Medicine,
Ho Chi Minh City, Vietnam
e-mail: trancongtoai@pnt.edu.vn

T. D. Huynh
Department of Embryology, Genetics and Stem Cells,
Pham Ngoc Thach University of Medicine,
Ho Chi Minh City, Vietnam
e-mail: thao_huynhduy@pnt.edu.vn

L. Filgueira
Faculty of Science and Medicine, University of
Fribourg, Fribourg, Switzerland
e-mail: luis.filgueira@unifr.ch

V. Van Nhan
University of Pharmacy and Medicine,
Ho Chi Minh City, Vietnam

Nha Khoa Nham Tam, Policlinic and International
Dental Implant Center, Ho Chi Minh City, Vietnam

© Springer Nature Switzerland AG 2019
D. Duscher, M. A. Shiffman (eds.), *Regenerative Medicine and Plastic Surgery*,
https://doi.org/10.1007/978-3-030-19962-3_13

degeneration need the use of proper biomaterials with the right dimensions and anatomy shape that can fit into the damaged area. The cases of larger and more difficult defects need highly osteogenic scaffolds to promote and improve the bone tissue formation and regeneration [2].

The majority of scaffolds made of ceramics and metals or from polymers showed a weak osteogenic capability. Bone grafts are currently available in many different forms, such as allogeneic/autologous demineralized bone matrix or implants, growth factor-loaded microbeads, or bio-derivate as calcium hydroxyapatite, gels, ceramic derivate, sea corals, and metals [2, 3]. The main target is to reduce the incidence of collateral complications such as rejection, infection and inflammation, and donor-site morbidity and of course reduce the overall costs and related expenses such as frequent hospitalization [3–5]. Breakdowns, such as partial ruptures or complete collapse, are the major issues related to synthetic implants, generally due to quality of the material and subtle autoimmune responses that may also create ideal conditions for bacterial growth, inflammation, and rejection [6–9]. Metal implants, for example, may cause malfunction due to aseptic loosening with specific inflammatory and immune responses to metal-wear particles released during bio-corrosion which intensify the osteolytic activity of osteoclasts at the bone-implant interface, leading to a progressive loss of fixation [8, 9]. Therefore, an optimal biomaterial should possess specific bio-characteristics and qualities that should be biodegradable, tolerable, and safely absorbed by the body. This should happen without causing any kind of damaging event such as an inflammation or an immune reaction, capable of carrying and supporting tissue growth and proliferation, thus allowing bone regeneration [5, 9].

The latest generations of bio-implants have been created with the precise intent of functioning as cell carriers capable of reproducing human bone formation process. The newest generation of these types of scaffolds has been developed with materials that possess specific mechanical and structural properties that are compatible with the anatomical site into which they are to be inserted, with enough volume fraction and high surface area to carry an enough number of cells within the scaf-fold and the surrounding host tissues. This allows ingrowth and vascularization [5]. Therefore, the new bio-implants tend to replicate the process of the formation of new bone development or which physiologically takes place after an injury [10].

An inflammatory response takes place after an implantation of a biomaterial as a consequence of host immune response [10]. During this phase monocytes differentiate to tissue macrophages. However, presence of MSCs promotes an immune-modulatory activity on macrophage M1/M2 balance towards M2 commencing a favorable cascade of events where interleukins such as IL-10, IL-4, IL-13, and IL-6 and prostaglandin E2 initiate the first step of the repairing process [11–14]. Bone plays a key role in well functioning of immune system and it is the site that immune cells are created. In fact, autoimmune disorders often induce bone tissue damages and degeneration, an event that has been confirmed by an experiment where macrophage ablation leads to intramembranous bone defection and inhibiting of the healing process [14].

In effect, previous studies have shown that some biomaterials due to high similarity with human tissues are able to trigger physic-chemical signals leading to stem cell differentiation towards diverse cell phenotypes as osteoblasts [15, 16]. Results have shown that biomaterials based on calcium phosphate (CaP), a major constituent of native bone tissue, induce naïve stem cells towards osteogenic differentiation promoting in vivo bone tissue formation and augmentation [16, 17].

However, though CaP is quickly absorbed in vivo, the process often occurs preceding the formation of new bone tissue that results in an incongruence between the host's new bone and scaffold. Conversely, β-TCP seems to be better compatible as the absorption rate is slower with a steady release of both calcium (Ca^{2+}) and sulfate (SO_4^{2-}) ions [18].

In line with our published study, we can confirm that hPB contains the right amount of different subsets of pluripotent and multipotent stem cells such as MSCs, HSCs, NSCs, and ESCs capable of differentiating into cells of different lineages such as osteoblasts [19]. In this current study, we have noted that part of hPB-SCs were induced to differentiate to active osteoblasts under the direct influence of β-TCP granules within a period of 7–10 days without the need of osteo-inductor medium in cell culture flask. Therefore we have hypothesized that the adjunct of autologous hPB-SCs together with β-TCP embedded in fibrin matrix gel could enhance both the correct time fraction for endogenous bone formation and the quality of the bone tissue itself in both in vitro and in vivo condition.

13.2 Material and Methods

13.2.1 Beta-Tricalcium Phosphate

The biomaterial composed of β-TCP (GUIDOR Calc-i-Oss) granules with diameters of 3×0.5ml $500-1000$ μM were supplied by Sunstar Degradable Solution AG Co., AS. Switzerland. The granules were irregular in shape, and each had interconnection diameters ranging from 70 to 200 μm. The porous structure of β-TCP was characterized using a scanning electron microscope (SEM, Zeiss-Sigma, USA). The porosity and interconnections of the porous as well as the composition and weight percentage of β –TCP granules were calculated using an analysis software package performed at Sunstar Degradable Solution AG laboratories. Guidor Calc-i-Oss consists of pure β-TCP with a purity of > 99% with a Ca/P molecule ratio of 1.5. Resorption will take place mostly parallel to bone regeneration. Depending on the regeneration potential of the host tissue, it will be completely resorbed within 9–15 months.

Peripheral blood 24 mL from each consent donor ($n = 10$) has been collected in four HSC Vacutainer[R] tube kit (1 red, 1 green, and 2 whites (Human-Stem Cells[R] provided by Silfradent, Italy)). The red cap tube consists of internal coating material made by silicon crystals in a coagulum and when separation/centrifugation is done at room temperature the final product is in a condensed form. White cap tube has no material and after separation the fibrin gel will be ready after 20 min. Green cap tube contains heparin and Ficoll-Paque PLUS (GE Healthcare Life Science-Uppsala, Sweden) to which blood is added in a ratio of 1:2.

The procedure to obtain Compact Bio-Bone[R] scaffold includes the use of β-TCP granules mixed together with different preparations obtained from the blood collected in four Vacutainer[R] tube kit. The blood needs to be centrifuged for 12 min at different variation speeds (Medifuge, Silfradent, Italy). In the first step, the clotted plasma-rich fibrin from the red cap tube is collected, separated from blood part, and squeezed to obtain fibrin serum, while the membrane was chopped into pieces and mixed with β-TCP granules and fibrin serum. Then, 2 mL plasma from the white cap tube without anticoagulant was used and mixed with β-TCP and fibrin, moved into Eppendorf conical tube 2 mL, and inserted in the APAG heater machine (Silfradent, Italy) for max 1 min at 74 °C. Meanwhile, peripheral blood stem cells (about $2 \times 10\,6 \pm 5 \times 10^5$ cells/mL) from the green cap tubes were isolated by using one part of Ficoll-Paque PLUS and two parts of blood, gently washed twice with PBS, and injected directly into the β-TCP and fibrin composite precedently obtained. The scaffold was put in flasks containing 5 mL serum-free medium (SFM, Gibco, Germany) and incubated at 37 °C with 5% CO_2; the medium was changed every 5 days.

13.2.2 Cytochemical Staining, Flow Cytometry Analysis, and RT-PCR

Mineral matrix deposits and bone nodules of osteoblasts, from human PB-SCs and es, were evaluated by staining cell cultures with alizarin red (AR), alkaline phosphatase (AP), and von Kossa (VK); flow cytometry analysis was used to confirm the expression of multipotent and pluripotent stem cells such as CD34, CD45, CD90, CD105, and SSEA3.

13.2.3 Alizarin Red Stain Procedure

The presence of calcium deposits was detected by washing cells with cold PBS and fixing them in NFB-neutral formalin buffer solution 10% for 30 min in a chemical hood. Cells were rinsed three times with distilled water and immersed in 2% solution of alizarin red for 5 min. Cells were rinsed 2–3 times in distilled water and checked under inverse microscope and photographed.

13.2.4 Alkaline Phosphatase Stain Procedure

The presence of alkaline phosphates was detected by washing cells with cold PBS and fixing them in NFB-neutral formalin buffer solution 10% for 30 min in a chemical hood. Cells were then stained with solution naphthol As-MX-PO$_4$ (Sigma) and Fast red violet LB salt (Sigma) for 45 min in the dark at room temperature. Cells were rinsed three times in distilled water and checked by inverse microscope and photographed.

13.2.5 Von Kossa Stain Procedure

The presence of calcium deposits was detected by washing cells with cold PBS and fixing them in NFB-neutral formalin buffer solution 10% for 30 min in a chemical hood. They were then stained with 2.5% silver nitrate (Merck, Germany) for 30 min in the dark. Cells were rinsed three times with distilled water, checked by inverse microscope, and photographed.

13.2.6 Cytochemical Staining of CFU-Fs

After being cultured in serum-free medium (SFM) for 7 days, PB-SCs were fixed with cold 10% neutral-buffered formalin (30 min at 4 °C) and then assayed for colony-forming unit fibroblasts (CFU-Fs). Briefly, the substrate solution was prepared by removing flasks from the incubator. We removed off cell culture medium and washed the flask with 10–15 mL of PBS. We added 10 mL of a 0.5% crystal violet solution (Sigma) made with methanol and we incubated the dishes, on the bench at room temperature, for 30 min. Then, the crystal violet solution was removed and the flasks were carefully rinsed 4× with 10–15 mL D-PBS. Then, each flask was gently rinsed 1× with tap water. All remained tap water was accurately pipetted off and let dry. The colonies were enumerated by microscope.

13.2.7 Immunophenotyping by Flow Cytometry Analysis

Immunophenotyping by flow cytometry analysis was performed to evaluate the presence of a set of markers CD34, CD45, CD90, CD105, and SSEA3. Cell samples were washed twice in Dulbecco's PBS containing 1% BSA (Sigma-Aldrich). Cells were stained for 30 min at 4 °C with anti-CD14-fluorescein isothiocyanate, anti-CD34-fluorescein isothiocyanate, anti-CD133-fluorescein isothiocyanate, anti-CD44-phycoerythrin, anti-CD45-fluorescein isothiocyanate, anti-CD90-phycoerythrin, anti-CD105-fluorescein isothiocyanate mAb, anti-Nestin, anti-Tra1, and anti-SSEA3 (BD Biosciences, Franklin Lakes, NJ, USA). Stained cells were analyzed by a FACSCalibur flow cytometer (BD Biosciences). Isotype controls were used for all analyses.

13.2.8 Real-Time-Polymerase Chain Reaction Procedure

RT-PCR for the expression of OCT-4, Sox-2, osteocalcin, Nanog, Nestin, DMP, and GAPDH was performed on human PB-SC-derived osteoblasts. Results were confirmed using a negative control procedure. Total RNA extraction of the tested cell culture was carried out including the positive control using the Trizol LS reagent (Invitrogen): (1) 150 μL of cell culture was added to one biopure Eppendorf tube containing 450 μL Trizol LS (Invitrogen) and homogenized by mixing up and down several times. (2) The homogenized solution was incubated at 30 °C for 10 min. (3) 120 μL of chloroform was added, then rapidly and carefully shaken for 15 s, kept at 30 °C for 10 min, and centrifuged in a refrigerated centrifuge (2–8 °C) at 1300 × g for 15 min. (4) 300 μL of the supernatant was carefully transferred into another biopure Eppendorf tube. While taking care not to disturb the interface, 300 μL isopropanol was added, and the tube was shaken gently up and down for a few seconds, then kept at 30 °C for 10 min, and centrifuged in the refrigerated

Table 13.1 Number of patient donors recruited for the study and it was analyzed the CFU-s

No. of patients	Age	Sex	Blood volume (mL)	Number of CFU
1	24	Female	9	38
2	32	Male	9	4
3	58	Male	9	9
4	55	Male	9	14
5	59	Male	9	3
6	53	Male	9	2
7	58	Male	9	144
8	48	Female	9	6
9	47	Male	9	5
10	53	Male	9	33

The number of CFU-s is not strictly corresponding to the donor's gender or age

centrifuge at $1300 \times g$ for 20 min. (5). The supernatant was removed without losing the precipitated RNA (sometimes invisible) at the bottom or on one side of the bottom of the tube. (6) About 1 mL ethanol 80% was added without any shaking, and then centrifuged in the refrigerated centrifuge at $800 \times g$ for 5 min; all of the supernatant was removed using a vacuum pump. (7) The pellet was dried at 55 °C for 10–15 min; 40 µL Q water is added to the dried pellet and then kept at 56 °C for 10 min. The isolated RNA was stored at −20 °C prior to RT-PCR. The cDNA synthesis was carried out using the iScript cDNA synthesis kit (BioRad): (1) In one tube PCR 0.2, 1 µL reverse transcriptase was added, 4 µL RT mix, and 15 µL of the isolated RNA; this was gently mixed by pipetting up and down several times. (2) The cDNA synthesis was carried out in the thermal cycler by the following thermal cycle, 25 °C/5 min, 42 °C/30 min, and 85 °C/5 min, and then kept at 4 °C until PCR (Table 13.1).

13.2.9 Primers

DMP

GCAGAGTGATGACCCAGAG **sense primer (3′ to 5′)**

GCTCGCTTCTGTCATCTTCC **antisense primer (5′ to 3′)**

200 **Primer expected size (bps)**

OCN

CAAAGGTGCAGCCTTTGTGTC **sense primer (3′ to 5′)**

TCACAGTCCGGATTGAGCTCA **antisense primer (5′ to 3′)**

150 **Primer expected size (bps)**

Nestin

AGAGGGGAATTCCTGGAG **sense primer (3′ to 5′)**

CTGAGGACCAGGACTCTCTA **antisense primer (5′ to 3′)**

496 **Primer expected size (bps)**

RUNX2

GGTTAATCTCCGCAGGTCACT **sense primer (3'to 5′)**

CACTGTGCTGAAGAGGCTGTT **antisense primer (5′ to 3′)**

203 **Primer expected size (bps)**

GAPDH

CCCATCACCATCTTCCAGGA **sense primer (5′ to 3′)**

TTGTCATACCAGGAAATGAGC **antisense primer (5′ to 3′)**

94 **Primer expected size (bps)**

13.3 Discussion

The improved understanding of the tissue microenvironment where the replacements had to be done resulted in better quality of the biomaterials used for the generation of bone implants. Early graft substitutes were conceptually made respecting the mechanical properties of the affected area; therefore biomaterial should have matched the physical properties of the replaced structure while keeping integrity with the surrounding environment [20]. Metals, ceramics, and polymers were the main materials used in these procedures. The collateral was the presence of consistent idiopathic immune responses probably due to the formation of fibrous tissues at the biomaterial-tissue interface that led to aseptic loosening. The persistent inflammatory response induced the generation of fibrotic connective tissue that encapsulated the foreign structure repairing it from the attack of immune system. This generated a deep friction between the implant and the surrounding tissues that eventually led to the total removal of the implant [20].

Thus second generation of bone graft substitutes were studied with the intent of inducing specific biological responses improving osteoconduction and vascularization to avoid the formation of this fibrous layer by using bioactive and biodegradable coatings. The procedure based on the use of bio-components that promote the integration with the surrounding tissues such as bioactive ceramics to include HA, β-TCP, or bioactive glass that closely mimics the bone tissue microenvironment [20]. These bio-compounds were designed to allocate stem cells and growth factors enhancing bone repair and regeneration and providing the necessary cellular and molecular support for vascularization and enough nutritional support [20].

We have presented through this chapter a new bioscaffold that meets the basic needs essential to enhance the repair, regeneration, and growth of host bone tissue, easy to manipulate and insert in in vivo procedures as presented in (Figs. 13.14, 13.15, and 13.16). This new scaffold due to the combined presence of autologous fibrin gel matrix, β-TCP, and autologous PB-SC-derived osteoblasts closely mimics the bone tissue microenvironment. The in vitro phase showed few prominent points that allow us to predict that this new bioscaffold may eventually match the biomaterial degradation rate with the bone regeneration rate and also a proper control of stem cells growth over their release kinetics and it may induce enough vascularization on the site.

The presence of pluripotent and multipotent PB-SCs both adherent and floating was tested by flow cytometry analysis and resulted positive for markers like CD 34-45-90-105 and SSEA-3 but negative for SSEA-1 (Figs. 13.18, 13.19, 13.20, 13.21, 13.22, and 13.23). The autologous PB-SCs obtained by consent donors were seeded with β-TCP embedded in a gel of fibrin gel obtained by donor's blood and were named the Compact Bio-Bone. Compact Bio-Bone. Compact Bio-Bone was positively tested by specific immune stain like alizarin red (AR, Figs. 13.1, 13.2, and 13.3), alkaline phosphatase (AP, Figs. 13.4, 13.5, and 13.6), and von Kossa (VK, Figs. 13.7 and 13.8); the Compact Bio-Bone cells were then tested by RT-PCR for the expression of osteo-matrix producing genes and the outcomes were positive for Runx2 (203 Kb), OCN (150 kb), DMP (200 Kb) and Nestin (496 Kb) (Fig. 13.13); eventually the scaffold was pictured in high resolution by SEM (Figs. 13.9, 13.10, 13.11 and 13.12). These results though obtained from in vitro showed highly promising outcomes especially if one considers the use in bone reconstructive procedures; the presence of mature osteoblast-like cells in a relatively short period of 7–12 days not only prefigures the possibility of a faster resorption rate once applied in vivo, but also definitively opens up the chance to resolve the formation of better endogenous bone formation. The presence of Nestin detected within Compact Bio-Bone cells, as recently confirmed by other authors though from bone marrow-derived stromal cells, showed to contribute to the normal osteoblast lineage cell turnover in the adult bone [21] (Figs. 13.19, 13.20, 13.21, 13.22, and 13.23).

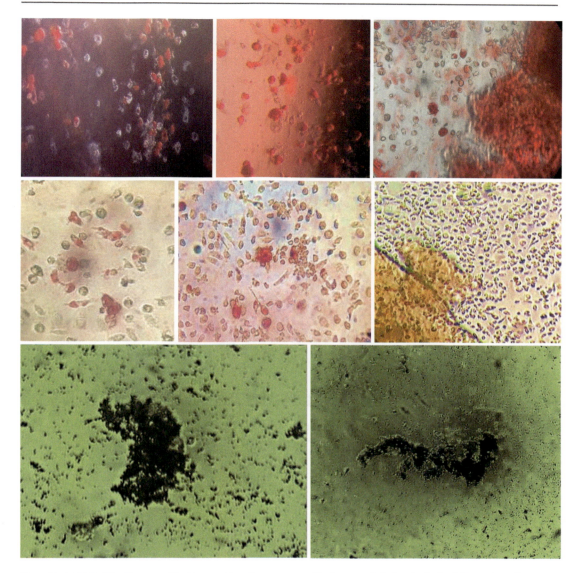

Figs. 13.1–13.8 The Compact Bio Bone with cells was incubated with SFM in an specific cell incubator at 37 °C with 5% CO_2 and at day 7 the scaffold with the fibrin sheet was stained by AR (Figs. 13.1–13.3) cells results positive to the red stain, it's possible to see under the microscope (X 100) the presence of osteoblast like cells; the scaffold was then stained by AP at day 7 and the cells positively revealed the typical violet color of mature osteoblasts, it's possible to visualize under microscope the presence of osteoblast cells expressing the AP within the fibrin compartment (Figs. 13.4–13.6- X 100); cells positively revealed the expression of VK typical of osteo cell expressing calcium matrix in a black color (Figs. 13.7 and 13.8)

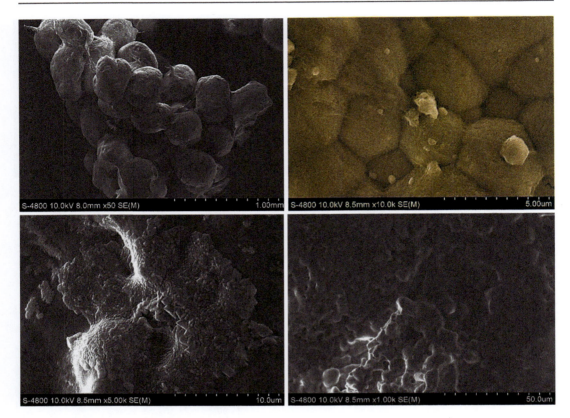

Figs. 13.9–13.12 The SEM analysis showed the ingrowth newly osteoblasts like cells within the Compact Bio Bone scaffold. The cells were shown to be able to interact with their micro-environment and change it, typical filaments and osteo-composit material are seen in all pictures

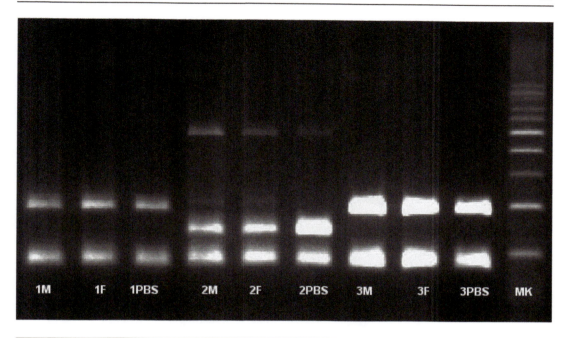

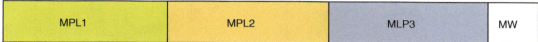

Fig. 13.13 RT-PCR analysis for osteoblast from PB-SCs seeded in beta tricalcium phosphate scaffold. The results show the expression of typical bone matrix gene such as Runx2 (203 Kb), OCN (150 kb), DMP (200 Kb) and Nestin (496 Kb) an intermediate filament protein that is commonly expressed in the neural or glial progenitors. PB-derived Nestin+ stem cells showed self-renewal capacity contributing to the regular osteoblast lineage cell turnover in the adult bone

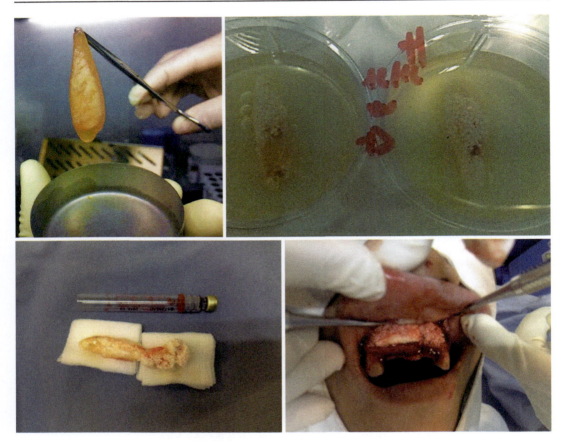

Figs. 13.14–13.17 Preparation of Compact Bio Bone with freshly harvested PB-SCs embedded in a autologous fibrin gel (15); the Compact Bio Bone could be bultured in FSM up to 12 days in cell incubator 37 °C with 5% CO_2; the Compact Bio Bone is ready to be used by dentist and inserted on patient (17-17)

13 Innovative Scaffold Solution for Bone Regeneration Made of Beta-Tricalcium Phosphate Granules... 177

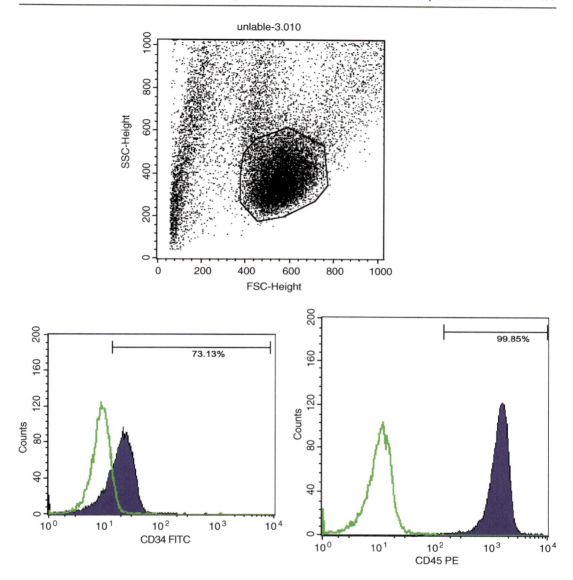

Figs. 13.18–13.23 Both adherent and floating autologous hPB-SCs were resulted positive for stem cell pluripotent and multipotent CD markers by Flow-cytometry analysis CD 34-45-90-105 SSEA3 but negative for SSEA1

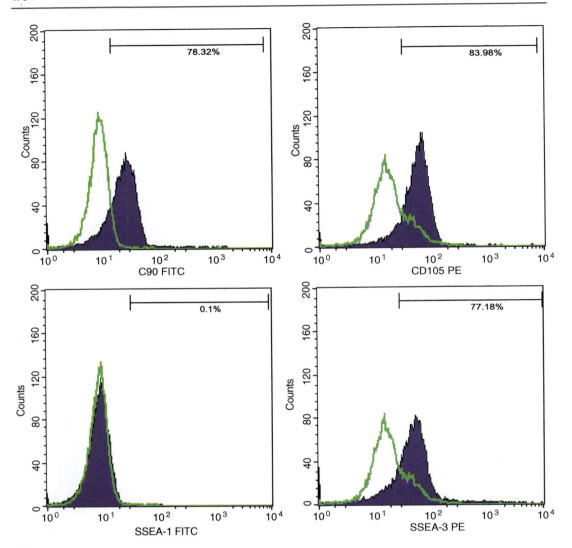

Figs. 13.18–13.23 (continued)

13.4 Conclusion

The big majority of researches in the development of these types of biocompatible bone graft substitutes are still in the *experimental* testing phase and very few have been conducted on clinical scale. A large *clinical* phase testing will be a necessity as it will open a new vision about the complexity of physiological-molecular processes that are involved in the clinical scenario. In addition, there is still much to learn about the strict relationship between bone tissue microenvironment and the current endocrine-metabolic condition of the patient to assess a proper individualized treatment and procedure. All of these questions surely will need a close interdisciplinary approach between all the different branches of science as medicine, chemistry, physics, and engineering.

References

1. Mistry AS, Mikos AG. Tissue engineering strategies for bone regeneration. Adv Biochem Eng Biotechnol. 2005;94:1–22.
2. Ling JL, Liu N, Shi JG, Liu Q, et al. Osteogenic scaffolds for bone reconstruction. BioResearch Open Access. 2012;1(3):137–44.
3. Tran CT, Gargiulo C, Thao HD, Tuan HM, Filgueira L, Michael Strong D. Culture and differentiation of osteoblasts on coral scaffold from human bone marrow mesenchymal stem cells. Cell Tissue Bank. 2011;12(4):247–61.
4. Fisher JN, Peretti GM, Scotti C. Stem cells for bone regeneration: from cell-based therapies to decellularised engineered extracellular matrices. Stem Cell Int. 2016;2016:1–15.
5. O'Brien FJ, Farrell E, Waller MA, Connell I, et al. Scaffolds and cells: preliminary biomechanical analysis and results for the use of a collagen gag scaffold for bone tissue engineering. Topic Bio-Mech Eng. 2004:167–83.
6. Einhorn TA. Enhancement of fracture healing. J Bone Joint Surg. 1995;77:940–56.
7. Puska M, Aho AJ, Vallittu P. Polymer composites for bone reconstruction. Adv Compos Mater Anal Nat Man-Made Mater. 2009;3:55–74.
8. Carlsson AS, Magnusson B, Moller H. Metal sensitivity in patients with metal-to-plastic total hip arthroplasties. Acta Orthop Scand. 1980;51(1):57–62.
9. Thomas P, Thomsen M. Allergy diagnostics in implant intolerance. Orthopedics. 2008;37(2):131–5.
10. Gamblin AL, Brennan AM, Renaud A, Yagita H, Lezot F, et al. Bone tissue formation with human mesenchymal stem cells and biphasic calcium phosphate ceramics: the local implication of osteoclasts and macrophages. Biomaterials. 2014;35:9660–7.
11. Grage-Griebenow E, Flad HD, Ernst M. Heterogeneity of human peripheral blood monocyte subsets. J Leukoc Biol. 2001;69:11e20.
12. Adutler-Lieber S, Ben-Mordechai T, Naftali-Shani N, Asher E, Loberman D, et al. Human macrophage regulation via interaction with cardiac adipose tissue-derived mesenchymal stromal cells. J Cardiovasc Pharmacol Ther. 2013;18:78e86.
13. Nemeth K, Leelahavanichkul A, Yuen PST, Mayer B, Parmelee A, et al. Bone marrow stromal cells attenuate sepsis via prostaglandin E(2)-dependent reprogramming of host macrophages to increase their interleukin-10 production. Nat Med. 2009;15:42e9.
14. Alexander KA, Chang MK, Maylin ER, Kohler T, et al. Osteal macrophages promote in vivo intramembranous bone healing in a mouse tibial injury model. J Bone Min Res. 2011;26:1517–32.
15. Benoit DS, Schwartz MP, Durney AR, Anseth KS. Small functional groups for controlled differentiation of hydrogel-encapsulated human mesenchymal stem cells. Nat Mater. 2008;7(10):816–23.
16. Shih YRV, Hwang YS, Phadke A, Kang H, et al. Calcium phosphate-bearing matrices induce osteogenic differentiation of stem cells through adenosine signaling. PNAS. 2014;111(3):990–5.
17. Peng G, Haoqiang Z, Yun L, Bo F, et al. Beta-tricalcium phosphate granules improve osteogenesis *in vitro* and establish innovative osteo-regenerators for bone tissue engineering *in vivo*. Sci Rep. 2016;6:23367.
18. Chang YL, Stanford CM, Keller JC. Calcium and phosphate supplementation promotes bone cell mineralization: implications for hydroxyapatite (HA)-enhanced bone formation. J Biomed Mater Res. 2000;52(2):270–8.
19. Gargiulo C, Pham VH, Hai NT, Nguyen NCD, Pham VP, Abe K, Flores V, Shifman M. Isolation and characterization of multipotent and pluripotent stem cells from human peripheral blood. Stem Cell Discov. 2015;5(3):1–17.
20. Polo-Corrales L, Latorre-Esteves M, Ramirez-Vick JE. Scaffold design for bone regeneration. J Nanosci Nanotechnol. 2014;14(1):15–56.
21. Panaroni C, Tzeng YS, Saeed H, Wu JY. Mesenchymal progenitors and the osteoblast lineage in bone marrow hematopoietic niches. Curr Osteoporos Rep. 2014;12(1):22–32.

Ordinary and Activated Bone Substitutes

14

Ilya Y. Bozo, R. V. Deev, A. Y. Drobyshev, and A. A. Isaev

14.1 Introduction

Bone grafting is one of the frequently used surgical procedures in traumatology, orthopedics, and oral and maxillofacial surgery. The treatment of patients with skeletal bone pathology requires often the use of bone substitutes, which temporarily replace the lost volume of bone tissue and accelerate reparative osteogenesis. The high rate of such operations is associated with the prevalence and variety of pathological conditions that result in the formation of bone defects. The specific group of indications for bone grafts and substitutes in traumatology and orthopedics consists of degenerative diseases of spine and major

---

I. Y. Bozo (✉)
Department of Maxillofacial Surgery, A.I. Burnazyan Federal Medical Biophysical Center of FMBA of Russia, Moscow, Russia

"Histograft", LLC, Moscow, Russia

R. V. Deev
"Histograft", LLC, Moscow, Russia

Department of General Pathology, I.P. Pavlov Ryazan State Medical University, Ryazan, Russia

PJSC "Human Stem Cells Institute", Moscow, Russia

A. Y. Drobyshev
Department of Maxillofacial and Plastic Surgery, A.I. Evdokimov Moscow State University of Medicine and Dentistry, Moscow, Russia

A. A. Isaev
PJSC "Human Stem Cells Institute", Moscow, Russia

joints, and in dental and maxillofacial surgery the atrophy of the alveolar ridges of the upper and lower jaws [1].

In the USA, according to the National Center for Health Statistics, in 2010, 4,392,000 bone and joint surgeries were made. Approximately one million of them involved cranial bones, extremities, ribs, and sterna affected by injuries, postsurgical deformations, and oncological and inflammatory diseases, and 1,394,000 more were joint replacements of the lower extremities (with regard to revision surgeries). Bone grafting materials were required at least in 20–25% of the cases. There were 500,000 spine fusions (including 27,000 reoperations), which usually utilized bone substitutes, and 21,000 cases of arthrodesis. In other words, the total number of surgeries with bone grafting was at least 1.3–1.5 million. As the total number of autogenic bone harvesting procedures did not exceed 207,000, the need for approved bone substitutes is evident.

Bone grafting is also required for one of every four dental implant placement. According to the estimate of Straumann (Germany) in 2016, the total number of implants annually placed among 19 countries with highest level of dental implant penetration is not less than 14.2 million. The demand for bone substitutes exceeds 3.5 million units in this category of indications alone.

More than 200 bone grafting materials have been approved for clinical use all over the world. A larger number of them are investigated in experi-

---

© Springer Nature Switzerland AG 2019
D. Duscher, M. A. Shiffman (eds.), *Regenerative Medicine and Plastic Surgery*,
https://doi.org/10.1007/978-3-030-19962-3_14

mental and clinical studies. Such a variety of products is the result of not only high demand but also the absence of a universal bone grafting material that could be effective in most clinical cases. Even with a correctly chosen treatment plan and an optimal surgical technique with advanced medical equipment, the bone substitute may often predetermine the unpredictability and, in some cases, unacceptability of the clinical outcome.

The array of bone substitutes that have been implemented in clinical practice and evaluated in different studies should be systematized. For this purpose, a row of classifications based on nature, chemical composition, physical properties, and other parameters have been described [2]. Moreover, we proposed chronological classification dividing all developed bone substitutes into five generations: xeno-, allo-, and autogenic bone fragments not specifically processed; preserved allogenic bone materials; bone matrix analogues of synthetic and natural origin, including items with growth factors; tissue-engineered bone grafts; and gene-activated bone substitutes [3]. All of these systems are logical, but have only theoretical relevance that is not associated with therapeutic indications and, accordingly, do not aid in the selection of the most optimal variant of material in a particular clinical situation. For this reason, we described applied classification that unified both theoretical aspects significant for biomaterial specialists and practical aspects that physicians needed [1]. The sense of this classification is presented in the chapter title and will be explained below. But before that we need to detail the key aspects of reparative osteogenesis regulation that formed the background for "ordinary and activated bone substitute conception."

14.2 Features of Reparative Osteogenesis Regulation: Local Osteoinductive Factors and "Osteogenic Insufficiency"

Reparative osteogenesis is difficultly regulated process which requires coherent cooperation of different cell types belonging to the lines of osteocytes, osteoclasts, fibroblasts, endotheliocytes, and leukocytes interacted between each other with cytokines, growth factors, and direct cell-to-cell and cell-to-matrix contacts. Each kind of cells on every stage of development has special role that is realized via proliferation, further differentiation, production of extracellular matrix components, bone matrix resorption, etc. Local osteoinductive factors (Table 14.1) that cells secrete within the bone defect area are critically important to make this "orchestra" work correctly for the sake of complete bone restoration.

All factors involved in reparative osteogenesis regulation have different impacts on bone formation, but some of them are dramatically important.

Bone morphogenetic proteins. BMP are members of the transforming growth factor family discovered in the second half of the twentieth century whose biological effect is not limited by bone tissue. They are so referred to because they were first discovered in demineralized and lyophilized bone matrixes that were implanted into rabbit muscles and showed osteoinductive properties [44]. Among all members of the BMP family, BMP-2, -4, -6, -7, and -9 have the largest impact on cells of osteoblastic differon (lowest—BMP-3, -5, -8, -10–15) [13].

Binding BMP with specific membrane tyrosine kinase receptors (types 1 and 2) results in the phosphorylation of intracellular proteins Smad-1, -5, and -8, which after activation form a "transport" complex, with Smad-4 translocating them to cell nuclei. In the nuclei, the Smad receptor proteins increase the expression of genes encoding key transcription factors responsible for the activation of the "osteoblastic phenotype" in cells (Fig. 14.1) [45–50]. Such transcription factors include Runx2 (runt-related transcription factor 2) [50, 51], Msx2 [52], and Dlx 5 and 6 [53]. By interacting with each other and with other transcription factors, such as Osx (osterix) [54, 55], they affect target genes. As a result, they increase the proliferative activity of progenitor cells (mainly Msx2 [52]) and differentiation to osteoblasts, as well as the production of the components of the bone intracellular matrix (osteocalcin,

14 Ordinary and Activated Bone Substitutes

Table 14.1 The main local osteoinductive factors

Factor	Effect on osteogenesis	Effect on angiogenesis
BMP-2,4	Activation of proliferation, differentiation, synthesis of components of bone intercellular matrix and growth factors (VEGF, bFGF, etc.) [4, 5]. Biological action is decreased by impact of BMP-3 [6]	Influence on EPC. Stimulation of migration, proliferation, and formation of capillary-like structures; increase of VEGF and ANG-1 receptor expression; no effect on cell differentiation and survival [7–9]
BMP-3	Suppression of differentiation; decrease of osteogenic activity [6]	–
BMP-6	Decrease of proliferative activity of MMSCs and activation of their differentiation [10] (to a greater extent than the other BMPs [11])	Activation of EPC proliferation, organization of capillary-like structures [12]
BMP-7	Activation of proliferation, differentiation, and synthesis of components of bone intercellular matrix [13]	Increase of endothelial cell proliferation, production of VEGF receptors, induction of capillary-like structure formation [14]
BMP-9	Increase of bone intercellular matrix production without negative regulation by BMP-3 [15]	Activation of endothelial cell proliferation, including production of angiogenic factor receptors (VEGF and ANG-1) [7]

(continued)

bone sialoprotein, alkaline phosphatase, and collagens of types III and I) [54]. Interestingly, two other Smad types, 6 and 7, have an inhibitory

Table 14.1 (continued)

Factor	Effect on osteogenesis	Effect on angiogenesis
Vascular endothelial growth factor (VEGF)	Increase of proliferative activity, differentiation, and chemotaxis induction by gradient of concentration [16–19]	Stimulation of proliferation, differentiation, migration, formation of capillary-like structures, inhibition of endothelial cell apoptosis [20–22]
Stromal derived factor-1 (SDF-1)	Induction of cambial cell homing by concentration gradient, inhibition of differentiation [23]	Activation of migration, proliferation, adhesion, and differentiation of EPCs [24]
Angiopoietin-1, 2	–	Activation of differentiation, intercellular contact formation of endothelial cells in vessel wall (vascular stabilization) [25, 26]
Erythropoietin	Stimulation of MMSC differentiation to osteoblasts, monocytes—to osteoclasts, without increase of their activity [27]; increase of chondrocyte proliferation [28]	Stimulation of endothelial cell proliferation [29] and NO production [30]
Basic fibroblast growth factor	Increase of proliferation and suppression of differentiation [31]	Increase of proliferation and suppression of EPC differentiation [32]
Hepatocyte growth factor	Activation of differentiation and synthesis of bone intercellular matrix components [33]	Activation of proliferation and migration [34], inhibition of apoptosis, decrease of endothelial permeability [35]

(continued)

Table 14.1 (continued)

Factor	Effect on osteogenesis	Effect on angiogenesis
Insulin-like growth factor-1	Increase of mechanic sensitivity of specialized cells, induction of differentiation, and synthesis of bone intercellular matrix components in response to physical exercise [36]	Activation of migration, proliferation, and differentiation of endothelial cells, induction of capillary-like structure formation [37]
PDGF-AA	Insignificant increase of proliferation and differentiation; chemotaxis activation (to lesser extent than when exposed to PDGF-BB) [38]; increase of IGF-1 production [39]	–
PDGF-BB	Activation of cell proliferation and migration [40]	Induction of pericyte migration, adhesion and incorporation to walls of forming vessels, activation of EPC migration [40]
TGF-β1	Increase of proliferative activity, decrease of differentiation, and synthesis of bone intercellular matrix components [41]	Activation, migration, proliferation, and formation of capillary-like structures [42]

(continued)

Table 14.1 (continued)

Factor	Effect on osteogenesis	Effect on angiogenesis
Angiogenin	–	Release of endothelial cells from vascular vessels and their activation, stimulation of migration, and proliferation [43]

EPC endothelial progenitor cells

effect on Smad-mediated BMP action [56]. The intracellular Smad signal pathway is not solely for BMP [4], and the list of Smad activators and transcription factors is not limited to the trans-forming growth factor family (BMP and TGF-β) (Fig. 14.1).

Loss-of-function mutations in genes encoding BMP-2 or key intracellular proteins (Runx2, Msx2, Dxl5, 6; Osx, etc.) providing transduction of its signals result in the development of severe disorders that are non-survivable in homozygote status. Therefore, genetically mediated BMP-2 deficiency leads to increased bone fragility, disturbance of enchondral osteogenesis, and mineralization of the bone matrix [57, 58]. Hereby, only BMP-2 function could not be compensated by the activities of other proteins: the selective knockout of other BMPs (4, 7) does not have a significant effect on the histophysiology of skeletal bone, although it is accompanied by pathological symptoms from other organs and systems (urinary, cardiovascular, etc.) [59, 60]. Loss of function of the α-subunit of Runx2 due to mutation, if the identical β-subunit is preserved (Runx2$^{+/-}$), results in the formation of cleidocranial dysplasia (dysostosis) [61, 62], whereas the Runx2$^{-/-}$ genotype is unsurvivable [63]. Autosomal-dominant craniosynostosis is based on Msx2 gene mutations [64].

Vascular endothelial growth factor. VEGF is a family of biologically active proteins first isolated by Folkman et al. in 1971 [65] that comprises the main auto- and paracrine regulation factors of vasculo-, angio- (VEGF-A, B; PIGF), and lymphogenesis (VEGF-C, D); they are produced by cells of all body tissues including epithelial.

In postnatal period of human development, VEGF-A (isoforms 121, 145, 148, 165, 183, 189, 206) [66] has the greatest impact on the formation

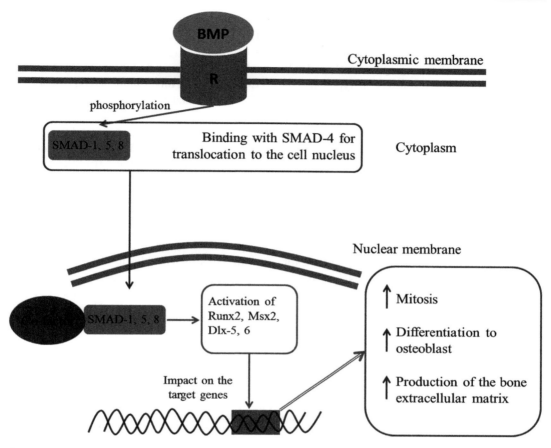

Fig. 14.1 Scheme of intracellular SMAD-mediated transduction pathway for BMP signals. *BMP* bone morphogenetic protein, *R* receptor

of blood vessels. There are three types of VEGF receptors. Types 1 and 2 are involved in angiogenesis, and type 3 is involved in the formation of lymphatic vessels. Hereby, type 1 receptor has a greater affinity to VEGF, but its tyrosine kinase activity is much lower than that of type 2 receptor, which is considered one of the regulatory mechanisms preventing excessive VEGF activity. Correspondingly, VEGF effects are implemented via the type 2 receptor [66]. After VEGF interaction with a specific type 2 receptor, the intracellular tyrosine sites of its kinase and carboxy-terminal domains undergo autophosphorylation (Y951, 1054, 1059, 1175, 1214) which, in turn, activate several intracellular proteins such as phospholipases Cγ, Cβ3, adapter proteins SRK, NCK, SHB, and SCK, which are the first complex cascades of signal transduction that change the morphofunctional state of target cells (mainly endothelial). In particular, phospholipase Cγ hydrolyzes membrane phospholipid PIP_2 by forming diacylglycerol and inositol-1,4,5-triphosphate, which increases the intracellular calcium levels that activate protein kinase C, which, in turn, initiates the subsequent activation of signalling pathway RAS-ERK leading to mitosis induction. As a result, the proliferative activity of endothelial cells is increased [66]. Phospholipase Cβ3 is involved in actin polymerization and the formation of stress fibrils that provide migration and motor cell activity. VEGF suppresses apoptosis via activation of the "phosphoinositide-3-kinase—protein kinase B (PI3K/AKT)" signalling pathway, inhibiting caspases 3, 7, and 9 and increasing cell survival. Moreover, axis PI3K/AKT along with calcium ions modulates the activity of endothelial NO synthase, which is accompanied by a rise in NO production and an increase in vascular permeability, which leads to angiogenesis (Fig. 14.2) [67, 68]. Thus, VEGF via a specific type 2 receptor induces activation, migration, proliferation, and differentiation of endotheliocytes and their progenitor cells, increasing cell survival which, combined with the modulation of intracellular interactions and increase in vascular permeability, are essential prerequisites for the formation of capillary-like structures and subsequent remodelling into mature vessels [69]. Because in both primary and secondary osteogenesis vessels sprouting into fibrous or cartilaginous tissues, respectively, provide the necessary conditions for the differentiation of resident cells into osteoblasts, as well as the migration of cambial reserves (perivascularly and with blood flow), VEGF-A may be considered an indirect osteoinductive factor.

Along with the angiogenesis-mediated effect, VEGF also has a direct influence on osteoblastic differon cells that not only produce VEGF [70] but also express its type 1 and 2 receptors both in embryogenesis [71] and postnatal period of development [72]. It is shown that the proliferation of cambial cells of bone tissue exposed to VEGF significantly increases (up to 70%), and the migration of osteogenic cells is activated by the gradient of VEGF concentration [16].

More recently, apart from the canonical, a receptor, mechanism of VEGF action on the progenitor cells of osteoblastic differon, data on a fundamentally different "intracrine" mechanism is available. Its existence is confirmed by results

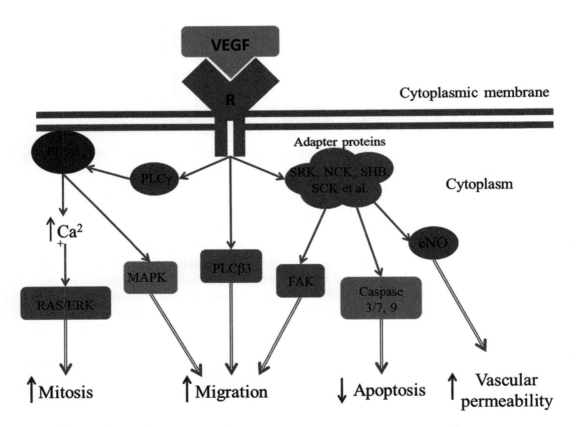

Fig. 14.2 Scheme of the intracellular cascade pathway of VEGF signals. *VEGF* vascular endothelial growth factor, *R* receptor, *PIP-2* phosphatidylinositol biphosphate, *PLCγ* phospholipases Cγ; *PLCβ* phospholipases Cβ, *SRK, NCK, SHB, SCK* group of adapter proteins, *MAPK* mitogen-activated protein kinase, *ERK* complex of extracellular signal-regulated kinase, *FAK* focal adhesion kinase, *eNO* endothelial NO synthase

showing that progenitor cells committed to osteoblasts (expressing Osx) synthesized VEGF not only for "export" but also to differentiate themselves to osteoblasts [73]. Liu et al. [74] investigated bone marrow MMSC cultures obtained from healthy mice (control) and animals with a "loss-of-function" mutation of the gene encoding VEGF. It appeared that cells of the experimental group underwent osteogenic differentiation to a lesser extent than that of the control; hereby, their adipogenic potential was increased. The addition of recombinant VEGF to the culture medium of the "mutant" cells did not result in normalization of osteogenic differentiation, and the addition of antibodies blocking the VEGF receptors in the control was not accompanied by negative effects. However, after transfection of cells in the experimental group by a retroviral vector with the *vegf* gene to compensate for the knockout, an increase in the intracellular concentration of VEGF proteins was observed, which led to normalization of osteogenic differentiation and a simultaneous decrease of adipogenic potential [74].

Therefore, VEGF has a wide spectrum of action on cells of endothelial and mesenchymal cellular differons involved in reparative osteogenesis having both an angiogenesis-mediated stimulatory effect and a direct inducing impact on osteoblastic cells via the receptor and intracrine mechanisms.

Stromal derived factor-1. SDF-1 (CXCL12) is a protein from the chemokine group represented by two forms derived from alternative splicing, SDF-1α (89 amino acids) and SDF-1β (93 amino acids) [75]. Both are produced by cells of the bone marrow, fibroblastic and osteoblastic differons, and perivasculocytes.

The main SDF-1 receptor is CXCR4. After formation of a complex with the ligand, an intracellular G-protein that consists of three subunits (α, β, γ) is activated by separating into a heterodimer Gβ/γ and monomer Gα (4 isoforms) possessing both different and common intracellular pathways of signal transduction. Gβ/γ activates phospholipase Cγ which, as mentioned above, increases the release of calcium ions from intracellular depots and via subsequent chains activates mitogen-activated protein kinases (MAPKs)

which, in turn, initiate chemotaxis (Fig. 14.3) [76]. Moreover, the cell impulse to migration is also provided via PI3K activation, and p38 provides the impulse to proliferation [77]. Transcription factor NF-kappa B, which level is increased under exposure to SDF-1, has a wide range of action due to its increase of expression by more than 200 target genes encoding proteins involved in the regulation of cell proliferation, differentiation, and migration [78]. It should be noted that recent data indicates that NF-kappa B, in general, inhibits osteogenesis via suppression of osteoblastic cambial cell differentiation. In this regard, SDF-1 secreted by osteoblasts from the lesion area [79] may bring a bifacial effect: inducing the homing of progenitor cells including MMSC to the target area [23] and inhibiting their differentiation to osteoblasts. However, there are reasons to suppose that the effects, "undesirable" during a certain period of time, may be eliminated by other factors. In particular, it is shown that activated Smad proteins (1, 5) may inhibit SDF-1 production of osteoblasts [80]. Hence, in the inflammatory phase when activity of BMP proteins is reduced, osteoblasts actively secrete SDF-1 to attract additional cambial reserves, including endothelial precursors to the bone defect by a gradient of the chemokine concentration. With the transition of the recovery process to the phase of regeneration, with the increase of BMP-2 and -7 levels and under their action (through Smad-1, -5), bone cells cease producing SDF-1, which inhibits differentiation of the migrated progenitor cells to osteoblasts. During the late stages of regeneration and as part of the remodelling of newly formed bone tissue when cancellous bone should be populated with cells of bone marrow, the BMP level is decreased, whereas SDF-1 is again increased, which provides a homing of hematopoietic stem cells (HSC). The same mechanism allows cells of the osteoblastic line to hold HSC within bone marrow niches formed by both MMSCs and osteoblasts [81].

Thus, collected and systematized current data on cell populations, systemic and local osteoinductive factors (Table 14.1), and regulatory pathways involved in reparative osteogenesis made a

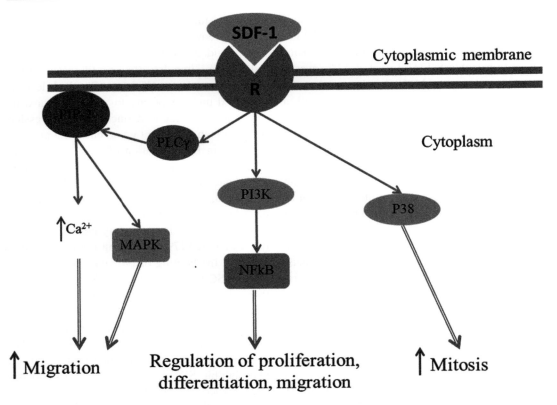

Fig. 14.3 Scheme of intracellular cascade pathway of SDF-1 signal transduction. *SDF-1* stromal derived factor-1, *R* receptor, *PIP-2* phosphatidylinositol biphosphate, *PLCγ* phospholipases Cγ; *MAPK* mitogen-activated protein kinase, *NFkB* nucleic factor-kappa B, *PI3K* phosphoinositide-3-kinase

rationale for the concept of **osteogenic insufficiency** that simply clarified why small and large bone defects healed differently. Osteogenic insufficiency is a pathological condition associated with the low activity of osteoinductive factors and (or) a decreased level of cambial cell in the bone lesion area, so that the natural process of reparative osteogenesis may not provide complete histo- and organotypic bone recovery [82].

Causes of osteogenic insufficiency may be divided into local and general; the former includes defect size, geometry, number of walls, damaging factor (high- and low-energy injuries), presence of pathological inflammatory processes and related factors, and low density of functional blood vessels in the bone defect area, and the latter involves age, coexisting disorders (diabetes mellitus, osteoporosis), social habits (smoking), and administration of drugs that negatively affect osteogenesis (cytostatic agents and possibly bisphosphonates, although a meta-analysis by Xue et al. [83] did not show any negative effects of these drugs on fracture healing time [83]).

It is therefore reasonable to divide all bone defects into two groups based on the absence/presence of osteogenic insufficiency. The first group is characterized by high activity of natural reparative processes, so optimization of bone regeneration alone is sufficient to decrease the treatment term and derive a larger volume of newly formed bone tissue. In the latter, defects with osteogenic insufficiency are determined by poor intensity of osteogenesis, and accordingly they require not optimization but rather induction and maintenance of reparative processes on a high level that may be achieved by introducing additional growth factors, substances increasing their synthesis, or cells that are able to produce them. In these terms, each kind of the defects are specific in their demand to bone substitute's mechanism of action that found practical embodiment in ordinary and activated bone substitute's conception.

14.3 Ordinary Bone Substitutes

This category includes the majority of bone grafting materials approved for clinical applications that do not contain biologically active components standardized by qualitative and quantitative parameters. The present category, which we referred to as "ordinary materials," integrates allogenic and xenogenic bone matrixes from various processing technologies (demineralized, deproteinized, etc.); calcium phosphates (β-tricalcium phosphate, octacalcium phosphate [84], etc.); natural or synthetic hydroxyapatites; synthetic (PLGA, etc.) and natural (collagen, chitosan) organic polymers; silicates; and composite products of the abovementioned materials.

It is well known that bone substitutes may possess different properties that have specific effects on reparative osteogenesis. Such properties include *osteoconduction, osteoprotection, osteoinduction*, and *osteogenicity* [85]. The majority of ordinary materials have mainly osteoconduction. Some of them (e.g., demineralized bone matrixes derived using different processing technologies and calcium phosphates) are additionally characterized by a moderate osteoinductive effect, most likely due to optimal physical and chemical properties and (or) the presence of certain biologically active substances in the matrix that are not standardized by qualitative and qualitative parameters. Their principal mechanism of action is to guide bone regeneration coming from natural osteogenic sources (preserved bone fragments) to the central part of a defect, and their range of ultimate effectiveness is limited by biological properties of damaged bone tissue.

In an attempt to improve ordinary bone substitute properties the researchers correct their composition, structure, surface macro- and microrelief, and bioresorption rate to get more similarities with natural bone matrix. There are some data indicating that material's surface topography is capable to modulate cells' morphofunctional features [86] that theoretically can make the mechanism of ordinary bone substitute action more beneficial through some direct influence on the cells involved in reparative osteogenesis.

Overall ordinary materials are appropriate for bone grafting in cases with high activity of native osteoinductive factors and (or) levels of cambial cells, but not enough for large bone defect repair. In other words, ordinary substitutes are ineffective in bone defects with osteogenic insufficiency, as they cannot modulate the effects of factors regulating osteogenesis. For that, ordinary materials as the scaffolds are combined with cells, growth factors, or gene constructs encoding them to make complex products, the "activated bone substitutes" containing biologically active components standardized by qualitative and quantitative parameters. Based on the nature of osteoinductive components, these materials may be divided into three main groups: tissue engineered, and protein and gene activated (Fig. 14.4).

14.4 Activated Bone Substitutes

The "ideal prototype" and the main competitor of activated bone substitutes is autogenic bone tissue, a "gold standard" for bone grafting that could be used in any clinical situation. In some cases during a surgery it is possible to harvest certain amount of autobone applicable to be grafted (either alone or being mixed with some other materials in various ratios) in the corresponding step of the same operation. However, majority of cases require expanded or additional surgical approach that increased duration of surgery, complications rate, and associated supplementary trauma. Large bone autograft harvesting including those with vascular supplies preserved for restoration of defect with osteogenic insufficiency is especially complicated, traumatic, and long-term. Moreover, this approach is not absolutely effective because there is a high rate of subsequent graft resorption (up to 50% and more

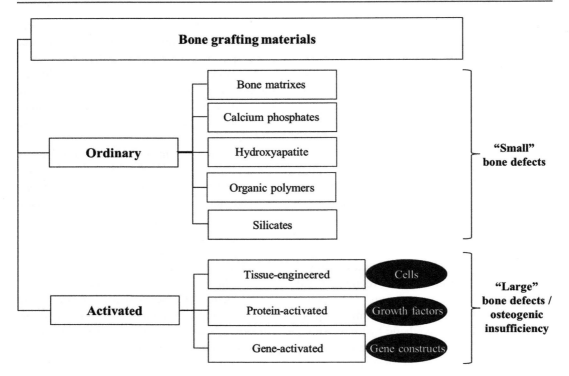

Fig. 14.4 Applied classification of current bone grafting materials according to composition, mechanism of action, and theoretical range of clinical indications

of primary volume lost in case of free iliac crest flap) [86], as well as a risk of microvascular anastomosis failure [87]. Thus, the activated bone substitutes are intended to replace bone autografts in clinical practice.

14.5 Tissue-Engineered Bone Substitutes

This group of materials includes those containing two main components, a bioresorbable scaffold and live (auto- or allogenic) cells. The principal idea of the approach is to replace lost cambial reserves and increase the concentration of osteoinductive factors in the area of material implantation. With high survival rates, cells transplanted into the recipient site may have a beneficial therapeutic effect due to two mechanisms of action: direct—differentiation to specialized cells of impaired tissues (mostly indicated for autogenous cells)—as well as indirect, a paracrine effect, the modulation of morphofunctional activity of other cells by the production of biologically active substances that are the factors of local osteogenesis regulation (Table 14.1). According to many authors, the paracrine activity of cells of the tissue-engineered bone substitutes, in particular, is their main mechanism of action [88]. Among the most significant factors for reparative osteogenesis produced by transplanted cells, bone morphogenetic protein (BMP), vascular endothelial growth factor (VEGF), and stromal derived factor (SDF-1) should be specified. It is of interest that ten million bone marrow multipotent mesenchymal stromal cells (MMSC) in vitro produce approximately 750 pg/mL VEGF, 1030 pg/mL of TGFβ1 [89], and 220 pg/mL of SDF-1 per day [90], whereas ten million osteogenic periosteal cells secrete up to 40 ng/mL of BMP-2 and 200 pg/mL of VEGF per day [91].

Cells for the development of tissue-engineered bone substitutes may be either expanded by cultural technologies or used immediately as a "fresh population" after harvesting from a tissue source. The main types of cells exposed to in vitro processing are MMSCs, osteogenic cells and osteoblasts, as well as their combination. For that purpose, some investigators use endotheliocytes as independent or additional cell components and even induced pluripotent stem cells. Uncultured cell populations include bone marrow cells (a mixture of MMSCs, fibroblasts, endothelial progenitor cells, HSC and definitive blood cells, etc.) and the stromal-vascular fraction of adipose tissue (SVF-AT) (MMSCs, endotheliocytes and endothelial progenitor cells, smooth muscle cells, fibroblasts, pre-adipocytes, and immunocompetent cells).

Multiple preclinical studies have shown the safety and efficacy of different variants of developed tissue-engineered bone substitutes that formed the basis for clinical translation. Moreover, several bone substitutes containing live cells have already been registered and approved for clinical application:

1. Allogenic: "Osteocel plus" (NuVasive, USA) (2005), "Trinity Evolution" (Orthofix, USA) (2009), "AlloStem" (AlloSource) (2011), "Cellentra VCBM" (BioMet, USA) (2012), and "OvationOS" (Osiris Therapeutics, USA) (2013)
2. Autogenous ("cell service," which consists of harvesting a primary cell population, cultivation, combination with an appropriate scaffold, and transfer to the clinic for use): "BioSeed-Oral Bone" (BioTissue Technologies, Germany) (2001), "Osteotransplant DENT" (co.don, Germany) (2006)

The large number of registered tissue-engineered bone substitutes proves both the safety and the efficacy of the approach for certain bone defects. The majority of the registered products (Trinity Evolution, AlloStem, Osteocel Plus, Cellentra VCBM) are indicated for particular variants of spondylo- and arthrodesis, and others are for substitution of jaw defects (BioSeed-Oral Bone, Osteotransplant DENT). "Osteocel Plus" is one of the first and the most successful tissue-engineered bone substitutes. The product is allogenous spongy bone tissue with live cells

$(5.25 \times 10^5 \pm 4.6 \times 10^3$ in 5 mL [92]) that are preserved in the composition due to a special "gentle" processing technology for cadaveric material with immunodepletion. Osteocel Plus is intended for spine surgery, including cervical spinal fusion. Separate clinical studies were performed on each of the five surgery types in which the product was indicated, a total number of 384 patients were enrolled, and adverse events were not reported. Most patients (over 90%) achieved complete fusion at 5–6 months after surgery [93]. The successful results of pilot clinical studies were published on the use of Osteocel Plus for other indications, such as augmentation of the alveolar ridge and arthrodesis of the lower extremities [94]. The developers associate the mechanism of action of Osteocel Plus and other allogenous tissue-engineered bone substitutes with the osteoinductive effect of the matrix, as well as the paracrine activity of cells producing BMP, VEGF, SDF-1, and other growth factors [92]. A certain concentration of biologically active substances is also contained in the bone matrix.

None of these tissue-engineered bone substitutes has been registered in Russia as a medical product approved for clinical application, although successful results in the area have been obtained since the 1980s [95, 96]. This is mainly due to the absence of approved legal regulations on the registration of medical products consisting of cells. Nevertheless, pilot and initiative clinical studies under local authorities, beyond registration, have been ongoing [97–101]. In particular, successful results were obtained with the use of tissue-engineered bone substitutes consisting of autogenous adipose-derived MMSCs and two types of matrices (hydroxyapatite and a composite material of hydroxyapatite and collagen) in the treatment of patients with the atrophy of the alveolar bone of the upper jaw and the alveolar part of the lower jaw at the A.I. Evdokimov Moscow State University of Medicine and Dentistry [98]. A pilot clinical trial on the safety and efficacy of tissue-engineered bone substitutes of tricalcium phosphate and autogenous gingiva-derived MMSCs with sinus lifting has been initiated at the A.I. Burnazyan Federal

Medical Biophysical Center (NCT02209311) [99].

However, several negative aspects of tissue-engineered bone substitutes should be mentioned:

- Lack of efficacy for large bone defects due to the death of most cells shortly after the transplantation of the tissue-engineered bone substitutes (cells require an active blood supply which is crucially minimized in a large lesion area) [100]
- High self-cost and complexity of technological process (cellular service) for making tissue-engineered bone substitutes in accordance with GMP and GTP standards
- Impossibility to organize full-scale batch production of the most effective personalized (containing autogenous cells) tissue-engineered products
- Special storage conditions that are not always available at medical institutions (e.g., temperatures below −80 °C)
- Complicated legal regulation and registration of medical products containing live cells

Thus, the tissue-engineered approach to the development of activated bone substitutes allows the creation of safe medical products that are effective for certain indications. However, there are some problems that limit the implementation of tissue-engineered products to routine clinical practice that predetermines the development of alternative approaches.

14.6 Bone Substitutes with Growth Factors

This group includes bone grafting materials consisting of a scaffold and growth factors (one or a few) that provide an osteoinductive effect; this is the most successful trend considering the precedents of clinical translation. Numerous products have already been registered and approved for clinical use such as "Emdogain" (Straumann, Germany)—a material with enamel matrix proteins (1997); "OP-1" (Stryker Biotech, USA)—

with recombinant BMP-7 (2001); "Infuse" (Medtronic, USA) (2002, 2004, 2007)—with recombinant BMP-2; "GEM21S," "Augment bone graft" (BioMimetic Therapeutics Inc., USA)—with recombinant PDGF-BB (2005, 2009); and "i-Factor Putty" (Cerapedalloics, USA)—with protein P-15 (ligand for integrin α2β1 expressed by cells of an osteoblastic line) (2008) [1].

"Infuse" was approved by the FDA for interbody spinal fusion in 2002, for bone grafting in shin bone fractures in 2004 (in combination with intramedullary fixation), and for sinus lifting and augmentation of the alveolar ridge in defects related to tooth extraction in 2007 [102]. The product is manufactured as a set consisting of a collagen matrix and recombinant BMP-2, which should be combined immediately prior to use. For spine surgery, because of the suboptimal biomechanical properties of the material, it should be implanted in a complex with special metallic cages. Two hundred and seventy patients were enrolled in the first clinical study and underwent anterior lumbar interbody spinal fusion. Of these patients, 143 had surgery with Infuse, and the others had surgery with a bone autograft of the iliac crest. During the 2-year follow-up, adequate safety was shown, as well as high efficacy of treatment, with fusion rates of 94.5% and 88.7% in the clinical and control groups, respectively (the differences were not statistically significant). Only in patients of the control group (5.9%) were adverse events related to autograft withdrawal identified [103]. Subsequently, several postmarketing clinical studies were performed; the results were published, and a systematic analysis revealed the safety and efficacy of Infuse to be equal to those of bone autografts [104–106].

However, critical articles were also published that emphasized the complications and adverse events of Infuse, as well as their concealment by the company developer [107, 108]. A special issue of the journal "Spine" was fully devoted to the problem, including the central review of the chief editor Carragee et al. [107]. The authors conducted a detailed analysis of 13 official clinical studies on Infuse, including reports submitted to FDA, on a total of 780 patients, and revealed that the rates of complications and adverse events (osteolysis with horizontal or vertical implant dislocation, lack of fusion, retrograde ejection, heterotopic ossification, radiculitis and infections) were approximately 10% for on-label application and up to 50% for thoracic or cervical spinal fusion [107]. Summarizing the results of multiple clinical studies on all three Infuse indications, it was stated that "on-label" use of the product is safe and effective in most cases, although there is a definite possibility of complications, as well as unsatisfactory results requiring re-surgery. The off-label application, for other variants of bone grafting, is accompanied with a significant increase in the risk of complications and adverse events [109]. The total number of surgeries with bone substitutes containing BMP alone (mainly Infuse) in 2010 was 107,000, and the total number of spine interventions with interbody cages was 206,000. This shows both the success of bone substitutes with growth factors and perhaps the high rates of off-label application of Infuse.

Some bone substitutes with growth factors are at different stages of experimental and clinical studies in Russia. For example, the results of the evaluation of bone substitutes with recombinant BMP-2 [110] or VEGF [111] have been published.

Advanced studies on the development of bone substitutes with growth factors focus on two main aspects, the combination of several factors including angiogenic and osteogenic, e.g., VEGF and BMP-2, in one scaffold [112] and providing the prolonged and controlled release of therapeutic proteins from the matrix structure, in particular due to the regulated dynamics of hydrogel matrix biodegradation [113–115] or the encapsulation of growth factors into microspheres made of organic polymers [116]. Some authors changed the structure of growth factors using special technologies (e.g., site-directed mutagenesis) to combine several factors, creating "mutant" molecules with higher efficacy in the activation of reparative osteogenesis. For example, Kasten et al. [117] modified growth/differentiation factor-5 (GDF-5) by adding BMP-2 sites to its sequence to enable it to bind with specific receptors. As a

result, molecule GDF-5 acquired properties typical of BMP-2 [117, 118].

Bone substitutes with growth factors also have shortcomings and problems that limit their efficacy. Firstly, protein molecules in surgical wounds (due to exudation and the high activity of proteolytic enzymes) undergo rapid biodegradation, making them short-lived, which does not allow the bone substitute to demonstrate its osteoinductive action to the fullest extent. Second, the amount of therapeutic protein is limited, and its action is short-term and difficult even with controlled and limited release. In other words, the low concentration of protein molecules that left the scaffold and preserved its biological activity reaches a target cell, interacts with specific receptors on its surface, and induces a biological effect. Hereby, the receptors are rapidly inactivated in the presence of the ligand as compensatory adaptation mechanism that protects cells from excessive stimulation. The biological effect of growth factors will cease, and the protein concentration will be exhausted.

Theoretically, tissue-engineered and gene-activated materials are devoid of such shortcomings. In the first case, surviving cells protractedly produce a range of biologically active substances that accurately react on microenvironment signals, and in the second they act more gently and long-term than bone substitutes with growth factors, as therapeutic proteins are produced for a certain period of time due to the expression of gene constructs delivered to target cells that can be regulated by the microenvironment.

14.7 Gene-Activated Bone Substitutes

The main active component of these products is gene constructs (nucleic acids). In this regard, the development of gene-activated bone substitutes is directly related to the advances of gene therapy, in which gene constructs are used as active substances of gene therapeutic drugs.

Since 1989, more than 1900 clinical studies on gene therapy have been already registered [119], which highlights the activity of the conducted studies. Moreover, several gene thera-peutic drugs have been already implemented into routine clinical practice: "Gendicine," "Oncorine" (SiBiono GeneTech, China), "Neovasculgen" (HSCI, Russia), and "Glybera" (uniQuro, The Netherlands). "Neovasculgen" is a Russian research product approved for clinical use in Russia and Ukraine [120, 121]. The experience of "Neovasculgen" development has been extrapolated to the first clinical trial of a gene-activated bone substitute (NCT02293031) [122] and subsequent clinical trial approved by Russian Federal Service for Surveillance in Healthcare (NCT03076138).

A gene-activated bone substitute is a complex "scaffold-nucleic acid" combined using methods such as "chemical binding" [123], adjuvants (e.g., gel biopolymers) [124], or direct incorporation of nucleic acids into the scaffold at a certain stage of the matrix synthesis. The total efficacy of the product is thereby determined with the total mechanism of action including both gene constructs (osteoinduction) and a scaffold (osteoconduction).

Two subsequent stages may be distinguished in the mechanism of the osteoinductive action of a gene-activated bone substitute, nonspecific and specific. The first is associated with the release of nucleic acids from the scaffold structure after implantation into the bone defect area, delivery to the cells of recipient area, and expression. This step is similar for any gene constructs, and the variability of transfection is provided mainly by transgene delivery systems. The second consists of the specific action of a protein regulatory molecule produced by transfected cells which act as "bioreactors of therapeutic proteins," synthesizing them for a certain period of time. In contrast to bone substitutes with growth factors, the main component of a gene-activated bone substitutes acts "gently," as mentioned above. In other words, transgene entry into the nucleus of target cell does not force obligatory expression of the therapeutic protein. The cell preserves its normal functional state and reaction to microenvironment stimuli, so that if the therapeutic protein is not needed at a particular period in time the transfected cell may decrease the mRNA of the trans-

gene by intracellular posttranscriptional mechanism regulated stability and the half-life of mRNA and thereby prevent protein production [125]. This mode of action of gene constructs significantly increases the efficacy of gene-activated bone substitutes in comparison with substitutes containing growth factors [126].

Gene constructs consist of a therapeutic gene (cDNA or RNA) and its intracellular delivery system (vector). Vectors are divided into two main groups, viral and nonviral. In the first case, a transgene is incorporated into a particle of a retro-, lenti-, adenovirus, or adeno-associated virus, and in the second case a transgene is incorporated into a plasmid, a circular molecule of nucleic acids containing several additional sequences providing transgene expression. Viral and nonviral delivery systems differ in their efficacy of transfection. 40% or more of viral gene constructs can enter target cells, and the rate of plasmid DNA uptake ("naked") does not exceed 1–2% due to its size and negative charge. Some approaches were proposed (physical and chemical) to increase the efficacy of plasmid DNA transfection up to 8–10% [127].

It should also be mentioned that several viral vectors (retro-, lentiviral, etc.) are incorporated into the genome. In other words, a transgene has an almost lifelong expression, and others, including plasmid DNA, are not integrated into the genome and therefore only temporarily express for 10–14 days. Considering that the production of a therapeutic protein encoded by a gene construct should not exceed the terms of complete reparative regeneration, retro- and lentiviral vectors are rarely used in making gene-activated bone substitutes; they are more often applied in the gene-cellular approach, wherein a cell culture is transfected ex vivo and then combined with a scaffold [128].

Hence, all gene-activated materials may be divided by the technology of scaffold and gene construct combinations, as well as by the compositions of their biologically active components: the nature of the vector or transgenes or the number of transgenes or various gene constructs in one product. However, it is evident that the main differences in the biological effect of a

14 Ordinary and Activated Bone Substitutes

Table 14.2 Compositions of gene constructs developed for induction of reparative osteogenesis (as components of gene-activated bone substitutes or gene-cellular products)

No.	Protein encoded by transgene	Vector	References
Genes encoding growth factors/hormones			
1	Angiopoietin-1		Cao et al. [129]
2	BMP-2	Plasmid DNA	Betz et al. [130]
		Adenoviral	Baltzer et al. [131]
		Lentiviral	Virk et al. [132]
		Liposomal	Lutz et al. [133]
3	BMP-4	Plasmid DNA	Chen et al. [134]
		Retroviral	Rose et al. [128]
4	BMP-6	Plasmid DNA	Sheyn et al. [135]
		Adenoviral	Bertone et al. [136]
		Adeno-accociated	Li et al. [137]
		Lentiviral	Sheyn et al. [135]
5	BMP7 (OP-1)	Plasmid DNA	Bright et al. [138]
		Adenoviral	Schek et al. [139]
		Adeno-associated	Song et al. [140]
		Retroviral	Breitbart et al. [141]
6	BMP-9	Plasmid DNA	Kimelman-Bleich et al. [142]
		Adenoviral	Abdelaal et al. [143]
7	BMP-12	Plasmid DNA	Kuroda et al. [144]
8	Cyclooxygenase-2 (Cox-2)	Retroviral	Rundle et al. [145]
9	Erythropoietin (EPO)	Adenoviral	Li et al. [146]
10	Epidermal growth factor (EGF)	Plasmid DNA	Wallmichrath et al. [147]
11	bFGF	Plasmid DNA	Guo et al. [148]
12	HGF	Adenoviral	Wen et al. [149]
13	HIF-1α	Lentiviral	Zou et al. [150]
14	IGF-1	Plasmid DNA	Shen et al. [151]
15	Integrin-α5	Lentiviral	Srouji et al. [152]
16	LIM mineralization protein-1 (LMP-1)	Retroviral	Strohbach et al. [153]
17	LMP-3	Adenoviral	Lattanzi et al. [154]
18	Nell-1	Adenoviral	Lu et al. [155]
19	Osterix	Retroviral	Tu et al. [156]
20	PDGF-A	Adenoviral	Jin et al. [157]
21	PDGF-B	Plasmid DNA	Elangovan et al. [158]
		Adenoviral	Jin et al. [157]
22	Parathyroid hormone (amino acids 1–34)	Plasmid DNA	Fang et al. [159]
23	TGF-β1	Nonviral vector (K)16GRGDSPC	Pan et al. [160]
24	VEGF-A	Plasmid DNA	Geiger et al. [161]
		Adenoviral	Tarkka et al. [162]
25	BMP2 + BMP7	Adenoviral	Koh et al. [163]
26	BMP-2 + BMP-6	Adenoviral	Menendez et al. [164]
27	BMP-2 + IHH	Adenoviral	Reichert et al. [165]
28	BMP-2+ VEGF	Adenoviral	Deng et al. [166]
29	BMP-2 + VEGF + IGF-1 + TGF-β1		Liu et al. [167]
30	BMP-7 + PDGF-b	Adenoviral	Zhang et al. [168]
31	RANKL + VEGF	Adeno-associated	Ito et al. [169]

(continued)

Table 14.2 (continued)

No.	Protein encoded by transgene	Vector	References
32	BMP-2/BMP-7	Plasmid DNA	Feichtinger et al. [170]
33	BMP-2/BMP-4	Liposomal	Wehrhan et al. [171]
34	BMP-6/BMP-9	Adenoviral	Die et al. [172]
35	BMP-6/VEGF	Adenoviral	Seamon et al. [173]
36	BMP-7/IGF-1	Adenoviral	Yang et al. [174]
37	BMP-7/OPG	Plasmid DNA	Liu et al. [175]
38	VEGF-A/SDF-1α	Plasmid DNA	Bozo et al. [122]
Genes encoding transcriptional factors			
39	Cbfa1	Lentiviral	Kim et al. [176]
		Adenoviral	Li et al. [177]
40	c-myb	Plasmid DNA	Bhattarai et al. [178]
41	Runx2	Adenoviral	Zhao et al. [179]
		Retroviral	Takahashi [180]
42	SOX9	Adeno-associated	Cucchiarini et al. [181]
43	caALK6 + Runx2	Plasmid DNA	Itaka et al. [182]

gene-activated bone substitutes are driven by the transgene. Nucleotide sequences encoding the main osteoinductive and osteoblast-specific transcription factors are, as expected, the most frequently used for the development of gene-activated bone substitutes (Table 14.2).

Among the transgenes most often selected for induction of recovery processes are *BMP*, especially encoding BMP-2 (Table 14.2), and *VEGFA*. The first studies were related to direct gene transfer; the method injected gene constructs into the soft tissues surrounding the bone defect as a solution, i.e., without immobilization on or into a scaffold [131, 170]. It is important that even in such a case, positive results were obtained, which proved the supreme importance of gene constructs in gene-activated materials. In particular, in the study by Baltzer et al. [131], complete consolidation was shown at 12 weeks after adenovirus administration (2×10^{10} particles) of the DNA encoding BMP-2 in the muscle around defects (1.3 cm) of the femur in rabbits. In the control, in which a gene encoding a fluorescent marker protein (luciferase) without osteoinductive activity was used as a transgene, a central part of the defect was preserved in all cases and filled by fibrous tissue. Until now, in vitro or in vivo direct gene transfer was mainly used in bone indications for the selective assessment of the biological effect of gene constructs chosen for the development of gene-activated bone substitutes.

Feichtinger [170] et al. developed a co-expressive plasmid DNA with genes encoding BMP-2 and BMP-7 that is subcutaneously injected as a solution (20 μg) and found that in 46% of cases induction of heterotopic osteogenesis resulted [170].

However, despite the published positive results of direct gene transfer, without mechanical filling of the bone defects with osteoconductive materials, especially in cases of large defects, complete recovery of the bone seems impossible. In this regard, gene-activated bone substitutes have become the logical "evolution" of direct gene transfer. The peak for the development of such products containing gene constructs with *BMP* occurred in 2004–2007, which may be related to the prior success of an alternative approach: the FDA approval and wide use in clinical practice of bone substitutes containing growth factors BMP-7 (OP-1, Stryker Biotech, USA) and BMP-2 (Infuse, Medtronic, USA) in 2001 and 2002, respectively.

Subsequently, the specification of the role of angiogenesis in bone regeneration, as well as a detailed description of the intracellular signal pathways regulating proliferation, differentiation, and morphofunctional activity of bone cells, formed a fundamental ground for an increasing number of investigators to use sequences encoding VEGF as transgenes and some transcription factors as well (Table 14.2).

Keeney et al. [183] developed a gene-activated bone substitute made of a collagen-calcium-phosphate matrix and plasmid DNA encoding VEGF-A165 (0.35 μg/mm^3 of the carrier). The item was implanted subcutaneously in mice and into the defects of the intercondyloid fossa of the femur (diameter 1 mm, length 7 mm). Although no signs of osteogenesis or a significant difference in the number of vessels appeared under heterotopic conditions, a significantly larger volume of bone was regenerated in the experimental group under orthotopic conditions than in the control (a scaffold with DNA encoding a marker gene) at 30 days after surgery [183]. However, the experimental model for the assessment of bone substitute efficacy could not be considered optimal due to minimum size of the defect.

In Russia, some variants of gene-activated bone substitutes have been already developed using *VEGFA* as a transgene and different scaffolds (xenogenic bone matrix, composite material of collagen and hydroxyapatite, octacalcium phosphate, etc.). The efficacy of the products was shown in a more complex model, with the substitution of bilateral cranial defects (diameter 10 mm) of parietal bones in rabbits [123].

Based on an analysis of the published study results associated with development of gene-activated bone substitutes (as well as the gene-cellular approach and direct gene transfer), we can conclude that most of them showed acceptable safety and high effectiveness in the experimental models, regardless of the vector type and scaffold. However, some difficulties remain for gene-activated materials in general: manufacture, sterilization, standardization of control for preservation of the specific activity of the gene constructs after the completion of the production cycle, and necessity of increasing the transfection level of nonviral gene constructs and enabling their prolonged release from the scaffold's structure after implantation. Delivery problem is the most acute for gene-activated materials containing plasmid DNA. But it was proposed and confirmed in some studies that calcium-consisting scaffold passively improved plasmid DNA delivery by realizing them in microscale complexes with scaffold fragment that was similar to well-known calcium-phosphate-mediated transfection method in vitro [123, 184].

In spite of all the advantages of gene-activated bone substitutes and a huge amount of experimental data the precedents of clinical studies are extremely rare. Below we can present detailed description of world's first **clinical case** [122].

The patient, aged 37 years, a female, was admitted to the department of maxillofacial surgery with complaints on the mandible mobility in the frontal and right distal regions when opening and closing the mouth, and impaired mastication due to partial mandibular edentulism on the right side.

In 2011 in a clinic at the place of the patient's residence she was diagnosed to have fibrous dysplasia of the mandible on the right side, for which the patient underwent the resection of the lower jaw from the frontal region to the right ramus with external approach. Later on she had two surgeries of microsurgical mandibular reconstruction with the use of vascularized fibular bone autografts carried out in the regional clinical center. Unfortunately, the autotransplants were removed in both cases due to vascular anastomosis failure. In May 2012 a reconstruction of the right mandible was done with a free non-vascularized rib autograft in our department. Within a year after the surgery, the patient had a retention of more than 90% of the autotransplant volume and the satisfactory function of mastication on the left side. However, no consolidation was observed and nonunions were diagnosed within the proximal and distal fixation areas that caused mobility and prevented any prosthetic treatment in the mandible on the right side. Therefore, reconstructive surgery with the resection of nonunions, bone grafting, and osteosynthesis was performed. Despite the surgical intervention a control clinical and instrumental examination in 0.5 year after the operation detected slight mobility in the frontal region and within the right ramus and no radiological evidence of consolidation (Fig. 14.5). The diastasis in the frontal mandible ranged from 5 mm on the upper edge to 14 mm on the lower one; the average tissue density between bone edges was

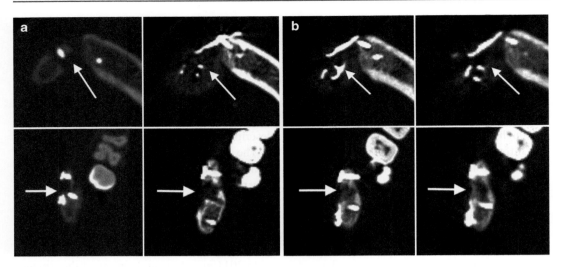

Fig. 14.5 CT scans of the patient's mandible in the frontal region (**a**) and within the ramus on the right side (**b**) prior to the operation and 3, 6, and 12 months after surgery. Arrows indicate nonunions (before operation) and sites of gene-activated bone substitute implantation

158.55 ± 116.29 HU; the separation between the transplant edge and the mandibular ramus on the vestibular surface achieved 9.2 mm, with the average tissue density in the nonunion area being 204.52 ± 97.84 HU [122].

Complex clinical cases with bone defects characterized by "osteogenic insufficiency" (large ones, nonunions, etc.) are the main indications for the use of activated bone substitutes as conventional ones will be suitable in less complicated settings. Therefore, to study the safety and efficacy of the gene-activated bone substitute we started with a very difficult clinical case when the main standard methods and materials used either resulted in specific complications or had limited efficacy. Four previous surgical interventions on the mandible resulted in scarring and impaired blood supply within the fixation areas of the bone autograft that predisposed to the development of osteogenic insufficiency and as a consequence—a formation of nonunions with high score according to Non-union Scoring System proposed by Calori et al. [184]. Moreover, one side of each nonunion was a non-vascularized bone autograft intended to be completely resorbed and simultaneously replaced by newly formed bone tissue. Such a feature has not been taken into account by current Non-union Scoring System but obviously significantly increased a complexity of the clinical case.

Taking into account a potentially poor blood supply within the fixation of the autotransplant and the mandibular fragments due to numerous operations previously performed including two failed microsurgical ones as well as a prolonged smoking experience (more than 15 years) the patient was offered to undergo a surgical treatment with the use of the gene-activated bone substitute. Considering the patient's characteristics and anamnesis the total score according to Non-union Scoring System [184] was estimated to be 31 that corresponded with high risk of nonunion relapse and required more specialized care. Additionally, the nonunions were complicated being formed by non-vascularized bone autograft intended to be resorbed and replaced by newly generated bone tissue. The voluntary written informed consent was obtained.

The gene-activated bone substitute we developed consists of the two components. The first one is the composite scaffold of bovine collagen and synthetic hydroxyapatite (granules with diameter of 500–1000 μm) registered as a bone substitute (CJSC Polystom, Russia) and approved for clinical use in Russia, and the second—a supercoiled naked plasmid DNA with cytomegalovirus promoter and gene encoding VEGF which is the active substance of "Neovasculgen." The pilot batch of the gene-activated bone substitutes in the form of dense elastic plates with sizes of 20 × 10 × 10 mm and weight of 200 ± 10 mg and that contained 0.2 mg of the gene constructs was produced for clinical study. Five plates were used for a surgical intervention (total amount of the scaffold—1000 mg, total dose of the plasmid DNA—1 mg).

The standard surgical protocol with metal construct removal, nonunion fibrous tissue excision, and approximated bone surface careful grinding was performed. Bone defects (5–14 mm in the frontal, 7–9 mm in the distal region of the mandible) between the rib autograft still present from previous interventions and mandibular fragments were filled in with the gene-activated bone substitute (Fig. 14.6). The autotransplant was fixed in the correct position with four straight miniplates and miniscrews.

In a postoperative period a soft diet and conservative therapy including antibiotics, analgetics, and desensitizing and anti-inflammatory agents were prescribed to the patient.

To evaluate the treatment results clinical and radiological diagnostic methods were used during the first 14 days of the postoperative period (in a hospital) and in 3, 6, and 12 months after surgery. A pain level in the postoperative region was rated with the use of the Visual Analog Scale, and edema was scored with the Numeric Rating Scale. A control panoramic radiograph was made on the next postoperative day, and dental CT was done on the other time points. A manual segmentation of the mandible was performed in the software 3DSlicer (Brigham, USA). The newly formed tissues within the bone substitute grafting were separately selected; their average density was calculated in Hounsfield units (HU) by using the "Label statistics" module. 3D bone reconstruction with volume rendering in the range of 250–2000 HU was made that complied with an optimal "bone window" with retention of spongy and lamellar bone in a model without metal constructs. A minimal size of diastases between mandibular fragments and rib autograft edges was determined with standard morphometry in the software Planmeca Romexis viewer (Planmeca Oy, Finland).

Neither adverse events (hypersensitive reaction, abnormal pain, edema, inflammatory complications, development of local vascular malformations, tumors, etc.) nor serious adverse events (life-threatening conditions) were observed. The postoperative pain score did not exceed 6 within the first 3 days after surgery; it was controlled with pain relievers; an average score for the following 4 days was 3.5, and no pain relief was required. Later on the patient did not notice any tenderness or discomfort within the postsurgical area. The maximal edema rated as 5 by the Numeric Rating Scale was observed on the third postoperative day. Then edema gradually decreased; its score was 3 by the end of the first week and remained at the same level for up to 14 days. These clinical data on safety were expected because each of the gene-activated bone substitute components, the

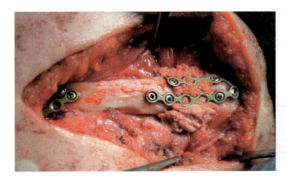

Fig. 14.6 Intraoperative view: nonunions are removed; mandible fragments and rib autograft are fixed with miniplates; bone defects filled with gene-activated bone substitute

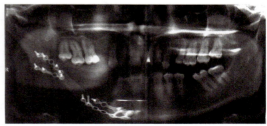

Fig. 14.7 Panoramic radiography, next day after surgery

collagen-hydroxyapatite scaffold and plasmid DNA, separately was registered and approved for clinical use previously in Russia as medical device for bone grafting and drug for CLI treatment, respectively.

Based on the panoramic radiograph data (Fig. 14.7) the autograft was fixed in a right position, the gene-activated bone substitute was located within bone defects, and its radiodensity was approximately twice as less as that of the bone autograft.

No inflammation sings, edema, or pain was observed in the postsurgical area for 12 months after surgery. Control CT showed that the rib autograft and metal constructs were correctly positioned.

Three months after surgery increased density regions were visualized in the zones of the distal and proximal autograft fixation and bone grafting (Fig. 14.5). The average density of these areas was 402.21 ± 84.40 in the frontal and 447.68 ± 106.75 HU within the distal fixation (Fig. 14.8). Newly formed tissues with an increased density equal to spongy bone were visualized within both zones of gene-activated bone substitute grafting. The volume of those tissues correlated with that of the substitute implanted. It is known from the previous experimental studies that the gene-activated bone substitute investigated has a baseline density equal to 130 ± 350 HU. Collagen comprising 60% of the scaffold mass undergoes rapid biodegradation (within 1 month). Starting from 45 days after the operation only single hydroxyapatite granules remain from the scaffold within the zone of a bone defect [123]. Therefore and considering a low standard deviation, bone tissue formation within the substitute implanted rather than the density of the remained scaffold fragments determined the high tissue density in the zone of bone grafting in 3 months.

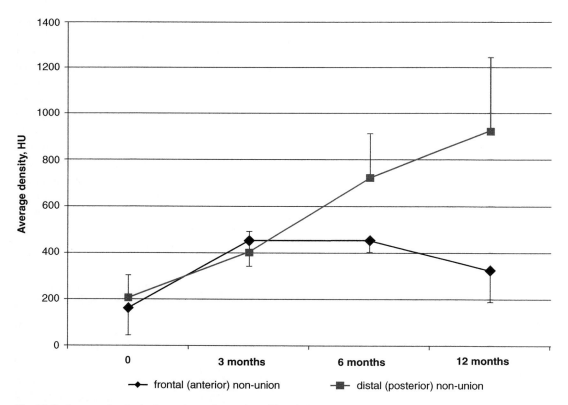

Fig. 14.8 Average density in the regions of nonunions (0) and the sites of gene-activated bone substitute implantation in 3, 6, and 12 months after surgery

The diastasis sizes between the bone fragments were 4.8 mm on the upper edge and 12.5 mm on the lower one in the frontal, and 6.2 mm on the vestibular surface and without dissociations on the lingual one within the distal region. No defects in the zones of proximal and distal fixation of the autograft were detected using 3D reconstruction. Heteromorphic newly formed tissues were seen in these areas; the tissues overstretched the bone borders of the reconstructed mandible outlining to a certain extent the substitute engrafted previously (Fig. 14.9).

The newly formed tissues with average density about 400 HU within gene-activated bone substitute implantation area were observed in the frontal region 6 and 12 months after surgery (Figs. 14.1 and 14.4). However, there was moderate partial resorption in the proximal edge of the bone autotransplant that prevented consolidation and maintained a diastasis. Clinical examination identified the appearance of minimal mandible mobility in the frontal region only 12 months after surgery that corresponded with CT results.

Meanwhile the distal edge of the rib autograft was completely integrated with adjacent mandibular ramus on both latest time points that did not allow to distinguish the borders between the mandible fragment, newly formed bone tissue, and rib autograft to segment these regions (Fig. 14.9). Normotrophic bone callus with no defects was formed 6 months after surgery and fully mineralized later on revealing the average density of 921.51 ± 321.89 on the last time point. Moreover, we found the completed remodeling of newly formed bone tissue with distinguished vestibular (1028.67 ± 169.77 HU) and lingual (1528.78 ± 81.53) cortical plates and spongy bone between them in 12 months. No mandibular mobility was detected in this region.

Thus, the main encouraging result we showed was that complete consolidation with normotrophic bone formation was achieved on the site of previously diagnosed distal nonunion. Bone remodeling process with cortical plate formation was completed in this area within 12 months after surgery. Unfortunately, the treatment of the other, more complex, nonunion was not successful: although the gene-activated bone substitute stimulated bone formation preserved in the frontal region during the 12 months of follow-up, the partial resorption of the adjacent autotransplant edge contributed to the proximal nonunion relapse. Such a resorption of the rib bone is normal but expectedly increased in response of the surgical intervention. Basically, any non-vascularized bone autograft undergoes continuous bioresorption to be completely replaced by newly formed bone tissue. No bone substitute is able to stop this natural process. The only option for gene-activated bone substitute to provide a consolidation was to accelerate the reparative osteogenesis so highly to make it outrunning the bioresorption rate. This bone formation activity was achieved in the distal nonunion site, but was not enough in the proximal one. That prevents any prosthetic treatment and requires additional surgery in the frontal region with bone autografting and more rigid fixation (for example, long custom-made titanium reconstructive plate).

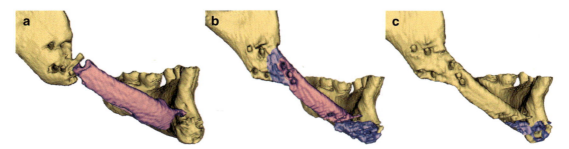

Fig. 14.9 The patient's mandible in different time points after bone grafting with the gene-activated bone substitute: (**a**) before; (**b**) 6 months; (**c**) 12 months. Dental CT, 3D reconstruction with volume rendering 250–2000 HU (titanium constructs are excluded)

Considering the results of the first clinical case we can conclude that more clinical trials and published data are needed to assess the treatment options and opportunities gene-activated bone substitutes could provide.

14.8 Conclusion

A detailed understanding of the regulation features of reparative osteogenesis, its dynamics, and results depending on the presence or absence of osteogenic insufficiency, as well as a comprehension of the modes of action characterized for various groups of bone substitutes that fall under two main technological trends, allowed us to propose the conception of "ordinary and activated bone substitutes."

The category of ordinary materials includes those which do not contain biologically active components standardized by qualitative and quantitative parameters. Osteoconduction and, in some cases, moderate osteoinduction allow these materials to optimize reparative regeneration for the promotion and increase in size of newly formed bone tissue. They are therefore intended for substitution of bone defects (and recovery of jaw atrophy) in the absence of osteogenic insufficiency. The main problem for this category of bone substitutes is low osteoinductive potential, which surgeons often try to overcome using an improvised, empirical activation of mixing the material with the patient's blood, autogenous bone (generally, in a ratio of 1:1), plasma enriched with thrombocytes, or plasma enriched with growth factors immediately prior to implantation [185].

Due to their biologically active components, activated materials have pronounced osteoinduction and (or) osteogenicity and are therefore able both to support the natural course of reparative osteogenesis and induce and provide high activity up to complete histotypical recovery. This quality makes them theoretically applicable for substitution of even large bone defects that are characterized by osteogenic insufficiency. Autogenous bone tissue is a prototype, a type of reference sample or a "gold standard" of materials for bone substitution [186]. The origin of their development lies in the necessity to develop effective alternatives for autogenous bone that may allow limiting or completely eliminating the use thereof.

The complex composition and mode of action of activated bone substitutes predetermine the necessity to carry out more comprehensive standardized preclinical studies with individual assessment of active components (growth factors, cells, or gene constructs) in terms of safety and biological action.

The practical value of the proposed classification of all bone substitutes as ordinary and activated, along with developing an understanding of osteogenic insufficiency, is to form a fundamental ground for physicians to make the most effective objective choice of bone graft for every particular clinical situation. However, before active using the presented system, some additional studies should be performed on methods for the quantitative evaluation of osteogenic insufficiency and the real clinical efficacy of all variants of activated bone substitutes.

Further development of ordinary and activated bone substitutes is directly related with 3D printing technologies that provide the opportunity to extrapolate all the advantages in personalized form that fully corresponds with the size and shape of a bone defect to be completely repaired. Being primarily devoted for large bone defects repair-activated bone substitutes are needed to be 3D printed; some methods have been already applied for this purpose [187, 188], and least of them were proposed for gene-activated bone substitutes [189].

References

1. Deev RV, Drobyshev AY, Bozo IY, Isaev AA. Ordinary and activated bone grafts: applied classification and the main features. Biomed Res Int. 2015;2015:365050.
2. Bhatt RA, Rozental TD. Bone graft substitutes. Hand Clin. 2012;28(4):457–68.
3. Deev RV, Bozo IY. Evolution of bone grafts. In: Muldashev ER, editor. Materials of the V Russian symposium with international participation. Ufa: Bashkortostan Publishing; 2012. p. 130–2.

4. Chen G, Deng C, Li YP. TGF-β and BMP signaling in osteoblast differentiation and bone formation. Int J Biol Sci. 2012;8(2):272–88.
5. Rivera JC, Strohbach CA, Wenke JC, et al. Beyond osteogenesis: an in vitro comparison of the potentials of six bone morphogenetic proteins. Front Pharmacol. 2013;4:125.
6. McMahon MS. Bone morphogenic protein 3 signaling in the regulation of osteogenesis. Orthopedics. 2012;35(11):920.
7. Suzuki Y, Ohga N, Morishita Y, et al. BMP-9 induces proliferation of multiple types of endothelial cells in vitro and in vivo. J Cell Sci. 2010;123(Pt 10):1684–92.
8. Finkenzeller G, Hager S, Stark GB. Effects of bone morphogenetic protein 2 on human umbilical vein endothelial cells. Microvasc Res. 2012;84(1):81–5.
9. Bai Y, Leng Y, Yin G, et al. Effects of combinations of BMP-2 with FGF-2 and/or VEGF on HUVECs angiogenesis in vitro and CAM angiogenesis in vivo. Cell Tissue Res. 2014;356(1):109–21.
10. Zhu F, Friedman MS, Luo W, et al. The transcription factor osterix (SP7) regulates BMP6-induced human osteoblast differentiation. J Cell Physiol. 2012;227(6):2677–85.
11. Friedman MS, Long MW, Hankenson KD. Osteogenic differentiation of human mesenchymal stem cells is regulated by bone morphogenetic protein-6. J Cell Biochem. 2006;98(3):538–54.
12. Glienke J, Schmitt AO, Pilarsky C, et al. Differential gene expression by endothelial cells in distinct angiogenic states. Eur J Biochem. 2000;267(9):2820–30.
13. Kang Q, Sun MH, Cheng H, et al. Characterization of the distinct orthotopic bone-forming activity of 14 BMPs using recombinant adenovirus-mediated gene delivery. Gene Ther. 2004;11(17):1312–20.
14. Akiyama I, Yoshino O, Osuga Y, et al. Bone morphogenetic protein 7 increased vascular endothelial growth factor (VEGF)-a expression in human granulosa cells and VEGF receptor expression in endothelial cells. Reprod Sci. 2014;21(4):477–82.
15. Lamplot JD, Qin J, Nan G, et al. BMP9 signaling in stem cell differentiation and osteogenesis. Am J Stem Cells. 2013;2:1): 1–21.
16. Mayr-Wohlfart U, Waltenberger J, Hausser H, et al. Vascular endothelial growth factor stimulates chemotactic migration of primary human osteoblasts. Bone. 2002;30(3):472–7.
17. D' Alimonte I, Nargi E, Mastrangelo F, et al. Vascular endothelial growth factor enhances in vitro proliferation and osteogenic differentiation of human dental pulp stem cells. J Biol Regul Homeost Agents. 2011;25(1):57–69.
18. Yang YQ, Tan YY, Wong R, et al. The role of vascular endothelial growth factor in ossification. Int J Oral Sci. 2012;4(2):64–8.
19. Wu Y, Cao H, Yang Y, et al. Effects of vascular endothelial cells on osteogenic differentiation of non-

contact co-cultured periodontal ligament stem cells under hypoxia. J Periodontal Res. 2013;48(1):52–65.
20. Matsumoto T, Bohman S, Dixelius J, et al. VEGF receptor-2 Y951 signaling and a role for the adapter molecule TSAd in tumor angiogenesis. EMBO J. 2005;24(13):2342–53.
21. Bhattacharya R, Kwon J, Li X, et al. Distinct role of PLCbeta3 in VEGF-mediated directional migration and vascular sprouting. J Cell Sci. 2009;122(Pt 7):1025–34.
22. Koch S, Claesson-Welsh L. Signal transduction by vascular endothelial growth factor receptors. Cold Spring Harb Perspect Med. 2012;2(7):a006502.
23. Marquez-Curtis LA, Janowska-Wieczorek A. Enhancing the migration ability of mesenchymal stromal cells by targeting the SDF-1/CXCR4 axis. Biomed Res Int. 2013;2013:561098.
24. Li B, Bai W, Sun P, et al. The effect of CXCL12 on endothelial progenitor cells: potential target for angiogenesis in intracerebral hemorrhage. J Interferon Cytokine Res. 2014. [Epub ahead of print].
25. Fagiani E, Christofori G. Angiopoietins in angiogenesis. Cancer Lett. 2013;328(1):18–26.
26. Herzog DP, Dohle E, Bischoff I, et al. Cell communication in a coculture system consisting of outgrowth endothelial cells and primary osteoblasts. Biomed Res Int. 2014;2014:320123.
27. Shiozawa Y, Jung Y, Ziegler AM, et al. Erythropoietin couples hematopoiesis with bone formation. PLoS One. 2010;5(5):e10853.
28. Wan L, Zhang F, He Q, et al. EPO promotes bone repair through enhanced cartilaginous callus formation and angiogenesis. PLoS One. 2014;9(7):e102010.
29. Buemi M, Donato V, Bolignano D. Erythropoietin: pleiotropic actions. Recenti Prog Med. 2010;101(6):253–67.
30. Cokic BB, Cokic VP, Suresh S, et al. Nitric oxide and hypoxia stimulate erythropoietin receptor via MAPK kinase in endothelial cells. Microvasc Res. 2014;92:34–40.
31. Park JB. Effects of the combination of fibroblast growth factor-2 and bone morphogenetic protein-2 on the proliferation and differentiation of osteoprecursor cells. Adv Clin Exp Med. 2014;23(3):463–7.
32. Sai Y, Nishimura T, Muta M, et al. Basic fibroblast growth factor is essential to maintain endothelial progenitor cell phenotype in TR-BME2 cells. Biol Pharm Bull. 2014;37(4):688–93.
33. Aenlle KK, Curtis KM, Roos BA, et al. Hepatocyte growth factor and p38 promote osteogenic differentiation of human mesenchymal stem cells. Mol Endocrinol. 2014;28(5):722–30.
34. Burgazli KM, Bui KL, Mericliler M, et al. The effects of different types of statins on proliferation and migration of HGF-induced human umbilical vein endothelial cells (HUVECs). Eur Rev Med Pharmacol Sci. 2013;17(21):2874–83.

35. Nakamura T, Mizuno S. The discovery of hepatocyte growth factor (HGF) and its significance for cell biology, life sciences and clinical medicine. Proc Jpn Acad Ser B Phys Biol Sci. 2010;86(6):588–610.

36. Sheng MH, Lau KH, Baylink DJ. Role of osteocyte-derived insulin-like growth factor i in developmental growth, modeling, remodeling, and regeneration of the bone. J Bone Metab. 2014;21(1):41–54.

37. Subramanian IV, Fernandes BC, Robinson T, et al. AAV-2-mediated expression of IGF-1 in skeletal myoblasts stimulates angiogenesis and cell survival. J Cardiovasc Transl Res. 2009;2(1):81–92.

38. Colciago A, Celotti F, Casati L, et al. In vitro effects of PDGF isoforms (AA, BB, AB and CC) on migration and proliferation of SaOS-2 osteoblasts and on migration of human osteoblasts. Int J Biomed Sci. 2009;5(4):380–9.

39. Levi B, James AW, Wan DC, et al. Regulation of human adipose-derived stromal cell osteogenic differentiation by insulin-like growth factor-1 and platelet-derived growth factor-alpha. Plast Reconstr Surg. 2010;126(1):41–52.

40. Wong VW, Crawford JD. Vasculogenic cytokines in wound healing. Biomed Res Int. 2013;2013:190486.

41. Palioto DB, Rodrigues TL, Marchesan JT, et al. Effects of enamel matrix derivative and transforming growth factor-β1 on human osteoblastic cells. Head Face Med. 2011;7:13.

42. Peshavariya HM, Chan EC, Liu GS, et al. Transforming growth factor-β1 requires NADPH oxidase 4 for angiogenesis in vitro and in vivo. J Cell Mol Med. 2014;18(6):1172–83.

43. Gao X, Xu Z. Mechanisms of action of angiogenin. Acta Biochim Biophys Sin (Shanghai). 2008;40(7):619–24.

44. Urist MR. Bone: formation by autoinduction. Science. 1965;150(698):893–9.

45. Omelyanenko NP, Slutsky LI, Connective Tissue MSP. Histophysiology, biochemistry, molecular biology. London: CRC Press; 2013.

46. Heldin CH, Miyazono K, ten Dijke P. TGF-beta signalling from cell membrane to nucleus through SMAD proteins. Nature. 1997;390(6659):465–71.

47. Zhou Z, Xie J, Lee D, et al. Neogenin regulation of BMP-induced canonical Smad signaling and endochondral bone formation. Dev Cell. 2010;19:90–102.

48. Bessa PC, Casal M, Reis RL. Bone morphogenetic proteins in tissue engineering: the road from laboratory to the clinic. Part I—Basic concepts. J Tissue Eng Regen Med. 2008;2(1):1–13.

49. Jonason JH, Xiao G, Zhang M, et al. Post-translational regulation of Runx2 in bone and cartilage. J Dent Res. 2009;88:693–703.

50. Hanai J, Chen LF, Kanno T, et al. Interaction and functional cooperation of PEBP2/CBF with Smads. Synergistic induction of the immunoglobulin germline Calpha promoter. J Biol Chem. 1999;274(44):31577–82.

51. Yoshida A, Yamamoto H, Fujita T, et al. Runx2 and Runx3 are essential for chondrocyte maturation, and Runx2 regulates limb growth through induction of Indian hedgehog. Genes Dev. 2004;18(8):952–63.

52. Cheng SL, Shao JS, Charlton-Kachigian N, et al. MSX2 promotes osteogenesis and suppresses adipogenic differentiation of multipotent mesenchymal progenitors. J Biol Chem. 2003;278(46):45969–77.

53. Merlo GR, Zerega B, Paleari L, et al. Multiple functions of Dlx genes. Int J Dev Biol. 2000;44(6):619–26.

54. Matsubara T, Kida K, Yamaguchi A, et al. BMP2 regulates Osterix through Msx2 and Runx2 during osteoblast differentiation. J Biol Chem. 2008;283(43):29119–25.

55. Liu TM, Lee EH. Transcriptional regulatory cascades in Runx2-dependent bone development. Tissue Eng Part B Rev. 2013;19(3):254–63.

56. Yano M, Inoue Y, Tobimatsu T. Smad7 inhibits differentiation and mineralization of mouse osteoblastic cells. Endocr J. 2012;59(8):653–62.

57. Tsuji K, Bandyopadhyay A, Harfe BD, et al. BMP2 activity, although dispensable for bone formation, is required for the initiation of fracture healing. Nat Genet. 2006;38:1424–9.

58. Shu B, Zhang M, Xie R, et al. BMP2, but not BMP4, is crucial for chondrocyte proliferation and maturation during endochondral bone development. J Cell Sci. 2011;124:3428–40.

59. Bandyopadhyay A, Tsuji K, Cox K, Harfe BD, et al. Genetic analysis of the roles of BMP2, BMP4, and BMP7 in limb patterning and skeletogenesis. PLoS Genet. 2006;2:e216.

60. Tsuji K, Cox K, Bandyop Adhyay A, Harfe BD, et al. BMP4 is dispensable for skeletogenesis and fracture-healing in the limb. J Bone Joint Surg Am. 2008;90(Suppl):14–8.

61. Cohen MM Jr. Biology of RUNX2 and cleidocranial dysplasia. J Craniofac Surg. 2013;24(1):130–3.

62. Roberts T, Stephen L, Beighton P. Cleidocranial dysplasia: a review of the dental, historical, and practical implications with an overview of the South African experience. Oral Surg Oral Med Oral Pathol Oral Radiol. 2013;115(1):46–55.

63. Otto F, Thornell AP, Cromptonetal T. Cbfa1, a candidate gene for cleidocranial dysplasia syndrome, is essential for osteoblast differentiation and bone development. Cell. 1997;89(5):765–71.

64. Ciurea AV, Toader C. Genetics of craniosynostosis: review of the literature. J Med Life. 2009;2(1):5–17.

65. Folkman J, Merler E, Abernathy C, et al. Isolation of a tumor factor responsible for angiogenesis. J Exp Med. 1971;133(2):275–88.

66. Goel HL, Mercurio AM. VEGF targets the tumour cell. Nat Rev Cancer. 2013;13(12):871–82.

67. Coultas L, Chawengsaksophak K, Rossant J. Endothelial cells and VEGF in vascular development. Nature. 2005;438(7070):937–45.

68. Olsson AK, Dimberg A, Kreuger J, et al. VEGF receptor signalling—in control of vascular function. Nat Rev Mol Cell Biol. 2006;7(5):359–71.

69. Arutyunyan IV, Kananihina YE, Makarov AV. Role of VEGF-A165 receptors in angiogenesis. Cell Transplant Tissue Eng. 2013;8(1):12–8.

70. Neve A, Cantatore FP, Corrado A, et al. In vitro and in vivo angiogenic activity of osteoarthritic and osteoporotic osteoblasts is modulated by VEGF and vitamin D3 treatment. Regul Pept. 2013;184:81–4.

71. Marini M, Sarchielli E, Toce M, et al. Expression and localization of VEGF receptors in human fetal skeletal tissues. Histol Histopathol. 2012;27(12):1579–87.

72. Tombran-Tink J, Barnstable CJ. Osteoblasts and osteoclasts express PEDF, VEGF-A isoforms, and VEGF receptors: possible mediators of angiogenesis and matrix remodeling in the bone. Biochem Biophys Res Commun. 2004;316(2):573–9.

73. Berendsen AD, Olsen BR. How vascular endothelial growth factor-A (VEGF) regulates differentiation of mesenchymal stem cells. J Histochem Cytochem. 2014;62(2):103–8.

74. Liu Y, Berendsen AD, Jia S, et al. Intracellular VEGF regulates the balance between osteoblast and adipocyte differentiation. J Clin Invest. 2012;122(9):3101–13.

75. Tashiro K, Tada H, Heilker R, et al. Signal sequence trap: a cloning strategy for secreted proteins and type I membrane proteins. Science. 1993;261(5121):600–3.

76. Mellado M, Rodríguez-Frade JM, Mañes S, et al. Chemokine signaling and functional responses: the role of receptor dimerization and TK pathway activation. Annu Rev Immunol. 2001;19:397–421.

77. Ward SG. T lymphocytes on the move: chemokines, PI 3-kinase and beyond. Trends Immunol. 2006;27(2):80–7.

78. Niederberger E, Geisslinger G. Proteomics and NF-κB: an update. Expert Rev Proteomics. 2013;10(2):189–204.

79. Jung Y, Wang J, Schneider A, et al. Regulation of SDF-1 (CXCL12) production by osteoblasts; a possible mechanism for stem cell homing. Bone. 2006;38(4):497–508.

80. Khurana S, Melacarne A, Yadak R, et al. SMAD signaling regulates CXCL12 expression in the bone marrow niche, affecting homing and mobilization of hematopoietic progenitors. Stem Cells. 2014;32(11):3012–22.

81. Christopher MJ, Liu F, Hilton MJ, et al. Suppression of CXCL12 production by bone marrow osteoblasts is a common and critical pathway for cytokine-induced mobilization. Blood. 2009;114(7):1331–9.

82. Gololobov VG, Deduh NV, Deev RV. Skeletal tissues and organs. In: Guidelines for histology, vol. 2, 2nd ed. Saint-Petersburg: Special Literature Publishing; 2011. p. 238–322.

83. Xue D, Li F, Chen G, et al. Do bisphosphonates affect bone healing? A meta-analysis of randomized controlled trials. J Orthop Surg Res. 2014;9:45.

84. Komlev VS, Barinov SM, Bozo II, et al. Bioceramics composed of octacalcium phosphate demonstrate enhanced biological behaviour. ACS Appl Mater Interfaces. 2014;6(19):16610–20.

85. Gololobov VG, Dulaev AK, Deev RB, et al. Morphofunctional organization, reactivity and regeneration of bone tissue. Saint-Petersburg: MMA; 2006.

86. Chiapasco M, Casentini P, Zaniboni M. Bone augmentation procedures in implant dentistry. Int J Oral Maxillofac Implants. 2009;24(Suppl):237–59.

87. Rogers GF, Greene AK. Autogenous bone graft: basic science and clinical implications. J Craniofac Surg. 2012;23(1):323–7.

88. Knight MN, Hankenson KD. Mesenchymal stem cells in bone regeneration. Adv Wound Care (New Rochelle). 2013;2(6):306–16.

89. Amable PR, Teixeira MV, Carias RB, et al. Protein synthesis and secretion in human mesenchymal cells derived from bone marrow, adipose tissue and Wharton's jelly. Stem Cell Res Ther. 2014;5(2):53.

90. Zhang M, Mal N, Kiedrowski M, et al. SDF-1 expression by mesenchymal stem cells results in trophic support of cardiac myocytes after myocardial infarction. FASEB J. 2007;21(12):3197–207.

91. Samee M, Kasugai S, Kondo H, et al. Bone morphogenetic protein-2 (BMP-2) and vascular endothelial growth factor (VEGF) transfection to human periosteal cells enhances osteoblast differentiation and bone formation. J Pharmacol Sci. 2008;108(1):18–31.

92. Neman J, Duenas V, Kowolik CM, et al. Lineage mapping and characterization of the native progenitor population in cellular allograft. Spine J. 2013;13(2):162–74.

93. Kerr EJ, Jawahar A, Wooten T, et al. The use of osteoconductive stem-cells allograft in lumbar interbody fusion procedures: an alternative to recombinant human bone morphogenetic protein. J Surg Orthop Adv. 2011;20(3):193–7.

94. Hollawell SM. Allograft cellular bone matrix as an alternative to autograft in hindfoot and ankle fusion procedures. J Foot Ankle Surg. 2012;51(2):222–5.

95. Osepyan IA, Chaylahyan RK, Garibyan ES. Treatment of non-union fractures, pseudarthroses, defects of long bones with transplantation of autologous bone marrow-derived fibroblasts grown in vitro and placed on the spongy bone matrix. Ortop Travmatol Protez. 1982;9:59.

96. Osepyan IA, Chaylahyan RK, Garibyan ES, et al. Transplantation of autologous bone marrow-derived fibroblasts in traumatology and orthopedics. Vestn Khir Im I I Grek. 1988;5:56.

97. Shchepkina EA, Kruglyakov PV, Solomin LN, et al. Transplantation of autologous multipotent mesenchymal stromal cells seeded on demineralized bone matrix in the treatment of pseudarthrosis of long bones. Cell Transplant Tissue Eng. 2007;2(3):67–74.

98. Drobyshev AY, Rubina KA, Sysoev VY, et al. Clinical trial of tissue-engineered construction based on autologous adipose-derived stromal cells in patients with alveolar bone atrophy of the upper and lower jaws. Bull Exp Clin Surg. 2011;IV(4):764–72.

99. Effectiveness and safety of method of maxilla alveolar process reconstruction using synthetic tricalcium phosphate and autologous MMSCs. https://clinicaltrials.gov/ct2/show/NCT02209311

100. Alekseev IS, Volkov AV, Kulakov AA, et al. Clinical and experimental study of combined cell transplant based on adipose-derived multipotent mesenchymal stromal cells in patients with severe bone tissue deficiency of jaws. Cell Transplant Tissue Eng. 2012;7(1):97–105.

101. Grudyanov AI, Zorin VL, Pereverzev RV, Zorina AI, Bozo IY. Efficiency of autofibroblasts in surgical treatment of parodontitis. Cell Transplant Tissue Eng. 2013;8(3):72–7.

102. McKay WF, Peckham SM, Badura JM. A comprehensive clinical review of recombinant human bone morphogenetic protein-2 (INFUSE Bone Graft). Int Orthop. 2007;31(6):729–34.

103. Burkus JK, Gornet MF, Dickman C, et al. Anterior lumbar interbody fusion using rhBMP-2 with tapered interbody cages. J Spinal Disord Tech. 2002;15(5):337–49.

104. Dimar JR, Glassman SD, Burkus JK, et al. Clinical and radiographic analysis of an optimized rhBMP-2 formulation as an autograft replacement in posterolateral lumbar spine arthrodesis. J Bone Joint Surg Am. 2009;91:1377–86.

105. Glassman SD, Carreon LY, Djurasovic M, et al. RhBMP-2 versus iliac crest bone graft for lumbar spine fusion: a randomized, controlled trial in patients over sixty years of age. Spine (Phila Pa 1976). 2008;33(26):2843–9.

106. Boden SD, Kang J, Sandhu H, et al. Use of recombinant human bone morphogenetic protein-2 to achieve posterolateral lumbar spine fusion in humans: a prospective, randomized clinical pilot trial: 2002 Volvo Award in clinical studies. Spine. 2002;27:2662–73.

107. Carragee EJ, Hurwitz EL, Weiner BK. A critical review of recombinant human bone morphogenetic protein-2 trials in spinal surgery: emerging safety concerns and lessons learned. Spine J. 2011;11(6):471–91.

108. Woo EJ. Adverse events reported after the use of recombinant human bone morphogenetic protein 2. J Oral Maxillofac Surg. 2012;70(4):765–7.

109. Epstein NE. Complications due to the use of BMP/INFUSE in spine surgery: the evidence continues to mount. Surg Neurol Int. 2013;4(Suppl 5):S343–52.

110. Chekanov AV, Fadeev IS, Akatov VS, et al. Quantitative effect of improving osteoinductive property of a material due to application of recombinant morphogenetic bone protein rhBMP-2. Cellular Transplant Tissue Eng. 2012;7(2):75–81.

111. Muraev AA, Ivanov SY, Artifexova AA, et al. Study of the biological properties of a new material based on osteoplastic nondemineralized collagen containing vascular endothelial growth factor in bone defects repair. Curr Technol Med. 2012;1:21–6.

112. Zhang W, Zhu C, Wu Y, et al. VEGF and BMP-2 promote bone regeneration by facilitating bone marrow stem cell homing and differentiation. Eur Cell Mater. 2014;27:1–11.

113. Holloway JL, Ma H, Rai R, et al. Modulating hydrogel crosslink density and degradation to control bone morphogenetic protein delivery and in vivo bone formation. J Control Release. 2014;191:63–70.

114. Lauzon MA, Bergeron E, Marcos B, et al. Bone repair: new developments in growth factor delivery systems and their mathematical modeling. J Control Release. 2012;162(3):502–20.

115. Yun YR, Jang JH, Jeon E, et al. Administration of growth factors for bone regeneration. Regen Med. 2012;7(3):369–85.

116. Chang PC, Dovban AS, Lim LP, et al. Dual delivery of PDGF and simvastatin to accelerate periodontal regeneration in vivo. Biomaterials. 2013;34(38):9990–7.

117. Kasten P, Beyen I, Bormann D, et al. The effect of two point mutations in GDF-5 on ectopic bone formation in a beta-tricalcium phosphate scaffold. Biomaterials. 2010;31:3878–84.

118. Kleinschmidt K, Ploeger F, Nickel J, et al. Enhanced reconstruction of long bone architecture by a growth factor mutant combining positive features of GDF-5 and BMP-2. Biomaterials. 2013;34(24):5926–36.

119. Gene therapy clinical trials worldwide. http://www.abedia.com/wiley/years.php.

120. Deev RV, Bozo IY, Mzhavanadze ND, et al. pCMV-vegf165 intramuscular gene transfer is an effective methomd of treatment for patients with chronic lower limb ischemia. J Cardiovasc Pharmacol Ther. 2015;20(5):473–82.

121. Deev R, Plaksa I, Bozo I, et al. Results of an international postmarketing surveillance study of pl-VEGF165 safety and efficacy in 210 patients with peripheral arterial disease. Am J Cardiovasc Drugs. 2017;17(3):235–42.

122. Bozo IY, Deev RV, Drobyshev AY, et al. World's first clinical case of gene-activated bone substitute application. Case Rep Dent. 2016;2016:8648949.

123. Deev RV, Drobyshev RV, Bozo IY, et al. Construction and biological effect evaluation of gene-activated osteoplastic material with human vegf gene. Cell Transplant Tissue Eng. 2013;8(3):78–85.

124. Wegman F, Bijenhof A, Schuijff L, et al. Osteogenic differentiation as a result of BMP-2 plasmid DNA based gene therapy in vitro and in vivo. Eur Cell Mater. 2011;21:230–42.

125. Baboo S, Cook PR. "Dark matter" worlds of unstable RNA and protein. Nucleus. 2014;5(4):281–6.

126. Evans CH. Gene delivery to bone. Adv Drug Deliv Rev. 2012;64(12):1331–40.

127. Grigorian AS, Schevchenko KG. Some possible molecular mechanisms of VEGF encoding plasmids functioning. Cell Transplant Tissue Eng. 2011;6(3):24–8.

128. Rose T, Peng H, Usas A, et al. Ex-vivo gene therapy with BMP-4 for critically sized defects and enhancement of fracture healing in an osteoporotic animal model. Unfallchirurg. 2005;108(1):25–34.

129. Cao L, Liu X, Liu S, et al. Experimental repair of segmental bone defects in rabbits by angiopoietin-1 gene transfected MSCs seeded on porous β-TCP scaffolds. J Biomed Mater Res B Appl Biomater. 2012;100(5):1229–36.
130. Betz VM, Betz OB, Glatt V, et al. Healing of segmental bone defects by direct percutaneous gene delivery: effect of vector dose. Hum Gene Ther. 2007;18(10):907–15.
131. Baltzer AW, Lattermann C, Whalen JD, et al. Genetic enhancement of fracture repair: healing of an experimental segmental defect by adenoviral transfer of the BMP-2 gene. Gene Ther. 2000;7(9):734–9.
132. Virk MS, Conduah A, Park SH, et al. Influence of short-term adenoviral vector and prolonged lentiviral vector mediated bone morphogenetic protein-2 expression on the quality of bone repair in a rat femoral defect model. Bone. 2008;42(5):921–31.
133. Lutz R, Park J, Felszeghy E, et al. Bone regeneration after topical BMP-2-gene delivery in circumferential peri-implant bone defects. Clin Oral Implants Res. 2008;19(6):590–9.
134. Chen JC, Winn SR, Gong X, et al. rhBMP-4 gene therapy in a juvenile canine alveolar defect model. Plast Reconstr Surg. 2007;120(6):1503–9.
135. Sheyn D, Kallai I, Tawackoli W, et al. Gene-modified adult stem cells regenerate vertebral bone defect in a rat model. Mol Pharm. 2011;8(5):1592–601.
136. Bertone AL, Pittman DD, Bouxsein ML, et al. Adenoviral-mediated transfer of human BMP-6 gene accelerates healing in a rabbit ulnar osteotomy model. J Orthop Res. 2004;22(6):1261–70.
137. Li JZ, Li H, Hankins GR, et al. Different osteogenic potentials of recombinant human BMP-6 adeno-associated virus and adenovirus in two rat strains. Tissue Eng. 2006;12(2):209–19.
138. Bright C, Park YS, Sieber AN, et al. In vivo evaluation of plasmid DNA encoding OP-1 protein for spine fusion. Spine (Phila Pa 1976). 2006;31(19):2163–72.
139. Schek RM, Hollister SJ, Krebsbach PH. Delivery and protection of adenoviruses using biocompatible hydrogels for localized gene therapy. Mol Ther. 2004;9(1):130–8.
140. Song K, Rao N, Chen M, et al. Construction of adeno-associated virus system for human bone morphogenetic protein 7 gene. J Huazhong Univ Sci Technolog Med Sci. 2008;28(1):17–21.
141. Breitbart AS, Grande DA, Mason J, et al. Gene-enhanced tissue engineering: applications for bone healing using cultured periosteal cells transduced retrovirally with the BMP-7 gene. Ann Plast Surg. 1999;42(5):488–95.
142. Kimelman-Bleich N, Pelled G, Zilberman Y, et al. Targeted gene-and-host progenitor cell therapy for nonunion bone fracture repair. Mol Ther. 2011;19(1):53–9.
143. Abdelaal MM, Tholpady SS, Kessler JD, et al. BMP-9-transduced prefabricated muscular flaps for the treatment of bony defects. J Craniofac Surg. 2004;15(5):736–41.
144. Kuroda S, Goto N, Suzuki M, et al. Regeneration of bone- and tendon/ligament-like tissues induced by gene transfer of bone morphogenetic protein-12 in a rat bone defect. J Tissue Eng. 2010;2010:891049.
145. Rundle CH, Strong DD, Chen ST, et al. Retroviral-based gene therapy with cyclooxygenase-2 promotes the union of bony callus tissues and accelerates fracture healing in the rat. J Gene Med. 2008;10(3):229–41.
146. Li C, Ding J, Jiang L, et al. Potential of mesenchymal stem cells by adenovirus-mediated erythropoietin gene therapy approaches for bone defect. Cell Biochem Biophys. 2014;70(2):1199–204.
147. Wallmichrath JC, Stark GB, Kneser U, et al. Epidermal growth factor (EGF) transfection of human bone marrow stromal cells in bone tissue engineering. J Cell Mol Med. 2009;13(8B):2593–601.
148. Guo X, Zheng Q, Kulbatski I, et al. Bone regeneration with active angiogenesis by basic fibroblast growth factor gene transfected mesenchymal stem cells seeded on porous beta-TCP ceramic scaffolds. Biomed Mater. 2006;1(3):93–9.
149. Wen Q, Zhou C, Luo W, et al. Pro-osteogenic effects of fibrin glue in treatment of avascular necrosis of the femoral head in vivo by hepatocyte growth factor-transgenic mesenchymal stem cells. J Transl Med. 2014;12:114.
150. Zou D, Zhang Z, He J, et al. Blood vessel formation in the tissue-engineered bone with the constitutively active form of HIF-1α mediated BMSCs. Biomaterials. 2012;33(7):2097–108.
151. Shen FH, Visger JM, Balian G, et al. Systemically administered mesenchymal stromal cells transduced with insulin-like growth factor-I localize to a fracture site and potentiate healing. J Orthop Trauma. 2002;16(9):651–9.
152. Srouji S, Ben-David D, Fromigué O, et al. Lentiviral-mediated integrin α5 expression in human adult mesenchymal stromal cells promotes bone repair in mouse cranial and long-bone defects. Hum Gene Ther. 2012;23(2):167–72.
153. Strohbach CA, Rundle CH, Wergedal JE, et al. LMP-1 retroviral gene therapy influences osteoblast differentiation and fracture repair: a preliminary study. Calcif Tissue Int. 2008;83(3):202–11.
154. Lattanzi W, Parrilla C, Fetoni A. Ex vivo-transduced autologous skin fibroblasts expressing human Lim mineralization protein-3 efficiently form new bone in animal models. Gene Ther. 2008;15(19):1330–43.
155. Lu SS, Zhang X, Soo C, et al. The osteoinductive properties of Nell-1 in a rat spinal fusion model. Spine J. 2007;7(1):50–60.
156. Tu Q, Valverde P, Li S, et al. Osterix overexpression in mesenchymal stem cells stimulates healing of critical-sized defects in murine calvarial bone. Tissue Eng. 2007;13(10):2431–40.
157. Jin Q, Anusaksathien O, Webb SA, et al. Engineering of tooth-supporting structures by delivery of PDGF gene therapy vectors. Mol Ther. 2004;9(4):519–26.

158. Elangovan S, D'Mello SR, Hong L, et al. The enhancement of bone regeneration by gene activated matrix encoding for platelet derived growth factor. Biomaterials. 2014;35(2):737–47.

159. Fang J, Zhu YY, Smiley E, et al. Stimulation of new bone formation by direct transfer of osteogenic plasmid genes. Proc Natl Acad Sci U S A. 1996;93(12):5753–8.

160. Pan H, Zheng Q, Yang S, et al. A novel peptide-modified and gene-activated biomimetic bone matrix accelerating bone regeneration. J Biomed Mater Res A. 2014;102(8):2864–74.

161. Geiger F, Bertram H, Berger I, et al. Vascular endothelial growth factor gene-activated matrix (VEGF165-GAM) enhances osteogenesis and angiogenesis in large segmental bone defects. J Bone Miner Res. 2005;20(11):2028–35.

162. Tarkka T, Sipola A, Jämsä T, et al. Adenoviral VEGF-A gene transfer induces angiogenesis and promotes bone formation in healing osseous tissues. J Gene Med. 2003;5(7):560–6.

163. Koh JT, Zhao Z, Wang Z, et al. Combinatorial gene therapy with BMP2/7 enhances cranial bone regeneration. J Dent Res. 2008;87(9):845–9.

164. Menendez MI, Clark DJ, Carlton M, et al. Direct delayed human adenoviral BMP-2 or BMP-6 gene therapy for bone and cartilage regeneration in a pony osteochondral model. Osteoarthr Cartil. 2011;19(8):1066–75.

165. Reichert JC, Schmalzl J, Prager P, et al. Synergistic effect of Indian hedgehog and bone morphogenetic protein-2 gene transfer to increase the osteogenic potential of human mesenchymal stem cells. Stem Cell Res Ther. 2013;4(5):105.

166. Deng Y, Zhou H, Yan C, et al. In vitro osteogenic induction of bone marrow stromal cells with encapsulated gene-modified bone marrow stromal cells and in vivo implantation for orbital bone repair. Tissue Eng Part A. 2014;20(13-14):2019–29.

167. Liu J, Xu L, Li Y, et al. Temporally controlled multiple-gene delivery in scaffolds: a promising strategy to enhance bone regeneration. Med Hypotheses. 2011;76(2):173–5.

168. Zhang Y, Cheng N, Miron R, et al. Delivery of PDGF-B and BMP-7 by mesoporous bioglass/silk fibrin scaffolds for the repair of osteoporotic defects. Biomaterials. 2012;33(28):6698–708.

169. Ito H, Koefoed M, Tiyapatanaputi P, et al. Remodeling of cortical bone allografts mediated by adherent rAAV-RANKL and VEGF gene therapy. Nat Med. 2005;11(3):291–7.

170. Feichtinger GA, Hofmann AT, Slezak P, et al. Sonoporation increases therapeutic efficacy of inducible and constitutive BMP2/7 in vivo gene delivery. Hum Gene Ther Methods. 2014;25(1):57–71.

171. Wehrhan F, Amann K, Molenberg A, et al. Critical size defect regeneration using PEG-mediated BMP-2 gene delivery and the use of cell occlusive barrier membranes—the osteopromotive principle revisited. Clin Oral Implants Res. 2013;24(8):910–20.

172. Die X, Luo Q, Chen C, et al. Construction of a recombinant adenovirus co-expressing bone morphogenic proteins 9 and 6 and its effect on osteogenesis in C3H10 cells. Nan Fang Yi Ke Da Xue Xue Bao. 2013;33(9):1273–9.

173. Seamon J, Wang X, Cui F, et al. Adenoviral delivery of the VEGF and BMP-6 genes to rat mesenchymal stem cells potentiates osteogenesis. Bone Marrow Res. 2013;2013:737580.

174. Yang L, Zhang Y, Dong R, et al. Effects of adenoviral-mediated coexpression of bone morphogenetic protein-7 and insulin-like growth factor-1 on human periodontal ligament cells. J Periodontal Res. 2010;45(4):532–40.

175. Liu JZ, Hu YY, Ji ZL. Co-expression of human bone morphogenetic protein-2 and osteoprotegerin in myoblast C2C12. Zhongguo Xiu Fu Chong Jian Wai Ke Za Zhi. 2003;17(1):1–4.

176. Kim MJ, Park JS, Kim S, et al. Encapsulation of bone morphogenic protein-2 with Cbfa1-overexpressing osteogenic cells derived from human embryonic stem cells in hydrogel accelerates bone tissue regeneration. Stem Cells Dev. 2011;20(8):1349–58.

177. Li J, Zhao Q, Wang E, et al. Transplantation of Cbfa1-overexpressing adipose stem cells together with vascularized periosteal flaps repair segmental bone defects. J Surg Res. 2012;176(1):e13–20.

178. Bhattarai G, Lee YH, Lee MH, et al. Gene delivery of c-myb increases bone formation surrounding oral implants. J Dent Res. 2013;92(9):840–5.

179. Zhao Z, Wang Z, Ge C, et al. Healing cranial defects with AdRunx2-transduced marrow stromal cells. J Dent Res. 2007;86(12):1207–11.

180. Takahashi T. Overexpression of Runx2 and MKP-1 stimulates transdifferentiation of 3T3-L1 preadipocytes into bone-forming osteoblasts in vitro. Calcif Tissue Int. 2011;88(4):336–47.

181. Cucchiarini M, Orth P, Madry H. Direct rAAV SOX9 administration for durable articular cartilage repair with delayed terminal differentiation and hypertrophy in vivo. J Mol Med (Berl). 2013;91(5):625–36.

182. Itaka K, Ohba S, Miyata K, et al. Bone regeneration by regulated in vivo gene transfer using biocompatible polyplex nanomicelles. Mol Ther. 2007;15(9):1655–62.

183. Keeney M, van den Beucken JJ, van der Kraan PM, et al. The ability of a collagen/calcium phosphate scaffold to act as its own vector for gene delivery and to promote bone formation via transfection with VEGF(165). Biomaterials. 2010;31(10):2893–902.

184. Calori GM, Phillips M, Jeetle S, Tagliabue L, Giannoudis PV. Classification of non-union: need for a new scoring system? Injury. 2008;39(Suppl 2):S59–63.

185. Anitua E, Alkhraisat MH, Orive G. Perspectives and challenges in regenerative medicine using plasma rich in growth factors. J Control Release. 2012;157(1):29–38.

186. Shaw RJ, Brown JS. Osteomyocutaneous deep circumflex iliac artery perforator flap in the reconstruc-

tion of midface defect with facial skin loss: a case report. Microsurgery. 2009;29(4):299–302.

187. Duarte Campos DF, Blaeser A, Buellesbach K, et al. Bioprinting organotypic hydrogels with improved mesenchymal stem cell remodeling and mineralization properties for bone tissue engineering. Adv Healthc Mater. 2016;5(11):1336–45.

188. Shim JH, Kim SE, Park JY, et al. Three-dimensional printing of rhBMP-2-loaded scaffolds with long-term delivery for enhanced bone regeneration in a rabbit diaphyseal defect. Tissue Eng Part A. 2014;20(13–14):1980–92.

189. Bozo IY, Komlev VS, Drobyshev AY, et al. Method for creating a personalized gene-activated implant for regenerating bone tissue. EP3130342, US20170209626 A1. Priority date Feb. 10; 2015.

Absorbable Bone Substitute Materials Based on Calcium Sulfate as Triggers for Osteoinduction and Osteoconduction

15

Dominik Pförringer and Andreas Obermeier

15.1 Introduction

In times of rising numbers of infections and growing populations bone defects caused by trauma, infection, or tumor pose a challenge for reconstructive surgery. Bone grafting is used as an established method, both autologous and hetero- and xenologous. In addition bone grafts can be antibiotically loaded [1] to prevent or treat infections. The final fate of a bone graft is influenced by local phenomena such as osteocon- and -induction as well as osteogenesis [2]. Autologous grafts are regarded as a near-ideal solution for defect reconstruction. However, their use is not without restrictions, harvesting sometimes complicated and availability limited. An alternative approach for defect reconstruction can be the use of allografts. Research activities have been focused on attempts to create ideal artificial bone grafts. Calcium sulfate (CS) poses a readily available and inexpensive solution with high biocompatibility [3]. In addition CS has the potential for incorporation of therapeutic substances and thus can function as a carrier for drugs, extending its usability in the field of bone reconstruction [4, 5].

Furthermore it can serve as an osteoconductive structure for ingrowing bone, through a process in which newly formed host bone replaces the CS material simultaneously without inducing a significant inflammatory reaction [6]. CS generally has no intrinsic osteoinductive or osteogenic capacity, and its resorption generally happens more rapidly than bone formation [7]. One way to antagonize this is the addition of osteoinductive and/or osteogenic components such as hydroxyapatite (HA) which has proven to deliver favorable results [8, 9]. However, the main limitation for the use of HA ceramics is the inherent brittleness and difficulty in processing [10]. Another promising additive is calcium carbonate. It has been shown that resorption of the CS can leave porosities enhancing the ingrowth of bone. In this study, we analyzed the biological remodeling in response to implantation of new formulations of CS augmented with antibiotics and calcium carbonate/tripalmitin in rabbit tibiae. Bone ingrowth and biomaterial performance were evaluated using optical microscopy, fluorescence microscopy, conventional radiography, and micro-computed tomography (micro-CT). Specifically, the effect of calcium carbonate/tripalmitin formulations of CS on resorption period was examined. Hence, the provision of temporary structural support during the remodeling process, and thus synchronization of the ingrowing bone with biomaterial resorption, as well as preservation of the bone graft's initial volume

D. Pförringer (✉) · A. Obermeier
Klinik und Poliklinik für Unfallchirurgie, Klinikum rechts der Isar, Technische Universität München, Munich, Germany
e-mail: Dominik.Pfoerringer@mri.tum.de; aobermeier@tum.de

© Springer Nature Switzerland AG 2019
D. Duscher, M. A. Shiffman (eds.), *Regenerative Medicine and Plastic Surgery*,
https://doi.org/10.1007/978-3-030-19962-3_15

were investigated. New formulations for delayed resorption and osteoconduction of artificial bone substitute materials may facilitate the bone regeneration process in an antimicrobial environment. Herewith, trauma and orthopedic surgeons may find a future bone substitute material for complex bone regeneration as infect prophylaxis, especially in case of methicillin-resistant staphylococcus aureus (MRSA) infections.

15.2 Materials and Methods

15.2.1 Implants

The resorbable bone substitute materials based on CS formulations used in the experiment consisted of CS dihydrate, gentamicin, and tripalmitin (group 1: Herafill®-G), CS dihydrate, vancomycin, and tripalmitin (group 2: CaSO4-V), as well as commercially available tobramycin-loaded CS hemihydrate (group 3: Osteoset®). The specific beads varied in size and consequently in implanted quantity to match overall implanted mass of approximately 500 mg. For this purpose, two units of group 1 implants (500 mg), 14 units of group 2 implants (490 mg), and five units of group 3 (537.5 mg) were used to implant inside an artificial rabbit tibial bone defect for investigation.

Group 1 implants at 6.0 mm diameter and 250 mg weight per unit consist of calcium sulfate dihydrate (71.6%), calcium carbonate (17.9%), tripalmitin (8.8%), and gentamicin sulfate (1.7%). Group 2 implants at 3.0 mm diameter and 35 mg per unit consist of calcium sulfate dihydrate (72.0%), calcium carbonate (18.0%), tripalmitin (8.9%), and vancomycin hydrochloride (1.1%). Group 3 implants at 4.8 mm diameter and 107.5 mg weight per unit consist of a hemihydrate modification of calcium sulfate (96.0%) and tobramycin sulfate (4.0%). Composition of bead implants is given in contents per weight.

15.2.2 Animal Study: Protocol

The study was approved by the Animal Experimentation Ethics Committee of Bavaria (Reg. No. 209.1/211-2531.2-22/05) and conducted with reference to the OECD Principles of Good Laboratory Practice. Fifty-four female New Zealand white rabbits (Charles River Laboratories, Sulzfeld, Germany) with a mean body weight (BW) of 4.5 kg (range: 4.3–4.7 kg) underwent surgery. For acclimatization purposes, the animals were delivered to the facility at least 2 weeks prior to surgical intervention. Animals were housed in cages (2–3 animals) at normal room temperature and daylight illumination with ad libitum access to food and water. The animals were divided into three groups according to the test implant groups (Fig. 15.1).

15.2.3 Animal Study: Surgical Procedure

The surgical procedure was performed under general anesthesia utilizing weight-adapted intramuscular injection of medetomidine 0.25 mg/kg BW (Domitor®, Pfizer Inc., Germany) and ketamine 17 mg/kg BW (S-Ketanest®, Parke-Davis GmbH). Analgesia during surgery was obtained through intravenous application of metamizole 30 mg/kg BW (Novaminsulfon®, Ratiopharm GmbH, Germany) at the beginning of surgery. The left hind leg prior to surgery was shaved, and the skin cleaned antiseptically (Cutasept®, Bode Chemie, Germany). The skin incision was placed laterally of the tibial tuberosity; consequently bone manipulation was performed at the medial side. This local skin displacement serves the purpose of infection prevention through the skin wound. A water-cooled surgical diamond fraise was employed to drill an 8 mm spherical bone cylinder to open the medullary cavity. Using a sterile forceps beads were inserted into the proximal medullary cavity. Specific quantities of implants were implanted according to varying size of beads to match overall implanted mass of 500 mg (±8%) (group 1: 2 units, group 2: 14 units, and group 3: 5 units). After implantation, the cavity was closed using the initially removed bone cylinder while surgical site was irrigated using sterile saline. Subcutaneous tissues were readapted (3-0 Vicryl®, Ethicon GmbH, Germany), and skin closed (3-0

15 Absorbable Bone Substitute Materials Based on Calcium Sulfate as Triggers for Osteoinduction... 213

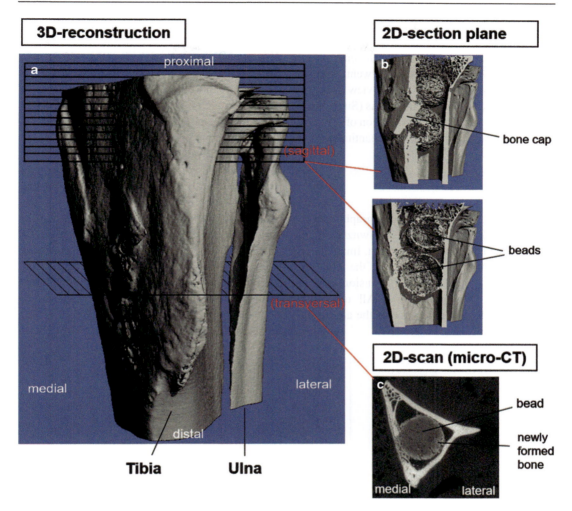

Fig. 15.1 Explanation of micro-CT planes used for tibia of a sacrificed animal (group 1 after 6 weeks). Exemplary declaration of micro-CT planes used in sacrificed rabbits from group 1 after 6 weeks: (**a**) 3-D reconstruction was carried out for gross overview; 2-D section plane in the sagittal direction (**b**) and the transversal plane from micro-CT scan (**c**) were analyzed for bone ingrowth and implant degradation. Reproduced with permission

Prolene®, Ethicon GmbH, Germany). Thereafter, a spray bandage (Opsite®, Smith & Nephew PLC, England) was applied. Anesthesia was antagonized using atipamezole 0.25 mg/kg (Antisedan®, Pfizer Inc., Germany).

Buprenorphine (0.03 mg/kg BW s.c., Temgesic®, Essex Pharma GmbH, Germany) was administered for analgesia towards the end of the surgical procedure immediately before wound closure was completed and postoperatively every 8 h for 4 days. In addition, carprofen 4 mg/kg s.c. (Rimadyl®, Pfizer Inc., Germany) was given for 7 days every 12 h. Animals underwent daily control examinations considering general condition, body temperature, and surgical site of the operated leg. Blood was obtained from the ear vein towards the end of the surgical procedure and weekly until sacrifice of the animal to determine blood count, and calcium/alkaline phosphatase levels.

Animals were euthanized after 4, 6, 8, and 12 weeks according to different testing groups employing an overdose of pentobarbital-sodium (Narcoren®, Merial GmbH, Germany 50 mg/kg). The tibiae were dissected. Then, all soft tissue was stripped from the bone, which was consecutively stored in 100% methanol.

15.2.4 Radiographic Evaluation

Immediately after surgery and at the end of each observation period, the tibiae underwent X-ray via contact radiography in a lateral view using an X-ray generator at 44 kV and 4 mAs (Super 80CP, Philips GmbH, Germany). Evaluation of the imaging changes was performed semiquantitatively.

15.2.5 Micro-CT

In addition, micro-computed tomography (micro-CT 80, Fa. Scanco, Brüttisellen, Switzerland) of the explanted tibiae was carried out. Images were obtained as transverse planes of the tibia and were further processed as 3-dimensional reconstructions to get an overview. All micro-CTs were performed after sacrifice of the animals on dissected tibiae.

15.2.6 Histology

Histomorphological analysis (Fig. 15.2) was performed to evaluate bone growth and bone vitality around the implants using two kinds of staining. In vivo fluorochrome substances were injected postoperatively at fixed time intervals to achieve a fluorescent labeling of newly formatted bone. Additionally, a classical histomorphological staining of specimen surfaces was performed after polymerization and down polishing of tibiae ex vivo.

Histological specimens were prepared by embedding with methyl-methacrylate resin (Technovit® 4004, HeraeusKulzer GmbH, Germany) and dissecting mid-tibiae sagittally utilizing a diamond saw (Exakt 300CL, Exakt GmbH, Germany), followed by grinding down with a polishing machine (Planopol-V and Pedemax-2, Struers GmbH, Germany) to a

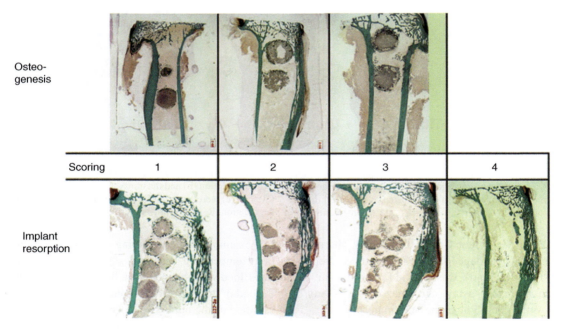

Fig. 15.2 Histological scoring system (modified Giemsa stain) for osteogenesis (group 1) and implant resorption (group 2). Histological scoring system (modified Giemsa stain) for osteogenesis (*1*: no osteogenesis, *2*: <50% of implant circumference with signs of osteogenesis, *3*: >50% of implant circumference with signs of osteogenesis) and implant resorption (*1*: <10% of implant degraded, *2*: 10–50% of implant degraded, *3*: >50% of implant degraded, *4*: <10% of implant remnants detectable). Reproduced with permission

thickness of 40–120 μm with diamond discs to get histological slides.

Approximately 100 μm thick tibia sections of every group at 4, 6, 8, and 12 weeks were then histomorphologically stained with a modified Grünwald-Giemsa staining called "K2." This staining is suitable to evaluate the newly formed bone, with newly formed osteoid tissue stained yellowish-brown, old bone stained light-green/blue, connective tissue stained orange-red, and calcifications stained bright-greenish-blue. The surface-stained specimens were then investigated via photo macroscopy (Wild M400, Wild Heerbrugg AG, Switzerland) at ×2 magnification to get an overview and using light microscopy (DMRB, Leica Microsystems GmbH, Germany) for more detail at ×25 and up to ×250 magnifications.

All implants and surrounding tissue were evaluated for osteogenesis (criteria: amount of trabecular bone formation, number of osteoblasts, amount of osteoid, general impression) and amount of implant resorption (criteria: lysis of implant surface, presence and number of surrounding giant cells, general inspection of morphology). Osteogenesis was graded on a scale from 1 to 3 (little to maximum), and resorption of implants from 1 to 4 (none to complete).

To measure bone apposition rates after surgery (i.e., the rate of bone ingrowth or bone deposition) within and around the implants, five separate fluorochrome labels were administered subcutaneously. For this purpose different fluorochrome solutions were injected at the following days postoperatively: 21 and 22 days (tetracycline, 30 mg/kg BW), 35 and 36 days (alizarin complexone, 30 mg/kg BW), 49 and 50 days (calcein green, 30 mg/kg BW), 63 and 64 days (xylenol orange, 90 mg/kg BW), and 77 and 78 days (calcein blue, 30 mg/kg BW). Fluorochrome labels are bound to sites of active bone deposition shortly after administration [11]. Therefore, the time of bone formation can be related to the individual fluorescence color of each fluorochrome label administered chronologically.

Later on, specimens were investigated with their fluorochrome labeling, because of new bone formation using a light microscope (DMRB,

Table 15.1 Correlation of fluorescence color and bone formation time

Substance	Time of bone formation (postoperative week)	Fluorescence color (UV excitation)
Tetracycline	3	Yellow/green
Alizarin complexone	5	Red
Calcein green	7	Green
Xylenol orange	11	Orange

Correlation of fluorescence color of newly formed bone dependent on substance solution and time of administration used. Resulting in visually distinguishable stains of newly formed bone at different time periods by defined colors

Leica Microsystems GmbH, Germany) with an ultraviolet light source and at ×25 and up to ×250 magnifications. Polychrome sequence marking was performed during the time of postoperative observation, exemplarily for one animal of each group. The fluorescence evaluation was performed at weeks 8 and 12.

Fluorescence colors of bone tissue specimens indicate the time of bone formation, correlating with the time of fluorescence stain application in vivo. Thus, time of bone formation can be assigned to individual label colors.

15.2.7 Statistical Analysis (Table 15.1)

Experimental investigations via X-ray, micro-CT, and classical histology were conducted on a minimum of four animals. One animal for each group was sacrificed additionally at weeks 8 and 12, to achieve specimen for fluorescence microscopy after polychrome sequence labeling. Scores for X-rays and classical histological (modified Giemsa stain) pictures were generated to collect semiquantitative information on the progress of implant resorption and osteogenesis.

15.3 Outcome and Scientific Literature Context

Rapid and complete reconstruction of bone defects through autologous or allogeneic bone or bone-related tissue can be crucial in trauma and orthopedic surgery. Calcium sulfate (CS), also known as plaster of paris, has been used for multiple implications for more than a century [12] and its use in the surgical setting has been proposed [2]. Key features of this bone graft substitute include its biocompatibility as well as its features as a binding and stabilizing agent. Additionally it can serve osteoconductive purposes potentially leading to bone ingrowth [13, 14]. On the other hand, rapid resorption rate limits its use as a structural allograft. In this context, some authors found unsatisfactory results for CS if used as a bone graft [15]. We had anticipated that a combination of CS with calcium carbonate/tripalmitin can purposely delay the degradation process, providing temporary structural support during remodeling, synchronizing the ingrowing bone with biomaterial resorption, as well as preserving the bone grafts' initial volume. This study aimed to visualize the mechanisms of bone formation in the presence of novel antibiotic-impregnated CS implants in a rabbit model. With the use of radiographic and histologic methods, none of the three tested bead formulations (Herafill®-G (1), CaSO4-V (2), and Osteoset® (3)) displayed any adverse reactions, while all of them were undergoing at least partial decomposition within the tibial bone.

In the underlying research the surgical procedure was well tolerated by all rabbits as well as the follow-up period. The animals were immediately after surgery fully weight bearing. No local or generalized signs of infection were observed.

These three bone substitutes composed of CS; differing side groups consisting of dihydrate (1, 2) and hemihydrate (3) in combination with differing antibiotics (gentamicin (1), vancomycin (2), and tobramycin (3)) were used as study implants. Beads of group 3 consisted of a hemihydrate formulation of CS and are commercially available (Osteoset®). In contrast, beads of groups 1 and 2 are novel preparations consisting of CS dihydrates with calcium carbonate and tripalmitin. Also, group 1 (Herafill®-G) is a commercial product, whereas group 2was at the time of testing still in the development stage to treat MRSA infections in

the future. The rationale for the use of calcium carbonate as an additive is its ability to delay degradation of CS and buffer the pH value in the implant region, thus counteracting healing problems of the tissue [16, 17]. Tripalmitin as a fatty acid influences the release kinetics of the antibiotics gentamicin and vancomycin to allow the beads to act potentially antimicrobially over a longer period of time [18, 19].

The tibiae were analyzed by different methods, in which drug release kinetics had already been described [12]. Conventional radiographs allowed a statement on the rate of resorption of implants during the study period. Further examination by micro-CT enabled a more precise insight into the processes of bone formation and implant resorption. While the 3-D reconstructions gave an overview of the processes in the interior of the entire bone, the main disadvantage is the influence on the image by adjusting black-and-white values. On the other hand, the transversal 2-D images reflect the unaltered facts and ongoing visible mechanisms at the bone-implant interface. However, the amount of pictures is not suitable for an overview. Therefore, the combination of both views is essential for analysis of the mechanisms acting around the implants.

15.4 Resulting Radiographic Imaging

Conventional X-ray imaging showed differences of biodegradation between the implants over time. Initially, 4 weeks after implantation, all beads were radiographically detectable at the level of the initial insertion sites. Evaluation of changes in radio-opaqueness revealed no relevant changes from the 4th to the 12th weeks, with well-distinguishable beads at 12 weeks in group 1. Radiolucency of the beads increased over the study period in groups 2 and 3. In group 2 radiolucency increased uniformly starting from the 4th week with completely disbanded implants at 12 weeks. In group 3 the implants were only faintly detectable after the 4th week and had already disappeared radiologically after 6 weeks.

15.4.1 Micro-CT

In group 1 all investigated implants showed radiopaque cores and localized mineralized surface changes after 4 weeks. Two of four animals showed beads with well-distinguishable trabecular structures between the implant's surface and the host bone cortex. After 6 weeks the radiopaque core of all implants had disappeared, and the resorption left a homogenously disorganized surface. In addition, cortical bone bridges reached all implants. After 8 and 12 weeks, bone bridging on the implants had progressed, with no change in resorption of the implants.

Almost complete resorption for groups 2 and 3 was detected within 12 and 6 weeks, respectively. The implants were almost completely resorbed in all animals. Varying degrees of mineralization at the periphery of the implant bed were detected. Group 2 indicated a delayed resorption until week 12 and superficial trabecular structures, compared to pure calcium sulfate. Bone bridges from the animals' cortex to the mineralized areas were found in a smaller quantity in group 2, but not for group 3 implants.

Resorption of the beads between the groups of this study differed considerably. The implants of group 3 consisted only of CS and tobramycin and had already shown peripheral and central signs of resorption in the 4th week. In the 6th and especially the 8th weeks after implantation, beads had almost fully degraded. Beads of group 2 were no longer detectable radiologically after the 12th week, while the micro-CT scans and histological slices showed remnants of the beads as well as low amounts of bone formation around the former implant sites. This observation was supported by the data of the polychrome sequence marking. The beads of group 1 (Herafill®-G) were well detectable radiographically after 4 weeks. In the 6th week increasing radiolucency of the beads was detectable, which increased up to week 8. At week 12 no further changes had shown. These observations suggest that both the size and composition of the implants influence the degradation rate. For Osteoset®, CS alone (group 3), a faster degradation was observed than for the compositions containing tripalmitin and

calcium carbonate (groups 1 and 2 ($CaSO_4$-V)). Within the latter group, the size of the implants seems to play an important role, since the small beads of group 2 were absorbed more rapidly than the large beads of group 1. These findings are in accordance to reported data from literature, where different degradation times for CS are reported for different sizes of implants: 3–4 weeks [20], 4–10 weeks [21], 6 weeks [22], 5–7 weeks [23], 45–72 days [3, 24], 8 weeks [25], 2–4 months [16], and 6 months [26]. This shows that size correlates directly with resorption time. Additionally, the localization of implantation influences degradation time as presented on beads in the radius [3], tibia [27], femur [27], or maxilla and mandible [28, 29]. Lastly, the used animal model can also influence the degradation as already investigated in the dog [3, 28, 29] or rabbit [20].

15.5 Histological Results

In group 1, trabecular bone formation, invasion of osteoblasts, and erosion were detectable on all implants' surfaces after 4 weeks. These mechanisms protruded up to 12 weeks with still detectable implants and surrounding giant cells at the time of sacrifice. In group 2, changes described above for group 1 were visible after 4 weeks as well. Implant resorption was more pronounced compared to group 1 with almost fully degraded implants after 12 weeks. Giant cells were not detectable from week 8 on. Group 3 showed similar osteogenesis as group 2 with invasion of osteoblasts and production of osteoid starting from week 4. Resorption was very pronounced with almost fully resorbed implants from week 6 on. No more beads were recognizable after 12 weeks.

Analysis of the histological specimen by means of polychrome sequence marking after 8 and 12 weeks revealed the peri-implant bone as newly formed vital bone with enhancement of applied fluorochromes (Fig. 15.3). In the newly formed bone, no evidence of foreign-body response to the biomaterials (e.g., presence of inflammatory cells or fibrous tissue) was observed.

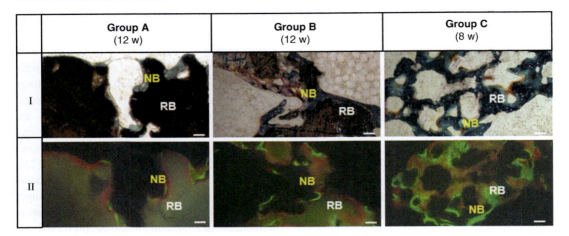

Fig. 15.3 Histological slices and fluorochrome-labeled bone at the implants in tibia of each group. Histological slices (*I*) and fluorochrome-labeled bone (*II*) of tibia of each group 1–3 after 12 weeks (1: Herafill®-G and 2: CaSO4-V), as well as after 8 weeks (3: Osteoset®). Corresponding to histological pictures fluorochromes were deposited around and inside the beads (*NB* new bone, *RB* remnants of beads). (*II*) Fluorescence colors indicate newly formed bone at different postsurgical periods (yellow/green (3w): tetracycline; red (5w): alizarin complexone; green (7w): calcein green; orange (11w): xylenol orange). The white scale bars represent 100 µm length. Reproduced with permission

By means of histological staining in the present study, we were able to find deep lacunae in the peri-implant bone of all three formulations as clear signs of cellular degradation. Additionally, the presence of multinucleated giant cells on or in the implant material confirms the presence of cellular degradation. Regarding the nature of CS degradation contradicting mechanisms have been described. Bell et al. and Tay et al. argue that CS degrades through the process of dissolution [23, 25]. This supports other investigators, who explain accelerated bone formation by provision of calcium ions during the degradation process [20]. Walsh et al. [30] conclude that the pH reduction leads to a demineralization of the adjacent bone, which subsequently leads to a release of BMP, which in turn stimulates bone regeneration. Sidqui et al. [31] also noted in their study that the activity of osteoclasts was excited by an acidic pH. Other authors however argue that CS is degraded by cellular phagocytosis [3]. Orsini et al. [32] report in their study a combination of both dissolution and cell-mediated degradation of CS [32]. However, whether CS additionally is degraded by dissolution cannot be stated with the findings of the present study.

The osteoconductive potential of CS was clearly detected in this study. All three formulations showed invading osteoblasts with active production of osteoid at the implant sites and walled osteocytes with surrounding newly formed bone at the level of the implants' surfaces and in the resorption lacunae. Group 3 showed osteogenesis up to the 8th week, with only remnants of osteoid at the level of the former implant site up to the 12th week (Fig. 15.4). Similar events were found in group 2, with active osteogenesis from the 4th week starting and decreasing over time. Group 1 beads seemed to induce osteogenesis at a steady-state level from the 4th up to the 12th weeks. Here, fine cortical bone trabeculae bridged the implant-bone interface at the day of sacrifice. Possibly, the osteoconductive effect of group 2 may be improved to the level of group 1 by using larger bead dimensions. In summary, osteogenesis was found to decrease over time in groups 2 and 3 almost parallel to degradation of the beads. On the other hand, high levels of osteogenesis in group 1 were detected over 12 weeks. These mechanisms were also supported by polychrome sequence marking after 12 weeks. In all three formulations, clear osteogenic signs detecting bone formation in direct contact to the degrading implants were found.

One limitation of this study is the use of different sizes for implants employed. The weights of the beads in sum were comparable accordingly

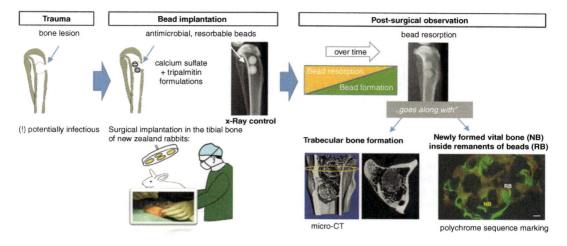

Fig. 15.4 Graphical abstract: experimental method overview. Reproduced with permission

using different numbers of beads per type. To allow for an ideally comparable situation, beads of exact same size and antibiotic content should be used. However, this study delivers valid results for resorption and bone formation over time, as the use of real market-ready material was compared. Molecular biologic methods in future studies might help in further evaluation of the newly formed bone inside and around the implants during resorption.

In summary, this study supported the biocompatibility of novel antibiotic-impregnated CS beads and specified the mechanisms in a rabbit model. The addition of calcium carbonate/tripalmitate delayed the degradation of the implants and resulted in synchronization of the ingrowing bone with biomaterial resorption but did not have negative effects on biocompatibility. Further in vivo studies are needed, in order to estimate time for complete resorption of these new formulations of CS beads and to determine responses of bone cells to this biomaterial.

15.6 Outlook

In a rabbit model novel formulations of absorbable bone substitute materials based on calcium sulfate exhibit delayed time of resorption, while osteoconduction is facilitated. Moreover, a sustaining osteogenic effect is determined over 12 weeks after implantation for Herafill®-G beads. The use of novel calcium sulfate-based bone substitute materials, incorporated with calcium carbonate/tripalmitin and antibiotics, can stimulate the bone regeneration process in an antimicrobial environment. Herewith, trauma and orthopedic surgeons may receive a future bone substitute material for complex bone regeneration as well as infection prophylaxis.

Acknowledgments At first, we would like to thank Mr. Dr. H. Büchner and Mr. Dr. S. Vogt (Heraeus Medical GmbH, Wehrheim, Germany) for their kind supply of bone substitute materials (Herafill®-G, as well as CaSO4-V). Second, many thanks to the central preclinical research division (ZPF) of the Klinikum rechts der Isar at the Technical University of Munich for their excellent support in performing the animal study. Especially, many thanks to Mrs. Dr. M. Rößner and Prof. Dr. H. Gollwitzer for their guidance in surgical procedure. Also, many thanks to Mrs. Dr. S. Kerschbaumer for generating and interpreting histological slices. Moreover, special thanks to Prof. Dr. P. Augat (Department of Biomechanics at the Unfallklinik Murnau) for his kind support in micro-CT investigations. Our special gratitude goes to Dr. Meredith Kiokekli for co-conduction of the experiments as well as to Mr. F. Seidl (M.A. Interpreting and Translating, MBA) for his kind support due to his perfect command of scientific English.

Disclaimer: Parts of this scientific article have by the authors previously been published in Journal of Material Science Materials in Medicine (Springer International Publishing AG). Reproduction was permitted by Springer.

References

1. Lalidou F, Kolios G, Drosos GI. Bone infections and bone graft substitutes for local antibiotic therapy. Surg Technol Int. 2014;24:353–62.
2. Cortez PP, Silva MA, Santos M, Armada-da-Silva P, Afonso A, Lopes MA, Santos JD, Maurício AC. A glass-reinforced hydroxyapatite and surgical-grade calcium sulfate for bone regeneration: in vivo biological behavior in a sheep model. J Biomater Appl. 2012;27(2):201–17.
3. Peltier LF. The use of plaster of paris to fill large defects in bone. Am J Surg. 1959;97(3):311–5.
4. Helgeson MD, Potter BK, Tucker CJ, Frisch HM, Shawen SB. Antibiotic-impregnated calcium sulfate use in combat-related open fractures. Orthopedics. 2009;32(5):323.
5. Beuerlein MJ, McKee MD. Calcium sulfates: what is the evidence? J Orthop Trauma. 2010;24(Suppl 1):S46–51.
6. Thomas MV, Puleo DA. Calcium sulfate: properties and clinical applications. J Biomed Mater Res B Appl Biomater. 2009;88(2):597–610.
7. Slater N, Dasmah A, Sennerby L, Hallman M, Piattelli A, Sammons R. Back-scattered electron imaging and elemental microanalysis of retrieved bone tissue following maxillary sinus floor augmentation with calcium sulphate. Clin Oral Implants Res. 2008;19(8):814–22.
8. Parsons JR, Ricci JL, Alexander H, Bajpai PK. Osteoconductive composite grouts for orthopedic use. Ann N Y Acad Sci. 1988;523:190–207.
9. Stubbs D, Deakin M, Chapman-Sheath P, Bruce W, Debes J, Gillies RM, Walsh WR. Biomaterials. 2004;25(20):5037–44.
10. Fan X, Ren H, Luo X, Wang P, Lv G, Yuan H, Li H, Yan Y. Mechanics, degradability, bioactivity, in vitro, and in vivo biocompatibility evaluation of poly(amino acid)/hydroxyapatite/calcium sulfate composite for potential load-bearing bone repair. J Biomater Appl. 2016;30(8):1261–72.
11. Frost HM. Tetracycline-based histological analysis of bone remodeling. Calcif Tissue Res. 1969;3(3):211–37.
12. Pforringer D, Obermeier A, Kiokekli M, Büchner H, Vogt S, Stemberger A, Burgkart R, Lucke M. Antimicrobial formulations of absorbable bone substitute materials as drug carriers based on calcium sulfate. Antimicrob Agents Chemother. 2016;60(7):3897–905.
13. Borrelli J Jr, Prickett WD, Ricci WM. Treatment of nonunions and osseous defects with bone graft and calcium sulfate. Clin Orthop Relat Res. 2003;411:245–54.
14. Evaniew N, Tan V, Parasu N, Jurriaans E, Finlay K, Deheshi B, Ghert M. Use of a calcium sulfate-calcium phosphate synthetic bone graft composite in the surgical management of primary bone tumors. Orthopedics. 2013;36(2):e216–22.
15. Glazer PA, Spencer UM, Alkalay RN, Schwardt J. In vivo evaluation of calcium sulfate as a bone graft substitute for lumbar spinal fusion. Spine J. 2001;1(6):395–401.
16. Coetzee AS. Regeneration of bone in the presence of calcium sulfate. Arch Otolaryngol. 1980;106(7):405–9.
17. Coraca-Huber D, Hausdorfer J, Fille M, Nogler M, Kuhn KD. Calcium carbonate powder containing gentamicin for mixing with bone grafts. Orthopedics. 2014;37(8):e669–72.
18. Coraca-Huber DC, Putzer D, Fille M, Hausdorfer J, Nogler M, Kuhn KD. Gentamicin palmitate as a new antibiotic formulation for mixing with bone tissue and local release. Cell Tissue Bank. 2014;15(1):139–44.
19. Obermeier A, Matl FD, Schwabe J, Zimmermann A, Kühn KD, Lakemeier S, von Eisenhart-Rothe R, Stemberger A, Burgkart R. Novel fatty acid gentamicin salts as slow-release drug carrier systems for anti-infective protection of vascular biomaterials. J Mater Sci Mater Med. 2012;23(7):1675–83.
20. Lebourg L, Biou C. The imbedding of plaster of paris in surgical cavities of the maxilla. Sem Med Prof Med Soc. 1961;37:1195–7.
21. Geldmacher J. Therapy of enchondroma with a plaster implant—renaissance of a treatment principle. Handchir Mikrochir Plast Chir. 1986;18(6):336–8.
22. Petruskevicius J, Nielsen S, Kaalund S, Knudsen PR, Overgaard S. No effect of Osteoset, a bone graft substitute, on bone healing in humans: a prospective randomized double-blind study. Acta Orthop Scand. 2002;73(5):575–8.
23. Bell WH. Resorption characteristics of bone and bone substitutes. Oral Surg Oral Med Oral Pathol. 1964;17:650–7.
24. Lillo R, Peltier LF. The substitution of plaster of Paris rods for portions of the diaphysis of the radius in dogs. Surg Forum. 1956;6:556–8.
25. Tay BK, Patel VV, Bradford DS. Calcium sulfate- and calcium phosphate-based bone substitutes. Mimicry of the mineral phase of bone. Orthop Clin North Am. 1999;30(4):615–23.
26. Kelly CM, Wilkins RM, Gitelis S, Hartjen C, Watson JT, Kim PT. The use of a surgical grade calcium sulfate as a bone graft substitute: results of a multicenter trial. Clin Orthop Relat Res. 2001;382:42–50.
27. Blaha JD. Calcium sulfate bone-void filler. Orthopedics. 1998;21(9):1017–9.
28. Calhoun NR, Greene GW Jr, Blackledge GT. Plaster: a bone substitute in the mandible of dogs. J Dent Res. 1965;44(5):940–6.
29. McKee JC, Bailey BJ. Calcium sulfate as a mandibular implant. Otolaryngol Head Neck Surg. 1984;92(3):277–86.

30. Walsh WR, Morberg P, Yu Y, Yang JL, Haggard W, Sheath PC, Svehla M, Bruce WJ. Response of a calcium sulfate bone graft substitute in a confined cancellous defect. Clin Orthop Relat Res. 2003;406:228–36.

31. Sidqui M, Collin P, Vitte C, Forest N. Osteoblast adherence and resorption activity of isolated osteoclasts on calcium sulphate hemihydrate. Biomaterials. 1995;16(17):1327–32.

32. Orsini G, Ricci J, Scarano A, Pecora G, Petrone G, Iezzi G, Piattelli A. J Biomed Mater Res B Appl Biomater. 2004;68(2):199–208.

Perivascular Progenitor Cells for Bone Regeneration

16

Carolyn Meyers, Paul Hindle, Winters R. Hardy,
Jia Jia Xu, Noah Yan, Kristen Broderick,
Greg Asatrian, Kang Ting, Chia Soo, Bruno Peault,
and Aaron W. James

16.1 Introduction

The perivascular niche has long been hypothesized to house a mesenchymal progenitor cell population. Over the last 9 years, the MSC-like identity of perivascular progenitor cells has become well defined [1]. First described in pericytes of the tunica intima [1], it is clear that at least two and likely several multipotent perivascular cell populations exist. The tunica adventitia (outermost layer of the blood vessel wall) also houses progenitor cells in man [2] and mouse [3]. Specific combined cell surface markers on each perivascular progenitor cell population allow for prospective isolation using antibodies, rather than retrospective culture-based isolation techniques which are more commonly employed in the MSC field. Since their initial identification, multiple research groups have confirmed the MSC characteristics of perivascular progenitor cells, and sought to leverage these cell types for tissue regeneration—both within and outside of the orthopedic field. In this review, we summarize what is known regarding the identity of perivascular progenitor cells, studies to date in bone tissue engineering, and evolving concepts of diversity within the perivascular niche.

C. Meyers · J. J. Xu · N. Yan · A. W. James (✉)
Department of Pathology, Johns Hopkins University,
Baltimore, MD, USA
e-mail: cmeyer30@jhmi.edu; nyan5@jhu.edu;
Awjames@jhmi.edu

P. Hindle
Department of Trauma and Orthopaedic Surgery,
The University of Edinburgh, Edinburgh, UK

W. R. Hardy
Orthopedic Hospital Research Center,
University of California, Los Angeles, CA, USA

K. Broderick
Department of Surgery, Johns Hopkins University,
Baltimore, MD, USA

G. Asatrian · K. Ting
School of Dentistry, University of California,
Los Angeles, CA, USA
e-mail: kting@dentistry.ucla.edu

C. Soo
Orthopedic Hospital Research Center,
University of California, Los Angeles, CA, USA

Department of Surgery, University of California,
Los Angeles, Los Angeles, CA, USA
e-mail: bsoo@ucla.edu

B. Peault
Orthopedic Hospital Research Center,
University of California, Los Angeles, CA, USA

BHF Center for Vascular Regeneration and MRC
Center for Regenerative Medicine, University of
Edinburgh, Edinburgh, UK
e-mail: bpeault@mednet.ucla.edu

© Springer Nature Switzerland AG 2019
D. Duscher, M. A. Shiffman (eds.), *Regenerative Medicine and Plastic Surgery*,
https://doi.org/10.1007/978-3-030-19962-3_16

Table 16.1 Summary of perivascular progenitor cell markers

Surface marker	Expression in pericytes	Expression in adventitial cells	Pericyte reference	Adventitial cell reference
CD10	+	+	[1]	[29]
CD13	+	+	[1]	[30]
CD14	−	−	[31–33]	[31]
CD29	+	Unknown	[32]	Unknown
CD31	−	−	[1, 31–36])	[2, 34–37]
CD33	−	−	[31]	[31, 38]
CD34	−	+	[1, 31, 34–36]	[2, 12, 13, 39]
CD44	+	+	[1, 31–34]	[2, 31, 37]
CD45	−	−	[1, 31, 32, 34]	[2, 31, 36, 37]
CD56	−	Unknown	[1, 34]	Unknown
CD73	+	+	[1, 31, 32, 34]	[2, 31]
CD90	+	+	[1, 31–36]	[2, 24, 31, 37]
CD105	+	+	[1, 32–34]	[2]
CD106	−	−	[1, 33]	Unknown
CD117	−	−	[36]	[36, 37]
CD133	−		[1]	Unknown
CD140b	+	+	[1, 31, 32, 34, 35]	[30]
CD146	+	−	[1, 31–36]	[2, 31, 36, 37]
ASMA	+	−	[1, 34, 35]	[2, 35, 37]
NG2	+	−	[1, 32, 34, 35]	[2, 37]
STRO-1	+	Unknown	[33, 34]	Unknown
Nestin	+	Unknown	[37]	Unknown
Vimentin	+	+	[37]	[37]

16.2 Introduction to Pericytes and Perivascular Stem Cells

16.2.1 Identification of Perivascular Progenitors

The definitive in situ identification of pericytes as a tissue-resident MSC population was first reported by Crisan et al. in 2008 [1]. Prior to this, scattered reports had described the possible multipotentiality of CD146+ pericytes, including an ability to undergo osteogenic, chondrogenic, fibrogenic, and adipogenic differentiation [4–6] and high expression of cytokines and differentiation factors [7, 8]. Crisan et al. [1] used a combination of immunofluorescent tissue section staining and flow cytometry to reveal the MSC characteristics of pericytes, including expression of canonical MSC markers such as CD44, CD73, CD90, and CD105. Pericytes can be identified using a set of characteristic markers (such as NG2, CD146, PDGFRB, and αSMA), along with co-staining for endothelial markers (such as CD31, CD34, or VWF von Willebrand

factor) for exclusion of intimal cells and to confirm a perivascular residence. Additional pericyte markers are listed in Table 16.1. Across human organs, pericytes were ubiquitously identified around arterioles, capillaries, and venules. Next, Crisan et al. used fluorescent-activated cell sorting (FACS) to purify a CD146 + CD34-CD45-CD56- pericyte population. Subsequent studies have refined this to identify a CD146 + CD34-CD31-CD45-CD56- cell population [9]. Crisan et al. [1] identified the multilineage differentiation potential of FACS-purified pericytes, including an ability to differentiate down bone, fat, skeletal muscle, and cartilage lineages. Since this time, multiple investigators have confirmed the pericytic/perivascular location of presumptive MSC [10].

A second perivascular location of MSC-like cells was soon identified in the tunica adventitia of larger vessels. Hu et al. [11] identified a CD117+, Sca-1+ (stem cell antigen-1) cell population within the mouse aorta with MSC characteristics. Likewise, investigators have isolated CD34+ mesenchymal progenitors from human arteries and veins [12, 13]. These findings were formalized by

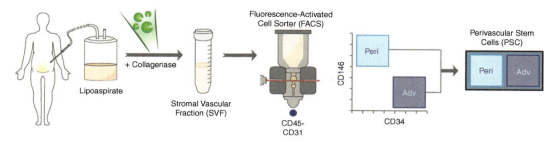

Fig. 16.1 Overview of perivascular stem/stromal cell isolation (PSC). From left to right: Perivascular stem/stromal cells are most commonly derived from liposuction of human subcutaneous adipose tissue, yielding lipoaspirate. After an enzymatic digestion using collagenase, the cell pellet is obtained, consisting of an unpurified stromal cell population, most commonly referred to as "stromal vascular fraction" or SVF. Fluorescence-activated cell sorting is performed to obtain CD146+CD34-CD45-CD31- pericyte (peri) and CD146-CD34+CD45-CD31- adventitial cell (adv) population. Collectively, these perivascular progenitor cell populations are termed "perivascular stem/stromal cell" or PSC

Corselli et al. [2], in which the tunica adventitia of human larger arteries and veins was observed to have MSC characteristics, including typical MSC marker expression (in situ and after purification) as well as multilineage differentiation potential. Under permissive culture conditions, these so-called adventitial cells can adopt pericyte markers, suggesting that this represents a less differentiated progenitor cell subset. In combination, pericytes and adventitial cells (collectively termed perivascular stem/stromal cells or PSC) comprise approximately 40% of lipoaspirate stromal tissues [14]. Analyzing the frequency of viable PSC purified by FACS from 60 consecutive donors of lipoaspirate, James and colleagues reported a mean yield of over 15 million purified PSC per 100 mL of lipoaspirate [15]. These results were the first study to demonstrate that PSC could be derived in clinically meaningful numbers for orthopedic applications. To date, it is the combination of human CD146+ pericytes and CD34+ adventitial cells that has been most often studied for use in bone tissue engineering (Fig. 16.1).

16.2.2 Relationship of Pericytes and Adventitial Cells

Although features of multipotency and perivascular locale are shared between pericytes and adventitial cells, the exact relationship between these two perivascular cell populations is unclear. As mentioned, under certain culture conditions, adventitial cells may adopt the expression of pericyte markers. These data led to the hypothesis of adventitial cells being a "less differentiated" stromal cell population. In order to more thoroughly understand these distinct cell populations, single-cell gene expression was performed on individual adventitial cells ($n = 67$) or pericytes ($n = 73$) from a single patient's adipose tissue sample [16]. Cells were transcriptionally profiled using Fluidigm single-cell quantitative PCR. To these single-cell data, we applied differential expression and principal component and clustering analysis, as well as an original gene co-expression network reconstruction algorithm. Despite the stochasticity at the single-cell level, an analysis of covariance patterns in gene expression yielded multiple network connectivity parameters suggesting that adventitial cells are the "least differentiated" while pericytes are the "more differentiated" cell populations. ALDH staining, a functional marker of primitivity was able to further subclassify perivascular progenitor cells into discreet subclasses. These findings are important from a tissue engineering standpoint, as most studies to date have used a combined pericyte + adventitial cell preparation for bone regeneration. The functional differences, if any, in bone forming potential remain unknown.

16.2.3 Perivascular Progenitor Cells in Ectopic Bone Formation

Proof-of-concept studies in PSC applications in bone tissue engineering were first conducted in an ectopic (intramuscular) model. This model examined the bone-forming potential of isolated

PSC as compared to the unpurified stromal vascular fraction (SVF) from the same patient's adipose sample [14]. Here, a demineralized bone matrix putty was used as an additional osteoinductive stimulus. Results showed that PSC intramuscular implantation resulted in larger bone particles and increased quantitative metrics of bone formation by micro-CT, and increased bone matrix accumulation by histologic analysis. In addition to increased bone formation among PSC-treated groups, an increase in vascularity of the implant site, associated with increased production of vascular endothelial growth factor (VEGF), was observed [17]. Implanted PSC were also observed to return to their perivascular location once engrafted [17], a finding that has also been observed in cardiac implantation models [18]. These "proof-of-concept" studies established for the first time that a purified perivascular cell therapy was superior to an unpurified stromal cell population across all outcomes of bone formation.

16.2.4 Perivascular Progenitor Cells in Orthotopic Models of Bone Formation

The bone-forming advantage of purified PSC has been examined across three clinically relevant models of bone regeneration/repair, including a (1) mouse calvarial defect model, (2) rat spinal fusion model, and (3) rat fibrous nonunion model. Importantly, all studies have used different application techniques and different scaffold carriers—highlighting the versatility of the cell therapy.

The first application of PSC for bone defect healing was studied in a murine calvarial defect model. Here, equal numbers of unpurified SVF or purified PSC from the same patient's adipose sample were implanted in a nonhealing calvarial defect in the mid-aspect of the parietal bone. Radiographic and histologic analysis showed that PSC induced a significant increase in bony regenerate within the defect site, with significant bone defect healing over the 8-week study period. In comparison, unpurified SVF from the same patient had no sta-

tistically significant benefit in comparison to an acellular scaffold control. Thus, across both ectopic and orthotopic bone models, PSC showed significantly greater potential for bone formation in comparison to unpurified stroma.

Spinal fusion models must withstand biomechanical strain and represent a more functionally demanding environment for a bone graft substitute. Chung and colleagues examined the effects of PSC application in a rat posterolateral lumbar spinal fusion model [19]. Briefly, human PSC implantation was performed across three cellular densities in the rat spinal fusion model, using a demineralized bone matrix putty as a cell carrier. Results showed that PSC showed a dose-dependent increase in endochondral ossification, increase in bone deposition, increase in measurements of bone strength, and complete fusion (100%) between spinal segments. PSC-induced bone formation was observed to be a result of both direct ossification of PSC and indirect paracrine effects exerted by PSC. By species-specific MHC class I immunohistochemical staining, the indirect effects on host osteoprogenitor cells seemed to be the principal mechanism of action.

To test the use of pericyte cell therapies for fibrous nonunion, investigators examined a well-established model of rat tibial atrophic nonunion [20, 21]. Pericytes were percutaneously injected 3 weeks after the establishment of fibrous nonunion [22]. Results showed that pericyte injection increased fracture callus size, increased mineralization, and resulted in increased bone union. Three differences should be noted between this and past studies. First, pericytes were injected as a cell suspension—as opposed to being mixed with a scaffold material. Second, pericyte injection in nonunion did not show significant persistence of implanted cells when defect sites were interrogated by species-specific immunohistochemistry. These data are at odds with prior studies in which human cell persistence was seen. Third and finally, pericytes did not outcompete BMSC cells in terms of primary outcomes when transplanted in equal numbers. These data suggest that differences across model systems and cell applications methods are critical when assessing interstudy comparability.

16.3 Emerging Topics: Regional Specification of Pericytes

Despite the ubiquitous distribution of pericytes and perivascular progenitor cells in all vascularized organs, their tissue-specific functions remain shrouded in mystery. Data to date supports the notion that pericytes have organ-specific functions in tissue repair while maintaining MSC-like properties. For example, pericytes derived from the fat pad adjacent to the cartilage of the knee show MSC-like multilineage differentiation potential, but with an unusual predilection for forming cartilage [23]. Likewise, dental pulp-derived CD146+ pericytes form a mixture of fibrous pulp and dentin once transplanted, in effect recapitulating their native environment with marked fidelity [24]. Similarly, brain-derived pericytes are primed to participate in neural repair [25, 26]. In order to directly compare the bone-forming potential of pericytes by anatomic location, Sacchetti et al. [27] purified and cultured four preparations of human CD146+ pericytes from either bone marrow, periosteum, skeletal muscle, or cord blood. Fascinatingly, in an ectopic bone formation model, vastly different quantities of bone and cartilage were produced depending on the anatomic site of origin [27]. Major differences in the transcriptome of each pericyte reservoir were associated with these differences in bone-forming potential. In a striking example of tissue specification, we recently demonstrated that pericytes within the kidney produce renin with functional enzyme activity, thereby participating in blood volume control [28]. Importantly, these tissue-specific pericytes retained a canonical feature of all pericytes—capability for multilineage differentiation [28]. In sum, the concept is emerging of pericytes as a reservoir of tissue-specific progenitor MSCs.

16.4 Discussion

In summary, the application of perivascular progenitor cells for orthopedic indications is a significant and growing field of study. There exists significant overlap in the use of PSC as compared to other adipose-derived cell therapies. However, there are significant risks and benefits of using a purified perivascular cell population. A cell sorting approach bypasses the time, costs, and risks associated with cell culture, and increases the homogeneity of the cell therapy. Precise cell identity may also speed regulatory approval of a cellular therapy. Conversely, there are several drawbacks for the use of a purified cellular constituent. First, the isolation of PSC is a complex procedure which currently uses FACS. GMP facilities with FACS capability are uncommon, and the associated costs will be high. Nevertheless, a purified cell product that bypasses any culture requirement represents a significant advance in the field. Additional, direct head-to-head comparisons of PSC to other MSC sources are needed to clarify the most optimum cell source for orthopedic indications.

Acknowledgements This work was supported by the NIH/NIAMS (grants R01 AR061399, R01 AR066782, K08 AR068316), the Musculoskeletal Transplant Foundation, and Orthopaedic Research and Education Foundation with funding provided by the Musculoskeletal Transplant Foundation.

Disclosure/Conflict of Interest.

K.T., B.P., and C.S. are inventors of perivascular stem cell-related patents filed from UCLA. K.T and C.S. are founders of Scarless Laboratories, Inc. which sublicenses perivascular stem cell-related patents from the UC Regents, and who also hold equity in the company. C.S. is also an officer of Scarless Laboratories, Inc.

References

1. Crisan M, Yap S, Casteilla L, Chen CW, Corselli M, Park TS, Andriolo G, Sun B, Zheng B, Zhang L, Norotte C, Teng PN, Traas J, Schugar R, Deasy BM, et al. A perivascular origin for mesenchymal stem cells in multiple human organs. Cell Stem Cell. 2008;3(3):301–13.
2. Corselli M, Chen CW, Sun B, Yap S, Rubin JP, Péault B. The tunica adventitia of human arteries and veins as a source of mesenchymal stem cells. Stem Cells Dev. 2012;21(8):1299–308.
3. Kramann R, Goettsch C, Wongboonsin J, Iwata H, Schneider RK, Kuppe C, Kaesler N, Chang-Panesso M, Machado FG, Gratwohl S, Madhurima K, Hutcheson JD, Jain S, Aikawa E, Humphreys BD. Adventitial MSC-like cells are progenitors of vascular smooth muscle cells and drive vascular calcification in chronic kidney disease. Cell Stem Cell. 2016;19(5):628–42.

4. Collett GD, Canfield AE. Angiogenesis and pericytes in the initiation of ectopic calcification. Circ Res. 2005;96(9):930–8.
5. Doherty MJ, Canfield AE. Gene expression during vascular pericyte differentiation. Crit Rev Eukaryot Gene Expr. 1999;9(1):1–17.
6. Farrington-Rock C, Crofts NJ, Doherty MJ, Ashton BA, Griffin-Jones C, Canfield AE. Chondrogenic and adipogenic potential of microvascular pericytes. Circulation. 2004;110(15):2226–32.
7. Invernici G, Emanueli C, Madeddu P, Cristini S, Gadau S, Benetti A, Ciusani E, Stassi G, Siragusa M, Nicosia R, Peschle C, Fascio U, Colombo A, Rizzuti T, Parati E, Alessandri G. Human fetal aorta contains vascular progenitor cells capable of inducing vasculogenesis, angiogenesis, and myogenesis in vitro and in a murine model of peripheral ischemia. Am J Pathol. 2007;170(6):1879–92.
8. Howson KM, Aplin AC, Gelati M, Alessandri G, Parati EA, Nicosia RF. The postnatal rat aorta contains pericyte progenitor cells that form spheroidal colonies in suspension culture. Am J Physiol Cell Physiol. 2005;289(6):C1396–407.
9. West CC, Hardy WR, Murray IR, James AW, Corselli M, Pang S, Black C, Lobo SE, Sukhija K, Liang P, Lagishetty V, Hay DC, March KL, Ting K, Soo C, Péault B. Prospective purification of perivascular presumptive mesenchymal stem cells from human adipose tissue: process optimization and cell population metrics across a large cohort of diverse demographics. Stem Cell Res Ther. 2016;7:47.
10. Murray IR, West CC, Hardy WR, James AW, Park TS, Nguyen A, Tawonsawatruk T, Lazzari L, Soo C, Péault B. Natural history of mesenchymal stem cells, from vessel walls to culture vessels. Cell Mol Life Sci. 2014;71(8):1353–74.
11. Hu Y, Zhang Z, Torsney E, Afzal AR, Davison F, Metzler B, Xu Q. Abundant progenitor cells in the adventitia contribute to atherosclerosis of vein grafts in ApoE-deficient mice. J Clin Invest. 2004;113(9):1258–65.
12. Campagnolo P, Cesselli D, Al Haj Zen A, Beltrami AP, Krankel N, Katare R, Angelini G, Emanueli C, Madeddu P. Human adult vena saphena contains perivascular progenitor cells endowed with clonogenic and proangiogenic potential. Circulation. 2010;121(15):1735–45.
13. Pasquinelli G, Tazzari PL, Vaselli C, Foroni L, Buzzi M, Storci G, Alviano F, Ricci F, Bonafè M, Orrico C, Bagnara GP, Stella A, Conte R. Thoracic aortas from multiorgan donors are suitable for obtaining resident angiogenic mesenchymal stromal cells. Stem Cells. 2007;25(7):1627–34.
14. James AW, Zara JN, Zhang X, Askarinam A, Goyal R, Chiang M, Yuan W, Chang L, Corselli M, Shen J, Pang S, Stoker D, Wu B, Ting K, Péault B, Soo C. Perivascular stem cells: a prospectively purified mesenchymal stem cell population for bone tissue engineering. Stem Cells Transl Med. 2012;1(6):510–9.

15. James AW, Zara JN, Corselli M, Askarinam A, Zhou AM, Hourfar A, Nguyen A, Megerdichian S, Asatrian G, Pang S, Stoker D, Zhang X, Wu B, Ting K, Péault B, Soo C. An abundant perivascular source of stem cells for bone tissue engineering. Stem Cells Transl Med. 2012;1(9):673–84.
16. Hardy WR, Moldovan NI, Moldovan L, Livak KJ, Datta K, Goswami C, Corselli M, Traktuev DO, Murray IR, Péault B, March K. Transcriptional networks in single perivascular cells sorted from human adipose tissue reveal a hierarchy of mesenchymal stem cells. Stem Cells. 2017;35(5):1273–89.
17. Askarinam A, James AW, Zara JN, Goyal R, Corselli M, Pan A, Liang P, Chang L, Rackohn T, Stoker D, Zhang X, Ting K, Péault B, Soo C. Human perivascular stem cells show enhanced osteogenesis and vasculogenesis with Nel-like molecule I protein. Tissue Eng Part A. 2013;19(11–12):1386–97.
18. Chen CW, Okada M, Proto JD, Gao X, Sekiya N, Beckman SA, Corselli M, Crisan M, Saparov A, Tobita K, Péault B, Huard J. Human pericytes for ischemic heart repair. Stem Cells. 2013;31(2):305–16.
19. Chung CG, James AW, Asatrian G, Chang L, Nguyen A, Le K, Bayani G, Lee R, Stoker D, Pang S, Zhang X, Ting K, Péault B, Soo C. Human perivascular stem cell-based bone graft substitute induces rat spinal fusion. Stem Cells Transl Med. 2015;4(5):538.
20. Reed AA, Joyner CJ, Isefuku S, Brownlow HC, Simpson AH. Vascularity in a new model of atrophic nonunion. J Bone Joint Surg Br. 2003;85(4):604–10.
21. Tawonsawatruk T, Kelly M, Simpson H. Evaluation of native mesenchymal stem cells from bone marrow and local tissue in an atrophic nonunion model. Tissue Eng Part C Methods. 2014;20(6):524–32.
22. Tawonsawatruk T, West CC, Murray R, Soo C, Peault B, Simpson AHRW. Adipose derived pericytes rescue fractures from a failure of healing-nonunion. Sci Rep. 2016;6(1):22779.
23. Hindle P, Khan N, Biant L, Péault B. The infrapatellar fat pad as a source of perivascular stem cells with increased chondrogenic potential for regenerative medicine. Stem Cells Transl Med. 2017;6(1):77–87.
24. Shi S, Gronthos S. Perivascular niche of postnatal mesenchymal stem cells in human bone marrow and dental pulp. J Bone Miner Res. 2003;18(4):696–704.
25. Tavazoie M, Van der Veken L, Silva-Vargas V, Louissaint M, Colonna L, Zaidi B, Garcia-Verdugo JM, Doetsch F. A specialized vascular niche for adult neural stem cells. Cell Stem Cell. 2008;3(3):279–88.
26. Cai W, Liu H, Zhao J, Chen LY, Chen J, Lu Z, Hu X. Pericytes in brain injury and repair after ischemic stroke. Transl Stroke Res. 2017;8(2):107–21.
27. Sacchetti B, Funari A, Remoli C, Giannicola G, Kogler G, Liedtke S, Cossu G, Serafini M, Sampaolesi M, Tagliafico E, Tenedini E, Saggio I, Robey PG, Riminucci M, Bianco P. No identical "mesenchymal stem cells" at different times and sites: human committed progenitors of distinct origin and differentia-

tion potential are incorporated as adventitial cells in microvessels. Stem Cell Rep. 2016;6(6):897–913.

28. Stefanska A, Kenyon C, Christian HC, Buckley C, Shaw I, Mullins JJ, Péault B. Human kidney pericytes produce renin. Kidney Int. 2016;90(6):1251–61.

29. Kim JH, Hwang SE, Yu HC, Hwang HP, Katori Y, Murakami G, Cho BH. Distribution of CD10-positive epithelial and mesenchymal cells in human mid-term fetuses: a comparison with CD34 expression. Anat Cell Biol. 2014;47(1):28–39.

30. Baer PC. Adipose-derived mesenchymal stromal/stem cells: an update on their phenotype in vivo and in vitro. World J Stem Cells. 2014;6(3):256–65.

31. Zimmerlin L, Donnenberg VS, Rubin JP, Donnenberg AD. Mesenchymal markers on human adipose stem/progenitor cells. Cytometry A. 2013;83(1):134–40.

32. Dar A, Domev H, Ben-Yosef O, Tzukerman M, Zeevi-Levin N, Novak A, Germanguz I, Amit M, Itskovitz-Eldor J. Multipotent vasculogenic pericytes from human pluripotent stem cells promote recovery of murine ischemic limb. Circulation. 2012;125(1):87–99.

33. Zannettino AC, Paton S, Kortesidis A, Khor F, Itescu S, Gronthos S. Human multipotential mesenchymal/stromal stem cells are derived from a discrete subpopulation of STRO-1bright/CD34/CD45(−)/glycophorin-A-bone marrow cells. Haematologica. 2007;92(12):1707–8.

34. Psaltis PJ, Harbuzariu A, Delacroix S, Holroyd EW, Simari RD. Resident vascular progenitor cells—diverse origins, phenotype, and function. J Cardiovasc Transl Res. 2011;4(2):161–76.

35. Tallone T, Realini C, Böhmler A, Kornfeld C, Vassalli G, Moccetti T, Bardelli S, Soldati G. Adult human adipose tissue contains several types of multipotent cells. J Cardiovasc Transl Res. 2011;4(2):200–10.

36. Zimmerlin L, Donnenberg VS, Pfeifer ME, Meyer EM, Péault B, Rubin JP, Donnenberg AD. Stromal vascular progenitors in adult human adipose tissue. Cytometry A. 2010;77(1):22–30.

37. Crisan M, Corselli M, Chen WC, Péault B. Perivascular cells for regenerative medicine. J Cell Mol Med. 2012;16(12):2851–60.

38. Corselli M, Parekh C, Giovanna E, Montelatici A, Sahghian A, Wang W, Ge S, Scholes J, Codrea F, Lazzari L, Crooks GM, Peault B. Vascular pericytes sustain hematopoietic stem cells. Blood. 2011;118:2394.

39. Corselli M, Crisan M, Murray IR, West CC, Scholes J, Codrea F, Khan N, Péault B. Identification of perivascular mesenchymal stromal/stem cells by flow cytometry. Cytometry A. 2013;83(8):714–20.

Bone Repair and Regeneration Are Regulated by the Wnt Signaling Pathway

17

Khosrow Siamak Houschyar, Dominik Duscher, Zeshaan N. Maan, Malcolm P. Chelliah, Mimi R. Borrelli, Kamran Harati, Christoph Wallner, Susanne Rein, Christian Tapking, Georg Reumuth, Gerrit Grieb, Frank Siemers, Marcus Lehnhardt, and Björn Behr

17.1 Introduction

Bone tissue is capable of spontaneous scarless self-repair, generating new tissue that is indistinguishable from surrounding uninjured bone [1]. The process of fracture healing in the adult skeleton recapitulates embryogenic bone development and can be considered a form of tissue regeneration [2]. Fracture healing is a complicated metabolic process that follows specific regenerative patterns and involves changes in the expression of several thousand genes [3]. If these factors are inadequate or interrupted, healing is delayed or impaired, resulting in nonunions [4]. The exact etiology of nonunions and delayed fracture healing are unknown [4], but numerous pre-, intra-, and postoperative factors have been found to be associated with impaired bone healing. Excessive periosteal strip-

---

K. S. Houschyar (✉)
Department of Plastic Surgery, BG University Hospital Bergmannsheil, Ruhr University Bochum, Bochum, Germany

Burn Unit, Department for Plastic and Hand Surgery, Trauma Center Bergmannstrost Halle, Halle (Saale), Germany

K. Harati · C. Wallner · M. Lehnhardt · B. Behr
Department of Plastic Surgery, BG University Hospital Bergmannsheil, Ruhr University Bochum, Bochum, Germany
e-mail: christoph.wallner@bergmannsheil.de; marcus.lehnhardt@bergmannsheil.de; bjorn.behr@rub.de

D. Duscher
Department for Plastic Surgery and Hand Surgery, Division of Experimental Plastic Surgery, Technical University of Munich, Munich, Germany

Z. N.Maan · M. P. Chelliah · M. R. Borrelli
Division of Plastic and Reconstructive Surgery, Department of Surgery, Stanford School of Medicine, Stanford, CA, USA
e-mail: zmaan@stanford.edu; mchelliah@stanford.edu; mimib@stanford.edu

S. Rein
Department of Plastic and Hand Surgery, Burn Center-Clinic St. Georg, Leipzig, Germany

C. Tapking
Department of Surgery, Shriners Hospital for Children-Galveston, University of Texas Medical Branch, Galveston, TX, USA

Department of Hand, Plastic and Reconstructive Surgery, Burn Trauma Center, BG Trauma Center Ludwigshafen, University of Heidelberg, Heidelberg, Germany
e-mail: chtapkin@utmb.edu

G. Reumuth
Department of Plastic and Hand Surgery, Burn Unit, Trauma Center Bergmannstrost Halle, Halle, Germany

G. Grieb · F. Siemers
Department of Plastic Surgery and Hand Surgery, Gemeinschaftskrankenhaus Havelhoehe, Teaching Hospital of the Charité Berlin, Berlin, Germany

---

© Springer Nature Switzerland AG 2019
D. Duscher, M. A. Shiffman (eds.), *Regenerative Medicine and Plastic Surgery*,
https://doi.org/10.1007/978-3-030-19962-3_17

ping, damage to surrounding soft tissue, inadequate postoperative immobilization, repeated manipulations, and excessive early motion of a fracture all worsen the outcomes of fracture healing [4].

Fracture repair is regulated by a variety of growth factors [5]. The canonical Wnt signaling pathway plays a central role in bone development, homeostasis, as well as bone repair and regeneration following injury [6]. In most cases, Wnt ligands promote bone growth, which has led to speculation that Wnt factors could be used to stimulate bone healing [7]. As a result of the increasing appreciation for Wnts in bone biology, several small molecules and biologics that enhance canonical Wnt signaling have been

tested in preclinical models, and some are entering clinical trials [8].

Here, we present a summation of data, which provide strong evidence on components of canonical Wnt signaling likely to be targeted by future treatment regimens to augment fracture healing. An overview of the Wnt pathway will be provided, followed by discussion of specific canonical Wnt signaling molecules that are favorable targets for facilitating fracture repair.

Table 17.1 Key molecules and cells involved in bone repair

Key factors	Function	In vivo and in vitro effects
Extracellular messengers		
IL-1, IL6, TNFα	Elicit inflammation and migration	In vitro inhibit osteoblastic differentiation, but in vivo TNFα is crucial for bone repair; role of IL-6 is controversial (anti- or pro-osteogenic probably, depending on soluble IL-6 receptor)
TGFβ	Mitogenic factor, osteogenic factor	Can induce osteoblast differentiation at the early stage of immature cells but can also inhibit osteogenesis in committed cells
BMP2	Osteogenic factor	Osteochondrogenic factor; might initiate bone formation and bone healing and can induce expression of other BMPs
BMP4	Osteogenic factor	Osteochondrogenic factor in vivo and in vitro
BMP7	Osteogenic factor	Osteogenic factor in vivo and in vitro; active on more mature osteoblasts
SDF1	Chemotactic factor	Allows MSC homing both in vitro and in vivo
Noggin	BMP2, 4, and 7 specific inhibitor	Suppresses osteoblastic differentiation
FGFb	Angiogenic and mitogenic factor, osteogenic factor (controversial)	Mutations induce chondrodysplasia and craniosynostosis; can stimulate Sox9; might be a negative regulator of postnatal
IGF-I, II	Mitogenic factors, osteogenic factors	Stimulates growth plate formation, endochondrate ossification, and bone formation by osteoblasts
PlGF	Angiogenic and vasculogenic factor	Induces proliferation and osteogenic differentiation of MSCs; crucial for vascularization
VEGF	Angiogenic and vasculogenic factor	Most potent angiogenic and vasculogenic factor; crucial at the onset of bone formation
PDGF	Mitogenic and chemotactic factor	Highly mitogenic factor for MSCs and chemotactic for MSCs, osteoblasts, and perivascular cells
Wnts	Mitogenic and osteogenic factors	Depending on Wnt type, crucial for osteoprogenitor proliferation; can also inhibit final osteoblast maturation
DKK1	Inhibitor of Wnt signaling	Strongly inhibits osteogenesis of MSC and osteoprogenitor cells; can stimulate terminal maturation
Ihh	Osteochondrogenic factor	Pivotal role for growth plate and endochondral formation; can inhibit osteoblast differentiation; might induce PTHrP expression
PTHrP	Osteochondrogenic factor	Pivotal role for growth plate and endochondral formation; can induce or inhibit osteogenesis
OPG	Decoy receptor of RANKL, inhibition of RANKL	Strongly inhibits bone resorption and has a pivotal role in bone remodeling

17 Bone Repair and Regeneration Are Regulated by the Wnt Signaling Pathway

Table 17.1 (continued)

Key factors	Function	In vivo and in vitro effects
RANKL	Induces osteoclastogenesis	Strongly stimulates bone resorption and has a pivotal role in bone remodeling
M-CSF	Induces osteoclastogenesis	Crucial for osteoclastogenesis
Gastrointestinal serotonin	Neurotransmitter inhibiting osteogenesis	Expressed by enterochromatin cells, inhibits bone formation and repressed by Lrp5
Intracellular messengers		
PKA/CREB	Transduce osteogenic signaling	Can transduce osteogenic signaling (still controversial); possible indirect effect
MAPKs	Transduce osteogenic signaling by phosphorylation	Crucial for regulation of intracellular signaling induced by osteogenic factors (still controversial)
β-Catenin	Osteogenic transducer factor	Pivotal role in transducing osteogenic signal from Wnt and is negatively regulated by GSK3β
Runx2	Early osteogenic transcription factor	Master regulator of early osteogenesis; runx2 mice died, with no bone formation
Osterix	Late osteogenic transcription factor	Master regulator of late osteogenesis, inhibiting chondrogenesis
Dlx5	Osteogenic homeobox protein	Induces osteoblast maturation but inhibits osteocyte formation
Msx2	Osteogenic homeobox protein	Induces proliferation of immature cells; responses depend on Dlx5 quantity
NF-kB	Inflammation transducer factor, inhibits osteogenesis	Inhibits the differentiation of MSCs and committed osteoblastic cells
Cells		
MSCs	Origin of osteoblasts	Can form bone in vivo and osteoblasts in vitro
Osteoblasts	Osteogenic professional cells	Generate bone formation
Adipose tissue-derived stromal cells	Multipotential cells	Can give rise to bone in vivo and in vitro but are less effective than bone marrow MSCs

Table 17.2 Clinical relevance of key factors in bone repair

Key factors tested	Observations
BMP2	Used for spine fusion, bone nonunion, and bone defects; clinically efficient for bone repair and regeneration; some adverse effects observed (osteolysis and ectopic bone formation)
BMP7	Used for spine fusion and bone nonunion; clinically efficient for bone repair
PTHrP/PTH	Used for osteoporosis; efficient for increasing bone mass when intermittently administered
Wnt/β-catenin	LiCl used as a specific inhibitor of GSK3β to increase bone mass postfracture and to diminish fracture risk bortezomib, proteasome inhibitor used in treatment of multiple myeloma (MM); also increases bone mass anti-DKK1 monoclonal antibody (BHQ880) used to inhibit osteolysis in MM or to increase BMD anti-sclerostin antibody used to increase bone mass
RANKL/OPG	Targeting RANKL to treat osteoporosis; e.g., denosumab (anti-RANKL antibody), which can be used with biphosphonates
Biphosphonates	Widely used for osteoporosis, bone necrosis, osteogenesis imperfecta, and some osteolytic tumors (MM) (zoledronate, alendronate, risedronate); some adverse effects noted (osteonecrosis, inhibition of osteogenesis)
TGFβ	Used as a bone nonunion marker
Platelet-rich plasma	Used in maxillofacial surgery and for bone defects with or without biomaterials with or without osteoregenerative cells
MSCs or osteoblasts	In vitro-expanded MSCs (or osteoblasts) used for bone defects, osteonecrosis, immune rejection; randomized controlled clinical trials are required

17.2 Formation of Bone During Embryological Development

Bone biology basically depends on molecular pivots (Table 17.1) that represent disease targets as well as tools for new treatments (Table 17.2). Skeletal formation involves synchronized integration of genetic programs governing the specification, proliferation, differentiation, programmed cell death, remodeling of the extracellular matrix, and vasculogenesis [9]. These same cellular and extracellular events occur during adult bone repair and regeneration, leading us and others to propose that the molecular machinery responsible for fetal skeletogenesis also plays a role in the process of skeletal repair [10].

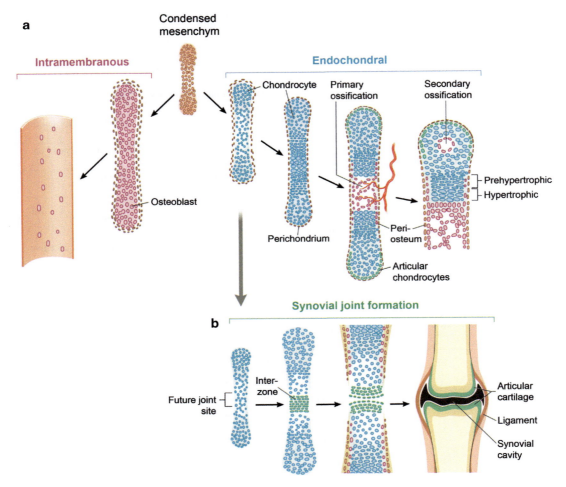

Fig. 17.1 Mechanisms of skeleton formation. (**a**) Bones can form by either intramembranous or endochondral ossification. Both processes are initiated by the condensation of mesenchymal cells. During intramembranous ossification, mesenchymal cells differentiate directly into osteoblasts and deposit bone. During endochondral ossification, mesenchymal cells differentiate into chondrocytes and first make a cartilage intermediate. Chondrocytes initiate a growth plate in the center of the bone, stop proliferating, and undergo hypertrophy. Hypertrophic chondrocytes mineralize their matrix and undergo apoptosis, attracting blood vessels and osteoblasts that remodel the intermediate into bone. (**b**) The first histologic sign of synovial joint formation is the gathering and flattening of cells, forming the interzone. Cavitation occurs within the presumptive joint separating the two cartilaginous structures. Remodeling and maturation proceed to give rise to the mature synovial joint. Wnt signaling plays a significant role in controlling almost all aspects of skeleton formation. Osteoblasts (purple); chondrocytes (blue); osteochondroprogenitor cells (brown)

Cartilage and bone define the skeleton and are produced by chondrocytes and osteoblasts, respectively. During embryological development, bone is formed by two distinct processes: intramembranous and endochondral ossification (Fig. 17.1). The developing skeletal elements are often segmented to form joints, which are required to support mobility. Synovial joints, which allow movement via smooth articulation between bony fronts, form when chondrogenic cells in newly formed cartilage undergo a program of dedifferentiation and flattening to form an interzone.

17.3 Mechanism of Bone Repair and Regeneration

Fracture healing is a complex and well-orchestrated regenerative process, initiated in response to injury, and culminating in optimal skeletal repair and restoration of skeletal function [11]. Unlike other adult tissues, which generate scar tissue at the site of an injury, the skeleton heals by forming new bone that is indistinguishable from adjacent, uninjured tissue [12]. Fracture healing occurs by two general mechanisms, direct and indirect repair, which mimic early developmental processes [13]. Direct or primary repair takes place when there is contact between adjacent bone cortices [8]. Osteoprogenitor cells, osteoclasts, and undifferentiated mesenchymal stem/stromal cells (MSCs) are recruited to the fracture site and contribute to formation of new bone in a mechanism similar to formation of bone during intramembranous ossification in the skull and clavicles [14]. Direct repair usually occurs when the injury is treated surgically by means of stable fixation. During indirect or secondary healing, bone formation is akin to endochondral ossification, the developmental process that produces long bones [15]. Following injury, a soft callus forms in indirect healing, which develops into a cartilaginous template that undergoes calcification into a hard callus and eventually is replaced by new woven bone [16]. Woven bone is slowly remodeled into lamellar bone, the final stage of a process requiring several months before the afflicted bone is able to support normal load bearing [17]. Most fractures heal with temporary immobilization, surgical fixation, or both, but 3–10% of fractures fail to heal and result in the formation of a fibrous or nonunion. Therapies aimed at inducing bone formation at the break point may increase the chances of a successful bone union but also decrease the time required for normal fracture healing.

17.4 Three Wnt Signaling Pathways

The Wnt signaling pathway is an evolutionarily conserved pathway that regulates crucial aspects of cell polarity, cell migration, cell fate determination, primary axis formation, organogenesis, and stem cell renewal during embryonic development [18]. A dysregulation of Wnt signaling has been implicated in the pathogenesis of many disease types, including autoimmune diseases and cancer [19]. The name Wnt originates from the fusion of *wingless,* the Drosophila segment polarity gene, and *integrated or int-1,* the name of the vertebrate homolog [20]. In mammals, complexity and specificity in Wnt signaling are in part achieved through 19 Wnt ligands, which are cysteine-rich proteins of approximately 350–400 amino acids that contain an N-terminal signal peptide for secretion [6]. As the signaling pathways that play crucial role during embryogenesis are tightly regulated, the expressions of Wnt proteins and antagonists are exquisitely controlled during development, both temporally and spatially [18].

Intracellular Wnt signaling diversifies into at least three main branches: (1) the β-catenin pathway (canonical Wnt pathway), which activates target genes in the nucleus; (2) the planar cell polarity (PCP) pathway, which involves jun N-terminal kinase (JNK), and cytoskeletal rearrangements; and (3) the Wnt/Ca^{2+} pathway [21]. In the canonical Wnt signaling pathway, glycogen synthase kinase 3 (GSK-3)-mediated β-catenin ubiquitination and degradation are inhibited by binding of Wnt to the receptor-coreceptor complex, frizzled/LRP [22]. In the planar cell polarity pathway, Wnt signaling activates JNK and directs

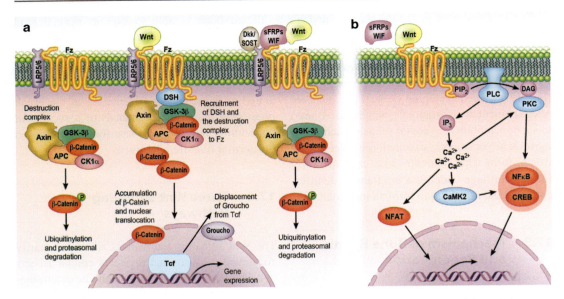

Fig. 17.2 Wnt signaling cascades. (**a**) The canonical Wnt signaling cascade is dependent on the intracellular signaling molecule β-catenin. In the absence of Wnt binding to Fz receptors, β-catenin is sequestered into a destruction complex composed of Axin, CK1α, APC, and GSK3β, phosphorylated, ubiquitinylated, and subsequently degraded by the proteasome. Upon binding of Wnt to Fz receptors and LRP5/6 coreceptors, DSH interacts with the receptor complex and recruits the destruction complex to the cell membrane allowing newly synthesized β-catenin to accumulate within the cytoplasm and to translocate to the nucleus. Nuclear β-catenin can displace the transcriptional co-repressor Groucho from TCF transcription factors and promote activation of a gene transcription program. Both Wnt-binding antagonists (sFRPs/WIF) and Wnt receptor antagonists (Dkk/SOST) inhibit the canonical cascade. (**b**) In the noncanonical Wnt signaling cascade, different phosphorylation cascades are activated by specific ligand–receptor interactions, seemingly without engagement of the LRP coreceptors. Many of these cascades are triggered by an increase in intracellular Ca^{2+} concentrations secondary to PLC and DAG production. Subsequently, PKC and CaMKII can activate transcription factors like NFκB and CREB; mediated IP3 and calmodulin are involved in the activation of NFAT. Only the Wnt-binding antagonists inhibit the noncanonical cascade. Abbreviations: APC, adenomatous polyposis coli; CaMKII, calcium/calmodulin-dependent protein kinase type II; CK1α, casein kinase 1-α; CREB, cyclic AMP-responsive element-binding protein; DAG, diacylglycerol; Dkk, Dickkopf; DSH, dishevelled; GSK3β, glycogen synthase kinase-3 β; IP3, inositol 1,4,5-triphosphate; LRP, low-density lipoprotein receptor-related protein; NFAT, nuclear factor of activated T cells; NFκB, nuclear factor κB; PIP2, phosphatidylinositol 4,5-bisphosphate; PKC, protein kinase C; PLC, phospholipase C; sFRPs, secreted frizzled-related proteins; SOST, sclerostin; WIF, Wnt inhibitory factor

asymmetric cytoskeletal organization and coordinated polarization of cell morphology within the plane of epithelial sheets [23]. This pathway branches from the canonical Wnt pathway downstream of frizzled, at the level of dishevelled, and involves downstream components like the small GTPase Rho and a kinase cascade including Misshapen, JNK kinase, and JNK [24]. Several members of the canonical Wnt signaling pathway, including GSK-3 and APC, have also been implicated in spindle orientation and asymmetric cell division of *C. elegans* and Drosophila [25]. In previous studies, Wnt has been shown to play a role in the release of intracellular calcium, possibly mediated via G proteins [26, 27]. This pathway involves activation of PLC, PKC, and calmodulin-dependent kinase II and is implicated in Xenopus ventralization and in regulation of convergent extension movements [28]. Here, we focus on canonical/ß-catenin-dependent Wnt signaling, which is the most extensively studied Wnt signaling pathway and is strongly implicated in skeletal tissue regeneration and repair (Fig. 17.2).

17.4.1 Canonical Wnt Signaling Pathway

In recent years, novel insights into the various levels of canonical Wnt signaling have refined the model of how this pathway is regulated [29]. At least 7 of 19 Wnt proteins, including Wnt-1, Wnt-2, Wnt-3, Wnt-3b, Wnt-4, Wnt-8, and Wnt-10b, have been reported to activate this pathway [30]. The hallmark of canonical Wnt signaling is the accumulation and translocation of adherens junction-associated protein, β-catenin, into the nucleus [31]. β-Catenin has been shown to perform two apparently unrelated functions: cell-cell adhesion and signaling in the Wnt/wg pathway [32]. Accumulation of intranuclear β-catenin results in transcriptional activation of specific target genes during development [33]. Dysregulation of β-catenin signaling plays a role in the genesis of a number of malignancies, suggesting an important role in the control of cellular proliferation or cell death [34]. In the absence of appropriate Wnt ligands, cytoplasmic β-catenin is degraded by a β-catenin destruction complex, comprised of Axin, adenomatous polyposis coli (APC), GSK3, and casein kinase 1 (CK1) [35]. CK1 and GSK3 phosphorylate β-catenin in the NH2-terminal degradation box, targeting it for ubiquitination. The phosphorylated β-catenin is polyubiquitinated by bTRCP1 (a component of ubiquitin E3 ligase) or bTRCP2 complex for the following proteasome-mediated degradation by the multi-protein complex [36].

The canonical Wnt pathway is initiated by binding of appropriate Wnt ligands to the Fzs and LRP-5/6 coreceptor [37, 38]. When an appropriate Wnt ligand binds the Wnt receptor complex, the intracellular protein, dishevelled (Dvl), is activated and transduces this signal from the receptor complex [39]. Activated Dvl inhibits GSK-3b, resulting in the collapse of the multi-protein complex [40]. β-Catenin cannot be targeted for degradation and it accumulates and translocates to the nucleus, where, in concert with members of the T-cell factor/lymphoid enhancer factor (TCF/LEF) family, it activates the transcription of a wide range of genes, including c-myc and cyclin D1 [41]. The complexity of Wnt intracellular signaling pathways parallels the complexity observed in the diversity of Wnt receptors [32]. To date, there are ten human Fz receptors. It should be emphasized that, although the role of Fz in acting as a receptor for Wnts is long established, LRP-5 and its closely related homolog, LRP-6, are two important molecules which mediate Wnt/β-catenin signaling. Both LRP-5 and LRP-6 act as coreceptors for Wnt proteins, and this canonical Wnt pathway can be antagonized by secreted proteins from the Dickkopf (Dkk) family that bind with high affinity to LRP-5 or LRP-6 and thereby directly prevent Wnt binding [42].

17.4.2 Noncanonical Wnt Signaling Pathway

The noncanonical pathway, often referred to as the β-catenin-independent pathway, is divided into two distinct branches: the planar cell polarity pathway (PCP pathway) and the Wnt/Ca^{2+} pathway [29]. The defining feature of the PCP pathway is its role in the regulation of actin cytoskeleton for polarized organization of structures and directed migration [18]. Moreover, this pathway appears to function independently of transcription. The PCP pathway emerged from genetic studies in Drosophila in which mutations in Wnt signaling components, including those in frizzled and Dvl, were found to randomize the orientation of epithelial structures such as cuticle hairs and sensory bristles [18].

The Wnt/Ca^{2+} noncanonical Wnt signaling pathway shares a number of components of the PCP pathway, but is sufficiently distinct to be considered separate. Its unique roles center around the modulation of both the canonical signaling for dorsal axis formation and the PCP signaling for gastrulation cell movements [43]. The Wnt/Ca^{2+} pathway is dependent on G-proteins, and its discovery was founded on the observation that some Wnts and Fz receptors can stimulate intracellular Ca^{2+} release from endoplasmic reticulum (ER) [44]. Wnt5a, Wnt11, and rat Fz2

(RFz-2) are capable of intracellular Ca^{2+} release, without affecting β-catenin stabilization [18]. The role of the Wnt/Ca^{2+} pathway during embryogenesis is diverse and includes negative regulation of dorsal axis formation, promotion of ventral cell fate, regulation of tissue separation, and convergent extension movements during gastrulation, and later in heart formation [45].

17.5 Mesenchymal Stem/Stromal Cells (MSCs) and Wnt Signaling in Bone Development and Homeostasis

MSCs are multipotent progenitors that retain the capacity to differentiate into multiple types of tissues, including bone, cartilage, fat, tendon, and muscle [46]. MSCs carry vast therapeutic potential in regenerative medicine due to their relative abundance in various locations in the patient's body, especially bone marrow, and their impressive differentiation capacity [47]. Induction of MSCs along the osteogenic lineage may serve as an effective therapy to promote bone formation in osteogenic disorders [48].

Commitment of MSCs into a single-cell lineage is regulated by a variety of growth factors, but current understanding of the process influencing cell fates is still limited [49]. Nonetheless, it is relatively well established that the Wnt signaling pathway plays an important role in promoting osteogenic differentiation of MSCs [49]. In addition to its role in pushing skeletal stem cells into the osteogenic lineage, Wnt ligands also stimulate osteoblast proliferation and support osteoblast maturation (Fig. 17.3) [50]. The Wnt signaling pathway is involved in both intramembranous and endochondral ossification [51]. Minear et al. [52] demonstrated for the first time that Wnt signaling, and perhaps the application of Wnt protein, could be used to stimulate bone healing. They used a mouse strain with increased cellular response to Wnt Axin2LacZ/LacZ, and demonstrated that these mice had accelerated fracture healing as a result of a more robust proliferation and earlier differentiation of skeletal stem cells/progenitor cells. In the same article, the investigators packaged purified Wnt3a protein into liposomal vesicles, and delivered these liposomal vehicles into

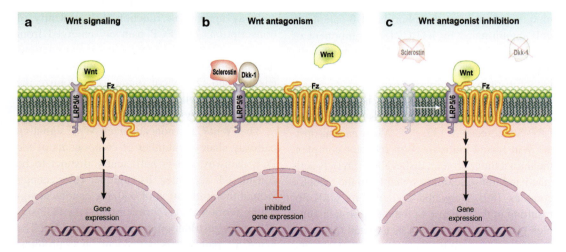

Fig. 17.3 Wnt signaling in osteoblasts. (**a**) When Wnt binds its receptor, frizzled, and coreceptors LRP5 and LRP6, its signaling pathway is activated, leading to gene expression (and ultimately protein synthesis and formation of bone). (**b**) Wnt antagonists sclerostin and Dkk-1 bind LRP5 and LRP6, preventing their interaction with frizzled and resulting in inhibition of gene expression. (**c**) Inhibition of Wnt antagonism promotes gene expression. This may occur as a result of a loss-of-function mutation in a gene that encodes for a Wnt antagonist, or by pharmacological engagement of the antagonist with an inhibitory molecule such as an antibody

the skeletal defect, and reported that Wnt3a stimulated proliferation of skeletal progenitor cells and accelerated their differentiation into osteoblasts. This biochemical strategy highlights the therapeutic potential of a protein being able to increase the duration and strength of Wnt signaling at the site of injury and thus enhance skeletal healing. This exciting finding opens new directions for research within a system known to substantially alter human phenotypes. Previous studies have shown that β-catenin both promotes the progression of MSCs from osteoblastic precursor cells into more mature osteoblasts and suppresses the differentiation of MSCs into adipogenic and chondrogenic lineages [53, 54]. Specifically, the canonical Wnt pathway inhibits the expression of the major adipogenic inducers PPARγ and CCAAT/enhancer-binding protein α to suppress adipogenic differentiation while upregulating the osteogenic regulators Runx2, Dlx5, and Osterix [55]. In addition, noncanonical Wnt signaling can induce osteogenic differentiation, albeit through a different mechanism. The noncanonical ligand Wnt 5a suppresses PPARγ through the inactivation of chromatins rather than through β-catenin action [56]. Although the interplay between the two independent mechanisms induced by Wnt ligands is still unclear, it is evident that Wnt signaling regulates the osteogenic differentiation of MSCs.

It is important to understand that cross talk exists between the different Wnt pathways and signaling pathways regulating osteogenic differentiation of MSCs. For example, bone morphogenic proteins (BMPs) have been shown to either enhance or antagonize osteogenic differentiation induced by Wnt signaling [49]. BMPs, particularly BMP2, 6, and 9, are major osteogenic growth factors that induce osteogenic differentiation in MSCs [49]. Studies have demonstrated that Wnt signaling and BMP signaling pathways have common targets which induce the osteogenic differentiation of MSCs; one such target is connective growth tissue factor [57, 58]. Furthermore, functional Wnt signaling was demonstrated to be required for BMP-induced osteogenic differentiation of MSCs [57, 58]. Wnt 3a enhanced the osteogenic effects of BMP9 while

β-catenin knockdown or overexpression of anFzd antagonist, FrzB, inhibited the osteogenic effects of BMP9 [59]. Similarly, conditional β-catenin knockout or Dkk-1 overexpression inhibit BMP2-induced ectopic bone formation [60]. It was suggested that BMP2 stimulates LRP5 expression and downregulates β-Trcp, leading to stabilization of β-catenin and its signaling to promote osteogenic differentiation [61].

Mutations in the Wnt signaling cascade can lead to excessive bone growth or resorption. The first indication of a link between bone biology and canonical Wnt signaling was discovered over a decade ago [62]. Loss-of-function mutation of the coreceptor LRP5 causes syndrome characterized by low bone mass, accompanied with frequent bone fractures. Gain-of-function mutations of LRP5 receptor, on the other hand, lead to high bone mass [63]. The essential role of LRP5 in the regulation of bone mass in humans is further underscored by the association of SNPs of the LRP5 gene with decreased bone mineral density and an increased risk of osteoporotic fractures [51]. The mechanism by which LRP5 regulates bone mass is not fully understood, but LRP5 and LRP6 are known to transduce Wnt signaling in vitro and indicated overlapping roles during in vivo skeletal patterning [64].

Gene variation in Wnt16 has recently been associated with bone mineral density and osteoporotic fractures. Wnt16 knockout mice demonstrate a substantial decrease in bone thickness and strength, highlighting its crucial role in bone biology [65]. The initial phase of skeletal tissue repair or active bone remodeling is similar to that occurring during skeletal embryogenesis as skeletal stem cells are shuttled to either the osteogenic or the chondrogenic route [14]. One report of Wnt involvement in fracture repair identified upregulation of Wnt5A, β-catenin, FZD, and numerous target genes during the process [8]. A later follow-up study demonstrated upregulation of additional Wnt related markers such as Wnt4, Wnt5B, LRP5, dishevelled (Dvl), TCF1, and peroxisome proliferator-activated receptor delta (PPARD) [37]. In contrast, the transcription factor LEF1 was repressed during the initial phases of bone repair and subsequent maximal bone formation [66]. However, since it is recognized that LEF1 inhibits RUNX2-dependent activation of OCN in osteoblasts, and the fact that RUNX2 is the transcription factor required for osteoblast development, it seems likely that decreased LEF1 expression is necessary for bone repair to occur [67]. β-Catenin appears to have various roles at different stages of bone repair; early in the process β-catenin regulates the ratio of osteoblast versus chondrocytes that arises from pluripotent MSCs [68, 69]. Later in the bone-healing process, β-catenin induces differentiation of osteoblasts and boosts their matrix production. Upregulation of LRP5 gene expression is shown in fracture callus, and β-catenin expression has been observed in callus and in proliferating chondrocytes, osteoblasts, and periosteal osteoprogenitor cells [70]. This indicates that canonical Wnt signaling pathway is active both in endochondral and in intramembranous ossification. The involvement of the Wnt signaling pathway during intramembranous ossification has also been shown in recent studies [71, 72]. Fractured femurs of LRP5 knockout mice have been reported to be smaller, less mineralized, and biomechanically inferior to those from wild-type littermates [70]. The study further showed that DKK1 antibody administration increased the size, mineralization, and biomechanical properties of fractured tissue, demonstrating that deletion of LRP5 delays the reestablishment of biomechanical integrity during fracture repair. The LRP5-mediated canonical Wnt signaling seems to be less important to mineral accumulation at the fracture site than for the restoration of proper tissue structural arrangement. This identifies LRP5 and canonical Wnt pathway as key components of fracture repair, although the noncanonical Wnt pathways have recently been implicated in bone formation during both intramembranous and endochondral fracture healing.

Wnt5a, a classical noncanonical Wnt, was recently reported as a critical component of BMP2-mediated osteogenic differentiation [49]. Further studies have also shown that BMPs can downregulate Wnt signaling in osteogenic differentiation via sclerostin and Dkk-1 [6, 73]. Knocking out BMP receptor type 1 in osteoblasts led to downregulation of sclerostin and Dkk-1 and an increased bone mass phenotype in mice [74]. As an explanation for the Wnt-antagonizing effects of BMP, it was suggested that Smad1 may form a complex with Dvl, thereby sequestering Dvl from the canonical Wnt pathway [75]. However, these seemingly conflicting findings on the cross talk between BMPs and Wnts remain unresolved.

Activation of the Notch pathway has been shown to inhibit the osteogenic differentiation induced by the Wnt/β-catenin signaling [60]. Overexpression of Notch intracellular domain, both in vivo and in vitro, led to the downregulation of Wnt signaling and impaired osteoblastogenesis [72]. The Hedgehog (Hh) pathway has been reported to work upstream of Wnt signaling in a sequential manner to induce osteogenic differentiation of MSCs [76]. Inhibition of Wnt signaling was shown to reduce Hh-induced osteogenic activity both in vitro and in vivo. It has been suggested that Hh signaling regulates the early stages of osteogenic differentiation of MSC followed by the Wnt signaling further downstream [77].

Wnt signaling also cross talks with inflammatory signaling processes during bone formation [78]. Tumor necrosis factor (TNF)-α has been shown to induce Dkk-1, an endogenous regulator of Wnt signaling, thus blocking osteoblast differentiation [79]. Overexpression of TNFα in mice leads to joint destruction without proper bone repair, mimicking rheumatoid arthritis [80]. However, neutralizing Dkk-1 with anti-Dkk-1 antibodies in TNFα transgenic mice inhibited the joint destruction and resulted in osteophyte formation, indicative of active bone repair [81]. In other words, the intricate balance between bone formation and bone resorption is maintained by the cross talk between the Wnt signaling and the TNFα-induced inflammatory process.

Knowledge of the cross talk between Wnt and other signaling pathways continues to expand. Recently, it was shown that the Wnt signaling pathway reciprocally regulates progranulin growth factor in frontotemporal dementia [82]. Meanwhile, progranulin, also known as proepithelin, is one of the newly identified growth factors that promotes chondrogenic differentiation of MSCs and endochondral ossification [83]. More details of the cross talk between Wnts and progranulin in bone formation will need to be elucidated by further investigation.

MicroRNA (miRNA) represents another category of elements that interact with Wnt signaling to regulate osteogenic differentiation [84]. Numerous miRNAs have been reported to either promote or inhibit osteogenic differentiation of MSCs [85]. Different miRNAs intervene at different locations and stages of osteogenic differentiation, interacting with extracellular growth factors and intracellular transcriptional factors such as Runx2 and osterix [86]. Several miRNAs have been reported to specifically interact with Wnt signaling molecules to affect osteogenesis [87]. miR-27 inhibits APC, thus activating canonical Wnt signaling to promote bone formation [88]. miR-335-5p was shown to downregulate Dkk-1, thus enhancing osteogenic differentiation [89].

Overall, a complex regulatory network exists between the Wnt signaling pathway and other signaling pathways during osteogenic differentiation. A complete characterization of all the interactions between Wnts and these other pathways is far from complete. Nevertheless, a better understanding of the intricate cross talks between Wnt and other signaling pathways in osteogenesis will prove to be essential in discovering therapeutic interventions to effectively manipulate these signaling pathways and treat osteogenic disorders.

17.6 Opportunities for Therapeutics

The ability to control the fate of skeletal stem cell, between self-renewal, proliferation, and differentiation, could lead to the possibility of expanding a limited population of adult progenitor cells and inducing their timely differentiation to restore the function of skeletal and cartilaginous tissue. Bone regeneration for fracture repair and defect healing may be the first major attempted procedure in orthopedic surgery.

Although internal fixation devices that can successfully achieve short-term stabilization at virtually all orthotopic sites have already been developed, long-term stability still requires bone fusion or bone augmentation [90]. Autogenous bone graft is a commonly used approach for promoting bone repair, especially for large-sized defects [91]. However, it can only be performed on a limited scale, and its harvesting can involve substantial donor-site morbidity. Allograft bone has the potential for antigenicity and disease transmission [92]. Biomaterials have increased infection rate and poor biomechanical properties [93]. Currently, BMP-2 and BMP-7, also known as osteogenic protein 1 (OP-1), are being increasingly employed in multiple clinical trials, in which they have been shown to be extremely effective in enhancing bone formation [94]. However, the requirement of large doses, short half-life and thus short-term bioavailability of BMPs, and lack of a practical and suitable method for sustained delivery of these exogenous proteins have greatly limited the application of BMPs in humans [95]. The most ideal treatment for bone regeneration would be a pharmacologic agent that is cost effective and does not require the addition of invasive procedures. Fortunately, BMP signaling is not the only pathway that can lead to an anabolic effect on the skeleton; over the past decade, the Wnt/β-catenin signaling pathway has emerged as central to the formation of bone.

In terms of skeletal homeostasis and bone repair, the Wnt pathway is among the most attractive targets for such therapeutic intervention. There is now a substantial literature supporting a role of Wnt signaling in skeletogenesis and a growing appreciation for the functions of Wnt in regulating stem cell and skeletal cell behavior [6]. The ability to target the Wnt signaling pathway as a means to enhance skeletal healing has not been lost on the pharmaceutical industry. At least two major drug companies have begun programs to develop products that inhibit Wnt antagonists; one of these may soon be engaged in early-phase clinical trials involving patients with long bone fractures. The challenge is to take advantage of these opportunities, not only by developing effective new therapies, but also by doing so within an economic framework that would lead to affordable products. A move from the use of recombinant gene technology to produce a therapeutic peptide to the development of a high-specificity antibody to inhibit a ligand or receptor may assist in achieving this goal. The interactions between Wnt receptors and coreceptors are reasonable targets for such strategies. Although exciting, manipulation of the Wnt signaling cascade should be performed with caution as it regulates numerous diverse pathological processes, including the development of cancer. An additional challenge is the hydrophobic, and therefore insoluble, nature of Wnt proteins. Wnt, however, has been successfully purified and packaged into liposomes, circumventing this delivery challenge. Additionally, other molecules that act on different components of the canonical Wnt signaling pathway may offer therapeutic potential. Lithium, for example, inhibits GSK3 and can thereby increase β-catenin, with promising effects on bone healing. Further investigation may elucidate additional molecules able to potentiate the bone-healing effects of the Wnt signaling pathway.

17.7 Conclusions and Future Perspectives

The ability of adult bone to scarlessly self-repair and regenerate following injury is a fact that most individuals learn and appreciate at an early age. However, in a small percentage of severe and disabling fractures, repair never occurs and a fibrous or nonunion is the result. The signaling molecules that have been developed as therapeutics for promoting bone regeneration have focused on factors that either enhance MSCs in the fracture or are osteoinductive. A wide spectrum of signaling factors influence the fracture-healing process and continuing to study these factors and their mechanisms will lead to promising new clinical treatments to repair bone. The Wnt signaling pathway has effects on cell proliferation, differentiation, stem cell maintenance, and tissue homeostasis and plays a fundamental role during the embryological

development of a number of tissues and organs including bone and cartilage. In the adult skeleton, Wnt signaling is critical for bone homeostasis, repair, and regeneration. Mutations in Wnt genes, receptors, and inhibitors of the Wnt signaling pathway can have deleterious effects on normal bone formation and turnover, resulting in skeletal abnormalities and contributing to the pathophysiology of some bone-related disorders or cancers. Recent advances in our understanding of the crucial roles that Wnt/β-catenin signaling plays in the development and maturation of osteoblast lineage cells have generated new opportunities to treat nonunions and perhaps to accelerate repair. Despite these rapid and measurable accomplishments, much remains to be learned about the effects of Wnts and Wnt antagonists on skeletal physiology and regeneration. Meanwhile, clinical trials will test the effectiveness of current Wnt pathway drugs on a variety of endocrine and orthopedic conditions and advanced genome sequencing technologies will point us in new directions.

References

1. Fisher JN, Peretti GM, Scotti C. Stem cells for bone regeneration: from cell-based therapies to decellularised engineered extracellular matrices. Stem Cells Int. 2016;2016:9352598.
2. Marsell R, Einhorn TA. The biology of fracture healing. Injury. 2011;42:551–5.
3. Mountziaris PM, Mikos AG. Modulation of the inflammatory response for enhanced bone tissue regeneration. Tissue Eng Part B Rev. 2008;14:179–86.
4. Victoria G, Petrisor B, Drew B, Dick D. Bone stimulation for fracture healing: what's all the fuss? Indian J Orthop. 2009;43:117–20.
5. Barnes GL, Kostenuik PJ, Gerstenfeld LC, Einhorn TA. Growth factor regulation of fracture repair. J Bone Miner Res. 1999;14:1805–15.
6. Wang Y, Li YP, Paulson C, Shao JZ, Zhang X, Wu M, Chen W. Wnt and the Wnt signaling pathway in bone development and disease. Front Biosci (Landmark Ed). 2014;19:379–407.
7. Chen T, Li J, Córdova LA, Liu B, Mouraret S, Sun Q, Salmon B, Helms J. A WNT protein therapeutic improves the bone-forming capacity of autografts from aged animals. Sci Rep. 2018;8:119.
8. Secreto FJ, Hoeppner LH, Westendorf JJ. Wnt signaling during fracture repair. Curr Osteoporos Rep. 2009;7:64–9.

9. Ingber DE, Levin M. What lies at the interface of regenerative medicine and developmental biology? Development. 2007;134:2541–7.
10. Gadjanski I, Spiller K, Vunjak-Novakovic G. Time-dependent processes in stem cell-based tissue engineering of articular cartilage. Stem Cell Rev. 2012;8:863–81.
11. Einhorn TA, Gerstenfeld LC. Fracture healing: mechanisms and interventions. Nat Rev Rheumatol. 2015;111:45–54.
12. Dimitriou R, Jones E, McGonagle D, Giannoudis PV. Bone regeneration: current concepts and future directions. BMC Med. 2011;9:66.
13. Cameron JA, Milner DJ, Lee JS, Cheng J, Fang NX, Jasiuk IM. Employing the biology of successful fracture repair to heal critical size bone defects. Curr Top Microbiol Immunol. 2013;367:113–32.
14. Arvidson K, Abdallah BM, Applegate LA, Baldini N, Cenni E, Gomez-Barrena E, Granchi D, Kassem M, Konttinen YT, Mustafa K, Pioletti DP, Sillat T, Finne-Wistrand A. Bone regeneration and stem cells. J Cell Mol Med. 2011;15:718–46.
15. Kostenuik P, Mirza FM. Fracture healing physiology and the quest for therapies for delayed healing and nonunion. J Orthop Res. 2017;35:213–23.
16. Scammell BE, Roach HI. A new role for the chondrocyte in fracture repair: endochondral ossification includes direct bone formation by former chondrocytes. J Bone Miner Res. 1996;11:737–45.
17. Panetta NJ, Gupta DM, Longaker MT. Bone regeneration and repair. Curr Stem Cell Res Ther. 2010;5:122–8.
18. Komiya Y, Habas R. Wnt signal transduction pathways. Organogenesis. 2008;4:68–75.
19. Shi J, Chi S, Xue J, Yang J, Li F, Liu X. Emerging role and therapeutic implication of Wnt signaling pathways in autoimmune diseases. J Immunol Res. 2016;2016:9392132.
20. Liu H, Liu Q, Zhou X, Huang Y, Zhang Z. Genome editing of Wnt-1, a gene associated with segmentation, via CRISPR/Cas9 in the pine caterpillar moth, Dendrolimus punctatus. Front Physiol. 2016;17:666.
21. Houschyar KS, Momeni A, Pyles MN, Maan ZN, Whittam AJ, Siemers F. Wnt signaling induces epithelial differentiation during cutaneous wound healing. Organogenesis. 2015;11:95–104.
22. Verheyen EM, Gottardi CJ. Regulation of Wnt/beta-catenin signaling by protein kinases. Dev Dyn. 2010;239:34–44.
23. Geetha-Loganathan P, Nimmagadda S, Scaal M. Wnt signaling in limb organogenesis. Organogenesis. 2008;4:109–15.
24. Habas R, Dawid IB. Dishevelled and Wnt signaling: is the nucleus the final frontier? J Biol. 2005;4:2.
25. Wu M, Herman MAA. A novel noncanonical Wnt pathway is involved in the regulation of the asymmetric B cell division in C. elegans. Dev Biol. 2006;293:316–29.
26. Huelsken J, Behrens J. The Wnt signalling pathway. J Cell Sci. 2002;115:3977–8.
27. Lu D, Carson DA. Spiperone enhances intracellular calcium level and inhibits the Wnt signaling pathway. BMC Pharmacol. 2009;9:13.
28. Kestler HA, Kuhl M. From individual Wnt pathways towards a Wnt signalling network. Philos Trans R Soc Lond B Biol Sci. 2008;363:1333–47.
29. Zhan T, Rindtorff N, Boutros M. Wnt signaling in cancer. Oncogene. 2017;36:1461–73.
30. Choi HJ, Park H, Lee HW, Kwon YG. The Wnt pathway and the roles for its antagonists, DKKS, in angiogenesis. IUBMB Life. 2012;64:724–31.
31. Enzo MV, Rastrelli M, Rossi CR, Hladnik U, Segat D. The Wnt/beta-catenin pathway in human fibrotic-like diseases and its eligibility as a therapeutic target. Mol Cell Ther. 2015;3:1.
32. Clevers H. Wnt/beta-catenin signaling in development and disease. Cell. 2006;127:469–80.
33. Cong F, Schweizer L, Chamorro M, Varmus H. Requirement for a nuclear function of beta-catenin in Wnt signaling. Mol Cell Biol. 2003;23:8462–70.
34. Tarapore RS, Siddiqui IA, Mukhtar H. Modulation of Wnt/beta-catenin signaling pathway by bioactive food components. Carcinogenesis. 2012;33:483–91.
35. Stamos JL, Weis WI. The beta-catenin destruction complex. Cold Spring Harb Perspect Biol. 2013;5:a007898.
36. Gao C, Xiao G, Hu J. Regulation of Wnt/beta-catenin signaling by posttranslational modifications. Cell Biosci. 2014;4:3.
37. Mohammed MK, et al. Wnt/beta-catenin signaling plays an ever-expanding role in stem cell self-renewal, tumorigenesis and cancer chemoresistance. Genes Dis. 2016;3:11–40.
38. Tauriello DV, Maurice MM. The various roles of ubiquitin in Wnt pathway regulation. Cell Cycle. 2010;9:3700–9.
39. Sethi JK, Vidal-Puig A. Wnt signalling and the control of cellular metabolism. Biochem J. 2010;427:1–17.
40. Voronkov A, Krauss S. Wnt/beta-catenin signaling and small molecule inhibitors. Curr Pharm Des. 2013;19:634–64.
41. Quarto N, Wan DC, Kwan MD, Panetta NJ, Li S, Longaker MT. Origin matters: differences in embryonic tissue origin and Wnt signaling determine the osteogenic potential and healing capacity of frontal and parietal calvarial bones. J Bone Miner Res. 2010;25:1680–94.
42. MacDonald BT, He X. Frizzled and LRP5/6 receptors for Wnt/beta-catenin signaling. Cold Spring Harb Perspect Biol. 2012;4:a007880.
43. Gomez-Orte E, Saenz-Narciso B, Moreno S, Cabello J. Multiple functions of the noncanonical Wnt pathway. Trends Genet. 2013;29:545–53.
44. Kuhl M, Sheldahl LC, Park M, Miller JR, Moon RT. The Wnt/Ca2+ pathway: a new vertebrate Wnt signaling pathway takes shape. Trends Genet. 2000;16:279–83.
45. De A. Wnt/Ca2+ signaling pathway: a brief overview. Acta Biochim Biophys Sin (Shanghai). 2011;43:745–56.

46. Garcia-Castro J, Trigueros C, Madrenas J, Pérez-Simón JA, Rodriguez R, Menendez P. Mesenchymal stem cells and their use as cell replacement therapy and disease modelling tool. J Cell Mol Med. 2008;12:2552–65.

47. Patel DM, Shah J, Srivastava AS. Therapeutic potential of mesenchymal stem cells in regenerative medicine. Stem Cells Int. 2013;2013:496218.

48. Undale AH, Westendorf JJ, Yaszemski MJ, Khosla S. Mesenchymal stem cells for bone repair and metabolic bone diseases. Mayo Clin Proc. 2009;84:893–902.

49. Kim JH, Liu X, Wang J, Chen X, Zhang H, Kim SH, Cui J, Li R, Zhang W, Kong Y, Zhang J, Shui W, Lamplot J, Rogers MR, Zhao C, Wang N, Rajan P, Tomal J, Statz J, Wu N, Luu HH, Haydon RC, He TC. Wnt signaling in bone formation and its therapeutic potential for bone diseases. Ther Adv Musculoskelet Dis. 2013;5:13–31.

50. Regard JB, Zhong Z, Williams BO, Yang Y. Wnt signaling in bone development and disease: making stronger bone with Wnts. Cold Spring Harb Perspect Biol. 2012;4:a007997.

51. Krishnan V, Bryant HU, Macdougald OA. Regulation of bone mass by Wnt signaling. J Clin Invest. 2006;116:1202–9.

52. Minear S, Leucht P, Jiang J, Liu B, Zeng A, Fuerer C, Nusse R, Helms JA. Wnt proteins promote bone regeneration. Sci Transl Med. 2010;2:29ra30.

53. Ullah I, Subbarao RB, Rho GJ. Human mesenchymal stem cells—current trends and future prospective. Biosci Rep. 2015;35:e00191.

54. Case N, Rubin J. Beta-catenin—a supporting role in the skeleton. J Cell Biochem. 2010;110:545–53.

55. Kang S, Bennett CN, Gerin I, Rapp LA, Hankenson KD, Macdougald OA. Wnt signaling stimulates osteoblastogenesis of mesenchymal precursors by suppressing CCAAT/enhancer-binding protein alpha and peroxisome proliferator-activated receptor gamma. J Biol Chem. 2007;282:14515–24.

56. Takada I, Mihara M, Suzawa M, Ohtake F, Kobayashi S, Igarashi M, Youn MY, Takeyama K, Nakamura T, Mezaki Y, Takezawa S, Yogiashi Y, Kitagawa H, Yamada G, et al. A histone lysine methyltransferase activated by non-canonical Wnt signalling suppresses PPAR-gamma transactivation. Nat Cell Biol. 2007;9:1273–85.

57. Beederman M, Lamplot JD, Nan G, Wang J, Liu X, Yin L, Li R, Shui W, Zhang H, Kim SH, Zhang W, Zhang J, Kong Y, Denduluri S, Rogers MR, et al. BMP signaling in mesenchymal stem cell differentiation and bone formation J Biomed Sci Eng. 2013;6:32–52.

58. Tang N, Song WX, Luo J, Luo X, Chen J, Sharff KA, Bi Y, He BC, Huang JY, Zhu GH, Su YX, Jiang W, et al. BMP-9-induced osteogenic differentiation of mesenchymal progenitors requires functional canonical Wnt/beta-catenin signalling. J Cell Mol Med. 2009;13:2448–64.

59. Yang K, Wang X, Zhang H, Wang Z, Nan G, Li Y, Zhang F, Mohammed MK, Haydon RC, Luu HH, Bi Y, He TC. The evolving roles of canonical WNT signaling in stem cells and tumorigenesis: implications in targeted cancer therapies. Lab Invest. 2016;96:116–36.

60. Lin GL, Hankenson KD. Integration of BMP, Wnt, and notch signaling pathways in osteoblast differentiation. J Cell Biochem. 2011;112:3491–501.

61. Zhang M, Yan Y, Lim YB, Tang D, Xie R, Chen A, Tai P, Harris SE, Xing L, Qin YX, Chen D. BMP-2 modulates beta-catenin signaling through stimulation of Lrp5 expression and inhibition of beta-TrCP expression in osteoblasts. J Cell Biochem. 2009;108:896–905.

62. Yavropoulou MP, Yovos JG. The role of the Wnt signaling pathway in osteoblast commitment and differentiation. Hormones (Athens). 2007;6:279–94.

63. Semenov MV, He X. LRP5 mutations linked to high bone mass diseases cause reduced LRP5 binding and inhibition by SOST. J Biol Chem. 2006;281:38276–84.

64. Johnson ML. LRP5 and bone mass regulation: where are we now? Bonekey Rep. 2012;1:1.1.

65. Zheng HF, Tobias JH, Duncan E, Evans DM, Eriksson J, Paternoster L, Yerges-Armstrong LM, Lehtimäki T, Bergström U, Kähönen M, Leo PJ, et al. WNT16 influences bone mineral density, cortical bone thickness, bone strength, and osteoporotic fracture risk. PLoS Genet. 2012;8:e1002745.

66. Shahi M, Peymani A, Sahmani M. Regulation of bone metabolism. Rep Biochem Mol Biol. 2017;5:73–82.

67. Huang W, Yang S, Shao J, Li YP. Signaling and transcriptional regulation in osteoblast commitment and differentiation. Front Biosci. 2007;12:3068–92.

68. Xu H, Duan J, Ning D, Li J, Liu R, Yang R, Jiang JX, Shang P. Role of Wnt signaling in fracture healing. BMB Rep. 2014;47:666–72.

69. Bao Q, Chen S, Qin H, Feng J, Liu H, Liu D, Li A, Shen Y, Zhao Y, Li J, Zong Z. An appropriate Wnt/beta-catenin expression level during the remodeling phase is required for improved bone fracture healing in mice. Sci Rep. 2017;7:695.

70. Komatsu DE, Mary MN, Schroeder RJ, Robling AG, Turner CH, Warden SJ. Modulation of Wnt signaling influences fracture repair. J Orthop Res. 2010;28:928–36.

71. Zhong Z, Ethen NJ, Williams BO. WNT signaling in bone development and homeostasis. Wiley Interdiscip Rev Dev Biol. 2014;3:489–500.

72. Rahman MS, Akhtar N, Jamil HM, Banik RS, Asaduzzaman SM. TGF-beta/BMP signaling and other molecular events: regulation of osteoblastogenesis and bone formation. Bone Res. 2015;3:15005.

73. Zhang W, Xue D, Yin H, Wang S, Li C, Chen E, Hu D, Tao Y, Yu J, Zheng Q, Gao X, Pan Z. Overexpression of HSPA1A enhances the osteogenic differentiation of bone marrow mesenchymal stem cells via activa-

tion of the Wnt/beta-catenin signaling pathway. Sci Rep. 2016;6:27622.

74. Pinzone JJ, Hall BM, Thudi NK, Vonau M, Qiang YW, Rosol TJ, Shaughnessy JD Jr. The role of Dickkopf-1 in bone development, homeostasis, and disease. Blood. 2009;113:517–25.

75. Guo X, Wang XF. Signaling cross-talk between TGF-beta/BMP and other pathways. Cell Res. 2009;19:71–88.

76. Chen G, Deng C, Li YP. TGF-beta and BMP signaling in osteoblast differentiation and bone formation. Int J Biol Sci. 2012;8:272–88.

77. James AW. Review of signaling pathways governing MSC osteogenic and adipogenic differentiation. Scientifica (Cairo). 2013;2013:684736.

78. Gao Y, Huang E, Zhang H, Wang J, Wu N, Chen X, Wang N, Wen S, Nan G, Deng F, Liao Z, Wu D, Zhang B, Zhang J, Haydon RC, Luu HH, Shi LL, He TC. Crosstalk between Wnt/beta-catenin and estrogen receptor signaling synergistically promotes osteogenic differentiation of mesenchymal progenitor cells. PLoS One. 2013;8:e82436.

79. Hiyama A, Yokoyama K, Nukaga T, Sakai D, Mochida J. A complex interaction between Wnt signaling and TNF-alpha in nucleus pulposus cells. Arthritis Res Ther. 2013;15:R189.

80. Baum R, Gravallese EM. Impact of inflammation on the osteoblast in rheumatic diseases. Curr Osteoporos Rep. 2014;12:9–16.

81. Wehmeyer C, Pap T, Buckley CD, Naylor AJ. The role of stromal cells in inflammatory bone loss. Clin Exp Immunol. 2017;189:1–11.

82. Rosen EY, Wexler EM, Versano R, Coppola G, Gao F, Winden KD, Oldham MC, Martens LH, Zhou P, Farese RV Jr, Geschwind DH. Functional genomic analyses identify pathways dysregulated by progranulin deficiency, implicating Wnt signaling. Neuron. 2011;71:1030–42.

83. Zhao YP, Tian QY, Frenkel S, Liu CJ. The promotion of bone healing by progranulin, a downstream molecule of BMP-2, through interacting with TNF/TNFR signaling. Biomaterials. 2013;34:6412–21.

84. Wang C, Liao H, Cao Z. Role of Osterix and MicroRNAs in bone formation and tooth development. Med Sci Monit. 2016;22:2934–42.

85. Kang H, Hata A. The role of microRNAs in cell fate determination of mesenchymal stem cells: balancing adipogenesis and osteogenesis. BMB Rep. 2015;48:319–23.

86. Zhang Y, Xie RL, Croce CM, Stein JL, Lian JB, van Wijnen AJ, Stein GS. A program of microRNAs controls osteogenic lineage progression by targeting transcription factor Runx2. Proc Natl Acad Sci U S A. 2011;108:9863–8.

87. Song JL, Nigam P, Tektas SS, Selva E. microRNA regulation of Wnt signaling pathways in development and disease. Cell Signal. 2015;27:1380–91.

88. Guo D, Li Q, Lv Q, Wei Q, Cao S, Gu J. MiR-27a targets sFRP1 in hFOB cells to regulate proliferation, apoptosis and differentiation. PLoS One. 2014;9:e91354.

89. Zhang J, Tu Q, Bonewald LF, He X, Stein G, Lian J, Chen J. Effects of miR-335-5p in modulating osteogenic differentiation by specifically downregulating Wnt antagonist DKK1. J Bone Miner Res. 2011;26:1953–63.

90. Buser D, Dula K, Lang NP, Nyman S. Long-term stability of osseointegrated implants in bone regenerated with the membrane technique. 5-year results of a prospective study with 12 implants. Clin Oral Implants Res. 1996;7:175–83.

91. Yu X, Tang X, Gohil SV, Laurencin CT. Biomaterials for bone regenerative engineering. Adv Healthc Mater. 2015;4:1268–85.

92. Oryan A, Alidadi S, Moshiri A, Maffulli N. Bone regenerative medicine: classic options, novel strategies, and future directions. J Orthop Surg Res. 2014;9:18.

93. Hunter JD 3rd, Cannon JA. Biomaterials: so many choices, so little time. What are the differences? Clin Colon Rectal Surg. 2014;27:134–9.

94. Roberts TT, Rosenbaum AJ. Bone grafts, bone substitutes and orthobiologics: the bridge between basic science and clinical advancements in fracture healing. Organogenesis. 2012;8:114–24.

95. Di Marco M, Shamsuddin S, Razak KA, Aziz AA, Devaux C, Borghi E, Levy L, Sadun C. Overview of the main methods used to combine proteins with nanosystems: absorption, bioconjugation, and encapsulation. Int J Nanomedicine. 2010;5:37–49.

Part III

Cartilage Regeneration

Cartilage Tissue Engineering: Role of Mesenchymal Stem Cells, Growth Factors, and Scaffolds

18

Mudasir Bashir Gugjoo, Hari Prasad Aithal, Prakash Kinjavdekar, and Amarpal

18.1 Introduction

The articular cartilage, a connective tissue with characteristic structural, biochemical, and metabolic features, furnishes an exceptional resiliency and almost frictionless movement to the diarthrodial joints [1]. The average articular cartilage thickness is at the most a few millimeters with knee thickness being 0.3 mm in rabbits, 0.4–0.5 mm in sheep, 0.6–1.3 mm in dog, 0.7–1.5 mm in goats, and 1.5–2.0 mm and 2.2–2.5 mm in humans. Among the commonly used animals for preclinical studies, horse knee cartilage thickness has closest approximation to human knee cartilage followed by goats [2]. Its composition as well as thickness even vary from joint to joint and with age among species [3]. In general, articular cartilage constitutes three layers/zones with the deep zone separated from subchondral bone by a wavy calcified zone known as tidemark (Fig. 18.1). The three zones bear unique arrange-

M. B. Gugjoo
Faculty of Veterinary Sciences and Animal Husbandry, SKUAST-Kashmir,
Jammu & Kashmir, India

H. P. Aithal (✉)
Training and Education Centre, ICAR-Indian Veterinary Research Institute, College of Agriculture Campus, Pune, Maharashtra, India

P. Kinjavdekar · Amarpal
Division of Surgery, ICAR-Indian Veterinary Research Institute, Izatnagar, Uttar Pradesh, India

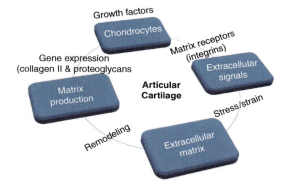

Fig. 18.1 Articular cartilage constitutes three layers/zones with the deep zone separated from subchondral bone by a wavy calcified zone known as tidemark

ment of matrix and cells. In the superficial zone, the cells are flattened disc-like, while in deeper zones the cells appear more rounded. Collagen arrangement appears parallel to the surface in superficial zone, while it becomes random in middle zone and perpendicular in deep zone. The main proteoglycan, aggrecan, content in superficial zone is limited, while in deeper zone it constitutes a major portion. The tissue ingredients in decreasing order of their concentration include water (approximately 75%), collagen especially type II (15%), proteoglycans (10%), and chondrocytes (<2%) [4]. The collagen provides the tissue strength while the proteoglycans provide functional resistance against compression [5]. The resident cells, chondrocytes that reside in lacunae singly or in groups (cell nests), occupy

© Springer Nature Switzerland AG 2019
D. Duscher, M. A. Shiffman (eds.), *Regenerative Medicine and Plastic Surgery*,
https://doi.org/10.1007/978-3-030-19962-3_18

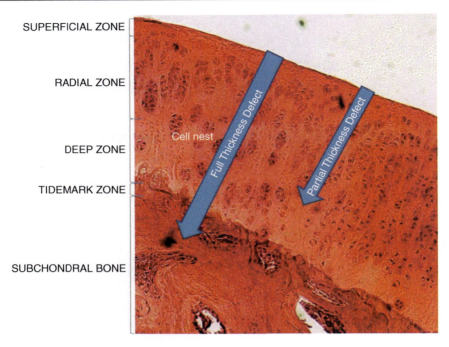

Fig. 18.2 Articular cartilage growth factors. The cells maintain tissue homeostasis through mechanical links generated from extracellular matrix (ECM) via cell surface receptors known as integrins

less than 10% of the tissue. The cells maintain tissue homeostasis through mechanical links generated from extracellular matrix (ECM) via cell surface receptors known as integrins [6]. In addition, the growth factors/cytokines act upon chondrocytes and/synovial cells to secrete proteinases such as aspartic/cysteine/serine and metalloproteinases for tissue homeostasis (Fig. 18.2). Among various proteinases, currently, matrix metalloproteinases that degrade all elements of ECM are considered to carry arthritic degeneration potential [7].

Cartilage is a highly differentiated tissue devoid of any direct blood, lymph, or nerve supply and with a scarce number of less proliferative chondrocytes [8, 9]. Articular cartilage upon damage carries limited regeneration potential. The injury in the form of defects is generally divided into partial- and full-thickness defects with the former confined to the tissue itself and the latter penetrating subchondral bone [10]. Partial-thickness defects do not heal spontaneously as the lesion remains devoid of fibrin clot and thus reparative stem cells. The defects are analogous to fissures or clefts seen in early stages of osteoarthritis [11]. Full-thickness defects though heal spontaneously but with a fibrous tissue that is weaker in structural and mechanical competence [11–15]. Osteoarthritis is a progressive erosion of articular cartilage with about 21.4% of the humans [16] and 20% of dogs [17] affected. The exact pathophysiological basis of osteoarthritis is still disputed but the cardinal signs include inflammation and pain, and the pathognomonic radiological features include articular cartilage thinning characterized by decreased joint space, sclerosis, and osteophyte formation [18, 19]. The pain and subsequent loss of functional activity that arise from an insult to the cartilage and its advancement into osteoarthritis demand advanced techniques for better cartilage rehabilitation [11, 12, 14, 20, 21].

To date no repair procedure has been able to heal the cartilage defects to a satisfactory level. Immediately post-injury the local death of cells hampers matrix production that may integrate with the native tissue. The main aim remains to repair the defects by true hyaline cartilage that has seamless local integration. Numerous invasive procedures such as microfracture [22], subchon-

18.2 Mesenchymal Stem Cells

dral bone drilling [23], lavage, debridement and perichondral arthroplasty [24], periosteal arthroplasty [25], autologous osteochondral transplantation [26], autologous chondrocyte implantation [12, 27, 28], and application of autogenic cancellous bone graft [29, 30] have been attempted for cartilage rehabilitation. The techniques, however, lack true hyaline cartilage repair potential besides being limited to small/medium focal sized osteochondral defects [31]. Autologous chondrocyte implantation (ACI), currently better among the lot, has drawbacks in the form of limited chondrocyte source availability, proneness of the cells to dedifferentiate to fibroblasts, and degeneration in pre-damaged cartilage [32, 33]. In addition, the ageing chondrocytes show declining mitotic and synthetic activity, and synthesize smaller and less uniform aggrecan molecules bearing less functional link proteins [34].

Currently, tissue engineering is being employed to achieve better cartilage rehabilitation. For successful cartilage tissue engineering, various components are required such as cells, growth factors, and three-dimensional matrices. Appropriate cells like autologous chondrocytes or autologous or allogenic stem cells may be implanted. Most of the cell-based therapies currently utilize chondrocytes (approx. 80%), while stem cells constitute only 15% [35]. The limitations associated with ACI mentioned above demand other cell types like stem cells, which are considered to be immunosuppressive. Growth factors incorporated by either viral/nonviral vectors, nucleofection, or direct delivery may regulate directed differentiation. However, the growth factors such as bone morphogenetic proteins (BMPs) direct both bone and cartilage formation and thus need to be regulated at particular step towards chondrogenic lineage [36]. The cells should be implanted on three-dimensional matrices that support the growth and prevent hazardous effect of local environment [10]. Scaffolds, either natural or synthetic, however, bear limitations like early degradation, lack of sufficient porosity, and non-supportive cell growth, and thus the scaffolds that mimic the desired properties of both and exclude the limitations are in demand.

Stem cells (SCs), characterized by the properties of self-renewal, multiplication, immunomodulation, and multi-lineage differentiation potential, are present in almost all the adult tissues of an individual to maintain normal cells, and thus tissue matrix turnover [10, 37]. The stem cells are of various types such as pluripotent (embryonic SCs, and induced pluripotent SCs) or multipotent (mesenchymal stem cells) based upon their potential to differentiate (Fig. 18.3). Pluripotent stem cells carry extended potential to act multipurpose research and clinical tools to understand and model diseases, develop and screen candidate drugs, and deliver cell replacement in regenerative medicine including cartilage [38]. However, limitations in the form of uncontrolled forced expression (iPSCs), and teratogenic effects and ethical issues (iPSCS/ESCs), have restricted their clinical applications [39, 40]. Currently, mesenchymal stem cells (MSCs) carry maximum share among all stem cells both in preclinical and clinical settings in human and veterinary medicine. The cells are easily available, are capable to differentiate, and secrete certain factors that modulate inflammation and promote healing, and in comparison to pluripotent stem cells they have minimal teratogenic and ethical issues associated [39, 41]. The cells are differentiated as per the available local niche/microenvironment and thus contribute to tissue repair or regeneration. Mesenchymal stem cells implanted into osteochondral defects differentiate into chondrocytes [42–44], while MSC-derived cartilage pellets if implanted subcutaneously either disappear [45] or calcify upon vascular invasion [32]. This indicates the role of microenvironment plausibly through cell-surface receptor stimulation by growth factors, extracellular matrix, or direct interaction with surface receptors of other resident cells (chondrocytes) [46–48]. Currently, MSCs are believed to largely act therapeutically by releasing a diverse array of cytokines, growth factors, chemokines, and immunomodulatory proteins, though they may also achieve terminal differentiation [49]. Despite the studies that show immunomodulatory potential of MSCs, two

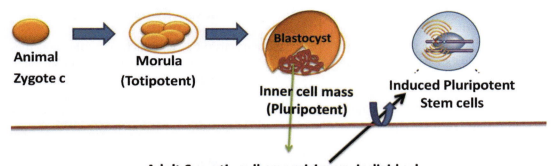

Fig. 18.3 Stem cell sources in animals and humans

recent studies in equines demonstrated development of allo-MSC antibody [50, 51]. One of the studies even showed that the MHC-mismatched MSCs underwent targeted death due to the activation of complement-dependent cytotoxicity. Thus, cautioning about some potential adverse effects that may ensue in addition to the reduced therapeutic efficacy on application of allogenic MSCs [50]. Lack of in-depth understanding in the area demands further steps that need to be deliberated to understand the mechanism(s) behind such differentiation and thereby controlled cell applications.

MSCs that carry maximum share in therapeutics may be derived from almost all the adult tissues (Fig. 18.3) including bone marrow, adipose tissue, embryonic tissue, synovial fluid and membrane, umbilical and peripheral blood, umbilical cord vein, Wharton's jelly, periosteum, muscle, heart, dental pulp, gingiva, periodontal ligament, and mammary tissue [52], each of which carries the potential to differentiate into chondrogenic lineage [36]. Among all the above mentioned sources the most commonly utilized stem cell sources for therapeutics so far have been bone marrow and adipose tissue [53].

Chondrogenic potential of MSCs was first evaluated under in vitro conditions in 1998 employing transforming growth factor-β (TGF-β) and dexamethasone [54]. Further investigations employing various other growth factors such as bone morphogenetic proteins (BMPs), insulin-like growth factor-1 (IGF-1), and parathyroid hormone-related peptide (pTHRP) showed enhanced MSC chondrogenesis [54–57]. However, the in vitro micromass culture method used in such studies may not produce tissue comparable to the native one as the process does not mimic the developmental sequences that actually occur during fetal development. A thorough understanding of embryonic development of the concerned tissue and biological features of the implanted cells is a must-learn criterion for successful cartilage tissue engineering [10]. Recently, under in vitro conditions cartilage tissue was generated approaching hyaline cartilage in physiologic stratification and biomechanical features. This could only be done after recapitulating various developmental processes of mesenchymal condensation via TGF-β1 [58, 59]. The various processes involved include MSC condensation into cellular bodies and condensed mesenchymal cell bodies (CMBs) followed by chondrogenic differentiation that leads to cartilaginous tissue formation. The CMBs under in vitro conditions have been able to generate tissue comparable to native cartilage on osseous tissue surface and also developed mechanically strong cartilage-to-cartilage tissue interface with complete integration [60].

Variations in MSCs' chondrogenic potential have been observed with respect to their source, culture periods, and age of the donors [53]. Among MSCs from various sources, synovial derived MSCs had better chondrogenic potential and led to formation of a large and heavy cartilage pellet compared to BM-MSCs, AD-MSC, Periosteal-MSC and M-MSCs [61]. In another study that compared BM-MSCs and AD-MSCs, the frequency of colony-forming units reportedly had been three times in the latter compared to the former [62]. In elderly patients, the differentiation potential and proliferation capacity of MSCs are reduced and may affect the healing outcome. The immunomodulation property of MSCs may allow allogenic cells to be used [63, 64]. MSCs are able to maintain their differentiation potential for limited periods with long ex vivo-cultured MSCs manifesting reduced chondrogenic matrix formation, undesired mineralization, and rapid cell death after implantation [32, 65]. The reduced cell population may be compensated by implantation of higher cell density for better cartilage healing as reported in some studies [66, 67]. But it may be noted that higher cell density has chances of more cell apoptosis and thus more inflammation at the site.

18.3 Growth Factors

In healthy cartilage environment various growth factors work either individually or in combination to complement each other for maintenance of cartilage homeostasis [68]. The main roles played by the growth factors are to promote MSC differentiation towards chondrogenic lineage, stimulate chondrocytic matrix synthesis, and decrease catabolic effect of MMPs and cytokines such as interleukin-1 [10, 69–71]. The factors act either at earliest phases to promote chondrocyte prolifera-

tion and differentiation like TGF-β [72] or at later stages to promote chondrocyte differentiation rather than initiation of maturation like BMP-2, BMP-4, BMP-6, and TGFβ-3 [73, 74]. To promote MSC differentiation towards chondrogenic lineage, BMP-2 appears superior but has the tendency to promote differentiation towards hypertrophy and osteogenesis characterized by type X collagen and Runx2 expression [72]. Similarly, high intraarticular doses of TGF-β1 have been reported to induce chemotaxis and activation of inflammatory cells tending towards fibrosis and osteophyte formation [72]. To address this issue, combinations of the growth factors have been used either to reduce the activity of each other at certain stage or to complement each other's physiological function. One of the proposals is to co-treat cells with BMP-2 and TGF-β as the latter may potentially prevent differentiation of MSCs into osteogenic lineage [75]. BMP-7 has been reported to inhibit MSC proliferation but does allow proliferation in the presence of TGF-β [76, 77]. Further, growth factors may complement each other and work in synergism. BMP-7 and IGF-1 lead to an enhanced cartilage matrix synthesis [78]. Similarly, IGF-1, IGF-2, and TGF-β regulate each other's gene expression and thus protein production [79]. Further, combination of IGF-1 and TGF-β has better healing potential compared to individual effect as the former is involved in protection of synovium and reduces the synovial thickening depicting lack of chronic inflammation [80]. Limitations in the form of osteogenic synthesis [72], synovial thickening [81, 82], and osteophyte formation [71, 83] as mentioned above may be managed by using growth factors in right combinations and dosages [72, 80, 84].

18.4 Scaffolds

Another criterion for successful cartilage tissue engineering is availability of three-dimensional matrices, as evidences have shown that two-dimensional culture system hardly supports MSCs' chondrogenic differentiation. The micromass culture system as mentioned earlier has failed to recapitulate the cartilage developmental stages, besides express hypertrophic marker, collagen type X [85]. For cartilage rehabilitation most of the investigators prefer MSC application along with scaffold. This allows cellular growth and prevents them against deleterious effects of local environment. In addition, the cells are retained in situ at the desired locations avoiding the common problem of cell leakage [10]. Selected scaffold is supposed to bear features of biocompatibility, support cellular growth and expansion, and facilitate diffusion and movement, yet maintain adequate mechanical strength and properties till tissue is regenerated and integrated [10, 86–88]. In osteochondral lesions, survival time of scaffold is critical as the neocartilage that replaces it should have preformed subchondral bone to survive in addition to its integration with surrounding native cartilage [89]. Usually the cartilage islands that form during healing fail to survive unless not integrated with the adjacent native cartilage [11].

The scaffold design in cartilage tissue engineering is aimed at maintaining the physical (scaffold architecture, mechanical function, and degradation) and biochemical (relevant to cellular behavior and activity) properties [89]. The matrices evaluated include natural fibrin [43, 90–93], agarose and alginate [86], collagen [94–97], hyaluronan [47, 98–100] as well as synthetic polylactic acid [101–103], polyglycolic acid [104], and polylactic and polyglycolic acid [105, 106]. Natural scaffolds that bear desired biocompatibility, better cell attachment, and differentiation have limitations in the form of availability, ease of fabrication, mesh properties, and controllable biodegradability, in addition to immunological reactions and disease transmission [10]. Synthetic scaffolds in comparison though are modified chemically for desired fabrication, and have better versatility, suitable mesh properties, and controllable degradability, but again fall short with respect to cyto-compatibility and may elicit host response upon release of toxic by-products [86, 87]. To overcome such impediments, hybrid scaffolds have been developed incorporating solid polymer scaffold and hydrogel [10]. The former provides mechanical strength and the latter supports cell delivery resembling the biphasic (solid and liquid phases) nature of cartilage. The cells in hydrogel are maintained in three-dimensional stages and are homogenously distributed in solid polymer scaffold pores [107].

In order to utilize such scaffolds in clinical practice, both in vitro and in vivo studies need to be conducted especially in relation to their biocompatibility and mechanical strength. Apart from the above mentioned scaffold designs, two other types including biomimetic zonal and non-fibrous/nanoporous scaffolds have been developed based on the concept to provide microenvironment comparable to that of native cartilage for the cells [10]. Biomimetic zonal scaffold comprises different zones like that of cartilage in order to mimic the physical properties. The implanted cells thus secrete matrix based on the available environment [108]. Nonfibrous/nanoporous scaffolds constitute nano-size matrix that mimics physicochemical and biological properties of cartilage matrix, and thus tends to develop relevant signals for cellular differentiation (MSCs) and matrix synthesis (from MSCs and chondrocytes) [109]. For creating such scaffolds, numerous fabrication techniques (electro-spinning, chemical etching, particulate clumping, 3D printing, and phase separation) may be employed [10]. Preclinical studies that encapsulated cells in nanofibrous scaffolds by electro-spinning have failed to maintain cell homogeneity and have resulted in cell clumping [110]. 3D printing is currently seen to carry the potential to replicate the cartilage structure. The cells are delivered in a suspension or with a gel as an ink in layer-by-layer process creating an appropriate pericellular environment for the cells located in each cartilage zone [111, 112]. One of the impediments in utilizing the technology in tissue engineering is the need to integrate vascular network for proper nutrient and gas supply. Cartilage, however, being devoid of direct blood, lymph, and nerve supply may act as a good candidate for 3D bioprinting [113]. Direct bioprinting into an ex vivo cartilage defect has resulted in some level of integration into native cartilage and mechanical competence [114]. This demands a detailed analysis of the

18.5 Clinical Trials

The successful outcome in clinical settings is the ultimate aim of cartilage tissue engineering. So far the aim is unmet both in veterinary practice and in human medicine though the reports appear promising. The application in animals may provide the basis for human stem cell therapy. In veterinary practice, canines and equines comprise majority of the clinical application studies.

Stem cell therapy in canines has been instituted both in preclinical [115–117] and in clinical settings [118–123]. A single-time, local implantation of the cells in all the studies has been made barring a single study wherein cells were implanted at acupoints [122]. The cells were either applied directly without employing the vehicle [122, 123] or implanted with platelet-rich plasma [120] or hyaluronic acid [124]. All these studies have reported improved healing (pain, visual analog scale, and range of motion) on MSC application with follow-up varying from 1 month [122, 124] and 6 months [120, 121] to 5 years [119–122, 124]. Two comparative studies were conducted involving AD-MSCs versus platelet-rich growth factors (PRGF) [121] and AD-MSC versus stromal vascular fraction (SVF) [122]. In both the studies improved results have been reported with MSCs; however, in the former study MSCs showed better results at 6 months compared to PRGF, while in the latter SVF had better results than MSCs. In another comparative study, vascular endothelial growth factor transgenic BM-MSCs were shown to improve early healing in comparison to simple MSCs [117].

In equines, most of the studies so far have been unable to fetch positive results for better cartilage repair in osteoarthritis patients [71, 125, 126]. Some of the studies, clinical as well as experimental, though have shown beneficial effects in cartilage repair but are mainly on the basis of reduction in pain perception [127–129]. In a clinical study of 40 horses having joint affections treated with BM-MSCs, 77% of the patients returned to work; among them 38% were able to work to the previous condition or exceeded [125]. Currently, the stem cell being implanted is at 2×10^7 concentration in hyaluronan scaffold (22 mg of Hyvisc) (hyaluronate sodium, 3×10^6 Da, Anika Therapeutics, Woburn, MA) [130], prior to which NSAIDs were recommended to reduce joint flare [131].

In human medicine numerous cartilage-related clinical trials implanting stem cells have been registered at http://www.clinicaltrial.gov/. Among them some are completed, while some are in progress. The cells have been injected either locally (intra-articularly) or implanted surgically. All the registered studies located were uncontrolled. The stem cell reported studies are either case series [66, 132–137], case reports [138–144], or comparative [66, 141–148] type. The cell types employed in such studies have been AD-MSCs, bone marrow concentrate, and BM-MSCs with or without the scaffolds. The patient number in case series studies ranged from 4 to 48. The follow-up period of at least 3 months and a maximum of 5 years has been made. An overall improvement in the clinical parameters (Visual Analog Score, Improved Knee and Osteoarthritis Outcome Score, and International Knee Documentation Score), MRI, and histological score in the patients has been reported with no major adverse effect observed on cell application. With respect to the formation of the healing tissue, the variability in outcome was reported. Some of the patients had hyaline-like tissue [135, 140, 145, 149], while others had combination of the hyaline/fibrocartilage [135] or mainly fibrocartilaginous tissue [141]. In a study that compared MSCs versus ACI with equal patient number of 36 in each group, the clinical results were comparable except for improvement in physical functioning of patients in BM-MSC groups [146]. In a study that evaluated dose-dependent healing potential of MSCs, the group of patients that received higher dose (1.0×10^8) had better clinical scores and reduced pain compared to those patients that received lower dose of AD-MSCs (1.0×10^7 and 5.0×10^7) [67].

In clinical settings, variability in lesion type, site, duration of existence, age of the patient, cell culture techniques, and cell application methods and their number, besides addition of growth factors and scaffolds, have bearing on the outcome, and thus demand controlled studies [10].

18.6 Conclusions and Future Perspectives

Articular cartilage upon damage carries limited regeneration potential. Currently, tissue engineering, employing cells, growth factors, and scaffolds are considered to have the potential to support regeneration and integration of neocartilage with the surrounding native tissue. MSCs especially BM-MSCs and AD-MSCs carry maximum share among all stem cells in cartilage tissue engineering. There is a need to investigate cell source to find out whether only autogenous cells or both autogenic and allogenic/xenogenic cells can be utilized. The cell survival posttransplantation and integration of regenerated tissue matrix with the host native tissue remain the major causes of concern. One of the promising technologies to develop mechanically strong cartilage-to-cartilage interface includes the mesenchymal condensation into cellular bodies under the influence of growth factors. However, more research especially under in vivo conditions is desired in the area to evaluate its actual clinical application. Growth factors form an indispensable part of the tissue engineering and demand further evaluation on the basis of their individual properties as well as combinations including dosages. Scaffold that affects the desired chondrogenesis remains to be elucidated. Newer fabrication technologies that appear promising need to be evaluated and compared against the conventional technologies especially in relation to the maintenance of scaffold mechanical and biological properties. Tissue engineering that appears promising needs to be evaluated with respect to the cell sources; culture methods; concentration; implantation methods; growth factors, their combinations, doses, and frequency; and scaffolds, their sources, design, and type, before it becomes a clinical reality.

References

1. Mankin HJ. Synovium and cartilage in health and disease. In: Newton CD, Nunamaker DM, editors. Textbook of small animal orthopaedics. Philadelphia: JB. Lippincott Company; 1984. p. 90.
2. Frisbie DD, Cross MW, McIlwraith CW. A comparative study of articular cartilage thickness in the stifle of animal species used in human pre-clinical studies compared to articular cartilage thickness in the human. Vet Comp Orthop Traumatol. 2006;19(3):142–6.
3. Athanasiou AK, Agarwal A, Muffoletto A. Biomechanical properties of hip cartilage in experimental animal models. Clin Orthop. 1995;316:254–66.
4. Poole AR. Cartilage in health and disease. In: Koopman WJ, editor. Arthritis and allied conditions. Philadelphia: Lippincott Williams and Wilkins; 2001. p. 226–84.
5. Maroudas A. Physicochemical properties of articular cartilage. In: Freeman M, editor. Adult articular cartilage. London: Pitman Medical; 1979. p. 215–90.
6. Jeffrey AK, Blunn GW, Archer CW. Three-dimensional collagen architecture in bovine articular cartilage. J Bone Joint Surg. 1991;73:795–801.
7. Woessner JF, Nagase H. Matrix Metalloproteinases and TIMPs. Oxford, UK: Oxford University Press; 2002.
8. Kinner B, Capito RM, Spector M. Regeneration of articular cartilage. Adv Biochem Eng Biotechnol. 2005;94:91–123.
9. Duarte Campos DF, Drescher W, Rath B, Tingart M, Fischer H. Supporting biomaterials for articular cartilage repair. Cartilage. 2012;3:205–21.
10. Gugjoo MB, Amarpal, Sharma GT, Kinjavdekar P, Aithal HP, Pawde AM. Cartilage tissue engineering: role of mesenchymal stem cells along with growth factors and scaffolds. Indian J Med Res. 2016;144:339–47.
11. Hunziker EB. Biologic repair of articular cartilage. Defect models in experimental animals and matrix requirements. Clin Orthop Relat Res. 1999;367:S135–46.
12. Breinan HA, Minas T, Hsu HP, Nehrer S, Sledge CB, Spector M. Effect of cultured autologous chondrocytes on repair of chondral defects in a canine model. J Bone Joint Surg Am. 1997;79:1439–51.
13. Arican M, Koylu O, Uyaroglu A, Erol M, Calim KN. The effect of (Hylan G-F 20) on bone metabolism in dogs with experimental osteochondral defects. J Turk Vet Surg. 2006;12:20–3.
14. Günes T, Sen C, Erdem M, Köseoglu RD, Filiz NO. Combination of microfracture and periosteal transplantation techniques for the treatment of full-thickness cartilage defects. Acta Orthop Traumatol Turc. 2006;40:315–23.

15. Tiwary R, Amarpal, Aithal HP, Kinjavdekar P, Pawde AM, Singh R. Effect of IGF-1 and uncultured autologous bone-marrow-derived mononuclear cells on repair of osteochondral defect in rabbits. Cartilage. 2014;5:43–54.

16. Barbour KE, Helmick CG, Boring M. Prevalence of doctor diagnosed arthritis at state and county levels-United States, 2014. MMWR—Morbid Mortal Week Report. 2016;65(19):489–94.

17. Johnston SA. Osteoarthritis. Joint anatomy, physiology, and pathobiology. Vet Clin North Am Small Anim Pract. 1997;27:699–723.

18. Brandt KD, Dieppe P, Radin E. Etiopathogenesis of osteoarthritis. Med Clin North Am. 2009;93(1):1e24.

19. Hugle T, Geurts J. What drives osteoarthritis?—synovial versus subchondral bone pathology. Rheumatology (Oxford). 2016;56(9):1461–71.

20. Bilgili H, Yildiz C, Kurum B, Soysal Y, Bahce M. Repair of osteochondral defects with autologous chondrocyte implantation: clinical study on the stifle joint of 9 dogs. Ankara Univ J Vet Fac. 2006;53:103–9.

21. Juneau C, Paine R, Chicas E, Gardner E, Bailey L, McDermott J. Current concepts in treatment of patellofemoral osteochondritis dissecans. Int J Sports Phys Ther. 2016;11(6):903–25.

22. Steadman JR, Briggs KK, Rodrigo JJ, Kocher MS, Gill TJ, Rodkey WG. Outcomes of microfracture for traumatic chondral defects of the knee: average 11-year follow-up. Art Ther. 2003;19:477–84.

23. Sgaglione NA, Miniaci A, Gillogly SD, Carter TR. Update on advanced surgical techniques in the treatment of traumatic focal articular cartilage lesions in the knee. Art Ther. 2002;18:9–32.

24. O'Driscoll SW. The healing and regeneration of articular cartilage. J Bone Joint Surg Am. 1998;80:1795–812.

25. Tsai CL, Liu TK, Fu SL, Perng JH, Lin AC. Preliminary study of cartilage repair with autologous periosteum and fibrin adhesive system. J Formos Med Assoc. 1992;91:S239–45.

26. Outerbridge HK, Outerbridge AR, Outerbridge RE. The use of a lateral patellar autologous graft for the repair of a large osteochondral defect in the knee. J Bone Joint Surg Am. 1995;77:65–72.

27. Grande DA, Halberstadt C, Naughton G, Schwartz R, Manji R. Evaluation of matrix scaffolds for tissue engineering of articular cartilage grafts. J Biomed Mater Res. 1997;34:211–20.

28. Tins BJ, McCall IW, Takahashi T, Cassar-Pullicino V, Roberts S, Ashton B, Richardson J. Autologous chondrocyte implantation in knee joint: MR imaging and histologic features at 1-year follow-up. Radiology. 2005;234:501–8.

29. van Dyk GE, Dejardin LM, Flo G, Johnson LL. Cancellous bone grafting of large osteochondral defects: an experimental study in dogs. Art Ther. 1998;14:311–20.

30. Gunay C, Sagliyan A, Unsaldi E, Yaman M. Repair of experimentally induced osteochondral defects of dog knee joint with cancellous autograft. Firat Univ J Health Sci. 2005;19:107–13.

31. Reddy S, Pedowitz DI, Parekh SG, Sennett BJ, Okereke E. The morbidity associated with osteochondral harvest from asymptomatic knees for the treatment of osteochondral lesions of the talus. Am J Sports Med. 2007;35:80–5.

32. Pelttari K, Winter A, Steck E, Goetzke K, Hennig T, Ochs BG, Aigner T, Richter W. Premature induction of hypertrophy during in vitro chondrogenesis of human mesenchymal stem cells correlates with calcification and vascular invasion after ectopic transplantation in SCID mice. Arthritis Rheum. 2006;54:3254–66.

33. Punwar S, Khan WS. Mesenchymal stem cells and articular cartilage repair: clinical studies and future direction. Open Orthop J. 2011;5:296–301.

34. Adkisson HD, Gillis MP, Davis EC, Maloney W, Hruska KA. In vitro generation of scaffold independent neocartilage. Clin Orthop Relat Res. 2001;391(Suppl):280–94.

35. Fraser JK, Wulur I, Alfonso Z, Hedrick MH. Fat tissue: an underappreciated source of stem cells for biotechnology. Trends Biotechnol. 2006;24(4):150–4.

36. Kessler MW, Grande DA. Tissue engineering and Cartilage. Organogenesis. 2008;4(1):28–32.

37. Gade NE, Pratheesh MD, Nath A, Dubey PK, Amarpal, Sharma B, Saikkumar G, Taru Sharma G. Molecular and cellular characterization of buffalo bone marrow-derived mesenchyme stem cells. Reprod Domest Anim. 2013;48(3):358–67.

38. Guzzo RM, Scanlon V, Sanjay A, Xu RH, Drissi H. Establishment of human cell type-specific iPS cells with enhanced chondrogenic potential. Stem Cell Rev Rep. 2014;10(6):820–9.

39. Zuk P, Zhu M, Mizuno H, Huang J, Futrell J, Katz A, Behhaim P, Lorenz HP, Hedrick MH. Multilineage cells from human adipose tissue: implications for cell-based therapies. Tissue Eng. 2001;7:211–28.

40. Wang M, Yuan Z, Ma N, Hao C, Guo W, Zou G, Zhang Y, Chen M, Gao S, Peng J, Wang A, Wang Y, Sui X, Xu W, Lu S, Liu S, Guo Q. Advances and prospects in stem cells for cartilage regeneration. Stem Cells Int. 2017;2017:Article ID 4130607.

41. Singh A, Singh A, Sen D. Mesenchymal stem cells in cardiac regeneration: a detailed progress report of the last 6 years (2010–2015). Stem Cell Res Ther. 2016;7(1):82.

42. Wakitani S, Goto T, Pineda SJ, Young RG, Mansour JM, Caplan AI, Goldberg VM. Mesenchymal cell-based repair of large, full-thickness defects of articular cartilage. J Bone Joint Surg Am. 1994;76:579–92.

43. Wang F, Li Z, Tamama K, Sen CK, Guan J. Fabrication and characterization of prosurvival growth factor releasing, anisotropic scaffolds for enhanced mesenchymal stem cell survival/growth and orientation. Biomacromolecules. 2009;10:2609–18.

44. Kazemi D, Asenjan SK, Dehdilani N, Para H. Canine articular cartilage regeneration using mesenchymal stem cells seeded on platelet rich fibrin. Bone Joint Res. 2017;6(2):98–107.

45. De Bari C, Dell 'Accio F, Luyten FP. Failure of in vitro differentiated mesenchymal stem cells from the synovial membrane to form ectopic stable cartilage in vivo. Arthritis Rheum. 2004;50:142–50.

46. Csaki C, Schneider PR, Shakibaei M. Mesenchymal stem cells as a potential pool for cartilage tissue engineering. Ann Anat. 2008;190:395–412.

47. Solchaga LA, Penick KJ, Welter JF. Chondrogenic differentiation of bone marrow-derived mesenchymal stem cells: tips and tricks. Methods Mol Biol. 2011;698:253–78.

48. Gugjoo MB, Amarpal, Ahmed AA, Kinjavdekar P, Aithal HP, Pawde AM, Kumar GS, Sharma GT. Mesenchymal stem cells with IGF-1 and TGF-β1 in laminin gel for osteochondral defects in rabbits. Biomed Pharmacother. 2017;93:1165–74.

49. Stewart MC, Stewart AA. Mesenchymal stem cells: characteristics, sources, mechanisms of action. Vet Clin North Am Equine Pract. 2011;27:243–61.

50. Berglund AK, Schnabel LV. Allogeneic major histocompatibility complex-mismatched equine bone marrow-derived mesenchymal stem cells are targeted for death by cytotoxic anti-major histocompatibility complex antibodies. Equine Vet J. 2017;49(4):539–44.

51. Owens SD, Kol A, Walker NJ, Borjesson DL. Allogeneic mesenchymal stem cell treatment induces specific allo-antibodies in horses. Stem Cells Int. 2016;2016:Article ID 5830103.

52. Mafi R, Hindocha S, Mafi P, Griffin M, Khan WS. Sources of adult mesenchymal stem cells applicable for musculoskeletal applications - a systematic review of the literature. Open Orthop J. 2011;5:242–8.

53. Lee WYW, Wang BW. Cartilage repair by mesenchymal stem cells: clinical trial update and perspectives. J Orthop Translation. 2017;9:76–88.

54. Johnstone B, Hering TM, Caplan AI, Goldberg VM, Yoo JU. In vitro chondrogenesis of bone marrow-derived mesenchymal progenitor cells. Exp Cell Res. 1998;238:265–72.

55. Sekiya I, Colter DC, Prockop DJ. BMP-6 enhances chondrogenesis in a subpopulation of human marrow stromal cells. Biochem Biophys Res Commun. 2001;284:411–8.

56. Kim YJ, Kim HJ, Im GI. PTHrP promotes chondrogenesis and suppresses hypertrophy from both bone marrow-derived and adipose tissue-derived MSCs. Biochem Biophys Res Commun. 2008;373:104–8.

57. Pei M, He F, Vunjak-Novakovic G. Synovium-derived stem cell-based chondrogenesis. Differentiation for cartilage repair: monitoring its success by magnetic resonance imaging and histology. Arthritis Res Ther. 2008;5:R60–3.

58. DeLise AM, Fischer L, Tuan RS. Cellular interactions and signalling in cartilage development. Osteoarthr Cartil. 2000;8:309–34.

59. Hall BK, Miyake T. All for one and one for all: condensations and the initiation of skeletal development. Bioessays. 2000;22:138–47.

60. Bhumiratana S, Eton RE, Oungoulian SR, Wan LQ, Ateshian GA, Vunjak-Novakovic G. Large, stratified, and mechanically functional human cartilage grown in vitro by mesenchymal condensation. Proc Natl Acad Sci U S A. 2014;111:6940–5.

61. Shirasawa S, Sekiya I, Sakaguchi Y, Yagishita K, Ichinose S, Muneta T. In vitro chondrogenesis of human synovium derived mesenchymal stem cells: optimal condition and comparison with bone marrow-derived cells. J Cell Biochem. 2006;97(1):84–97.

62. Mitchell JB, Mcintosh K, Zvonic S, Garrett S, Floyd ZE, Kloster A, Di Halvorsen Y, Storms RW, Goh B, Kilroy G, Wu X, Gimble JM. Immunophenotype of human adipose-derived cells: temporal changes in stromal-associated and stem cell–associated markers. Stem Cells. 2006;24(2):376–85.

63. Steinert AF, Ghivizzani SC, Rethwilm A, Tuan RS, Evans CH, Noth U. Major biological obstacles for persistent cell-based regeneration of articular cartilage. Arthritis Res Ther. 2007;9(3):213.

64. Roobrouck VD, Ulloa-Montoya F, Verfaillie CM. Self-renewal and differentiation capacity of young and aged stem cells. Exp Cell Res. 2008;314(9):1937–44.

65. van der Bogt KE, Schrepfer S, Yu J, Sheikh AY, Hoyt G, Govaert JA, Velotta JB, Contag CH, Robbins RC, Wu JC. Comparison of transplantation of adipose tissue- and bone marrow-derived mesenchymal stem cells in the infarcted heart. Transplantation. 2009;87(5):642–52.

66. Koga H, Muneta T, Ju YJ, Nagase T, Nimura A, Mochizuki T, Ichinose S, von der Mark K, Sekiya I. Synovial stem cells are regionally specified according to local microenvironments after implantation for cartilage regeneration. Stem Cells. 2007;25:689–96.

67. Jo CH, Lee YG, Shin WH, Kim H, Chai JW, Jeong EC, Kim JE, Shim H, Shin JS, Shin IS, Ra JC, Oh S, Yoon KS. Intra-articular injection of mesenchymal stem cells for the treatment of osteoarthritis of the knee: a proof-of-concept clinical trial. Stem Cells. 2014;32:1254–66.

68. Goldring MB, Tsuchimochi K, Ijiri K. The control of chondrogenesis. J Cell Biochem. 2006;97:33–4.

69. Middleton J, Manthey A, Tyler J. Insulin-like growth factor (IGF) receptor, IGF-I, interleukin-1 beta (IL-1 beta), and IL-6 mRNA expression in osteoarthritic and normal human cartilage. J Histochem Cytochem. 1996;44:133–41.

70. Gouttenoire J, Valcourt U, Ronzière MC, Aubert-Foucher E, Mallein-Gerin F, Herbage D. Modulation of collagen synthesis in normal and osteoarthritic cartilage. Biorheology. 2004;41:535–42.

71. Wilke MM, Nydam DV, Nixon AJ. Enhanced early chondrogenesis in articular defects following arthroscopic mesenchymal stem cell implantation in an equine model. J Orthop Res. 2007;25:913–25.

72. Baugé C, Girard N, Lhuissier E, Bazille C, Boumediene K. Regulation and role of TGFβ signaling pathway in aging and osteoarthritis joints. Aging Dis. 2014;5(6):394–405.

73. Kameda T, Koike C, Saitoh K, Kuroiwa A, Iba H. Analysis of cartilage maturation using micromass cultures of primary chondrocytes. Dev Growth Differ. 2000;42(3):229–36.

74. Spagnoli A. Mesenchymal stem cells and fracture healing. Orthopedics. 2008;31(9):855–6.

75. Mehlhorn AT, Schmal H, Kaiser S, Lepski G, Finkenzeller G, Stark GB, Südkamp NP. Mesenchymal stem cells maintain TGF-β-mediated chondrogenic phenotype in alginate bead culture. Tissue Eng. 2006;12(6):1393–403.

76. Goodrich LR, Hidaka C, Robbins PD, Evans CH, Nixon AJ. Genetic modification of chondrocytes with insulin-like growth factor-1 enhances cartilage healing in an equine model. J Bone Joint Surg Br. 2007;89:672–85.

77. Elshaier AM, Hakimiyan AA, Rappoport L, Rueger DC, Chubinskaya S. Effect of interleukin-1beta on osteogenic protein 1-induced signaling in adult human articular chondrocytes. Arthritis Rheum. 2009;60:143–54.

78. Loeser RF, Pacione CA, Chubinskaya S. The combination of insulin-like growth factor 1 and osteogenic protein 1 promotes increased survival of and matrix synthesis by normal and osteoarthritic human articular chondrocytes. Arthritis Rheum. 2003;48:2188–96.

79. Shi S, Mercer S, Eckert GJ, Trippel SB. Growth factor regulation of growth factors in articular chondrocytes. J Biol Chem. 2009;284:6697–704.

80. Davies LC, Blain EJ, Gilbert SJ, Caterson B, Duance VC. The potential of IGF-1 and TGFbeta1 for promoting "adult" articular cartilage repair: an in vitro study. Tissue Eng Part A. 2008;14:1251–61.

81. Bakker AC, van de Loo FA, van Beuningen HM, Sime P, van Lent PL, van der Kraan PM, Richards CD, van den Berg WB. Overexpression of active TGF-beta-1 in the murine knee joint: evidence for synovial layer-dependent chondro-osteophyte formation. Osteoarthr Cartil. 2001;9:128–36.

82. Boehm AK, Seth M, Mayr KG, Fortier LA. Hsp90 mediates insulin-like growth factor 1 and interleukin-1beta signaling in an age-dependent manner in equine articular chondrocytes. Arthritis Rheum. 2007;56:2335–43.

83. Miyakoshi N, Kobayashi M, Nozaka K, Okada K, Shimada Y, Itoi E. Effects of intraarticular administration of basic fibroblast growth factor with hyaluronic acid on osteochondral defects of the knee in rabbits. Arch Orthop Trauma Surg. 2005;125:683–92.

84. Mierisch CM, Cohen SB, Jordan LC, Robertson PG, Balian G, Diduch DR. Transforming growth factor-beta in calcium alginate beads for the treatment of articular cartilage defects in the rabbit. Art Ther. 2002;18:892–900.

85. Johnstone B, Alini M, Cucchiarini M, Dodge GR, Eglin D, Guilak F. Tissue engineering for articular cartilage repair—the state of the art. Eur Cell Mater. 2013;25:248–67.

86. Lu L, Zhu X, Valenzuela RG, Currier BL, Yaszemski MJ. Biodegradable polymer scaffolds for cartilage tissue engineering. Clin Orthop Relat Res. 2001;391(Suppl):251–70.

87. Risbud MV, Sittinger M. Tissue engineering: advances in in vitro cartilage generation. Trends Biotechnol. 2002;20:351–6.

88. Frenkel SR, Di Cesare PE. Scaffolds for articular cartilage repair. Ann Biomed Eng. 2004;32:26–34.

89. Hutmacher DW. Scaffold design and fabrication technologies for engineering tissues—state of the art and future perspectives. J Biomater Sci Polym Ed. 2001;12:107–24.

90. Hendrickson DA, Nixon AJ, Grande DA, Todhunter RJ, Minor RM, Erb H, Lust G. Chondrocyte-fibrin matrix transplants for resurfacing extensive articular cartilage defects. J Orthop Res. 1994;12:485–97.

91. Brittberg M, Sjögren-Jansson E, Lindahl A, Peterson L. Influence of fibrin sealant (Tisseel) on osteochondral defect repair in the rabbit knee. Biomaterials. 1997;18:235–42.

92. Fortier LA, Mohammed HO, Lust G, Nixon AJ. Insulin like growth factor-I enhances cell-based repair of articular cartilage. J Bone Joint Surg Br. 2002;84:276–88.

93. Fortier LA, Nixon AJ, Lust G. Phenotypic expression of equine articular chondrocytes grown in three-dimensional cultures supplemented with supraphysiologic concentrations of insulin-like growth factor-1. Am J Vet Res. 2002;63:301–5.

94. Grande DA, Pitman MI, Peterson L, Menche D, Klein M. The repair of experimentally produced defects in rabbit articular cartilage by autologous chondrocyte transplantation. J Orthop Res. 1989;7:208–18.

95. Nehrer S, Breinan HA, Ramappa A, Shortkroff S, Young G, Minas T, Sledge CB, Yannas IV, Spector M. Canine chondrocytes seeded in type I and type II collagen implants investigated in vitro. J Biomed Mater Res. 1997;38:95–104.

96. Wakitani S, Imoto K, Yamamoto T, Saito M, Murata N, Yoneda M. Human autologous culture expanded bone marrow mesenchymal cell transplantation for repair of cartilage defects in osteoarthritic knees. Osteoarthr Cartil. 2002;10:199–206.

97. Lee CR, Grodzinsky AJ, Hsu HP, Spector M. Effects of a cultured autologous chondrocyte-seeded type II collagen scaffold on the healing of a chondral defect in a canine model. J Orthop Res. 2003;21:272–81.

98. Knudson W, Casey B, Nishida Y, Eger W, Kuettner KE, Knudson CB. Hyaluronan oligosaccharides perturb cartilage matrix homeostasis and induce chondrocytic chondrolysis. Arthritis Rheum. 2000;43:1165–74.

99. Gao J, Dennis JE, Solchaga LA, Goldberg VM, Caplan AI. Repair of osteochondral defect with tissue-engineered two-phase composite material of injectable calcium phosphate and hyaluronan sponge. Tissue Eng. 2002;8:827–37.

100. Solchaga LA, Gao J, Dennis JE, Awadallah A, Lundberg M, Caplan AI, Goldberg VM. Treatment of osteochondral defects with autologous bone marrow in a hyaluronan-based delivery vehicle. Tissue Eng. 2002;8:333–47.

101. Chu CR, Dounchis JS, Yoshioka M, Sah RL, Coutts RD, Amiel D. Osteochondral repair using perichondrial cells. A 1-year study in rabbits. Clin Orthop Relat Res. 1997;340:220–9.

102. Dounchis JS, Bae WC, Chen AC, Sah RL, Coutts RD, Amiel D. Cartilage repair with autogenic perichondrium cell and polylactic acid grafts. Clin Orthop Relat Res. 2000;377:248–64.

103. Frenkel SR, Chang J, Maurer S, Baitner A, Wright K. Bone protein in a grafton flex carrier for articular cartilage repair. Trans Am Acad Orthop Surg. 2001;26:356.

104. Liu Y, Chen F, Liu W, Cui L, Shang Q, Xia W, Wang J, Cui Y, Yang G, Liu D, Wu J, Xu R, Buonocore RD, Cao Y. Repairing large porcine full-thickness defects of articular cartilage using autologous chondrocyte-engineered cartilage. Tissue Eng. 2002;8:709–21.

105. Niederauer GG, Slivka MA, Leatherbury NC, Korvick DL, Harroff HH, Ehler WC, Dunn CJ, Kieswetter K. Evaluation of multiphase implants for repair of focal osteochondral defects in goats. Biomaterials. 2000;21:2561–74.

106. Cohen SB, Meirisch CM, Wilson HA, Diduch DR. The use of absorbable co-polymer pads with alginate and cells for articular cartilage repair in rabbits. Biomaterials. 2003;24:2653–60.

107. Caterson EJ, Nesti LJ, Li WJ, Danielson KG, Albert TJ, Vaccaro AR, Tuan RS. Three-dimensional cartilage formation by bone marrow-derived cells seeded in polylactide/alginate amalgam. J Biomed Mater Res. 2001;57:394–03.

108. Klein TJ, Schumacher BL, Schmidt TA, Li KW, Voegtline MS, Masuda K, Thonar EJ, Sah RL. Tissue engineering of stratified articular cartilage from chondrocyte subpopulations. Osteoarthritis Cartilage. 2003;11:595–602.

109. Zhang L, Webster TJ. Nanotechnology and nanomaterials: promises for improved tissue regeneration. Nano Today. 2009;4:66–80.

110. Li WJ, Jiang YJ, Tuan RS. Cell-nanofiber-based cartilage tissue engineering using improved cell seeding, growth factor, and bioreactor technologies. Tissue Eng Part A. 2008;14:639–48.

111. Cohen DL, Lipton JI, Bonassar LJ, Lipson H. Additive manufacturing for in situ repair of osteochondral defects. Biofabrication. 2010;2(3):12.

112. Fedorovich NE, Schuurman W, Wijnberg HM, Prins HJ, van Weeren PR, Malda J, Alblas J, Dhert WJ. Biofabrication of osteochondral tissue equivalents by printing topologically defined, cell-laden hydrogel scaffolds. Tissue Eng Part C Methods. 2012;18(1):33–44.

113. Ozbolat IT, Yu Y. Bioprinting toward organ fabrication: challenges and future trends. IEEE Transactions Biomed Eng. 2013;60(3):691–9.

114. Cui XF, Breitenkamp K, Finn MG, Lotz M, D'Lima DD. Direct human cartilage repair using three-dimensional bioprinting technology. Tissue Eng Part A. 2012;18(11–12):1304–12.

115. Mokbel A, El-Tookhy O, Shamaa AA, Sabry D, Rashed L, Mostafa A. Homing and efficacy of intraarticular injection of autologous mesenchymal stem cells in experimental chondral defects in dogs. Clin Exp Rheumatol. 2011;29:275–84.

116. Yang Q, Peng J, Lu SB, Guo QY, Zhao B, Zhang L, Wang AY, Xu WJ, Xia Q, Ma XL, Hu YC, Xu BS. Evaluation of an extracellular matrix-derived acellular biphasic scaffold/cell construct in the repair of a large articular high-load-bearing osteochondral defect in a canine model. Chin Med J. (Engl). 2011;124:3930–8.

117. Hang D, Wang Q, Guo C, Chen Z, Yan Z. Treatment of osteonecrosis of the femoral head with VEGF165 transgenic bone marrow mesenchymal stem cells in mongrel dogs. Cells Tissues Organs. 2012;195:495–506.

118. Black LL, Gaynor J, Adams C, Dhupa S, Sams AE, Taylor R, Harman S, Gingerich DA, Harman R. Effect of intraarticular injection of autologous adipose-derived mesenchymal stem and regenerative cells on clinical signs of chronic osteoarthritis of the elbow joint in dogs. Vet Ther. 2008;9:192–200.

119. Yoon HY, Lee JH, Jeong SW. Long-term follow-up after implantation of autologous adipose tissue derived mesenchymal stem cells to treat a dog with stifle joint osteoarthrosis. J Vet Clin. 2012;29:82–6.

120. Vilar JM, Morales M, Santana A, Spinella G, Rubio M, Cuervo B, Cugat R, Carrillo JM. Controlled, blinded force platform analysis of the effect of intraarticular injection of autologous adipose-derived mesenchymal stem cells associated to PRGF-Endoret in osteoarthritic dogs. BMC Vet Res. 2013;9:131.

121. Cuervo B, Rubio M, Sopena J, Dominguez JM, Vilar J, Morales M, Cugat R, Carrillo JM. Hip osteoarthritis in dogs: a randomized study using mesenchymal stem cells from adipose tissue and plasma rich in growth factors. Int J Mol Sci. 2014;15:13437–60.

122. Marx C, Silveira MD, Selbach I, Da Silva AS, De Macedo Braga LMG, Camassola M, Nardi NB. Acupoint injection of autologous stromal vascular fraction and allogeneic adipose-derived stem cells to treat hip dysplasia in dogs. Stem Cells Int. 2014;2014:391274.

123. Vilar JM, Batista M, Morales M, Santana A, Cuervo B, Rubio M, Cugat R, Sopena J, Carrillo JM. Assessment of the effect of intraarticular injection of autologous adipose-derived mesenchymal stem cells in osteoarthritic dogs using a double

blinded force platform analysis. BMC Vet Res. 2014;10:143.

124. Guercio A, Di Marco P, Casella S, Cannella V, Russotto L, Purpari G, Di Bella S, Piccione G. Production of canine mesenchymal stem cells from adipose tissue and their application in dogs with chronic osteoarthritis of the humeroradial joints. Cell Biol Int. 2012;36:189–94.

125. Frisbie DD, Kisiday JD, Kawcak CE, Werpy NM, McIlwraith CW. Evaluation of adipose-derived stromal vascular fraction or bone marrow-derived mesenchymal stem cells for treatment of osteoarthritis. J Orthop Res. 2009;27:1675–80.

126. McIlwraith CW, Frisbie DD, Rodkey WG, Kisiday JD, Werpy NM, Kawcak CE, Steadman JR. Evaluation of intraarticular mesenchymal stem cells to augment healing of microfractured chondral defects. Art Ther. 2011;27:1552–61.

127. Ferris D, Frisbie DD, Kisiday J, McIlwraith CW, Hague B, Major MD, Schneider RK, Zubrod CJ, Watkins JJ, Kawcak CE, Goodrich LR. Clinical evaluation of bone marrow-derived mesenchymal stem cells in naturally occurring joint disease. Regen Med. 2009;4(Suppl 2):16.

128. Raheja LF, Galuppo LD, Bowers-Lepore J, Dowd JP, Tablin F, Yelowley CE. Treatment of bilateral medial femoral condyle articular cartilage fissures in a horse using bone marrow-derived multipotent mesenchymal stromal cells. J Equine Vet Sci. 2011;31:147–54.

129. Yamada ALM, Carvalho AD, Moroz A, Deffune E, Watanabe MJ, Hussni CA, Rodrigues CA, Alves ALG. Mesenchymal stem cell enhances chondral defects healing in horses. Stem Cell Disc. 2013;3(4):218–25.

130. Schnabel LV, Fortier LA, McIlwraith CW, Nobert KM. Therapeutic use of stem cells in horses: which type, how, and when? Vet J. 2013;197:570–7.

131. Ferris DJ, Frisbie DD, Kisiday JD, McIlwraith CW, Hague BA, Major MD, Schneider RK, Zubrod CJ, Kawcak CE, Goodrich LR. Clinical follow up of thirty-three horses treated for stifle injury with bone marrow derived mesenchymal stem cells intraarticularly. Vet Surg. 2014;43(3):255–65.

132. Haleem AM, Singergy AA, Sabry D, Atta HM, Rashed LA, Chu CR, El Shewy MT, Azzam A, Abdel Aziz MT. The clinical use of human culture-expanded autologous bone marrow mesenchymal stem cells transplanted on platelet-rich fibrin glue in the treatment of articular cartilage defects: a pilot study and preliminary results. Cartilage. 2010;1(4):253–61.

133. Buda R, Vannini F, Cavallo M, Grigolo B, Cenacchi A, Giannini S. Osteochondral lesions of the knee: a new one-step repair technique with bone-marrow-derived cells. J Bone Joint Surg Am. 2010;92(2):2–11.

134. Davatchi F, Abdollahi BS, Mohyeddin M, Shahram F, Nikbin B. Mesenchymal stem cell therapy for knee osteoarthritis. Preliminary report of four patients. Int J Rheum Dis. 2011;14(2):211–5.

135. Gigante A, Calcagno S, Cecconi S, Ramazzotti D, Manzotti S, Enea D. Use of collagen scaffold and autologous bone marrow concentrate as a one-step cartilage repair in the knee: histological results of second-look biopsies at 1 year follow-up. Int J Immunopathol Pharmacol. 2011;24(2):69–72.

136. Emadedin M, Aghdami N, Taghiyar L, Fazeli R, Moghadasali R, Jahangir S, Farjad R, Baghaban EM. Intraarticular injection of autologous mesenchymal stem cells in six patients with knee osteoarthritis. Arch Iran Med. 2012;15(7):422–8.

137. Mardones R, Jofre CM, Tobar L, Minguell JJ. Mesenchymal stem cell therapy in the treatment of hip osteoarthritis. J Hip Preservation Surg. 2017;4(2):159–63.

138. Wakitani S, Mitsuoka T, Nakamura N, Toritsuka Y, Nakamura Y, Horibe S. Autologous bone marrow stromal cell transplantation for repair of full-thickness articular cartilage defects in human patellae: two case reports. Cell Transplant. 2004;13(5):595–600.

139. Adachi N, Ochi M, Deie M, Ito Y. Transplant of mesenchymal stem cells and hydroxyapatite ceramics to treat severe osteochondral damage after septic arthritis of the knee. J Rheumatol. 2005;32(8):1615–8.

140. Kuroda R, Ishida K, Matsumoto T, Akisue T, Fujioka H, Mizuno K, Ohgushi H, Wakitani S, Kurosaka M. Treatment of a full-thickness articular cartilage defect in the femoral condyle of an athlete with autologous bone-marrow stromal cells. Osteoarthr Cartil. 2007;15(2):226–31.

141. Wakitani S, Nawata M, Tensho K, Okabe T, Machida H, Ohgushi H. Repair of articular cartilage defects in the patellofemoral joint with autologous bone marrow mesenchymal cell transplantation: three case reports involving nine defects in five knees. J Tissue Eng Regen Med. 2007;1(1):74–9.

142. Centeno CJ, Busse D, Kisiday J, Keohan C, Freeman M, Karli D. Increased knee cartilage volume in degenerative joint disease using percutaneously implanted, autologous mesenchymal stem cells. Pain Physician. 2008;11(3):343–53.

143. Kasemkijwattana C, Hongeng S, Kesprayura S, Rungsinaporn V, Chaipinyo K, Chansiri K. Autologous bone marrow mesenchymal stem cells implantation for cartilage defects: two cases report. J Med Assoc Thai. 2011;94(3):395–400.

144. Pak J. Regeneration of human bones in hip osteonecrosis and human cartilage in knee osteoarthritis with autologous adipose-tissue-derived stem cells: a case series. J Med Case Reports. 2011;5:296.

145. Giannini S, Buda R, Vannini F, Cavallo M, Grigolo B. One-step bone marrow-derived cell transplantation in talar osteochondral lesions. Clin Orthop Relat Res. 2009;467(12):3307–20.

146. Nejadnik H, Hui JH, Feng Choong EP, Tai BC, Lee EH. Autologous bone marrow-derived mesenchymal stem cells versus autologous chondrocyte implantation: an observational cohort study. Am J Sports Med. 2010;38(6):1110–6.

147. Varma HS, Dadarya B, Vidyarthi A. The new avenues in the management of osteoarthritis of knee-stem cells. J Indian Med Assoc. 2010;108:583–5.

148. Koh YG, Choi YJ. Infrapatellar fat pad-derived mesenchymal stem cell therapy for knee osteoarthritis. Knee. 2012;19(6):902–7.

149. Giannini S, Buda R, Cavallo M, Ruffilli A, Cenacchi A, Cavallo C, Vannini F. Cartilage repair evolution in post-traumatic osteochondral lesions of the talus: from open field autologous chondrocyte to bone-marrow-derived cells transplantation. Injury. 2010;41(11):1196–203.

Sox9 Potentiates BMP2-Induced Chondrogenic Differentiation and Inhibits BMP2-Induced Osteogenic Differentiation

19

Junyi Liao, Ning Hu, Nian Zhou, Chen Zhao, Xi Liang, Hong Chen, Wei Xu, Cheng Chen, Qiang Cheng, and Wei Huang

19.1 Introduction

From degenerative disorders to traumatic injuries, cartilaginous pathologies present a very significant clinical challenge to the medical fraternity especially due to its lack of regenerative capabilities [1]. To overcome this drawback, many surgical interventions were applied. Of these methods, the two important surgical methods designed to promote cartilage repair were the bone marrow stimulation techniques and restoration techniques [2–4]. But the bone marrow stimulation techniques such as microfracture and drilling produce fibrocartilage with insufficient long-term effects [5, 6]. Restoration techniques such as autologous chondrocyte implantation and osteochondral allograft were limited by insufficient cell supply, damage to the donor site, and immunological reactions [7, 8]. As a consequence, stem cell-based and gene-enhanced tissue engineering cartilage is considered to be more promising in the treatment of cartilaginous pathologies [7, 9].

Mesenchymal stem cells (MSCs) can undergo self-replenishment and have the potential to differentiate into multiple lineages, including osteo-genic, chondrogenic, and adipogenic lineages [9–12]. Due to the abundant source, easy isolation, and stable expression of the exogenous genes, MSCs have been regarded as ideal seed cells for scientific research on cartilage tissue engineering [12]. Many studies have reported that different growth factors such as BMPs [13], FGFs [14, 15], IGF1 [16], and TGF-β [17, 18] have been identified for their ability to direct MSCs towards the chondrocyte phenotypes [19]. However, the use of these growth factors is still disputable, because of their limited ability towards the synthesis of specific cartilage matrix components. Hence, optimizing a chondrogenic growth factor and amplifying its specific chondrogenic ability are among the most crucial and key steps in the process of cartilage tissue engineering.

Bone morphogenetic protein 2 (BMP2), belonging to the transforming growth factor beta (TGF-β) superfamily, is known to induce human bone mesenchymal stem cells (hBMSCs) [14], adipose-derived stem cells (ADSCs) [16], mouse embryonic fibroblasts (MEFs) [20], and chondrogenic differentiation, and it promotes MSC condensation, chondrogenic differentiation, chondrocyte proliferation, and hypertrophic differentiation [13]. BMP2 has a greater potential to induce chondrogenic differentiation of MSCs compared with other growth factors such as TGF-β and IGF1 [21, 22]. However, BMP2 is also known to induce MSC osteogenic differentiation and stimulates endochondral ossification [23, 24]. Thus, potentiating

J. Liao · N. Hu · N. Zhou · C. Zhao · X. Liang
H. Chen · W. Xu · C. Chen · Q. Cheng · W. Huang (✉)
Department of Orthopaedic Surgery, The First Affiliated Hospital of Chongqing Medical University, Chongqing, China
e-mail: huangwei68@263.net

© Springer Nature Switzerland AG 2019
D. Duscher, M. A. Shiffman (eds.), *Regenerative Medicine and Plastic Surgery*,
https://doi.org/10.1007/978-3-030-19962-3_19

BMP2-induced MSC chondrogenic differentiation and inhibiting BMP2-induced MSC osteogenic differentiation may play a vital role in chondrogenesis and cartilage formation.

In this study, we made an attempt to investigate whether overexpression of Sry-related transcription factor Sox9 could potentiate BMP2-induced chondrogenic differentiation of MSCs, and inhibit osteogenic differentiation of MSCs. Sox9 is a transcription factor belonging to the Sry-related high-mobility group box (Sox) protein family [25]. It is essential for chondrogenesis of MSCs [26–28], since it is considered as the key transcription factor for BMP2-induced chondrogenesis [20]. It is also reported that Sox9 inhibits the transactivation of Runt-related transcription factor 2 (Runx2) [29], which is a key transcription factor for osteogenesis and endochondral ossification [30, 31]. However, it is not clear whether Sox9-mediated inhibition of osteogenic differentiation signaling plays any role in the BMP2-induced differentiation of MSCs. Nevertheless, we found that BMP2-induced Sox9 expression was transient and relatively at a lower level during the early stages of MSC differentiation. Exogenous overexpression of Sox9 enhanced the BMP2-induced chondrogenic differentiation and marker expression, and inhibited BMP2-induced osteogenic differentiation and marker expression. In stem cell implantation studies, Sox9 was shown to potentiate BMP2-induced cartilage formation, and inhibit endochondral ossification during ectopic bone/cartilage formation. Through perinatal limb explant culture, we demonstrated that Sox9 and BMP2 synergistically promoted chondrocyte condensation and proliferation. However, Sox9 inhibited BMP2-induced chondrocyte hypertrophy, and ossification. Our findings strongly suggest that overexpression of Sox9 in BMP2-induced MSC differentiation may become a new strategy for cartilage tissue engineering.

19.2 Materials and Methods

19.2.1 Ethics Statement

The experimental protocols were approved by the Ethical Committee of the First Affiliated Hospital of Chongqing Medical University. All animal protocols were approved by Ethical Committee of the First Affiliated Hospital of Chongqing Medical University. All surgery was performed under sodium pentobarbital anesthesia, and all efforts were made to minimize suffering.

19.2.2 Cell Culture and Chemicals

The HEK 293 and C3H10T1/2 cell lines were obtained from ATCC (Manassas, VA). Cell lines were preserved in complete Dulbecco's modified Eagle's medium (DMEM, Hyclone, China), supplemented with 10% fetal bovine serum (FBS, Gibco, Australia), 100 U/mL penicillin, and 100 mg/mL streptomycin, maintained at 37 °C in a humidified 5% carbon dioxide (CO_2) atmosphere. Unless indicated otherwise, all chemicals were purchased from Sigma-Aldrich or Corning.

19.2.3 Recombinant Adenoviruses Expressing GFP, BMP2, and Sox9

Recombinant adenoviruses were generated using AdEasy technology as described previously [32, 33]. The coding regions of GFP, BMP2, and Sox9 were amplified with PCR, and cloned into adenoviral shuttle vectors. Then the vectors were used to generate recombinant adenoviruses in HEK 293 cells. The resulting adenoviruses were designated as AdGFP, AdBMP2, and AdSox9. AdGFP was used as a vector control.

19.2.4 Chondrogenic Differentiation of MSCs in Micromass Culture

The C3H10T1/2 cells transduced with AdGFP, AdBMP2, and/or AdSox9 were cultured in 100 mm dishes. At 80% confluence, cells were harvested and resuspended in the high-density (10^5 cells in a 10 μL drop of media) culture medium. Then the medium (50 μL in each well) was added at the center of each well in the 12-well plates. The plates were then carefully transferred to CO_2 incubators and incubated for 2 h. About 2–3 mL of the medium was added to each well. Fresh medium was added to the wells every 4–5 days. Chondrogenic assays were carried out at desired time points.

19.2.5 Osteogenic Differentiation of MSCs in Monolayer Culture

The C3H10T1/2 cells were seeded in 6-well or 24-well plates, at 40–50% confluence. Cells were infected with AdBMP2 and/or AdSox9, and AdGFP was used as control. Fresh medium was added to the wells every 4–5 days. Cells were harvested at desired time points for subsequent analysis.

19.2.6 Alcian Blue Staining for Micromass Pellet

Micromass cell pellet was washed with phosphate-buffered saline (PBS), treated with 4% paraformaldehyde for 30 min, and again washed with PBS. The pellet was stained with 0.5% Alcian blue in 0.1 M HCl (pH 1.0) for 12 h, and washed with distilled water. The pellet was then photographed with Nikon microscope. The Alcian blue-stained cultures were extracted at room temperature using the 6 M guanidine hydrochloride. Optical density (OD) of the extracted dye was measured at 630 nm in a microplate reader (Bio-Rad, America). All the experiments were performed in triplicate. All the results are represented as mean ± standard deviation (SD).

19.2.7 RNA Isolation and Semiquantitative RT-PCR

The total RNA of the cells was isolated using TRIZOL reagent (Invitrogen, USA) according to the manufacturer's instructions. Reverse transcription reactions were performed using PrimeScript RT reagent kit (Takara, Dalian, China) according to the manufacturer's instructions. The first-strand cDNA products were further diluted ten times and used as PCR templates. Semiquantitative PCR was performed using TaKaRa Ex Taq (Takara, Dalian, China), a touchdown cycling program, which is as follows: 95 °C for 3 min for 1 cycle; 95 °C for 30 s, 58 °C for 30 s, and 72 °C for 13 cycles and decreasing 0.5 °C per cycle; and then at 95 °C for 30 s, 58 °C for 30 s, and 72 °C for 30 s for 20–25 cycles; 72 °C for 7 min; and finally hold at 4 °C. PCR products were resolved on 1.5% agarose gels. All sample values were normalized to GAPDH expression. The primer sequences used for this analysis are listed in Table 19.1.

19.2.8 Real-Time PCR

The total RNA of the cells was isolated using TRIZOL reagent (Invitrogen, USA) according to the manufacturer's instructions. Reverse transcription reactions were performed using iScript cDNA synthesis kit (Bio-Rad, USA) according to the manufacturer's instructions. Real-time PCR was performed using SsoAdvanced SYBR Green Supermix (Bio-Rad, USA). The conditions maintained for real-time PCR are as follows: 95 °C for 3 min for 1 cycle; 95 °C for 10 s and 58 °C for 5 s for 40 cycles. Dissociation stage was applied at the end of the amplification procedure. The dissolve curve did not determine any nonspecific amplification. All sample values were normalized to GAPDH expression by using the $2^{-\triangle\triangle Ct}$ method. The primer sequences used for this analysis are listed in Table 19.1.

19.2.9 Immunocytochemical Staining

The C3H10T1/2 cells were infected with AdGFP, AdBMP2, and/or AdSox9. At desired time points, the cells were treated with 4% paraformaldehyde for 30 min at room temperature. After washing with PBS, the fixed cells were treated with 1% NP-40 and 10% goat serum, to render the cells more permeable. Following this step, the cells were incubated with primary antibodies against osteopontin (OPN) (Santa Cruz Biotechnology, USA) at 4 °C overnight. These cells were again

Table 19.1 Primer oligonucleotide sequences used for PCR

Gene	Forward primer (5′-3′)	Reverse primer (5′-3′)	Product size
Aggrecan	TGGCTTCTGGAGACAGGACT	TTCTGCTGTCTGGGTCTCCT	188 bp
BMP2	ACCAGACTATTGGACACCAG	AATCCTCACATGTCTCTTGG	174 bp
Col2a1	CAACACAATCCATTGCGAAC	TCTGCCCAGTTCAGGTCTCT	159 bp
GAPDH	CTACACTGAGGACCAGGTTGTCT	TTGTCATACCAGGAAATGAGCTT	123 bp
Osteocalcin	CCTTCATGTCCAAGCAGGA	GGCGGTCTTCAAGCCATAC	161 bp
Osteopontin	CCTCCCGGTGAAAGTGAC	CTGTGGCGCAAGGAGATT	124 bp
Runx2	CCGGTCTCCTTCCAGGAT	GGGAACTGCTGTGGCTTC	122 bp
Sox9	AGCTCACCAGACCCTGAGAA	TCCCAGCAATCGTTACCTTC	200 bp

washed with PBS, and were incubated with biotin-containing secondary antibodies for 30 min, followed by incubation with streptavidin-labeled horseradish peroxidase for 15 min at room temperature. The presence of the expected protein was visualized by DAB staining. Stains with IgG were used as negative controls. The results were repeated in at least three independent experiments.

19.3 Western Blot Analysis

The cell lysates were prepared using cell lysis buffer containing a protease inhibitor PMSF (Beyotime, Shanghai, China). About 60 μg of total protein for each sample was loaded onto 10% SDS-PAGE and transferred to PVDF membrane. The membrane was incubated overnight with antibodies against Sox9 (Santa Cruz Biotechnology, USA), collagen type II alpha 1 (Col2a1) (Santa Cruz Biotechnology, USA), OPN (Santa Cruz Biotechnology, USA), osteocalcin (OC) (Santa Cruz Biotechnology, USA), and β-actin (Bioworld Technology, USA) at a dilution of 1:500 or 1:1000, respectively. Following this, the membrane was again incubated with a secondary antibody conjugated with horseradish peroxidase (Earthox, USA). Immune-reactive signals were detected using ECL kit (Millipore, USA).

19.4 Alkaline Phosphatase (ALP) Assay

The ALP activities were assessed using the modified Great Escape SEAP chemiluminescence assay (BD Clontech) and/or histochemical stain-ing, as described previously [34, 35]. For ALP histochemical staining, the cells were induced for osteogenic differentiation using AdGFP, AdBMP2 and/or Sox9, and DMSO (solvent control) as control. Infected cells were fixed with fixative solution (two volumes of citrate working solution to three volumes of acetone) at room temperature for 30 s. After washing with distilled water, cells were stained subjected to histochemical staining with a mixture of 0.1 mg/mL of naphthol AS-MX phosphate and 0.6 mg/mL of Fast Blue BB salt. Histochemical staining was recorded using bright-light microscopy.

For the chemiluminescence assays, each assay condition was performed in triplicate, and the results were repeated in at least three independent experiments. The ALP activity was normalized by total cellular protein concentrations among the samples.

19.5 Matrix Mineralization Assay (Alizarin Red S Staining)

The C3H10T1/2 cells were seeded in 24-well plates and infected with AdGFP, AdBMP2, and/or AdSox9. Infected cells were cultured in the presence of ascorbic acid (50 mg/mL) and β-glycerophosphate (10 mM). On day 14 after infection, mineralized matrix nodules were stained for calcium precipitation by means of alizarin red S staining, as described previously [34–36]. Briefly, cells were treated with 4% paraformaldehyde for 30 min. After washing with PBS, cells were incubated with 2% alizarin red S for 30 min, followed by extensive washing with distilled water. The staining of calcium mineral deposits was recorded using bright-light microscopy.

19.6 Subcutaneous Stem Cell Implantation (Ectopic Cartilage/Bone Formation Assay)

The C3H10T1/2 cells were infected with AdGFP, AdBMP2, and/or AdSox9. Twenty-four hours after infection, cells were harvested and resuspended in PBS containing 300 U/mL penicillin, and 300 mg/mL streptomycin for subcutaneous injection (5×10^6 per injection) into the flanks of athymic nude (nu/nu) mice (three animals per group, 4- to 6-week-old males, Experimental Animal Center, Chongqing Medical University, Chongqing, China). Animals were euthanized, and the implantation sites were retrieved for histological and other staining evaluations at 5 and 8 weeks, respectively.

19.7 Mouse Fetal Limb Explant Culture

The forelimbs of mouse embryos (E18.5) were dissected under sterile conditions and incubated in DMEM (Hyclone, China) containing 0.5% FBS (Gibco, Australia), 50 mg/mL ascorbic acid, 1 mM β-glycerophosphate, 100 U/mL penicillin, and 100 µg/mL streptomycin at 37 °C in humidified air with 5% CO_2 for up to 14 days, as described previously [37, 38]. The limbs were infected by AdGFP, AdBMP2, and/or AdSox9 directly, 12 h after dissection. About 50% of the medium was replaced every 2–3 days. Cultured tissues were observed at different time points under the microscope to confirm the survival of tissue cells and the expression of fluorescence markers.

19.8 Histologic Evaluation: Hematoxylin and Eosin, Masson's Trichrome, Alcian Blue, and Safranin O-Fast Green Staining

Retrieved and cultured tissues were treated with 10% formalin, decalcified, and embedded in paraffin. Paraffin-embedded sections were deparaffinized and then rehydrated in a graduated fashion. The deparaffinized samples were subjected to antigen retrieval and fixation. The sections were stained with hematoxylin and eosin (H&E), Masson's trichrome, Alcian blue, and Safranin O-fast green. Histological evaluation was performed using a light microscope (Nikon).

19.9 Statistical Analysis

All quantitative experiments were performed in triplicate and/or repeated three times. Data were expressed as mean ± standard deviation (SD). The one-way analysis of variance was used to analyze statistical significance. A value of $P < 0.05$ was considered statistically significant.

19.10 Results

19.10.1 Gene Transduction of MSCs and Expression of Sox9 in Each Treatment Groups

The C3H10T1/2 cells were infected by recombinant adenovirus expression of GFP, BMP2, and/or Sox9 successfully both in monolayer and micromass culture (Fig. 19.1). The BMP2 and Sox9 mRNA were upregulated at 36 h after BMP2 and Sox9 transduction using semiquantitative RT-PCR analysis. Expression of Sox9 after infection was evaluated by western blot technique on days 2, 5, and 7 after incubation in a normal medium without any supplemented growth factors in micromass culture. Some studies have reported that Sox9 could be upregulated by BMP2 [13, 20]. We found that Sox9 expression induced by BMP2 was time dependent and showed a highest level on day 5. When combined with Sox9, it showed a significant earlier onset of overexpression of Sox9 (day 2) and maintained significant values at high level from days 2 to 7. These results suggest an efficient and relatively sustained overexpression of Sox9 in BMP2 and Sox9 coinfected C3H10T1/2 cells.

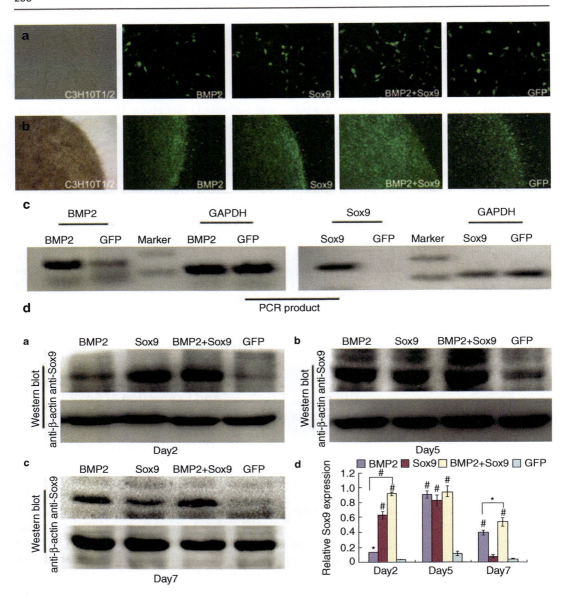

Fig. 19.1 Gene transduction and Sox9 expression in each treatment group. (**A, B**) Bright-light and fluorescence microscope examination showed the transduction efficiency of recombinant adenoviruses in monolayer culture (24 h after transduction, ×100) and micromass culture (3 days after transduction, ×40), respectively. (**C**) Recombinant adenovirus-mediated overexpression of BMP2 and Sox9 mRNA was evaluated by semiquantitative RT-PCR analysis using GAPDH as a housekeeping gene. (**D**) Sox9 expression was evaluated by western blot analysis in each treatment group at days 2, 5, and 7 after transduction (**a** to **c**) and relative Sox9 expression was analyzed by Quantity One software using β-actin as controls (**d**); the results are expressed as mean ± SD of triplicate experiments, $*P < 0.05$, $^{\#}P < 0.01$

19.10.2 Sox9 Potentiates BMP2-Induced Chondrogenic Differentiation of MSCs in Micromass Cultures In Vitro

Since cell density is necessary for chondrogenic differentiation of MSCs, micromass culture could provide cell-to-cell contact in 3-dimensions (3D), which is similar to MSC condensation to induce chondrogenesis in vivo [39]. So, micromass cultures were used to evaluate the influence of exogenous overexpression of Sox9 in BMP2-induced chondrogenic differentiation of MSCs in vitro.

Glycosaminoglycans and collagen type II alpha 1 (Col2a1) are the two main markers of chondrocyte and formed cartilage matrix. Therefore, we evaluated the two markers in micromass culture in vitro. As shown in Fig. 19.2, aggrecan (ACAN) mRNA was upregulated by BMP2 ($P < 0.05$), and Sox9 enhanced this effect ($P < 0.01$) on days 7 and 14, respectively. Alcian blue staining was used to detect the sulfated glycosaminoglycans on days 7 and 14. Gross observation and microscopic examination are shown in Fig. 19.2. Alcian blue staining quantification was also used to evaluate the chondrogenic differentiation of MSCs.

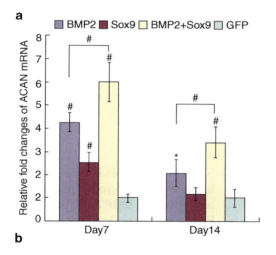

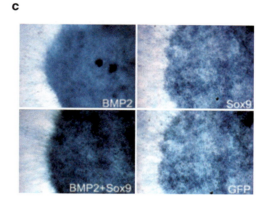

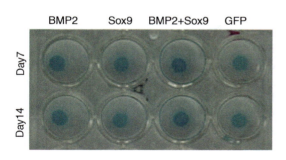

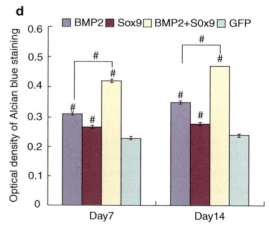

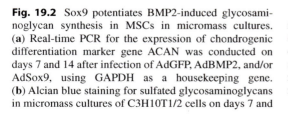

Fig. 19.2 Sox9 potentiates BMP2-induced glycosaminoglycan synthesis in MSCs in micromass cultures. (**a**) Real-time PCR for the expression of chondrogenic differentiation marker gene ACAN was conducted on days 7 and 14 after infection of AdGFP, AdBMP2, and/or AdSox9, using GAPDH as a housekeeping gene. (**b**) Alcian blue staining for sulfated glycosaminoglycans in micromass cultures of C3H10T1/2 cells on days 7 and 14 after transduction of indicated recombinant adenoviruses, gross observation. (**c**) Microscope examination (×40). (**d**) Alcian blue staining quantifying: cells were extracted with 6 M guanidine hydrochloride; optical density of the extracted dye was measured at 630 nm. The results were expressed as mean ± SD of triplicate experiments, *$P < 0.05$, #$P < 0.01$

The synergistic effect of Sox9 on BMP2-induced chondrogenic differentiation was observed on days 7 and 14, respectively ($P < 0.01$).

Col2a1 is one of the most important molecular markers for chondrogenesis. We evaluated the expression of Col2a1 on both mRNA and protein levels at continuous time points (Fig. 19.2). While exogenous overexpression of Sox9 alone did not exert any significant effect on Col2a1 mRNA and protein expression, Sox9 was shown to exhibit a synergistic effect on BMP2-induced Col2a1 expression on both mRNA and protein levels (Fig. 19.3). Interestingly, BMP2 failed to upregulate Col2a1 mRNA and protein expression at early stage (days 3 and 5) of MSC differentiation ($P > 0.05$). However, when combined with Sox9, they showed significant earlier onset of expression of Col2a1 (day 3), and they maintained their significant values at high level at all time points (from days 3 to 14) on both mRNA and protein levels ($P < 0.01$).

These results show that exogenous overexpression of Sox9 potentiates BMP2-induced chondrogenic differentiation of MSCs in vitro persistently.

19.10.3 Sox9 Inhibits BMP2-Induced Osteogenic Differentiation of MSCs In Vitro

BMP2 also induces MSC osteogenic differentiation and upregulates osteogenic markers at the late stage of chondrogenesis. We studied the effect of exogenous overexpression of Sox9 on BMP2-induced MSC osteogenic differentiation. ALP activity is an early marker of osteogenic differentiation. ALP activity was measured on days 7 and 9 after transduction. As expected, BMP2 induced a significant increase in ALP activity ($P < 0.01$), especially on day 9 (Fig. 19.4). However, when coinfected with Sox9, ALP activity decreased dramatically both on days 7 and 9 ($P < 0.01$). This indicates that Sox9 inhibits BMP2-induced early osteogenic differentiation.

Moreover, we analyzed the effect of Sox9 on late osteogenic markers OPN using immunocytochemical staining, and found that Sox9 inhibited BMP2-induced OPN expression on day 11 after transduction (Fig. 19.4). We also demonstrated that Sox9 inhibited BMP2-induced matrix mineralization using alizarin red staining on day 14 after transduction. These indicate that Sox9 inhibits BMP2-induced late osteogenic differentiation of MSCs.

We further determined the effect of Sox9 on BMP2-induced osteogenic gene and protein expression. Runx2 is the key transcription factor of osteogenesis, and functions at the early stage of osteogenic differentiation. We found that Sox9 inhibited Runx2 mRNA expression at early stage (days 2 and 3) of MSC differentiation (Fig. 19.5), and similar results were detected on protein level ($P < 0.01$). We also detected that Sox9 inhibited BMP2-induced late osteogenic differentiation markers OPN and OC at both mRNA and protein levels on days 7 and 14 ($P < 0.01$).

These results show that Sox9 inhibits BMP2-induced early and late osteogenic differentiation of MSCs in vitro.

19.10.4 Sox9 Potentiates BMP2-Induced Cartilage Formation and Inhibits BMP2-Induced Endochondral Ossification in MSC Implantation In Vivo

While the in vitro studies established that exogenous overexpression of Sox9 potentiates BMP2-induced chondrogenesis and inhibits BMP2-induced osteogenesis, it was imperative to demonstrate if overexpression of Sox9 plays such a role in vivo. Using our previously established stem cell implantation assay [35, 37, 40], we injected C3H10T1/2 cells infected with AdGFP, AdBMP2, and/or AdSox9 at the same infection ratio subcutaneously into the flanks of athymic nude (nu/nu) mice for 5 weeks and 8 weeks, respectively. The cells transduced with AdGFP or AdSox9 alone failed to form any detectable masses (data not shown). The BMP2 and Sox9 coinfected cells formed cartilaginous/bony masses, which were noticeably smaller than those formed by cells infected by BMP2 alone (Fig. 19.6). On histological examination, masses

19 Sox9 Potentiates BMP2-Induced Chondrogenic Differentiation and Inhibits BMP2-Induced... 271

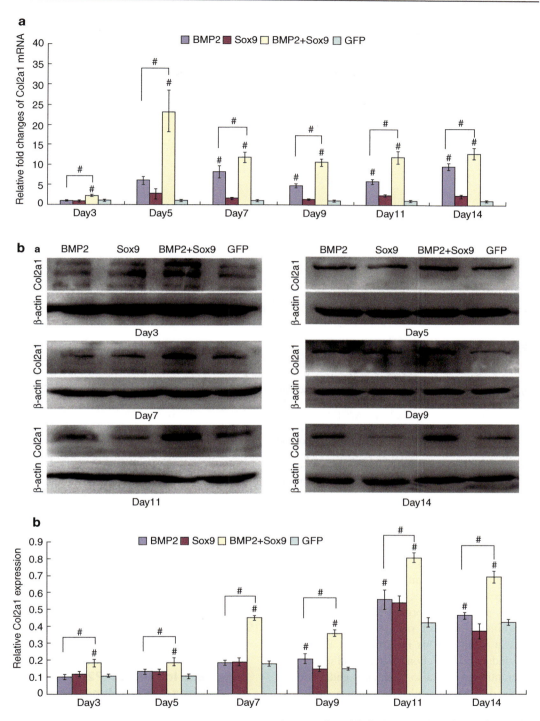

Fig. 19.3 Sox9 potentiates BMP2-induced Col2a1 synthesis in MSCs in micromass cultures. (**A**) Real-time PCR for the expression of chondrogenic differentiation marker gene Col2a1 was conducted at continuous time points (from days 3 to 14) after infection of AdGFP, AdBMP2, and/or AdSox9, using GAPDH as a housekeeping gene. (**B**) Western blot for the expression of Col2a1 was conducted at continuous time points (from days 3 to 14) after transduction of indicated recombinant adenoviruses (**a**); quantitatively, relative Sox9 expression was analyzed by Quantity One software using β-actin as controls (**b**). The results are expressed as mean ± SD of triplicate experiments, ∗$P < 0.05$, #$P < 0.01$

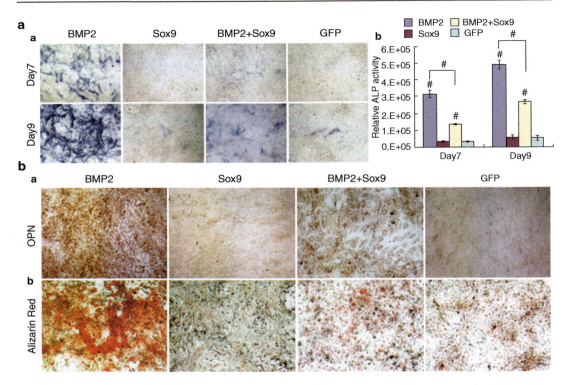

Fig. 19.4 Sox9 inhibits BMP2-induced early and late osteogenic differentiation of MSCs in vitro. (**A**) C3H10T1/2 cells were infected with AdGFP, AdBMP2, and/or AdSox9. The ALP activities were measured on days 7 and 9 using ALP histochemical staining (**a**), and chemiluminescent assays (**b**). (**B**) C3H10T1/2 cells were infected with indicated recombinant adenoviruses. On day 11 after infection, the expression of osteopontin (OPN) was assayed by immunocytochemical staining using anti-OPN antibody (**a**). For matrix mineralization, C3H10T1/2 cells were infected with indicated recombinant adenoviruses and cultured in mineralization medium. Alizarin red staining was conducted on day 14 after infection (**b**)

Each assay was done in triplicate. ALP assay results are expressed as mean ± SD, $*P < 0.05$, $^{\#}P < 0.01$

formed in BMP2 transduced groups showed both bony and cartilaginous component. Masses in BMP2 transduced cell groups showed some mature bone matrices and trabeculae with the presence of a significant number of chondrocytes at week 5. The trabeculae turned thicker and more bone matrices formed with the presence of a significant number of hypertrophy chondrocytes at week 8. This indicates that BMP2 not only induces MSC chondrogenic and osteogenic differentiation, but also stimulates endochondral ossification. On the other hand, masses formed by BMP2 and Sox9 coinfected cells showed a large number of chondrocytes with no obvious trabecula formation both at weeks 5 and 8. This indicates that Sox9 inhibits BMP2-induced osteogenesis and endochondral ossification. Masson's trichrome, Alcian blue, and safranin O-fast green staining were also used to analyze the components of the masses, which confirmed that Sox9 potentiates BMP2-induced cartilage formation and inhibits BMP2-induced osteogenesis and endochondral ossification.

19.10.5 Sox9 Promotes Expansion of the Proliferating Chondrocyte Zone and Inhibits BMP2-Induced Chondrocyte Hypertrophy and Ossification in Fetal Limb Explant Cultures

After the in vitro and in vivo tests, we also explored the effect of Sox9 on skeletal development using the fetal limb culture assay. The skinned fetal limbs

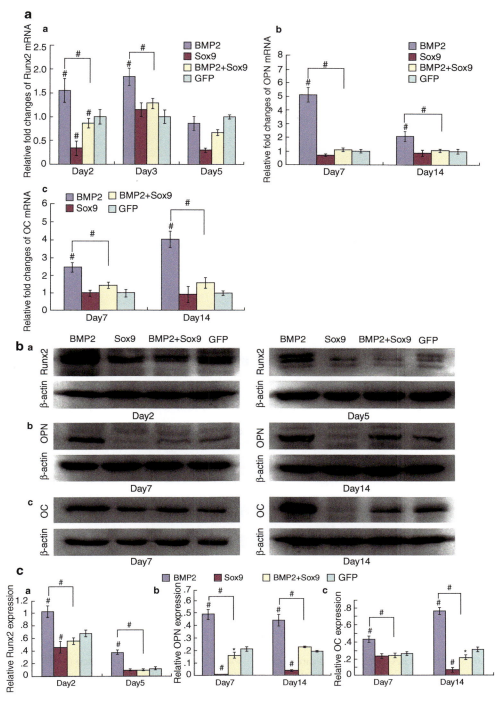

Fig. 19.5 Sox9 inhibits BMP2-induced osteogenic marker expression in MSCs in vitro. (**A**) Real-time PCR for the expression of early osteogenic differentiation gene Runx2 (**Aa**), late osteogenic gene osteocalcin (OC), and osteopontin (OPN) (**b, c**) was conducted at indicated time points after infection with AdGFP, AdBMP2, and/or AdSox9, using GAPDH as a housekeeping gene. (**B**) Western blot for the expression of Runx2 (**a**), OPN (**b**), and OC (**c**) was conducted at indicated time points after transduction of indicated recombinant adenoviruses, respectively. (**C**) Relative protein expression was analyzed by Quantity One software using β-actin as controls, respectively. The results are expressed as mean ± SD of triplicate experiments, *$P < 0.05$, #$P < 0.01$

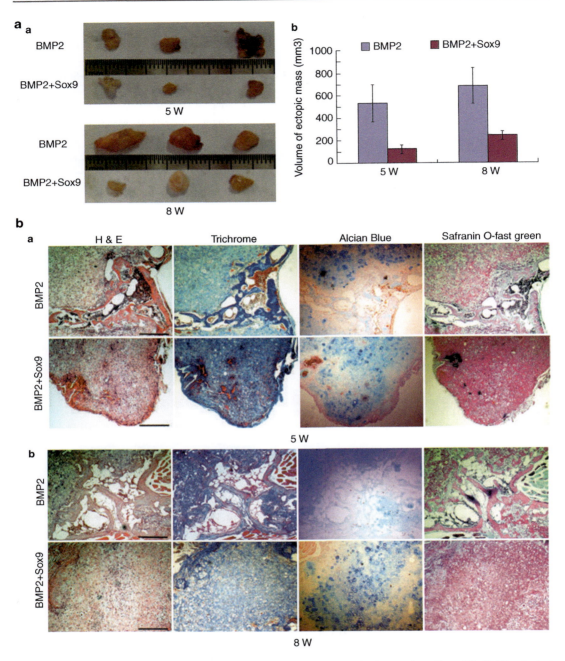

Fig. 19.6 Sox9 potentiates BMP2-induced cartilage formation and inhibits BMP2-induced endochondral ossification in MSC implantation in vivo. (**A**) Macrographic images of ectopic masses. BMP2 or BMP2 and Sox9 coinfected C3H10T1/2 cells were implanted subcutaneously to the flanks of nude mice. Ectopic masses were retrieved at 5 and 8 weeks (**a**). The volume of the masses was determined using vernier calipers (**b**). (**B**) Histological analysis of the retrieved samples. The retrieved samples were fixed, decalcified, paraffin embedded, and subjected to H&E, Masson's trichrome, Alcian blue, and safranin O-fast green staining. Representative images are shown, magnification ×100, scale bar = 1 mm

were isolated from mouse E18.5 perinatal embryos and cultured in the organ culture medium in the presence of AdGFP, AdBMP2, and/or AdSox9 for 14 days. The limbs were infected with indicated recombinant adenoviruses effectively at day 5 (Fig. 19.7). On histological examination, both BMP2 and Sox9 induced chondrocyte proliferation and condensation. However, only BMP2 induced chondrocyte hypertrophy and ossification. When the limbs were coinfected with AdBMP2 and AdSox9, the proliferating chondrocyte zone was expanded with no obvious expansion of hypertrophic chondrocyte zone. Quantitative analysis of the histologic data also indicated combined treatment of BMP2 and Sox9 had the largest length of proliferating chondrocyte zone ($P < 0.01$), while BMP2 alone exhibited the largest length of hypertrophic chondrocyte zone ($P < 0.01$). These results suggest that BMP2 and Sox9 act synergistically to induce chondrocyte/chondroblast proliferation and condensation, and Sox9 inhibits BMP2-induced chondrocyte hypertrophy and ossification in fetal limb explant culture.

19.11 Discussion

Gene-enhanced tissue engineering cartilage is a promising strategy for cartilaginous pathologies. However, it is very difficult to acquire seed cells with sufficient biological activity [41, 42]. We demonstrated here that exogenous overexpression of Sox9 significantly enhanced BMP2-induced chondrogenic differentiation and cartilage formation, while it also inhibited BMP2-induced osteogenic differentiation and endochondral ossification. Thus, overexpression of Sox9 in BMP2-induced chondrogenic differentiation may result in stable chondrogenic phenotype of MSCs.

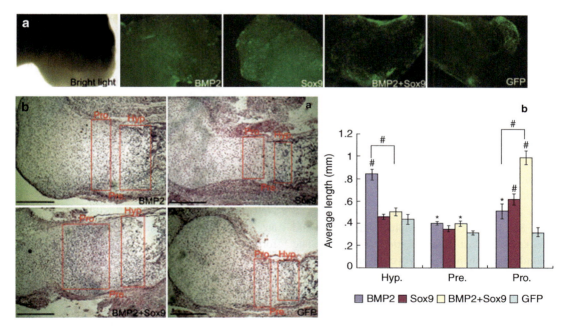

Fig. 19.7 Sox9 promotes expansion of the proliferating chondrocyte zone and inhibits BMP2-induced chondrocyte hypertrophy and ossification in organ cultures. (**A**) Mouse E18.5 forelimbs ($n = 4$ each group) were harvested and transduced with AdGFP, AdBMP2, and/or AdSox9. The forelimbs were cultured in organ culture medium and the transduction efficiency was visualized under bright-light and fluorescence microscope (×40). (**B**) Histological analysis of the cultured forelimbs. The forelimbs were fixed, decalcified, paraffin embedded, and subjected to H&E staining. Representative images are shown (**a**), magnification ×100. The average lengths of the hypertrophic zones, prehypertrophic zones, and proliferating zones were also determined by using ImageJ software (**b**). *Hyp* hypertrophic chondrocyte zone, *Pre* prehypertrophic chondrocyte zone, *Pro* proliferating chondrocyte zone. ∗$P < 0.05$, #$P < 0.01$, scale bar = 1 mm

Recombinant human bone morphogenetic protein 2 (rhBMP-2) has been approved for treating acute, open tibial shaft fractures by FDA [43]. BMP2 also exhibits high chondrogenic activity at early stage of endochondral ossification of MSCs. However, BMP2-induced osteogenic activity and endochondral ossification affect the maintenance of hyaline cartilage phenotype [44, 45]. Such limitations can be overcome by inhibiting BMP2-induced osteogenic differentiation and endochondral ossification of MSCs through directing BMP2-stimulated MSCs towards chondrogenic lineage with chondrogenic factors, such as Sox9. We found that BMP2 showed both high chondrogenic and osteogenic differentiation activity in vitro. Interestingly, overexpression of Sox9 enhanced BMP2-induced chondrogenesis and inhibited BMP2-induced osteogenesis (Figs. 19.2, 19.3, 19.4, and 19.5). We further evaluate the effect of overexpression of Sox9 on BMP2-induced MSC differentiation in vivo; the MSC subcutaneous implantation assays showed that BMP2 induced MSC chondrogenic differentiation, and mediated endochondral ossification in a time-dependent manner, which is in agreement with the previous studies [45–47]. When transduction with Sox9 occurred, endochondral ossification was inhibited, and hyaline cartilage was maintained (Fig. 19.6). Finally, we explored the effect of Sox9 on skeletal development in fetal limb culture assay. Fetal limb culture could mimic the progress of chondrogenesis and endochondral ossification in vitro [37, 38, 48]. Similar with the effect of Sox9 in vivo, which showed that Sox9 directs hypertrophic maturation and blocks osteoblast differentiation of growth plate chondrocyte, using a doxycycline-inducible Cre transgene and Sox9 conditional null alleles in the mouse [49], our data showed that Sox9 acted synergetically with BMP2 to expend proliferating chondrocyte zone and inhibited BMP2-induced chondrocyte hypertrophy, and ossification. Based on these results, it was concluded that Sox9 can effectively potentiate BMP2-induced chondrogenic differentiation of MSCs, as well as inhibit BMP2-induced osteogenic differentiation, and endochondral ossification of MSCs both in vitro and in vivo. Therefore, exogenous overexpression of Sox9 in BMP2-induced differentiation of MSCs may be an efficient way for cartilage tissue engineering.

Sox family was originally identified in Sry proteins, the male sex-determination transcription factor, a gene localized on the Y chromosome. It has 12 groups, and along with Sox8 and Sox10 Sox9 belongs to group E. The Sox5, Sox6, and Sox13 belong to group D [25]. Sox transcription factors act as architecture organizers. Sox5, Sox6, and Sox9, which are known as Sox trio, worked in coordination during the chondrogenesis [25, 50, 51]. Venkatesan et al. [52] reported that Sox9 gene transfer through replication-defective recombinant adeno-associated virus (rAAV) vectors can induce human MSC chondrogenic differentiation and decrease the expression of osteogenic differentiation markers for 21 days. Cucchiarini et al. [53] also showed a process of cartilage defect repair in rabbits' knee joints using rAAV as a gene transfer tool. Cao et al. [54] showed that implantation of Sox9 modifying MSCs in a polyglycolic acid (PGA) scaffold resulted in better repair of knee osteochondral defect in rabbit using recombinant adenovirus-mediated gene transfer. However, it is also reported that Sox9 alone was insufficient to induce MSC chondrogenic differentiation, but required other growth factors, such as Sox5, Sox6, IGF1, FGF, or TGF-β [20, 50, 55, 56]. Our study indicated that transient overexpression of Sox9 using adenovirus vector was insufficient to induce chondrogenic differentiation of MSCs. However, overexpression of Sox9 could potentiate BMP2-induced chondrogenesis and cartilage formation.

Runx2 is a transcription factor essential for BMP2-induced bone formation, endochondral ossification, and vascular invasion [10, 23, 24, 31, 57]. However, Sox9 represses Runx2 expression by promoting transcriptional repressor Bapx1 expression [30]. In agreement with the other previous reports regarding the effect of Sox9 on Runx2 expression, we found that exogenous overexpression of Sox9 in BMP2-induced osteogenic differentiation of MSCs showed a significant decrease in the levels of Runx2 expression, sequentially with delayed osteogenic differentiation, and endochondral ossification. Apart from overexpression of Sox9, silencing or removing Runx2 might achieve a similar outcome in BMP2-induced MSC differentiation. Kawato et al. [58] found that Nkx3.2-induced suppression of Runx2 is crucial for the maintenance of chondrocyte phenotypes. Lin et al. [59] showed that endochondral bone formation would be inhibited by silencing Runx2 in trauma-induced heterotopic ossification. Yoshida et al. [60] showed that chondrocyte differentiation was inhibited depending on the dosages of Runx2, and Runx2 (−/−) mice showed a complete absence of chondrocyte maturation. Also noteworthy, it is reported that Runx2 is essential for chondrogenesis. Kim et al. [61] found that Runx2 plays an important role in Ihh signaling, which induces early chondrogenesis that consists of mesenchymal cell condensation, proliferation, and differentiation into chondrocytes at the early stage of embryogenesis. Therefore, it may be a more efficient way to suppress the expression of Runx2 rather than completely removing it for cartilage tissue engineering.

Articular cartilage is an avascular, aneural tissue and lacks lymphatic drainage, which is composed of chondrocytes and cartilage matrix. It is essential for cartilage tissue engineering to retain the hyaline cartilage phenotype. However, chondrogenesis and endochondral ossification are tightly coupled and well coordinated during bone and cartilage formation [62]. The activation of endochondral ossification results in failure of maintaining the hyaline cartilage phenotype. The TGF-β, BMPs, and FGFs have been reported for their ability to direct MSCs towards the chondrocyte lineage. Yet these growth factors led to undesirable endochondral ossification or ectopic ossification [13, 14, 16, 43, 63, 64]. Sox9 as a master transcription factor for chondrogenesis also delayed BMP2-induced bone formation of MSCs, and endochondral ossification through repressed Runx2 expression, and thus plays a crucial role in cartilage formation and cartilaginous pathology healing [49, 53]. We confirmed that Sox9-mediated inhibition of osteogenic differentiation plays an important role in BMP2-induced cartilage formation and keeping hyaline cartilage phenotype.

Although it is well known that chondrogenesis and endochondral ossification are tightly coupled in cartilage and bone formation, it is unclear how these processes are linked in BMP2-induced MSC differentiation. Our study demonstrated that exogenous overexpression of Sox9 potentiates BMP2-induced MSC chondrogenesis and cartilage formation, as well as inhibits BMP2-induced MSC osteogenesis and endochondral ossification. Thus, exogenous overexpression of Sox9 in BMP2-induced MSC differentiation can be considered as a new strategy for cartilage tissue engineering.

Acknowledgments We are grateful for the Department of Clinical Hematology, Third Military Medical University, and all persons in this department. We greatly thank Pro. Tong-Chuan He (Molecular Oncology Laboratory, Department of Surgery, University of Chicago Medical Center) for the use of AdBMP2, AdSox9, and AdGFP.

References

1. Steinert AF, Noth U, Tuan RS. Concepts in gene therapy for cartilage repair. Injury. 2008;39(Suppl 1):S97–113.
2. Benthien JP, Schwaninger M, Behrens P. We do not have evidence based methods for the treatment of cartilage defects in the knee. Knee Surg Sports Traumatol Arthrosc. 2011;19:543–52.
3. Rodriguez-Merchan EC. Regeneration of articular cartilage of the knee. Rheumatol Int. 2013;33(4):837–45.
4. Kao YJ, Ho J, Allen CR. Evaluation and management of osteochondral lesions of the knee. Phys Sportsmed. 2011;39:60–9.
5. Chen H, Sun J, Hoemann CD, Lascau-Coman V, Ouyang W, McKee MD, Shive MS, Buschmann MD. Drilling and microfracture lead to different bone structure and necrosis during bone-marrow stimulation for cartilage repair. J Orthop Res. 2009;27:1432–8.
6. Dhinsa BS, Adesida AB. Current clinical therapies for cartilage repair, their limitation and the role of stem cells. Curr Stem Cell Res Ther. 2012;7:143–8.
7. Matricali GA, Dereymaeker GP, Luyten FP. Donor site morbidity after articular cartilage repair procedures: a review. Acta Orthop Belg. 2010;76:669–74.
8. Longo UG, Petrillo S, Franceschetti E, Berton A, Maffulli N, Denaro V. Stem cells and gene therapy for cartilage repair. Stem Cells Int. 2012;2012:168385.

9. Deng ZL, Sharff KA, Tang N, Song WX, Luo J, Chen J, Bennett E, Reid R, Manning D, Xue A, Montag AG, Luu HH, Haydon RC, He TC. Regulation of osteogenic differentiation during skeletal development. Front Biosci. 2008;13:2001–21.

10. Pelttari K, Steck E, Richter W. The use of mesenchymal stem cells for chondrogenesis. Injury. 2008;39(Suppl 1):S58–65.

11. Augello A, Kurth TB, De Bari C. Mesenchymal stem cells: a perspective from in vitro cultures to in vivo migration and niches. Eur Cell Mater. 2010;20:121–33.

12. Yoon BS, Lyons KM. Multiple functions of BMPs in chondrogenesis. J Cell Biochem. 2004;93:93–103.

13. Cucchiarini M, Ekici M, Schetting S, Kohn D, Madry H. Metabolic activities and chondrogenic differentiation of human mesenchymal stem cells following recombinant adeno-associated virus-mediated gene transfer and overexpression of fibroblast growth factor 2. Tissue Eng Part A. 2011;7:1921–33.

14. Handorf AM, Li WJ. Fibroblast growth factor-2 primes human mesenchymal stem cells for enhanced chondrogenesis. PLoS One. 2011;6(7):e22887.

15. An C, Cheng Y, Yuan Q, Li J. IGF-1 and BMP-2 induces differentiation of adipose-derived mesenchymal stem cells into chondrocytes-like cells. Ann Biomed Eng. 2010;38:1647–54.

16. Shintani N, Siebenrock KA, Hunziker EB. TGF-β1 enhances the BMP-2-induced chondrogenesis of bovine synovial explants and arrests downstream differentiation at an early stage of hypertrophy. PLoS One. 2013;8(1):e53086.

17. Keller B, Yang T, Chen Y, Munivez E, Bertin T, Zabel B, Lee B. Interaction of TGFbeta and BMP signaling pathways during chondrogenesis. PLoS One. 2011;6(1):e16421.

18. Danisovic L, Varga I, Polak S. Growth factors and chondrogenic differentiation of mesenchymal stem cells. Tissue Cell. 2012;44:69–73.

19. Pan Q, Yu Y, Chen Q, Li C, Wu H, Wan Y, Ma J, Sun F. Sox9, a key transcription factor of bone morphogenetic protein-2-induced chondrogenesis, is activated through BMP pathway and a CCAAT box in the proximal promoter. J Cell Physiol. 2008;217:228–41.

20. Mahmoudifar N, Doran PM. Chondrogenic differentiation of human adipose-derived stem cells in polyglycolic acid mesh scaffolds under dynamic culture conditions. Biomaterials. 2010;31:3858–67.

21. Kurth T, Hedbom E, Shintani N, Sugimoto M, Chen FH, Haspl M, Martinovic S, Hunziker EB. Chondrogenic potential of human synovial mesenchymal stem cells in alginate. Osteoarthr Cartil. 2007;15:1178–89.

22. Lian JB, Stein GS, Javed A, van Wijnen AJ, Stein JL, Montecino M, Hassan MQ, Gaur T, Lengner CJ, Young DW. Networks and hubs for the transcriptional control of osteoblastogenesis. Rev Endocr Metab Disord. 2006;7:1–16.

23. Nishimura R, Hata K, Ono K, Amano K, Takigawa Y, Wakabayashi M, Takashima R, Yoneda T. Regulation of endochondral ossification by transcription factors. Front Biosci (Landmark Ed). 2012;17:2657–66.

24. Lefebvre V, Dumitriu B, Penzo-Mendez A, Han Y, Pallavi B. Control of cell fate and differentiation by Sry-related high-mobility-group box (Sox) transcription factors. Int J Biochem Cell Biol. 2007;39:2195–214.

25. Akiyama H. Control of chondrogenesis by the transcription factor Sox9. Mod Rheumatol. 2008;18:213–9.

26. Guerit D, Philipot D, Chuchana P, Toupet K, Brondello JM, Mathieu M, Jorgensen C, Noël D. Sox9-regulated miRNA-574-3p inhibits chondrogenic differentiation of mesenchymal stem cells. PLoS One. 2013;8(4):e62582.

27. Cairns DM, Liu R, Sen M, Canner JP, Schindeler A, Little DG, Zeng L. Interplay of Nkx3.2, Sox9 and Pax3 regulates chondrogenic differentiation of muscle progenitor cells. PLoS One. 2012;7(7):e39642.

28. Cheng A, Genever PG. SOX9 determines RUNX2 transactivity by directing intracellular degradation. J Bone Miner Res. 2010;25:2680–9.

29. Yamashita S, Andoh M, Ueno-Kudoh H, Sato T, Miyaki S, Asahara H. Sox9 directly promotes Bapx1 gene expression to repress Runx2 in chondrocytes. Exp Cell Res. 2009;315:2231–40.

30. Ding M, Lu Y, Abbassi S, Li F, Li X, Song Y, Geoffroy V, Im HJ, Zheng Q. Targeting Runx2 expression in hypertrophic chondrocytes impairs endochondral ossification during early skeletal development. J Cell Physiol. 2012;227:3446–56.

31. He TC, Zhou S, da Costa LT, Yu J, Kinzler KW, Vogelstein B. A simplified system for generating recombinant adenoviruses. Proc Natl Acad Sci U S A. 1998;95:2509–14.

32. Luo J, Deng ZL, Luo X, Tang N, Song WX, et al. A protocol for rapid generation of recombinant adenoviruses using the AdEasy system. Nat Protoc. 2007;2:1236–47.

33. Cheng H, Jiang W, Phillips FM, Haydon RC, Peng Y, Zhou L, Luu HH, An N, Breyer B, Vanichakarn P, Szatkowski JP, Park JY, He TC. Osteogenic activity of the fourteen types of human bone morphogenetic proteins (BMPs). J Bone Joint Surg Am. 2003;85-A(8):1544–52.

34. Kang Q, Sun MH, Cheng H, Peng Y, Montag AG, Deyrup AT, Jiang W, Luu HH, Luo J, Szatkowski JP, Vanichakarn P, Park JY, Li Y, Haydon RC, He TC. Characterization of the distinct orthotopic bone-forming activity of 14 BMPs using recombinant adenovirus-mediated gene delivery. Gene Ther. 2004;11:1312–20.

35. Sharff KA, Song WX, Luo X, Tang N, Luo J, Chen J, Bi Y, He BC, Huang J, Li X, Jiang W, Zhu GH, Su Y, He Y, Shen J, Wang Y, Chen L, Zuo GW, Liu B, Pan X, Reid RR, Luu HH, Haydon RC, He TC. Hey1 basic helix-loop-helix protein plays an important role in mediating BMP9-induced osteogenic differentiation of mesenchymal progenitor cells. J Biol Chem. 2009;284:649–59.

36. Chen L, Jiang W, Huang J, He BC, Zuo GW, Zhang W, Luo Q, Shi Q, Zhang BQ, Wagner ER, Luo J, Tang M, Wietholt C, Luo X, Bi Y, Su Y, Liu B, et al. Insulin-like growth factor 2 (IGF-2) potentiates BMP-9-induced osteogenic differentiation and bone formation. J Bone Miner Res. 2010;25:2447–59.

37. Wang JH, Liu YZ, Yin LJ, Chen L, Huang J, Liu Y, Zhang RX, Zhou LY, Yang QJ, Luo JY, Zuo GW, Deng ZL, He BC. BMP9 and COX-2 form an important regulatory loop in BMP9-induced osteogenic differentiation of mesenchymal stem cells. Bone. 2013;57:311–21.

38. Lengner CJ, Lepper C, van Wijnen AJ, Stein JL, Stein GS, Lian JB. Primary mouse embryonic fibroblasts: a model of mesenchymal cartilage formation. J Cell Physiol. 2004;200:327–33.

39. Hu N, Jiang D, Huang E, Liu X, Li R, Liang X, Kim SH, Chen X, Gao JL, Zhang H, Zhang W, Kong YH, Zhang J, Wang J, Shui W, Luo X, Liu B, et al. BMP9-regulated angiogenic signaling plays an important role in the osteogenic differentiation of mesenchymal progenitor cells. J Cell Sci. 2013;126:532–41.

40. Hollander AP, Dickinson SC, Kafienah W. Stem cells and cartilage development: complexities of a simple tissue. Stem Cells. 2010;28:1992–6.

41. Prockop DJ, Kota DJ, Bazhanov N, Reger RL. Evolving paradigms for repair of tissues by adult stem/progenitor cells (MSCs). J Cell Mol Med. 2010;14:2190–9.

42. Woo EJ. Adverse events after recombinant human BMP2 in nonspinal orthopaedic procedures. Clin Orthop Relat Res. 2013;471:1707–11.

43. Jang WG, Kim EJ, Kim DK, Ryoo HM, Lee KB, Kim SH, Choi HS, Koh JT. BMP2 protein regulates osteocalcin expression via Runx2-mediated Atf6 gene transcription. J Biol Chem. 2012;287:905–15.

44. Yu YY, Lieu S, Lu C, Colnot C. Bone morphogenetic protein 2 stimulates endochondral ossification by regulating periosteal cell fate during bone repair. Bone. 2010;47:65–73.

45. Wang Y, Zheng Y, Chen D, Chen Y. Enhanced BMP signaling prevents degeneration and leads to endochondral ossification of Meckel's cartilage in mice. Dev Biol. 2013;381:301–11.

46. Yu L, Han M, Yan M, Lee J, Muneoka K. BMP2 induces segment-specific skeletal regeneration from digit and limb amputations by establishing a new endochondral ossification center. Dev Biol. 2012;372:263–73.

47. Hu N, Wang C, Liang X, Yin L, Luo X, Liu B, Zhang H, Shui W, Nan G, Wang N, Wu N, Chen X, He Y, Wen S, Deng F, Zhang H, Liao Z, Luu HH, Haydon RC, He TC, Huang W. Inhibition of histone deacetylases potentiates BMP9-induced osteogenic signaling in mouse mesenchymal stem cells. Cell Physiol Biochem. 2013;32:486–98.

48. Dy P, Wang W, Bhattaram P, Wang Q, Wang L, Ballock RT, Lefebvre V. Sox9 directs hypertrophic maturation and blocks osteoblast differentiation of growth plate chondrocytes. Dev Cell. 2012;22:597–609.

49. Yang HN, Park JS, Woo DG, Jeon SY, Do HJ, Lim HY, Kim SW, Kim JH, Park KH. Chondrogenesis of mesenchymal stem cells and dedifferentiated chondrocytes by transfection with SOX trio genes. Biomaterials. 2011;32:7695–704.

50. Ikeda T, Kamekura S, Mabuchi A, Kou I, Seki S, Takato T, Nakamura K, Kawaguchi H, Ikegawa S, Chung UI. The combination of SOX5, SOX6, and SOX9 (the SOX trio) provides signals sufficient for induction of permanent cartilage. Arthritis Rheum. 2004;50:3561–73.

51. Venkatesan JK, Ekici M, Madry H, Schmitt G, Kohn D, Cucchiarini M. SOX9 gene transfer via safe, stable, replication-defective recombinant adeno-associated virus vectors as a novel, powerful tool to enhance the chondrogenic potential of human mesenchymal stem cells. Stem Cell Res Ther. 2012;3:22.

52. Cucchiarini M, Orth P, Madry H. Direct rAAV SOX9 administration for durable articular cartilage repair with delayed terminal differentiation and hypertrophy in vivo. J Mol Med (Berl). 2012;91(5):625–36.

53. Cao L, Yang F, Liu G, Yu D, Li H, Fan Q, Gan Y, Tang T, Dai K. The promotion of cartilage defect repair using adenovirus mediated Sox9 gene transfer of rabbit bone marrow mesenchymal stem cells. Biomaterials. 2011;32:3910–20.

54. Park JS, Yang HN, Woo DG, Jeon SY, Do HJ, Lim HY, Kim JH, Park KH. Chondrogenesis of human mesenchymal stem cells mediated by the combination of SOX trio SOX5, 6, and 9 genes complexed with PEI-modified PLGA nanoparticles. Biomaterials. 2011;32:3679–88.

55. Garza-Veloz I, Romero-Diaz VJ, Martinez-Fierro ML, Marino-Martinez IA, Gonzalez-Rodriguez M, Martinez-Rodriguez HG, Espinoza-Juarez MA, Bernal-Garza DA, Ortiz-Lopez R, Rojas-Martinez A. Analyses of chondrogenic induction of adipose mesenchymal stem cells by combined co-stimulation mediated by adenoviral gene transfer. Arthritis Res Ther. 2013;15:R80.

56. Nishimura R, Hata K, Matsubara T, Wakabayashi M, Yoneda T. Regulation of bone and cartilage development by network between BMP signalling and transcription factors. J Biochem. 2012;151:247–54.

57. Kawato Y, Hirao M, Ebina K, Tamai N, Shi K, Hashimoto J, Yoshikawa H, Myoui A. Nkx3.2-induced suppression of Runx2 is a crucial mediator of hypoxia-dependent maintenance of chondrocyte phenotypes. Biochem Biophys Res Commun. 2011;416:205–10.

58. Lin L, Shen Q, Leng H, Duan X, Fu X, Yu C. Synergistic inhibition of endochondral bone formation by silencing Hif1alpha and Runx2 in trauma-induced heterotopic ossification. Mol Ther. 2011;19:1426–32.

59. Yoshida CA, Yamamoto H, Fujita T, Furuichi T, Ito K, Inoue K, Yamana K, Zanma A, Takada K, Ito Y, Komori T. Runx2 and Runx3 are essential for chondrocyte maturation, and Runx2 regulates limb growth through induction of Indian hedgehog. Genes Dev. 2004;18:952–63.

60. Kim EJ, Cho SW, Shin JO, Lee MJ, Kim KS, Jung HS. Ihh and Runx2/Runx3 signaling interact to coordinate early chondrogenesis: a mouse model. PLoS One. 2013;8(2):e55296.

61. Chun JS, Oh H, Yang S, Park M. Wnt signaling in cartilage development and degeneration. BMB Rep. 2008;41:485–94.

62. Bakker AC, van de Loo FA, van Beuningen HM, Sime P, van Lent PL, van der Kraan PM, Richards CD, van den Berg WB. Overexpression of active TGF-beta-1 in the murine knee joint: evidence for synovial-layer-dependent chondro-osteophyte formation. Osteoarthr Cartil. 2001;9:128–36.

63. Mi Z, Ghivizzani SC, Lechman E, Glorioso JC, Evans CH, Robbins PD. Adverse effects of adenovirus-mediated gene transfer of human transforming growth factor beta 1 into rabbit knees. Arthritis Res Ther. 2003;5:R132–9.

64. Steinert AF, Ghivizzani SC, Rethwilm A, Tuan RS, Evans CH, Nöth U. Major biological obstacles for persistent cell-based regeneration of articular cartilage. Arthritis Res Ther. 2007;9:213.

Stem Cells and Ear Regeneration

20

Hamid Karimi, Seyed-Abolhassan Emami, and Ali-Mohammad Karimi

20.1 History and Introduction

Reconstruction of ears after trauma or due to congenital malformation (microtia-anotia) is a heavy and time-consuming procedure and it needs several sessions of surgery. In spite of obvious progress in reconstruction techniques during the last six decades, still some problems have not been solved in this issue (Fig. 20.1).

20.2 Etiology of Missed Ears

Besides the microtia or anotia, there are other categories of patients who have lost their external ear due to trauma, accident, frost bite, cancer removal, animal or human bites, or burns [1–5].

Congenital malformation of ear or microtia is a deformity that there is underdeveloped ear. A completely missed pinna is called anotia. Because microtia and anotia have the same origin, it is usually referred to as microtia-anotia. Microtia can be unilateral or bilateral. Incidence of microtia is 1 out of about 8000–10,000 births.

In unilateral microtia, the right ear is most commonly affected.

20.2.1 Classification

There are four grades of microtia [1]:

1. Grade I: A less than complete development of the external ear with minimal structures and a small external ear canal.
2. Grade II: A partially developed ear (usually the top portion is underdeveloped) with a closed (stenotic) external ear canal with conductive hearing loss.
3. Grade III: Absence of the external ear with a small peanut-like vestige cartilage and an absence of the external ear canal and eardrum. The most common form of microtia.
4. Grade IV: Absence of the total ear or anotia.

The pathophysiology of microtia is still unknown, but most probably is multifactorial, and it happens during weeks 8–12 of gestation period, when the external ear is forming [1].

H. Karimi (✉) · S.-A. Emami
Department of Plastic and Reconstructive Surgery,
Hazrat Fatemeh Hospital, Iran University of Medical
Sciences, Tehran, Iran

A.-M. Karimi
School of Medicine, Iran University of Medical
Sciences, Tehran, Iran

© Springer Nature Switzerland AG 2019
D. Duscher, M. A. Shiffman (eds.), *Regenerative Medicine and Plastic Surgery*,
https://doi.org/10.1007/978-3-030-19962-3_20

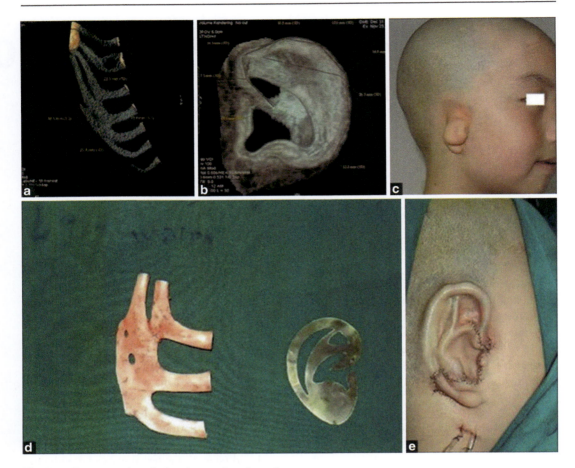

Fig. 20.1 Reconstruction of microtia ear using rib cartilage

20.3 Traditional Methods for Ear Reconstruction

For reconstruction of ear several methods have been advised (Table 20.1):

1. Reconstruction with rib costal cartilage
2. Reconstruction with Medpor (porous polyethylene) and placing a skin flap (or multiple flaps) over it
3. Using an external ear prosthesis

One of the most popular one is reconstruction with costal cartilage; some authors recommend this technique at the age of 7 (before entering the primary school) and some others recommend ages 8–10 as the ear in this age is completely developed and reached to its maximum size. The major advantage of this surgery is that the patient's own tissue is used for the reconstruction. But the technique involves from two to four stages depending on the surgeon's preferred method. The main disadvantages of this method are that it is a heavy and time-consuming procedure and it needs several sessions of surgery (at least three sessions of operations), removal of costal cartilage (donor-site morbidity) (Table 20.2), lack of enough cartilage elasticity in adults (calcification of cartilage in adults), and shortage of and small-size costal cartilage in some patients, remaining of scar in the chest, hypertrophic scar in the chest, some respiratory problems after the surgery specially during exercise, lack of enough cartilage for both sides

20 Stem Cells and Ear Regeneration

Table 20.1 Summary of total autologous auricular reconstructive techniques

Surgeon	Technique	Pros	Cons
Tanzer	Four stages: 1. Rotation of the lobule into a transverse position 2. Fabrication and placement of a costal cartilage framework 3. Elevation of the ear from the side of the head 4. Construction of a tragus and conchal cavity	– First stepwise total auricular reconstruction – Good results	– Multiple operations – Transposing lobule first poses risk of vascular compromise of skin flap
Brent	Four stages: 1. Rib cartilage framework fabrication and placement 2. Lobule transposition 3. Elevation of framework and creation of a retroauricular sulcus 4. Conchal excavation and tragus construction	– Good contour – Postoperative drain limits complications of bolster dressings	– Multiple operations – Lack definition of conchal bowl – Composite skin/cartilage tragal grafts can contract
Nagata	Two stages: 1. Fabrication of costal cartilage framework including the tragus, conchal excavation and rotation of the lobule 2. Elevation of framework, placement of cartilage graft in auriculocephalic sulcus, covered with temporoparietal fascial flap and skin graft	– Less operations – High-definition framework to create a good tragus	– More cartilage needed – Detailed framework so long learning curve – Minimum age 10 years – Partial necrosis of posterior flap – Wire sutures increase extrusion

From: Z. M. Jessop, M. Javed, I. A. Otto, E. J. Combellack, S. Morgan, C.C. Breugem, et al.; Combining regenerative medicine strategies to provide durable reconstructive options: auricular cartilage tissue engineering; Stem Cell Research & Therapy 2016 **7**:19, **DOI:** 10.1186/s13287-015-0273-0 **Published:** 28 January 2016. [The article is distributed under the terms of the Creative Commons Attribution 4.0 International License (http://creativecommons.org/licenses/by/4.0/), which permits unrestricted use, distribution, and reproduction in any medium, provided you give appropriate credit to the original author(s) and the source, provide a link to the Creative Commons license]

Table 20.2 Donor site morbidity associated with total autologous auricular reconstruction

	Donor site morbidity	Incidence	Total number of patients per study
Early	Pneumothorax	3 (1%)	270
		19 (22%)	88
	Atelectasis	4 (22%)	18
		7 (8%)	88
	Pleural effusion	–	–
Delayed	Persistent pain	6 (14%)	42
	Thoracic scoliosis	4 (25%)	16
	Seroma	9 (8%)	108, rhinoplasty group
	Clicking	3 (7%)	42
	Abnormal scarring	0 (0%)	42
		3 (2.7%)	110
		12 (14%)	88
		14 (5.3%)	264
		21 (6.5%)	322
	Contour deformity	3 (7%)	42
		16 (50%)	32
		22 (25%)	88

From: Z. M. Jessop, M. Javed, I. A. Otto, E. J. Combellack, S. Morgan, C.C. Breugem, et al.; Combining regenerative medicine strategies to provide durable reconstructive options: auricular cartilage tissue engineering; Stem Cell Research & Therapy 2016 **7**:19, **DOI:** 10.1186/s13287-015-0273-0 **Published:** 28 January 2016. [The article is distributed under the terms of the Creative Commons Attribution 4.0 International License (http://creativecommons.org/licenses/by/4.0/), which permits unrestricted use, distribution, and reproduction in any medium, provided you give appropriate credit to the original author(s) and the source, provide a link to the Creative Commons license]

(bilateral missed ears), lack of reserve cartilage for a third surgery (re-sculpturing of a new costal cartilage for secondary reconstruction), shape and configuration of the reconstructed ear sometimes not completely perfect, and low rate of resemblance to normal ear [6–11]. Some of the normal configurations of normal ears cannot be reproduced and some of them have not long durability (Table 20.3).

Reconstruction with Medpor has some advantages, like one-stage procedure, no donor-site morbidity, and good shape and size (Figs. 20.2 and 20.3). But disadvantages are using of a synthetic material, infection, extrusion of framework, inflammatory reactions, inability to grow with patient, and so on [12].

For external prosthesis only one stage of fixation is enough, but some inflammatory response to the screws or fixation site, easy dropping of loose prosthesis, incompatibility of other children in school (they may make a joke by removing the ear and playing with it in the class), changing of the color during the time, changing of quality during the time, need for buying a new one after some years, and inability to grow with patient's age (Fig. 20.4). Therefore these problems promote some surgeons to find other options.

20.4 New Concepts for Ear Reconstruction

For reconstruction of ear two different tissues have to be restored, skin envelope and cartilaginous ear framework: skin and ear framework.

20.4.1 Skin Reconstruction

The skin can be repaired as follows:

1. Using the postauricular skin + skin graft.
2. It can be reproduced with a flap from superficial temporal fascia which covers the framework as a sandwich and total skin graft of it.
3. Free fascial flaps (such as radial forearm flap) and total skin graft of it.

4. Placing tissue expander in postauricular skin and expansion of skin [12].
5. Using multiple flaps.

20.4.2 Reconstruction of Ear Framework

For reconstruction of framework using new methods, surgeons from 15 years ago started to use cell expansion techniques and cell cultures.

One of the first steps in this regard was using the chondrocytes of the patient from normal side. The scientists extracted the chondrocytes and cultured it in vitro and after expansion they used it over a scaffold, placed it in vitro or in vivo, and tried to build a cartilaginous framework. The more the scaffold resembles to the patient's ear, the more the reconstructed ear resembles to the normal ear. Although it is a one-stage procedure and has no morbidity in chest, it needs another operation for harvesting the chondrocytes.

20.4.2.1 Animal Chondrocytes

In one of the reports from Japan in 2004, the authors used bovine articular chondrocytes and expended it and seeded over PLLAEC (poly(L-lactic acid-epsilon-caprolactone) copolymer scaffold (which resembles to human ear). Then they inserted it in the back of mouse for 40 weeks. The resulted ear was very similar to human ears [13].

In 2007 from the USA, the authors used chondrocytes of joint, septum, or ear of the rabbit and reconstructed a new cartilage over a scaffold. They found high proteoglycan, collagen type II, and glycosaminoglycan in the cartilage and used them for reconstruction of trachea cartilage. They concluded that only ear cartilage chondrocytes maintained their biomechanical characteristics [14].

In a report from Germany in 2003, the authors used septal chondrocytes and expended them and seeded over hyaluronic acid scaffold (Hyaff 11) and in 4 weeks they reconstructed a new cartilage [15].

In 2003 there is a report that authors used bovine chondrocytes with scaffold and acrylic internal support. And in 12 weeks in rats they

20 Stem Cells and Ear Regeneration

Table 20.3 Long-term limitations of autologous auricular reconstruction

Long-term limitations		Reasons
Stiffness		1. Different biomechanical properties of fibrocartilage donor 2. Heterotopic calcification
Extrusion		1. Skin flap necrosis 2. Wire sutures to assemble cartilage framework 3. Wound infection or pressure dressings
Projection loss		1. Effacement of postauricular sulcus due to contraction of skin grafts
Distortion		1. Constriction of skin and soft tissue overlying the construct due to scarring or ischemia 2. Cartilage degradation and resorption leading to loss of definition

From: Z. M. Jessop, M. Javed, I. A. Otto, E. J. Combellack, S. Morgan, C.C. Breugem, et al.; Combining regenerative medicine strategies to provide durable reconstructive options: auricular cartilage tissue engineering; Stem Cell Research & Therapy 2016 **7**:19, **DOI:** 10.1186/s13287-015-0273-0 **Published:** 28 January 2016. [The article is distributed under the terms of the Creative Commons Attribution 4.0 International License (http://creativecommons.org/licenses/by/4.0/), which permits unrestricted use, distribution, and reproduction in any medium, provided you give appropriate credit to the original author(s) and the source, provide a link to the Creative Commons license]

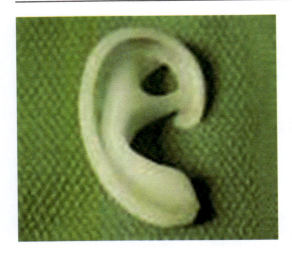

Fig. 20.2 Medpor ear framework

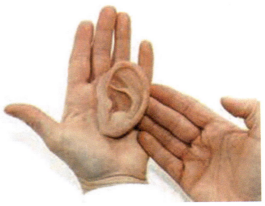

Fig. 20.4 Artificial ear prosthesis

Fig. 20.3 Medpor ear framework

reconstructed new human shape ear and human nasal tip cartilages which have good rigidity [16] (Fig. 20.5). In 2002 there was a report that stated that rabbit chondrocytes (autograft or allograft) and a rabbit decellularized cartilage can be cultured together (they used decellularized cartilage as a framework or scaffold). They found that new ear has viable chondrocytes [17]. For evaluation of scaffolds a study in 2014 was done. The authors used rabbit articular chondrocytes with swine articular scaffolds and used the new cartilage in a defect in rabbit trachea. They concluded that allogenic chondrocytes with xenogenic scaffolds would result in a cartilage that has good shape and function [18].

These reports have shown us that a new cartilage can be regenerated from xeno-scaffolds. But there are reports that the new cartilages are not durable [19] or stable [20] although still other reports emphasized on rigidity [16], biocompatibility [21], and proper biomechanical characteristics of new cartilages [14].

Then researches focused on the origin of chondrocytes. It was found that ear chondrocytes will produce elastic cartilage and articular chondrocytes will produce articular cartilages [22]. Besides it is found that costal chondrocytes will become calcified during the long term and the best chondrocytes for tissue engineering are those extracted from ear, septum, and joints [23].

20.4.2.2 Human Chondrocytes

Gradually after successful results with animal chondrocytes, the studies went for the culturing of human chondrocytes. There are reports that used human chondrocytes and cultured them in

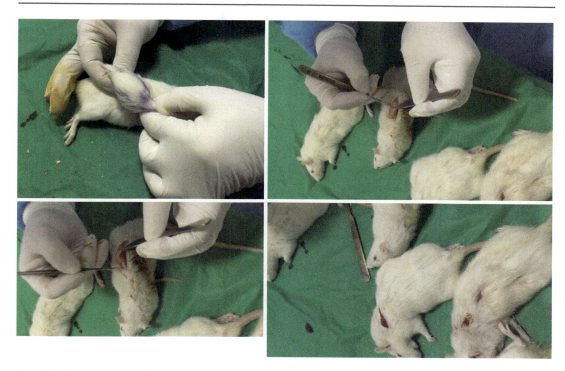

Fig. 20.5 Reconstruction of human ear cartilages in the back of the rats, in vivo

the lower abdomen. The result is a block of cartilage that can be used for sculpturing and reconstruction of ear framework [24, 25]. The authors stated that the cartilage has a neo-perichondrium and it can also be used as chondro-fat composite graft too. They reported that the cartilages were stable for 1–5 years [4]. These works are suitable as there is no donor site in the chest and rib cartilages. And in some other studies normal human chondrocytes were cultured in special media and used for reproduction of ear cartilage in lab [5] (Fig. 20.6).

Also there is another report from the USA that children ear chondrocytes can be cultured for framework reconstruction. The cells have a better speed for doubling and it has longer biocompatibility [26]. There is also another report from the USA that human chondrocytes were seeded over PGA/PLA (polyglycolic acid/poly lactic acid) scaffolds and the regenerated cartilages had no differences with normal ones regarding the mechanical and histological properties, unless

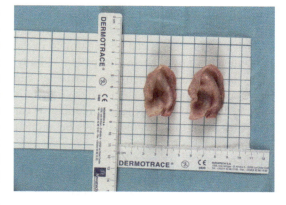

Fig. 20.6 Reconstructed ears with chondrocytes

the cells were smaller and there was a fibrous capsule around the cartilage [27]. In a report from Switzerland in 2014, the authors mentioned that they used human septal chondrocytes and after 4 weeks the results were proper hyaline and fibrous cartilages which remained stable and functional for 1 year [28].

20.4.2.3 Chondrocytes from Microtia Ear

But problem of taking chondrocytes from a normal ear has led the scientists to another option. They used chondrocytes from microtia ear, as they are expandable and with no donor morbidity and no morbidity for normal ears. These new cartilages had proper physical characteristics and abundant amount of proteoglycans and collagen type II [29]. It is reported that for human chondrocytes, FGF-2 promotes more proliferation and OP-1 has induced and maintained the phenotypic characteristics of ear chondrocytes [30]. So in this option, surgery is one stage with no donor morbidity in normal ear and finds a way of use for microtia cartilage.

20.4.2.4 Stem Cell Regeneration

After microtial chondrocytes, another step was to culture and expand the stem cells and differentiate them into the chondroblasts. Obviously the researches began with animal stem cells. Stem cells have the ability to increase into number of millions and they can also be kept in liquid nitrogen for further culture and use. And after multiplication, they can differentiate into target cell, i.e., cartilage cells. These processes can be performed in the lab and have no stage in the patient's body. Many stem cells and many ear cartilages can be regenerated from one stem cell, so if there is any need for re-sculpturing and redoing of ear reconstruction, there would always be another cartilage for this purpose.

Many sources for stem cells have been used in this regard: bone marrow stem cells (BMSC) [22, 31–55], adipose-derived stem cells (ADSC) [52, 56–59], mesenchymal stem cells (MSC) [32, 60–63], dental pulp stem cells [5, 56], perichondrial stem cells [64–66], chondrocyte-derived progenitor cells (CDPC) [67], and fetal cartilage-derived progenitor cells (FCPC) [68].

The stem cells from these origins are harvested, cultured, and expanded and then they will be set in a chondrogenic culture media in order to produce the chondroblasts [38]. Then the new chondroblasts are placed in vitro and with the help of external molds or internal scaffolds they will gain the shape of a human ear.

The chondrogenic culture media (like DMEM/F12 media) should have special characteristics and it should have the ability to induce stem cells into chondroblasts [38]. These inducing media are very specific for the type of cells; for example in a report in 2013, ADSC and ear chondrocytes of rat were cultured separately in the same media. ADSC produced proper cartilage while ear chondrocyte did not produce proper cartilage [59]. Some specific cell mediators and growth factors should be added to this media for promoting the stem cells. Some of them are for increasing the number of stem cells such as FGF-2 [30] and some of them are necessary for changing into chondroblasts such as TGF-B3 [33], OP-1 [30], PTH [36], b FGF [69], and PRP [44] and some are needed to maintain the phenotype of cartilage like chondromodulin-1 [45].

Problems with a pre-induced media have led the scientists to a new innovation. Some of the authors used mature chondrocytes as a co-culture for stem cells and chondrogenic microenvironment that was produced and had an induction effect on the stem cells and after induction the chondroblasts and chondrocytes were produced and could be seeded to the scaffolds. It has been shown that co-culturing of ADSC with osteocytes can produce osteoblasts and co-culturing with chondrocytes can produce chondroblasts [56]. Some of the authors used xenograft chondrocytes [32, 61, 62] and some others used allograft human chondrocytes [37]; some others used autograft chondrocytes with good results [34, 41, 42, 44, 46, 49, 54, 56, 57].

About the time of induction in the induction media or co-culture several studies have been done. Some scientists advise from 7 to 10 days and some others up to 3 weeks and others even 8–12 weeks. It is obvious that there is no consensus about this issue [37, 38].

What should be the ratio or the percentage of stem cells versus chondrocytes in co-culture media. There are many reports from 1:2 to 1:1 to 7:3 to 75:25 to 80:20. But most of the authors advised 75:25 ratio [32, 40–42, 54, 57].

20.4.2.5 Neonatal Chondrocytes Versus Adult Chondrocytes

There is a very good study from the UK in which the authors compared the quality of cartilage after co-culturing of BMSC with human neonatal chondrocytes versus adult chondrocytes. They found that there were no differences and even the quality of cartilage in adult cartilages in some issues was better than neonatal cartilages. So it seems that power of reproduction in chondrocytes themselves is not an important issue and only chondrogenic effect is important [37].

20.4.2.6 Choosing the Stem Cells

There was a very good study in 2010 and authors reported that among BMSC, ADSC, and chondrocytes, the best cartilages were obtained from BMSCs. Therefore it seems that the choice for regeneration of ear cartilage is BMSCs [52].

20.4.2.7 Scaffolds

There are some few reports that did not use the scaffolds and still reported the good results. Although the longevity of these cartilages and quality of them are in question, they stated the proper results while some of them reported that the cartilages were not stable [20, 29, 36]. Some of the authors who did not use scaffolds used fibrin sealant in order to put the regenerated cells together. This sealant may have some minor effects of scaffolds, but the regenerated cartilages would be with the shape of a block [29, 36, 39, 47].

For using a stable structure that puts every cells together, some authors used external molds and more authors advised internal scaffolds. There are reports about externals molds that will shape the cartilage into a human ear cartilage [70, 71]. These authors reported good and proper results after 8–12 weeks of in vivo implantation.

But the difficulty of using an external mold for a long time in an animal model has led the scientists to use an internal scaffold that would shape the regenerative cartilage. There are several kinds of scaffolds: absorbable, nonabsorbable, double scaffolds, and so on. These scaffolds can help scientists to produce a 3D cartilage which resembles the ear framework. Several characteristics have been written for a good and proper scaffold; it should provide good surface for adhesion of the stem cells, it should provide good structure for cell support, it should have mechanical properties similar to the ideal cartilage, scaffolds and its degradation should not be toxic to the host, scaffold has to have the ability for 3D reproduction, it should have high (90%) porosity for cell-to-polymer interaction, it should have enough space for matrix production, it should help to prevent loss of phenotype, it is better to be absorbable after regeneration of cartilage, degraded scaffolds should not be toxic to stem cells, and it should be easily replaced by the new cartilage tissue. The scaffolds should have the ability for cell induction in early steps and ability to support cell differentiation in the next stages (Table 20.4).

Absorbable scaffolds are PLA (poly-lactic acid), PLLA (poly-L-lactic acid) [35], PLA/PGA [22, 27, 40, 62], PGA (polyglycolic acid) [34], collagen scaffolds [65], collagen type I hydrogel [34], PLCG (poly-lactic-co-glycolide) [37], PLLAEC poly(L-lactic acid-epsilon-caprolactone) copolymer [13], collagen type I glycosaminoglycan [38], Hyaff 11 (a hyaluronic acid derivative) [15], swine scaffold PCS [18], hexanol acetone/carbonate, decellularized ear cartilage (only collagen and elastin) [48], cadaver ear framework [55], acellular dermal matrix (ADM) [49], PLA/PLEC poly(L-lactic acid) and poly(L-lactide-epsilon-caprolactone) [23], silk fibroin [43], GT/PCL (gelatin/polycaprolactone) [42], PCL (poly-e-caprolactone) [61], and poly-coprolactone-based polyurethane [72].

Nonabsorbable scaffolds include chitosan nonwoven [31], polyethylene [73], silk polymer [60], acrylic internal support [16], and porous coral scaffold [46].

Double scaffolds are also numerous such as chitosan nonwoven/PLGA (poly (dl-lactide-co-glycolide)) [31], PLAEC/PLA (poly(L-lactic acid-epsilon-caprolactone)/poly-lactic acid) [23], alginate and silk polymer [60], porous collagen/titanium wire [69], and silk fibroin/chitosan [43]. In some mixed scaffolds both scaffolds are absorbable; in others one is absorbable and nour-

Table 20.4 Different type of scaffolds

Absorbable	Poly-L-lactic acid Polyglycolic acid Poly galactin Polyurethan	Dexon Vicryl
Non-absorbable	Poly vinyl alcohol Polytetrafluoroethylene Polyethylene Nylon Polyester Carbon fiber meshwork	Porous sponge Teflon Dacron
Bio-materials non-synthetic	Collagen type I sponge Decalcified bone Fibrin polymer Hyaluronic acid Meniscus ADM Decellularized cartilage Cadaver cartilage	Our previous report (Karimi H, Emami SA, Olad-Gobad MK)

From: Z. M. Jessop, M. Javed, I. A. Otto, E. J. Combellack, S. Morgan, C.C. Breugem, et al.; Combining regenerative medicine strategies to provide durable reconstructive options: auricular cartilage tissue engineering; Stem Cell Research & Therapy 2016 7:19, **DOI:** 10.1186/s13287-015-0273-0 **Published:** 28 January 2016. [The article is distributed under the terms of the Creative Commons Attribution 4.0 International License (http://creativecommons.org/licenses/by/4.0/), which permits unrestricted use, distribution, and reproduction in any medium, provided you give appropriate credit to the original author(s) and the source, provide a link to the Creative Commons license]

ishing for the cells and the other is nonabsorbable and provides heavy and stable structure for a long time and in other less frequent types both are nonabsorbable.

Yu Liu et al. [5] also used double-mesh framework, one as structure and the other as nourishing mesh, and have done it in vitro. After 12 weeks they had good results for elasticity and shape. In a study by Nayyer-Leila et al. [74], they used human stem cells and collagen–polyester mesh for reconstruction of ear. But they had done it in only one patient and this type of study needs further cases and researches.

Selection of scaffolds depends on the experience of surgeon or scientists, type of stem cells that they want to use, and type of regenerative cartilages that are needed to be regenerated. For example for reconstruction of ear most frequent scaffolds that are used include PLA/PGA and PLA.

Scaffolds are widely used to reconstruct cartilage. Yet, the fabrication of a scaffold with a highly organized microenvironment that closely resembles native cartilage remains a major challenge. Some authors suggested that scaffolds derived from acellular extracellular matrices are able to provide such a microenvironment. There is a report specifically on decellularization of full-thickness ear cartilage. In this study, decellularized ear cartilage scaffolds were prepared. The authors removed cells and cell remnants from elastic cartilage and the obtained scaffolds retained their native collagen and elastin contents as well as their architecture and shape. High-magnification scanning electron microscopy showed no obvious difference in matrix density after decellularization. However, glycosaminoglycan content was significantly reduced. Then the authors used BMSC over this scaffold and the new cartilages had good and proper characteristics [48]. In another report from our center, we used ear framework of cadaver with BMSC. The resulted ears have very good elastic and biomechanical properties (Figs. 20.7–20.11). And the shape, weight, and size of the ears were retained. This was the first time in the published literature that the weight of ears was measured and with BMSC new cartilages were formed with proper shape, weight, and size. In that study we concluded that the using

20 Stem Cells and Ear Regeneration

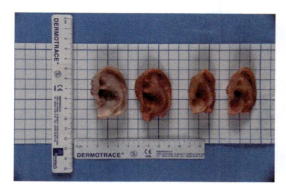

Fig. 20.7 Reconstruction of human ears with BMSC and cadaver framework

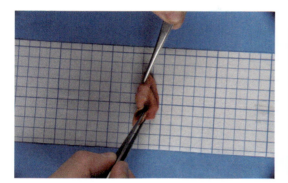

Fig. 20.8 Testing the elasticity of reconstructed ear that was regenerated from the patients' own BMSC and cadaver framework

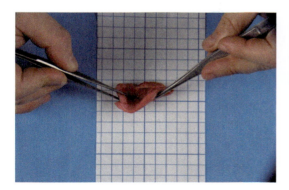

Fig. 20.9 Testing the flexibility of reconstructed ear that was regenerated from the patients' own BMSC and cadaver framework

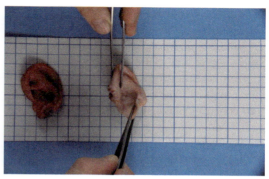

Fig. 20.10 Testing the flexibility of reconstructed ear that was regenerated from the patients' own BMSC and cadaver framework

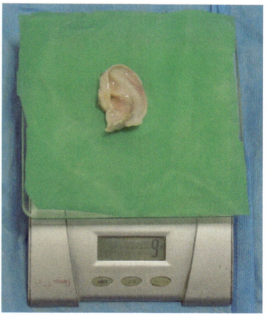

Fig. 20.11 Measuring the weight of frameworks in two groups after 12 weeks in the back of the rat, in vivo. Our results showed that BMSC and cadaver framework will maintain the weight and shape and elasticity of the ear framework and after 12 weeks, the reconstructed ear can be used for reconstruction of missed ear in the patients

of ear cadaver framework seeded with bone marrow stem cells for reconstruction of ear is a feasible, fast, one-stage technique and the elasticity, shape, size, and weight of the framework would be preserved. Using the patient stem cells also would provide the most HLA compatibility for reconstructed ear [55]. There is another report in 2015 that used human cadaver ear framework for reconstruction of ear [48].

20.4.2.8 Growth Factors

For culturing and expanding of the stem cells, it is needed to have a chondrogenic media and provide some growth factors and cytokines [40, 56] to facilitate transformation of stem cells into chondroblasts. The stem cells can differentiate into chondrogenic cells by the help of chondrogenic media with growth factors or co-culturing with chondrocytes or chondrogenic matrix. Some of these growth factors are:

1. FGF-2: for promoting cell proliferation
2. OP-1: for promoting cell phenotypes into chondroblasts
3. TGF-β3: for transforming into chondrocytes
4. βGF: for cell proliferation
5. Chrondromodulin-1: for stabilization of phenotype and prevention of calcification of cartilage
6. TGF-β: for cell differentiation and matrix formation and transformation to chondroblast; it works synergistically with IGF-1
7. Dexamethasone: for transformation to chondroblast
8. Ascorbic acid: as adjuvant
9. TGF-β3: for transforming into chondrocytes
10. IGF-1: for transforming into chondrocytes and matrix formation
11. TGF-β1: for chondrogenicity
12. BMP: bone morphologic protein
13. Basic FGF: FGF-2 for cell proliferation
14. PDGF: platelet-derived growth factor; for cell proliferation [75–88]

20.5　Maturation Process

It is proven that tissue-specific stem cells can produce a new auricular cartilage, but this new cartilage structure is immature phenotypically; advantage of this issue is that it is highly active metabolically and it can grow and develop into mature adult ear cartilage, but disadvantage is that it may be liable to resorption. In the first and highly innovative study of auricular tissue engineering by Vacanti et al. [6, 16], a new ear-shaped cartilage was made from bovine chondrocytes and biocompatible scaffolds that were xenografted into a nude mouse [89]. The shape of the cartilage was supported by an externally fixed mold, but after removal of the mold the cartilage deformed and started to shrink. So the new cartilage was not stable in long term (Fig. 20.12). Newer studies have shown that the process of resorption may be due to an intrinsic property of new cartilage, or due to extrinsic factors, e.g., cell mediators or inflammatory mediators or cell-to-cell interactions.

Advantage of an internal permanent support, like coiled wire or titanium wire or nylon scaffolds, is that it is shown in animal studies that it helps to prevent or to reduce shrinkage of the neo-cartilage [19, 75], but disadvantage is that

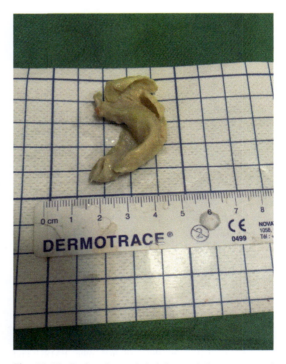

Fig. 20.12 Deformity and shrinkage of the regenerated ear framework

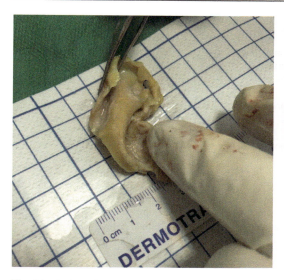

Fig. 20.13 The shape and flexibility of the framework has been lost

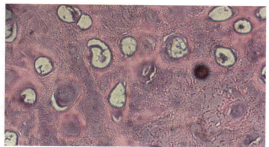

Fig. 20.14 Microscopic examination of cartilage after shrinkage

implanted synthetic materials may be extruded [75–80].

The extracellular matrix of cartilage has collagen and elastin framework and can last forever, as it has extensive chemical cross-linking that stabilizes the structural of the cartilage. The maturation process of cross-linking starts after the birth and in 15 years will be completed. It is called functional adaptation [81–83] (Figs. 20.13 and 20.14).

Gradual increasing of the number of cross-linking would increase the biomechanical strength of cartilage [84]; in this way a durable cartilage will be produced [85, 86]. Until now there were no solutions for promoting maturation in a new cartilage. The lack of maturation is one of the major causes of the failure for production of a durable tissue-engineered ear cartilage. There is a marvelous study that fibroblast growth factor (FGF)-2 and transforming growth factor beta-1 (TGF-β1) may induce maturation in the cartilage in immature articular cartilage. No one knows to what extent does auricular cartilage undergo tissue maturation. But the goal of newer studies is to accelerate the maturation process and have a functional and durable cartilage [87, 88].

20.6 Achieved Results

Up to now several progressions have been made by scientists to produce ear cartilages by tissue engineering. These cartilages can be formed in vivo or in vitro. If you want to use in vivo option, there would be needs for a two-stage operation for the patients. And if you want to use in vitro option, there would be one-stage operation but the mechanical properties of the new cartilages are weaker than the in vivo one.

The 3D manufacturing of the scaffolds is a new concept that has many successful results and the regenerated ear would have a natural 3D configuration very similar to a normal ear.

20.7 Horizons of Research

The new research can be focused over reconstructing cartilage in fewer days, promoting faster maturation for new cartilages, having strong mechanical properties, and regenerating ear framework with bilayer skin over both sides of the framework (Table 20.5; Fig. 20.15).

Table 20.5 Potential future benefits and challenges of combining regenerative medicine with additive manufacturing

	Feature	Benefits	Challenges
Bioprinting	Control over macrostructure and microstructure of tissue produced	Replicate anatomical form Reduce surgical technique learning curve	Biomechanical properties of bioinks Effect of printing on cells Printing resolution
	Patient-specific macrostructure from image acquisition (CT/MRI)	Reduce variability in surgical outcomes	Macrostructure may alter during bioreactor maturation
	Manufacture ex vivo	Avoid donor site morbidity Reduce operating time	Potential for contamination Regulatory constraints
Regenerative medicine	Tissue-specific stem cells to improve quality and functionality of engineered tissue	True "like for like" replacement Restoring native anisotropy allows improved matching of mechanical properties	Genetic stability and differentiation capacity of cells after prolonged expansion in culture
	Tissue maturation utilizing growth factors	Reduce degradation and constriction	Optimal growth factor combinations and temporal effects

CT computed tomography, *MRI* magnetic resonance imaging
From: Z. M. Jessop, M. Javed, I. A. Otto, E. J. Combellack, S. Morgan, C.C. Breugem, et al.; Combining regenerative medicine strategies to provide durable reconstructive options: auricular cartilage tissue engineering; Stem Cell Research & Therapy 2016 **7**:19, **DOI:** 10.1186/s13287-015-0273-0 **Published:** 28 January 2016. [The article is distributed under the terms of the Creative Commons Attribution 4.0 International License (http://creativecommons.org/licenses/by/4.0/), which permits unrestricted use, distribution, and reproduction in any medium, provided you give appropriate credit to the original author(s) and the source, provide a link to the Creative Commons license]

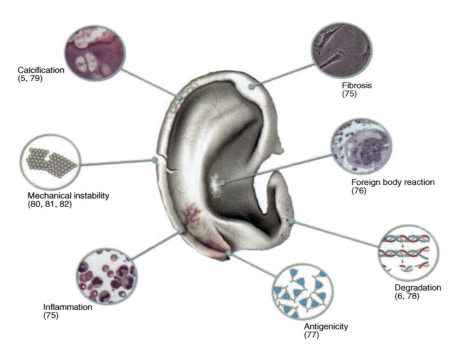

Fig. 20.15 Limitations of current tissue-engineered auricular cartilage constructs. (From: Z. M. Jessop, M. Javed, I. A. Otto, E. J. Combellack, S. Morgan, C.C. Breugem, et al.; Combining regenerative medicine strategies to provide durable reconstructive options: auricular cartilage tissue engineering; Stem Cell Research & Therapy 2016 **7**:19, **DOI:** 10.1186/s13287-015-0273-0 **Published:** 28 January 2016. [The article is distributed under the terms of the Creative Commons Attribution 4.0 International License (http://creativecommons.org/licenses/by/4.0/), which permits unrestricted use, distribution, and reproduction in any medium, provided you give appropriate credit to the original author(s) and the source, provide a link to the Creative Commons license])

References

1. Nagata S. A new method of total reconstruction of the auricle for microtia. Plast Reconstr Surg. 1993;92:187–201.
2. De La Cruz A, Kesser BW. Management of the unilateral atretic ear. In: Pensak M, editor. Controversies in otolaryngology—head and neck surgery. New York: Thieme Medical Publishers; 1999. p. 381–5.
3. Kountakis SE, Helidonis E, Jahrsdoerfer RA. Microtia grade as an indicator of middle ear development in aural atresia. Arch Otolaryngol Head Neck Surg. 1995;121(8):885–6.
4. Vrabec JT, Lin JW. Inner ear anomalies in congenital aural atresia. Otol Neurotol. 2010;31:1421.
5. Liu Y, Zhang L, et al. In vitro engineering of human ear-shaped cartilage assisted with CAD/CAM technology. Biomaterials. 2010;31:2176–83.
6. Tanzer RC. Total reconstruction of the external ear. Plast Reconstr Surg. 1959;23:1–15.
7. Brent B. Technical advances with autogenous rib cartilage grafts—a personal review of 1,200 cases. Plast Reconstr Surg. 1999;104(2):319–34.
8. Brent B. Auricular repair with autogenous rib cartilage grafts: two decades of experience with 600 cases. Plast Reconstr Surg. 1992;90(3):355–74.
9. Firmin F. Microtie Reconstruction par la Technique de Brent. Ann Chir Plast Esthet. 1992;1:119.
10. Nagata S. Modification of the stages in total reconstruction of the auricle: Part I. Grafting the three-dimensional costal cartilage framework for lobule-type microtia. Plast Reconstr Surg. 1994;93(2):221–30.
11. Brent B. The team approach to treating the microtia-atresia patient. Otolaryngol Clin N Am. 2000;33(6):1353–65.
12. Song C, Jiao F, Zhuang H. [Clinical study on external ear reconstruction using expanded postauricular flap and medpor framework]. Zhongguo Xiu Fu Chong Jian Wai Ke Za Zhi. 2007;21(1):40–3.
13. Isogai N, Asamura S, Higashi T, Ikada Y, Morita S, Hillyer J, Jacquet R, Landis WJ. Tissue engineering of an auricular cartilage model utilizing cultured chondrocyte-poly(L-lactide-epsilon-caprolactone) scaffolds. Tissue Eng. 2004;10(5–6):673–87.
14. Henderson JH, Welter JF, Mansour JM, Niyibizi C, Caplan AI, Dennis JE. Cartilage tissue engineering for laryngotracheal reconstruction: comparison of chondrocytes from three anatomic locations in the rabbit. Tissue Eng. 2007;13(4):843–53.
15. Naumann A, Aigner J, Staudenmaier R, Seemann M, Bruening R, Englmeier KH, Kadegge G, Pavesio A, Kastenbauer E, Berghaus A. Clinical aspects and strategy for biomaterial engineering of an auricle based on three-dimensional stereolithography. Eur Arch Otorhinolaryngol. 2003;260(10):568–75.
16. Kamil SH, Kojima K, Vacanti MP, Bonassar LJ, Vacanti CA, Eavey RD. In vitro tissue engineering to generate a human-sized auricle and nasal tip. Laryngoscope. 2003;113(1):90–4.
17. Kelley TF, Sutton FM, Wallace VP, Wong BJ. Chondrocyte repopulation of allograft cartilage: a preliminary investigation and strategy for developing cartilage matrices for reconstruction. Otolaryngol Head Neck Surg. 2002;127(4):265–70.
18. Shin YS, Lee BH, Choi JW, Min BH, Chang JW, Yang SS, Kim CH. Tissue-engineered tracheal reconstruction using chondrocyte seeded on a porcine cartilage-derived substance scaffold. Int J Pediatr Otorhinolaryngol. 2014;78(1):32–8.
19. Sterodimas A, de Faria J, Correa WE, Pitanguy I. Tissue engineering and auricular reconstruction: a review. J Plast Reconstr Aesthet Surg. 2009;62(4):447–52.
20. Hong HJ, Lee JS, Choi JW, Min BH, Lee HB, Kim CH. Transplantation of autologous chondrocytes seeded on a fibrin/hyaluronan composite gel into tracheal cartilage defects in rabbits: preliminary results. Artif Organs. 2012;36(11):998–1006.
21. Komura M, Komura H, Otani Y, Kanamori Y, Iwanaka T, Hoshi K, Tsuyoshi T, Tabata Y. The junction between hyaline cartilage and engineered cartilage in rabbits. Laryngoscope. 2013;123(6):1547–51.
22. Kang N, Liu X, Cao Y, Xiao R. [Comparison study of tissue engineered cartilage constructed with chondrocytes derived from porcine auricular and articular cartilage]. Zhonghua Zheng Xing Wai Ke Za Zhi. 2014;30(1):33–40.
23. Kusuhara H, Isogai N, Enjo M, Otani H, Ikada Y, Jacquet R, Lowder E, Landis WJ. Tissue engineering a model for the human ear: assessment of size, shape, morphology, and gene expression following seeding of different chondrocytes. Wound Repair Regen. 2009;17(1):136–46.
24. Yanaga H, Imai K, Fujimoto T, Yanaga K. Generating ears from cultured autologous auricular chondrocytes by using two-stage implantation in treatment of microtia. Plast Reconstr Surg. 2009;124(3):817–25.
25. Yanaga H, Imai K, Tanaka Y, Yanaga K. Two-stage transplantation of cell-engineered autologous auricular chondrocytes to regenerate chondrofat composite tissue: clinical application in regenerative surgery. Plast Reconstr Surg. 2013;132(6):1467–77.
26. Rodriguez A, Cao YL, Ibarra C, Pap S, Vacanti M, Eavey RD, Vacanti CA. Characteristics of cartilage engineered from human pediatric auricular cartilage. Plast Reconstr Surg. 1999;103(4):1111–9.
27. Park SS, Jin HR, Chi DH, Taylor RS. Characteristics of tissue-engineered cartilage from human auricular chondrocytes. Biomaterials. 2004;25(12):2363–9.
28. Fulco I, Miot S, Haug MD, Barbero A, Wixmerten A, Feliciano S, Wolf F, Jundt G, Marsano A, Farhadi J, Heberer M, Jakob M, Schaefer DJ, Martin I. Engineered autologous cartilage tissue for nasal reconstruction after tumour resection: an observational first-in-human trial. Lancet. 2014;384(9940):337–46.
29. Ishak MF, See GB, Hui CK, Abdullah AB, Saim LB, Saim AB, Idrus RB. The formation of human auricular cartilage from microtic tissue: an in vivo study. Int J Pediatr Otorhinolaryngol. 2015;79(10):1634–9.

30. Shasti M, Jacquet R, McClellan P, Yang J, Matsushima S, Isogai N, Murthy A, Landis WJ. Effects of FGF-2 and OP-1 in vitro on donor source cartilage for auricular reconstruction tissue engineering. Int J Pediatr Otorhinolaryngol. 2014;78(3):416–22.

31. Cheng Y, Cheng P, Xue F, Wu KM, Jiang MJ, Ji JF, Hang CH, Wang QP. Repair of ear cartilage defects with allogenic bone marrow mesenchymal stem cells in rabbits. Cell Biochem Biophys. 2014;70(2):1137–43.

32. Pleumeekers MM, Nimeskern L, Koevoet WL, Karperien M, Stok KS, van Osch GJ. Cartilage regeneration in the head and neck area: combination of ear or nasal chondrocytes and mesenchymal stem cells improves cartilage production. Plast Reconstr Surg. 2015;136(6):762e–74e.

33. Wei B, Jin C, Xu Y, Tang C, Hu W, Wang L. [Effect of bone marrow mesenchymal stem cells-derived extracellular matrix scaffold on chondrogenic differentiation of marrow clot after microfracture of bone marrow stimulation in vitro]. Zhongguo Xiu Fu Chong Jian Wai Ke Za Zhi. 2013;27(4):464–74.

34. Zhang L, He A, Yin Z, Yu Z, Luo X, Liu W, Zhang W, Cao Y, Liu Y, Zhou G. Regeneration of human-ear-shaped cartilage by co-culturing human microtia chondrocytes with BMSCs. Biomaterials. 2014;35(18):4878–87.

35. Wang M, Xia Y, Wang S. Experimental study on repair of articular cartilage defects with homograft of marrow mesenchymal stem cells seeded onto poly-l-lactic acid/gelatin. Zhongguo Xiu Fu Chong Jian Wai Ke Za Zhi. 2007;21(7):753–8.

36. Chen Y, Chen Y, Zhang S, Du X, Bai B. Parathyroid hormone-induced bone marrow mesenchymal stem cell chondrogenic differentiation and its repair of articular cartilage injury in rabbits. Med Sci Monit Basic Res. 2016;22:132–45.

37. Saha S, Kirkham J, Wood D, Curran S, Yang XB. Informing future cartilage repair strategies: a comparative study of three different human cell types for cartilage tissue engineering. Cell Tissue Res. 2013;352(3):495–507.

38. Xiang Z, Hu W, Kong Q, Zhou H, Zhang X. Preliminary study of mesenchymal stem cells-seeded type I collagen-glycosaminoglycan matrices for cartilage repair. Zhongguo Xiu Fu Chong Jian Wai Ke Za Zhi. 2006;20(2):148–54.

39. Lü CW, Hu YY, Bai JP, Liu J, Meng GL, Lü R. Three dimensional induction of autologous mesenchymal stem cell and the effects on depressing long-term degeneration of tissue-engineering cartilage. Zhonghua Wai Ke Za Zhi. 2007;45(24):1717–21.

40. Zhou GD, Miao CL, Wang XY, Liu TY, Cui L, Liu W, Cao YL. [Experimental study of in vitro chondrogenesis by co-culture of b0one marrow stromal cells and chondrocytes]. Zhonghua Yi Xue Za Zhi. 2004;84(20):1716–20.

41. Qing C, Wei-ding C, Wei-min F. Co-culture of chondrocytes and bone marrow mesenchymal stem cells in vitro enhances the expression of cartilaginous extracellular matrix components. Braz J Med Biol Res. 2011;44(4):303–10.

42. He X, Feng B, Huang C, Wang H, Ge Y, Hu R, Yin M, Xu Z, Wang W, Fu W, Zheng J. Electrospun gelatin/polycaprolactone nanofibrous membranes combined with a coculture of bone marrow stromal cells and chondrocytes for cartilage engineering. Int J Nanomedicine. 2015;10:2089–99.

43. Deng J, She R, Huang W, Dong Z, Mo G, Liu B. A silk fibroin/chitosan scaffold in combination with bone marrow-derived mesenchymal stem cells to repair cartilage defects in the rabbit knee. J Mater Sci Mater Med. 2013;24(8):2037–46.

44. Ba R, Wei J, Li M, Cheng X, Zhao Y, Wu W. Cell-bricks based injectable niche guided persistent ectopic chondrogenesis of bone marrow-derived mesenchymal stem cells and enabled nasal augmentation. Stem Cell Res Ther. 2015;6:16.

45. Chen Z, Wei J, Zhu J, Liu W, Cui J, Li H, Chen F. Chm-1 gene-modified bone marrow mesenchymal stem cells maintain the chondrogenic phenotype of tissue-engineered cartilage. Stem Cell Res Ther. 2016;7(1):70.

46. Zheng YH, Su K, Kuang SJ, Li H, Zhang ZG. [New bone and cartilage tissues formed from human bone marrow mesenchymal stem cells derived from human condyle in vivo]. Zhonghua Kou Qiang Yi Xue Za Zhi. 2012;47(1):10–3.

47. Ge W, Jiang W, Li C, You J, Qiu L, Zhao C. [Conduction of injectable cartilage using fibrin sealant and human bone marrow mesenchymal stem cells in vivo]. Zhongguo Xiu Fu Chong Jian Wai Ke Za Zhi. 2006;20(2):139–43.

48. Utomo L, Pleumeekers MM, Nimeskern L, Nürnberger S, Stok KS, Hildner F, van Osch GJ. Preparation and characterization of a decellularized cartilage scaffold for ear cartilage reconstruction. Biomed Mater. 2015;10(1):015010.

49. Qi H, Jie Y, Chen L, Jiang L, Gao X, Sun L. Preparation of acellular dermal matrix as a kind of scaffold for cartilage tissue engineering and its biocompatibility. Zhongguo Xiu Fu Chong Jian Wai Ke Za Zhi. 2014;28(6):768–72.

50. Sato Y, Wakitani S, Takagi M. Xeno-free and shrinkage-free preparation of scaffold-free cartilage-like disc-shaped cell sheet using human bone marrow mesenchymal stem cells. J Biosci Bioeng. 2013;116(6):734–9.

51. Liu Y, He L, Tian J. [Effect of basic fibroblast growth factor and parathyroid hormone-related protein on early and late chondrogenic differentiation of rabbit bone marrow mesenchymal stem cells induced by transforming growth factor beta 1]. Zhongguo Xiu Fu Chong Jian Wai Ke Za Zhi. 2013;27(2):199–206 (in Chinese).

52. Jakobsen RB, Shahdadfar A, Reinholt FP, Brinchmann JE. Chondrogenesis in a hyaluronic acid scaffold: comparison between chondrocytes and MSC from bone marrow and adipose tissue. Knee Surg Sports Traumatol Arthrosc. 2010;18(10):1407–16. https://doi.org/10.1007/s00167-009-1017-4. Epub 2009 Dec 18. Erratum in: Knee Surg Sports Traumatol Arthrosc. 2014 Jul;22(7):1711-4.

53. Wang H, Li Y, Chen J, Wang X, Zhao F, Cao S. [Chondrogenesis of bone marrow mesenchymal stem cells induced by transforming growth factor beta3 gene in Diannan small-ear pigs]. Zhongguo Xiu Fu Chong Jian Wai Ke Za Zhi. 2014;28(2):149–54.

54. Kang N, Liu X, Yan L, Wang Q, Cao Y, Xiao R. Different ratios of bone marrow mesenchymal stem cells and chondrocytes used in tissue-engineered cartilage and its application for human ear-shaped substitutes in vitro. Cells Tissues Organs. 2013;198(5):357–66.

55. Karimi H, Emami SA, Olad-Gubad MK. Bone marrow stem cells and ear framework reconstruction. J Craniofac Surg. 2016;27(8):2192–6.

56. Carbone A, Valente M, Annacontini L, Castellani S, Di Gioia S, Parisi D, Rucci M, Belgiovine G, Colombo C, Di Benedetto A, Mori G, Lo Muzio L, Maiorella A, Portincasa A, Conese M. Adipose-derived mesenchymal stromal (stem) cells differentiate to osteoblast and chondroblast lineages upon incubation with conditioned media from dental pulp stem cell-derived osteoblasts and auricle cartilage chondrocytes. J Biol Regul Homeost Agents. 2016;30(1):111–22.

57. Cai Z, Pan B, Jiang H, Zhang L. Chondrogenesis of human adipose-derived stem cells by in vivo co-graft with auricular chondrocytes from Microtia. Aesthet Plast Surg. 2015;39(3):431–9.

58. Bahrani H, Razmkhah M, Ashraf MJ, Tanideh N, Chenari N, Khademi B, Ghaderi A. Differentiation of adipose-derived stem cells into ear auricle cartilage in rabbits. J Laryngol Otol. 2012;126(8):770–4.

59. Meric A, Yenigun A, Yenigun VB, Dogan R, Ozturan O. Comparison of chondrocytes produced from adipose tissue-derived stem cells and cartilage tissue. J Craniofac Surg. 2013;24(3):830–3.

60. Sterodimas A, de Faria J. Human auricular tissue engineering in an immunocompetent animal model. Aesthet Surg J. 2013;33(2):283–9.

61. Dahlin RL, Kinard LA, Lam J, Needham CJ, Lu S, Kasper FK, Mikos AG. Articular chondrocytes and mesenchymal stem cells seeded on biodegradable scaffolds for the repair of cartilage in a rat osteochondral defect model. Biomaterials. 2014;35(26):7460–9.

62. Liu K, Zhou GD, Liu W, Zhang WJ, Cui L, Liu X, Liu TY, Cao Y. The dependence of in vivo stable ectopic chondrogenesis by human mesenchymal stem cells on chondrogenic differentiation in vitro. Biomaterials. 2008;29(14):2183–92.

63. Raghunath J, Sutherland J, Salih V, Mordan N, Butler PE, Seifalian AM. Chondrogenic potential of blood acquired mesenchymal progenitor cells. J Plast Reconstr Aesthet Surg. 2010;63:841–7.

64. Xu J-W, Shane Johnson T, et al. Tissue-engineered flexible ear-shaped cartilage. Plast Reconstr Surg. 2005;115:1633.

65. Togo T, Utani A, Naitoh M, Ohta M, Tsuji Y, Morikawa N, Nakamura M, Suzuki S. Identification of cartilage progenitor cells in the adult ear perichondrium: utilization for cartilage reconstruction. Lab Invest. 2006;86(5):445–57.

66. Kagimoto S, Takebe T, Kobayashi S, Yabuki Y, Hori A, Hirotomi K, Mikami T, Uemura T, Maegawa J, Taniguchi H. Autotransplantation of monkey ear perichondrium-derived progenitor cells for cartilage reconstruction. Cell Transplant. 2016;25(5):951–62.

67. Jiang Y, Cai Y, Zhang W, Yin Z, Hu C, Tong T, Lu P, Zhang S, Neculai D, Tuan RS, Ouyang HW. Human cartilage-derived progenitor cells from committed chondrocytes for efficient cartilage repair and regeneration. Stem Cells Transl Med. 2016;5(6):733–44. https://doi.org/10.5966/sctm.2015-0192. Epub 2016 Apr 29. PubMed PMID: 27130221; PubMed Central PMCID: PMC4878331.

68. Choi WH, Kim HR, Lee SJ, Jeong N, Park SR, Choi BH, Min BH. Fetal cartilage-derived cells have stem cell properties and are a highly potent cell source for cartilage regeneration. Cell Transplant. 2016;25(3):449–61.

69. Pomerantseva I, Bichara DA, Tseng A, Cronce MJ, Cervantes TM, Kimura AM, Neville CM, Roscioli N, Vacanti JP, Randolph MA, Sundback CA. Ear-shaped stable auricular cartilage engineered from extensively expanded chondrocytes in an immunocompetent experimental animal model. Tissue Eng Part A. 2016;22(3–4):197–207.

70. Liao HT, Zheng R, Liu W, Zhang WJ, Cao Y, Zhou G. Prefabricated, ear-shaped cartilage tissue engineering by scaffold-free porcine chondrocyte membrane. Plast Reconstr Surg. 2015;135(2):313e–21e.

71. Neumeister MW, Wu T, Chambers C. Vascularized tissue-engineered ears. Plast Reconstr Surg. 2006;117(1):116–22.

72. von Bomhard A, Veit J, Bermueller C, Rotter N, Staudenmaier R, Storck K. The HN. Prefabrication of 3D cartilage constructs: towards a tissue engineered auricle—a model tested in rabbits. PLoS One. 2013;8(8):e71667. https://doi.org/10.1371/journal.pone.0071667. eCollection 2013. Erratum in: PLoS One. 2014;9(11):e113017

73. Ruszymah BH, Chua KH, Mazlyzam AL, Aminuddin BS. Formation of tissue engineered composite construct of cartilage and skin using high density polyethylene as inner scaffold in the shape of human helix. Int J Pediatr Otorhinolaryngol. 2011;75(6):805–10.

74. Nayyer L, Patel K, et al. Revolution and challenge in auricular cartilage reconstruction. Plast Reconstr Surg. 2012;129(5):1123–37.

75. Cronin TD. Use of a silastic frame for total and subtotal reconstruction of the external ear: preliminary report. Plast Reconstr Surg. 1966;37:399.

76. Cronin TD, Greenberg RL, Brauer RO. Follow-up study of silastic frame for reconstruction of external ear. Plast Reconstr Surg. 1968;42:522.

77. Cronin TD, Ascough BM. Silastic ear construction. Clin Plast Surg. 1978;5:367.

78. Reinisch J. Microtia reconstruction using a polyethylene implant: an eight year surgical experience. In: Paper presented at the 1999 annual meeting of the American Association of Plastic Surgeons. Colorado Springs, CO; 1999 May 5.

79. Reinisch JF, Lewin S. Ear reconstruction using a porous polyethylene framework and temporoparietal fascia flap. Facial Plast Surg. 2009;25(3):181–9.

80. Romo T 3rd, Morris LG, Reitzen SD, Ghossaini SN, Wazen JJ, Kohan D. Reconstruction of congenital microtia-atresia: outcomes with the Medpor/bone-anchored hearing aid-approach. Ann Plast Surg. 2009;62(4):384–9.

81. Eyre DR, Dickson IR, Van Ness KP. Collagen cross-linking in human bone and articular cartilage. Age-related changes in the content of mature hydroxypyridinium residues. Biochem J. 1988;252:495–500.

82. Bank RM, Bayliss FPJG, Lafeber AM, Tekoppele J. Ageing and zonal variation in post-translational modification of collagen in normal human articular cartilage. Biochem J. 1998;330:345–51.

83. Brommer H, Brama PAJ, Laasanen MS, Helminen HJ, van Weeren PR, Jurvelin JS. Functional adaptation of articular cartilage from birth to maturity under the influence of loading: a biomechanical analysis. Equine Vet J. 2005;37(2):148–54.

84. Williamson AK, Chen AC, Masuda K, Thonar EJ, Sah RL. Tensile mechanical properties of bovine articular cartilage: variations with growth and relationships to collagen network components. J Orthop Res. 2003;21(5):872–80.

85. Bichara DA, O'Sullivan NA, Pomerantseva I, Zhao X, Sundback CA, Vacanti JP, et al. The tissue-engineered auricle: past, present, and future. Tissue Eng Part B Rev. 2012;18:51–61.

86. Zopf DA, Flanagan CL, Nasser HB, Mitsak AG, Huq FS, Rajendran V, et al. Biomechanical evaluation of human and porcine auricular cartilage. Laryngoscope. 2015;125(8):E262–8.

87. Khan IM, Evans SL, Young RD, Blain EJ, Quantock AJ, Avery N, et al. Fibroblast growth factor 2 and transforming growth factor $\beta 1$ induce precocious maturation of articular cartilage. Arthritis Rheum. 2011;63(11):3417–27.

88. Khan IM, Francis L, Theobald PS, Perni S, Young RD, Prokopovich P, et al. In vitro growth factor-induced bio engineering of mature articular cartilage. Biomaterials. 2013;34(5):1478–87.

89. Cao Y, Vacanti JP, Paige KT, Upton J, Vacanti CA. Transplantation of chondrocytes utilizing a polymer-cell construct to produce tissue-engineered cartilage in the shape of a human ear. Plast Reconstr Surg. 1997;100:297–302.

Part IV

Muscle and Tendon Regeneration

Muscle Fiber Regeneration in Long-Term Denervated Muscles: Basics and Clinical Perspectives

21

Ugo Carraro, Helmut Kern, Sandra Zampieri, Paolo Gargiulo, Amber Pond, Francesco Piccione, Stefano Masiero, Franco Bassetto, and Vincenzo Vindigni

21.1 Introduction

In recent years, basic research into the use of electrostimulation to provoke muscle plasticity after spinal cord injury (SCI) has prompted clinical trials which have investigated the use of long impulse electrical stimulation as a treatment for long-term permanently denervated human muscles [1, 2]. These studies were aimed at improving muscle trophism and increasing external muscle power to a level sufficient to restore an impulse-assisted ability to perform "stand-up" exercises [3–9]; that is, the muscle mass and strength would be increased to the point that individuals could actually stand up with assistance from direct electrical stimulation of the denervated muscles. These treatments usually were initiated late after denervation because of clinical constraints and/or outdated beliefs that electrical stimulation is ineffective and may actually interfere with ensuing myofiber reinnervation. These ideas are still maintained by many experts despite recent evidence that electrostimulation may indeed enhance nerve growth and appropriate muscle reinnervation [10–12]. Here we discuss works exploring the effects of combining spontaneous or induced aneural myogenesis with delayed long-term electrical stimulation.

Permanent denervation of skeletal muscle results in early loss of function and subsequent tissue wasting that is believed to lead to muscle fiber death and ultimately to substitution of contractile tissue with adipocytes and collagenous sheets [8, 9]. More specifically, permanent mus-

U. Carraro
IRCCS Fondazione Ospedale San Camillo, Venezia-Lido, Italy

Department of Neurorehabilitation, Foundation San Camillo Hospital, I.R.C.C.S., Venice, Italy
e-mail: ugo.carraro@unipd.it

H. Kern
Physiko- und Rheumatherapie, St. Poelten, Austria
e-mail: helmut@kern-reha.at

S. Zampieri · S. Masiero · F. Bassetto
Interdepartmental Research Centre of Myology, University of Padova, Padova, Italy
e-mail: sanzamp@unipd.it; stef.masiero@unipd.it; franco.bassetto@unipd.it

P. Gargiulo
Clinical Engineering and Information Technology, Landspitali—University Hospital, Reykjavik, Iceland
e-mail: paologar@landspitali.is

A. Pond
Anatomy Department, Southern Illinois University School of Medicine, Carbondale, IL, USA
e-mail: apond@siumed.edu

F. Piccione
IRCCS Fondazione Ospedale San Camillo, Venezia-Lido, Italy
e-mail: francesco.piccione@ospedalesancamillo.net

V. Vindigni (✉)
Unit of Plastic Surgery, Department of Neuroscience, University of Padova, Padova, Italy
e-mail: vincenzo.vindigni@unipd.it

© Springer Nature Switzerland AG 2019
D. Duscher, M. A. Shiffman (eds.), *Regenerative Medicine and Plastic Surgery*,
https://doi.org/10.1007/978-3-030-19962-3_21

cle denervation in mammals produces a long-term period of denervation atrophy followed by many months later by severe atrophy accompanied with lipodystrophy and fibrosis, and additionally by non-compensatory myogenic events (regeneration of muscle fibers). In the rodent life span (3–4 years), these events begin 24 months after permanent denervation and persist the entire lifetime; however differences between species can be significant and at this point are poorly outlined (see below). Nonetheless, all of these events, regardless of time frame, constitute the concept of muscle plasticity, that is, the wide range of adaptive responses of muscle tissue to increased or diminished use [13].

The mechanisms leading to cell death in rodent skeletal muscle undergoing post-denervation atrophy have been described in detail [14–16]. Briefly, as time elapses from the denervation event, ultrastructural characteristics very similar to those considered as markers of apoptosis are noted along with clear morphological manifestations of muscle cell death, starting from progressive destabilization of the differentiated phenotype of muscle cells, as evidenced by spatial disorganization of myofibrils and formation of myofibril-free zones. Dead muscle fibers are observed, those being typically surrounded by a folded intact basal lamina and having an intact sarcolemma and highly condensed chromatin and sarcoplasm. The numbers of nuclei displaying abnormal morphology exceed the numbers of nuclei positive for apoptosis by 30–40-fold.

Muscle degeneration (we will not use the word dystrophy to avoid confusion with genetic muscle diseases) is an even later effect of denervation during which a significant portion of the muscle mass is substituted by different cells (mainly fibroblasts) and collagen sheets which surround myofibers (endomysial fibrosis) followed by degeneration or lipodystrophy [17]. These later alterations appear to occur in the long-term denervated muscle of rat after a heavy reduction in the number of capillaries per myofiber [18].

The fully differentiated pattern of fast and slow myosins becomes established in normal adult skeletal muscles and acute denervation has very little influence on the type of contractile proteins synthesized in the early atrophying muscle fibers. Nonetheless, a small net change in fiber type appears to be a typical feature of the early phases of denervation. This relative imbalance in fiber type is attributed to the preferential atrophy of fast fibers followed by atrophy of slow fibers [19, 20]. However, after several months of permanent denervation in rats, there is an almost complete transformation of mixed muscles into nearly pure fast muscles [21–23], with only a small amount of residual slow myosin present with the fast myosin. Analyses of denervated and aneurally regenerated muscles suggest that in long-term denervated rat soleus the slow-to-fast transformation is mainly the result of repeated cycles of cell death and regeneration [24–26]. Such a slow myosin disappearance is less pronounced in other species [27], but it is not known if this means that post-denervation myofiber regeneration is less pronounced.

21.2 Spontaneous Myogenesis in Denervated Skeletal Muscles

There is a general consensus that in mature mammalian skeletal muscles "satellite cells" are the source of new myonuclei, in particular in regenerative processes after trauma or myotoxic injuries [28, 29]. Despite some doubt [30], in mammalian skeletal muscle there is also some evidence suggesting that satellite cell proliferation is necessary to support the process of compensatory hypertrophy. Specifically, when elevated radiation doses are used in rodents to prevent satellite cell proliferation prior to initiation of skeletal muscle functional overloading, the hypertrophy response is nearly absent [31].

Evidence of myogenic events is also observed in atrophying denervated skeletal muscles. In contrast with older reports, light and electron microscopy studies reveal that long-term denervated muscle maintains a steady-state severe atrophy for the animal life span. Further, some morphological and molecular features indicate that events of aneural regeneration occur continuously [21, 32]. New muscle fibers are present as early as 1 month after nerve section and reach a maximum between 2 and 4 months in rat leg muscles following denervation. Myogenesis gradually decreases with progressive post-denervation degeneration, although myogenic events continue to occur and are not secondary to muscle reinnervation [33]. Muscle partial denervation or reinnervation is a major technical problem in the study of long-term denervated muscle in rodents. The small size of rat legs and the high capacity for peripheral nerve regeneration necessitate the development and execution of careful surgical approaches in establishing the experimental model [18]. Furthermore, minimal residual innervation is the most common event in clinical cases. Research on this "disturbing" variable in long-term denervated human muscle management is needed.

The myogenic response in long-term denervated rat muscle is biphasic and includes two distinct processes. The first process, which dominates during the first 2 months post-denervation, resembles the formation of secondary and tertiary generations of myotubes which occurs during normal muscle development. The activation of this type of myogenic response (myofiber generation, which truly increases the number of myofibers present in an anatomically defined muscle, i.e., hyperplasia) does not depend on cell death and degenerative processes [17].

A second type of myogenesis is a typical regenerative reaction that occurs mainly within the spaces surrounded by the basal lamina of dead muscle fibers [18]. Myofibers of varying sizes are vulnerable to degeneration and death, which indicates that cell death does not correlate with levels of muscle cell atrophy in denervated muscle [17]. These regenerative processes frequently result in the development of abnormal muscle cells that branch or form small clusters surrounded by two layers of basal lamina (the old layer and the new one which was secreted by the new myofiber) [17, 21, 32, 33]. Spontaneous myofiber regeneration in long-term denervation has been quantified and shown to be non-compensatory and to result in the reduction of

satellite cell pools [17, 24, 25, 34–39]. The turnover of myonuclei (but not necessarily of regenerating myofibers) in adult rats, studied by continuous infusion of 5-bromo-2-deoxyuridine (BRDU), occurs at a rate of 1–2% per week at most [38, 40]. Evidence of satellite cell depletion in denervated muscle raises some questions about the long-term potential effects of regenerative myogenesis [35, 36]; however, myotoxin-induced myogenesis suggests that a long-lasting effect could occur(see below).

Interspecies differences raise additional doubts, since there are significant differences in post-denervation effects, even in rate of atrophy [41]. In humans, denervation atrophy progresses at a relatively slower rate in comparison to rat, but we have observed that the myogenic reactions to denervation in human muscle are very similar to those well described in rodents; this is indeed encouraging as to the potential application of our work in clinical rehabilitation.

21.3 Induced Myogenesis in Long-Term Denervation

In addition to traumatic events, regeneration of muscle fibers has been studied after induction of muscle damage and regeneration by vitamin E deprivation [42], autografting [17, 18], and as induced by myotoxin treatment [43, 44]. Studies demonstrate that permanent denervation does not prevent induced muscle regeneration [32] and a long-term retention of this capability has been demonstrated. For example, 4 months after denervation, rats treated with bupivacaine develop massive and synchronous myofiber regeneration in both fast and slow muscles within a few days of treatment [26, 45]. Additionally, when rat muscles are denervated for 7 months and then the rats are treated with marcaine or notexin, autografting of the muscles is followed by substitution of old fibers by new fibers [32, 34, 36, 46, 47]. Specifically, 2 weeks after autotransplantation, the aneurally regenerated myofibers increased in size up to 25% of the normal size of innervated muscle fibers, and then they decreased in size to almost one-tenth of their normal adult size [32]. Evidence that some of the proliferating satellite cells reenter an undifferentiated, stem cell-like state, being capable of myogenesis after further injuries, was provided by a study consisting of a series of myotoxic exposures which provided time for regeneration to occur between each toxic injury. The results revealed that new muscle fibers form after insult for up to four treatments [48, 49]. Further, satellite cell proliferation and myofiber regeneration are enhanced in long-term denervated muscles when they are treated with electrical stimulation [50]. This is not surprising because satellite cell proliferation is increased in normal muscle by increased physical activity, in particular strength training [51], but also by massage [52].

21.4 What Is Feasible and What Will Be Possible

21.4.1 h-bFES of Long-Term Denervated Human Muscles

Over the last few decades there has been an increased interest in the use of home-based functional electrical stimulation (h-bFES) to restore movement to the limbs of immobilized spinal cord injury (SCI) patients (i.e., upper motor neuron lesion, spastic paralysis) [1–9, 53–55]. However, there is another group of SCI patients whose issues are more difficult to treat. Patients living with complete conus and cauda equina syndrome present with paralysis and severe secondary medical problems because of the marked atrophy of denervated muscles and the associated loss of bone mass and skin dystrophy [1]. In these patients, injury also causes irreversible loss of the nerve supply to some or all the muscle fibers of the affected limbs, resulting in flaccid paralysis subsequent to lower motor neuron lesion. It is technically more difficult to treat these patients because the absence of functional nerve fibers makes it more difficult to recruit the population of myofibers necessary to regain functional movements at an acceptable force level using surface electrodes. Thus, the electrical energy required to stimulate these muscles directly is greater than that which can be delivered by commercially available stimulation devices used for innervated muscles.

Despite these difficulties, pilot studies of the functional clinical application of FES to denervated muscles have been published. One study demonstrated that direct FES of the denervated tibialis anterior muscle could result in gait correction [56]. Other papers [3–9, 57], contrary to widely accepted opinion, have shown that electrical stimulation of even long-term denervated muscles can produce muscle contractions strong enough (i.e., tetanic contractions) to restore muscle mass and force production. Indeed, through the successful EU Program RISE [Use of electrical stimulation to restore standing in paraplegics with long-term denervated degenerated muscles (QLG5-CT-2001-02191), we have demonstrated that h-bFES therapy improves the muscle condition of mobility-impaired persons, even in extreme cases in which post-denervation muscle degeneration has occurred [3–9, 57]. Complete conus and cauda equina syndrome, for example, is an SCI sequela in which the leg muscles are completely disconnected from the nervous system. In this condition, affected muscles undergo sequential stages of loss of function and then, finally, complete loss of skeletal muscle tissue. We analyzed muscle biopsies from RISE patients at different time points after SCI and discovered that (a) within months of injury the loss of stimulation-induced contractility produced ultrastructural disorganization; (b) progressive atrophy persisted for up to 2 years after injury; and (c) substantial loss of myofibers appeared more than 3 years after SCI. Importantly, we have shown that h-bFES of denervated muscles can inhibit muscle loss and also recover muscle from degeneration. However, even though h-bFES substantially improves muscle mass and strength, excitability of the treated muscles never reaches the level of normal innervated muscles. Indeed, we reported that long-term discontinuation of h-bFES resulted in loss of the improvements that came with the previous period of treatment [58, 59]. Furthermore, with results produced by a 2-year longitudinal prospective study of 25 patients with complete conus/cauda equina lesions, we showed that the improvements produced by h-bFES can be maintained over time when treatment is continued [9]. Specifically, in this study, denervated leg muscles were stimulated by h-bFES using the prototype of a custom-designed stimulator and large surface electrodes for long-term denervated muscle [3, 5–9, 45, 53, 57] that are now commercially available (the Stimulette Den2x, Dr. Schuhfried Medizintechnik GmbH, Vienna, Austria). Samples were harvested both before and after 2 years of h-bFES and muscle mass, force, and structure were determined using (a) computed tomography; (b) measurements of knee torque during stimulation; and (c) muscle biopsies analyzed by histology and electron microscopy. Twenty out of 25 patients completed the 2-year h-bFES program. The data demonstrated that treatment resulted in (a) a 35% increase in cross section in an area of the quadriceps ($P < 0.001$); (b) a 75% increase in mean diameter of muscle fibers ($P > 0.001$); and (c) improvements of the ultrastructural organization of contractile material. Further, an exciting 1187% increase in force output during electrical stimulation ($P > 0.001$) was achieved. The recovery of quadriceps force was sufficient to allow 25% of the subjects to perform FES-assisted stand-up exercises, demonstrating that h-bFES of denervated muscle is an effective home therapy capable of rescuing tetanic contractility and muscle mass [9]. Important benefits for the patients included the improved cosmetic appearance of lower extremities and the enhanced cushioning effect for seating.

Functional data highly correlate to histomorphometric results [9]. Interestingly, immunohistochemistry for anti-embryonic MHC revealed that regenerating myofibers were present in all of the muscle biopsies. Some myofibers had central nuclei, a feature suggesting that they had regenerated no more than 10 days before muscle biopsy harvesting. Frequency distribution of myofibers according to their minimum diameter in semithin sections showed that about 50% were severely atrophic (i.e., having a minimum diameter smaller than 10 μm), but a large proportion of myofibers were eutrophic (i.e., with a mini-

mum diameter larger than 40 μm). The results were substantiated by structure-to-function correlations and by an advanced clinical muscle imaging technique, i.e., quantitative muscle color-computed tomography (QMC-CT) [9].

Medical imaging, a vital field of research for diagnostic and investigative assessment, is of particular interest here as a tool to recapitulate and quantify internal and external tissue morphologies in a noninvasive manner. The most current research aspires to improve instrument design, image processing software, data acquisition methodology, and computational modeling. In particular, visually simplistic imaging methods with high resolution for assessing diseased or damaged tissues are a strategic priority in translational myology research. The follow-up of muscle atrophy/degeneration in neuro-muscularly traumatized people is difficult because of the lack of adequate imaging analyses and also the practical and ethical constraints on harvesting the muscle biopsies necessary to monitor the efficacy of the therapy/rehabilitation strategies. More sensitive methods for quantitative clinical imaging of skeletal muscle are needed. False color computed tomography is popular in cardiology [60], but it is still a novelty for quantitative analyses of total volume and quality of anatomically defined skeletal muscles. QMC-CT is a highly sensitive quantitative imaging analysis recently developed to monitor skeletal muscle and perform follow-up examinations of muscle affected by wasting conditions [61–63]. QMC-CT uses CT numbers, i.e., Hounsfield units (HU), for tissue characterization. It allows for discrimination of soft tissues as follows: subcutaneous fat, intramuscular fat, low-density muscle, normal muscle, and fibrous-dense connective tissue. To evaluate this data, pixels within the defined interval of HU values (or, more generally, gray values when these data are not from CT scans) are selected and highlighted, while others with HU values outside the threshold remain black. Specific soft-tissue areas are colored as follows: subcutaneous fat (yellow: −200 to −10 HU), intramuscular fat (orange: −200 to −10 HU), low-density muscle (cyan: −9 to 40 HU), normal muscle (red: 41–70 HU), and fibrous-dense connective tissue (gray: 71–150

HU). Thus, soft-tissue discrimination and its quantitation in a leg can be achieved. The Hounsfield values for the entirety of the lower limbs can be plotted on a histogram to display the tissue profiles [58, 59, 61–63].

21.5 Perspectives

After many years of basic research concerning electrostimulation-induced muscle plasticity, we conclude that functional electrical stimulation using long biphasic impulses is able to restore muscle mass, force production, and movement in humans even after years of complete denervation. Patients suffering from flaccid paraplegia (denervation of lower extremity muscles, e.g., conus and cauda equina syndrome) are especially good candidates for these approaches [9].

In the long term, we may consider the development and application of implantable devices as alternatives to the approaches based on surface electrodes; however, because electrical stimulation can elicit pain in patients for whom residual innervation is functional, a better knowledge and control of stimulation-induced muscle trophism must be achieved. Once this control is obtained, artificial synapses (i.e., a pool of miniaturized electrodes which contact each of the surviving or regenerated myofibers in the denervated muscle) would have to be designed and developed. Considering the powerful angiogenesis of regenerating muscle [52], one might consider that the sacrifice of some of the new vascular branches could be worth the development of the ability to deliver sufficient current to new myofibers by means of nano-fabricated electrodes.

Whether procedures based on in vivo protocols such as the induction of muscle damage/regeneration by injection of anesthetics in local anoxic conditions [26, 32, 34, 45, 46, 52, 55] or ex vivo techniques such as the proliferation of autologous myoblasts derived from patient's muscle biopsies (i.e., from their own "satellite cells" or from conversion of autologous fibroblasts to myogenic cells) will develop into applied methods is still open to preclinical and clinical research. A promising aim would be to replenish the degenerated skeletal muscles of long-term SCI persons with these methods before or at the same time h-bFES for denervated muscles is applied. This option could give SCI patients the opportunity to train muscles which have been denervated for more than 10 years.

For now, based upon pilot human studies and application of existing experimental knowledge, it can be anticipated that FES of long-term denervated muscles [9] and atrophic muscles in aging people [64–68] using surface electrodes may improve mobility with substantial reductions in the risk and the severity of secondary medical problems, resulting in less frequent hospitalizations and a reduced burden on public health services. Integration of h-bFES with good nutrition and exercise where possible (e.g., volitional in-bed gym [68, 69]) will allow SCI patients to look forward to improved health, independence, and quality of life, and the prospects of better professional and social integration.

Acknowledgements This work was supported by Italian Ministero per l'Università e la RicercaScientifica eTecnologica (M.U.R.S.T.), the EU Commission Shared Cost Project RISE (Contract no. QLG5-CT-2001-02191), the Austrian Ministry of Science, and the Ludwig Boltzmann Society (Vienna). U. Carraro thanks the IRCCS Fondazione Ospedale San Camillo, Venice (Italy), for scientific support and hospitality.

References

1. Kern H, Carraro U. Home-based Functional Electrical Stimulation (h-b FES) for long-term denervated human muscle: History, basics, results and perspectives of the Vienna Rehabilitation Strategy. Eur J Transl Myol. 2014;24:27–40.
2. Carraro U, Kern H, Gava P, Hofer C, Loefler S, Gargiulo P, Edmunds K, Árnadóttir D, Zampieri S, Ravara B, Gava F, Nori A, Gobbo V, Masiero S, Marcante A, Baba A, Piccione F, Schils S, Pond A, Mosole S. Recovery from muscle weakness by exercise and FES: lessons from Masters, active or sedentary seniors and SCI patients. Aging Clin Exp Res. 2017;29(4):579–90.
3. Kern H, Boncompagni S, Rossini K, Mayr W, Fanò G, Zanin ME, Podhorska-Okolow M, Protasi F, Carraro U. Long-term denervation in humans causes degeneration of both contractile and excitation-contraction coupling apparatus that can be reversed by functional electrical stimulation (FES). A role for

myofiber regeneration? J Neuropathol Exp Neurol. 2004;63:919–31.

4. Kern H, Rossini K, Carraro U, Mayr W, Vogelauer M, Hoellwarth U, Hofer C. Muscle biopsies show that FES of denervated muscles reverses human muscle degeneration from permanent spinal motoneuron lesion. J Rehabil Res Dev. 2005;42:43–53.

5. Carraro U, Rossini K, Mayr W, Kern H. Muscle fiber regeneration in human permanent lower motoneuron denervation: relevance to safety and effectiveness of FES-training, which induces muscle recovery in SCI subjects. Artif Organs. 2005;29:187–91.

6. Kern H, Salmons S, Mayr W, Rossini K, Carraro U. Recovery of long-term denervated human muscles induced by electrical stimulation. Muscle Nerve. 2005;31(1):98–101.

7. Boncompagni S, Kern H, Rossini K, Hofer C, Mayr W, Carraro U, Protasi F. Structural differentiation of skeletal muscle fibers in the absence of innervation in humans. Proc Natl Acad Sci U S A. 2007;104:19339–44.

8. Kern H, Carraro U, Adami N, Hofer C, Loefler S, Vogelauer M, Mayr W, Rupp R, Zampieri S. One year of home-based Functional Electrical Stimulation (FES) in complete lower motor neuron paraplegia: recovery of tetanic contractility drives the structural improvements of denervated muscle. Neurol Res. 2010;32:5–12.

9. Kern H, Carraro U, Adami N, Biral D, Hofer C, Forstner C, Mödlin M, Vogelauer M, Pond A, Boncompagni S, Paolini C, Mayr W, Protasi F, Zampieri S. Home-based functional electrical stimulation rescues permanently denervated muscles in paraplegic patients with complete lower motor neuron lesion. Neurorehabil Neural Repair. 2010;24:709–21.

10. Zealear DL, Rodriguez RJ, Kenny T, Billante MJ, Cho Y, Billante CR, Garren KC. Electrical stimulation of a denervated muscle promotes selective reinnervation by native over foreign motoneurons. J Neurophysiol. 2002;87:2195–9.

11. Willand MP. Electrical stimulation enhances reinnervation after nerve injury. Eur J Transl Myol. 2015;25(4):243–8.

12. Willand MP, Rosa E, Michalski B, Zhang JJ, Gordon T, Fahnestock M, Borschel GH. Electrical muscle stimulation elevates intramuscular BDNF and GDNF mRNA following peripheral nerve injury and repair in rats. Neuroscience. 2016;334:93–104.

13. Pette D, Vrbová G. The contribution of neuromuscular stimulation in elucidating muscle plasticity revisited. Eur J Transl Myol. 2017;27(1):33–9.

14. Carraro U, Franceschi C. Apoptosis of skeletal and cardiac muscles and physical exercise. Aging (Milano). 1997;9(1-2):19–34.

15. Carraro U, Sandri M. Apoptosis of skeletal muscles during development and disease. Int J Biochem Cell Biol. 1999;31(12):1373–90.

16. Borisov AB, Carlson BM. Cell death in denervated skeletal muscle is distinct from classical apoptosis. Anat Rec. 2000;258(3):305–18.

17. Borisov AB, Dedkov EI, Carlson BM. Interrelations of myogenic response, progressive atrophy of muscle fibers, and cell death in denervated skeletal muscle. Anat Rec. 2001;264:203–18.

18. Carlson BM, Borisov AI, Dekov EI, Dow D, Kostrominova TY. The biology and restorative capacity of long-term denervated skeletal muscle. Basic Appl Myol. 2002;12:247–54.

19. Jakubiec-Puka A, Kordowska J, Catani C, Carraro U. Myosin heavy chain isoform composition in striated muscle after denervation and self-reinnervation. Eur J Biochem. 1990;193:623–8.

20. Talmadge RJ, Roy RR, Bodine-Fowler SC, Pierotti DJ, Edgerton VR. Adaptations in myosin heavy chain profile in chronically unloaded muscles. Basic Appl Biol. 1995;5:119–34.

21. Carraro U, Morale D, Mussini I, Lucke S, Cantini M, Betto R, Catani C, Dalla Libera L, Danieli-Betto D, Noventa D. Chronic denervation of rat diaphragm: maintenance of fiber heterogeneity with associated increasing uniformity of myosin isoforms. J Cell Biol. 1985;100:161–74.

22. Carraro U, Catani C, Biral D. Selective maintenance of neurotrophically regulated proteins in denervated rat diaphragm. Exp Neurol. 1979;63(3):468–75.

23. Carraro U, Catani C, DallaLibera L. Myosin light and heavy chains in rat gastrocnemius and diaphragm muscles after chronic denervation or reinnervation. Exp Neurol. 1981;72(2):401–12.

24. Lewis DM, Schmalbruch H. Contractile properties of a neurally regenerated compared with denervated muscles of rat. J Muscle Res Cell Motil. 1994;15(3):267–77.

25. Schmalbruch H, Lewis DM. A comparison of the morphology of denervated with aneurally regenerated soleus muscle of rat. J Muscle Res Cell Motil. 1994;15(3):256–66.

26. Carraro U, Catani C, Degani A, Rizzi C. Myosin expression in denervated fast-and slow-twitch muscles: fiber modulation and substitution. In: Pette D, editor. The dynamic state of muscle fibers. Berlin: Walter de Gruyter; 1990. p. 247–62.

27. Bacou F, Rouanet P, Barjot C, Janmot C, Vigneron P, d'Albis A. Expression of myosin isoforms in denervated, cross-reinnervated, and electrically stimulated rabbit muscles. Eur J Biochem. 1996;236(2):539–47.

28. Mauro A. Satellite cell of skeletal muscle fibers. J Biophys Biochem Cytol. 1961;9:493–5.

29. Yin H, Price F, Rudnicki MA. Satellite cells and the muscle stem cell niche. Physiol Rev. 2013;93:23–67.

30. Gundersen K, Bruusgaard JC. Nuclear domains during muscle atrophy: nuclei lost or paradigm lost? J Physiol. 2008;586(Pt 11):2675–81.

31. Yablonka-Reuveni Z. The skeletal muscle satellite cell: still young and fascinating at 50. J Histochem Cytochem. 2011;59(12):1041–59.

32. Mussini I, Favaro G, Carraro U. Maturation, dystrophic changes and the continuous production of fibers in skeletal muscle regenerating in the absence of nerve. J Neurophatol Exp Neurol. 1987;46:315–31.

33. Allen DL, Monke SR, Talmadge RJ, Roy RR, Edgerton VR. Plasticity of myonuclear number in hypertrophied and atrophied mammalian skeletal muscle fibers. J Appl Physiol. 1995;78:1969–76.

34. Billington L, Carlson BM. The recovery of long-term denervated rat muscles after Marcaine treatment and grafting. Anat Rec. 1996;144(1-2):147–55.

35. Dedkov AR, Kostrominova TY, Borisov AB, Carlson BM. Reparative myogenesis in long-term denervated skeletal muscles of adult rats results in a reduction of the satellite cell population. Anat Rec. 2001;263:139–54.

36. Jejurikar SS, Marcelo CL, Kuzon WM Jr. Skeletal muscle denervation increases satellite cell susceptibility to apoptosis. Plast Reconstr Surg. 2002;110(1):160–8.

37. Lewis DM, Schmalbruch H. Effects of age on aneural regeneration of soleus muscle in rat. J Physiol (Lond). 1995;488(2):483–92.

38. Rodrigues AC, Schmalbruch H. Satellite cells and myonuclei in long-term denervated rat muscle. Anat Rec. 1995;243(4):430–7.

39. Yoshimura K, Harii K. A regenerative change during muscle adaptation to denervation in rats. J Surg Res. 1999;81(2):139–46.

40. Schmalbruch H, Lewis DM. Dynamics of nuclei of muscle fibers and connective tissue cells in normal and denervated rat muscles. Muscle Nerve. 2000;23(4):617–26.

41. Hnik P. Rate of denervation muscle atrophy. In: Gutmann E, editor. The denervated muscle. Prague: Publishing House of Czechoslovak Academy of Science; 1962. p. 341–71.

42. Gallucci V, Novello F, Margreth A, Aloisi M. Biochemical correlates of discontinuous muscle regeneration in rat. Br J Exp Pathol. 1966;47:215–27.

43. Carraro U, DallaLibera L, Catani C. Myosin light and heavy chains in muscle regenerating in absence of the nerve: transient appearance of the embryonic light chains. Exp Neurol. 1983;79:106–17.

44. Carraro U, Catani C. A sensitive SDS PAGE method separating heavy chain isoforms of rat skeletal muscles reveals the heterogeneous nature of the embryonic myosin. Biochem Biophys Res Commun. 1983;116:793–802.

45. Carraro U, Rossini K, Zanin ME, Rizzi C, Mayr W, Kern H. Induced myogenesis in long-term permanent denervation: perspective role in functional electrical stimulation of denervated legs in humans. Basic Appl Myol. 2002;12(2):53–64.

46. Gulati AK. Long-term retention of regenerative capacity after denervation of skeletal muscle, and dependency of late differentiation on innervation. Anat Rec. 1988;220:429–34.

47. Lu DX, Huang SK, Carlson BM. Electron microscopic study of long-term denervated rat skeletal muscle. Anat Rec. 1997;248(3):355–6.

48. Gross JG, Morgan JE. Muscle precursor cells injected into irradiated mdx mouse muscle persist after serial injury. Muscle Nerve. 1999;22(2):174–85.

49. Gross JG, Bou-Gharios G, Morgan JE. Potentiation of myoblast transplantation by host muscle irradiation is dependent on the rate of radiation delivery. Cell Tissue Res. 1999;298(2):371–5.

50. Putman CT, Dusterhoft S, Pette D. Satellite cell proliferation in low frequency-stimulated fast muscle of hypothyroid rat. Am J Physiol Cell. 2000;279(3):C682–C90.

51. Kadi F, Schjerling P, Andersen LL, Charifi N, Madsen JL, Christensen LR, Andersen JL. The effects of heavy resistance training and detraining on satellite cells in human skeletal muscles. J Physiol. 2004;558(Pt 3):1005–12.

52. Best TM, Gharaibeh B, Huard J. Stem cells, angiogenesis and muscle healing: a potential role in massage therapies? Br J Sports Med. 2013;47(9):556–60.

53. Kern H, Hofer C, Mödlin M, Forstner C, Raschka-Höger D, Mayr W, Stöhr H. Denervated muscles in humans: limitations and problems of currently used functional electrical stimulation training protocols. Artif Organs. 2002;26(3):216–8.

54. Franceschini M, Cerrel Bazo H, Lauretani F, Agosti M, Pagliacci MC. Age influences rehabilitative outcomes in patients with spinal cord injury (SCI). Aging Clin Exp Res. 2011;23(3):202–8.

55. Huang H, Sun T, Chen L, Moviglia G, Chernykh E, von Wild K, Deda H, Kang KS, Kumar A, Jeon SR, Zhang S, Brunelli G, Bohbot A, Soler MD, Li J, Cristante AF, Xi H, Onose G, Kern H, Carraro U, Saberi H, Sharma HS, Sharma A, He X, Muresanu D, Feng S, Otom A, Wang D, Iwatsu K, Lu J, Al-Zoubi A. Consensus of clinical neurorestorative progress in patients with complete chronic spinal cord injury. Cell Transplant. 2014;23(Suppl 1):S5–17.

56. Valencic V, Vodovnik L, Stefancic M, Jelnikar T. Improved motor response due to chronic electrical stimulation of denervated tibialis anterior muscle in humans. Muscle Nerve. 1986;9:612–7.

57. Hofer C, Mayr W, Stöhr H, Unger E, Kern H. A stimulator for functional activation of denervated muscles. Artif Organs. 2002;26(3):276–9.

58. Edmunds KJ, Árnadóttir Í, Gíslason MK, Carraro U, Gargiulo P. Nonlinear trimodal regression analysis of radiodensitometric distributions to quantify sarcopenic and sequelae muscle degeneration. Comput Math Methods Med. 2016;2016:8932950.

59. Edmunds KJ, Gargiulo P. Imaging approaches in functional assessment of implantable myogenic biomaterials and engineered muscle tissue. Eur J Transl Myol. 2015;25(2):4847.

60. Nieman K, Hoffmann U. Cardiac computed tomography in patients with acute chest pain. Eur Heart J. 2015;36:906–14.

61. Gargiulo P, Reynisson PJ, Helgason B, Kern H, Mayr W, Ingvarsson P, Helgason T, Carraro U. Muscle, tendons, and bone: structural changes during denervation and FES treatment. Neurol Res. 2011;33:750–8.

62. Carraro U, Edmunds KJ, Gargiulo P. 3D false color computed tomography for diagnosis and follow-up of permanent denervated human muscles submitted to home-based functional electrical stimulation. Eur J Transl Myol. 2015;25:129–40.

63. Edmunds KJ, Gíslason MK, Arnadottir ID, Marcante A, Piccione F, Gargiulo P. Quantitative computed tomography and image analysis for advanced muscle assessment. Eur J Transl Myol. 2016;26(2):6015.

64. Kern H, Barberi L, Löfler S, Sbardella S, Burggraf S, Fruhmann H, Carraro U, Mosole S, Sarabon N, Vogelauer M, Mayr W, Krenn M, Cvecka J, Romanello V, Pietrangelo L, Protasi F, Sandri M, Zampieri S, Musaro A. Electrical stimulation counteracts muscle decline in seniors. Front Aging Neurosci. 2014;6:189.

65. Zampieri S, Pietrangelo L, Loefler S, Fruhmann H, Vogelauer M, Burggraf S, Pond A, Grim-Stieger M, Cvecka J, Sedliak M, Tirpáková V, Mayr W, Sarabon N, Rossini K, Barberi L, De Rossi M, Romanello V, Boncompagni S, Musarò A, Sandri M, Protasi F, Carraro U, Kern H. Lifelong physical exercise delays age-associated skeletal muscle decline. J Gerontol A Biol Sci Med Sci. 2015;70(2):163–73.

66. Zampieri S, Mosole S, Löfler S, Fruhmann H, Burggraf S, Cvečka J, Hamar D, Sedliak M, Tirptakova V, Šarabon N, Mayr W, Kern H. Physical exercise in aging: nine weeks of leg press or electrical stimulation training in 70 years old sedentary elderly people. Eur J Transl Myol. 2015;25(4):237–42.

67. Zampieri S, Mammucari C, Romanello V, Barberi L, Pietrangelo L, Fusella A, Mosole S, Gherardi G, Höfer C, Löfler S, Sarabon N, Cvecka J, Krenn M, Carraro U, Kern H, Protasi F, Musarò A, Sandri M, Rizzuto R. Physical exercise in aging human skeletal muscle increases mitochondrial calcium uniporter expression levels and affects mitochondria dynamics. Physiol Rep. 2016;4(24):e13005.

68. Kern H, Hofer C, Loefler S, Zampieri S, Gargiulo P, Baba A, Marcante A, Piccione F, Pond A, Carraro U. Atrophy, ultra-structural disorders, severe atrophy and degeneration of denervated human muscle in SCI and Aging. Implications for their recovery by Functional Electrical Stimulation, updated 2017. Neurol Res. 2017;13:1–7.

69. Carraro U, Gava K, Baba A, Piccione F, Marcante A. Fighting muscle weakness in advanced aging by take-home strategies: Safe anti-aging full-body in-bed gym and functional electrical stimulation (FES) for mobility compromised elderly people. Biol Eng Med. 2016;1:1–4.

Rejuvenating Stem Cells to Restore Muscle Regeneration in Aging

22

Eyal Bengal and Maali Odeh

22.1 Introduction

Skeletal muscles that comprise 40% of the body weight are necessary to generate mechanical force and body movement. Muscle is composed of contractile multinucleated myofibers made of organized sarcomeres. Sarcopenia is the age-associated process of muscle mass loss which is particularly severe after the seventh decade of life [1, 2]. Weakening of the muscle at advanced age increases the likelihood of injury in old people; it reduces their mobility and lowers their quality of life. Therefore, it is important to find strategies that would slow down or even reverse sarcopenia. In parallel to the gradual loss of muscle mass and function, declining muscle fiber repair in elderly people prevents muscle recovery after injury. The capacity of muscle to regenerate relies on a population of adult stem cells, known as satellite cells (SCs) which reside between the muscle sarcolemma and the basal lamina of each muscle fiber [3–5]. These cells are responsible for the robust regenerative capacity of skeletal muscle, as was demonstrated by the complete lack of regeneration in adult skeletal muscle depleted of Pax7-expressing satellite cells [6–9]. The causes of the decline in satellite cell number and function in aged model animals, mostly in rodents, have been studied extensively in the recent years with the ultimate goal of rejuvenating satellite cell functionality in old people.

22.2 Satellite Cell Biology

The population of adult mononucleated stem cells, known as satellite cells, was first described by Mauro [10]. In resting muscles, satellite cells are quiescent and express the transcription factor Pax7. Following injury, some satellite cells become activated and express the muscle regulatory factors (MRFs), MyoD and Myf5, and proliferate. A subset of the proliferating cells commit to differentiation following the expression of other MRFs, myogenin and MRF4, and fuse with damaged fibers, while another subset of the cells that downregulates the expression of MRFs self-renew and reinstate quiescence, thus preserving a pool of stem cells for future regeneration. The number of satellite cells that remains constant even after multiple injuries indicates that there must be a mechanism that balances quiescent, self-renewal, and commitment that preserves the number of functional stem cells throughout the lifetime of the organism. The present perception is that the population of satellite cells is heterogeneous, being composed of subpopulations of cells that are more committed to the myogenic

E. Bengal (✉) · M. Odeh
Department of Biochemistry, The Ruth and Bruce
Rappaport Faculty of Medicine, Technion-Israel
Institute of Technology, Haifa, Israel
e-mail: bengal@technion.ac.il

© Springer Nature Switzerland AG 2019
D. Duscher, M. A. Shiffman (eds.), *Regenerative Medicine and Plastic Surgery*,
https://doi.org/10.1007/978-3-030-19962-3_22

lineage and other cells that are more prone towards self-renewal and preserving stemness [11–14].

22.3 Origin and Heterogeneity of Satellite Cells

During embryo development, the progenitors of skeletal muscle arise from the dorsal part of the myotome, the dermomyotome [15]. These embryonic progenitors express two paired box proteins Pax3 and Pax7 [16, 17]. Most of these cells undergo muscle differentiation resulting in the formation of muscle compartments within the myotome, while a subset of myogenic progenitors that do not express MRFs migrate to a niche positioned between the primitive basal lamina structure and the myotome at mouse embryonic days 16.5–18.5 [18]. It is believed that these cells give rise to the satellite cell population in the adult muscle which lies between the fiber sarcolemma and the basal lamina. Although these quiescent cells do not express MRFs in the adult, lineage-tracing reporter allele studies suggest that they did express those genes at earlier embryonic stages. One study indicated that all adult satellite cells transcribed MyoD prenatally [19]. However, other studies suggested that a small subpopulation consisting of 10% of adult SCs never expressed Myf5 [12]. In muscle transplantation experiments, this cell subpopulation repopulated the satellite cell niche more efficiently than cells that had been expressing Myf5 at earlier stages. This result suggested the existence of two populations of adult satellite cells, a Myf5-negative population that possesses high self-renewal capability and a Myf5-positive population that tends to commit to the myogenic lineage. Another criterion to distinguish between different subpopulations of satellite cells was the level of Pax7 expression; two populations of SCs were separated, one expressing high levels of Pax7 protein and a second that expresses low levels of Pax7 [13]. Cells expressing higher levels of Pax7 cycled more slowly, their metabolism was slower, and they engrafted better in transplantation experiments than cells that expressed lower levels of Pax7. Segregation of satellite cells based on their proliferative history revealed that cells that underwent less replication cycles better retained their self-renewal potential than those that underwent more replication cycles and that were more committed to differentiation [20]. This finding was in agreement with other studies that demonstrated that slow-dividing cells had better self-renewal capacity [21, 22]. The above studies demonstrate that satellite cells consist of heterogeneous populations of cells that are more or less committed to differentiation. Active research effort is aiming to explore how each population is maintained and whether these cellular states are interchangeable.

22.4 Cell Cycle Regulation of Satellite Cells

Muscle regeneration involves exit of satellite cells from the quiescent state, followed by activation, proliferation, commitment to myogenic differentiation or self-renewal, and return to the quiescent state.

22.4.1 Satellite Cell Quiescence

This state is characterized by expression of Pax7 and absence of MRF expression [18, 23–25]. The quiescent state is also characterized by expression of genes involved in the inhibition of cell proliferation, adhesion to the niche, transport, and lipid processing necessary for quiescent cell metabolism [26–28]. Epigenetic analysis indicates that many silent genes are marked by "active chromatin" suggesting that they are in "poised" state that enables fast release from quiescence to the activated state [29, 30]. Notch signaling plays a key role in maintaining satellite cell quiescence [31, 32]. Notch activity in satellite cells inhibits MyoD expression and induces Pax7 expression that further reduces MyoD protein stability [33, 34]. Deletion of Notch target genes, Hey1 and HeyL, in the Pax7 lineage resulted in spontaneous activation of quiescent satellite cells and impairment of self-renewal that blunted muscle regeneration following muscle injury [31, 32, 35]. Interestingly, the transcription factor Forkhead box protein O3 (FoxO3), that was also found to be required for quiescence reentry, induces Notch signaling by increasing the expression of Notch receptors [36]. It is also highly likely that the quiescent state involves additional transcription regulators, noncoding RNAs, epigenetic regulators, and signaling molecules yet to be explored.

22.4.2 Satellite Cell Activation

Satellite cells are activated upon muscle injury by growth factors and cytokines released by muscle-resident cells and infiltrating inflammatory cells [37–42]. The resident cells include fibroblasts and mesenchymal stem cells, which include the fibro-adipogenic progenitors (FAPs), resident macrophages, and myofiber while the infiltrating inflammatory cells include neutrophils, eosinophils, and macrophages. Together these cells induce a pro-inflammatory phase that promotes cell cycle activation and is followed by an anti-inflammatory phase that facilitates differentiation [43–45]. The first cell division of SCs is much slower than subsequent cell cycles, indicating that the exit from quiescence is a slow process and involves intermediate cell states [46]. The environmental signals induce the immediate expression of MRFs MyoD and Myf5 that control the transcriptional program of activated satellite cells [47–49]. Transcriptome analysis of activated SCs reveals upregulation of genes implicated in cell cycle progression, metabolic processes, and immune system [28, 30]. Unlike the quiescent state, many of the silent genes are associated with repressive chromatin state that is possibly needed to restrict commitment of this stage to the myogenic fate [50–53]. Like other stem cells, SCs undergo both symmetric and asymmetric cell divisions. In asymmetric cell division, one daughter cell is destined to self-renewal and replenishment of the quiescent SC pool while the other is directed to differentiation. The mode of cell division is dependent on several parameters, including spindle orientation relative to the attached myofiber, environmental signaling

events, distribution of cellular components of Notch signaling, and their patterns of expression [54, 55]. It is becoming evident that asymmetric cell cycle divisions are affected by the heterogeneity of satellite cells. For example, satellite cells that have never expressed Myf5, as determined by lineage tracing, represent a population of stem cells that divides asymmetrically, while those cells that had previously expressed Myf5 divide symmetrically and both daughter cells are destined to myogenic differentiation [12].

22.4.3 Cell Cycle Exit and Return to Quiescence

Cell cycle exit and return to quiescence require the upregulation of specific cyclin-dependent inhibitors such as p27^{Kip1} [20]. One major signaling pathway that is downregulated during return of SCs to quiescence is the ERK mitogen-activated protein kinase (MAPK) pathway [56]. The induced expression of the tyrosine kinase inhibitor, Sprouty1, is partially responsible for ERK MAPK reduced activity and exit of SCs from the cell cycle. Notch signaling is another crucial pathway for the return of SCs to quiescence. Activation of Notch in one of the daughter cells following asymmetric cell division promotes its return to quiescence [12]. Notch signaling inhibits MyoD while promoting Pax7 expression that maintains the stem cell identity [34, 57]. Notch signaling is also necessary for the targeting of satellite cells into their proper niche under the basal lamina, by inducing the expression of specific adhesion molecules and extracellular matrix (ECM) proteins [57]. Balanced fate decisions throughout most of the organism lifetime maintain the stem cell pool while repairing the damaged myofibers.

22.5 The Decline of Satellite Cells During Aging

Skeletal muscle mass and strength gradually decline with aging in a process known as sarcopenia [58]. This decline occurs in parallel to regression in the regenerative capacity as a consequence of functional impairment of SCs [59–61]. The regenerative impairment of SCs at the old age is attributed to their inability to maintain or return to quiescence. As a result, their self-renewal capacity is diminished, and they display higher commitment to differentiation, and enter into a pre-senescence state [60–62]. In addition, satellite cell number decreases at highly advanced "geriatric" age (≥ 26 months in mice) due to increased apoptotic cell death [63, 64]. Investigations in the recent years revealed that regression of SCs during aging is due to extrinsic/environmental changes as well as due to cell-intrinsic/autonomous changes (Fig. 22.1).

22.5.1 Intrinsic Changes

Results of recent studies support the notion that SCs from aged muscle accumulate intrinsic changes that lead to a reduction in self-renewal and regenerative capacity in cell transplantation experiments. SCs isolated from aged mice that were injected into young recipient muscles performed poorly in regenerating the injured muscle relative to SCs isolated from young mice [61, 65–67]. These experiments indicate that extrinsic regulators of the "young environment" cannot sufficiently reverse the intrinsic accumulated lesions of the "old satellite cell."

Cell-intrinsic changes that are typical to aged SCs include genomic instability, DNA damage, oxidative damage, and deteriorated mitochondrial function. In spite of the fact that SCs are relatively resistant to DNA damage [68], they accumulate such damage due to their low turnover and long lifetime as quiescent cells that are unable to dilute the accumulated DNA damage [69]. The damage to DNA entails change in quiescent state metabolism that drives entry to cell cycle division and imbalance between commitment to the differentiated state and to cell self-renewal, all of which leading to cell senescence or cell death.

The age-associated functional defects observed in SCs may reflect alterations in epigenetic and transcriptional programs. Transcriptional changes could explain reduced antioxidant activity, changes in protein folding, reduced myogenic differentiation, and tendency of these cells to adopt fibroblastic and adipogenic fates [70]. The altered transcriptional program is at least partly associated with the dramatic changes that occur in the epigenetic landscape, such as changes in DNA methylation and posttranslational histone modifications [30]. Recent studies demonstrated a progressive increase in DNA methylation in the aging muscle [71–73]. In general, de novo DNA methylation of CpG islands recruits polycomb repressive complex 2 (PRC2) to gene promoters, and SCs isolated from aged mice show elevated levels and altered distribution of the H3K27me3 repression mark [30]. These changes affect gene expression that could be involved in dysregulation of signaling pathways necessary for an efficient regenerative response. One pathway that is chronically active in aged SCs is the p38 MAPK (reviewed in [74–76]). It remains unclear whether high p38 MAPK activity in SCs is induced by intracellular signal transduction/transcriptional changes (intrinsic), by environmental stress (extrinsic), or by both. High p38 MAPK activity was reported to reduce proliferative activity [66] and to decrease asymmetric cell divisions [65], ultimately decreasing the number of self-renewed satellite cells and increasing percentage of cells that exit the cell cycle. Self-renewal and regenerative capacity of "old satellite cells" can be restored by ex vivo treatment with a small-molecule p38 MAPK inhibitor [44]. Another gene whose expression is affected by epigenetic changes is *Cdkn2a*, which encodes the cell-cycle inhibitor p16[INK4A] that is thought to drive cellular senescence [34]. In young SCs, p16[INK4A] is silenced by the PRC1-mediated repressive histone H2AK119Ub modification which is significantly reduced in SCs isolated from geriatric mice, resulting in p16[INK4A] expression and their entry to a senescent state [61]. Interestingly, p38 MAPK may induce cellular senescence by activating p16[INK4A] [77]. Therefore, high intrinsic p38 activity may affect aged SCs by reducing asymmetric cell division and self-renewal and also by activating p16[INK4A] expression, driving these cells to a pre-senescent state.

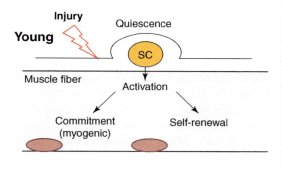 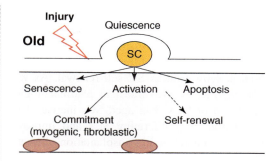

Intrinsic	Extrinsic	Intrinsic	Extrinsic
P38 MAPK ↓	FGF2 ↓	P38 MAPK ↑	FGF2 ↑
JAK/STAT3 ↓	Notch ↑	JAK/STAT3 ↑	Notch ↓
P16 INK ↓	IL6 ↓	P16 INK ↑	IL6 ↑
Autophagy ↑	TGFβ ↓	Autophagy ↓	TGFβ ↑
NAD+ ↑	Oxytocin ↑	NAD+ ↓	Oxytocin ↓
Spry1 ↑	Wnt ↓	Spry1 ↓	Wnt ↑
β1 Integrin ↑	Fibronaction ↑	β1 Integrin ↓	Fibronaction ↓
ROS ↓	Matrix stiffness ↓	ROS ↑	Matrix stiffness ↑
FoXO3 ↑		FoXO3 ↓	

Fig. 22.1 Intrinsic and extrinsic factors that are dysregulated during aging disrupt the stem cell functions of satellite cells. Muscle stem cells (satellite cells) reside between the muscle fiber and the basal lamina in a quiescent state. In this state, the satellite cell expresses Notch3 receptor, β1 integrin, and Sprouty1, a FGF signaling inhibitor. The myofiber secretes the Notch ligand, Delta1, into the satellite cell niche, activating Notch signaling in the satellite cell. In addition, the quiescent state is characterized by autophagy flux and sufficient amounts of nicotinamide adenine dinucleotide (NAD+); both sustain basal energy metabolism and mitochondrial integrity. With aging, the satellite cells, the niche, and the systemic environment undergo changes that weaken the satellite cell quiescent state. These changes include reduced expression of Delta1 by the myofiber and increased expression of FGF2 and TGFβ by niche resident cells. The niche extracellular matrix (ECM) composition is modified and becomes more rigid, affecting the satellite cell interactions with the ECM. The circulation is modified, with increased levels of TGFβ, Wnt ligands, and pro-inflammatory cytokines (such as Il-6) and decreased levels of oxytocin and reduction of the provision of fibronectin to the niche. The "aged quiescent cell" also presents elevated p38 mitogen-activated kinase (MAPK) and JAK/STAT activities, all of which leads to weakening of the quiescent state and consequently its function in muscle repair. In a more advance geriatric stage, the p16INK4a locus becomes derepressed and the expression of p16[INK4a] is induced provoking a switch of the satellite cell into a pre-senescent state. Upon injury, satellite cells of the "young muscle" exit quiescence and enter the cell cycle. Asymmetric cell cycle division with one daughter cell expressing p38 MAPK and committed to myogenesis, and the other daughter cell not expressing p38 MAPK and that self-renews to expand the population of stem cells that reinstate the niche. A balance between satellite cells that are competent for myogenic differentiation and cells that self-renew ensures an efficient muscle repair and preservation of sufficient number of stem cells for future muscle regeneration. In "old muscle," FGF2 levels are increased in the niche and p38 MAPK signaling is elevated in satellite cells, and as a result they lose asymmetric cell division; self-renewal is impaired while increased number of progenies commit to differentiation and also undergo apoptotic cell death. Self-renewal is also affected by elevated JAK/STAT3 levels and reduced deposition of fibronectin in the niche that weakens satellite cell interaction with ECM via β1 integrin. High levels of TGFβ antagonize Notch signaling and lead to the induction of CDK inhibitors that reduce the self-renewal capabilities of satellite cells, and promote a switch towards a fibroblastic fate. In addition, prolonged exposure to systemic pro-inflammatory cytokines and Wnt ligands results in aberrant satellite cell activation and loss of quiescence. At geriatric age, muscle injury and regenerative pressure drive the entry of pre-senescent satellite cells into full senescence. This process is accelerated by reduced autophagy flux that leads to accumulation of damaged mitochondria and increased levels of reactive oxygen (ROS) that fix the stem cells into the terminal senescent state. In summary, a variety of intrinsic and extrinsic factors that are modified during aging diminish the regenerative capacity of satellite cells. Rejuvenation of satellite cells should consider the intrinsic and extrinsic factors and the gradual accumulation of lesions generating a very heterogeneous population of satellite cells. Approaches that will preserve/restore satellite cell quiescence are expected to increase the functionality of muscle stem cells at the old age

SCs of old mice also have chronically elevated activity of the JAK-STAT pathway [67, 78]. STAT3 drives the expression of MyoD and commitment to myogenic differentiation, and thus its high activity reduces SC self-renewal. As with p38 MAPK, ex vivo transient pharmacological inhibition of STAT3 in aged SCs increased the population of proliferating SCs and improved muscle regeneration in transplantation of treated SCs into injured muscles [78].

Another cell-intrinsic change observed in old and geriatric SCs is unbalanced proteostasis (protein homeostasis) [62]. SCs from geriatric mice are characterized by low baseline autophagy (a quality-control mechanism whereby intracellular proteins and organelles are degraded within the lysosome), resulting in accumulation of damaged proteins, dysfunctional mitochondria, and oxidative stress that lead to the senescent state [62]. Consistent with this, SC senescence in old mice is driven by a decline in the level of oxidized cellular nicotinamide adenine dinucleotide (NAD+) that impairs mitochondrial activity. Treatment with the NAD+ precursor nicotinamide riboside rejuvenates SC function [79].

22.5.2 Extrinsic Changes

SCs are affected by the local microenvironment (niche) as well as the circulation, both of which undergo aging-associated alterations. **A. The niche:** Satellite cells are located in a protected membrane-enclosed niche between the basal lamina and the plasma membrane of the mature myofiber. This niche is affected by secreted factors originating from the myofiber and other resident cells. Expression of several extracellular ligands increases during aging of the niche, compromising SC quiescence and reducing their regenerative potential. Niche FGF signaling is elevated with aging, due to release of FGF2 by myofibers and decreased expression of Sprouty1, an inhibitor of FGF signaling in the SCs. These events lead to chronic activity of ERK MAPK resulting in loss of quiescence and a subsequent reduction in SC number. Indeed, deletion of the Spry1 gene (encoding Sprouty1) in SCs of mice

led to persistent ERK MAPK activation that impaired their self-renewal [56]. Yet, transient FGF signaling is needed for SC proliferation occurring following muscle injury [80, 81], suggesting that timely ERK MAPK signaling is beneficiary for the regeneration process while the same signal at the wrong timing impairs the same process. Increased activities of two other signaling molecules, TGFβ and canonical Wnt, in the aging niche, were implicated in the suppression of SC stemness and in their trans-differentiation from a myogenic to a fibrogenic lineage [70, 82]. Yet, trans-differentiation of SCs into other cell types, such as fibroblastic or adipogenic cells, may constitute rather infrequent events during aging or in dystrophic muscle [70, 82–85]. In contrast, Notch signaling, required to maintain the quiescent state, is reduced in the aged niche. The important role of Notch in maintaining the regenerative potential of SCs was demonstrated by the finding that inhibition of Notch signaling in young SCs causes regenerative defects while its activation in aged SCs restored their regeneration capacity [86–88]. Reduced Notch signaling in the "aged niche" was caused by insufficient expression of the Notch ligand, Delta1, by the adjacent myofibers [11]. Increased TGFβ signaling in SCs directly antagonized the Notch signaling pathway via the effector Smad3, indicating interconnection between the different signaling events that affect the niche [31, 82]. Recent studies provided evidence for the requirement of physical interactions between the SC and the niche matrix for the maintenance of regenerative capacity of these cells [89, 90]. The expression of the cell surface receptor β1-integrin and the extracellular matrix (ECM) protein fibronectin was altered in old SCs and their niche, respectively [91, 92]. Moreover, changes in the interactions between SCs and ECM during aging as a result of the modified tissue stiffness and topography may also alter SC regenerative functions [90–95]. The elasticity of the matrix is sensed by stem cell integrin-focal adhesion complexes. These complexes are engaged with the actin-myosin network, causing alteration in RhoAGTPase activity that transmits signals into cell nuclei and changes gene expression and cell phenotype [96]. Increased ECM rigidity in the aged niche may, therefore, contribute to disruption of SC homeostasis and reduced quiescence and failure to regenerate the injured muscle. **B. Systemic regulation:** The influence of the circulation on SCs was demonstrated in heterochronic whole-muscle transplant experiments [97–101] and heterochronic parabiosis, wherein two mice are surgically joined such that they shared the same circulatory system [63, 87, 102–104]. Interestingly, joining young and aged mice improved the regenerative response to muscle injury in the aged partner [86, 87], indicating that young blood contains "rejuvenating factors," and a major effort has been directed at identifying these molecules. One candidate is oxytocin, a hypothalamic hormone that declines with age in the blood and whose receptor is downregulated in SCs of aged mice [105]. Administration of oxytocin to aged mice enhanced SC proliferation and differentiation and improved overall regenerative potential after muscle injury [105]. The effect of another candidate rejuvenating factor, GDF11, is under debate. This protein is a member of the TGFβ family that shows structural and functional homology to myostatin (GDF8)whose expression was previously shown to decrease muscle mass and interfere with muscle repair [106]. While one report described a decline in the levels of GDF11 in the blood of aged animals and humans and also showed that administration of recombinant GDF11 to old mice improved SC regeneration [104], another study yielded opposite results: levels of GDF11 were increased in the circulation of aged mice and the administration of recombinant GDF11 to old mice had no beneficial effects, and even worsened regeneration after muscle injury in young mice [107]. More recent investigations supported the latter study in finding no evidence that GDF11 could rejuvenate old stem cells or extend life span in animal models of progeria [108–111]. Therefore, the role of GDF11 in muscle regeneration remains controversial and will certainly be a subject of future research.

Distinct cell types residing in the niche or infiltrating the injured muscle have been shown to influence SC functions by releasing growth factors and cytokines which may act at different stages of the regeneration process. These cell types include FAPs and other resident progenitor cells; several immune cell types such as macrophages, eosinophils, and T lymphocytes; and neurons or endothelial cells [112–125]. Since these cells also experience age-related alterations, it is likely that aging will affect their cross talk with the SCs, and thus provoke consequences on the repair process.

22.6 Conclusions and Future Perspective

The significant advances in the understanding of SC aging open up real possibilities for improving SC regenerative potential as a possible treatment for aging and diseased muscles. Emerging evidence indicates that the functional and numerical loss of SCs is a progressive process occurring throughout the lifetime of the organism. The long-lived quiescent SC accumulates many lesions that affect its transcriptional program, alter its metabolism, and impair homeostasis. These changes are a consequence of the aging of the SCs itself, and of age-related changes in the immediate environment of the niche and the circulation. Although this process is gradual, it is accelerated in advanced old age to the extent that SCs become practically nonfunctional due to senescence or apoptosis. In this context, disputes about which factors, intrinsic or extrinsic, are more dominant in dictating the fate of old SCs seem misplaced since both are likely to make important contributions to SC functional decline with aging. In fact, in many cases, it is difficult to discern between intrinsic and extrinsic effects. For example, the interaction of the SC with the ECM is affected by both the SC expressing adhesion molecules and the secreted molecules that constitute the ECM and resident cells of the niche that also affect the nature of the ECM. Resolving whether intrinsic or extrinsic events are dominant in the loss of regenerative capacity will require the reconstruction of the niche with biomimetic tools in which the effect of the matrix and SCs could be monitored independently [126].

A degree of success has been obtained in restoring the regenerative capacity of old muscle both in parabiosis experiments (extrinsic effect) and with transplantation of ex vivo-rejuvenated SCs into old animals (intrinsic effect). The simplest explanation for these effects is the heterogeneous nature and the plasticity of SCs. SC heterogeneity increases at advanced age with the rise of different subpopulations that have variable degrees of accumulated damage to the cells. One can suggest that beyond a certain level of damage, cells lose their ability to be rejuvenated. Beyond a certain age, the majority of SCs have accumulated damage beyond a level that could be corrected, yet a small number of cells that had suffered less damage can potentially be rejuvenated. Future therapy should target the small subpopulation of functional SCs with limited accumulated damage that can be still reversed. Proof of principle of this strategy was provided in experiments in which isolated populations of SCs were treated by ex vivo pharmacological inhibition of stress pathways such as p38 MAPK or JAK/STAT3 and transplanted back to old mice with proven improvement of injured muscles [66, 78]. This type of approach is based on two assumptions; one is that the "damaged" cells still maintain a certain level of plasticity that enables them to convert back into a more stemlike state. A second assumption is that at the "stem cell" state these cells have a tremendous proliferative capability that allows them to reinstate the "old niche." The second assumption was elegantly proved when single bioluminescent-labeled SCs that were transplanted into injured muscles of mice were shown to give rise to thousands of progenies that could differentiate and self-renew [14]. In an alternative strategy, health and fitness of old SCs could be increased by refueling "cleanup" activities such as autophagy (which also declines with age) to eliminate damage, thus improving SC regenerative capacity after muscle injury and in transplantation procedures. Reactivation of autophagy was expected to remove damaged proteins and mitochondria and restore the regenerative functions of old SCs [62]. Indeed, rapamycin (inhibitor of mTOR) treatment of aged mice that was sufficient to restore autophagy flux in SCs prevented their entry to senescence and increased their regenerative activity.

The evidence that was outlined in this chapter indicates a number of directions for future research. The key finding that the SC pool enters a state of irreversible senescence at a geriatric age [61] implies that any treatment to rejuvenate endogenous stem cells should be implemented before this point of no return. It is also important to consider the link between SC regenerative potential and quiescence. It is generally agreed that the more quiescent a stem cell is, the higher its regenerative capacity is. It has also become clear that somatic stem cell populations are heterogeneous, with cells showing differing levels of quiescence, particularly at an old age [127]. Subpopulations of quiescent SCs with distinct regenerative capacities have been identified based on differential expression of markers such as Pax7, CD34, Myf5, and M-cadherin [12–14, 24, 128]. Highly quiescent subpopulations probably change with aging to become less quiescent and therefore to have reduced regenerative capacity. SC heterogeneity should therefore be better demarcated with the aim of deciphering the molecular basis of quiescence. Understanding the quiescent state will allow early intervention aimed at preserving the highly regenerative quiescent subpopulations throughout life. Likewise, reversion of age-associated muscle regenerative loss should also gain from expansion of relevant subpopulations of resident progenitor cells in the SC niche. Another unresolved issue is the interplay between the various events that contribute to the loss of SC regenerative potential with aging. Research needs to focus on determining which events are causative and which are consequential. For example, DNA damage may induce loss of baseline autophagy flux in old SCs, or alternatively DNA damage may be the consequence of oxidative stress resulting from the loss of autophagy flux. Defining the hierarchy of events leading to SC deterioration should enable targeting of upstream events in order to achieve more efficient rejuvenation of SCs. Last but not least, in a low-turnover tissue such as muscle, much of the damage to the quiescent SC is the result of the gradual decline (aging) of the niche composition and the systemic system. Future efforts to rejuvenate the regenerative potential of SCs should thus adopt holistic view of the SC and its supportive environment. Hence, treatment should target both the SC and its niche. Improvement of SC-niche interactions could potentially be affected by delivery of bioengineered molecules that will reconstitute the niche [129]. Additional approaches should aim at preventing the progressive processes that disrupt tissue homeostasis in the course of the organism's lifetime. For example, it is well understood that increasing levels of circulating inflammatory cytokines in aged individuals have multiple deleterious effects on homeostasis of different tissues. Prevention of systemic chronic inflammation is expected, therefore, to greatly reduce the deterioration of many tissues and prevent diseases that are typical to the advanced age.

Current efforts to rejuvenate SCs in aged mice include genetic and pharmacological inhibition of p16^{INK4a} [61], STAT3 [67, 90], and p38 MAPK [66]; augmentation of autophagic flux [62]; NAD+ repletion [79]; and administration of rejuvenating hormones like oxytocin [105]. While these approaches hold great promise, their translation from the mouse to humans will require significant technological advances that will eliminate or minimize the potentially broad side effects and overcome the huge size difference between these organisms. Interestingly, SC activity has been found to increase in response to simple lifestyle changes that modify cell metabolism, such as adopting a low-calorie diet [130]. Similarly, exercise has been shown to enhance SC numbers and function and to promote better muscle regeneration in rodents [131–134]. This serves as a reminder that we should consider not only sophisticated methods but also simpler approaches.

References

1. Dutta C, Hadley EC, Lexell J. Sarcopenia and physical performance in old age: overview. Muscle Nerve Suppl. 1997;5:S5–9.
2. Evans WJ. What is sarcopenia? J Gerontol A Biol Sci Med Sci. 1995;50(Spec No):5–8.
3. Sambasivan R, Tajbakhsh S. Adult skeletal muscle stem cells. Results Probl Cell Differ. 2015;56:191–213.
4. Scharner J, Zammit PS. The muscle satellite cell at 50: the formative years. Skelet Muscle. 2011;1:28.
5. Yin H, Price F, Rudnicki MA. Satellite cells and the muscle stem cell niche. Physiol Rev. 2013;93:23–67.
6. Gunther S, Kim J, Kostin S, Lepper C, Fan CM, Braun T. Myf5-positive satellite cells contribute to Pax7-dependent long-term maintenance of adult muscle stem cells. Cell Stem Cell. 2013;13:590–601.
7. Lepper C, Partridge TA, Fan CM. An absolute requirement for Pax7-positive satellite cells in acute injury-induced skeletal muscle regeneration. Development. 2011;138:3639–46.
8. Sambasivan R, Yao R, Kissenpfennig A, Van Wittenberghe L, Paldi A, Gayraud-Morel B, Guenou H, Malissen B, Tajbakhsh S, Galy A. Pax7-expressing satellite cells are indispensable for adult skeletal muscle regeneration. Development. 2011;138:3647–56.
9. von Maltzahn J, Jones AE, Parks RJ, Rudnicki MA. Pax7 is critical for the normal function of satellite cells in adult skeletal muscle. Proc Natl Acad Sci U S A. 2013;110:16474–9.
10. Mauro A. Satellite cell of skeletal muscle fibers. J Biophys Biochem Cytol. 1961;9:493–5.
11. Collins CA, Olsen I, Zammit PS, Heslop L, Petrie A, Partridge TA, Morgan JE. Stem cell function, self-renewal, and behavioral heterogeneity of cells from the adult muscle satellite cell niche. Cell. 2005;122:289–301.
12. Kuang S, Kuroda K, Le Grand F, Rudnicki MA. Asymmetric self-renewal and commitment of satellite stem cells in muscle. Cell. 2007;129:999–1010.
13. Rocheteau P, Gayraud-Morel B, Siegl-Cachedenier I, Blasco MA, Tajbakhsh S. A subpopulation of adult skeletal muscle stem cells retains all template DNA strands after cell division. Cell. 2012;148:112–25.
14. Sacco A, Doyonnas R, Kraft P, Vitorovic S, Blau HM. Self-renewal and expansion of single transplanted muscle stem cells. Nature. 2008;456:502–6.
15. Bentzinger CF, Wang YX, Rudnicki MA. Building muscle: molecular regulation of myogenesis. Cold Spring Harb Perspect Biol. 2012;4(2):a008342.
16. Bober E, Franz T, Arnold HH, Gruss P, Tremblay P. Pax-3 is required for the development of limb muscles: a possible role for the migration of dermomyotomal muscle progenitor cells. Development. 1994;120:603–12.
17. Relaix F, Rocancourt D, Mansouri A, Buckingham M. Divergent functions of murine Pax3 and Pax7 in limb muscle development. Genes Dev. 2004;18:1088–105.
18. Kassar-Duchossoy L, Giacone E, Gayraud-Morel B, Jory A, Gomes D, Tajbakhsh S. Pax3/Pax7 mark a novel population of primitive myogenic cells during development. Genes Dev. 2005;19:1426–31.
19. Kanisicak O, Mendez JJ, Yamamoto S, Yamamoto M, Goldhamer DJ. Progenitors of skeletal muscle

19. satellite cells express the muscle determination gene. MyoD, Dev Biol. 2009;332:131–41.
20. Chakkalakal JV, Christensen J, Xiang W, Tierney MT, Boscolo FS, Sacco A, Brack AS. Early forming label-retaining muscle stem cells require p27kip1 for maintenance of the primitive state. Development. 2014;141:1649–59.
21. Ono Y, Masuda S, Nam HS, Benezra R, Miyagoe-Suzuki Y, Takeda S. Slow-dividing satellite cells retain long-term self-renewal ability in adult muscle. J Cell Sci. 2012;125:1309–17.
22. Schultz E. Satellite cell proliferative compartments in growing skeletal muscles. Dev Biol. 1996;175:84–94.
23. Gros J, Manceau M, Thome V, Marcelle C. A common somatic origin for embryonic muscle progenitors and satellite cells. Nature. 2005;435: 954–8.
24. Relaix F, Montarras D, Zaffran S, Gayraud-Morel B, Rocancourt D, Tajbakhsh S, Mansouri A, Cumano A, Buckingham M. Pax3 and Pax7 have distinct and overlapping functions in adult muscle progenitor cells. J Cell Biol. 2006;l172:91–102.
25. Seale P, Sabourin LA, Girgis-Gabardo A, Mansouri A, Gruss P, Rudnicki MA. Pax7 is required for the specification of myogenic satellite cells. Cell. 2000;102:777–86.
26. Fukada S, Ma Y, Ohtani T, Watanabe Y, Murakami S, Yamaguchi M. Isolation, characterization, and molecular regulation of muscle stem cells. Front Physiol. 2013;4:317.
27. Koopman R, Ly CH, Ryall JG. A metabolic link to skeletal muscle wasting and regeneration. Front Physiol. 2014;5:32.
28. Pallafacchina G, Francois S, Regnault B, Czarny B, Dive V, Cumano A, Montarras D, Buckingham M. An adult tissue-specific stem cell in its niche: a gene profiling analysis of in vivo quiescent and activated muscle satellite cells. Stem Cell Res. 2010;4:77–91.
29. Guenther MG, Levine SS, Boyer LA, Jaenisch R, Young RA. A chromatin landmark and transcription initiation at most promoters in human cells. Cell. 2007;130:77–88.
30. Liu L, Cheung TH, Charville GW, Hurgo BM, Leavitt T, Shih J, Brunet A, Rando TA. Chromatin modifications as determinants of muscle stem cell quiescence and chronological aging. Cell Rep. 2013;4:189–204.
31. Bjornson CR, Cheung TH, Liu L, Tripathi PV, Steeper KM, Rando TA. Notch signaling is necessary to maintain quiescence in adult muscle stem cells. Stem Cells. 2012;30:232–42.
32. Mourikis P, Sambasivan R, Castel D, Rocheteau P, Bizzarro V, Tajbakhsh S. A critical requirement for notch signaling in maintenance of the quiescent skeletal muscle stem cell state. Stem Cells. 2012;30:243–52.
33. Olguin HC, Olwin BB. Pax-7 up-regulation inhibits myogenesis and cell cycle progression in satellite cells: a potential mechanism for self-renewal. Dev Biol. 2004;275:375–88.
34. Wen Y, Bi P, Liu W, Asakura A, Keller C, Kuang S. Constitutive Notch activation upregulates Pax7 and promotes the self-renewal of skeletal muscle satellite cells. Mol Cell Biol. 2012;32:2300–11.
35. Fukada S, Yamaguchi M, Kokubo H, Ogawa R, Uezumi A, Yoneda T, Matev MM, Motohashi N, Ito T, Zolkiewska A, Johnson RL, Saga Y, Miyagoe-Suzuki Y, Tsujikawa K, Takeda S, Yamamoto H. Hesr1 and Hesr3 are essential to generate undifferentiated quiescent satellite cells and to maintain satellite cell numbers. Development. 2011;138:4609–19.
36. Gopinath SD, Webb AE, Brunet A, Rando TA. FOXO3 promotes quiescence in adult muscle stem cells during the process of self-renewal. Stem Cell Reports. 2014;2:414–26.
37. Allen RE, Boxhorn LK. Regulation of skeletal muscle satellite cell proliferation and differentiation by transforming growth factor-beta, insulin-like growth factor I, and fibroblast growth factor. J Cell Physiol. 1989;138:311–5.
38. Chen SE, Gerken E, Zhang Y, Zhan M, Mohan RK, Li AS, Reid MB, Li YP. Role of TNF-{alpha} signaling in regeneration of cardiotoxin-injured muscle. Am J Physiol Cell Physiol. 2005;289:C1179–87.
39. Chen SE, Jin B, Li YP. TNF-alpha regulates myogenesis and muscle regeneration by activating p38 MAPK. Am J Physiol Cell Physiol. 2007;292:C1660–71.
40. Mourkioti F, Rosenthal N. IGF-1, inflammation and stem cells: interactions during muscle regeneration. Trends Immunol. 2005;26:535–42.
41. Sheehan SM, Allen RE. Skeletal muscle satellite cell proliferation in response to members of the fibroblast growth factor family and hepatocyte growth factor. J Cell Physiol. 1999;181:499–506.
42. Tatsumi R, Anderson JE, Nevoret CJ, Halevy O, Allen RE. HGF/SF is present in normal adult skeletal muscle and is capable of activating satellite cells. Dev Biol. 1998;194:114–28.
43. Bosurgi L, Manfredi AA, Rovere-Querini P. Macrophages in injured skeletal muscle: a perpetuum mobile causing and limiting fibrosis, prompting or restricting resolution and regeneration. Front Immunol. 2011;2:62.
44. Chazaud B. Macrophages: supportive cells for tissue repair and regeneration. Immunobiology. 2014;219:172–8.
45. Tidball JG. Mechanisms of muscle injury, repair, and regeneration. Compr Physiol. 2011;1:2029–62.
46. Siegel AL, Kuhlmann PK, Cornelison DD. Muscle satellite cell proliferation and association: new insights from myofiber time-lapse imaging. Skelet Muscle. 2011;1:7.
47. Blum R, Vethantham V, Bowman C, Rudnicki M, Dynlacht BD. Genome-wide identification of enhancers in skeletal muscle: the role of MyoD1. Genes Dev. 2012;26:2763–79.

48. Cao Y, Yao Z, Sarkar D, Lawrence M, Sanchez GJ, Parker MH, MacQuarrie KL, Davison J, Morgan MT, Ruzzo WL, Gentleman RC, Tapscott SJ. Genome-wide MyoD binding in skeletal muscle cells: a potential for broad cellular reprogramming. Dev Cell. 2010;18:662–74.

49. Cooper RN, Tajbakhsh S, Mouly V, Cossu G, Buckingham M, Butler-Browne GS. In vivo satellite cell activation via Myf5 and MyoD in regenerating mouse skeletal muscle. J Cell Sci. 1999;112(Pt 17):2895–901.

50. Boonsanay V, Zhang T, Georgieva A, Kostin S, Qi H, Yuan X, Zhou Y, Braun T. Regulation of skeletal muscle stem cell quiescence by suv4-20h1-dependent facultative heterochromatin formation. Cell Stem Cell. 2016;18:229–42.

51. Dilworth FJ, Blais A. Epigenetic regulation of satellite cell activation during muscle regeneration. Stem Cell Res Ther. 2011;2:18.

52. Moresi V, Marroncelli N, Coletti D, Adamo S. Regulation of skeletal muscle development and homeostasis by gene imprinting, histone acetylation and microRNA. Biochim Biophys Acta. 1849;2015:309–16.

53. Segales J, Perdiguero E, Munoz-Canoves P. Epigenetic control of adult skeletal muscle stem cell functions. FEBS J. 2015;282:1571–88.

54. Conboy MJ, Karasov AO, Rando TA. High incidence of non-random template strand segregation and asymmetric fate determination in dividing stem cells and their progeny. PLoS Biol. 2007;5:e102.

55. Shinin V, Gayraud-Morel B, Gomes D, Tajbakhsh S. Asymmetric division and cosegregation of template DNA strands in adult muscle satellite cells. Nat Cell Biol. 2006;8:677–87.

56. Shea KL, Xiang W, LaPorta VS, Licht JD, Keller C, Basson MA, Brack AS. Sprouty1 regulates reversible quiescence of a self-renewing adult muscle stem cell pool during regeneration. Cell Stem Cell. 2010;6:117–29.

57. Brohl D, Vasyutina E, Czajkowski MT, Griger J, Rassek C, Rahn HP, Purfurst B, Wende H, Birchmeier C. Colonization of the satellite cell niche by skeletal muscle progenitor cells depends on Notch signals. Dev Cell. 2012;23:469–81.

58. Frontera WR, Hughes VA, Fielding RA, Fiatarone MA, Evans WJ, Roubenoff R. Aging of skeletal muscle: a 12-yr longitudinal study. J Appl Physiol (1985). 2000;88:1321–6.

59. Brack AS, Bildsoe H, Hughes SM. Evidence that satellite cell decrement contributes to preferential decline in nuclear number from large fibres during murine age-related muscle atrophy. J Cell Sci. 2005;118:4813–21.

60. Chakkalakal JV, Jones KM, Basson MA, Brack AS. The aged niche disrupts muscle stem cell quiescence. Nature. 2012;490:355–60.

61. Sousa-Victor P, Gutarra S, Garcia-Prat L, Rodriguez-Ubreva J, Ortet L, Ruiz-Bonilla V, Jardi M, Ballestar E, Gonzalez S, Serrano AL, Perdiguero E, Munoz-Canoves P. Geriatric muscle stem cells switch reversible quiescence into senescence. Nature. 2014;506:316–21.

62. Garcia-Prat L, Martinez-Vicente M, Perdiguero E, Ortet L, Rodriguez-Ubreva J, Rebollo E, Ruiz-Bonilla V, Gutarra S, Ballestar E, Serrano AL, Sandri M, Munoz-Canoves P. Autophagy maintains stemness by preventing senescence. Nature. 2016;529:37–42.

63. Brack AS, Rando TA. Intrinsic changes and extrinsic influences of myogenic stem cell function during aging. Stem Cell Rev. 2007;3:226–37.

64. Garcia-Prat L, Sousa-Victor P, Munoz-Canoves P. Functional dysregulation of stem cells during aging: a focus on skeletal muscle stem cells. FEBS J. 2013;280:4051–62.

65. Bernet JD, Doles JD, Hall JK, Kelly Tanaka K, Carter TA, Olwin BB. p38 MAPK signaling underlies a cell-autonomous loss of stem cell self-renewal in skeletal muscle of aged mice. Nat Med. 2014;20:265–71.

66. Cosgrove BD, Gilbert PM, Porpiglia E, Mourkioti F, Lee SP, Corbel SY, Llewellyn ME, Delp SL, Blau HM. Rejuvenation of the muscle stem cell population restores strength to injured aged muscles. Nat Med. 2014;20:255–64.

67. Price FD, von Maltzahn J, Bentzinger CF, Dumont NA, Yin H, Chang NC, Wilson DH, Frenette J, Rudnicki MA. Inhibition of JAK-STAT signaling stimulates adult satellite cell function. Nat Med. 2014;20:1174–81.

68. Vahidi Ferdousi L, Rocheteau P, Chayot R, Montagne B, Chaker Z, Flamant P, Tajbakhsh S, Ricchetti M. More efficient repair of DNA double-strand breaks in skeletal muscle stem cells compared to their committed progeny. Stem Cell Res. 2014;13:492–507.

69. Fulle S, Di Donna S, Puglielli C, Pietrangelo T, Beccafico S, Bellomo R, Protasi F, Fano G. Age-dependent imbalance of the antioxidative system in human satellite cells. Exp Gerontol. 2005;40:189–97.

70. Brack AS, Conboy MJ, Roy S, Lee M, Kuo CJ, Keller C, Rando TA. Increased Wnt signaling during aging alters muscle stem cell fate and increases fibrosis. Science. 2007;317:807–10.

71. Day K, Waite LL, Thalacker-Mercer A, West A, Bamman MM, Brooks JD, Myers RM, Absher D. Differential DNA methylation with age displays both common and dynamic features across human tissues that are influenced by CpG landscape. Genome Biol. 2013;14:R102.

72. Ong ML, Holbrook JD. Novel region discovery method for Infinium 450K DNA methylation data reveals changes associated with aging in muscle and neuronal pathways. Aging Cell. 2014;13:142–55.

73. Zykovich A, Hubbard A, Flynn JM, Tarnopolsky M, Fraga MF, Kerksick C, Ogborn D, MacNeil L, Mooney SD, Melov S. Genome-wide DNA methylation changes with age in disease-free human skeletal muscle. Aging Cell. 2014;13:360–6.

74. Li YP, Niu A, Wen Y. Regulation of myogenic activation of p38 MAPK by TACE-mediated TNFalpha release. Front Cell Dev Biol. 2014;2:21.

75. Madaro L, Latella L. Forever young: rejuvenating muscle satellite cells. Front Aging Neurosci. 2015;7:37.

76. Segales J, Perdiguero E, Munoz-Canoves P. Regulation of muscle stem cell functions: a focus on the p38 MAPK signaling pathway. Front Cell Dev Biol. 2016;4:91.

77. Munoz-Espin D, Serrano M. Cellular senescence: from physiology to pathology. Nat Rev Mol Cell Biol. 2014;15:482–96.

78. Tierney MT, Aydogdu T, Sala D, Malecova B, Gatto S, Puri PL, Latella L, Sacco A. STAT3 signaling controls satellite cell expansion and skeletal muscle repair. Nat Med. 2014;20:1182–6.

79. Zhang H, Ryu D, Wu Y, Gariani K, Wang X, Luan P, D'Amico D, Ropelle ER, Lutolf MP, Aebersold R, Schoonjans K, Menzies KJ, Auwerx J. NAD(+) repletion improves mitochondrial and stem cell function and enhances life span in mice. Science. 2016;352:1436–43.

80. Shefer G, Van de Mark DP, Richardson JB, Yablonka-Reuveni Z. Satellite-cell pool size does matter: defining the myogenic potency of aging skeletal muscle. Dev Biol. 2006;294:50–66.

81. Lefaucheur JP, Sebille A. Basic fibroblastic growth factor promotes in vivo muscle regeneration in murine muscular dystrophy. Neurosci Lett. 1995;202:121–4.

82. Carlson ME, Hsu M, Conboy IM. Imbalance between pSmad3 and Notch induces CDK inhibitors in old muscle stem cells. Nature. 2008;454:528–32.

83. Biressi S, Miyabara EH, Gopinath SD, Carlig PM, Rando TA. A Wnt-TGFbeta2 axis induces a fibrogenic program in muscle stem cells from dystrophic mice. Sci Transl Med. 2014;6:267ra176.

84. Pessina P, Kharraz Y, Jardi M, Fukada S, Serrano AL, Perdiguero E, Munoz-Canoves P. Fibrogenic cell plasticity blunts tissue regeneration and aggravates muscular dystrophy. Stem Cell Reports. 2015;4:1046–60.

85. Phelps M, Stuelsatz P, Yablonka-Reuveni Z. Expression profile and overexpression outcome indicate a role for betaKlotho in skeletal muscle fibro/adipogenesis. FEBS J. 2016;283:1653–68.

86. Conboy IM, Conboy MJ, Smythe GM, Rando TA. Notch-mediated restoration of regenerative potential to aged muscle. Science. 2003;302:1575–7.

87. Conboy IM, Conboy MJ, Wagers AJ, Girma ER, Weissman IL, Rando TA. Rejuvenation of aged progenitor cells by exposure to a young systemic environment. Nature. 2005;433:760–4.

88. Wagers AJ, Conboy IM. Cellular and molecular signatures of muscle regeneration: current concepts and controversies in adult myogenesis. Cell. 2005;122:659–67.

89. Tierney MT, Gromova A, Sesillo FB, Sala D, Spenle C, Orend G, Sacco A. Autonomous extracellular matrix remodeling controls a progressive adaptation in muscle stem cell regenerative capacity during development. Cell Rep. 2016;14:1940–52.

90. Tierney MT, Sacco A. The role of muscle stem cell-niche interactions during aging. Nat Med. 2016;22:837–8.

91. Lukjanenko L, Jung MJ, Hegde N, Perruisseau-Carrier C, Migliavacca E, Rozo M, Karaz S, Jacot G, Schmidt M, Li L, Metairon S, Raymond F, Lee U, Sizzano F, Wilson DH, Dumont NA, Palini A, Fassler R, Steiner P, Descombes P, Rudnicki MA, Fan CM, von Maltzahn J, Feige JN, Bentzinger CF. Loss of fibronectin from the aged stem cell niche affects the regenerative capacity of skeletal muscle in mice. Nat Med. 2016;22:897–905.

92. Rozo M, Li L, Fan CM. Targeting beta1-integrin signaling enhances regeneration in aged and dystrophic muscle in mice. Nat Med. 2016;22:889–96.

93. Kragstrup TW, Kjaer M, Mackey AL. Structural, biochemical, cellular, and functional changes in skeletal muscle extracellular matrix with aging. Scand J Med Sci Sports. 2011;21:749–57.

94. Wood LK, Kayupov E, Gumucio JP, Mendias CL, Claflin DR, Brooks SV. Intrinsic stiffness of extracellular matrix increases with age in skeletal muscles of mice. J Appl Physiol (1985). 2014;117:363–9.

95. Trappmann B, Gautrot JE, Connelly JT, Strange DG, Li Y, Oyen ML, Cohen Stuart MA, Boehm H, Li B, Vogel V, Spatz JP, Watt FM, Huck WT. Extracellular-matrix tethering regulates stem-cell fate. Nat Mater. 2012;11:642–9.

96. Halder G, Dupont S, Piccolo S. Transduction of mechanical and cytoskeletal cues by YAP and TAZ. Nat Rev Mol Cell Biol. 2012;13:591–600.

97. Carlson BM, Faulkner JA. Muscle transplantation between young and old rats: age of host determines recovery. Am J Physiol. 1989;256:C1262–6.

98. Grounds MD. Age-associated changes in the response of skeletal muscle cells to exercise and regeneration. Ann N Y Acad Sci. 1998;854:78–91.

99. Lee AS, Anderson JE, Joya JE, Head SI, Pather N, Kee AJ, Gunning PW, Hardeman EC. Aged skeletal muscle retains the ability to fully regenerate functional architecture. Bioarchitecture. 2013;3:25–37.

100. Shavlakadze T, McGeachie J, Grounds MD. Delayed but excellent myogenic stem cell response of regenerating geriatric skeletal muscles in mice. Biogerontology. 2010;11:363–76.

101. Smythe GM, Shavlakadze T, Roberts P, Davies MJ, McGeachie JK, Grounds MD. Age influences the early events of skeletal muscle regeneration: studies of whole muscle grafts transplanted between young (8 weeks) and old (13–21 months) mice. Exp Gerontol. 2008;43:550–62.

102. Villeda SA, Luo J, Mosher KI, Zou B, Britschgi M, Bieri G, Stan TM, Fainberg N, Ding Z, Eggel A, Lucin KM, Czirr E, Park JS, Couillard-Despres S, Aigner L, Li G, Peskind ER, Kaye JA, Quinn JF, Galasko DR, Xie XS, Rando TA, Wyss-Coray T. The ageing systemic milieu negatively regu-

lates neurogenesis and cognitive function. Nature. 2011;477:90–4.

103. Conboy MJ, Conboy IM, Rando TA. Heterochronic parabiosis: historical perspective and methodological considerations for studies of aging and longevity. Aging Cell. 2013;12:525–30.

104. Sinha M, Jang YC, Oh J, Khong D, Wu EY, Manohar R, Miller C, Regalado SG, Loffredo FS, Pancoast JR, Hirshman MF, Lebowitz J, Shadrach JL, Cerletti M, Kim MJ, Serwold T, Goodyear LJ, Rosner B, Lee RT, Wagers AJ. Restoring systemic GDF11 levels reverses age-related dysfunction in mouse skeletal muscle. Science. 2014;344:649–52.

105. Elabd C, Cousin W, Upadhyayula P, Chen RY, Chooljian MS, Li J, Kung S, Jiang KP, Conboy IM. Oxytocin is an age-specific circulating hormone that is necessary for muscle maintenance and regeneration. Nat Commun. 2014;5:4082.

106. McPherron AC, Huynh TV, Lee SJ. Redundancy of myostatin and growth/differentiation factor 11 function. BMC Dev Biol. 2009;9:24.

107. Egerman MA, Cadena SM, Gilbert JA, Meyer A, Nelson HN, Swalley SE, Mallozzi C, Jacobi C, Jennings LL, Clay I, Laurent G, Ma S, Brachat S, Lach-Trifilieff E, Shavlakadze T, Trendelenburg AU, Brack AS, Glass DJ. GDF11 increases with age and inhibits skeletal muscle regeneration. Cell Metab. 2015;22:164–74.

108. Freitas-Rodriguez S, Rodriguez F, Folgueras AR. GDF11 administration does not extend lifespan in a mouse model of premature aging. Oncotarget. 2016;7(35):55951–6.

109. Hinken AC, Powers JM, Luo G, Holt JA, Billin AN, Russell AJ. Lack of evidence for GDF11 as a rejuvenator of aged skeletal muscle satellite cells. Aging Cell. 2016;15:582–4.

110. Rinaldi F, Zhang Y, Mondragon-Gonzalez R, Harvey J, Perlingeiro RC. Treatment with rGDF11 does not improve the dystrophic muscle pathology of mdx mice. Skelet Muscle. 2016;6:21.

111. Rodgers BD, Eldridge JA. Reduced circulating GDF11 is unlikely responsible for age-dependent changes in mouse heart, muscle, and brain. Endocrinology. 2015;156:3885–8.

112. Uezumi A, Fukada S, Yamamoto N, Takeda S, Tsuchida K. Mesenchymal progenitors distinct from satellite cells contribute to ectopic fat cell formation in skeletal muscle. Nat Cell Biol. 2010;12:143–52.

113. Joe AW, Yi L, Natarajan A, Le Grand F, So L, Wang J, Rudnicki MA, Rossi FM. Muscle injury activates resident fibro/adipogenic progenitors that facilitate myogenesis. Nat Cell Biol. 2010;12:153–63.

114. Arnold L, Henry A, Poron F, Baba-Amer Y, van Rooijen N, Plonquet A, Gherardi RK, Chazaud B. Inflammatory monocytes recruited after skeletal muscle injury switch into anti-inflammatory macrophages to support myogenesis. J Exp Med. 2007;204:1057–69.

115. Perdiguero E, Sousa-Victor P, Ruiz-Bonilla V, Jardi M, Caelles C, Serrano AL, Munoz-Canoves P. p38/ MKP-1-regulated AKT coordinates macrophage transitions and resolution of inflammation during tissue repair. J Cell Biol. 2011;195:307–22.

116. Heredia JE, Mukundan L, Chen FM, Mueller AA, Deo RC, Locksley RM, Rando TA, Chawla A. Type 2 innate signals stimulate fibro/adipogenic progenitors to facilitate muscle regeneration. Cell. 2013;153:376–88.

117. Castiglioni A, Corna G, Rigamonti E, Basso V, Vezzoli M, Monno A, Almada AE, Mondino A, Wagers AJ, Manfredi AA, Rovere-Querini P. FOXP3+ T cells recruited to sites of sterile skeletal muscle injury regulate the fate of satellite cells and guide effective tissue regeneration. PLoS One. 2015;10:e0128094.

118. Zhang J, Xiao Z, Qu C, Cui W, Wang X, Du J. CD8 T cells are involved in skeletal muscle regeneration through facilitating MCP-1 secretion and Gr1(high) macrophage infiltration. J Immunol. 2014;193:5149–60.

119. Burzyn D, Kuswanto W, Kolodin D, Shadrach JL, Cerletti M, Jang Y, Sefik E, Tan TG, Wagers AJ, Benoist C, Mathis D. A special population of regulatory T cells potentiates muscle repair. Cell. 2013;155:1282–95.

120. Kuswanto W, Burzyn D, Panduro M, Wang KK, Jang YC, Wagers AJ, Benoist C, Mathis D. Poor repair of skeletal muscle in aging mice reflects a defect in local, interleukin-33-dependent accumulation of regulatory T cells. Immunity. 2016;44:355–67.

121. Villalta SA, Rosenthal W, Martinez L, Kaur A, Sparwasser T, Tidball JG, Margeta M, Spencer MJ, Bluestone JA. Regulatory T cells suppress muscle inflammation and injury in muscular dystrophy. Sci Transl Med. 2014;6:258ra142.

122. Abou-Khalil R, Mounier R, Chazaud B. Regulation of myogenic stem cell behavior by vessel cells: the "menage a trois" of satellite cells, periendothelial cells and endothelial cells. Cell Cycle. 2010;9:892–6.

123. Christov C, Chretien F, Abou-Khalil R, Bassez G, Vallet G, Authier FJ, Bassaglia Y, Shinin V, Tajbakhsh S, Chazaud B, Gherardi RK. Muscle satellite cells and endothelial cells: close neighbors and privileged partners. Mol Biol Cell. 2007;18:1397–409.

124. Liu W, Wei-LaPierre L, Klose A, Dirksen RT, Chakkalakal JV. Inducible depletion of adult skeletal muscle stem cells impairs the regeneration of neuromuscular junctions. eLife. 2015;4:e09221.

125. Carraro U, Boncompagni S, Gobbo V, Rossini K, Zampieri S, Mosole S, Ravara B, Nori A, Stramare R, Ambrosio F, Piccione F, Masiero S, Vindigni V, Gargiulo P, Protasi F, Kern H, Pond A, Marcante A. Persistent muscle fiber regeneration in long term denervation. past, present, future. Eur J Transl Myol. 2015;25:4832.

126. Blau HM, Cosgrove BD, Ho AT. The central role of muscle stem cells in regenerative failure with aging. Nat Med. 2015;21:854–62.

127. Goodell MA, Nguyen H, Shroyer N. Somatic stem cell heterogeneity: diversity in the blood, skin and intestinal stem cell compartments. Nat Rev Mol Cell Biol. 2015;16:299–309.

128. Montarras D, Morgan J, Collins C, Relaix F, Zaffran S, Cumano A, Partridge T, Buckingham M. Direct isolation of satellite cells for skeletal muscle regeneration. Science. 2005;309:2064–7.

129. Gilbert PM, Havenstrite KL, Magnusson KE, Sacco A, Leonardi NA, Kraft P, Nguyen NK, Thrun S, Lutolf MP, Blau HM. Substrate elasticity regulates skeletal muscle stem cell self-renewal in culture. Science. 2010;329:1078–81.

130. Cerletti M, Jang YC, Finley LW, Haigis MC, Wagers AJ. Short-term calorie restriction enhances skeletal muscle stem cell function. Cell Stem Cell. 2012;10:515–9.

131. Shefer G, Rauner G, Stuelsatz P, Benayahu D, Yablonka-Reuveni Z. Moderate-intensity treadmill running promotes expansion of the satellite cell pool in young and old mice. FEBS J. 2013;280:4063–73.

132. Shefer G, Rauner G, Yablonka-Reuveni Z, Benayahu D. Reduced satellite cell numbers and myogenic capacity in aging can be alleviated by endurance exercise. PLoS One. 2010;5:e13307.

133. Joanisse S, Nederveen JP, Baker JM, Snijders T, Iacono C, Parise G. Exercise conditioning in old mice improves skeletal muscle regeneration. FASEB J. 2016;30:3256–68.

134. Pietrangelo T, Di Filippo ES, Mancinelli R, Doria C, Rotini A, Fano-Illic G, Fulle S. Low intensity exercise training improves skeletal muscle regeneration potential. Front Physiol. 2015;6:399.

Silk Fibroin-Decorin Engineered Biologics to Repair Musculofascial Defects

23

Lina W. Dunne, Nadja Falk, Justin Hubenak, Tejaswi S. Iyyanki, Vishal Gupta, Qixu Zhang, Charles E. Butler, and Anshu B. Mathur

23.1 Introduction

Musculofascial defects of the anterior abdominal wall, such as ventral hernias, are a challenging surgical problem, with hernia recurrence being one of the most morbid complications [1]. The use of synthetic mesh (e.g., polypropylene [PP] mesh) for repair has reduced 10-year recurrence rates from 63 to 32%[8] but has increased the risk of complications such as bowel adhesions and mesh infection. A bioprosthetic mesh (e.g., human acellular dermal matrix, small intestinal submucosa, or porcine acellular dermal matrix [PADM]) provides an effective alternative for abdominal wall reconstruction, with good tensile strength, fewer bowel adhesions, and a lower risk of infection. Unlike the scar tissue and fibrous capsule response associated with synthetic meshes, recellularization, revascularization, and cellular remodeling and reintegration occur in the surrounding host tissue after repair with a bioprosthetic mesh, allowing tissue regeneration [2–4]. The limitations of bioprosthetic meshes include high cost, long remodeling times, limited fascial interface strength compared to synthetic mesh, limited control over initial mechanical properties and architecture for patient-specific needs, and risk of material laxity and bulging. The ideal implant materials for ventral hernia repair would bear the initial mechanical strength of the repaired mechanically loaded host site and provide patient-specific control of tissue regeneration at the same time.

With advancements in the multidisciplinary field of tissue engineering, engineered biomimetic scaffolds using biomaterials provide an alternative to overcome the shortcomings of current hernia repair materials. In addition to traditional biocompatibility and biodegradability, biomimetic materials mimic the spatial, physical, and biochemical cues that cells would sense in the native extracellular matrix (ECM). Natural polymers, such as collagen [5], chitosan, and silk fibroin (SF) [24], have good biocompatibility and have been successfully used as scaffold materials for tissue engineering [6]. *Bombyx mori* SF has been investigated for in vivo repair [7–11] because of its permeability to oxygen and water, biocompatibility, relatively low thrombogenicity, low inflammatory response, midrange degradation kinetics, and high tensile strength and flexibility [12–15]. Since natural elements of the in vivo basement membrane include other biochemical components, such as glycosaminoglycans [12, 16, 17] (e.g., chitosan) and proteoglycans (e.g., decorin), molecules such as chitosan are combined with SF to mimic the in vivo scenario [12].

In this study, decorin was combined with SF to fabricate scaffolds and matrices that mimic the

---

L. W. Dunne · N. Falk · J. Hubenak · T. S. Iyyanki
V. Gupta · Q. Zhang · C. E. Butler · A. B. Mathur (✉)
Department of Plastic Surgery, Unit 602, The University of Texas MD Anderson Cancer Center, Houston, TX, USA
e-mail: nfalk1@tulane.edu;
QZhang5@mdanderson.org;
cbutler@mdanderson.org; amathur@mdanderson.org

© Springer Nature Switzerland AG 2019
D. Duscher, M. A. Shiffman (eds.), *Regenerative Medicine and Plastic Surgery*,
https://doi.org/10.1007/978-3-030-19962-3_23

properties of native ECM. Decorin is a small leucine-rich proteoglycan with a core protein of ~40 kDa. It is made up of three domains: an N-terminal region, which possesses a single chondroitin/dermatan sulfate side chain and a distinct pattern of Cys residues; a central region, which is composed of ten leucine-rich repeats that are believed to interact with other proteins, including collagen and transforming growth factor-β; and another Cys-rich C-terminal region [18, 19]. Small leucine-rich proteoglycan affects collagen fibrillogenesis, growth factor modulation, and cellular growth regulation [18, 20–23].

Decorin binds to multiple collagen types, including types I, II, III, VI, and XIV. Weis et al. [23] reported that decorin has a role in collagen fibrillogenesis, and Danielson et al. [20] found that collagen morphological characteristics were abnormal, with coarser and more irregular fiber outlines in the absence of decorin. Bulges formed along the collagen fibrils, increasing the fibril diameter. Non-covalent interactions between the surface of the collagen fibril and the decorin core bind the two molecules at 67 nm intervals along the collagen fibril. The maximum binding force between collagen I and the decorin core protein was found along the sequence [-Gly-Pro-Ala-Gly-Ala-Arg-Gly-Pro-Ala-Gly-Pro-Gln-]$_3$ at an equilibrium distance of 0.642 nm [24]. The structural sequence of SF consists of amino acids similar to collagen that form its crystalline domain or a heavy chain that is mostly composed of glycine, alanine, serine, and tyrosine amino acid repeats (GAGAGSGAAG[SG(AG)2]$_8$Y) and short amorphous or light-chain domains consist of bulkier side-chain amino acids, such as aspartic acid [25]. The amino acid sequence enables SF to form an antiparallel β-pleated sheet secondary structure that leads to fibril formation similar to collagen.

The aim of this study was to develop a new SF-based scaffold with improved initial mechanical strength for ventral hernia repair. The SF-decorin (SFD) scaffolds were compared with PADM (degradable biologic control) and PP mesh (nondegradable control) in a guinea pig incisional ventral hernia repair model using an inlay repair technique [5, 7]. Composites of SFD-PADM were developed to determine whether they imparted or enhanced the initial strength of the repair site. This in vivo study evaluated adhesions to the bio-material, cell infiltration, vascularization, inflammatory markers, muscle differentiation, and mechanical properties of the repair site.

23.2 Materials and Methods

23.2.1 Scaffold Preparation and Characterization

23.2.1.1 Scaffold Preparation

The preparation of pure SF (donated by Dr. Samuel M. Hudson, TECS, North Carolina State University, Raleigh, NC) was described in detail by Gobin et al. [12]. Decorin (Sigma-Aldrich, St. Louis, MO), at a concentration of 28.6 µg/mL, was added to SF to make SFD. The solutions were frozen at −80 °C overnight in an ethanol bath. Frozen samples were then lyophilized for at least 2 days. The dry SFD samples were crystallized with 50:50 (v/v) water to methanol for 15 min, washed with phosphate-buffered saline (PBS) (15 min × 2), and immersed in PBS overnight. The dry SFCS-decorin samples were crystallized with 50:50 (v/v) methanol to sodium hydroxide (1 N) for 15 min and then washed with PBS; the solution was changed every 4 h until the pH had equilibrated to 7. Five scaffolds were prepared for each condition to evaluate structural and mechanical properties.

23.2.1.2 3D Architecture Characterization by Scanning Electron Microscopy

Scaffold samples (0.5 cm diameter circular punch) were coated under a vacuum to a thickness of 25 nm using a Balzer MED 010 evaporator (Technotrade International, Manchester, NH) with platinum alloy. The samples were placed in the chamber of the sputter coater at a distance of 5 cm from the platinum alloy target and coated under a vacuum for 180 s. Platinum yields a smaller grain size than does gold. The samples were examined using a JSM-5910 scanning electron microscope (JEOL, USA, Inc., Peabody, MA) at an accelerating voltage of 5 kV. Images were analyzed with ImageJ software. Three samples were imaged, and 5–10 images from each group were analyzed with ImageJ to determine scaffold structure properties such as fiber size and pore size.

23.3 Application of SFD Scaffolds for Ventral Hernia Repair

23.3.1 PP Mesh, PADM, SFD, and SFD-PADM Composite Implant Preparation

Before implantation, PP mesh (Prolene, Ethicon, Inc., Somerville, NJ), PADM (Strattice; LifeCell Corp., Branchburg, NJ), and SFD samples (4.58 ± 0.11 mm heterogeneous thickness) were cut into 2 × 4 cm² elliptical shapes (Fig. 23.1). The SFD scaffolds were sutured with 2 × 4 cm² elliptical PADM with 4-0 Vicryl sutures to create the SFD-PADM composite. All implants were sterilized by being immersed in 70% ethanol for 3 h and then washed with PBS (15 min × 2). Samples were left in PBS with 1% (v/v) penicillin and streptomycin sulfate (Invitrogen, Carlsbad, CA) until implantation. Seven samples were prepared for each condition for the in vivo test.

23.3.2 In Vivo Ventral Hernia Repair Model

A well-established acute ventral hernia model in guinea pigs was used in this study [5, 7, 26].

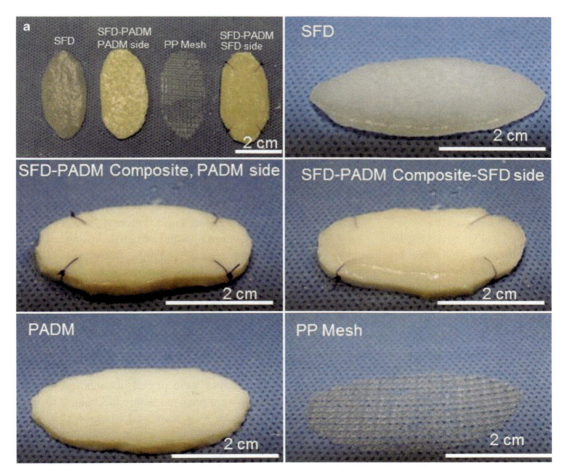

Fig. 23.1 (a) Four implants for guinea pig ventral hernia repair that were cut into 2 × 4 cm² elliptical shapes: silk fibroin-decorin blend (SFD), porcine acellular dermal matrix (PADM), polypropylene mesh (PP mesh), SFD-PADM composite-PADM side, and SFD-PADM composite-SFD side. (b) Ventral hernia repair with SFD, PADM, and PP mesh showing inlay repair of a 1 × 3 cm² defect with 0.5 cm elliptical sutured inlay

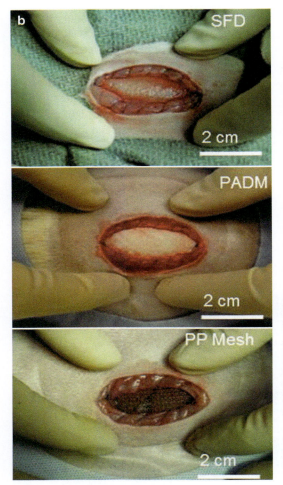

Fig. 23.1 (continued)

Prolene suture. This repair resulted in an elliptical musculofascial defect (1 × 3 cm^2) that was bridged only by the implant materials and a 0.5 cm peripheral zone of implant-musculofascial layer overlap (Fig. 23.1). The skin was closed with both 4-0 resorbable subcuticular sutures (Vicryl) and stainless steel clips. Animals were monitored until they fully recovered from anesthesia.

23.4 Gross Evaluation

Guinea pigs were euthanized by isoflurane inhalant anesthetic and subsequent IV or IC overdose injection of thiopental at week 4. The entire abdominal wall, including the repair site, was circumferentially incised to widely expose the repair site for analysis without disrupting any abdominal adhesions. Photographs were taken and the evaluation was performed in a blinded fashion by three independent observers. Each repair site underwent gross observation of the adhesion structures and evaluation of adhesion grade and strength according to the established Butler Adhesion Scale [7]. Adhesion structures included the omentum, bowel, stomach, and liver. The adhesion coverage area was quantified by assigning a percentage to the area occupied by adhesions per total visible implant area. Adhesions were graded from 0 to 3 with integrals of 0.5, where 0 = no adhesions, 1 = adhesions easily freed with gentle force, 2 = adhesions freed with blunt dissection, and 3 = adhesions requiring sharp dissection to be freed from the implant site. Evidence of seroma, hematoma, herniation, infection, perforation, bowel obstruction, or fistulization was also noted. Implant degradation, remodeling, and integration with surrounding tissues, as well as evidence of suture presence and vascularization, were all noted.

23.5 Mechanical Testing

The mechanical properties of all implants were measured before implantation and at week 4 after implantation by uniaxial tensile testing using an Enduratec ELF 3200 instrument (Bose, Minnetonka, MN). Samples were tested with a 225 N load cell (Honeywell Sensotec, Columbus, OH) at a 500 μm/s strain rate. E, ultimate tensile strength (UTS), and $\varepsilon_{failure}$ were calculated from stress-versus-strain curves. In brief, after the entire abdominal wall was removed at the appropriate time point, 1 cm wide strips were cut, and both sections were placed in saline on ice. All sutures were removed before mechanical testing; all samples were measured within 4 h of being harvested and tested at ambient temperature while moist. To test the interfacial strength between the muscle and the remodeled biomaterial, strips were cut in the middle. One clamp was placed on one end of the abdominal wall, and the other was placed over the implant material so that the interfacial area was between the two clamps. In the SFD group, the second strip was clamped at the junctional interface to determine the mechanical strength of the remodeled implant biomaterial. Clamps were placed at the muscle-implant interface on either side of the strip, and biaxial tension was applied. The normal abdominal wall was used as the control.

Cellular Infiltration, Cellular Differentiation, and Tissue Remodeling. One transverse section of full-thickness abdominal wall, including the repair site, adjacent abdominal wall, and attached viscera, was excised from each guinea pig [7]. Sections were fixed in 10% formalin, embedded in paraffin, and sectioned to 4 μm thick. Cellular infiltration was determined by hematoxylin staining. Infiltrated cells at the repair site in the SFD group were distinguished by immunohistological staining for macrophages at the pro-inflammatory stage (M1 profile) and remodeling stage (M2 profile) using surface markers CD80 and CD163, respectively. The differentiation of infiltrated cells into skeletal muscle was evaluated using specific staining for the skeletal muscle cell marker MyoD1. Remodeling of implants was studied by Movat pentachrome staining. In Movat staining, collagen appears yellow, ground substances blue, elastin black, and fibrin or fibronectin magenta red. Collagen deposition (both mature and immature) was studied by staining collagen type I and type III in the SFD group. Slides were imaged with an Olympus IX70 microscope (Olympus, Center Valley, PA).

23.6 Vascularization

Blood vessels were stained with anti-factor VIII immunolabeling (hematoxylin and von Willebrand factor). Vascularization was quantified by counting the stained vascular structures and expressed as vessels per cross-section area. The degree of vascularization of the SFD implants was determined for each of the three specific implant zones: interface, junction, and center.

23.7 Statistical Analysis

Data are presented as the mean ± the standard error of the mean. Data sets were analyzed using the ANOVA test in SigmaStat software. P values of less than 0.05 were considered statistically significant. A post hoc power analysis was used to detect the difference in adhesion coverage area (clinical measure for ventral hernia repair) between PP mesh and PADM, PADM composite with SFD-facing skin, composite with SFD-facing peritoneum, and SFD alone (94%, 95%, 86%, and 14%, respectively). The calculations were based on data from the study, using a two-sided t-test. The overall significance level was set as 0.05. The significance level for each test was set as 0.0125 when the Bonferroni approach was used to adjust for multiple comparisons. The post hoc power analysis was performed using nQuery 6.0. The power to detect the difference in adhesion coverage area between the use of PP mesh and that of SFD alone was only 14% when seven guinea pigs were used in each group. To obtain 70% or more power to detect the difference in adhesion coverage area between the use of PP mesh and that of SFD alone, 28 or more guinea pigs in each group were required (too many animals for an in vivo study). A post hoc analysis showed that the 14% power was due to a similar adhesion coverage area between PP mesh and SFD alone, although the type of adhesion was different for the two groups. The in vivo response of degradation and remodeling is driven by the type of adhesion, such as omentum and bowel for PP mesh and pure omentum for SFD.

23.8 Results

23.8.1 3D Architecture Characterization of SFD Scaffolds

SF scaffolds had a smooth surface texture, with a few fibril-like structures (fibril size = 3.07 ± 0.50 μm, n = 5) at the edge (Fig. 23.2). SFD scaffolds with a decorin concentration of 28.6 μg/mL had an entangled fibrillar

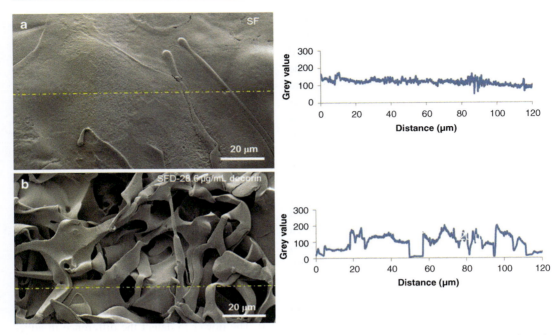

Fig. 23.2 Scanning electron microscope SEM images of SF and SFD scaffolds and ImageJ surface profile plots for each image. The yellow line in SEM images indicates where the surface profile plot was generated. The grey level changes reflect the variation in surface morphological characteristics. (**a**) SF. (**b**) SFD with a decorin concentration of 28.6 μg/mL. Scale bar = 20 μm

structure (fibril size = 5.92 ± 2.07 μm, $n = 6$) with porous structures (pore size = 21.65 ± 9.20 μm, $n = 7$) (Fig. 23.2).

23.9 Application of SFD Scaffolds for Ventral Hernia Repair

23.9.1 Gross Analysis

PP mesh, PADM, SFD, and SFD-PADM composites were tested in a well-established ventral hernia model in guinea pigs. One case of superficial wound dehiscence was found in each of the PP mesh and PADM groups due to skin closure versus three in both of the composite groups. The wound sites were repaired immediately after being discovered, and no infection was observed. If the wounds did not heal and the skin closures were not complete, the animals were euthanized according to approved protocols. One guinea pig in the SFD group died 10 days after surgery because of a bowel injury that occurred during suture placement.

The gross appearance of hernia defects at 4 weeks after repair with PP mesh, PADM, SFD, or composites was compared with the appearance of the uninjured abdominal wall (Fig. 23.3 and Table 23.1). The use of PP mesh resulted in omentum and bowel adhesions, whereas the use of other materials resulted in only omentum adhesions. The use of PP mesh resulted in a greater adhesion coverage area ($65.0 \pm 10.7\%$) than that of PADM ($4.5 \pm 1.5\%$), composite with SFD-facing skin ($2.3 \pm 0.9\%$), or composite with SFD-facing peritoneum ($12.6 \pm 6.3\%$) ($p < 0.05$); the use of SFD alone decreased the adhesion area ($35.8 \pm 13.3\%$) compared with the use of PP mesh. There was no significant difference among the SFD, PADM, and composite groups with respect to adhesion coverage area. The forces required to remove adhesions were significantly greater for PP mesh (2.1 ± 0.6) than for other groups, but the difference was significant only for PADM (0.3 ± 0.2) ($p < 0.05$). There was no significant difference in adhesion grade among the PADM, SFD, and SFD-PADM composite groups.

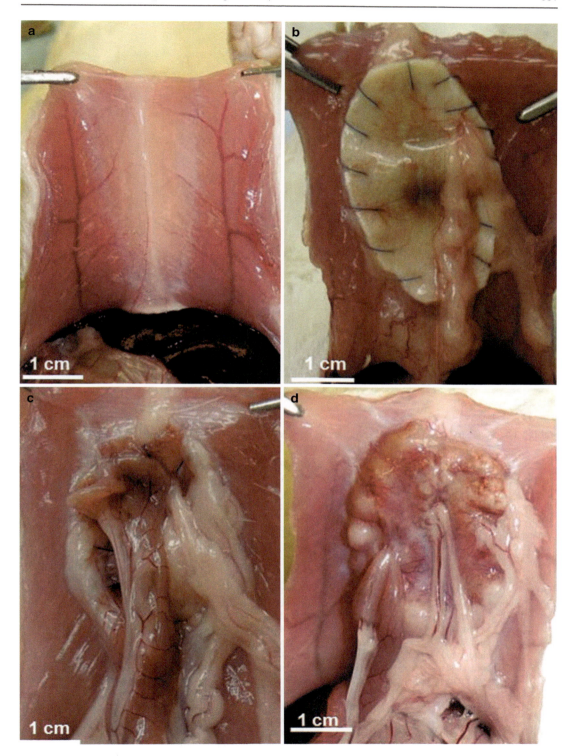

Fig. 23.3 (a) Normal uninjured abdominal wall compared with the following repair sites at 4 weeks after implantation. (b) PADM, (c) PP mesh, (d) SFD, (e) SFD-PADM composite with SFD facing the peritoneal side, and (f) SFD-PADM composite with SFD facing the skin side. PADM, SFD, and composite groups show omental adhesions, whereas the PP mesh group shows omental and bowel adhesions

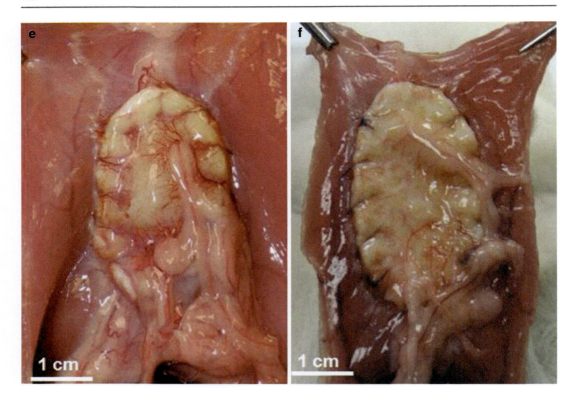

Fig. 23.3 (continued)

Table 23.1 Summary of gross evaluation of implant remodeling and integration at ventral hernia repair sites

Variable	SFD ($n = 6$)	PADM ($n = 6$)	SFD-PADM composite: SFD facing skin side ($n = 4$)	SFD-PADM composite: SFD facing peritoneal side ($n = 4$)	PP ($n = 6$)
Adhesion type	Omentum	Omentum	Omentum	Omentum	Omentum and bowel
Adhesion area (%)	35.8 ± 13.3	4.5 ± 1.5[a]	2.3 ± 0.9[a]	12.6 ± 6.3[a]	65.0 ± 10.7
Adhesions (grade)	1.2 ± 0.3	0.3 ± 0.2[b]	0.5 ± 0.0	0.5 ± 0.0	2.1 ± 0.6
Gross observation	Remodeled with observable blood vessels from the surface of implants, integrated with muscles, no suture observable	Clear separation between PADM and muscles, observable blood vessels on the PADM surface, observable sutures	Clear separation between SFD and PADM PADM: not degraded, not integrated with muscles with clear separation, observable blood vessels on the PADM surface at the peritoneal side, observable sutures SFD: not degraded, underwent remodeling	Clear separation between SFD and PADM SFD: not degraded, remodeled with observable blood vessels from the surface of implants at the peritoneal side, integrated with muscles, no suture observable PADM: not degraded	Embedded in scar tissues, covered by adhered tissues and organs

SFD silk fibroin decorin, *PADM* porcine acellular dermal matrix, *PP* polypropylene mesh
[a]$p < 0.05$ compared with PP mesh group for surface area
[b]$p < 0.05$ compared with PP mesh group for adhesion grade

PADM was present at the repair sites, with no obvious degradation in the PADM and composite groups at the end of 4 weeks (Fig. 23.3). Sutures could still be observed, and separation between the PADM and the surrounding muscle tissues could still be seen. In the composite and SFD groups, SFD scaffolds appeared to have been remodeled three dimensionally; the remodeled SFD site appeared to have been integrated into the adjacent abdominal wall at the SFD-fascial defect interface.

23.10 Mechanical Properties of Implants

The mechanical properties of the implants used in this study are summarized in Table 23.2. SFD implants had UTS and elastic modulus (E) values (UTS = 151.7 ± 7.7 kPa, E = 125.4 ± 19.3 kPa) similar to those of native abdominal wall tissue (UTS = 172.5 ± 42.0 kPa, E = 177.8 ± 2.5 kPa). The $\varepsilon_{failure}$ for SFD ($\varepsilon_{failure}$ = 0.75 ± 0.10) was lower ($\varepsilon_{failure}$ = 1.30 ± 0.04) ($p < 0.05$). Both PADM and PP mesh had $\varepsilon_{failure}$ values similar to those of the native abdominal wall; however, they had much higher UTS and E values ($p < 0.05$).

Figure 23.4 shows representative 1 cm wide strips of (a) SFD and (b) SFD-PADM composite-SFD facing the peritoneal side that were harvested and cut for mechanical testing. Figure 23.4 shows a schematic of the implant, interface, junction, and center zones with respect to the skin and peritoneal interfaces, where the mechanical properties of the center and the muscle interface were tested. Figure 23.5 shows representative stress-versus-strain curves for the native abdominal wall and remodeled SFD at the repaired defect site at week 4. The mechanical properties of the center and muscle interface, as measured using a uniaxial tensile test, are summarized in Table 23.3. The mechanical properties of the SFD-abdominal wall musculofascial interface (UTS = 157.7 ± 17.0 kPa, E = 118.0 ± 11.8 kPa, $\varepsilon_{failure}$ = 1.16 ± 0.15) were not significantly different from those of the native abdominal wall (UTS = 172.5 ± 42.0 kPa, E = 177.8 ± 2.5 kPa, $\varepsilon_{failure}$ = 1.30 ± 0.04). The use of composites did not affect mechanical properties such as E and $\varepsilon_{failure}$ at the implant-abdominal wall musculofascial interface, but it did compromise UTS at the interface compared with that for the SFD group and native abdominal wall tissues. UTS and E values for the PP-abdominal wall musculofascial interface were higher than those for the SFD-abdominal wall musculofascial interface and native abdominal wall ($p < 0.05$); however, the $\varepsilon_{failure}$ for the PP-abdominal wall musculofascial interface was significantly lower ($p < 0.05$). The UTS and E values for the PADM-abdominal wall musculofascial interface were similar to those for the SFD-abdominal wall musculofascial interface and native abdominal wall, but the $\varepsilon_{failure}$ values for the former interface were significantly lower ($p < 0.05$).

Table 23.2 Mechanical properties measured from stress-versus-strain curves obtained from implants via uniaxial tensile test

Property	PADM ($n = 6$)	PP ($n = 6$)	SFD ($n = 3$)	Native abdominal wall ($n = 2$)
UTS (kPa)	$1.39 \times 10^4 \pm 0.10 \times 10^4$	$1.41 \times 10^4 \pm 0.13 \times 10^4$	$1.52 \times 10^2 \pm 0.07 \times 10^{2a,b}$	$1.72 \times 10^2 \pm 0.42 \times 10^{2a,b}$
E (kPa)	$1.23 \times 10^4 \pm 0.17 \times 10^4$	$1.12 \times 10^4 \pm 0.19 \times 10^4$	$1.25 \times 10^2 \pm 0.19 \times 10^{2c,d}$	$1.18 \times 10^2 \pm 0.03 \times 10^{2c,d}$
$\varepsilon_{failure}$	1.11 ± 0.08^e	1.26 ± 0.08^e	0.75 ± 0.10	1.30 ± 0.04^e

PADM porcine acellular dermal matrix, *PP* polypropylene mesh, *SFD* silk fibroin decorin scaffolds, *UTS* ultimate tensile strength, *E* elastic modulus, $\varepsilon_{failure}$ strain at failure
[a]$p < 0.05$ compared with PP mesh group for UTS
[b]$p < 0.05$ compared with PADM group for UTS
[c]$p < 0.05$ compared with PP mesh group for E
[d]$p < 0.05$ compared with PADM group for E
[e]$p < 0.05$ compared with SFD group for $\varepsilon_{failure}$

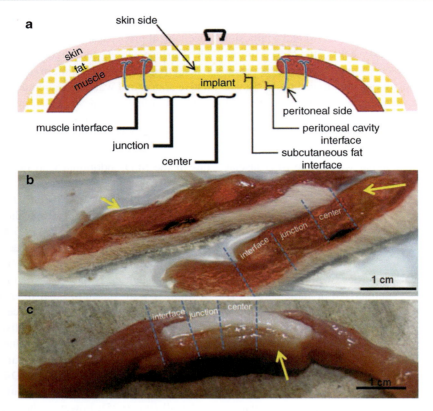

Fig. 23.4 (**a**) Cross section for the repair site. The implant is divided into three interfaces and two other zones. The three interfaces are where the implant contacts muscle, skin, and peritoneal cavity. Remodeling in the muscle interface, as well as in the junction and center zones, was noted. Representative cross-sectional views of (**b**) SFD and (**c**) SFD-PADM composite implant with SFD facing the peritoneal side 4 weeks after the repair. The SFD scaffold shows remodeling and integration at the muscle interface and across the junction and center, as indicated by the yellow arrows. SFD in the SFD-PADM composite shows a smooth surface facing the peritoneal cavity, as indicated by the yellow arrow (scale bar = 1 cm). Harvested tissues were also used to assess mechanical properties at the muscle interface and center zone of the implant and histologically to assess cellular and matrix deposition

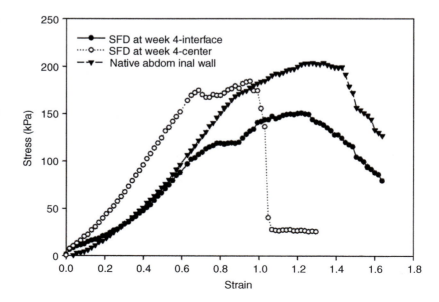

Fig. 23.5 Stress-versus-strain curves for the native abdominal wall and the regenerated tissue from the SFD scaffold 4 weeks after repair. The three curves (normal abdominal wall, SFD center, and SFD interface) indicate similar stress-versus-strain deformation profiles

Table 23.3 Mechanical properties measured from stress-versus-strain curves obtained at the abdominal wall interface of the implanted scaffolds after 4 weeks via uniaxial tensile test

Property	SFD ($n = 6$)	PADM ($n = 6$)	SFD-PADM composite: SFD facing skin side ($n = 4$)	SFD-PADM composite: SFD facing peritoneal side ($n = 4$)	PP ($n = 6$)	Native abdominal wall ($n = 2$)
UTS (kPa)	157.7 ± 17.0^a	129.5 ± 8.1^a	$75.8 \pm 12.4^{a,c}$	$63.7 \pm 7.6^{a-c}$	276.7 ± 25.33	172.5 ± 42.0^a
E (kPa)	118.0 ± 11.8^d	236.0 ± 36.5^d	92.7 ± 28.8^d	86.0 ± 10.6^d	552.2 ± 55.3	117.8 ± 2.5^d
$\varepsilon_{failure}$	$1.16 \pm 0.15^{e,f}$	0.52 ± 0.03	0.78 ± 0.10	0.87 ± 0.06	0.54 ± 0.01	$1.30 \pm 0.04^{e,f}$

SFD silk fibroin decorin, *PADM* porcine acellular dermal matrix, *PP* polypropylene mesh, *UTS* ultimate tensile strength, *E* elastic modulus, $\varepsilon_{failure}$ strain at failure

[a] $p < 0.05$ compared with PP mesh group for UTS
[b] $p < 0.05$ compared with native abdominal wall group for UTS
[c] $p < 0.05$ compared with SFD group for UTS
[d] $p < 0.05$ compared with PP mesh group for E
[e] $p < 0.05$ compared with PA DM group for $\varepsilon_{failure}$
[f] $p < 0.05$ compared with PP mesh group for $\varepsilon_{failure}$

The mechanical properties of remodeled SFD at the center of implant and at the implant-abdominal wall musculofascial interface were also measured and compared with those of the native abdominal wall. There was no significant difference in mechanical properties among SFD before implantation (UTS = 151.7 ± 7.7 kPa, $E = 125.4 \pm 19.3$ kPa, $\varepsilon_{failure} = 0.75 \pm 0.10$), SFD after implantation at the interface (UTS = 157.7 ± 17.0 kPa, $E = 118.0 \pm 11.8$ kPa, $\varepsilon_{failure} = 1.16 \pm 0.15$), SFD after implantation at the center (UTS = 168.4 ± 16.3 kPa, $E = 198.4 \pm 54.3$ kPa, $\varepsilon_{failure} = 0.82 \pm 0.07$), or native abdominal wall (UTS = 172.5 ± 42.0 kPa, $E = 117.8 \pm 2.5$ kPa, $\varepsilon_{failure} = 1.30 \pm 0.04$).

23.11 Histological Analysis

Implant remodeling was examined by Movat pentachrome staining of implant samples at the repair site at week 4 (Fig. 23.6 and Table 23.4). In Movat staining, collagen appears yellow, ground substance blue, elastin black, and fibrin and fibronectin magenta red. Since PP mesh is not degradable, the pores observed along the repair site in the PP mesh groups were due to the presence of PP fibers. Scar tissue formed around PP fibers all over the repair site: Fig. 23.6 shows "ghost" areas that were previously occupied by these fibers. The use of PP mesh resulted in fibrosis, and cells

were observed in the area surrounding the fibers. The cross sections of sites repaired with PADM still showed substantial quantities of PADM material after 4 weeks, indicating that PADM had undergone little remodeling (Fig. 23.6); partially remodeled PADM was observed close to the skin or peritoneal sides. In the PADM group, the implant was encapsulated and cells had infiltrated in a limited region close to the PADM-capsule layer interface (e.g., musculofascial interface, skin interface, or peritoneal interface). Away from these interfaces, few cells were observed within the PADM.

SFD was remodeled to different extents at different positions. SFD repair sites that interfaced with muscle, skin, or peritoneum were fully remodeled by densely packed connective tissue (such as collagen and ground substance) that was seamlessly integrated with the interfaces. The muscle interface was seamlessly integrated with the remodeled region at the interface and junction (Fig. 23.6) of the scaffold, as characterized by infiltrated cells and ECM intermingled with abdominal wall muscle. Positions away from these interfaces (center of the scaffold) were partially remodeled, as evidenced by the residual SFD scaffold fibrils, infiltrated cells, and deposited ECM.

Figure 23.6 shows cellular infiltration in SFD at the abdominal wall repair site at week 4.

There was homogeneous distribution of cells throughout the repair site (cell density$_{interface}$ = 1271 ± 49 cells/mm^2, cell density$_{junction}$ = 1351 ± 57 cells/mm^2, cell density$_{center}$ = 1402 ± 39 cells/mm^2). In the composite groups, cells infiltrated the PADM or SFD but in a limited region that was close to an interface (e.g., musculofascial interface, skin interface, and peritoneal interface); away from these interfaces, few cells were observed within the PADM or SFD. Clear separation of the PADM and SFD implants at the PADM-SFD interface was also observed in the composite groups.

Immunohistochemical staining for von Willebrand factor identified blood vessels at the repair site for all materials. In the PP group, blood vessels were observed around the scar tissues that formed around the fibrous mesh and the interface sites (e.g., musculofascial interface, skin interface, and peritoneal interface). Representative images of von Willebrand staining (Fig. 23.7) highlight vasculature in the remodeled SFD and SFD-PADM-SFD facing the peritoneal side. In the PADM group, blood vessels were observed in the capsular layer and within the PADM in limited local areas at interface sites (e.g., musculofascial interface, skin

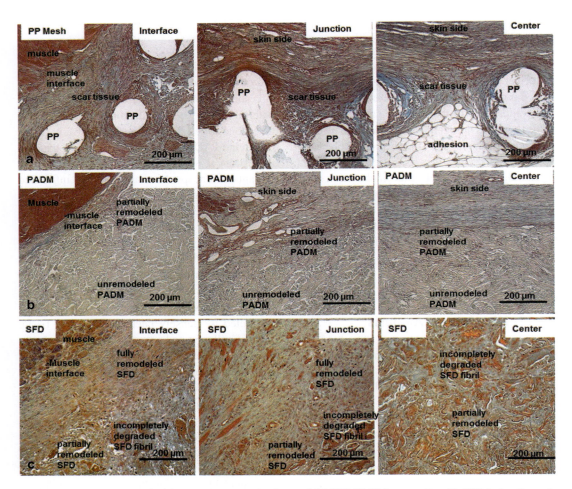

Fig. 23.6 Histological (MOVAT pentachrome stained) cross section at week 4 showing different ventral hernia repair sites for (**a**) PP mesh, (**b**) PADM, (**c**) SFD, and (**d**) SFD-PADM composites with SFD facing the skin side and (**e**) SFD-PADM composite with SFD facing the peritoneal side. MOVAT pentachrome staining shows collagen as yellow, ground substance as blue, elastin as black, and fibrin or fibronectin as magenta. Scale bar = 200 μm

23 Silk Fibroin-Decorin Engineered Biologics to Repair Musculofascial Defects

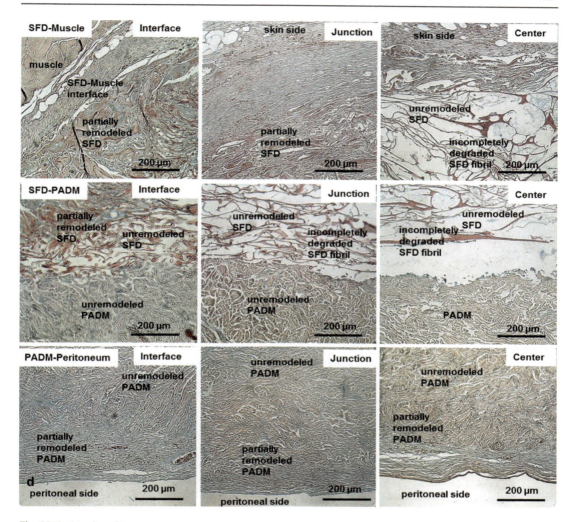

Fig. 23.6 (continued)

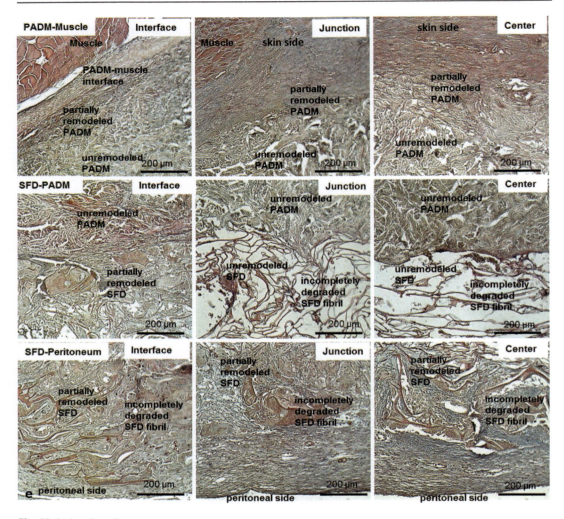

Fig. 23.6 (continued)

interface, and peritoneal interface) (Table 23.4). In sites repaired with SFD, blood vessels were uniformly distributed throughout the repair site. According to a histological analysis of neovascularization based on von Willebrand factor staining, vascular density was higher at the SFD-abdominal wall musculofascial interface site (9.7 ± 1.1 vascular structures/mm^2) and the junction site (7.7 ± 1.2 vascular structures/mm^2) than at the center site (4.3 ± 1.4 vascular structures/mm^2) ($p < 0.05$). In composite groups, blood vessels were observed in the capsular layer and within the SFD or PADM in limited local areas at these interface sites (e.g., musculofascial interface, skin interface, and peritoneal interface); few vessels were observed within either the PADM or the SFD that were not at these interfaces.

Collagen deposition in remodeled SFD and PADM (Fig. 23.8) was evaluated for each of the

Table 23.4 Summary of histological analysis of implants at week 4

Characteristic	SFD	PADM	SFD-PADM composite: SFD facing skin side	SFD-PADM composite: SFD facing peritoneal side
Cell infiltration	[a]	[b]Close to the muscle, skin, and peritoneal interfaces [c]Further away from these interfaces	[b]In **SFD**, close to the muscle and skin interfaces [c]In **SFD** and **PADM**, close to the SFD-PADM interface [b]In **PADM**, close to the peritoneal interfaces	[b]In **PADM**, close to the muscle and skin interfaces [c]In **SFD** and **PADM**, close to the SFD-PADM interface [b]In **SFD**, close to the peritoneal interfaces
Vascularization	[a]	[b]Close to the muscle, skin, and peritoneal interfaces [c]Further away from these interfaces	[b]In **SFD**, close to the muscle and skin interfaces [c]In SFD and PADM, close to the SFD-PADM interface [b]In PADM, close to the peritoneal interfaces	[b]In **PADM**, close to the muscle and skin interfaces [c]In **SFD** and **PADM**, close to the SFD-PADM interface [b]In **SFD**, close to the peritoneal interfaces
Implant remodeling	[d]Close to the muscle, skin, and peritoneal interfaces [e]Away from these interfaces	[e]Close to the muscle, skin, and peritoneal interfaces [f]Further away from these interfaces	[e]In **SFD**, close to the muscle and skin interfaces [f]In **SFD** and **PADM**, close to the SFD-PADM interface, clear separation between SFD and PADM [e]In **PADM**, close to the peritoneal interface	[e]In **PADM**, close to the muscle and skin interfaces [f]In **SFD** and **PADM**, close to the SFD-PADM interface, clear separation between SFD and PADM [e]In **SFD**, close to the peritoneal interface

SFD silk fibroin decorin, *PADM* porcine acellular dermal matrix

[a]Homogeneous distribution—cells or blood vessels were uniformly distributed throughout the implant

[b]Local distribution—cells or blood vessels were distributed in a limited fashion within part of the implant

[c]No distribution—no cell infiltration or blood vessels were found

[d]Fully remodeled—no signs of the implant, deposition of new extracellular matrix, or cell infiltration occurred

[e]Partially remodeled—some degradation of the implant and deposition of new extracellular matrix occurred

[f]Unremodeled—no degradation of the implant or cell infiltration occurred

three specific implant zones: interface, center, and junction. The muscle–SFD interface stained positive for collagen I. Collagen I was distributed all over SFD implants, whereas collagen III was observed mainly in areas close to the implant-peritoneum interface. In the composite groups, PADM underwent little remodeling, and the SFD part showed remodeling only in a limited area that was close to the interface. Away from the interface, SFD scaffolds in composite groups were barely remodeled.

The regeneration and remodeling of SFD were assessed at 4 weeks by marking the infiltrated cells in the SFD with macrophage M1 profile marker CD80, M2 profile marker CD163, and skeletal muscle marker MyoD1

(Fig. 23.9). Cells that stained positively for CD80 (marker for pro-inflammatory stage-M1 profile) were distributed around the remaining SFD fibril structures. A few cells stained positively for CD163 (marker for remodeling stage-M2 profile); these cells were observed only in the infiltrated cells that were distributed close to the peritoneal interface. The differentiation of infiltrated cells into skeletal muscle was evaluated using specific staining for skeletal muscle cell marker MyoD1. MyoD1 is a type of myogenic protein that is expressed early in skeletal muscle differentiation. MyoD1+ cells were observed in the infiltrated cells that were distributed close to the peritoneal interface.

Fig. 23.7 Hematoxylin- and von Willebrand factor-stained (**a**) SFD and (**b**) SFD-PADM composite with SFD facing the peritoneal side; blood vessels stained brown on a light blue background. SFD appears to be vascularized by week 4 compared to the SFD-PADM composite, which shows un-degraded scaffold fragments. The blood vessels are shown in dark brown and are marked by red arrows. Scale bar = 200 μm

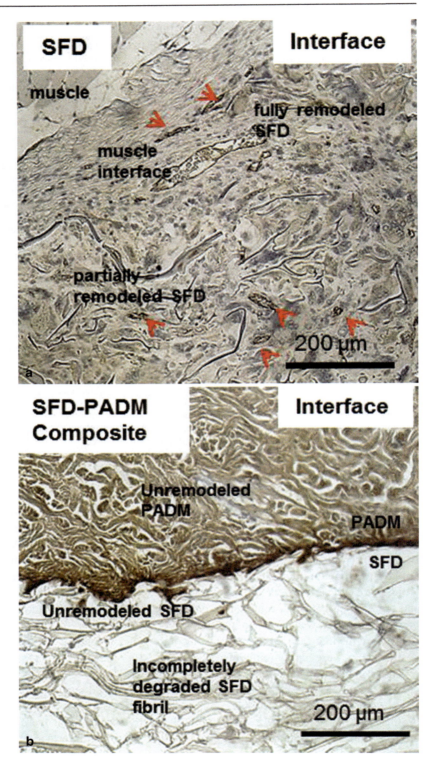

23 Silk Fibroin-Decorin Engineered Biologics to Repair Musculofascial Defects

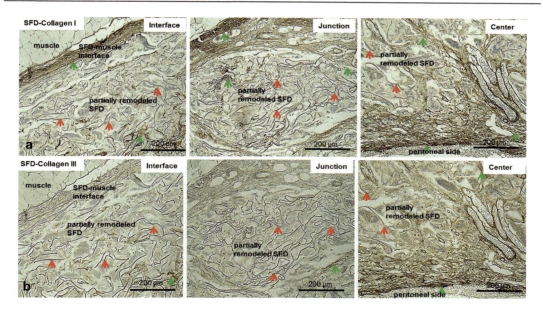

Fig. 23.8 Histological cross-sectional examination of SFD stained for (**a**) collagen I and (**b**) collagen III at week 4. Collagen deposition (brown staining, as indicated by green arrows) in remodeled SFD (silk fibril, as indicated by red arrows) was evaluated for each of the three specific implant zones: interface, center, and junction. Scale bar = 200 μm

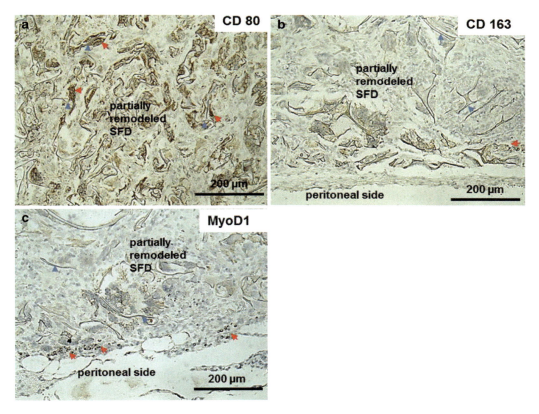

Fig. 23.9 Representative histological cross-sectional examination of SFD stained for (**a**) CD 80, (**b**) CD 163, and (**c**) MyoD1 at week 4. Positive staining appears brown and is indicated by a red arrow in the images. SFD residual fibrils are indicated by blue arrows. Scale bar = 200 μm

23.12 Discussion

In this study, we investigated the use of SF and decorin blends for ventral hernia repair in a guinea pig model. The presence of decorin induced fibril structures and affected the mechanical properties of the scaffold. SFD with a decorin concentration of 28.6 μg/mL resulted in a scaffold with mechanical properties similar to those of the native abdominal wall. By week 4, the remodeled scaffold had a gross appearance similar to that of the native abdominal wall musculofascial and integrated at the interface with adjacent native tissue. The SFD repair sites remained intact, and their mechanical properties were similar to those of the native abdominal wall. A histological analysis showed uniform cellular infiltration, vascularization, and deposition of new ECM at SFD repair sites, which may contribute to enhancement of mechanical strength at the implant-defect interface to prevent hernia recurrence. In all, this study revealed that a new biomaterial, SFD, supports tissue growth and promotes musculofascial regeneration in ventral hernia repair.

Blending decorin with SF provides a matrix with unique mechanical and structural properties. SFD scaffolds have 3D fibril-based structures, which differ from the sheet-based structures of SFCS scaffolds [12]. A fibrillar structure is an important native ECM property [27, 28], so it is encouraging that this new SFD scaffold system incorporated these fibrillar features to mimic the in vivo microenvironment. This fibrillar formation may be due to the collagen fibrillogenesis properties of decorin [24]. Differences in scaffolds' structural properties may lead to changes in their observed mechanical properties [29]. Our results showed that SFD with a decorin concentration of 28.6 μg/mL had a UTS and $\varepsilon_{failure}$ that were appropriate for ventral hernia repair.

It was previously shown in an in vivo guinea pig model that SFCS 70:30 scaffolds successfully repair abdominal wall musculofascial [7]. SFCS 70:30, a blend with a pre-implant UTS of 50 kPa, was used to repair a guinea pig native abdominal wall with a UTS of 172 kPa. After 4 weeks, the SFCS at the regenerated tissue sites had a UTS of 628 kPa [7]. This study showed that degradation of SFCS and deposition of new tissue resulted in a strengthened repair site. SFD scaffolds at a decorin concentration of 28.6 μg/mL have mechanical properties (UTS and E) higher than those of SFCS [12] and similar to those of native guinea pig abdominal wall musculofascial. These properties may be a function of the significant fibrillar structure of SFD scaffolds. These SFD scaffolds could provide enough mechanical support for ventral hernia repair. In addition, SFD scaffolds have similar initial mechanical properties to those of native tissue and facilitate and support tissue regeneration [30, 31]. Although decorin has previously been used in tissue engineering, such as bone engineering [32–34], this is the first study, to our knowledge, in which decorin-based scaffolds were investigated for use in ventral hernia repair.

Repair of the incisional ventral hernia with SFD resulted in only weak omentum adhesions (e.g., weak adhesion strength). The omentum consists of two mesothelial layers and central connective tissue, with abundant adipose tissue and macrophage and lymphocyte aggregates. The omentum detects, adheres to, and repairs injury and inflammation in the peritoneal cavity. It also reacts to foreign objects and materials by encapsulating them and provides an accessible and versatile source of growth factors and angiogenic factors [35–38]. The results of a recent study showed that the omentum contains stem cells for tissue regeneration and wound healing that are indistinguishable from mesenchymal stem cells [39]. Because of these regenerative properties, the omentum has been used by surgeons in reconstructive surgery [40–42]. In our study, omentum adhesions were observed in SFD implants, resulting in cell infiltration from the omentum to SFD and neovascularization within SFD implants that caused remodeling of the implant. Omentum adhesions may also explain the distribution of remodeling-stage marker CD163 (M2 profile) and skeletal muscle differentiation marker MyoD1 close to the peritoneal interface.

The mechanical properties of SFD were adjusted to mimic those of the native abdominal wall of guinea pigs in this study. The initial mechanical properties of SFD (28.6 µg/mL) were sufficient for surgical handling and sustained in vivo pressure during the following 4-week study. At week 4 after repair, the musculofascial-SFD interface and the remodeled center of the implant showed mechanical properties (UTS, E, and $\varepsilon_{failure}$) similar to those of native abdominal musculofascial tissues. Even though the mechanical properties of SFD implants are not as strong as those of other materials for clinical applications, SFD implants and remodeled SFD had similar mechanical properties to those of native guinea pig abdominal wall musculofascial. This result suggests that it is important to match the mechanical properties of implanted materials to host tissues at repair sites.

The PP mesh repair site had higher mechanical properties (UTS and E) at the musculofascial-PP mesh interface than did the PADM, SFD, or SFD-PADM composites at 4 weeks. However, the use of PP mesh causes strong bowel adhesions (e.g., a large adhesion area and strong adhesion strength), which can result in complications such as bowel obstruction, enterocutaneous fistulae, and pain [35–40]. PADM is gaining popularity as a replacement for PP mesh in clinical applications since it has similar mechanical properties but fewer and weaker adhesions. However, the PADM-musculofascial interface is weaker than the PP mesh [3, 7], PADM is not patient specific, and PADM scaffolds have a long remodeling time. SFD, an engineered biomaterial with a low cost that can be fabricated to meet specific mechanical and structural needs, provides another alternative for ventral hernia repair.

Remodeling is a process that consists of cellular infiltration, biomaterial degradation, and deposition of new ECM (e.g., collagen and ground substance). The SFD scaffold degraded and fully remodeled, with deposition of densely packed connective tissue where the scaffold interfaced with the muscle, skin, and peritoneal cavity. The muscle interface was seamlessly integrated with the remodeled region where the scaffold had originated, as characterized by infiltrated cells and ECM intermingled with abdominal wall muscle. Positions far away from these interfaces were partially remodeled with the combination of SFD scaffold, infiltrated cells, and deposited ECM. Mature collagen (i.e., collagen I) was distributed throughout the SFD implants. Scaffold remodeling and ECM deposition may contribute to the strong mechanical properties at the musculofascial-SFD interface as well as in the center of remodeled SFD implants. However, SFD scaffolds were not completely degraded. In our previous report on hernia repair with SFCS scaffolds, SFCS degraded with the deposition of new tissues and appeared to be fully remodeled by 4 weeks, both grossly and histologically, although a few scaffold fibrils were visible in the center of the implant where the center was still going through remodeling. The difference in wound healing may be due to a different chemical composition, degradation rate, microstructure, or thickness of the implanted materials [7].

PP mesh is a synthetic polymer and is nondegradable via in vivo enzymes and other lytic agents. At week 4, implanted PP mesh fibers were surrounded by cells and encapsulated by collagen and other ground substances that comprised the fibrotic scar tissue. Although it is biologically derived from an animal source and has biodegradable chemistry, PADM was associated with overall slow degradation, cellular infiltration in a limited region, remodeling at the outer layer of implants that interacted with host tissues, and insulated central regions of the implant [3, 4]. PADM has a condensed collagen fibril structure that may slow degradation and limit cell infiltration and new ECM deposition, leading to long-term remodeling.

Vascularization plays an important role in tissue remodeling and regeneration [43, 44]. Insufficient blood supply due to a lack of vascularization leads to nutrient deficiencies or hypoxia in the tissue, which affects its function [44]. In the sites repaired with SFD, blood vessels were uniformly distributed throughout the repair site, similar to the ubiquitous cellular infiltration, indicating the potential for remodeling and regeneration. Macrophages also play a dominant role in biological scaffold remodeling in vivo [45]. The macrophage phenotype can be characterized as pro-inflammatory or immunomodulatory (M1) or tissue remodeling (M2).

Infiltrated cells in SFD scaffolds were mainly M1 macrophages. However, a small population of M2 macrophages was observed on the peritoneal side. These results indicate that an active cellular response occurred within 4 weeks of repair and that SFD implants continued to undergo inflammation and remodeling phases within the wound-healing cycle during the 4 weeks after repair.

MyoD1 belongs to a family of proteins known as myogenic regulatory factors, which suggests that myogenic commitment occurs at an early stage of cellular differentiation. The function of MyoD1 in development is to commit mesodermal cells to a skeletal lineage and then regulate that process; it also plays a role in regulating muscle repair [46–48]. Myogenic proteins that are expressed early in skeletal muscle differentiation were also observed in infiltrated cells that were distributed at the peritoneal side, close to the peritoneal interface. The observation of skeletal muscle differentiation in the SFD group demonstrates that muscle regeneration had occurred by 4 weeks after the initial foreign-body response of SFD [49].

SFD was combined with PADM to develop a composite for hernia repair in this study. The SFD-PADM composite caused weaker omentum adhesion, with a reduced adhesion area compared to that observed in the SFD group, regardless of whether PADM or SFD was on the peritoneal side. However, UTS at the musculofascial-composite interface decreased significantly compared with that in the SFD group, which may be due to incomplete remodeling and weaker integration of composites with native surrounding tissues at the SFD side than in the SFD group alone at week 4. This study provides a proof of concept, using SFD-PADM composites to meet a clinical need. To translate this platform into a clinical application, more work is still needed to optimize the composite constructs (e.g., SFD thickness, spatial organization of SFD and PADM, and SFD-to-PADM ratio) and maximize the advantages of both SFD and PADM, such as minimum or omentum-based adhesions and a strong interface connection.

23.13 Conclusions

In this study, we evaluated a new SF-based platform by combining SF with decorin. The presence of decorin induced fibril structures, which affect the mechanical properties of the SFD scaffolds. SFD with a decorin concentration of 28.6 µg/mL had mechanical properties similar to those of the native musculofascial in guinea pigs. This new platform was used to repair the abdominal wall in an established rodent ventral hernia model. At week 4, the scaffolds had remodeled, with seamless integration at the interface with adjacent native tissue. The SFD repair sites remained intact, and their mechanical properties were similar to those of the native abdominal wall. A histological analysis showed uniform cellular infiltration, vascularization, and deposition of new ECM at SFD repair sites, which may contribute to enhancement of mechanical strength at the implant-defect interface to prevent hernia recurrence. In all, SFD shows promise in promoting musculofascial regeneration for ventral hernia repair. To be clinically relevant, novel techniques, such as processing SF-based scaffolds with dielectrophoresis to align molecular chains within SF and enhance mechanical properties [50], are being studied in our laboratory. The ideal implanted materials for ventral hernia repair are required to not only bridge musculofascial defect materials but also to function as active "engineered biologics" to induce and enhance tissue repair and regeneration.

Acknowledgements This study was supported by National Institutes of Health (NIH) and National Institute on Aging (NIA) via NIH/NIA grant R01AG034658. We thank Dr. Samuel M. Hudson (North Carolina State University) for the donation of raw silk and Carmen N. Rios and Victor L. Lam for technical assistance. In addition, we thank the High-Resolution Electron Microscopy Facility (HREMF; Cancer Center Core Grant CA16672) for scanning electron microscopy imaging.

References

1. Butler CE, Campbell KT. Minimally invasive component separation with inlay bioprosthetic mesh (MICSIB) for complex abdominal wall reconstruction. Plast Reconstr Surg. 2011;128:698–709.

2. Altman AM, Abdul Khalek FJ, Alt EU, Butler CE. Adipose tissue-derived stem cells enhance bioprosthetic mesh repair of ventral hernias. Plast Reconstr Surg. 2010;126:845–54.

3. Burns NK, Jaffari MV, Rios CN, Mathur AB, Butler CE. Non-cross-linked porcine acellular dermal matrices for abdominal wall reconstruction. Plast Reconstr Surg. 2010;125:167–76.

4. Butler CE, Burns NK, Campbell KT, Mathur AB, Jaffari MV, Rios CN. Comparison of cross-linked and non-cross-linked porcine acellular dermal matrices for ventral hernia repair. J Am Coll Surg. 2010;211:368–76.

5. Butler CE, Navarro FA, Orgill DP. Reduction of abdominal adhesions using composite collagen-GAG implants for ventral hernia repair. J Biomed Mater Res. 2001;58:75–80.

6. Mano JF, Silva GA, Azevedo HS, Malafaya PB, Sousa RA, Silva SS, Boesel LF, Oliveira JM, Santos TC, Marques AP, Neves NM, Reis RL. Natural origin biodegradable systems in tissue engineering and regenerative medicine: present status and some moving trends. J R Soc Interface. 2007;4:999–1030.

7. Gobin AS, Butler CE, Mathur AB. Repair and regeneration of the abdominal wall musculofascial defect using silk fibroin-chitosan blend. Tissue Eng. 2006;12:3383–94.

8. Gupta V, Mun GH, Choi B, Aseh A, Mildred L, Patel A, Zhang Q, Price JE, Chang D, Robb G, Mathur AB. Repair and reconstruction of a resected tumor defect using a composite of tissue flap-nanotherapeutic-silk fibroin and chitosan scaffold. Ann Biomed Eng. 2011;39:2374–87.

9. Rios CN, Skoracki RJ, Mathur AB. GNAS1 and PHD2 short-interfering RNA support bone regeneration in vitro and an in vivo sheep model. Clin Orthop Relat Res. 2012;470:2541–53.

10. Rios CN, Skoracki RJ, Miller MJ, Satterfield WC, Mathur AB. In vivo bone formation in silk fibroin and chitosan blend scaffolds via ectopically grafted periosteum as a cell source: a pilot study. Tissue Eng Part A. 2009;15:2717–25.

11. Zang M, Zhang Q, Davis G, Huang G, Jaffari M, Rios CN, Gupta V, Yu P, Mathur AB. Perichondrium directed cartilage formation in silk fibroin and chitosan blend scaffolds for tracheal transplantation. Acta Biomater. 2011;7(9):3422–31.

12. Gobin AS, Froude VE, Mathur AB. Structural and mechanical characteristics of silk fibroin and chitosan blend scaffolds for tissue regeneration. J Biomed Mater Res A. 2005;74:465–73.

13. Karageorgiou V, Tomkins M, Fajardo R, Meinel L, Snyder B, Wade K, Chen J, Vunjak-Novakovic G, Kaplan DL. Porous silk fibroin 3-D scaffolds for delivery of bone morphogenetic protein-2 in vitro and in vivo. J Biomed Mater Res A. 2006;78:324–34.

14. Lovett M, Cannizzaro C, Daheron L, Messmer B, Vunjak-Novakovic G, Kaplan DL. Silk fibroin microtubes for blood vessel engineering. Biomaterials. 2007;28:5271–9.

15. Mauney JR, Nguyen T, Gillen K, Kirker-Head C, Gimble JM, Kaplan DL. Engineering adipose-like tissue in vitro and in vivo utilizing human bone marrow and adipose-derived mesenchymal stem cells with silk fibroin 3D scaffolds. Biomaterials. 2007;28:5280–90.

16. She Z, Liu W, Feng Q. Self-assembly model, hepatocytes attachment and inflammatory response for silk fibroin/chitosan scaffolds. Biomed Mater. 2009;4:045014.

17. Silva SS, Motta A, Rodrigues MT, Pinheiro AF, Gomes ME, Mano JF, Reis RL, Migliaresi C. Novel genipin-cross-linked chitosan/silk fibroin sponges for cartilage engineering strategies. Biomacromolecules. 2008;9:2764–74.

18. Iozzo RV. Matrix proteoglycans: from molecular design to cellular function. Annu Rev Biochem. 1998;67:609–52.

19. Reed CC, Iozzo RV. The role of decorin in collagen fibrillogenesis and skin homeostasis. Glycoconj J. 2002;19:249–55.

20. Danielson KG, Baribault H, Holmes DF, Graham H, Kadler KE, Iozzo RV. Targeted disruption of decorin leads to abnormal collagen fibril morphology and skin fragility. J Cell Biol. 1997;136:729–43.

21. Ferdous Z, Grande-Allen KJ. Utility and control of proteoglycans in tissue engineering. Tissue Eng. 2007;13:1893–904.

22. Iwasaki S, Hosaka Y, Iwasaki T, Yamamoto K, Nagayasu A, Ueda H, Kokai Y, Takehana K. The modulation of collagen fibril assembly and its structure by decorin: an electron microscopic study. Arch Histol Cytol. 2008;71:37–44.

23. Weis SM, Zimmerman SD, Shah M, Covell JW, Omens JH, Ross J Jr, Dalton N, Jones Y, Reed CC, Iozzo RV, McCulloch AD. A role for decorin in the remodeling of myocardial infarction. Matrix Biol. 2005;24:313–24.

24. Vesentini S, Redaelli A, Montevecchi FM. Estimation of the binding force of the collagen molecule-decorin core protein complex in collagen fibril. J Biomech. 2005;38:433–43.

25. Asakura T, Ashida J, Yamane T, Kameda T, Nakazawa Y, Ohgo K, Komatsu K. A repeated beta-turn structure in poly(ala-Gly) as a model for silk I of Bombyx mori silk fibroin studied with two-dimensional spin-diffusion NMR under off magic angle spinning and rotational echo double resonance. J Mol Biol. 2001;306:291–305.

26. Butler CE, Prieto VG. Reduction of adhesions with composite AlloDerm/polypropylene mesh implants for abdominal wall reconstruction. Plast Reconstr Surg. 2004;114:464–73.

27. Frantz C, Stewart KM, Weaver VM. The extracellular matrix at a glance. J Cell Sci. 2010;123:4195–200.

28. Hubbell JA. Materials as morphogenetic guides in tissue engineering. Curr Opin Biotechnol. 2003;14:551–8.

29. Kim UJ, Park J, Li C, Jin HJ, Valluzzi R, Kaplan DL. Structure and properties of silk hydrogels. Biomacromolecules. 2004;5:786–92.

30. Discher DE, Janmey P, Wang YL. Tissue cells feel and respond to the stiffness of their substrate. Science. 2005;310:1139–43.

31. Engler AJ, Sen S, Sweeney HL, Discher DE. Matrix elasticity directs stem cell lineage specification. Cell. 2006;126:677–89.

32. Ameye L, Young MF. Mice deficient in small leucine-rich proteoglycans: novel in vivo models for osteoporosis, osteoarthritis, Ehlers-Danlos syndrome, muscular dystrophy, and corneal diseases. Glycobiology. 2002;12:107R–16R.

33. Douglas T, Hempel U, Mietrach C, Heinemann S, Scharnweber D, Worch H. Fibrils of different collagen types containing immobilised proteoglycans (PGs) as coatings: characterisation and influence on osteoblast behaviour. Biomol Eng. 2007;24:455–8.

34. Douglas T, Hempel U, Mietrach C, Viola M, Vigetti D, Heinemann S, Bierbaum S, Scharnweber D, Worch H. Influence of collagen-fibril-based coatings containing decorin and biglycan on osteoblast behavior. J Biomed Mater Res A. 2008;84:805–16.

35. Beelen RH, Fluitsma DM, Hoefsmit EC. The cellular composition of omentum milky spots and the ultrastructure of milky spot macrophages and reticulum cells. J Reticuloendothel Soc. 1980;28:585–99.

36. Collins D, Hogan AM, O'Shea D, Winter DC. The omentum: anatomical, metabolic, and surgical aspects. J Gastrointest Surg. 2009;13:1138–46.

37. Mandache E, Moldoveanu E, Savi G. The involvement of omentum and its milky spots in the dynamics of peritoneal macrophages. Morphol Embryol (Bucur). 1985;31:137–42.

38. Saqib NU, McGuire PG, Howdieshell TR. The omentum is a site of stromal cell-derived factor 1 alpha production and reservoir for CXC chemokine receptor 4-positive cell recruitment. Am J Surg. 2010;200:276–82.

39. Shah S, Lowery E, Braun RK, Martin A, Huang N, Medina M, Sethupathi P, Seki Y, Takami M, Byrne K, Wigfield C, Love RB, Iwashima M. Cellular basis of tissue regeneration by omentum. PLoS One. 2012;7:e38368.

40. Asai S, Kamei Y, Torii S. One-stage reconstruction of infected cranial defects using a titanium mesh plate enclosed in an omental flap. Ann Plast Surg. 2004;52:144–7.

41. Maloney CT Jr, Wages D, Upton J, Lee WP. Free omental tissue transfer for extremity coverage and revascularization. Plast Reconstr Surg. 2003;111:1899–904.

42. Shao ZQ, Kawasuji M, Takaji K, Katayama Y, Matsukawa M. Therapeutic angiogenesis with autologous hepatic tissue implantation and omental wrapping. Circ J. 2008;72:1894–9.

43. Nomi M, Atala A, Coppi PD, Soker S. Principals of neovascularization for tissue engineering. Mol Asp Med. 2002;23:463–83.

44. Rouwkema J, Rivron NC, van Blitterswijk CA. Vascularization in tissue engineering. Trends Biotechnol. 2008;26:434–41.

45. Badylak SF, Valentin JE, Ravindra AK, McCabe JP, Stewart-Akers AM. Macrophage phenotype as a determinant of biologic scaffold remodeling. Tissue Eng Part A. 2008;14:1835–42.

46. Charge SB, Rudnicki MA. Cellular and molecular regulation of muscle regeneration. Physiol Rev. 2004;84:209–38.

47. Grounds MD, Garrett KL, Lai MC, Wright WE, Beilharz MW. Identification of skeletal muscle precursor cells in vivo by use of MyoD1 and myogenin probes. Cell Tissue Res. 1992;267:99–104.

48. Megeney LA, Kablar B, Garrett K, Anderson JE, Rudnicki MA. MyoD is required for myogenic stem cell function in adult skeletal muscle. Genes Dev. 1996;10:1173–83.

49. Anderson JM, Rodriguez A, Chang DT. Foreign body reaction to biomaterials. Semin Immunol. 2008;20:86–100.

50. Gupta V, Davis G, Gordon A, Altman AM, Reece GP, Gascoyne PR, Mathur AB. Endothelial and stem cell interactions on dielectrophoretically aligned fibrous silk fibroin-chitosan scaffolds. J Biomed Mater Res A. 2010;94:515–23.

Skeletal Muscle Restoration Following Volumetric Muscle Loss: The Therapeutic Effects of a Biologic Surgical Mesh

24

Jenna L. Dziki, Jonas Eriksson, and Stephen F. Badylak

24.1 Introduction

Volumetric muscle loss (VML) as a result of tumor resection, disease, or trauma is a challenging problem with limited therapeutic options. Although skeletal muscle retains a limited capacity for regeneration following acute injury, the loss of 20% muscle mass or more at a single anatomic site overwhelms this inherent regenerative potential and results in scar tissue formation, significant loss of function, and in many cases limb amputation [1]. As described below, current reconstructive surgical approaches are fraught with limitations, and effective alternative strategies are needed to improve aesthetic and functional clinical outcomes.

J. L. Dziki · J. Eriksson
McGowan Institute for Regenerative Medicine, Pittsburgh, PA, USA

Department of Surgery, University of Pittsburgh School of Medicine, Pittsburgh, PA, USA
e-mail: dzikijl@upmc.edu; erikssonj@upmc.edu

S. F. Badylak (✉)
McGowan Institute for Regenerative Medicine, Pittsburgh, PA, USA

Department of Surgery, University of Pittsburgh School of Medicine, Pittsburgh, PA, USA

Department of Bioengineering, University of Pittsburgh, Pittsburgh, PA, USA
e-mail: badylaks@upmc.edu

Regenerative medicine approaches to skeletal muscle restoration following injury have largely been cell centric and, to date, less than satisfactory. The use of scaffold materials and/or bioactive signaling molecules, either alone or in combination with stem or progenitor cells, has been investigated. Recent reports show promising results with the use of a biologic scaffold material that facilitates formation of new functional skeletal muscle [2, 3]. Although advancements are being made, there is significant upside potential with these new strategies for skeletal muscle restoration. This chapter provides an overview of current invasive and noninvasive approaches to volumetric muscle loss, use of biologic surgical mesh materials as a therapeutic option, associated mechanisms by which these materials promote myogenesis, and factors that affect remodeling outcomes.

24.2 Therapeutic Strategies

Treatment options for volumetric muscle loss are limited and consist mainly of surgical attempts to translocate autologous skeletal muscle combined with extensive physical therapy. However, such approaches almost uniformly fail to restore adequate strength and function, and result in a lifetime of disability. Sixty-five percent of workplace disability in armed services personnel is associ-

© Springer Nature Switzerland AG 2019
D. Duscher, M. A. Shiffman (eds.), *Regenerative Medicine and Plastic Surgery*,
https://doi.org/10.1007/978-3-030-19962-3_24

ated with volumetric muscle loss and unsatisfactory available treatments [4]. Unfortunately, function of the affected limb continues to worsen over the lifetime of the patient [5].

Surgical procedures considered to be the standard for VML treatment include free and/or rotating flaps and autologous grafts or muscle transposition [6]. The rationale behind these approaches is that coverage of the wound site with fasciocutaneous and/or muscle tissue should reduce complications such as infection at the injury site [7]. Neither the flap procedures nor the muscle grafting replace lost functional muscle tissue with viable, innervated, and contractile muscle. The use of flaps and grafts is consistently associated with donor-site morbidity and often fails completely due to infection and necrosis, which, in some cases, leads to amputation of the affected limb [8]. Recent preclinical animal studies have attempted to utilize minced muscle grafts as a therapeutic strategy; however this approach has not yet resulted in appreciable restoration of functional tissue [9, 10]. Similarly, use of orthotics does not allow for full restoration of function, and orthotics are often not applicable for upper extremity injuries [1].

Physical therapy is a cornerstone of the care for VML patients. The objective of physical therapy is to strengthen the remaining injured muscle and improve the ability to perform tasks of daily living; however, physical therapy alone does not promote appreciable muscle regeneration. A growing body of literature suggests that mechanotransduction through both dynamic and static mechanical stimuli can alter the phenotype of cells within a skeletal muscle injury site and create a microenvironment conducive for repair and regeneration [11, 12]. Though complete tissue restoration is not attainable through physical therapy alone, the benefits of active load bearing across the defect site include immune cell modulation [13], increased vascularization [14], reduction of fibrosis, and promotion of myotube alignment and fusion [15, 16]. The diligent and informed use of physical therapy, in particular the timing of initiation of therapy and type of therapy (e.g., active vs. passive), can be a critical determinant of downstream outcomes [3].

24.3 Stem Cell-Centric Approaches

The limitations of current approaches to VML have been the impetus for investigating alternative strategies, including the use of stem/progenitor cells. The well-studied process of skeletal muscle regeneration involves the sequential steps of myogenic progenitor cell proliferation, mobilization, and differentiation, and the regulation of these processes by the immune system to form multinucleated myotubes. Satellite cells are the putative skeletal muscle progenitor cells that are activated following injury. Satellite cells and other stem/progenitor cells including perivascular stem cells, adult muscle-derived stem cells, embryonic stem cells, and induced pluripotent stem cells (iPSCs), among others, have been investigated as cell-based therapeutic approaches for reconstructing skeletal muscle tissue [20]. Results of preclinical studies involving stem cell administration for enhanced myogenesis have been widely variable, and such approaches have failed to gain traction in the clinical setting. VML injuries typically involve extensive tissue damage and an associated chronic inflammatory microenvironment. Direct injection of stem cells into such a microenvironment results in poor cell engraftment, low viability, and diminished differentiation potential. Recall that the extracellular matrix represents the native microenvironmental niche for tissue during development and beyond. Recent work by Webster et al. has shown that ultrastructural ECM units are required for guiding myogenic progenitor cells toward a regenerative phenotype [21]. Quarta et al. [22] utilized ECM hydrogels to maintain a favorable 3D environment for muscle-derived stem cells prior to implantation and demonstrated the ability of combination therapies, including physical therapy, to promote functional muscle restoration. The expense and regulatory hurdles associated with stem cell-based approaches have been a barrier to timely clinical translation of these and similar technologies [23, 24]. Attempts to combine myogenic stem/progenitor cells within synthetic or biologic scaffolds have yet to yield significant restoration of functional skeletal muscle and vasculature that translates to clinical improvement [17–19].

24.4 ECM Bioscaffold-Based Approaches

In addition to the technical challenges facing the clinical translation of cell-based therapies (described above), clinical use has also been limited by costly cell isolation techniques and steep regulatory hurdles. In contrast to some of the technical and nontechnical challenges with stem cell therapy as described above, cell-free approaches in the form of acellular biologic scaffolds have shown promising early results as a viable "off-the-shelf" option for patients suffering from VML and other severe musculotendinous injuries (Table 24.1).

Acellular biologic surgical meshes composed of allogenic or xenogeneic (typically porcine) extracellular matrix provide a template, or "microenvironmental niche," to initiate functional tissue reconstruction by mechanisms that include immunomodulation and stem cell differentiation. This acellular strategy is focused upon providing the complex and cytocompatible microenvironment required for skeletal muscle regeneration after a critical-size VML injury, rather than relying upon exogenous cells for repopulating the entirety of the stem/progenitor cell pool and ultimately promoting functional myogenesis. ECM meshes can obviate the need for exogenous stem cell delivery by stimulating recruitment of endogenous stem/progenitor cells. When source tissues from which the ECM surgical meshes are created are thoroughly decellularized and when care is taken to preserve the ECM ultrastructure and composition, the resulting ECM "microenvironmental niche" can facilitate the resolution of inflammation by release of specific signaling molecules during the process of mesh degradation. Pharmacologic approaches such as administration of osteoactivin or losartan, which purportedly prevent muscle atrophy and reduce fibrosis, have been associated with mixed

Table 24.1 Overview of novel treatment options for volumetric muscle loss

Desired result	ECM bioscaffolds	Exogenous stem cell delivery	Autologous minced muscle grafts	Pharmacologic approaches
Clinical translatability				
Stage of development	Clinical cohort study [1, 2]	Preclinical animal studies [15]	Preclinical animal studies	Preclinical animal studies
Hurdles	n/a	Significant technical, regulatory, and economic hurdles [12, 16, 17]	No advantage over current standard of care [21]	Side effects
Quality-of-life improvement				
Strength improvement	Yes [1–4, 6]	Partial [18]	Yes [21, 22]	No [24]
Functional improvement	Yes [1, 2]	Partial [18–20]	Partial [21]	No [24]
Histologic improvement				
Myofiber formation	Yes [1–5]	Yes	Yes [23]	No [25]
Fibrosis reduction	Yes [1, 2, 5]	Yes	Yes [21]	Yes [14]
Reinnervation			NT	NT
Histologic evidence	Yes [5, 7]	Yes [13, 21]		
Nerve conduction studies	Yes [2, 8]	NT		
Immune friendly				
Immunosuppression	Immunosuppression not required [5, 9, 10]	Immunosuppression required dependent upon cell source [17]	Immunosuppression not required	Immunosuppression not required
Resolution of inflammation	Yes [5, 6, 9, 10]	Partial [11]	No	NT

NT not tested

results in a severe inflammatory microenvironment such as VML, and may occasionally have a deleterious effect upon muscle function [25].

The efficient clinical translation of pharmacologic strategies, cell-based approaches, physical therapy, and combination therapies remains a challenge but is worthy of further pursuit. At the present time, the lower regulatory hurdles, lack of requirement for immunosuppression, relatively inexpensive production costs, and ability to stimulate endogenous functional muscle regeneration place the acellular surgical mesh approach at the forefront of novel therapies for VML.

24.5 Mechanisms by Which ECM Promotes Myogenesis

More than a superficial understanding of the mechanisms by which biologic surgical mesh materials can facilitate muscle regeneration is required to optimize the chances for a functional outcome. The rationale for use of an ECM-based surgical mesh can be best explained by first reviewing current knowledge of the default response to skeletal muscle injury in mammals. Immediately following muscle injury, damaged myocytes release cytokines that induce a proinflammatory phenotype in neutrophils and macrophages, which in turn secrete a profile of signaling molecules (interleukin-1β, interleukin-6, tumor necrosis factor alpha) that induce mobilization and proliferation of resident stem and progenitor cells (e.g., satellite cells). The subsequent transition of macrophages to a regulatory, pro-remodeling phenotype occurs as a result of as yet poorly defined signals. However, this phenotype transition is required for resolution of the inflammatory process and subsequent regeneration of lost or injured myocytes [26–29]. The pro-remodeling macrophage phenotype (M2-like) is associated with the expression of TGF-β, a profibrotic cytokine, increased connective collagen deposition, and arginase 1 expression which can promote fibroblast proliferation and increased connective tissue production [30, 31]. Inhibition of either the pro-inflammatory or the pro-remodeling macrophage phenotype severely limits skeletal muscle regeneration [32].

ECM surgical meshes have repeatedly been shown to facilitate constructive remodeling of the target tissue by activating macrophages toward the anti-inflammatory, and pro-remodeling macrophage phenotype. Likewise, adaptive immune B and T cells are recruited to the ECM scaffold remodeling site, and T cells activate to a Th2-like (pro-remodeling) phenotype [33]. The in vivo degradation of ECM mesh materials is associated with the release of low-molecular-weight matricryptic oligopeptides and other bioactive products [34, 35] that influence macrophage phenotype and recruit and influence the fate of endogenous progenitor cells. For example, during the course of mesh degradation, perivascular stem cells mobilize from their perivascular niche and differentiate in response to microenvironmental factors such as oxygen concentration, mechanical forces, and cytokines such as vascular endothelial growth factor (VEGF), bone morphogenetic protein (BMP), fibroblast growth factor (FGF), epidermal growth factor (EGF), platelet-derived growth factor (PDGF), and as previously mentioned TGF-β, among others [36]. Any chemical cross-linking agents such as glutaraldehyde or carbodiimide used during the production of ECM-based surgical mesh will inhibit this necessary functional degradation of the scaffold and will mitigate the pro-remodeling response [37].

Functional skeletal muscle regeneration is dependent upon a number of physiologic events including vascularization and innervation of the injury site [2, 38]. Preclinical studies that involve the implantation of ECM surgical mesh in rat abdominal wall and canine esophagus show neovascularization and evidence of nerve cells and Schwann cells in the acute and subacute postsurgical period [39]. The same study also showed the in vitro migration of Schwann cells when exposed to matricryptic peptides in a concentration-dependent fashion. A separate study that investigated the restoration of canine gastrocnemius musculotendinous junction by ECM surgical mesh implantation showed a 48% increase in contractile force and the presence of vascularized, innervated skeletal muscle [40].

A cohort clinical report of patients with volumetric muscle loss who were treated with ECM surgical meshes and given early-onset mechanical loading (i.e., physical therapy) was recently published [2]. Reinnervation of the wound area was evaluated by nerve conduction and electromyography [41]. The nerve conduction studies showed an increase in compound muscle action potential in four out of eight participants while electromyography showed that two participants had improvement in target muscle innervation. In vitro studies have repeatedly described the chemoattractant effect of ECM degradation products upon endothelial cells [34, 42, 43] and the proliferation and differentiation effects upon neuronal cells [44, 45]. Although the specific molecular mechanisms of neovascularization and innervation during skeletal muscle regeneration as a result of ECM mesh implantation are not known, the previous studies provide a useful database for futures studies.

24.6 Additional Factors That Affect Clinical Outcome

The use of acellular bioscaffolds (i.e., surgical meshes) to promote functional skeletal muscle replacement, as opposed to the use of skeletal muscle stem/progenitor cells, is not necessarily intuitive. When considering the mechanisms (described above) by which these materials induce a constructive host (patient) response, the positive results observed in many preclinical and early clinical studies become more plausible. However, there are many patient-, surgical-, and material-related factors that also influence the clinical outcome.

The ability of the host to interact with an implanted biologic scaffold composed of ECM is dependent upon variables such as the age of the patient, comorbidities such as diabetes and obesity, nutritional status, and anatomic site of implantation.

Surgical technique is an obvious and important factor in any reconstructive surgical procedure. When biologic meshes are used, surgical technique is even more important and includes an understanding of the mechanisms of remodeling (described above). Removal of scar tissue to create a fresh wound, assurance of close contact between mesh and adjacent healthy tissue, avoidance of seroma formation, and placement of the surgical mesh under tension, among others, can all be critical determinants of a successful outcome.

The composition of the biologic surgical mesh varies depending upon the source tissue (e.g., small intestine, urinary bladder, dermis) because the extracellular matrix for each tissue and organ is distinctive. Mechanical properties, rate of degradation, and composition vary among the surgical meshes derived from different tissues and can be an important factor in functional remodeling, especially in skeletal muscle where mechanical loading and weight bearing are critical.

The age of the animal from which the source tissue is harvested, nutritional status of the animal, and genetic variability of the herd from which the tissue is collected, among other factors, can introduce a degree of variability that must be taken into account when selecting an ECM-based surgical mesh for the treatment of volumetric muscle loss. In addition, the processing of the material including methods of decellularization, presence or absence of the use of chemical cross-linking agents, and method of terminal sterilization are important determinants of clinical success.

Although these ECM materials are regulated at surgical mesh devices, an understanding of the biology during the remodeling process and factors which affect outcomes is essential for optimizing the chances of a satisfactory clinical result. While biologic surgical mesh development has provided significant advancement in the treatment for VML, room for improvement remains, specifically to promote vascularized, innervated, and functional muscle tissue that fully restores strength and function. It is likely that a combination of cell-based therapies and physical therapy with acellular biologic mesh materials will provide a superior treatment option for VML patients.

References

1. Grogan BF, Hsu JR. Skeletal trauma research, volumetric muscle loss. J Am Acad Orthop Surg. 2011;19(Suppl 1):S35–7.
2. Dziki JB, Badylak S, Yabroudi M, Sicari B, Ambrosio A, Stearns K, Turner N, Wyse A, Boninger ML, Brown EHP, Rubin JP. An acellular biologic scaffold treatment for volumetric muscle loss: results of a 13-patient cohort study. NPJ Regen Med. 2016;1:16008.
3. Sicari BM, Rubin JP, Dearth CL, Wolf MT, Ambrosio F, Boninger M, Turner NJ, Weber DJ, Simpson TW, Wyse A, Brown EH, Dziki JL, Fisher LE, Brown S, Badylak SF. An acellular biologic scaffold promotes skeletal muscle formation in mice and humans with volumetric muscle loss. Sci Transl Med. 2014;6(234):234–58.
4. Corona BT, Rivera JC, Owens JG, Wenke JC, Rathbone CR. Volumetric muscle loss leads to permanent disability following extremity trauma. J Rehabil Res Dev. 2015;52(7):785–92.
5. Corona BT, Wenke JC, Ward CL. Pathophysiology of volumetric muscle loss injury. Cells Tissues Organs. 2016;202(3–4):180–8.
6. Lin CH, Lin YT, Yeh JT, Chen CT. Free functioning muscle transfer for lower extremity posttraumatic composite structure and functional defect. Plast Reconstr Surg. 2007;119(7):2118–26.
7. Paro J, Chiou G, Sen SK. Comparing muscle and fasciocutaneous free flaps in lower extremity reconstruction—does it matter? Ann Plast Surg. 2016;76(Suppl 3):S213–5.
8. Patzkowski JC, Owens JG, Blanck RV, Kirk KL, Hsu JR. Deployment after limb salvage for high-energy lower-extremity trauma. J Trauma Acute Care Surg. 2012;73(2 Suppl 1):S112–5.
9. Corona BT, Henderson BE, Ward CL, Greising SM. Contribution of minced muscle graft progenitor cells to muscle fiber formation after volumetric muscle loss injury in wild-type and immune deficient mice. Physiol Rep. 2017;5(7):e13249.
10. Ward CL, Ji L, Corona BT. An autologous muscle tissue expansion approach for the treatment of volumetric muscle loss. Biores Open Access. 2015;4(1):198–208.
11. Valdivia M, Vega-Macaya F, Olguin P. Mechanical control of myotendinous junction formation and tendon differentiation during development. Front Cell Dev Biol. 2017;5:26.
12. Qin YX, Hu M. Mechanotransduction in musculoskeletal tissue regeneration: effects of fluid flow, load-ing, and cellular-molecular pathways. Biomed Res Int. 2014;2014:863421.
13. Dziki JL, Giglio RM, Sicari BM, Wang DS, Gandhi RM, Londono R, Dearth CL, Badylak SF. The effect of mechanical loading upon extracellular matrix bioscaffold-mediated skeletal muscle remodeling. Tissue Eng Part A. 2018;24(1–2):34–46.
14. Ambrosio F, Kadi F, Lexell J, Fitzgerald GK, Boninger ML, Huard J. The effect of muscle loading on skeletal muscle regenerative potential: an update of current research findings relating to aging and neuromuscular pathology. Am J Phys Med Rehabil. 2009;88(2):145–55.
15. Palermo AT, Labarge MA, Doyonnas R, Pomerantz J, Blau HM. Bone marrow contribution to skeletal muscle: a physiological response to stress. Dev Biol. 2005;279(2):336–44.
16. Ambrosio F, Ferrari RJ, Fitzgerald GK, Carvell G, Boninger ML, Huard J. Functional overloading of dystrophic mice enhances muscle-derived stem cell contribution to muscle contractile capacity. Arch Phys Med Rehabil. 2009;90(1):66–73.
17. Corona BT, Ward CL, Baker HB, Walters TJ, Christ GJ. Implantation of in vitro tissue engineered muscle repair constructs and bladder acellular matrices partially restore in vivo skeletal muscle function in a rat model of volumetric muscle loss injury. Tissue Eng Part A. 2014;20(3–4):705–15.
18. Corona BT, Machingal MA, Criswell T, Vadhavkar M, Dannahower AC, Bergman C, Zhao W, Christ GJ. Further development of a tissue engineered muscle repair construct in vitro for enhanced functional recovery following implantation in vivo in a murine model of volumetric muscle loss injury. Tissue Eng Part A. 2012;18(11–12):1213–28.
19. Rossi CA, Flaibani M, Blaauw B, Pozzobon M, Figallo E, Reggiani C, Vitiello L, Elvassore N, De Coppi P. In vivo tissue engineering of functional skeletal muscle by freshly isolated satellite cells embedded in a photopolymerizable hydrogel. FASEB J. 2011;25(7):2296–304.
20. Collins CA, Olsen I, Zammit PS, Heslop L, Petrie A, Partridge TA, Morgan JE. Stem cell function, self-renewal, and behavioral heterogeneity of cells from the adult muscle satellite cell niche. Cell. 2005;122(2):289–301.
21. Webster MT, Manor U, Lippincott-Schwartz J, Fan CM. Intravital imaging reveals ghost fibers as architectural units guiding myogenic progenitors during regeneration. Cell Stem Cell. 2016;18(2):243–52.
22. Quarta M, Cromie M, Chacon R, Blonigan J, Garcia V, Akimenko I, Hamer M, Paine P, Stok M, Shrager JB, Rando TA. Bioengineered constructs combined with exercise enhance stem cell-mediated treatment of volumetric muscle loss. Nat Commun. 2017;8:15613.
23. Montarras D, Morgan J, Collins C, Relaix F, Zaffran S, Cumano A, Partridge T, Buckingham M. Direct isolation of satellite cells for skeletal muscle regeneration. Science. 2005;309(5743):2064–7.

24. Mueller GM, O'Day T, Watchko JF, Ontell M. Effect of injecting primary myoblasts versus putative muscle-derived stem cells on mass and force generation in mdx mice. Hum Gene Ther. 2002;13(9):1081–90.

25. Garg K, Corona BT, Walters TJ. Losartan administration reduces fibrosis but hinders functional recovery after volumetric muscle loss injury. J Appl Physiol (1985). 2014;117(10):1120–31.

26. Fadok VA, Bratton DL, Konowal A, Freed PW, Westcott JY, Henson PM. Macrophages that have ingested apoptotic cells in vitro inhibit proinflammatory cytokine production through autocrine/paracrine mechanisms involving TGF-beta, PGE2, and PAF. J Clin Invest. 1998;101(4):890–8.

27. Fadok VA, Bratton DL, Guthrie L, Henson PM. Differential effects of apoptotic versus lysed cells on macrophage production of cytokines: role of proteases. J Immunol. 2001;166(11):6847–54.

28. Arnold L, Henry A, Poron F, Baba-Amer Y, van Rooijen N, Plonquet A, Gherardi RK, Chazaud B. Inflammatory monocytes recruited after skeletal muscle injury switch into anti-inflammatory macrophages to support myogenesis. J Exp Med. 2007;204(5):1057–69.

29. Mounier R, Théret M, Arnold L, Cuvellier S, Bultot L, Göransson O, Sanz N, Ferry A, Sakamoto K, Foretz M, Viollet B, Chazaud B. AMPKalpha1 regulates macrophage skewing at the time of resolution of inflammation during skeletal muscle regeneration. Cell Metab. 2013;18(2):251–64.

30. Roberts AB, Sporn MB, Assoian RK, Smith JM, Roche NS, Wakefield LM, Heine UI, Liotta LA, Falanga V, Kehrl JH, et al. Transforming growth factor type beta: rapid induction of fibrosis and angiogenesis in vivo and stimulation of collagen formation in vitro. Proc Natl Acad Sci U S A. 1986;83(12):4167–71.

31. Mills CD, Kincaid K, Alt JM, Heilman MJ, Hill AM. M-1/M-2 macrophages and the Th1/Th2 paradigm. J Immunol. 2000;164(12):6166–73.

32. Ruffell D, Mourkioti F, Gambardella A, Kirstetter P, Lopez RG, Rosenthal N, Nerlov C. A CREB-C/EBPbeta cascade induces M2 macrophage-specific gene expression and promotes muscle injury repair. Proc Natl Acad Sci U S A. 2009;106(41):17475–80.

33. Sadtler K, Estrellas K, Allen BW, Wolf MT, Fan H, Tam AJ, Patel CH, Luber BS, Wang H, Wagner KR, Powell JD, Housseau F, Pardoll DM, Elisseeff JH. Developing a pro-regenerative biomaterial scaffold microenvironment requires T helper 2 cells. Science. 2016;352(6283):366–70.

34. Reing JE, Zhang L, Myers-Irvin J, Cordero KE, Freytes DO, Heber-Katz E, Bedelbaeva K, McIntosh D, Dewilde A, Braunhut SJ, Badylak SF. Degradation products of extracellular matrix affect cell migration and proliferation. Tissue Eng Part A. 2009;15(3):605–14.

35. Dietz GW, Heppel LA. Studies on the uptake of hexose phosphates. II. The induction of the glucose 6-phosphate transport system by exogenous but not by endogenously formed glucose 6-phosphate. J Biol Chem. 1971;246(9):2885–90.

36. Karalaki M, Fili S, Philippou A, Koutsilieris M. Muscle regeneration: cellular and molecular events. In Vivo. 2009;23(5):779–96.

37. Valentin JE, Stewart-Akers AM, Gilbert TW, Badylak SF. Macrophage participation in the degradation and remodeling of extracellular matrix scaffolds. Tissue Eng Part A. 2009;15(7):1687–94.

38. Tonnesen MG, Feng X, Clark RA. Angiogenesis in wound healing. J Investig Dermatol Symp Proc. 2000;5(1):40–6.

39. Agrawal V, Brown BN, Beattie AJ, Gilbert TW, Badylak SF. Evidence of innervation following extracellular matrix scaffold-mediated remodelling of muscular tissues. J Tissue Eng Regen Med. 2009;3(8):590–600.

40. Turner NJ, Yates AJ Jr, Weber DJ, Qureshi IR, Stolz DB, Gilbert TW, Badylak SF. Xenogeneic extracellular matrix as an inductive scaffold for regeneration of a functioning musculotendinous junction. Tissue Eng Part A. 2010;16(11):3309–17.

41. Han N, Yabroudi MA, Stearns-Reider K, Helkowski W, Sicari BM, Rubin JP, Badylak SF, Boninger ML, Ambrosio F. Electrodiagnostic evaluation of individuals implanted with extracellular matrix for the treatment of volumetric muscle injury: case series. Phys Ther. 2016;96(4):540–9.

42. Vorotnikova E, McIntosh D, Dewilde A, Zhang J, Reing JE, Zhang L, Cordero K, Bedelbaeva K, Gourevitch D, Heber-Katz E, Badylak SF, Braunhut SJ. Extracellular matrix-derived products modulate endothelial and progenitor cell migration and proliferation in vitro and stimulate regenerative healing in vivo. Matrix Biol. 2010;29(8):690–700.

43. Li F, Li W, Johnson S, Ingram D, Yoder M, Badylak S. Low-molecular-weight peptides derived from extracellular matrix as chemoattractants for primary endothelial cells. Endothelium. 2004;11(3–4):199–206.

44. Ghuman H, Massensini AR, Donnelly J, Kim SM, Medberry CJ, Badylak SF, Modo M. ECM hydrogel for the treatment of stroke: characterization of the host cell infiltrate. Biomaterials. 2016;91:166–81.

45. Crapo PM, Tottey S, Slivka PF, Badylak SF. Effects of biologic scaffolds on human stem cells and implications for CNS tissue engineering. Tissue Eng Part A. 2014;20(1–2):313–23.

Principles of Tendon Regeneration

25

Jacinta Leyden, Yukitoshi Kaizawa, and James Chang

25.1 Introduction

Tendons are connective tissues that connect muscle to bone. They have a unique protein arrangement, are hypocellular, and have a limited blood supply. Tendons can withstand significant tensile loads and confer the ability to move by transferring forces that muscle contractions create.

The United States spends nearly $30 billion on musculoskeletal injuries each year, 45% of which are tendon and ligament injuries [1]. Tendon injuries account for nearly 300,000 repair surgeries and a significant amount of morbidity and loss of productivity, in both the United States and across the globe [2].

Surgical interventions involve basic suture repair, tendon transfers, and tendon grafts. Due to its hypocellularity and hypovascularity, the tendon has low healing potential, and scar tissue formation easily occurs. Complications of these techniques involve high failure rates with a significant risk of recurrence of rupture, fibrous

J. Leyden
Stanford University School of Medicine,
Stanford, CA, USA
e-mail: leydenj@stanford.edu

Y. Kaizawa · J. Chang (✉)
Division of Plastic and Reconstructive Surgery,
Stanford University Medical Center,
Palo Alto, CA, USA
e-mail: kai0812@stanford.edu;
jameschang@stanford.edu

adhesions, and poor long-term functional outcomes [3, 4]. Even when surgically repaired, tendon injuries remain notoriously difficult to treat and manage, and both function and structure often remain compromised even 1 year following injury [5].

Tissue engineering plays a promising role in optimizing tendon healing and regeneration. Current methods have explored the use of various techniques to augment the native healing and regenerative capacity of tendons, and there have been great advancements in the development of allogenic and xenogeneic tendon grafts and other artificial materials to repair tendon defects.

This chapter describes an overview of tendon anatomy and structure, native tendon-healing cascade, and tissue engineering principles and advances for tendon regeneration.

25.2 Tendon Anatomy

Tendons are primarily white in color and possess a fibroelastic texture. They have different connective tissue layers that enable tendon collagen fibers to glide along each other with minimal friction when they experience tensile forces along the tendon axis. These layers carry blood vessels, lymphatics, and nerve channels throughout the tendon [6]. The epitenon, the outermost connective tissue layer, is the loose connective tissue that covers the entire tendon and extends deep between the tertiary bundles all

© Springer Nature Switzerland AG 2019
D. Duscher, M. A. Shiffman (eds.), *Regenerative Medicine and Plastic Surgery*,
https://doi.org/10.1007/978-3-030-19962-3_25

Fig. 25.1 Tendon structure (adapted from Sharma and Maffulli 2006 [16])

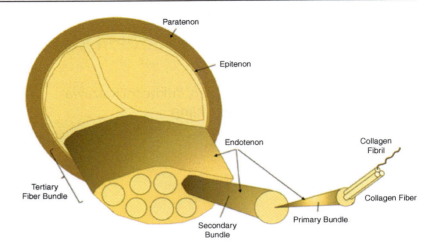

the way to the fascicles. The endotenon is a thin, reticular structure and a continuation of the more superficial epitenon that extends between the primary bundles [7]. The paratenon (or tendon sheath) surrounds the epitenon superficially and consists of type I and type III collagen fibrils and elastic fibrils [8]. Tendons are either sheathed or unsheathed, with the former having a more delicate blood supply system and an inner layer of synovial cells derived from the parietal and visceral membranes [9, 10]. This extra layer in between the sheath and paratenon provides additional lubrication to these tendons and allows the tendon to handle greater mechanical stress [11]. Examples of unsheathed tendons include the patellar and calcaneal tendons, which have larger vessels that pass through their respective vincula [12]. Sheathed tendons, like hand flexor tendons, are less vascularized with more delicate vincula and have avascular areas that receive nutrition by diffusion [13].

The smallest building block of the tendon is the collagen fibril. Collagen fibrils consist of tropocollagen, a triple-helix formation of proteins that cross-link together [14]. Fibrils collect into fibers (primary bundles). Fibers then collect into fascicles (secondary bundles), which connect into tertiary bundles that make up the mature tendon (Fig. 25.1) [15, 16].

25.3 Tendon Composition

Tendons have low cellularity and consist primarily of water (70%), with a variety of compounds comprising their dry weight, including collagen (86%). Type I collagen constitutes 97–98% of the overall collagen profile and gives tendons the ability to resist tension. Type II collagen is present in cartilaginous zones, while type III collagen is present in the reticulum fibers of vascular walls [15]. Elastin, a protein that contributes to tissue flexibility, extensibility, and recoil, is 2% of a tendon's dry weight [17]. 1–5% of a tendon's dry weight consists of proteoglycans. The remaining dry weight consists of inorganic components like copper and magnesium [18].

The primary cell of the tendon is the tenocyte. Tenocytes are specialized fibroblastic cells with an elongated form that provide the tendon with its unique biomechanical properties. Tenocytes synthesize the proteins of the extracellular matrix, namely collagen, proteoglycans, and glycoproteins [15]. They also release signals that regulate collagen formation and development [19].

The substance of the extracellular matrix (ECM) network around the collagen core consists of proteoglycans, glycosaminoglycans (GAGs), glycoproteins, and other small molecules. All these components play specific roles in tendon development, nutrition, and healing [20]. Proteoglycans, with their hydrophilic structure, allow for rapid diffusion of water-soluble molecules and migration of cells into the tendon structure [21]. Glycoproteins help with the repair process, and the various glycoproteins within the ECM structure serve varying roles that are still unclear [22, 23].

25.4 Blood Supply

Tendons and ligaments consume 7.5 times less oxygen per cellular unit compared to skeletal muscle [24]. This low metabolic rate and anaerobic energy generation reduce the risk of ischemia and necrosis while maintaining tension and carrying heavy loads over prolonged periods [25]. Nonetheless, this low metabolic rate and limited blood supply contribute to slow healing after injury. Three separate regions supply blood to tendons: the myotendinous junction, the osteotendinous junction, and arteries within surrounding connective tissues. The vascular contribution of each of these areas varies between tendon types [7].

25.5 Tendon Biomechanics

Tendons withstand tension and absorb excessive forces along the axis to limit damage to the tendon and its surrounding structures [26]. A tendon's mechanical behavior depends significantly on the number and types of intramolecular and intermolecular bonds within the tendon, which indicates its overall health. The stress–strain curve in Fig. 25.2 demonstrates this process [27]. Stress–strain curves facilitate comparison of tendons to determine whether a repair technique is effective, and the ultimate load of failure is one of the most indicative parameters for tendon biomechanical evaluation (Fig. 25.2) [28, 29].

25.6 Tendon-Healing Cascade

Tendon healing divides into three overlapping phases: inflammation, proliferation, and remodeling.

Inflammation is the first reaction following tendon injury. During this period, cells such as neutrophils and erythrocytes migrate to the injury site, and monocytes and macrophages begin phagocytosis of necrotic material [30]. During this time, vasoactive and chemotactic factors initiate angiogenesis and tenocyte proliferation.

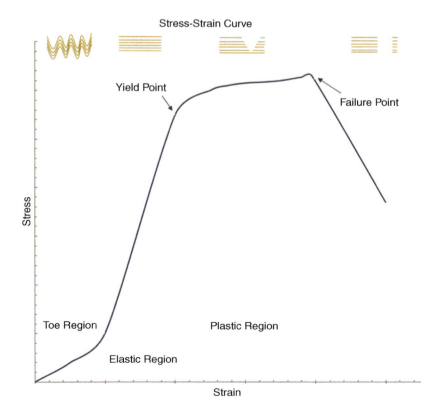

Fig. 25.2 Tendon stress–strain curve (adapted from Morales-Orcajo E [29])

Tenocytes start to synthesize type III collagen and other matrix components to rebuild the injured tendon. Several enzymes, like matrix metalloproteinases (MMPs), play a role in degrading the ECM at the injury site to increase cell turnover and further facilitate the healing process [31, 32].

Angiogenesis is a crucial component in the healing process during the inflammation phase. The new vascular network nourishes new tissues and removes cell debris. Some studies show that even the slightest decrease in blood flow during this period can have detrimental effects on healing [33]. Growth factors like vascular endothelial growth factor (VEGF), transforming growth factor-beta (TGF-β), and insulin-like growth factor 1 (IGF-1) mediate the inflammatory and angiogenic response [5, 34, 35].

Next, in the proliferative phase, there is a dramatic increase in cellularity and matrix production. There is a large spike in type III collagen production, which contributes to elasticity [36]. IGF-1, basic fibroblast growth factor (bFGF), platelet-derived growth factor (PDGF), TGF-β, and bone morphogenetic protein (BMP) are growth factors that play a critical role during this phase and promote tenogenic factors that contribute to recovery [32].

The final stage, remodeling, is further divided into the consolidation and maturation periods. Consolidation occurs within 6–10 weeks following the initial injury, during which the synthesis of collagen and GAGs decreases, and tenocytes and collagen fibers orient themselves in the direction of stress [36]. Cellularity also decreases during this stage as the tissue becomes even more fibrous. Type I collagen production increases, replacing type III collagen and GAGs built up during the previous phase [37]. Maturation occurs after 10 weeks. Collagen fibrils start to cross-link in greater numbers, increasing tissue stiffness [37]. Over the next year after injury, the tissue turns fibrous and displays more scar-like vascularity [38].

Many enzymes play critical roles in the tendon-healing process at the final healing stage, namely matrix metalloproteinases (MMPs). These enzymes play an important role in tendon

degradation and remodeling following injury by shaping and influencing the tendon ECM [39]. MMPs with collagenase activity, however, increase the risk of tendon rupture recurrence, as the presence of denatured collagen and the degradation of collagen fibrils from these enzymes weaken the tendon ECM [40]. This contributes to risk of tendon reinjury and degradation over time as MMP and cytokine release continues long after the initial tendon insult.

25.7 Surgical Tendon Repair and Regeneration

The main objective of surgical repair is to re-establish tendon alignment to maximize native tendon-healing abilities. Tendons were initially thought to have little to no regenerative ability given their low cellularity and limited blood supply, but as we continue to understand and recognize their intrinsic healing cascades, new technologies emerge that optimize and augment the native cues that contribute to tendon regeneration.

25.8 Principles of Tendon Tissue Engineering

Tissue engineering plays a promising role in promoting tendon regeneration and healing. Tissue engineering for tendon regeneration aims to create new, healthy tissue in place of injured or damaged space. The foundations of tissue engineering revolve around (1) establishing an appropriate scaffold to the physical space that a defect creates, (2) using cells to replace tenocytes and create tendon cellular and mechanical composition within the scaffold, and (3) creating an environment conducive to cell maturation and tenocyte survival within the scaffold.

25.8.1 Cells

Tissue engineering efforts involve investigating the ideal cell type to fill tendon defects. As tenocytes are the primary cells that make up the tendon and contribute to its mechanical and chemical properties, tenocytes are a common cell source.

Direct injection of autologous tenocytes into injury areas or within scaffolds can augment tendon healing [41, 42]. However, there are several limitations to tenocyte-based tendon healing, including difficulties with cell harvesting and expansion. Additionally, protocols for autologous tenocyte harvesting result in significant donor-site defects. In the context of these limitations, dermal fibroblasts present another option for cell harvesting [43].

The use of adult stem cells in tendon tissue engineering is associated with lower donor-site morbidity [44]. Adult stem cells are further subdivided into multipotent versus pluripotent lineage potential. Mesenchymal stem cells (MSCs) are multipotent cells with bone marrow and adipose cells as common sites for harvest. Their proliferative capacity is superior to tenocytes [45], and they are more easily expanded ex vivo for storage and banking [46]. Moreover, MSCs have homing abilities to areas of injury and exhibit anti-inflammatory and angiogenic effects within injury sites [45]. More recently, tissue engineering efforts have used tendon-derived stem cells (TDSCs) for tendon healing and regeneration. Differentiation properties of TDSCs are more favorable towards tenocyte differentiate than MSCs from other sources [47].

Tissue engineering efforts have also used pluripotent stems cells, like embryonic stem (ES) cells and induced pluripotent stem (iPS) cells [48]. Embryonic stem cells have a pluripotent differentiation capacity and a tenogenic differentiation profile that is more favorable than that of MSCs [49]. However, numerous ethical concerns surround the use and harvest of embryonic stem cells. Induced pluripotent stem cells do not present the same ethical concerns, and their differentiation capacity is similar to that of embryonic stem cells [50]. The quality of the iPS cells for tenogenic differentiation, however, remains variable and under-investigated.

There is no current consensus about the optimal cell source. Table 25.1 contains an overview of different cell types in tissue engineering [41, 51–57].

Table 25.1 Cells used in tendon tissue engineering

Cell type	Experimental design	Results
Tendon fibroblasts	• Lateral epicondylitis, direct injection [51] • PGA[a] scaffold, hen flexor digitorum profundus model [41]	• Improved clinical function and MRI tendinopathy scores • Increased tensile strength
Dermal fibroblasts	• Patellar tendinopathy, direct injection [52] • PGA scaffold, porcine flexor digital superficial tendon model [53]	• Improved clinical outcome scores • Similar tensile strength to tenocyte-seeded scaffolds
Bone marrow mesenchymal stem cells (BMSCs)	• Rabbit ACL[b], coating with BMSCs[c] [54] • PLGA[d] scaffold in vitro [55]	• Improved biomechanical testing • Tenogenic differentiation
Adipose-derived stem cells (ASCs)	• Rabbit Achilles tendon, injection with PRP [56]	• Enhanced primary tendon healing
Tendon-derived stem cells (TDSCs)	• Rat patellar tendon window defect, no scaffold [57]	• Increased collagen production and alignment
Induced pluripotent stem (iPS) cells	• No injury— in vitro differentiation to tenocytes [50]	• Ability to differentiate into tenocytes, but at lower rates than embryonic stem cells

[a]*PGA* poly-glycolic acid
[b]*ACL* anterior cruciate ligament
[c]*BMSCs* bone marrow-derived stem cells
[d]*PLGA* poly-L-lactic acid

25.8.2 Scaffolds

Scaffolds within tissue engineering provide biomechanical support for tissue healing until native healing channels have established enough matrix and cell production to maintain a regenerated structure on their own. The ideal scaffold for tissue regenerations should (1) be biocompatible, (2) support cell attachment and growth, (3) have high surface area, (4) promote cell differentiation to tenocytes or appropriate cell form, (5) minimize host immune inflammatory response, and (6) mimic native tendon architectures and mechanical properties if they are not biodegradable [36].

Scaffolds can consist of either biologically or synthetically derived materials. Biologic scaffolds include decellularized native tendon matrices or derivatives of naturally occurring proteins, like collagen. Synthetic scaffolds usually consist of polyester derivatives or other fabrics like silk.

25.8.2.1 Biologic Scaffolds
Biologic scaffolds are often created from decellularized allograft and xenografts, and this is one of the treatment options currently available for significant tendon defects. In theory, since tendon matrices are directly obtained from the same tissue type, they should retain native biomechanical and biochemical tendon properties. Nonetheless, in order to minimize disease transmission and an immune response against foreign tissue and cells, tissue allografts and xenografts must first be decellularized before implantation. Decellularization techniques allow for overall preservation of mechanical and chemical properties while preserving various proteins and growth factors within the tendon extracellular matrix [58, 59]. However, tendon implants still do not fully match the properties of native tendon when implanted from a source with exposure to different stress and strain parameters over time [60].

Biologic scaffolds also include biomaterials created from collagen or other tendon-related proteins. This presents a middle ground between mitigating the immune-related risks of a pure allograft or xenograft implantation and providing a cell signaling environment conducive to teno-cyte differentiation, proliferation, and survival. Collagen-based scaffolds are also highly biocompatible, as tendon ECMs consist mainly of type I collagen [7]. Biologic scaffolds in isolation can bridge large tendon defects or augment healing when seeded with various cell types.

25.8.2.2 Synthetic Scaffolds
Synthetic scaffolds are an alternative to biologically derived scaffolds. Most synthetic scaffolds are made of α-hydroxyl-polyesters, given their biocompatibility, biodegradability, and ability to support cell seeding [61, 62]. There are several limitations, however, with the use of synthetic scaffolds. Synthetically derived scaffolds do not retain the critical ECM properties and cell signaling abilities of native tissue. As a result, they do not properly regulate cell activity or tenocyte differentiation and have limited survival compared to biologic scaffolds, and are weaker than native tendon material [63]. The most common synthetic scaffolds are made of poly(glycolic acid) (PGA), poly(lactic-co-glycolic acid) (PGLA), and poly-L-lactic acid (PLLA). Table 25.2 summarizes an overview of their applications [41, 61, 62, 64–70].

25.9 Growth Factors

Growth factors play an important role in overall tendon healing. Therefore, the external application of these factors to tendon injuries can augment the native healing process. However, the timing and delivery of these growth factors in the setting of a healing tendon remain challenging. Table 25.3 summarizes the various growth factors investigated for tendon repair augmentation and the clinical implications of their delivery during the healing process [34, 55, 71–79].

25.10 Discussion

As our understanding of tendon development and healing continues to improve, we continue to tailor tissue engineering efforts towards augmenting the native healing process. While the

25 Principles of Tendon Regeneration

Table 25.2 Scaffolds used in tendon tissue engineering

Scaffold type	Experimental design	Results
Poly-glycolic acid (PGA)	• Hen flexor digitorum profundus model and autologous tenocytes [41] • Mouse skeletal muscle-derived cells in vitro [61]	• Increased tensile strength • Tendon structure formation with mature collagen fibrils
Poly-L-lactic acid (PLLA)	• Human tendon stem/progenitor cells in vitro [62]	• Tenocyte differentiation and cell survival
Poly lactic-co-glycolic acid (PLGA)	• Rat Achilles tendon model with BMSCs[a] [64]	• Increased tensile strength, increased collagen deposition compared to cell-free scaffold
Silk	• In vitro electrospun silk scaffold, seeded with BMSC[a] [65]	• Increased expression of tendon-related proteins
Collagen-silk	• In vitro scaffold seeded with tendon-derived stem cells [66]	• Increased collagen deposition and increased tensile strength
Collagen gels	• Rabbit patellar tendon model, BMSCs [67] • Collagen gel/sponge blend, rabbit patellar tendon, BMSCs [68]	• Increased tensile strength and neotissue formation compared to collagen gel alone or simple suture • Increased tensile strength
Tendon hydrogel (tHG)	• Rat Achilles tendon model, no cells [69] • Rat Achilles tendon model, ASCs[b], and PRP[c] [70]	• Increased ultimate failure load with gel alone compared to control • Increased ultimate failure load, increased ECM formation

[a]*BMSC* bone marrow-derived stem cell
[b]*ASCs* adipose-derived stem cells
[c]*PRP* platelet-rich plasma

Table 25.3 Growth factors in tendon healing

Growth factor	Experimental design	Results
Insulin-like growth factor 1 (IGF-1)	• Rat Achilles tendon model [71] • Rabbit flexor tendon model [34]	• Decreased inflammation-induced functional deficits • Increased proliferation and migration of fibroblasts to injury site, increased collagen, and ECM production
Transforming growth factor beta (TGF-β)	• Rabbit tenocytes in vitro [72] • Equine embryo-derived stem cells in vitro [73]	• Increased matrix synthesis in tendon healing • Induced tenocyte differentiation, increased tendon-specific gene markers
Bone morphogenetic protein (BMP)	• BMP-14, transected rat Achilles tendon [74]	• Increased tensile strength
Basic fibroblast growth factor (bFGF)	• Rat patellar tendon [75] • bFGF-releasing nanofibrous scaffold [55]	• Increased proliferation and expression of type III collagen. • Increased BMSC[a] proliferation, tenogenic differentiation, and collagen deposition within scaffold
Platelet-derived growth factor (PDGF)	• Rabbit tendon in vitro [76] • Rabbit medial collateral ligament [77]	• Stimulated collagen production and matrix production • Increased load to failure and elongation values
Platelet-rich plasma (PRP) (concentrated pool of growth factor: VEGF[a], PDGF, TGF, IGF-1, FGF)	• Rat patellar tendon model [78] • Human Achilles tendon model [79]	• Increased level of circulation-derived cells into injury site • Decreased time to previous range of motion and movement

[a]*VEGF* vascular endothelial growth factor

established pillars of tissue engineering have been well elucidated, the challenge remains on how to optimize the individual areas of tissue engineering in combination.

25.10.1 Current Problems of Cell Therapy

Tissue engineering efforts have utilized several different cell sources, as detailed in Table 25.1. However, there are limitations and risks associated with each of these cells, and many areas require further investigation.

First, the therapeutic effects of cell implantation require a large number of cells. Currently, harvested cells typically undergo ex vivo expansion within specific nutrient-rich mediums before achieving a sufficient quantity for implantation. Common feeder mediums, like fetal bovine serum (FBS), carry a risk of disease transmission and immune rejection [80]. A possible solution to this issue is to optimize the efficacy of harvesting cells and develop a method of culturing cells without FBS. Several trials have implanted MSCs in situ without ex vivo expansion with promising results, and these efforts should continue to be pursued [81]. Alternatively, other efforts have expanded iPS cells within feeder-free mediums before implantation, which reduces the infection risks of current feeder mediums [82].

Second, it is important to note that both multipotent and pluripotent stem cells carry a risk of tumorigenicity when applied clinically [83, 84, 99]. Undifferentiated cells, especially iPS cells, risk teratoma development after transplantation, although the rate of transformation potential is still unknown [85]. Ethical questions surrounding their use persist, as tendon injuries are not typically life threatening and do not confer an increased risk of mortality, though these treatment modalities may have lethal impacts on extreme outcomes.

Finally, optimal methods to keep viable cells around injury sites require further investigation. A significant percentage of implanted cells leave the injected site if injected in mediums like fetal bovine serum or other nutrient-rich mediums [86, 87]. Our laboratory developed a thermo-responsive, decellularized, human tendon-derived hydrogel (tHG) as a cell vehicle. Our tHG contains abundant natural cues for cellular attraction [88]. When tHG-containing, luciferase-transfected, adipose-derived stem cells were injected into a rat Achilles tendon defects, in vivo imaging system (IVIS) assays demonstrated that the implanted cells stayed around the repair site for up to 7 days following injection. Furthermore, the tHG-containing adipose-derived stem cells showed improvements in ultimate failure load 4 weeks after repair compared to a control surgical repair [70].

25.10.2 New Promising Techniques for Scaffold Development

A novel scaffold-creating technology, called electrospinning, creates scaffolds that mimic properties of native tendon matrices. Electrospun silk scaffolds, when seeded with bone marrow-derived stem cells (BMSCs), demonstrate enhanced differentiation of stem cells into tenocytes and increased expression of tendon- and ligament-specific proteins [65]. Other scaffold materials, like electrospun collagen-silk blends seeded with stem cells, also have increased collagen deposition and mechanical improvements [89].

Overall, the tensile strength, porosity, and elasticity of electrospun scaffolds are comparable to most biologic grafts [90]. However, micro- and nanoscale topographical differences between biomaterials influence an electrospun scaffold's ability to promote cell differentiation and survival [62]. Additional efforts to improve biomaterials for electrospinning will optimize the environment for cell seeding, differentiation, and survival.

25.10.3 Challenges for Clinical Application of Growth Factors

Various growth factors have been used to augment the tendon-healing process. However, the investigation of each of the factors in isolation is a notable limitation, given that the native tendon-healing process involves a coordinated mix of growth factors at specific time intervals. The timing and delivery of these growth factors during the healing process remain a significant clinical challenge.

Tissue engineering efforts can introduce growth factors to an injury at the time of repair, or as follow-up injections to the injury site at a later date [91]. The timing of this growth factor delivery is still unclear and requires further optimization as our understanding of the healing cascade improves. Future areas of research include investigating the effect of single application versus multiple applications of growth factors and the effect of immediate bolus versus continuous release of growth factor. Preliminary data describing the efficacy of cell and growth factor-eluting scaffolds suggests that this may be a promising vessel for growth factor delivery within tendon defects in the future. This scaffold was made from an electrospun nanofiber made of PLGA, and a unique heparin/fibrin-based delivery system (HBDS) allowed for a slow, controlled release of PDGF growth factor and ASCs. In vitro results demonstrated cell viability and sustained levels of growth factor across time [55, 92].

25.10.4 Biophysical Stimulation

Biophysical techniques can also augment tendon healing. These include mechanical, electrical, and sonographic stimulation. An in vitro study demonstrated that MSCs within a silk-collagen sponge scaffold that were stretched and subjected to mechanical stimulation throughout their seeding process demonstrated histologic findings comparable to morphological features of native tenocytes. Furthermore, these samples demonstrated greater levels of tendon-related gene markers compared to samples not subjected to mechanical stimulation [93].

Low-intensity pulsed ultrasound and pulsed electromagnetic fields or shockwaves can also improve tendon healing [94–96]. These methods have yielded similar results to mechanical stimulation, including increased failure load and elasticity.

Clinical practice has incorporated biophysical stimulation for musculocutaneous injuries for many years, but our understanding of their effects at the cellular level is still in its infancy [97]. With increased understanding of the effects these biophysical stimuli have at a microscopic level, we can optimize the timing and interfacing of this treatment modality clinically.

25.10.5 Drug Repositioning

Drug repositioning is another valuable area of exploration for tendon regeneration. Drug repositioning involves using an existing pharmacologic treatment and administering that treatment for an alternative purpose. In one study, erythropoietin, a drug typically used for anemia, showed increased ultimate failure loads and histological evidence of increased tenocyte proliferation within an injury site compared to a control repair [98]. Evidence surrounding its use, however, remains controversial, as other studies have failed to demonstrate any significant improvements to tendon healing with its use. Pharmacologic treatments for skeletal muscle, spinal cord, and other injuries may have secondary application for tendon injuries and merit further study.

25.11 Conclusions

Tissue engineering efforts have had significant advances in tendon regeneration and healing over the last decade. As our understanding of tendon regeneration improves, tissue engineering methods will yield more efficacious options to treat patients with tendon injuries.

References

1. Pietrzak WS. Musculoskeletal tissue regeneration: biological materials and methods. Totowa: Humana Press; 2008.

2. Pennisi E. Tending tender tendons. Science. 2002;295(5557):1011.

3. Klepps S, Bishop J, Lin J, Cahlon O, Strauss A, Hayes P, Flatow EL. Prospective evaluation of the effect of rotator cuff integrity on the outcome of open rotator cuff repairs. Am J Sports Med. 2004;32(7):1716–22.

4. Krueger-Franke M, Siebert CH, Scherzer S. Surgical treatment of ruptures of the Achilles tendon: a review of long-term results. Br J Sports Med. 1995;29(2):121–5.

5. Voleti PB, Buckley MR, Soslowsky LJ. Tendon healing: repair and regeneration. Annu Rev Biomed Eng. 2012;14:47–71.

6. Ackermann PW, Salo P, Hart DA. Tendon innervation. Adv Exp Med Biol. 2016;920:35–51.

7. Benjamin M, Kaiser E, Milz S. Structure-function relationships in tendons: a review. J Anat. 2008;212(3):211–28.

8. Fenwick SA, Hazleman BL, Riley GP. The vasculature and its role in the damaged and healing tendon. Arthritis Res. 2002;4(4):252–60.

9. Brockis J. The blood supply of the flexor and extensor tendons of the fingers in man. Bone Joint J. 1953;35-B(1):131–8.

10. Chaplin DM. The vascular anatomy within normal tendons, divided tendons, free tendon grafts and pedicle tendon grafts in rabbits. A microradioangiographic study. J Bone Joint Surg Br. 1973;55(2):369–89.

11. Ciatti R, Mariani PP. Fibroma of tendon sheath located within the ankle joint capsule. J Orthop Traumatol. 2009;10(3):147–50.

12. Rathbun JB, Macnab I. The microvascular pattern of the rotator cuff. J Bone Joint Surg Br. 1970;52(3):540–53.

13. Ochiai N, Matsui T, Miyaji N, Merklin RJ, Hunter JM. Vascular anatomy of flexor tendons. I. Vincular system and blood supply of the profundus tendon in the digital sheath. J Hand Surg Am. 1979;4(4):321–30.

14. Gohl KL, Listrat A, Béchet D. Hierarchical mechanics of connective tissues: integrating insights from nano to macroscopic studies. J Biomed Nanotechnol. 2014;10(10):2464–507.

15. Kannus P. Structure of the tendon connective tissue. Scand J Med Sci Sports. 2000;10(6):312–20.

16. Sharma P, Maffulli N. Biology of tendon injury: healing, modeling and remodeling. J Musculoskel Neuronal Interact. 2006;6(2):181–90.

17. Kielty CM, Stephan S, Sherratt MJ, Williamson M, Shuttleworth CA. Applying elastic fibre biology in vascular tissue engineering. Philos Trans R Soc Lond Ser B Biol Sci. 2007;362(1484):1293–312.

18. Weinreb JH, Sheth C, Apostolakos J, McCarthy MB, Barden B, Cote MP, Mazzocca AD. Tendon structure, disease, and imaging. Muscles Ligaments Tendons J. 2014;4(1):66–73.

19. Wall ME, Dyment NA, Bodle J, Volmer J, Loboa E, Cederlund A, Fox AM, Banes AJ. Cell signaling in tenocytes: response to load and ligands in health and disease. Adv Exp Med Biol. 2016;920:79–95.

20. Kannus P. Tendons—a source of major concern in competitive and recreational athletes. Scand J Med Sci Sports. 1997;7(2):53–4.

21. Jozsa L, Balint JB, Kannus P, Reffy A, Barzo M. Distribution of blood groups in patients with tendon rupture. An analysis of 832 cases. J Bone Joint Surg Br. 1989;71(2):272–4.

22. Bi Y, Ehirchiou D, Kilts TM, Inkson CA, Embree MC, Sonoyama W, Li L, Leet AI, Seo BM, Zhang L, Shi S, Young MF. Identification of tendon stem/progenitor cells and the role of the extracellular matrix in their niche. Nat Med. 2007;13(10):1219–27.

23. Chuen FS, Chuk CY, Ping WY, Nar WW, Kim HL, Ming CK. Immunohistochemical characterization of cells in adult human patellar tendons. J Histochem Cytochem. 2004;52(9):1151–7.

24. Vailas AC, Tipton CM, Laughlin HL, Tcheng TK, Matthes RD. Physical activity and hypophysectomy on the aerobic capacity of ligaments and tendons. J Appl Physiol Respir Environ Exerc Physiol. 1978;44(4):542–6.

25. Kubo K, Ikebukuro T, Tsunoda N, Kanehisa H. Changes in oxygen consumption of human muscle and tendon following repeat muscle contractions. Eur J Appl Physiol. 2008;104(5):859–66.

26. Wang JH, Guo Q, Li B. Tendon biomechanics and mechanobiology—a minireview of basic concepts and recent advancements. J Hand Ther. 2012;25(2):133–40.

27. Depalle B, Qin Z, Shefelbine SJ, Buehler MJ. Influence of cross-link structure, density and mechanical properties in the mesoscale deformation mechanisms of collagen fibrils. J Mech Behav Biomed Mater. 2015;52:1–13.

28. Barber FA, Herbert MA, Coons DA. Tendon augmentation grafts: biomechanical failure loads and failure patterns. Arthroscopy. 2006;22(5):534–8.

29. Morales-Orcajo E, de Bengoa Vallejo RB, Iglesias ML, Bayod J. Structural and material properties of human foot tendons. Clin Biomech. 2016;37:1–6.

30. D'Addona A, Maffulli N, Formisano S, Rosa D. Inflammation in tendinopathy. Surgeon. 2017;15(5):297–302.

31. Riley G. The pathogenesis of tendinopathy. A molecular perspective. Rheumatology (Oxford). 2004;43(2):131–42.

32. Praxitelous P, Edman G, Ackermann PW. Microcirculation after Achilles tendon rupture correlates with functional and patient-reported outcomes. Scand J Med Sci Sports. 2018;28(1):294–302.

33. Chang J, Thunder R, Most D, Longaker MT, Lineaweaver WC. Studies in flexor tendon wound healing: neutralizing antibody to TGF-beta1 increases postoperative range of motion. Plast Reconstr Surg. 2000;105(1):148–55.

34. Lyras DN, Kazakos K, Agrogiannis G, Verettas D, Kokka A, Kiziridis G, Chronopoulos E, Tryfonidis M. Experimental study of tendon healing early phase: is IGF-1 expression influenced by platelet

rich plasma gel? Orthop Traumatol Surg Res. 2010;96(4):381–7.

35. Abrahamsson SO. Similar effects of recombinant human insulin-like growth factor-I and II on cellular activities in flexor tendons of young rabbits: experimental studies in vitro. J Orthop Res. 1997;15(2):256–62.

36. Docheva D, Müller SA, Majewski M, Evans CH. Biologics for tendon repair. Adv Drug Deliv Rev. 2015;84:222–39.

37. Miyashita H, Ochi M, Ikuta Y. Histological and biomechanical observations of the rabbit patellar tendon after removal of its central one-third. Arch Orthop Trauma Surg. 1997;116(8):454–62.

38. Riley GP, Curry V, DeGroot J, van El B, Verzijl N, Hazleman BL, Bank RA. Matrix metalloproteinase activities and their relationship with collagen remodelling in tendon pathology. Matrix Biol. 2002;21(2):185–95.

39. Spiesz EM, Thorpe CT, Chaudhry S, Riley GP, Birch HL, Clegg PD, Screen HR. Tendon extracellular matrix damage, degradation and inflammation in response to in vitro overload exercise. J Orthop Res. 2015;33(6):889–97.

40. Wang A, Breidahl W, Mackie KE, Lin Z, Qin A, Chen J, Zheng MH. Autologous tenocyte injection for the treatment of severe, chronic resistant lateral epicondylitis: a pilot study. Am J Sports Med. 2013;41(12):2925–32.

41. Cao Y, Liu Y, Liu W, Shan Q, Buonocore SD, Cui L. Bridging tendon defects using autologous tenocyte engineered tendon in a hen model. Plast Reconstr Surg. 2002;110(5):1280–9.

42. Van Eijk F, Saris DB, Riesle J, Willems WJ, Van Blitterswijk CA, Verbout AJ, Dhert WJ. Tissue engineering of ligaments: a comparison of bone marrow stromal cells, anterior cruciate ligament, and skin fibroblasts as cell source. Tissue Eng. 2004;10(5–6):893–903.

43. Lui PP. Stem cell technology for tendon regeneration: current status, challenges, and future research directions. Stem Cells Cloning. 2015;8:163–74.

44. Zhang J, Wang JH. Characterization of differential properties of rabbit tendon stem cells and tenocytes. BMC Musculoskelet Disord. 2010;11:10.

45. Ben-David U, Benvenisty N. The tumorigenicity of human embryonic and induced pluripotent stem cells. Nat Rev Cancer. 2011;11(4):268–77.

46. Zachar L, Bačenková D, Rosocha J. Activation, homing, and role of the mesenchymal stem cells in the inflammatory environment. J Inflamm Res. 2016;9:231–40.

47. Guo J, Chan KM, Zhang JF, Li G. Tendon-derived stem cells undergo spontaneous tenogenic differentiation. Exp Cell Res. 2016;341(1):1–7.

48. Liu W, Yin L, Yan X, Cui J, Liu W, Rao Y, Sun M, Wei Q, Chen F. Directing the differentiation of parthenogenetic stem cells into tenocytes for tissue-engineered tendon regeneration. Stem Cells Transl Med. 2017;6(1):196–208.

49. Chen X, Song XH, Yin Z, Zou XH, Wang LL, Hu H, Cao T, Zheng M, Ouyang HW. Stepwise differentiation of human embryonic stem cells promotes tendon regeneration by secreting fetal tendon matrix and differentiation factors. Stem Cells. 2009;27(6):1276–87.

50. Bavin EP, Smith O, Baird AE, Smith LC, Guest DJ. Equine induced pluripotent stem cells have a reduced tendon differentiation capacity compared to embryonic stem cells. Front Vet Sci. 2015;2:55.

51. Wang A, Mackie K, Breidahl W, Wang T, Zheng MH. Evidence for the durability of autologous tenocyte injection for treatment of chronic resistant lateral epicondylitis: mean 4.5-year clinical follow-up. Am J Sports Med. 2015;43(7):1775–83.

52. Clarke AW, Alyas F, Morris T, Robertson CJ, Bell J, Connell DA. Skin-derived tenocyte-like cells for the treatment of patellar tendinopathy. Am J Sports Med. 2011;39(3):614–23.

53. Liu W, Chen B, Deng D, Xu F, Cui L, Cao Y. Repair of tendon defect with dermal fibroblast engineered tendon in a porcine model. Tissue Eng. 2006;12(4):775–88.

54. Lim JK, Hui J, Li L, Thambyah A, Goh J, Lee EH. Enhancement of tendon graft osteointegration using mesenchymal stem cells in a rabbit model of anterior cruciate ligament reconstruction. Arthroscopy. 2004;20(9):899–910.

55. Sahoo S, Ang LT, Cho-Hong Goh J, Toh SL. Bioactive nanofibers for fibroblastic differentiation of mesenchymal precursor cells for ligament/tendon tissue engineering applications. Differentiation. 2010;79(2):102–10.

56. Uysal AC, Mizuno H. Tendon regeneration and repair with adipose derived stem cells. Curr Stem Cell Res Ther. 2010;5(2):161–7.

57. Ni M, Lui PP, Rui YF, Lee YW, Lee YW, Tan Q, Wong YM, Kong SK, Lau PM, Li G, Chan KM. Tendon-derived stem cells (TDSCs) promote tendon repair in a rat patellar tendon window defect model. J Orthop Res. 2012;30(4):613–9.

58. Ning LJ, Jiang YL, Zhang CH, Zhang Y, Yang JL, Cui J, Zhang YJ, Yao X, Luo JC, Qin TW. Fabrication and characterization of a decellularized bovine tendon sheet for tendon reconstruction. J Biomed Mater Res A. 2017;105(8):2299–311.

59. Raghavan SS, Woon CY, Kraus A, Megerle K, Choi MS, Pridgen BC, Pham H, Chang J. Human flexor tendon tissue engineering: decellularization of human flexor tendons reduces immunogenicity in vivo. Tissue Eng Part A. 2012;18(7–8):796–805.

60. Yang G, Rothrauff BB, Tuan RS. Tendon and ligament regeneration and repair: clinical relevance and developmental paradigm. Birth Defects Res C Embryo Today. 2013;99(3):203–22.

61. Chen B, Wang B, Zhang WJ, Zhou G, Cao Y, Liu W. In vivo tendon engineering with skeletal muscle derived cells in a mouse model. Biomaterials. 2012;33(26):6086–97.

62. Yin Z, Chen X, Chen JL, Shen WL, Hieu Nguyen TM, Gao L, Ouyang HW. The regulation of tendon stem

62. cell differentiation by the alignment of nanofibers. Biomaterials. 2010;31(8):2163–75.

63. Wan Y, Chen W, Yang J, Bei J, Wang S. Biodegradable poly(L-lactide)-poly(ethylene glycol) multiblock copolymer: synthesis and evaluation of cell affinity. Biomaterials. 2003;24(13):2195–203.

64. Ouyang HW, Goh JC, Thambyah A, Teoh SH, Lee EH. Knitted poly-lactide-co-glycolide scaffold loaded with bone marrow stromal cells in repair and regeneration of rabbit Achilles tendon. Tissue Eng. 2003;9(3):431–9.

65. Teh TK, Toh SL, Goh JC. Aligned fibrous scaffolds for enhanced mechanoresponse and tenogenesis of mesenchymal stem cells. Tissue Eng Part A. 2013;19(11–12):1360–72.

66. Shen W, Chen J, Yin Z, Chen X, Liu H, Heng BC, Chen W, Ouyang HW. Allogenous tendon stem/progenitor cells in silk scaffold for functional shoulder repair. Cell Transplant. 2012;21(5):943–58.

67. Awad HA, Boivin GP, Dressler MR, Smith FN, Young RG, Butler DL. Repair of patellar tendon injuries using a cell-collagen composite. J Orthop Res. 2003;21(3):420–31.

68. Juncosa-Melvin N, Boivin GP, Gooch C, Galloway MT, West JR, Dunn MG, Butler DL. The effect of autologous mesenchymal stem cells on the biomechanics and histology of gel-collagen sponge constructs used for rabbit patellar tendon repair. Tissue Eng. 2006;12(2):369–79.

69. Kim MY, Farnebo S, Woon CY, Schmitt T, Pham H, Chang J. Augmentation of tendon healing with an injectable tendon hydrogel in a rat Achilles tendon model. Plast Reconstr Surg. 2014;133(5):645e–53e.

70. Chiou GJ, Crowe C, McGoldrick R, Hui K, Pham H, Chang J. Optimization of an injectable tendon hydrogel: the effects of platelet-rich plasma and adipose-derived stem cells on tendon healing in vivo. Tissue Eng Part A. 2015;21(9–10):1579–86.

71. Kurtz CA, Loebig TG, Anderson DD, DeMeo PJ, Campbell PG. Insulin-like growth factor I accelerates functional recovery from Achilles tendon injury in a rat model. Am J Sports Med. 1999;27(3):363–9.

72. Klein MB, Yalamanchi N, Pham H, Longaker MT, Chang J. Flexor tendon healing in vitro: effects of TGF-beta on tendon cell collagen production. J Hand Surg Am. 2002;27(4):615–20.

73. Barsby T, Guest D. Transforming growth factor beta3 promotes tendon differentiation of equine embryo-derived stem cells. Tissue Eng Part A. 2013;19(19–20):2156–65.

74. Forslund C, Aspenberg P. Tendon healing stimulated by injected CDMP-2. Med Sci Sports Exerc. 2001;33(5):685–7.

75. Chan BP, Fu S, Qin L, Lee K, Rolf CG, Chan K. Effects of basic fibroblast growth factor (bFGF) on early stages of tendon healing: a rat patellar tendon model. Acta Orthop Scand. 2000;71(5):513–8.

76. Yoshikawa Y, Abrahamsson S. Dose-related cellular effects of platelet-derived growth factor-BB differ in various types of rabbit tendons in vitro. Acta Orthop Scand. 2009;72(3):287–92.

77. Hildebrand K, Woo S, Smith D, Allen C, Deie M, et al. The effects of platelet-derived growth factor-BB on healing of the rabbit medial collateral ligament. Am J Sports Med. 2016;26(4):549–54.

78. Kajikawa Y, Morihara T, Sakamoto H, Matsuda K, Oshima Y, Yoshida A, Nagae M, Arai Y, Kawata M, Kubo T. Platelet-rich plasma enhances the initial mobilization of circulation-derived cells for tendon healing. J Cell Physiol. 2008;215(3):837–45.

79. Sánchez M, Anitua E, Azofra J, Andía I, Padilla S, Mujika I. Comparison of surgically repaired Achilles tendon tears using platelet-rich fibrin matrices. Am J Sports Med. 2017;35(2):245–51.

80. Bahn JJ, Chung JY, Im W, Kim M, Kim SH. Suitability of autologous serum for expanding rabbit adipose-derived stem cell populations. J Vet Sci. 2012;13(4):413–7.

81. Ito K, Aoyama T, Fukiage K, Otsuka S, Furu M, Jin Y, Nasu A, Ueda M, Kasai Y, Ashihara E, Kimura S, Maekawa T, Kobayashi A, Yoshida S, Niwa H, Otsuka T, Nakamura T, Toguchida J. A novel method to isolate mesenchymal stem cells from bone marrow in a closed system using a device made by nonwoven fabric. Tissue Eng Part C Methods. 2010;16(1):81–91.

82. Olmer R, Haase A, Merkert S, Cui W, Palecek J, Ran C, Kirschning A, Scheper T, Glage S, Miller K, Curnow EC, Hayes ES, Martin U. Long term expansion of undifferentiated human iPS and ES cells in suspension culture using a defined medium. Stem Cell Res. 2010;5(1):51–64.

83. Barkholt L, Flory E, Jekerle V, Lucas-Samuel S, Ahnert P, Bisset L, Büscher D, Fibbe W, Foussat A, Kwa M, Lantz O, Mačiulaitis R, Palomäki T, Schneider CK, Sensebé L, Tachdjian G, Tarte K, Tosca L, Salmikangas P. Risk of tumorigenicity in mesenchymal stromal cell-based therapies—bridging scientific observations and regulatory viewpoints. Cytotherapy. 2013;15(7):753–9.

84. Blum B, Benvenisty N. The tumorigenicity of human embryonic stem cells. Adv Cancer Res. 2008;100:133–58.

85. Walsh SK, Kumar R, Grochmal JK, Kemp SW, Forden J, Midha R. Fate of stem cell transplants in peripheral nerves. Stem Cell Res. 2012;8(2):226–38.

86. Jesuraj NJ, Santosa KB, Newton P, Liu Z, Hunter DA, Mackinnon SE, Sakiyama-Elbert SE, Johnson PJ. A systematic evaluation of Schwann cell injection into acellular cold-preserved nerve grafts. J Neurosci Methods. 2011;197(2):209–15.

87. Farnebo S, Farnebo L, Kim M, Woon C, Pham H, Chang J. Optimized repopulation of tendon hydrogel: synergistic effects of growth factor combinations and adipose-derived stem cells. Hand (N Y). 2017;12(1):68–77.

88. O'Brien F. Biomaterials & scaffolds for tissue engineering. Mater Today. 2011;14(3):88–95.
89. Baker SC, Atkin N, Gunning PA, Granville N, Wilson K, Wilson D, Southgate J. Characterisation of electrospun polystyrene scaffolds for three-dimensional in vitro biological studies. Biomaterials. 2006;27(16):3136–46.
90. Bucher TA, Ebert JR, Smith A, Breidahl W, Fallon M, Wang T, Zheng MH, Janes GC. Autologous tenocyte injection for the treatment of chronic recalcitrant gluteal tendinopathy: a prospective pilot study. Orthop J Sports Med. 2017;5(2):2325967116688866.
91. Manning CN, Schwartz AG, Liu W, Xie J, Havlioglu N, Sakiyama-Elbert SE, Silva MJ, Xia Y, Gelberman RH, Thomopoulos S. Controlled delivery of mesenchymal stem cells and growth factors using a nanofiber scaffold for tendon repair. Acta Biomater. 2013;9(6):6905–14.
92. Galloway MT, Lalley AL, Shearn JT. The role of mechanical loading in tendon development, maintenance, injury, and repair. J Bone Joint Surg Am. 2013;95(17):1620–8.
93. Ying ZM, Lin T, Yan SG. Low-intensity pulsed ultrasound therapy: a potential strategy to stimulate tendon-bone junction healing. J Zhejiang Univ Sci B. 2012;13(12):955–63.
94. Strauch B, Patel MK, Rosen DJ, Mahadevia S, Brindzei N, Pilla AA. Pulsed magnetic field therapy increases tensile strength in a rat Achilles' tendon repair model. J Hand Surg Am. 2006;31(7):1131–5.
95. Chow DH, Suen PK, Fu LH, Cheung WH, Leung KS, Wong MW, Qin L. Extracorporeal shockwave therapy for treatment of delayed tendon-bone insertion healing in a rabbit model: a dose-response study. Am J Sports Med. 2012;40(12):2862–71.
96. Ambrosio F, Wolf SL, Delitto A, Fitzgerald GK, Badylak SF, Boninger ML, Russell AJ. The emerging relationship between regenerative medicine and physical therapeutics. Phys Ther. 2010;90(12):1807–14.
97. Uslu M, Kaya E, Yaykaşlı KO, Oktay M, Inanmaz ME, Işık C, Erdem H, Erkan ME, Kandiş H. Erythropoietin stimulates patellar tendon healing in rats. Knee. 2015;22(6):461–8.
98. Bilal O, Guney A, Kalender AM, Kafadar IH, Yildirim M, Dundar N. The effect of erythropoietin on biomechanical properties of the Achilles tendon during the healing process: an experimental study. J Orthop Surg Res. 2016;11(1):55.
99. Majumdar MK, Thiede MA, Mosca JD, Moorman M, Gerson SL. Phenotypic and functional comparison of cultures of marrow-derived mesenchymal stem cells (MSCs) and stromal cells. J Cell Physiol. 1998;176(1):57–66.

Stem Cells and Tendon Regeneration

26

Hamid Karimi, Kamal Seyed-Forootan, and Ali-Mohammad Karimi

26.1 Introduction

Tendon injury and rotator cuff injury are very common in human population [1–8] and each year more than 30 million injuries have been reported from all over the world [2, 9, 10]. Some of the most famous injuries are rotator cuff injury, tendon-to-bone junction [11], and Achilles tendon injury. These injuries if not treated properly would lead to early osteoarthritis, pain, restriction of movements, excess pressure and force over other tendons and ligaments, adhesions, elongation of repair site (and impairment in tendon and muscle function), ectopic bone formation, significant impairment of joint and limb, joint weakness, wrong and erroneous position in the joint, joint stiffness, and eventually joint prosthesis (replacement) and a lower quality of life [2, 3, 12–15].

The tendons have very minimum amount of cells and vascularity, so when it is injured it has a minimum capacity to repair. It cannot self-repair and when it is injured the process of healing is very slow and usually results in fibrous tissue for-

H. Karimi (✉)
Department of Plastic and Reconstructive Surgery, Hazrat Fatemeh Hospital, Iran University of Medical Sciences, Tehran, Iran

K. Seyed-Forootan
Department of Plastic and Reconstructive Surgery, Iran University of Medical Sciences, Tehran, Iran

A.-M. Karimi
School of Medicine, Iran University of Medical Sciences, Tehran, Iran

mation [1–5, 7, 9, 15–22]. Therefore the tendon is hardly repaired and usually its function and strength would not be regained completely. Tendons normally repaired with fibrous scar or fibrovascular scar and sometimes with ectopic bone formation [1, 3–5, 7, 9, 15–22]. In the tendon bone junction, the specific fibrocartilage tissue will not be reproduced, so there is failure of attachment of tendon to the bone [4, 5, 19, 21, 23–26]. It is well known that tendons will not be healed completely, will not gain the previous force of normal tissue, and cannot stand the maximum loading and the function will not be recovered completely; therefore there is a chance for re-rupture and re-tear [7, 15, 21, 23, 24, 27, 28].

Nowadays usual methods for treatment of tendon and ligament injuries are not completely effective and have some rates of reinjury or re-tear, treatment failure, or re-rupture. Many of the physicians and authors tried to find newer techniques and methods to have a better outcome. Recently cell therapy, whether differentiated cells or stem cells, has provided new concepts and new hopes for complete and efficacious treatments.

In this chapter we discuss the normal anatomy and weak points of the tendons, methods of cell therapy, normal tenocyte therapy with advantages and disadvantages, stem cell therapy with advantages and disadvantages, and also different types of stem cells that have been used, growth hormones, scaffolds, specific media, limitations, results of treatments, and horizons of cell treatments.

© Springer Nature Switzerland AG 2019
D. Duscher, M. A. Shiffman (eds.), *Regenerative Medicine and Plastic Surgery*,
https://doi.org/10.1007/978-3-030-19962-3_26

26.2 Normal Anatomy of the Tendons

The tendons are composed of cells and extracellular matrix. The cells are including tenocytes, tenoblasts, and cells with capacity for growth and repairing the tendons. Tenocytes are the main cells in the tendons; they are very low in number, are of spindle shape, are situated parallel to the collagen fibers, and have a limited turnover. These cells are responsible for maintaining the extracellular matrix and microenvironment of the tendon. They are also responsible for collagen production. Since the cells are very sparse, the capacity of the tendons for repair is very low and they cannot heal the tendon completely [1–5, 7, 9, 20, 29] (Fig. 26.1).

The extracellular matrix is composed of glycose-amino-glycans (GAG) [proteo-glycan matrix]; collagen fibers types I, III, and V and elastin fibers; methyl metalloproteases (MMPs), tissue inhibitors of MMPs, and some blood vessels and nerve fibers [2, 10, 30].

The main structural fiber of tendon is collagen type I which is more than 65–80% weight of dry tendon weight and has mechanical function. Four collagen strands will form microfibrils and some microfibrils will form fibrils and then some fibrils together will form "fiber." Fibers are surrounded by endotenon and these fibers together will form primary sub-fascicles, then secondary fascicles, and tertiary fascicles; all of them are covered by endotenon which is a connective tissue and will provide blood vessels for them. A number of the collagen fibers will form tendon and in turn tendon is covered by epitenon. Some of the tendons have synovial sheaths over epitenons and some of them have paratenon over it. Both of them used to reduce the friction of tendons over other tissues specially bones [2, 3, 13, 17, 18, 29–31].

Collagen type III is in the ECM and has a regulatory function. Increasing of the collagen type III will result in decrease in diameter of collagen fibers type I. The same (increase in collagen type III) will be seen in healing of the tendon after injury. This may explain the reduction in maximum load of the tendons after injury and healing [2, 29].

Collagen type V has structural function in ECM and has no mechanical function [2, 29, 30]. Elastin fibers have a role in elasticity of the tendon but as they are few in number the tendons have a minimal elasticity and they strongly transfer the force of the muscles to the bones for movement. The elastin fibers composed of around 1–2% of dry weight of tendon. The tendon has a minimum elasticity and plasticity which are important for its function that are transferring the force of muscles to the bones. In this way the tendon is able to retract with high tensile strength [32]. MMPs have an important role in enzymatic decrease of the fibers and on the contrary tenocytes will produce an increase in the fibers, specially collagen fibers [30]. Tenoblasts are very sparse but can proliferate and can differentiate into mature tenocytes and are important in tendon repair, but hypocellularity and hypovascularity in the tendon will lead to incomplete repair and regeneration in the tendons [14, 17, 31]. Therefore the tendons have limited ability and limited capacity for self-regeneration [3, 14, 17, 31]. The change in microenvironment of tendon after injury will result in proliferation and differentiation of these cells into tenocytes and process of healing. Sometimes erroneous differentiation of tenoblasts will lead to tendon ossification [2, 30].

Tenocytes and in general all cells in the tendon will react to increase in the pressure, force, and/or tension in the tendon dynamically. Their phenotype will change according to mechanobiological changes in the tendon [31]. In this

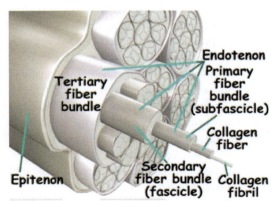

Fig. 26.1 Histology of tendons

regard it is believed that tenocytes in every tendon are different from tenocytes of other tendons. And each tendon has a specific tenocyte that differs with others. The metabolic activity of tenocytes is also different in the different tendons [2, 31, 33]. So it seems that tenocytes will respond to biophysical, biochemical, and biological signals to differentiate and to maintain their phenotypes [10]. It is said that the tendons are mechanosensitive tissues, so imposing stress and tension on the tendon would have a positive effect on tenocytes and regenerated tendon would have stronger characteristics [34].

Tendons have a few blood vessels which are mostly found in the endotenon and peritenon and in some mesos of tendons. In some special areas tendon has no vessels and these are the weak point of the tendons. The nervous system mostly is located on the periphery of the tendons and will regulate the blood flow of the vessels. There are also sensory nerves in the tendons. The insertion of the tendons has no vessels and nerves [2, 30, 31].

26.3 Classical (Traditional) Method for Tendon Repair

The most frequent methods for tendon repair are:

1. Conservative treatment
2. Surgical treatment

Both methods are not completely effective and the tendons healed with scar/fibrous scar, ectopic bones, and adhesions or lack of regeneration of fibrocartilage tissue in the junction of the tendon to the bones. The resulted tissue cannot stand the mechanical forces properly. The resulted complications will be re-injury and re-tear in the tendons, joint stiffness, limitation in the range of motion, pain, excess stress to the other tendons, wrong position of the joint, osteoarthritis, and eventually joint replacement. The rate of re-tear has been reported up to 28–32% of the cases especially in massive tears. Therefore the surgeons tried to find another way in treatment of the tendon injury.

26.4 Cell Therapy

Using the cells from other parts of body or from allogeneic sources and injecting them in the site of tendon injury have started from 15 to 20 years ago. Two types of cells can be used for this purpose: differentiated cells and stem cells. Differentiated cells include tenocyte (autologous or allogeneic) and fibroblast (mostly from the skin).

Tenocytes are very specialized and differentiated cells for producing and maintaining the microenvironment of the tendons. They can produce growth hormones and ECM for the tendon repair. So these cells have promising capabilities. They can be harvested from other parts of body and in vitro proliferate and increase in number. Then they will be injected in the site of injury (with surgical repair or even without repair with only percutaneous injection). But there are some problems in this regard. The tenocytes are very few and obtaining enough cells from normal tendons is very difficult and time consuming. It has also risk of injury to the normal tissues and tendons. Culturing of the tenocytes is very difficult too. Most of the cells after two or three passages will lose the characteristic features of the tenocytes; they become round (instead of spindle) shape and most of the capabilities of secretion of the ECM and collagen will be vanished. Therefore culturing of these cells will not result in tenocytes. It is needed to differentiate them again into the target cells. Sometimes this process is impossible and with very few resulted cells. Using allograft tenocytes has advantages that there is not any injury to the normal tissue of the patients (no donor-site morbidity), but still culturing is a big problem.

Tenocytes actually are very similar to fibroblasts, so we can use skin fibroblasts for culturing. Skin fibroblasts are numerous in number and a small piece of skin can produce enough fibroblasts for culture. And there is minimal injury into the normal skin of the patients. But again culturing has the same difficulties. Tenocytes will not produce teratoma and in this regard are better than stem cells [31, 35–40].

26.4.1 Stem Cells

Using stem cells are nowadays very popular in many fields of medicine and their capability for regeneration has many promising results in different tissues (Fig. 26.2). Stem cells have special characteristics including high ability for self-production, plasticity and proliferation, easily harvestable, ability to produce and differentiate into many lineages of the cells, ability to migrate into the site of injury (systemic migration, local migration), easy grafting into the site, secretion of bioactive molecules and cytokines, immunosuppression, compatibility of allogeneic stem cells for the patient's body, tissue healing and normal tissue regeneration, high synthetic activity, ability to differentiate into target cells, self-renewal, self-regeneration, autocrine secretion, paracrine secretion, angiogenesis, continuous production of the bioactive factors, production of specific proteins, and ability to differentiate according to microenvironment [2, 3, 5, 41–45].

26.4.2 Stem Cell Types That Are Used for Tendon Regeneration

1. Embryonic Stem Cells (ESC)

    These cells can highly proliferate but there is a risk of wrong differentiation and production of teratoma. So first they should be differentiated into mesenchymal stem cells (MSC) in vitro and then used for tendon repair. There is also an ethical issue for using embryonic tissues [2, 16, 29, 46, 47]. There is a report that embryonic stem cells in vitro can better differentiate into tenocytes than pluripotent stem cells [47]. And ESC have a longer half-life than pluripotent stem cells and bone marrow stem cells [2]. Also ESC better migrate (and to a longer distance) along the length of the tendon than bone marrow stem cells [2, 10].

2. (Human) Induced Pluripotent Stem Cells (hiPSC)

    These cells can be harvested from normal tissue of the patients and there is no ethical issue, but risk of teratoma still is present. They should be treated with numerous viral vectors and factors; these factors may be harmful for the tendon repair or harmful for differentiation of the cells. Again these cells should differentiate into MSC in vitro and then used for regeneration. Pretreatment of ESC and hiPSC is very important as this would reduce the risk of ectopic bone formation, reduce the risk of tumor formation, and improve the healing of the tendon. It is important for these two recent cells that only MSC be used for cell therapy and all of the under-differentiated stem cells should be removed from the treatment media; this process needs extra efforts for purification of the MSCs. Both cells are very low immunogenic, so they can be used as allogeneic without rejection [2, 29, 47, 48].

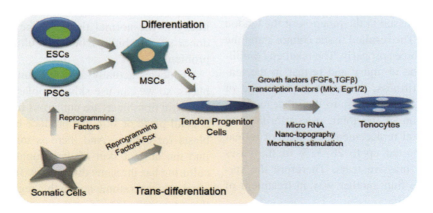

Fig. 26.2 Differentiation from stem cells to tenocytes

3. Mesenchymal Stem Cells (MSC)

   These cells are more differentiated. They are safer in teratoma production. They have no ethical issue. They can be easily harvested from the many tissues. They have no harmful immunologic induction, so can be used as allogeneic. Also they have some immunosuppressive effects. It is confirmed that they downregulate the inflammatory macrophages in order to produce less inflammation and adhesions and have better repair of the tissue [5, 6, 8, 12, 17, 27, 29, 32, 43–45, 49–60, 62–64]. The MSCs from the tear site in the rotator cuff injury have been harvested and shown to be multipotent. They can differentiate into bone, cartilage, and fat cells [58]. In symptomatic rotator cuff tears in human, it has been found that there is paucity of MSCs in the junction of tendon to bone [6] (Fig. 26.3).

4. Bone Marrow Stem Cell (BMSC)

   One of the most famous MSCs is bone marrow stem cell (BMSC) that has been used in many studies for tendon regeneration. They have been used for induction of healing in patella tendon, supraspinatus tendon, and Achilles. But sometimes they will produce bone, so ratio of these cells to collagen is very important for using them for tendon healing. So far BMSCs are the most frequent stem cells that have been used for tendon regeneration [10, 18, 19, 33, 61, 65–75]. The BMSCs have been used for patellar tendinopathy too [28].

5. Adipose-Derived Stem Cells (ADSC)

   These cells are easily harvested and proliferate into a large amount and have been used for tendon and ligament regeneration. Even allogeneic ADSC has been used for the treatment of epicondylosis. Therefore it seems that they have low immunologic reaction [11, 14, 22, 23, 25, 27, 28, 76–80]. There are some reports about using the adipose-derived stromal cells, probably with interstitial fluids that contain growth hormones and cytokines; one of the reports from the USA in 2016 mentioned that with some cytokines it produced more collagen and had better

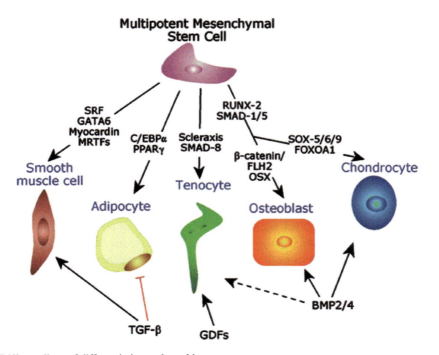

Fig. 26.3 Different lines of differentiation and cytokines

healing in tendon [81]. But in another report from the USA in 2015 adipose-derived stromal cells with scaffold had no effective results [21].

ADSC has been used successfully for the treatment of lateral epicondylar tendinopathy [28].

6. Tendon-Derived Stem Cells (TDSCs)

They are the most suitable stem cells for production of tenocytes. They are highly specialized for production of tendon. They have high capacity for proliferation, high ability for differentiation to the tenocytes, and low immunogenicity; in this way, they are ideal for tendon regeneration. Nowadays TDSCs are most popular stem cells that have been used as a choice [34] in repair of human tendons and BMSC are selective for repair of tendon in horses [4, 9, 10, 20, 24, 34, 66, 69, 74, 82–101]. It has been reported that TDSC had better results than BMSC for regeneration of tendons in rat, as they are endogenously designed to differentiate into tenocytes [20, 69]. The local TDSCs of the tendons are OCT ¾ (+) SC that has specific characteristics for regeneration [96].

7. Human Placenta-Derived Mesenchymal Progenitor (Stromal) Cells (HPMSC)

They are very effective in tendon regeneration and even human placental extract has been used for this purpose. Probably growth factors and cytokines have a very important role in this process [2, 51, 102, 103].

8. Bone Marrow Extract (Aspiration)

Bone marrow extract/aspiration has also been used for tendon regeneration. It would provide its effects most probably through stem cells that are located in the bone marrow or located in the blood. It has been used for regeneration of fibrocartilage tissue in the tendon to bone junction. It has been shown that it has positive effects on production of fibrocartilage tissue in the rotator cuff tear site [104].

9. Oral Stem Cells

These are of numerous types of stem cells and they have been used for repair of many tissue and for treatment of inflammatory diseases. Among them, periodontal ligament stem cells (PDLSC) and mesenchymal stem cells from gingiva (GMSC) are ideal for regeneration of tendons [29, 105, 106].

10. Peripheral Blood-MSC (PB-MSC)

Peripheral blood MSCs have been used with cytokines for induction of tenocytes and differentiation into tenocytes [107].

11. Umbilical Cord Blood-MSC (UC-MSC)

MSCs from umbilical cord have been used for several reasons and one of them is for repair of tendons and ligaments. These MSCs can repair the tendon and it is confirmed histologically [62, 108, 109].

Besides, there is a report that authors used human umbilical cord-MSC for conservative treatment of rotator cuff tear with only injection of MSCs [108]. MSC of umbilical cord had good results in race horse tendon also [62].

12. Perivascular-Derived Stem Cells (PDSC)

These cells have the capacity to heal and regenerate an injured tendon. They may also include neural crest stem cells (NCSC) that are present in the blood vessels in the peritenon and migrate to the interstitial area and produce ECM [26, 110].

13. Ligament Stem Cell (LSC)

Ligament stem cells have synergistic effects with umbilical cord-MSCs and together have been used for neovascularization and remodeling of the tear ligaments with good results [111].

14. Skeletal Muscle Progenitor Cells (SMPC)

In chronic tendon injuries, there would be muscle atrophy in the muscles around the joint. SMPCs can proliferate and help in the repair of the muscles, prevention of atrophy, and repair of tendon [15].

15. Trans-osseous Drilling of the Bone

In some papers it is shown that drilling or performing micro-fractures in the tendon-to-bone junction may bring stem cells and growth factors from the bone marrow and would help in regeneration of the tendon and its repair. In some other reports it is called bone marrow stimulation (BMS). BMS reduced significantly the ratio of re-tear from 23.9 to 9.1% [2, 112, 113].

26.4.3 Stem Cell Secretions (Paracrine)

The stem cells not only function as a cell progenitor but also play a role in secreting special growth factors and cytokines to prepare the microenvironment for induction, proliferation, and differentiation of the stem cells into tenocytes, to prevent immunogenicity, to have immune-modulatory effects, to reduce the inflammation, and in turn to reduce scar formation and adhesions, lessen lymphocyte infiltration, pass messages to the other cells, induce tenocytes for production of specialized ECM, facilitate migration of the cells, prevent overexpression of key transcriptors, and maintain the concentration of the cytokines in the media. As you know the half-life of the growth hormones is very short, around few minutes to few hours, so stem cells are needed to maintain the continuous production and secretion of the bioactive factors in order to have a complete healing in the tendon. During the process of healing, the body sometimes uses stem cells for delivery of gene to the site of injury. The stem cells sometimes directly secret the cytokines and sometimes produce micro-vesicles (secretomes) that contain the growth factors, cytokines, and miRNA and in this way transfer messages to the tissue and other cells. The main growth factors for proliferation and differentiation of the stem cells into tenocytes are hepatocyte growth factor, human platelet-derived growth factor-BB, intelukein-6, tumor growth factor beta (TGF-B), bone morphogenetic proteins (BMP), chemokine ligand-13 (CXCL-13), early growth response-1 (EGR-1), Mohawk (MKX) (specific transcription factor of the tendon), parathyroid hormone (PTH 1-34), inhibitors of TGF1, ascorbic acid, myostatin, human growth differentiation factor 5 (hGDF-5), TGF-B3, VEGF, nestin, TGF-B2, IL-4, MMPs, inhibitors of MMPs, PRP, connective tissue growth factor (CTGF), scleraxis, IL-10, basic fibroblast growth factor 2 (bFGF-2), growth and differenti-

ation factor-6 (GDF-6), tenogenic growth factor, leukocyte-rich PRP (L-PRP), and dexamethasone [1, 7, 9, 16, 18, 24, 30, 32, 49, 51, 54, 56, 57, 61, 65, 75, 77, 81–83, 86, 87, 93, 99, 107, 114–116].

Other than cytokines, ECM, fibrin glue, fibrin hydrogel, spongilization of the end of the tendon (tear tendon), tendon hydrogel, extracorporeal shock wave therapy, low-level laser therapy, extracellular matrix patch, and special scaffolds and tensile (mechanical) stress can act in the same way as bioactive factors and facilitate the differentiation of the stem cells into target cells (tenocytes) for tendon regeneration and produce better and stronger tendon. Proper tensile loading (mechanical loading) is a very important factor to maintain the composition and characteristics of the tendon and promote a strong tendon repair. It is shown that mechanical loading can help and promote BMSC and TDSC, which were seeded on a proper scaffold, to improve the healing of the tendon. The effect of stretching on BMSCs has been seen before. Cyclic stretching can upregulate the tendon proteins, can help in differentiation of stem cells into tenocytes, and can promote the tenocytes to maintain the mechanical characteristics of the tendon [23, 27, 31–33, 78, 89, 107, 116–118].

It has been shown that hypoxia promotes differentiation of ADSC into tenocytes [119].

There was a report from China in 2016 that in rats that had had previous practice and muscle tensile stress before injury, proliferation of TDSC would be better and would lead to better healing in the tendon [89].

26.4.4 Scaffolds

Specialized scaffolds usually are needed for stem cells to be seeded on it and induce stem cells to produce ECM and collagen molecules for regeneration of the tendon. These scaffolds should have special configuration and pores for stem cells (of every kind) to be seeded precisely and have enough space for secretion of cytokines and collagen and gradually to change it with the normal structure of a tendon. And in this way they produce a normal tendon with its biochemical and biomechanical characteristics that can stand against the tensile forces, transfer the force of muscles to the bone, and make movement possible.

The scaffolds may be absorbable or nonabsorbable or slow absorbable. These scaffolds should have good mechanical power in the early stages to support the tissue; should have few or no toxic materials for the stem cells and other tissues; should slowly degenerate; should have few toxic materials after degeneration; should promote and induce growth, proliferation, and differentiation of the stem cells; should promote significant induction of tenogenesis; should help in proper orientation of the collagen fibers and cells; should induce more tensile strength; should improve stiffness of the tendon; should promote regeneration of the tendon; and should improve production of the tendon proteins [1, 68, 121, 122] (Fig. 26.4).

Some of the most famous and frequently used scaffolds are polyhydroxyalkanoates (PHA), porcine small intestine submucosa (SIS), decellularized tendon tissue, decellularized slice of tendon, anisotropically aligned collagen biotextile (with 80% porosity), human tendon hydrogel, silk fibroin, poly-lactide-co-glycolic acid (PLGA), nanosized aligned fibers, biomimetic tendon extracellular matrix composite gradient scaffold, biphasic silk fibroin scaffold, vicryl mesh, ECM, electrospin silk fibroin mat (CSF), chitosan fibers, engineered tendon matrix (ETM), poly-e-caprolactone and methacrylated gelatin (PCL + mGLT), lyophilized tendon hydrogel (which is injectable), PLCGA, multilayer xenograft tendon, collagen I scaffold, and aligned electrospun multilayer collagen polymer scaffold

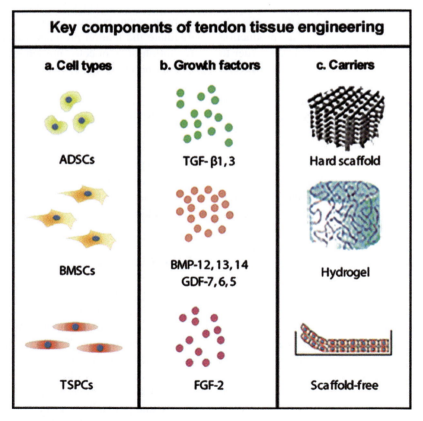

Fig. 26.4 Scafolds and stem cells

[2, 4, 7, 11, 21, 23, 52, 55, 56, 59, 60, 63, 67, 72, 74, 78–80, 116, 120, 121, 123] (Fig. 26.5).

26.4.5 Co-cultures

The other methods for promoting the differentiation of various stem cells into target cells are using the co-cultures in vitro. Co-culturing of stem cells with tendinous cells or tenoblast or other stem cells has resulted in more differentiation and more production of the tenoblasts and tenocytes and would help in regeneration of the new tendon. In this method one of the cells acts as inductor and with production of the bioactive factors and cytokines would induce other linages of cells to proliferate and differentiate into tenocytes. There was a report in 2016 that co-culturing of BMSC and TDSC would produce a good and normal tendon [66]. Other reports from the USA mentioned that co-culturing of MSC and TDSC is helpful for repair of tendon tear [85]. In another report from the USA, ADSC acted with tendon fibroblast to have a faster tendon healing [77]. In a study from the USA in 2016, co-culturing of adult stem cells with embryonic tendon cells with special growth factors resulted in differentiation and better healing [46]. In a study from China, co-culturing of TDSC and BMSC produced neo-tendon in 8 weeks [69]. In a study from Italy in 2015, the authors mentioned that co-culturing of tenocytes and ADSC resulted in differentiation into tenogenic lineage and more growth factor production [22]. Co-culturing of ligament stem cells with the umbilical cord stem cells has also been reported in 2015. The umbilical cord stem cells promoted neoangiogenesis and ligament stem cells differentiated into target cells [111].

26.4.6 Autologous or Allogeneic Stem Cells

Stem cells generally are multipotent cells and can produce cytokines to suppress the immune system or act as an immunomodulator. These cells are not strongly immunogenic, so they will not produce a strong immunity reaction. Therefore allogenic and autologous stem cells can be used safely. The only potential problem is the ESC and iHPSC. These cells may have immunologic reaction and should be used with caution and after HLA typing. BMSC, MSC, and ADSC have been used and compared as allogeneic and autologous with no differences in the results [2, 28, 30, 71].

26.4.7 Secretomes (Vesicular System) and PRPs

The stem cells particularly MSCs during the process of culture and healing will produce micro-vesicles or secretomes that contain

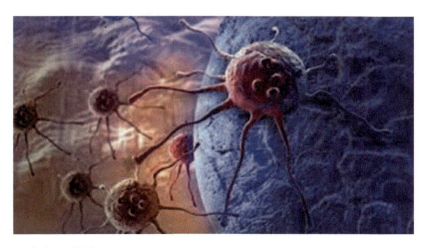

Fig. 26.5 Stem cells in scaffold

growth factors, bio-factors, cytokines, pigment epithelium-derived factor, follistatin, and miRNA. Stem cells use secretomes to deliver the factors to the other stem cells and neighbor cells. In this way they control the production of matrix and cytokines, migration, proliferation, and differentiation of the cells (and neighbor cells). So it has mechanochemical induction effects. It has been reported that secretomes can prevent fatty degeneration and muscle atrophy after tendon tears [2, 51, 54, 87].

Platelet-rich plasmas (PRPs) contain growth hormones and cytokines that in some studies have been reported to have good results for tendon repair. There was a study from China in 2016 in which authors used PRP with BMSC. They reported faster and better result in healing of tendon-to-bone junction [65]. In a study from the USA, they stated that TDSC with PRP can facilitate healing of tendon with numerous growth factors from PRP [86]. In a study from the USA, authors mentioned that PRP with IL-1 B, IL-6, TNFa, and PGE2 can induce proliferation of TDSCs and differentiate TDSCs into tenocytes [93]. In another report from the USA in 2015, PRP can induce differentiation of ADSCs [79]. In a study from the USA in 2015, authors used ADSCs and PRP with human cadaver tendon hydrogel in rat. They concluded that PRP would increase ECM and ADSCs proliferate and produce tenocytes [80]. There was a study from China in 2014 in which authors concluded that TDSC with PRP resulted in healing of tendon histologically and mechanically [9]. We have also conducted a study about PRP and healing of Achilles tendon. After 12 weeks we found two cases of ectopic cartilage metaplasia and PRP had no effect on strengthening of tendon repair [124].

26.4.8 Delivery Methods for Stem Cells

The stem cells can be delivered to the injury site by four methods: during the open surgery and after suturing, during arthroscopic surgery and after suturing, by percutaneous injection (directly or by the help of ultrasound), and by migration of endogenous stem cells into the injury site [2, 20, 28, 73, 96, 97, 108, 125–127].

There was a report from the USA in 2017 which stated that percutaneous injection of BMSC can prevent re-rupture and better healing [125]. In a study in 2015, the authors in vivo induced proliferation and differentiation of TDSC with connective tissue growth factors (CD146) (endogenous TDSC) and promoted healing of the tendon [97]. In another report from the USA in 2015, the authors stated that using granulocyte colony-stimulating factor (G-CSF) can increase cellularity of the injury site and promote migration of endogenous stem cells into the site [126]. In a study from South Korea in 2015, human umbilical cord MSC was injected in rabbit tendon percutaneously and resulted in better healing [108].

26.4.9 Cellular Markers for TDSC and Tenocytes

Whatever types of stem cells are used for the regeneration of the tendons, the end lineage of cells is the same. All of the stem cells have to go through TDSC and tenoblasts and tenocytes. It means that the target cells should be tenocytes.

For identifying the tenoblasts and tenocytes some of the markers have been found and developed for measuring the safety of the cell therapy and the cultures [49, 98, 100, 116]. Cell markers for tenoblasts are matrix protein, MMPs, and integrin [100]. The cell markers for tenocytes are collagen I, collagen III, decorin, sclerosis, tenascin-c, and tenomodulin [49, 116].

26.4.10 Number of Stem Cells

The number of stem cells that are needed for a good and proper induction of regeneration in the tendon is controversial. Some authors believe that between 10^6 and 10^7 cells are the needed cells. Some authors stated that there are no differences among 10^6 to 4×10^6 and yet some others believed that 4×10^6 is the optimum number [2, 10, 71, 128, 129].

26.4.11 Media

There are specific culture media for culturing or co-culturing of the stem cells. These media should have special composition for better proliferation of the stem cells and tenoblast. Proper osmolarity, pH, and oxygen have important effects on culturing. The media should not have toxic ingredient for growth of these lines of cells. And it should not have ingredients that oppose the action of cytokines and biomolecules [2, 10].

26.4.12 Eventual Fate of the Stem Cells During Cell Therapy

There is no unique consensus about the fate of the stem cells. Some authors believe that these cells will differentiate into tenocytes, will build the tendon, and will remain in the site. Some others believe that only 20% of stem cells will remain and the rest will leave the site. They think that the stem cells only have induction, angiogenesis, anti-inflammatory, and immune-modulatory role and promote other cells for healing [2, 10, 130, 131].

26.5 Horizons

Many aspects of stem cell therapy for tendon healing have not been clarified.

Some of them are as follows:

1. Most of the studies are on small animals and with short follow-up. Longer time studies are necessary in this regard.
2. Studies with bigger number of animals are needed.
3. Clinical studies are very few.
4. Several types of scaffolds are present. And many more studies are needed for testing these scaffolds.
5. Venous injection of the stem cells has not been studied. This is an open area for complex studies.
6. Percutaneous injection of stem cells in the site of tendon tear still needs more studies.

7. Cell density of cell solutions is another field for work.
8. It has been reported that more than 28% of tendon repairs will develop ectopic bone formation especially in BMSC therapy. Preventive measures for this issue should be studied more.

References

1. Yin Z, Guo J, Wu TY, Chen X, Xu LL, Lin SE, Sun YX, Chan KM, Ouyang H, Li G. Stepwise differentiation of mesenchymal stem cells augments tendon-like tissue formation and defect repair in vivo. Stem Cells Transl Med. 2016;5(8):1106–16.
2. Lui PP. Stem cell technology for tendon regeneration: current status, challenges, and future research directions. Stem Cells Cloning. 2015;8:163–74.
3. Neckař P, Syková E. Stem cells in orthopaedics. Cas Lek Cesk. 2015;154(3):107–9.
4. Zhang K, Asai S, Yu B, Enomoto-Iwamoto M. IL-1β irreversibly inhibits tenogenic differentiation and alters metabolism in injured tendon-derived progenitor cells in vitro. Biochem Biophys Res Commun. 2015;463(4):667–72.
5. Valencia Mora M, Ruiz Ibán MA, Díaz Heredia J, Barco Laakso R, Cuéllar R, García Arranz M. Stem cell therapy in the management of shoulder rotator cuff disorders. World J Stem Cells. 2015;7(4):691–9.
6. Hernigou P, Merouse G, Duffiet P, Chevalier N, Rouard H. Reduced levels of mesenchymal stem cells at the tendon-bone interface tuberosity in patients with symptomatic rotator cuff tear. Int Orthop. 2015;39(6):1219–25.
7. Shapiro E, Grande D, Drakos M. Biologics in Achilles tendon healing and repair: a review. Curr Rev Musculoskelet Med. 2015;8(1):9–17.
8. Bíró V. Use of tissue engineering in the reconstruction of flexor tendon injuries of the hand. Orv Hetil. 2015;156(6):216–20.
9. Chen L, Liu JP, Tang KL, Wang Q, Wang GD, Cai XH, Liu XM. Tendon derived stem cells promote platelet-rich plasma healing in collagenase-induced rat Achilles tendinopathy. Cell Physiol Biochem. 2014;34(6):2153–68.
10. Gaspar D, Spanoudes K, Holladay C, Pandit A, Zeugolis D. Progress in cell-based therapies for tendon repair. Adv Drug Deliv Rev. 2015;84:240–56.
11. Font Tellado S, Bonani W, Balmayor ER, Foehr P, Motta A, Migliaresi C, van Griensven M. Fabrication and characterization of biphasic silk fibroin scaffolds for tendon/ligament-to-bone tissue engineering. Tissue Eng Part A. 2017;23(15–16):859–72.
12. Linderman SW, Gelberman RH, Thomopoulos S, Shen H. Cell and biologic-based treatment

12. of flexor tendon injuries. Oper Tech Orthop. 2016;26(3):206–15.

13. Hsieh CF, Alberton P, Loffredo-Verde E, Volkmer E, Pietschmann M, Müller PE, Schieker M, Docheva D. Periodontal ligament cells as alternative source for cell-based therapy of tendon injuries: in vivo study of full-size Achilles tendon defect in a rat model. Eur Cell Mater. 2016;32:228–40.

14. Vieira MH, Oliveira RJ, Eça LP, Pereira IS, Hermeto LC, Matuo R, Fernandes WS, Silva RA, Antoniolli AC. Therapeutic potential of mesenchymal stem cells to treat Achilles tendon injuries. Genet Mol Res. 2014;13(4):10434–49.

15. Meyer GA, Farris AL, Sato E, Gibbons M, Lane JG, Ward SR, Engler AJ. Muscle progenitor cell regenerative capacity in the torn rotator cuff. J Orthop Res. 2015;33(3):421–9.

16. Dale TP, Mazher S, Webb WR, Zhou J, Maffulli N, Chen GQ, El Haj AJ, Forsyth NR. Tenogenic differentiation of human embryonic stem cells. Tissue Eng Part A. 2018;24(5–6):361–8.

17. Peach MS, Ramos DM, James R, Morozowich NL, Mazzocca AD, Doty SB, Allcock HR, Kumbar SG, Laurencin CT. Engineered stem cell niche matrices for rotator cuff tendon regenerative engineering. PLoS One. 2017;12(4):e0174789.

18. Bottagisio M, Lopa S, Granata V, Talò G, Bazzocchi C, Moretti M, Barbara LA. Different combinations of growth factors for the tenogenic differentiation of bone marrow mesenchymal stem cells in monolayer culture and in fibrin-based three-dimensional constructs. Differentiation. 2017;95:44–53.

19. Hao ZC, Wang SZ, Zhang XJ, Lu J. Stem cell therapy: a promising biological strategy for tendon-bone healing after anterior cruciate ligament reconstruction. Cell Prolif. 2016;49(2):154–62.

20. Chen L, Jiang C, Tiwari SR, Shrestha A, Xu P, Liang W, Sun Y, He S, Cheng B. TGIF1 gene silencing in tendon-derived stem cells improves the tendon-to-bone insertion site regeneration. Cell Physiol Biochem. 2015;37(6):2101–14.

21. Lipner J, Shen H, Cavinatto L, Liu W, Havlioglu N, Xia Y, Galatz LM, Thomopoulos S. In vivo evaluation of adipose-derived stromal cells delivered with a nanofiber scaffold for tendon-to-bone repair. Tissue Eng Part A. 2015;21(21–22):2766–74.

22. Veronesi F, Torricelli P, Della Bella E, Pagani S, Fini M. In vitro mutual interaction between tenocytes and adipose-derived mesenchymal stromal cells. Cytotherapy. 2015;17(2):215–23.

23. Orr SB, Chainani A, Hippensteel KJ, Kishan A, Gilchrist C, Garrigues NW, Ruch DS, Guilak F, Little D. Aligned multilayered electrospun scaffolds for rotator cuff tendon tissue engineering. Acta Biomater. 2015;24:117–26.

24. Tao X, Liu J, Chen L, Zhou Y, Tang K. EGR1 induces tenogenic differentiation of tendon stem cells and promotes rabbit rotator cuff repair. Cell Physiol Biochem. 2015;35(2):699–709.

25. Valencia Mora M, Antuña Antuña S, García Arranz M, Carrascal MT, Barco R. Application of adipose tissue-derived stem cells in a rat rotator cuff repair model. Injury. 2014;45(Suppl 4):S22–7.

26. Xu W, Sun Y, Zhang J, Xu K, Pan L, He L, Song Y, Njunge L, Xu Z, Chiang MY, Sung KL, Chuong CM, Yang L. Perivascular-derived stem cells with neural crest characteristics are involved in tendon repair. Stem Cells Dev. 2015;24(7):857–68.

27. Kim YS, Sung CH, Chung SH, Kwak SJ, Koh YG. Does an injection of adipose-derived mesenchymal stem cells loaded in fibrin glue influence rotator cuff repair outcomes? A clinical and magnetic resonance imaging study. Am J Sports Med. 2017;45(9):2010–8.

28. Pas HI, Moen MH, Haisma HJ, Winters M. No evidence for the use of stem cell therapy for tendon disorders: a systematic review. Br J Sports Med. 2017;51(13):996–1002.

29. Pillai DS, Dhinsa BS, Khan W. Tissue engineering in Achilles tendon reconstruction; the role of stem cells, growth factors and scaffolds. Curr Stem Cell Res Ther. 2017;12(6):506–12.

30. Hofer HR, Tuan RS. Secreted trophic factors of mesenchymal stem cells support neurovascular and musculoskeletal therapies. Stem Cell Res Ther. 2016;7(1):131.

31. Youngstrom DW, Barrett JG. Engineering tendon: scaffolds, bioreactors, and models of regeneration. Stem Cells Int. 2016;2016:3919030.

32. Zhang B, Luo Q, Halim A, Ju Y, Morita Y, Song G. Directed differentiation and paracrine mechanisms of mesenchymal stem cells: potential implications for tendon repair and regeneration. Curr Stem Cell Res Ther. 2017;12(6):447–54.

33. Qin TW, Sun YL, Thoreson AR, Steinmann SP, Amadio PC, An KN, Zhao C. Effect of mechanical stimulation on bone marrow stromal cell-seeded tendon slice constructs: a potential engineered tendon patch for rotator cuff repair. Biomaterials. 2015;51:43–50.

34. Liu Y, Xu J, Xu L, Wu T, Sun Y, Lee YW, Wang B, Chan HC, Jiang X, Zhang J, Li G. Cystic fibrosis transmembrane conductance regulator mediates tenogenic differentiation of tendon-derived stem cells and tendon repair: accelerating tendon injury healing by intervening in its downstream signaling. FASEB J. 2017;31(9):3800–15.

35. Zhang J, Wang JH. Characterization of differential properties of rabbit tendon stem cells and tenocytes. BMC Musculoskelet Disord. 2010;11:10.

36. Yao L, Bestwick CS, Bestwick LA, Maffulli N, Aspden RM. Phenotypic drift in human tenocyte culture. Tissue Eng. 2006;12(7):1843–9.

37. Schwarz R, Colarusso L, Doty P. Maintenance of differentiation in primary cultures of avian tendon cells. Exp Cell Res. 1976;102(1):63–71.

38. Jelinsky SA, Archambault J, Li L, Seeherman H. Tendon-selective genes identified from rat and

39. Wang A, Mackie K, Breidahl W, Wang T, Zheng MH. Evidence for the durability to autologous tenocyte injection for treatment of chronic resistant lateral epicondylitis: mean 4.5-year clinical follow-up. Am J Sports Med. 2015;43:1775–83.

40. Clarke AW, Alyas F, Morris T, Robertson CJ, Bell J, Connell DA. Skin-derived tenocyte-like cells for the treatment of patellar tendinopathy. Am J Sports Med. 2011;39(3):614–23.

41. Boháč M, Csöbönyeiová M, Kupcová I, Zamborský R, Fedeleš J, Koller J. Stem cell regenerative potential for plastic and reconstructive surgery. Cell Tissue Bank. 2016;17(4):735–44.

42. Dyment NA, Galloway JL. Regenerative biology of tendon: mechanisms for renewal and repair. Curr Mol Biol Rep. 2015;1(3):124–31.

43. Meyerrose T, Olson S, Pontow S, Kalomoiris S, Jung Y, Annett G, Bauer G, Nolta JA. Mesenchymal stem cells for the sustained in vivo delivery of bioactive factors. Adv Drug Deliv Rev. 2010;62(12):1167–74.

44. Caplan AI, Bruder SP. Mesenchymal stem cells: building blocks for molecular medicine in the 21st century. Trends Mol Med. 2001;7(6):259–64.

45. Bruder SP, Fink DJ, Caplan AI. Mesenchymal stem cells in bone development, bone repair, and skeletal regeneration therapy. J Cell Biochem. 1994;56(3):283–94.

46. Okech W, Kuo CK. Informing stem cell-based tendon tissue engineering approaches with embryonic tendon development. Adv Exp Med Biol. 2016;920:63–77.

47. Bavin EP, Smith O, Baird AE, Smith LC, Guest DJ. Equine induced pluripotent stem cells have a reduced tendon differentiation capacity compared to embryonic stem cells. Front Vet Sci. 2015;2:55.

48. Zhang C, Yuan H, Liu H, Chen X, Lu P, Zhu T, Yang L, Yin Z, Heng BC, Zhang Y, Ouyang H. Well-aligned chitosan-based ultrafine fibers committed teno-lineage differentiation of human induced pluripotent stem cells for Achilles tendon regeneration. Biomaterials. 2015;53:716–30.

49. Le W, Yao J. The effect of myostatin (GDF-8) on proliferation and tenocyte differentiation of rat bone marrow-derived mesenchymal stem cells. J Hand Surg Asian Pac Vol. 2017;22(2):200–7.

50. Tan EW, Schon LC. Mesenchymal stem cell-bearing sutures for tendon repair and healing in the foot and ankle. Foot Ankle Clin. 2016;21(4):885–90.

51. Julianto I, Rindastuti Y. Topical delivery of mesenchymal stem cells "secretomes" in wound repair. Acta Med Indones. 2016;48(3):217–20.

52. Yang G, Rothrauff BB, Lin H, Yu S, Tuan RS. Tendon-derived extracellular matrix enhances transforming growth factor-β3-induced tenogenic differentiation of human adipose-derived stem cells. Tissue Eng Part A. 2017;23(3-4):166–76.

53. Veronesi F, Salamanna F, Tschon M, Maglio M, Nicoli Aldini N, Fini M. Mesenchymal stem cells for tendon healing: what is on the horizon? J Tissue Eng Regen Med. 2017;11(11):3202–19.

54. Sevivas N, Teixeira FG, Portugal R, Araújo L, Carriço LF, Ferreira N, Vieira da Silva M, Espregueira-Mendes J, Anjo S, Manadas B, Sousa N, Salgado AJ. Mesenchymal stem cell secretome: a potential tool for the prevention of muscle degenerative changes associated with chronic rotator cuff tears. Am J Sports Med. 2017;45(1):179–88.

55. Nowotny J, Aibibu D, Farack J, Nimtschke U, Hild M, Gelinsky M, Kasten P, Cherif C. Novel fiber-based pure chitosan scaffold for tendon augmentation: biomechanical and cell biological evaluation. J Biomater Sci Polym Ed. 2016;27(10):917–36.

56. Hsieh CF, Alberton P, Loffredo-Verde E, Volkmer E, Pietschmann M, Müller P, Schieker M, Docheva D. Scaffold-free Scleraxis-programmed tendon progenitors aid in significantly enhanced repair of full-size Achilles tendon rupture. Nanomedicine (Lond). 2016;11(9):1153–67.

57. Aktas E, Chamberlain CS, Saether EE, Duenwald-Kuehl SE, Kondratko-Mittnacht J, Stitgen M, Lee JS, Clements AE, Murphy WL, Vanderby R. Immune modulation with primed mesenchymal stem cells delivered via biodegradable scaffold to repair an Achilles tendon segmental defect. J Orthop Res. 2017;35(2):269–80.

58. Nagura I, Kokubu T, Mifune Y, Inui A, Takase F, Ueda Y, Kataoka T, Kurosaka M. Characterization of progenitor cells derived from torn human rotator cuff tendons by gene expression patterns of chondrogenesis, osteogenesis, and adipogenesis. J Orthop Surg Res. 2016;11:40.

59. Tornero-Esteban P, Hoyas JA, Villafuertes E, Rodríguez-Bobada C, López-Gordillo Y, Rojo FJ, Guinea GV, Paleczny A, Lópiz-Morales Y, Rodriguez-Rodriguez L, Marco F, Fernández-Gutiérrez B. Efficacy of supraspinatus tendon repair using mesenchymal stem cells along with a collagen I scaffold. J Orthop Surg Res. 2015;10:124.

60. Younesi M, Islam A, Kishore V, Anderson JM, Akkus O. Tenogenic induction of human MSCs by anisotropically aligned collagen biotextiles. Adv Funct Mater. 2014;24(36):5762–70.

61. Tian F, Ji XL, Xiao WA, Wang B, Wang F. CXCL13 promotes the effect of bone marrow mesenchymal stem cells (MSCs) on tendon-bone healing in rats and in C3HIOT1/2 cells. Int J Mol Sci. 2015;16(2):3178–87.

62. Mohanty N, Gulati BR, Kumar R, Gera S, Kumar S, Kumar P, Yadav PS. Phenotypical and functional characteristics of mesenchymal stem cells derived from equine umbilical cord blood. Cytotechnology. 2016;68(4):795–807.

63. Zhang W, Yang Y, Zhang K, Li Y, Fang G. Weft-knitted silk-poly(lactide-co-glycolide) mesh scaffold combined with collagen matrix and

63. seeded with mesenchymal stem cells for rabbit Achilles tendon repair. Connect Tissue Res. 2015;56(1):25–34.
64. Ramdass B, Koka PS. Ligament and tendon repair through regeneration using mesenchymal stem cells. Curr Stem Cell Res Ther. 2015;10(1):84–8.
65. Teng C, Zhou C, Xu D, Bi F. Combination of platelet-rich plasma and bone marrow mesenchymal stem cells enhances tendon-bone healing in a rabbit model of anterior cruciate ligament reconstruction. J Orthop Surg Res. 2016;11(1):96.
66. Wu T, Liu Y, Wang B, Sun Y, Xu J, Yuk-Wai LW, Xu L, Zhang J, Li G. The use of cocultured mesenchymal stem cells with tendon-derived stem cells as a better cell source for tendon repair. Tissue Eng Part A. 2016;22(19-20):1229–40.
67. Zhi Y, Liu W, Zhang P, Jiang J, Chen S. Electrospun silk fibroin mat enhances tendon-bone healing in a rabbit extra-articular model. Biotechnol Lett. 2016;38(10):1827–35.
68. Degen RM, Carbone A, Carballo C, Zong J, Chen T, Lebaschi A, Ying L, Deng XH, Rodeo SA. The effect of purified human bone marrow-derived mesenchymal stem cells on rotator cuff tendon healing in an athymic rat. Art Ther. 2016;32(12):2435–43.
69. Kong X, Ni M, Zhang G, Chai W, Li X, Li Y, Wang Y. Application of tendon-derived stem cells and bone marrow-derived mesenchymal stem cells for tendon injury repair in rat model. Zhejiang Da Xue Xue Bao Yi Xue Ban. 2016;45(2):112–9.
70. Gao Y, Zhang Y, Lu Y, Wang Y, Kou X, Lou Y, Kang Y. TOB1 deficiency enhances the effect of bone marrow-derived mesenchymal stem cells on tendon-bone healing in a rat rotator cuff repair model. Cell Physiol Biochem. 2016;38(1):319–29.
71. He M, Gan AW, Lim AY, Goh JC, Hui JH, Chong AK. Bone marrow derived mesenchymal stem cell augmentation of rabbit flexor tendon healing. Hand Surg. 2015;20(3):421–9.
72. Omi R, Gingery A, Steinmann SP, Amadio PC, An KN, Zhao C. Rotator cuff repair augmentation in a rat model that combines a multilayer xenograft tendon scaffold with bone marrow stromal cells. J Shoulder Elbow Surg. 2016;25(3):469–77.
73. Havlas V, Kotaška J, Koníček P, Trč T, Konrádová Š, Kočí Z, Syková E. Use of cultured human autologous bone marrow stem cells in repair of a rotator cuff tear: preliminary results of a safety study. Acta Chir Orthop Traumatol Cech. 2015;82(3):229–34.
74. Ning LJ, Zhang YJ, Zhang Y, Qing Q, Jiang YL, Yang JL, Luo JC, Qin TW. The utilization of decellularized tendon slices to provide an inductive microenvironment for the proliferation and tenogenic differentiation of stem cells. Biomaterials. 2015;52:539–50.
75. Li J, Chen L, Sun L, Chen H, Sun Y, Jiang C, Cheng B. Silencing of TGIF1 in bone mesenchymal stem cells applied to the post-operative rotator cuff improves both functional and histologic outcomes. J Mol Histol. 2015;46(3):241–9.

76. Gelberman RH, Linderman SW, Jayaram R, Dikina AD, Sakiyama-Elbert S, Alsberg E, Thomopoulos S, Shen H. Combined administration of ASCs and BMP-12 promotes an m2 macrophage phenotype and enhances tendon healing. Clin Orthop Relat Res. 2017;475(9):2318–31.
77. Shen H, Kormpakis I, Havlioglu N, Linderman SW, Sakiyama-Elbert SE, Erickson IE, Zarembinski T, Silva MJ, Gelberman RH, Thomopoulos S. The effect of mesenchymal stromal cell sheets on the inflammatory stage of flexor tendon healing. Stem Cell Res Ther. 2016;7(1):144.
78. Crowe CS, Chattopadhyay A, McGoldrick R, Chiou G, Pham H, Chang J. Characteristics of reconstituted lyophilized tendon hydrogel: an injectable scaffold for tendon regeneration. Plast Reconstr Surg. 2016;137(3):843–51.
79. Crowe CS, Chiou G, McGoldrick R, Hui K, Pham H, Chang J. Tendon regeneration with a novel tendon hydrogel: in vitro effects of platelet-rich plasma on rat adipose-derived stem cells. Plast Reconstr Surg. 2015;135(6):981e–9e.
80. Chiou GJ, Crowe C, McGoldrick R, Hui K, Pham H, Chang J. Optimization of an injectable tendon hydrogel: the effects of platelet-rich plasma and adipose-derived stem cells on tendon healing in vivo. Tissue Eng Part A. 2015;21(9-10):1579–86.
81. Gelberman RH, Shen H, Kormpakis I, Rothrauff B, Yang G, Tuan RS, Xia Y, Sakiyama-Elbert S, Silva MJ, Thomopoulos S. Effect of adipose-derived stromal cells and BMP12 on intrasynovial tendon repair: a biomechanical, biochemical, and proteomics study. J Orthop Res. 2016;34(4):630–40.
82. Yin Z, Hu JJ, Yang L, Zheng ZF, An CR, Wu BB, Zhang C, Shen WL, Liu HH, Chen JL, Heng BC, Guo GJ, Chen X, Ouyang HW. Single-cell analysis reveals a nestin(+) tendon stem/progenitor cell population with strong tenogenic potentiality. Sci Adv. 2016;2(11):e1600874.
83. Hu C, Zhang Y, Tang K, Luo Y, Liu Y, Chen W. Downregulation of CITED2 contributes to TGFβ-mediated senescence of tendon-derived stem cells. Cell Tissue Res. 2017;368(1):93–104.
84. Chen H, Ge HA, Wu GB, Cheng B, Lu Y, Jiang C. Autophagy prevents oxidative stress-induced loss of self-renewal capacity and stemness in human tendon stem cells by reducing ROS accumulation. Cell Physiol Biochem. 2016;39(6):2227–38.
85. Leong DJ, Sun HB. Mesenchymal stem cells in tendon repair and regeneration: basic understanding and translational challenges. Ann N Y Acad Sci. 2016;1383(1):88–96.
86. Wang JH, Nirmala X. Application of tendon stem/progenitor cells and platelet-rich plasma to treat tendon injuries. Oper Tech Orthop. 2016;26(2):68–72.
87. Wang B, Guo J, Feng L, Suen CW, Fu WM, Zhang JF, Li G. MiR124 suppresses collagen formation of human tendon derived stem cells through targeting egr1. Exp Cell Res. 2016;347(2):360–6.

88. Wang JH, Komatsu I. Tendon stem cells: mechanobiology and development of tendinopathy. Adv Exp Med Biol. 2016;920:53–62.

89. Zhang J, Yuan T, Wang JH. Moderate treadmill running exercise prior to tendon injury enhances wound healing in aging rats. Oncotarget. 2016;7(8):8498–512.

90. Guo J, Chan KM, Zhang JF, Li G. Tendon-derived stem cells undergo spontaneous tenogenic differentiation. Exp Cell Res. 2016;341(1):1–7.

91. Lui PP, Wong OT, Lee YW. Transplantation of tendon-derived stem cells pre-treated with connective tissue growth factor and ascorbic acid in vitro promoted better tendon repair in a patellar tendon window injury rat model. Cytotherapy. 2016;18(1):99–112.

92. Chen W, Tang H, Liu X, Zhou M, Zhang J, Tang K. Dickkopf1 Up-Regulation induced by a high concentration of dexamethasone promotes rat tendon stem cells to differentiate into adipocytes. Cell Physiol Biochem. 2015;37(5):1738–49.

93. Zhou Y, Zhang J, Wu H, Hogan MV, Wang JH. The differential effects of leukocyte-containing and pure platelet-rich plasma (PRP) on tendon stem/progenitor cells - implications of PRP application for the clinical treatment of tendon injuries. Stem Cell Res Ther. 2015;6:173.

94. Al-Ani MKh XK, Sun Y, Pan L, Xu Z, Yang L. Study of bone marrow mesenchymal and tendon-derived stem cells transplantation on the regenerating effect of Achilles tendon ruptures in rats. Stem Cells Int. 2015;2015:984146.

95. Tokunaga T, Shukunami C, Okamoto N, Taniwaki T, Oka K, Sakamoto H, Ide J, Mizuta H, Hiraki Y. FGF-2 stimulates the growth of tenogenic progenitor cells to facilitate the generation of tenomodulin-positive tenocytes in a rat rotator cuff healing model. Am J Sports Med. 2015;43(10):2411–22.

96. Runesson E, Ackermann P, Karlsson J, Eriksson BI. Nucleostemin- and Oct 3/4-positive stem/progenitor cells exhibit disparate anatomical and temporal expression during rat Achilles tendon healing. BMC Musculoskelet Disord. 2015;16:212.

97. Lee CH, Lee FY, Tarafder S, Kao K, Jun Y, Yang G, Mao JJ. Harnessing endogenous stem/progenitor cells for tendon regeneration. J Clin Invest. 2015;125(7):2690–701.

98. Lui PP. Markers for the identification of tendon-derived stem cells in vitro and tendon stem cells in situ - update and future development. Stem Cell Res Ther. 2015;6:106.

99. Chen W, Tang H, Zhou M, Hu C, Zhang J, Tang K. Dexamethasone inhibits the differentiation of rat tendon stem cells into tenocytes by targeting the scleraxis gene. J Steroid Biochem Mol Biol. 2015;152:16–24.

100. Popov C, Burggraf M, Kreja L, Ignatius A, Schieker M, Docheva D. Mechanical stimulation of human tendon stem/progenitor cells results in upregulation of matrix proteins, integrins and MMPs, and activation of p38 and ERK1/2 kinases. BMC Mol Biol. 2015;16:6.

101. Liu J, Tao X, Chen L, Han W, Zhou Y, Tang K. CTGF positively regulates BMP12 induced tenogenic differentiation of tendon stem cells and signaling. Cell Physiol Biochem. 2015;35(5):1831–45.

102. Ilić N, Atkinson K. Manufacturing and use of human placenta-derived mesenchymal stromal cells for phase I clinical trials: establishment and evaluation of a protocol. Vojnosanit Pregl. 2014;71(7):651–9.

103. Petrou IG, Grognuz A, Hirt-Burri N, Raffoul W, Applegate LA. Cell therapies for tendons: old cell choice for modern innovation. Swiss Med Wkly. 2014;144:w13989.

104. Zong JC, Mosca MJ, Degen RM, Lebaschi A, Carballo C, Carbone A, Cong GT, Ying L, Deng XH, Rodeo SA. Involvement of Indian hedgehog signaling in mesenchymal stem cell-augmented rotator cuff tendon repair in an athymic rat model. J Shoulder Elbow Surg. 2017;26(4):580–8.

105. Xiao L, Nasu M. From regenerative dentistry to regenerative medicine: progress, challenges, and potential applications of oral stem cells. Stem Cells Cloning. 2014;7:89–99.

106. Górski B. Gingiva as a new and the most accessible source of mesenchymal stem cells from the oral cavity to be used in regenerative therapies. Postepy Hig Med Dosw (Online). 2016;70:858–71.

107. Gomiero C, Bertolutti G, Martinello T, Van Bruaene N, Broeckx SY, Patruno M, Spaas JH. Tenogenic induction of equine mesenchymal stem cells by means of growth factors and low-level laser technology. Vet Res Commun. 2016;40(1):39–48.

108. Park GY, Kwon DR, Lee SC. Regeneration of full-thickness rotator cuff tendon tear after ultrasound-guided injection with umbilical cord blood-derived mesenchymal stem cells in a rabbit model. Stem Cells Transl Med. 2015;4(11):1344–51.

109. Jang KM, Lim HC, Jung WY, Moon SW, Wang JH. Efficacy and safety of human umbilical cord blood-derived mesenchymal stem cells in anterior cruciate ligament reconstruction of a rabbit model: new strategy to enhance tendon graft healing. Arthroscopy. 2015;31(8):1530–9.

110. Takayama K, Kawakami Y, Mifune Y, Matsumoto T, Tang Y, Cummins JH, Greco N, Kuroda R, Kurosaka M, Wang B, Fu FH, Huard J. The effect of blocking angiogenesis on anterior cruciate ligament healing following stem cell transplantation. Biomaterials. 2015;60:9–19.

111. Jiang D, Yang S, Gao P, Zhang Y, Guo T, Lin H, Geng H. Combined effect of ligament stem cells and umbilical-cord-blood-derived CD34+ cells on ligament healing. Cell Tissue Res. 2015;362(3):587–95.

112. Weninger P, Wepner F, Kissler F, Enenkel M, Wurnig C. Anatomic double-bundle reinsertion after acute proximal anterior cruciate ligament injury using Knotless PushLock Anchors. Arthrosc Tech. 2015;4(1):e1–6.

113. Taniguchi N, Suenaga N, Oizumi N, Miyoshi N, Yamaguchi H, Inoue K, Chosa E. Bone marrow stimulation at the footprint of arthroscopic surface-holding repair advances cuff repair integrity. J Shoulder Elbow Surg. 2015;24(6):860–6.

114. Otabe K, Nakahara H, Hasegawa A, Matsukawa T, Ayabe F, Onizuka N, Inui M, Takada S, Ito Y, Sekiya I, Muneta T, Lotz M, Asahara H. Transcription factor Mohawk controls tenogenic differentiation of bone marrow mesenchymal stem cells in vitro and in vivo. J Orthop Res. 2015;33(1):1–8.

115. Lee DJ, Southgate RD, Farhat YM, Loiselle AE, Hammert WC, Awad HA, O'Keefe RJ. Parathyroid hormone 1-34 enhances extracellular matrix deposition and organization during flexor tendon repair. J Orthop Res. 2015;33(1):17–24.

116. Govoni M, Berardi AC, Muscari C, Campardelli R, Bonafè F, Guarnieri C, Reverchon E, Giordano E, Maffulli N, Della Porta G. An engineered multiphase three-dimensional microenvironment to ensure the controlled delivery of cyclic strain and human growth differentiation factor 5 for the tenogenic commitment of human bone marrow mesenchymal stem cells. Tissue Eng Part A. 2017;23(15–16):811–22.

117. Ficklscherer A, Serr M, Loitsch T, Niethammer TR, Lahner M, Pietschmann MF, Müller PE. The influence of different footprint preparation techniques on tissue regeneration in rotator cuff repair in an animal model. Arch Med Sci. 2017;13(2):481–8.

118. Leone L, Raffa S, Vetrano M, Ranieri D, Malisan F, Scrofani C, Vulpiani MC, Ferretti A, Torrisi MR, Visco V. Extracorporeal Shock Wave Treatment (ESWT) enhances the in vitro-induced differentiation of human tendon-derived stem/progenitor cells (hTSPCs). Oncotarget. 2016;7(6):6410–23.

119. Yu Y, Zhou Y, Cheng T, Lu X, Yu K, Zhou Y, Hong J, Chen Y. Hypoxia enhances tenocyte differentiation of adipose-derived mesenchymal stem cells by inducing hypoxia-inducible factor-1α in a co-culture system. Cell Prolif. 2016;49(2):173–84.

120. Chailakhyan RK, Shekhter AB, Ivannikov SV, Tel'pukhov VI, Suslin DS, Gerasimov YV, Tonenkov AM, Grosheva AG, Panyushkin PV, Moskvina IL, Vorob'eva NN, Bagratashvili VN. Reconstruction of ligament and tendon defects using cell technologies. Bull Exp Biol Med. 2017;162(4):563–8.

121. Chainani A, Little D. Current status of tissue-engineered scaffolds for rotator cuff repair. Tech Orthop. 2016;31(2):91–7.

122. Jiang D, Gao P, Zhang Y, Yang S. Combined effects of engineered tendon matrix and GDF-6 on bone marrow mesenchymal stem cell-based tendon regeneration. Biotechnol Lett. 2016;38(5):885–92.

123. Yang G, Lin H, Rothrauff BB, Yu S, Tuan RS. Multilayered polycaprolactone/gelatin fiber-hydrogel composite for tendon tissue engineering. Acta Biomater. 2016;35:68–76.

124. Seyed-Forootan K, Karimi H, Dayani AR. PRP and metaplasia in repaired tendon. J Acute Dis. 2014;3(4):284–9.

125. Kadakia AR, Dekker RG 2nd, Ho BS. Acute Achilles tendon ruptures: an update on treatment. J Am Acad Orthop Surg. 2017;25(1):23–31.

126. Ross D, Maerz T, Kurdziel M, Hein J, Doshi S, Bedi A, Anderson K, Baker K. The effect of granulocyte-colony stimulating factor on rotator cuff healing after injury and repair. Clin Orthop Relat Res. 2015;473(5):1655–64.

127. Lui PP. Identity of tendon stem cells—how much do we know? Identity of tendon stem cells—how much do we know? J Cell Mol Med. 2013;17(1):55–64.

128. Lee SY, Kim W, Lim C, Chung SG. Treatment of lateral epicondylosis by using allogeneic adipose-derived mesenchymal stem cells: a pilot study. Stem Cells. 2015;33:2995–3005.

129. Ilic N, Atkinson K. Manufacturing and use of human placenta-derived mesenchymal stromal cells for phase I clinical trials: establishment and evaluation of a protocol. Vojnosanit Pregl. 2014;71(7):651–9.

130. Lange-Consiglio A, Rossi D, Tassan S, Perego R, Cremonesi F, Parolini O. Conditioned medium from horse amniotic membrane-derived multipotent progenitor cells: immunomodulatory activity in vitro and first clinical application in tendon and ligament injuries in vivo. Stem Cells Dev. 2013;22(22):3015–24.

131. Manning CN, Martel C, Sakiyama-Elbert SE, Silva MJ, Shah S, Gelberman RH, Thomopoulos S. Adipose-derived mesenchymal stromal cells modulate tendon fibroblast responses to macrophage-induced inflammation in vitro. Stem Cell Res Ther. 2015;6:74.

Cell Therapies for Tendon: Treatments and Regenerative Medicine

27

Anthony Grognuz, Pierre-Arnaud Aeberhard, Murielle Michetti, Nathalie Hirt-Burri, Corinne Scaletta, Anthony de Buys Roessingh, Wassim Raffoul, and Lee Ann Laurent-Applegate

27.1 Introduction

Tendon is the tissue that is found between a muscle and a bone and it is thus implicated in the movements of our body. This role is very important in everyday life and many problems arise when it is compromised. Thus, the aim is to restore its function as rapidly as possible when degradation occurs.

Tendon afflictions can occur in multiple locations and in various forms. We can usually divide them into two categories depending on their chronic or acute nature. Chronic injuries are generally degenerative and mostly due to overuse and are referred under the general term "tendinopathy." On the other side, acute injuries are generally due to one specific traumatic event leading to a tear which can be partial or total. The two categories are not always so easy to separate as many spontaneous ruptures take place on a degenerative field [1, 2].

In many countries, the number of overuse injuries has augmented during the last decades, parallel with the democratization of sport practice [3, 4]. Tendinopathies are a very common condition and an epidemiologic study found that the incidence was 166.6/100,000 for males and 52.1/100,000 for females [5]. Therefore, recently, there is high concern to ever-increasing injuries related to the ageing population and rising sports practice [6]. While not directly life threatening, tendon afflictions have been found to have a great global disease burden with amazing high costs associated [7]. This phenomenon is in part related to the anatomic regions and disability caused. Acute hand tendon injuries are particularly highly prevalent in manual and office workers with the most commonly involved tendons being the extensor and flexor of the index and middle fingers [8]. Tendinopathies on the other hand also lead to disabilities and impairments, notably in the leg or arm movements, with the Achilles, the rotator cuff, or the common wrist extensors and flexors frequently concerned among others [3, 4].

The treatment of tendon pathologies conventionally extends from conservative care to surgical intervention depending on the gravity of the injury. Unfortunately the results are not always optimal and this leads researchers to look for new therapies. Therefore, cell therapies and tissue engineering have become more and more popular during the last decades.

A. Grognuz · P.-A. Aeberhard · M. Michetti
N. Hirt-Burri · C. Scaletta · A. de Buys Roessingh
W. Raffoul · L. A. Laurent-Applegate (✉)
Unit of Regenerative Therapy, Service of Plastic
and Reconstructive Surgery, Department
of Musculoskeletal Medicine, Lausanne University
Hospital, Epalinges, Switzerland
e-mail: Anthony.Grognuz@chuv.ch;
Pierre-Arnaud.Aeberhard@epfl.ch;
Murielle.Michetti@chuv.ch; Nathalie.Hirt@chuv.ch;
Corinne.Scaletta@chuv.ch;
Anthony.debuys-roessingh@chuv.ch;
Wassim.Raffoul@chuv.ch;
Lee.Laurent-Applegate@chuv.ch

© Springer Nature Switzerland AG 2019
D. Duscher, M. A. Shiffman (eds.), *Regenerative Medicine and Plastic Surgery*,
https://doi.org/10.1007/978-3-030-19962-3_27

27.2 Tendon Structure and Role

Tendons are a dense-connective tissue consisting of specialized fibroblastic cells called tenocytes which are scattered within an extracellular matrix. Since they have a relatively poor vascularization, tendons are portrayed as a white and shiny structure. They play an important role in our body by binding muscles to bones (Fig. 27.1) with only one exception where a muscle is attached to another muscle (rectus abdominis). Tendons permit to store the force created in the muscle and to transmit it to the bone. Collagen is highly organized in tendons and since they have low and mainly anaerobic metabolism, they are able to carry loads and maintain tension for long periods without risk of ischemia or necrosis [9]. Their elaborated structure not only allows the tendon to transmit forces but also to act as a buffer against tension and compression by absorbing external forces from various directions. Tendons are thus the major "protector" of muscles against tears [9–11]. Varying from flat-shaped to ribbon-like or even cylindrical, tendons differ considerably in shape and length depending on their location [9, 10, 12, 13]. For instance, muscles dedicated to create powerful forces are usually elongated by short and wide tendons (e.g., quadriceps) while muscles responsible for fine movements are usually associated to a long and thin tendon (e.g., hand flexor tendons) [9].

The transition zone from tendon to muscle is called the myotendinous junction and the location where the tendon inserts into bone is known as the enthesis (Fig. 27.1). The tissue structure can be either fibrous, with the same dense connective tissue structure as in tendon midsubstance, or fibrocartilaginous, with a transition composed of four zones from a noncalcified tendinous structure to a calcified cartilaginous bony tissue as in the enthesis [14–16].

In summary, tendons can best be described as a white poorly vascularized tissue connecting a muscle to a bone with the myotendinous junction being the transition from muscle to tendon and the enthesis being the region where tendon usually becomes fibrocartilaginous to insert into bone (Fig. 27.1).

Tendons, composed for 65–80% of dry mass of collagen, rely on a very hierarchical organization of this protein within the tissue [9, 10, 12, 17]. Tenocytes are tendon-specific fibroblastic cells interspersed throughout the dense tissue (Fig. 27.1), and are responsible for regulation of fibrillogenesis [13]. The manner to describe the aggregation and organization of collagen within tendon can vary (Fig. 27.1) [17, 18]. Collagen α-chains are rich in glycine (1/3 of amino acids) as well as proline and hydroxyproline and adopt a left-handed helical configuration. Three α-chains supercoil to form a right-handed super-helical collagen molecule (stabilized by interchain h-bonds) and provide the collagen footprint type [17]. Proteoglycans and glycosaminoglycans appear between the α-chains [9]. Collagen molecules assemble together to form collagen fibrils highly recognizable to their striated structure with a periodicity of 67 nm. Fibrils are bundled together to form a collagen fiber. Bundles of fibers, with elongated tenocytes in between, are organized as fascicles and are surrounded by a loose connective tissue known as the endotenon [17]. The collagen fibers are mainly oriented longitudinally, but fibers can also present transversal variations and this can give a wavy pattern to the tendon at rest [9, 17, 19]. To form the tendon, the different fascicles are then bundled together by the epitenon, another connective tissue contiguous to the endotenon [17]. These two sheaths permit gliding power between fascicles and can allow vascularization, lymphatics, and innervation deep into the tendon structure (Fig. 27.1) [9, 13, 20].

The tendon can further be surrounded by five different structural types which allow ideal gliding abilities [9]. Long tendons often go through canals covering grooves and notches which are called fibrous sheaths or retinacula such as in the flexor and extensor tendons of the hand and foot. These fibrous sheaths can be reinforced by a reflexion pulley which helps to maintain the tendon in its canal when it faces a change in direction. In some parts of the hand or of the foot where high lubrication is needed, synovial sheaths can be found around the tendon. Synovial fluid is located between the parietal and visceral

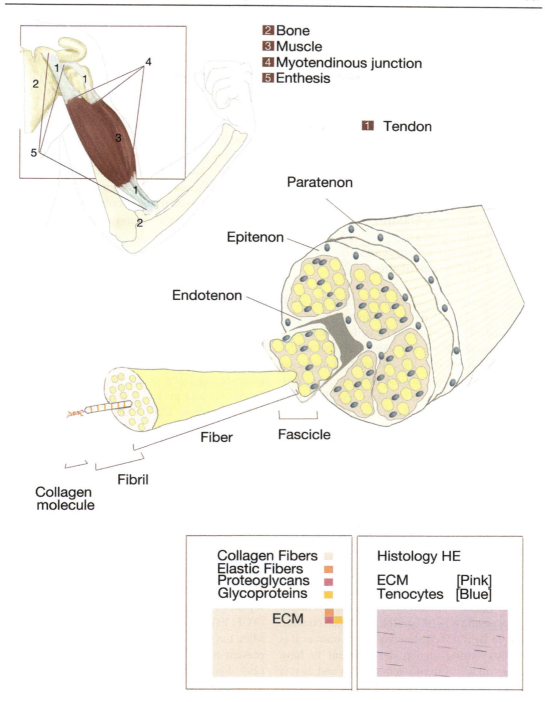

Fig. 27.1 Tendon location, structure, and composition. (*Top*) Tendon is part of the musculoskeletal system and it is found between muscle and bone. It allows transmitting the forces created in the muscles to the bone and thus permits movement. (*Center*) Tendon ultrastructure with hierarchical organization of collagen within the tissue, from collagen molecules to tendon. (*Bottom left*) The extracellular matrix of tendon is in great majority composed of collagen (represented in clear grey), but other proteins such as elastin, proteoglycans, and glycoproteins are also present in smaller quantities. (*Bottom right*) The histological structure shows elongated cells called tenocytes (in blue) interspersed within the dense matrix (in pink) with a hematoxylin-eosin (HE) stain. The cell/matrix ratio is low compared to other connective tissues

sheets to complete the composition of this structure. Paratenon is a loose connective tissue which is found in many tendons such as the Achilles tendon. Although it is not a real synovial sheath, it allows smooth gliding processes since it is an elastic sleeve. Finally, tendon bursae are found near bony prominences to reduce friction and protect the tendon [9].

27.3 Tendon Composition

27.3.1 Collagens, Elastin, Proteoglycans, Glycosaminoglycans, and Glycoproteins

Tendons are composed of approximately 70% water [10, 12] and have a cellular component composed of fibroblastic-type cells called tenocytes dispersed in the well-organized extracellular matrix (ECM) with the cell/matrix ratio being low. While collagens are the most frequent proteins and account for 60–80% of the dry mass, the organization of the tissue relies on efficient interaction of collagen I with other collagens as well as with many other matrix and cell proteins [9, 10, 12, 13, 17].

Different types of collagen coexist in tendon and arrange themselves in a different manner to give rise to various organizations. Collagens I, II, III, V, and XI are part of the fibril-forming class of collagens and are all found in tendons and in ligaments to different extents [17]. Collagen I, approximately 60% of dry mass of tendon and up to 90–95% of total collagen, represents by far the most abundant collagen [20] ahead of collagen III, representing 5–10% of total collagen content. Collagen III is nonetheless very important as it is necessary during tendon development to have regular collagen I fibrillogenesis [21] and as it is increased during the first phases of healing processes after tendon injuries [22]. Collagen II, which is more present in cartilage, is principally found at the enthesis zone [17, 23] but also around some gliding structures such as pulleys [12]. Collagens V and XI are present in very low amounts even though they can be very important

because their absence leads to altered collagen fibril development [17]. Collagens XII and XIV can be found both in tendons and ligaments and they are fibril-associated collagens [17] with collagen XIV shown to be highly expressed during fetal development (in a chick model), but with a net decrease after birth [24]. Collagen type IX interacts with collagen II and can be found in cartilaginous regions of tendon [12, 17]. Other collagens of low content include collagen IV (a basement membrane collagen implicated in the interface between tissues), collagen XIII (a transmembrane collagen found particularly in the myotendinous junction), and collagen VI (a beaded filament-forming collagen implicated in different aspects such as cell proliferation, migration, differentiation, and apoptosis). Even though in small quantities, collagen VI absence can lead to dysregulation in fibrillogenesis as it is also implicated in development, homeostasis, and repair of tendons [17, 25].

Elastin is another fibrous protein of the tendon representing 1–2% of the ECM dry weight, rich in glycine and proline like collagen, but not in hydroxyproline. Elastin gives flexibility to the tendon by being able to elongate to 70–150% of its length and by helping collagen fibers to recover their crimp after elongation [9, 12, 26].

Proteoglycans are composed of a core protein with chains of glycosaminoglycans (GAGs) and are encompassed within and between fibrils and fibers of collagen [9, 27]. Even though they represent less than 1% of dry weight of tendon, they are of great importance [12, 27] as small proteoglycans are particularly important in tendon development and organization regulating fibrillogenesis of collagen and growth factor signaling (TGF, EGF) and can influence cell proliferation. Mice having a deficiency in either decorin (most present proteoglycan in tendon) [28], biglycan [29], fibromodulin [30], or lumican [31] have shown abnormal collagen organization in tendons. Large proteoglycans are negatively charged and can trap a high content of water permitting a viscous environment and thus providing protection against compression, adapted environment for collagen fiber stretching, good diffusion of hydrosoluble molecules, and cell migration [27].

Among them, aggrecan and versican are particularly present in zones of compression.

GAGs are less present in tendons than in other connective tissues such as cartilage and their concentration varies from 0.2% dry mass in high-tension zones (i.e., dermatan sulfate) to 3.5–5% in pressure zones or enthesis (i.e., chondroitin sulfate, keratin sulfate). Heparan sulfate and heparin are found mostly in the myotendinous junction and hyaluronan constitutes around 6% of the total GAG amount in tendon and it differs from the other GAGs as it is larger and it is not bound to a core protein nor is it sulfated [27].

Glycoproteins represent less than 1% of the dry weight of tendon [9] and seem to be implicated in mechanical stability. Different adhesive glycoproteins are found within tendons, such as fibronectin, thrombospondin, tenascin-C, undulin, and laminin [9]. Fibronectin is found on collagen surfaces [20] but is not specific to tendon. It is important in adhesion and binds cells and many proteins of the ECM [12], increases in case of tendon injury, and helps to give strength to a wound by cross-linking with collagen [12]. Thrombospondin and tenomodulin are relatively specific to tendon and their genes have been identified as the best tendon-selective genes in both rat and human mature tendon [32]. The role of tenomodulin in tendon was probably the most studied among glycoproteins and was shown to be regulated by scleraxis, a vital transcription factor in tendon development [33]. The presence of tenomodulin is necessary for tendon maturation and tenocyte proliferation and its absence can lead to defects in tendon structure [34, 35]. Tenascin C, found frequently in tendon, is regulated by mechanical loading and could play a role in collagen alignment [1]. It has sometimes been proposed as a specific tendon marker but its expression is also elevated in other tissues such as cartilage [32]. Laminin is found in vascular regions and within the myotendinous junction [9].

In addition to all of the above, some inorganic molecules are implicated in tendon metabolism and can be found in small amounts within the extracellular matrix (less than 0.2% dry mass) [9]. Among them, calcium is the most present and it is found in higher amounts at the enthesis than in the mid-substance [9]. Its presence can be found intensified in the mid-substance in case of calcifying tendinopathy [36]. Other molecules also present include copper which is important in the formation of collagen cross-links and manganese which participates in enzymatic reactions during synthesis of ECM [9].

27.4 Tendon Injuries, Healing, and Treatments

Now that the basic structure and biology of tendons have been described, it is important to position what type of person is most affected by tendon pathologies and what are the alternatives for repairing or regenerating a tendon that malfunctions. The epidemiology of tendon injury is vast within a heterogeneous population affected. Tendon injuries occur within multiple anatomical sites and are frequently associated with debilitating pain and dysfunction. This creates functional and productivity problems in the workplace but also for recreational activities and sports. The healing process can be long and loss of mobility occurs frequently. Scar and adhesion formation due to non-organized repair processes are common, notably in the hand [37]. Injuries very often lead to re-ruptures [18, 38] and this is why multiple treatment plans have been proposed and have been evolving rapidly over the years.

Tendon injuries are principally divided into two categories whether they are acute or chronic. Acute injuries are those from traumatic partial or total tears while tendinopathies of chronic nature are typically associated with overuse degenerative processes. The separation between acute and chronic states is not always so easy as spontaneous ruptures of a tendon often occur following a degenerative status [1, 2]. As more and more people are practicing sports for leisure, the number of overuse injuries has dramatically increased parallel over the years. The most frequently affected tendons of lower extremities include the Achilles, patellar, biceps femoris (hamstrings), and tractus iliotibialis. For upper extremities, these are the rotator cuff, wrist extensors responsible for "tennis elbow," and wrist flexors responsible for "golfer's elbow" [3, 4].

Accidents with acute tendon laceration of the hand are particularly common as a total of 1775 new cases each year are reported in Switzerland alone with a mean insurance cost of 23,843 CHF (www.unfallstatistik.ch). Historically, the Swiss hand surgeon Claude Verdan was the first to introduce a system of classification for flexor tendon injuries and he separated these hand injuries into five anatomical zones [39].

As all of these conditions promote joint instability and arthritis, surgical intervention was the first-line therapy associated in management. As time and experience for tendon injury management progressed, other key elements on the pathophysiological level had to be considered in order to achieve optimal functional result including intrinsic tenocyte regenerative response, adequate nutrition, and prevention of adhesions. Conventional treatments depend therefore on the severity of the tendon injury and can range from conservative treatments including physical therapy or infiltration and extend to new cell therapy approaches. When these are not sufficient, surgical intervention is necessary such as tenotomy, suture, tendon transfer, graft, or in the extreme cases prosthetic implants.

Adhesions can be frequent despite early surgical treatment. Controlled mobilization is usually recommended in parallel to surgery, because mechanical stimulation leads to fewer adhesions and increased strength. However, caution is advised with injuries at the enthesis where mobilization inhibits tendon repair, and thus cast immobilization should be preferred when this location is injured [18, 22, 40, 41].

During the inflammatory phase directly after an injury, macrophages have a predominant role and release chemo-attractive substances notably to recruit and stimulate tenocytes. After a few days, the second step of healing (proliferative phase) is controlled by tenocytes and macrophages and there is a high deposition of matrix, mostly composed of type 3 collagen. After 1 or 2 months, a third phase (remodeling phase) takes place with the extracellular matrix (ECM) con-

tinuing to be produced, but with a higher amount of type 1 collagen. The tissue is reorganized and becomes more and more aligned. This process takes as long as 1 year or more, with the cell density and activity diminishing with time [10, 13].

Tendon regeneration is a time-consuming procedure and this can be associated, in part, because the adult tenocytes are characterized by a slow metabolism. Therefore, many recent cellular therapies have been proposed to stimulate tendon healing and have been integrated into the clinic already including minimally processed cells and tissues such as platelet-rich plasma (PRP), separated adipose stem cells (ASCs), and separated bone marrow stem cells (BM-MSCs) (Fig. 27.2). These cell therapies have been integrated rapidly into the clinic since the patient's own tissue and cells are only centrifuged or separated by gravitational means before injecting them back into the patient as soon as possible after isolation. However, variable patient relief from these proposed cell therapies depends highly on the biological product produced and the procedure chosen for isolation. The procedure should take into account that the "biological" could be sensitive to extreme mechanical conditions such as high centrifugation, passing the tissue through small needles and solutions that may come into contact with the biological material of interest. Overall, the sterility should be of high concern to assure the safety of the patient and closed systems should be implemented when possible and with a laminar flow safety cabinet (Fig. 27.2).

Other procedures which are technically more demanding (but have specific advantages of having more purified portions of cell fractions) are advancing also. These cell therapies are considered as standardized transplants and require varying degrees of good manufacturing processes (GMP) with highly specific infrastructure containing clean rooms for processing and procedures executed by a specially trained staff (Fig. 27.2).The cells are generally used within delivery systems (such as a biogel or scaffold) that must have been shown to be safe for the patient. All of these preparations should be prepared with closed system processing and/or a laminar flow hood to assure sterility of the prod-

Fig. 27.2 Regulatory processing for cellular therapy products. Different treatments require various procedures for final formulation within sterile biosafety laminar flow environment. Cellular transplants rely on simplified procedures for PRP, ASCs, and BM-MSCs using cell separation only and can be administered rapidly for patient treatment. Standardized transplants necessitate the implementation of cell culture procedures and specific infrastructure for GMP manufacturing are necessary for high safety standards worldwide. All of these procedures are under continuous monitoring and training of laboratory staff

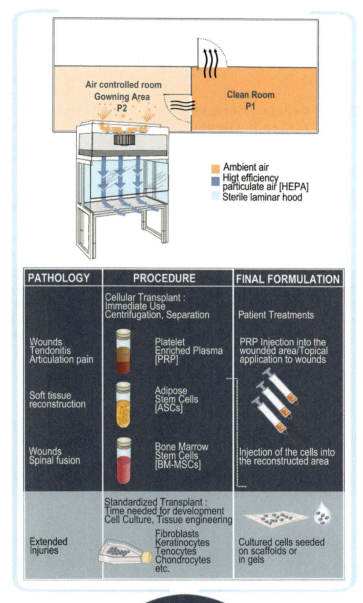

uct preparation. Different cell sources, isolation procedures, proliferation protocols, and stability of cells are all determining factors to bring new innovative therapies safely to the patient.

27.5 Cellular Therapies for Tendon Repair and Regeneration

As for any tissue engineering therapeutic strategy, the choice of the cell source and type is a main factor for success. A cell source destined for tendon repair/regeneration should be easy to collect and traceable (known origin) and easily and rapidly cultured and expanded. All of the procedures should be done with cells that possess a high tendon formation potential and with maximum stability of the desired phenotype to assure patient safety upon implantation. All processing should also assure a final product that will not trigger an immunological reaction of the recipient patient.

Figure 27.3 illustrates a continuum, based on development, for tissue sources that can provide cells that can be used in cellular therapies. Some of these tissues can be implemented for tendon repair/regeneration including the allogenic progenitor tenocytes derived from later stage development (fetal and adult) and for autologous repair (newborn and adult).

27.5.1 Embryonic Stem Cells

Embryonic stem cells (ESCs) could be used as a cell source for tendon repair and regeneration. The embryo stage is in development up to 8 weeks postfertilization (Fig. 27.3). During development, the cells composing the morula (around 96 h postfertilization) are "totipotent" and can differentiate into every cell type. The

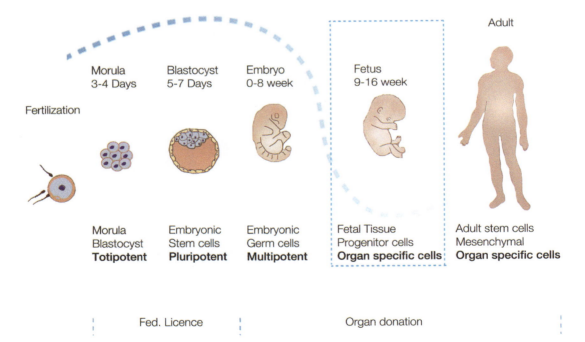

Fig. 27.3 Cell type potency available through development. Different cell types isolated at various stages of development can provide cell sources with variable differentiation advantages. In Switzerland like in many countries, cell capacity to differentiate into multiple cell types diminishes with development, as the cells become more "specialized." Early-stage development cells require a federal license for use whereas later stage development is under transplantation programs and organ donations. Human fetal progenitor tenocytes that are discussed in details in this chapter belong to the highlighted rectangle

cells then already begin to diverge when creating the blastocyst around 5–7 days postfertilization where they become pluripotent. Original tissue sources usually come from the inner cell mass found in the blastocyst at this early age and pluripotency means that the cells are not able to develop into another embryo. However, with proper stimulation and cocktails of growth factors, these cells can develop into every cell type.

In practice, due to their high potential to differentiate, the presence of factors like GDF5 and BMP2 in the wound could potentially lead to the formation of bone or cartilage tissues [42]. However, a rat tendon model successfully derived from ESCs has been developed with cells achieving maturation and differentiation into a tenocyte type 30 days postimplantation. Using the same cells in a fibrin gel led to better structural and functional results (than the controls with fibrin gel only) to treat rat patellar tendons. The cells survived for 4 weeks and activated the intrinsic tendon regeneration procedure with no presence of ectopic tissue or tumor formation at that time point. Injection of ESCs into an injured tendon of a horse with flexor tendinopathy (induced by collagenase gel—physical defect) led to an important clinical amelioration showing improved structural changes in the animals treated with ESCs over placebo using MRI as well as ultrasound analysis in a double-blind study [43]. Definition of the specific mechanism of amelioration through exogenous cell transplant, local cytokine, or immunological modulation or simply by the stimulation of environmental endogenous horse cells was suggested.

Cultures of these stem cell types begin with low quantities of material (<100 cells) and upscaling the stem cells in an undifferentiated state requires many growth factor supplements. In addition, ethical issues for the use of embryonic stem cells can lead to specific directives and legislation with even restrictive licensing to allow cell manipulation and research. Finally, attention should be given to the phenotype of these cells in tendon to be sure that with their high differentiation potential, they do not shift to undesired cell types. Cell encapsulation and cellular cloning could provide more specific techniques to assure

a better security to the patient and the delivery of a precise cell population. These cells could therefore provide an allogenic (cells not associated with the patient) cell source for product development.

27.5.2 Adult Cell Sources

Adult cell sources have been actively used to achieve tendon regeneration in the last years. In addition to the above-described ESCs of allogenic source, autologous cells have been proposed to assure safety and various adult cell populations have been advanced for tendon repair. Cell choice can be difficult as cell types are variable and each cell type possesses different advantages and disadvantages. The adult sources include the following:

1. Mesenchymal stem cells (MSCs) isolated by centrifugation or simple separation means or with complex culture procedures from bone marrow (BM-MSCs) adipose tissue (ASCs) or more recently from tendon (TSPCs)
2. Induced pluripotent stem cells (IPSCs)
3. Tenocytes or tendon sheath fibroblasts (fully differentiated cell lineages)
4. Blood derivatives such as platelets or plasma-enriched platelet preparations (PRP)

Fresh blood or bone marrow has been used extensively for over 40 years in transplantation medicine. Cells from fresh bone marrow or blood are accessible without much effort from most patients and could present immunological advantages for autologous use, but if this is not possible allotransplantation remains tedious as these transplantations could lead to an acute graft-versus-host disease [44, 45]. Therefore, modern cell-based therapeutic techniques focusing on the use of specific "purified" and culture-expanded lineages could have some advantages. To this end, a company specialized in stem cell therapies has shown that BM-MSCs isolated by density gradient, purified to eliminate non-MSC cell sources and expanded in cell culture, could be used as a secure allogenic cell source in the clinical field [46]. Clinical trials using allogenic BM-MSCs for burns and wounds were realized at the University of Miami as they were awarded a Department of Defense and Armed Forces Institute of Regenerative Medicine Grant (DOD-AFIRM) for tissue repair [47] (Miller School Physician/Scientists Receive $3M Grant to Treat Burn Wounds with Stem Cells, http://med.miami.edu/news/miller-school-physician-scientists-receive-3-million-defense-grant-to-treat Accessed 9/3/17). On the technical side, only 1 of 100,000 cells derived from bone marrow is a stem cell and therefore extensive separation protocols need to be developed with the least manipulation. ASCs are more abundant in adipose tissue (>100-fold compared to BM-MSC) and have better growth capacity in vitro.

Both ASCs and BM-MSCs have been used in animal models for tendon repair and regeneration potential including rats, horses, and sheep [48–52]. In most of the studies, slight improvements have been reported, but unfortunately the regeneration quality was not comparable to that seen intrinsically in fetal tendon [53–55]. MSCs have also been tested in a human clinical trial and the autotransplantation of mononuclear stem cells extracted from the iliac crest in humans has given encouraging results regarding safety and ability to enhance intrinsic tendon regeneration [56]. Nevertheless, there are still doubts about the stability of MSCs to function in a tenogenic manner for tendon repair. For example, the results of a study demonstrated improved biomechanical parameters of the tendon repaired with MSC-collagen composites, but calcifications in the repair site appeared in 28% of the grafted tendons versus 0% in the control group with natural repair [57]. It was also observed in another study that BM-MSCs transplanted into mice could form bone rather than tendon tissue and the authors warned that the use of such cells could potentially lead to calcifications. The development of calcifications remains an important obstacle to be avoided. In contrast to BM-MSCs and ASCs, the use of the more recently discovered tendon stem/progenitor cells (TSPCs) can avoid this problem of calcifications thanks to a phenotype more tuned towards tendon production. Unfortunately,

there is no easily available source for the harvesting of autologous TSPCs and their use thus does not seem possible for the clinics [58].

More recently, induced pluripotent stem cells (iPSCs) were tested in tendon and improved the outcomes of healing in a study [59], and iPSCs are inducible cells and thus could present a risk of dedifferentiation like ESCs or MSCs. Further studies would be needed to better evaluate their potential.

Cell choices from specific tendon environment have also been described such as tenocytes, and tendon sheath fibroblasts [53] and comparisons between ASC and BM-MSC were assessed for their qualities both in vitro and in vivo with all cell types being viable and depositing matrix. Unfortunately, like for TSPCs, there is no easily available source to collect such cells.

Another popular source for tendon regeneration is the "platelet." Platelet-rich plasma (PRP) can be used alone or with biocompatible scaffolds and there is still controversy regarding the real positive effect that this therapeutic agent can obtain most likely due to the highly variable treatments imposed on the biological components to elaborate the formulation to be administrated [60–62] but the isolation and preparation can be done rapidly and inexpensively [63].

27.5.3 Progenitor Cell Sources

Placenta, umbilical cord, amniotic liquid, and amniotic tissues contain various types of fetal progenitor cells that have been proposed for cellular therapies and tissue regeneration [64–66]. The latter cell types, fetal progenitor cells, are known for their qualities to promote scarless tissue repair. Most fetal cell research is based on specific material derived from the first trimester (11–14 weeks of gestation) since tissue-specific cells can be isolated and expanded to form Master and Working Cell Banks for long-term stocking and usage. These programs for clinical research normally are regulated in transplantation programs with strict guidelines or even laws and thus defining the tissue to be considered as an organ donation [67] providing legal structure and authorization in most countries.

Fetal progenitor cells were historically used in the development of the polio vaccine. The Nobel Prize for Medicine was awarded to American immunologists already in 1954, and these same cultured human fetal progenitor cells (known as MRC-5 and WI-38) are still used in contemporary vaccine development today. Following this pioneering work on vaccine production processing, use of fetal tissues or cells progressed mainly within the field of neurology and immunology. Transplantation of fetal neural cells has been used to treat a variety of conditions to date such as Huntington's [68, 69] or Parkinson's disease [70]. Fetal transplants have been used in other neurological situations with spinal cord affections or injuries and with encouraging results for recovery of motor function and reported procedure improvement in terms of security [71–74].

Human fetal liver cells have also been intensively studied and used for more than 30 years already to treat severe immunodeficiency, hematological disorders, and congenital disorders of metabolism [75]. Principally, liver failures and diabetes have been targeted as potential important medical conditions for fetal cell therapy strategies [76–79]. In these studies, human fetal liver cells were successfully isolated to treat end-stage liver disease and were shown to have improved the patients' state significantly within the first 18 months of follow-up. Overall, a better understanding of developmental embryology has helped for substantial technological progress and studies are continuing to reveal important factors that can play crucial roles from various stem and progenitor cell types.

Human fetal skin progenitor cells have been used to treat burns in children and also for chronic wounds such as ulcers in elderly patients [67, 80–82]. With this particular technique, cells developed within the Swiss Transplantation Platform were from one dedicated cell bank which could be expanded and stocked. Vials from the cell bank are able to produce over 35×10^9 three-dimensional biological bandages (~100 cm^2), providing an off-the-shelf cell-based therapy [80–82]. Under the same umbrella of the Transplantation Platform, human fetal bone cells [83] and chondro-progenitor cells [84] have been used as potential regenerative agents for human skeletal tissue, and depending on delivery systems the cells can be used either in injectable techniques for difficult-to-treat areas or on scaffolds for cavity filling [85]. Fetal progenitor tenocytes are also in this Transplantation Platform and have been proposed in an extensive evaluation [86, 87] for their potential use for cellular therapy and for tendon tissue engineering. The specific requirements and evaluation for these specific cells and for the production of a biocompatible neo-tendon are illustrated in Fig. 27.4.

27.5.4 Potential of Human Progenitor Tenocytes for Cell Therapy and Tendon Tissue Engineering Applications

Cell choice and technical specifications that are related to their collection, culture, expansion, storage, and stability are determining factors in successful cell therapies. Therapeutically, the cells should be capable of stimulating high tissue regenerative properties, and they should produce low or no immunological or inflammatory reactions. Figure 27.5 depicts the rapid culture process for a clinical cell bank of human progenitor tenocytes which can be developed in a very short time period from one single-organ donation of Achilles heel tendon (approximately 2 mm^3). All processes for the clinical cells were developed and registered in a Transplantation Platform in Switzerland as of 2007 for tenocytes. The developed process has provided a unique cell population with the associated cell deposits for patent application [88].

Human fetal progenitor tenocytes were isolated from the Achilles tendon of a male 14-week gestation organ donation according to a protocol approved by an ethics committee (University Hospital of Lausanne (CHUV), Ethics Committee Protocol #62/07: 14-week gestation organ donation, registered under the Federal Transplantation Program and its Biobank complying with the laws and regulations). In general,

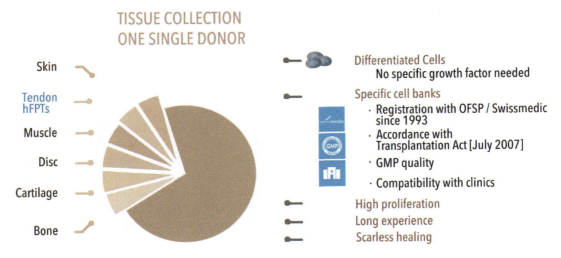

Fig. 27.4 Transplantation platform for progenitor cells. Specific tissues are used for developing cell sources which are stocked as cell banks (Master and Working Cell Banks) under GMP conditions. The processed cell banks are under registration in accordance with the Swiss Transplantation Act, the Department of Public Health, and Swissmedic. This program has existed since 1993 in the Swiss University Hospital of Lausanne and the resulting cells have been shown to have high proliferation characteristics over the years and are known in research to have scarless healing properties for multiple tissues. One cell bank is composed of hFPTs which are described in more details in this chapter

the parental primary culture can be produced in less than 11 days using a simple classic medium as nutrient (Dulbecco's modified Eagle's medium (DMEM) supplemented with 10% fetal bovine serum and 1% glutamine) or with other nonanimal media solutions. Master and Working Cell Bank vials (MCB and WCB) of 5×10^5–10^7 cells each can be stored at −165 °C in the vapor phase of liquid nitrogen for at least 10 years with no incidence on stability (Fig. 27.5). Other progenitor cell types have been shown to remain stable for around 30 years to date (i.e., skin) with cell concentrations from 0.5 to 10 million cells per vial. From the original 2 mm³ of tissue, it would be possible to develop around 100–200 vials of a MCB and each vial could be further expanded to an equivalent quantity of WCB vials. The overall potential of the one, unique organ donation can thus be illustrated for thousands of billions of cells and millions of treatments produced from the clinical cell bank. The cells produced could be delivered in different manners, either in hydrogels or within a scaffold, to provide variable treatment options for repair and regeneration. There can be enough cells produced from one clinical cell bank to be conserved for the long term in liquid nitrogen to assure all experimentation necessary, evaluate formulation, and eventually provide millions of treatments (Fig. 27.5).

27.5.4.1 Progenitor Tenocyte Characterization and Stability

The hFPTs have been characterized extensively for cell growth over passaging to determine that there is no decrease in cell growth rate until late passages, no morphological variations, and no genetic instability seen during expansion and processing. Human fetal progenitor tenocytes are grown easily under 2D culture conditions in tissue culture polystyrene flasks placed in cell culture incubators at 37 °C in a humidified atmosphere containing 5% CO_2. Standard growth medium (DMEM supplemented with 10% fetal bovine serum and 1% glutamine) free of antibiotic supplementation permits a good growth and is changed every 3–4 days. Seeding densities of 3000 and 6000 cells/cm² are generally used for banked cells and lower concentrations can also be

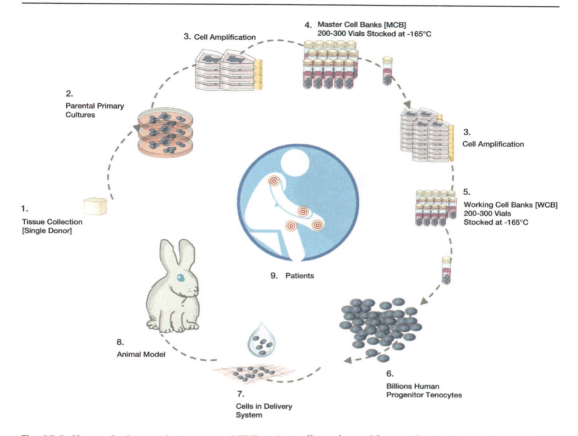

Fig. 27.5 Human fetal progenitor tenocyte (hFPT) cell bank production and use. From one organ donation, cells can be expanded to both Master and Working Cell Banks (MCB, WCB) with the resulting stocked cells in liquid nitrogen. This permits an off-the-shelf availability and the cells can be used for experiments, cell characterization, or formulation in a delivery system to prepare ready-to-use cellular products. A product must have presented in vivo efficacy and safety before it can finally be delivered to patients

used. Higher seeding density implementation would not necessarily accelerate the overall growth of cells. When cell growth and morphology of hFPTs in 2D standard culture conditions are analyzed, no growth difference or morphology anomalies are seen up to passage 12. Mean population doublings and doubling time can be determined for each passage to characterize and define the cell banking procedures for clinical use [86]. Concerning the morphology, hFPTs maintain very similar spindle-shaped morphology (Fig. 27.6) from low passages up to passage 12 with highly aligned configuration when the density increases. It is only in much higher passages that the cells are seen to be slightly larger and there is weaker cellular alignment when the density increases.

In addition, their phenotype is stable throughout the different passages in both 2D and 3D culture with specific tendon marker expression for collagen I, scleraxis, and tenomodulin being present (Fig. 27.6). HFPTs have the ability to form 3D pellets and in such a configuration they are able to deposit extracellular matrix during the creation of micro-tissues.

Population homogeneity and stability can also be evaluated at various passages by conventional karyotype and flow cytometry. Karyotype can be implemented in order to observe potential numerical or structural abnormalities and to identify low-degree mosaic conditions of hFPTs. It can be done at each passage to show that stability remains throughout processing. The surface markers of hFPTs seen with flow cytometry can

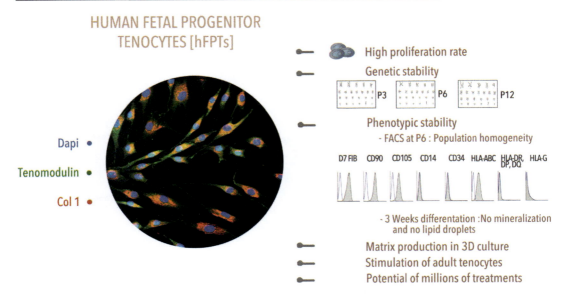

Fig. 27.6 Human fetal progenitor tenocyte (HFPT) characteristics. hFPTs have a high proliferation rate when grown in 2D and they produce the typical substances found in tendon such as collagen I, tenomodulin, or scleraxis. Cells grown in 3D are able to show matrix deposition potential and keep their capacity to secrete the same substances. They present a good genetic and phenotypic stability over passages which can be demonstrated by karyotype, FACS analysis, and differentiation assays. They are also able to stimulate the activity of older adult tenocytes when co-cultured with such a population

show proof of a homogenous cell population. For hFPTs, the overall profile is similar to fibroblastic adherent cells (positive for D7-fib and CD90). Potential contaminating subpopulations such as hematopoietic or endothelial cells (CD34 and CD14) are not present, which is important as it is known that the absence of such cells (for example T lymphocytes) prevents significant graft-versus-host disease reactions [89]. In addition there is no shift in the expression of the surface markers between low and high passages, indicating a good overall stability for banked hFPT cells. It is also interesting that these cells express CD105 as it has been shown to increase the stability of cells in the tenogenic phenotype and a lack of this protein in tenocytes leads to more chondrogeneration in healing tendon than when this protein is present [90].

Other markers tested for hFPTs can show proof of safety. As expected, a lack of MHC class II proteins (HLA-DR, DP, DQ) is shown for these cells while MHC class I (HLA-A, B, C) can be mildly present which is similar to the profile of MSCs [91]. T lymphocytes are known to be responsible in part for allograft rejection through HLA recognition. Therefore, the presence of MHC class I proteins on the surface of a cell could perhaps trigger an immune reaction from CD8+ T cells. However, there is evidence that MSCs, even if they present MHC class I antigens, are able to be grafted without any immunological response noted. Evidence has shown that they possess the ability to inhibit proliferation of T lymphocytes in vitro [91] and are well tolerated in vivo in MHC-mismatched primates [92, 93]. Some products used in patient treatment for wounds and oral mucosa repair have a similar MHC profile. It is the case of the first cover product Apligraft, made with neonatal foreskin cells, which has been shown to be well tolerated by patients [94]. Therefore, the mechanism is not fully understood as tolerance seen with these examples would indicate that some allogenic cell types presenting HLA proteins can be used for transplantation without frank immunosuppression. Fetal progenitor cells most likely have the same mechanism of action and do not illicit a negative immune reaction. The most important demonstration is during gestation, where it has been shown that even if paternal HLA-C is recognized during pregnancy, there is no harm to the fetus [95]. Past clinical trials and clinical experience with severe burn patients have also illustrated an excellent tolerance for fetal skin progenitor cells. Biologic bandages containing these specifically banked cells lead to improved outcomes in burn and wound healing and multiple applications have not initiated immune reactions [80, 81].

Phenotypic stability with osteogenic or adipogenic inducing conditions can be implemented to assure that the hFPTs remain in the "differentiation program" of tenocytes even under severe pressure for modifications by external growth factors. Typical conditions for phenotypic stability studies are represented as follows. An osteogenic induction is usually accomplished over 21 days using a medium composed of alpha-MEM, 10% FBS, 5.97 mM L-glut, 284 μM L-ascorbic acid, 5 mM β-glycerophosphate, and 100 nM dexamethasone. The inducing media is changed every 3–4 days. Cells would then be rinsed with deionized water and fixed in 4% formalin solution for 10 min at room temperature before Von Kossa staining or alizarin red (pH 9) staining to observe the production of mineralized matrix, with the first showing phosphate deposition and the second calcium deposition. An adipogenic induction is also realized over a 21-day period using a medium composed of DMEM, 5.97 mM L-glut, 1% insulin-transferrin-selenite (ITS), 1 μM dexamethasone, 100 μM indomethacin, and 100 μM 3-isobutyl-1-methylxanthine (IBMX). Similar to the method for osteogenic induction, cells are rinsed with deionized water and fixed in 4% formalin solution for 10 min at room temperature but are stained with Oil Red O, to highlight the presence of neutral lipids. Even with heavily induced conditions, the hFPTs remain as the same uniform population which was derived originally. In comparison, BM-MSCs and ASCs change their phenotype and deposit mineral matrix or accumulate lipids under such conditions.

In vitro cell culture models for tendon regeneration can provide general results on dose response. Such models may lead to crucial preclinical information for critical assessment of potential responses of injury in vivo. Importantly, hFPTs have shown the potential to promote adult tenocyte activity in a dose-dependant manner when these two cell types were grown together in a co-culture assay [86].

All of the above data show that the cell source provided by hFPTs is a unique stable cell supply that can be produced under stringent current GMP manufacturing. The possibility to create large cell banks to obtain off-the-shelf reserves would provide an ideal solution for cell therapy and patient treatments. As the cells are all derived from only one organ donation, extensive screening can be implemented and easily adaptable to GMP cell banking conditions. Overall, these cells have a remarkable genetic and phenotypic stability and are able to conserve their tenogenic nature even when high inductive conditions towards other phenotypes are imposed. Finally, their ability to stimulate the activity of adult tenocytes is of importance as it shows a proof of concept that could lead to a more rapid healing in vivo.

27.5.4.2 Formulations with Human Fetal Progenitor Tenocytes in Delivery Systems

Hydrogel Solutions

Human fetal progenitor tenocytes present suitable characteristics for the regeneration and repair of tendon. Nevertheless, a suitable delivery system is also of importance for the delivery of the cells in an appropriate and flexible manner to the injured tissue. A first solution could be proposed using hydrogels and those that are already for clinical use would give a beginning proof of concept for determining the stability of cells in a gel, the biocompatibility, and if the formulation could be stored for some period of time and under what conditions for transport. Commercial HA gels could be easily used to resuspend hFPTs which would allow for good delivery of the cells where no culture medium or growth supplement is used in the formulation in order to make it therapeutically dispensable. Upon analyzing multiple hydrogels, surprisingly different formulations could allow a good survival of hFPT banked cells for up to 3 days when stored at 4 °C (refrigerator stable). In the preparations that were stored for up to 3 days at 4 °C, the recovery of the cells was the same as cells that would have been kept under growth conditions [85]. This parameter is essential when defining final preparations for all of the logistics as it allows good flexibility of the formulation for clinical practice. Parallel to good survival, other parameters tested for such preparations would be a good physical stability of the preparation (with absence of sedimentation) through time and the cells should maintain ability to attach and proliferate (biocompatibility). Moreover, sufficient viscosity without causing cell death would be essential for the gel to remain in place if deposited on a wounded location. Even though hFPTs were able to survive and to recover within multiple commercially available gels, some products portrayed some advantages over others in terms of survival and viscosity (Fig. 27.7).

Matrix Solutions

Hydrogels may help for degenerated areas and small defects, but when facing extensively injured tissue material with sufficient biomechanical properties would be necessary and could permit to deliver the cells correctly and safely for tissue replacement. Therefore, good biocompatibility and sufficient mechanical properties are criteria of high importance among others for various types of scaffolds which can be used in tendon healing. Synthetic scaffolds are a first category of matrix which have been tested for tendon treatment [52] and they present some advantages such as a well-controlled process in formulation and manufacturing and easy adaptation for structural, physical, and mechanical properties. Some disadvantages are their poor biocompatibility and limited integration such as shown with polymers widely used for tendon repair such as polyglycolic acid (PGA) and poly-lactic glycolic acid (PLGA) presenting varying results [96, 97]. Therefore, biological scaffolds are an alternative to synthetic scaffolds. Among them, decellularized extracellular matrix (ECM) is prepared from

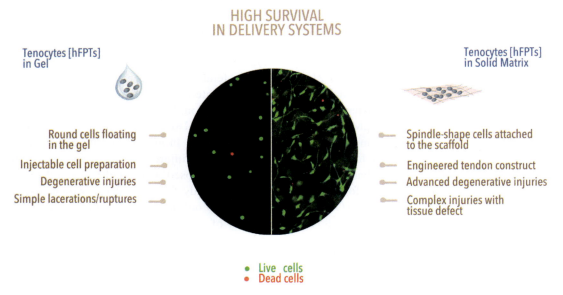

Fig. 27.7 hFPTs in delivery systems. hFPTs can be delivered into hydrogels or solid matrix with excellent survival. The fact to be able to work with different types of delivery systems is an advantage as it is possible to target variable injuries. Standard tendinopathies and simple lacerations or ruptures could be treated by injections. On the other hand, complex degenerative injuries or extensive acute injuries could benefit from engineered tendon constructs to fill the defects and bring biomechanical support

natural tissue and is processed to make them usable for allografts (not the same patient) or xenografts (animal to human use). As cell-associated immunogenic antigens could be involved in the development of an immune reaction by the recipient patient, it is important to have an entire decellularization of the tissue [97]. The overall treatment of the tissue must be strong enough to eliminate the cellular material, but mild enough to conserve the initial structure, composition, and mechanical properties of the tissue when possible.

Thereafter, the decellularization process can begin with the choice of various physical or chemical treatment regimens. Enzymatic treatments are sometimes used in complement to eliminate the unwanted cellular components. All these processes require extensive rinsing to eliminate the products employed to lyse cells as they could have deleterious effect on the seeded cells or even to the patient tissue. Sterilization is finally required to avoid a risk of contamination of the final preparation.

Many biological sources have been proposed and tested. Among them, the small intestinal submucosa (SIS) has probably been the most widely used and it was tested for ligament and tendon repair already since the 1990s [98, 99]. Since this time, there are commercial products available for tendon augmentation when rotator cuff is injured. Most of the products are not developed with tendon tissue but with the following:

1. Porcine SIS (Restore, Orthobiologic Implant, CuffPatch)
2. Human dermis (Graftjacket, Allopatch HD, ArthroFlex)
3. Bovine dermis (TissueMend, Bioblanket)
4. Porcine dermis (Zimmer, Conexa)
5. Equine pericardium (OrthoADAPT)
6. Human fascia lata (AlloPatch)

All of these products have been evaluated for their characteristics and efficacy [38, 100–103]. Unfortunately, the biomechanical characteristics of these products are far weaker to those found in native tendon except for AlloPatch which demonstrates good results [38]. Therefore, beginning

with tendon as a source could prepare the ideal scaffold with better overall mechanical results as tendon structure is indeed very specific and optimally adapted to its role to transmit forces from muscle to bone.

Many techniques have been tested for decellularization and strategies have been proposed to decellularize tendon of various species including rat [104, 105] (Cartmell JA and Dunn MG 2000), porcine [106–108], canine [109], rabbit [110–113], equine [114, 115], and even human [116–118]. Unfortunately, tendons of small animals would not be well adapted to humans and human cadaveric tendons would be of interest if their availability was not limited. Equine superficial digital flexor tendon (SDFT) presents an alternative interesting source due to its large dimension and availability through the food industry. As this industry is highly regulated, high quantities of tissue could be obtained from accredited and traceable sources and thus could be valorized.

This equine tendon source could be sliced to the desired size and decellularized with simple chemical treatments, like detergents that have the capacity to solubilize the cell membranes, and various methods can be proposed with a variety of actions such as

1. Sodium dodecyl sulfate (SDS) as an ionic detergent
2. T-octyl-phenoxypolyethoxyethanol (Triton X-100) as a nonionic detergent
3. Tri-n-butyl phosphate (TBP) as a zwitterionic detergent

SDS and Triton X-100 have been the detergents most widely reported for decellularization of different tissues and organs [119, 120], but TBP has also permitted good results on tendon in different studies [104, 105, 121].

The effectiveness of the decellularization must be controlled along with evaluating the impact of treatment on the structure (Fig. 27.8). Histological staining and DNA measurements can provide quality control measures. Hematoxylin-eosin (HE) is a simple staining procedure of histological sections of tissues allowing the detection of the cell nuclei which

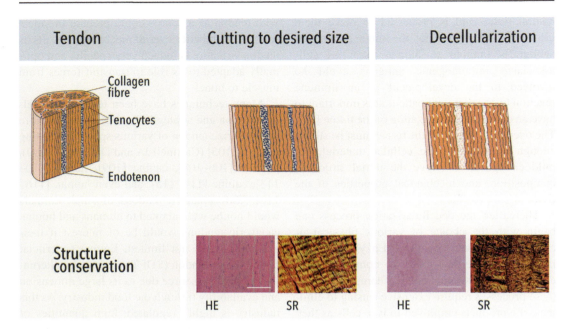

Fig. 27.8 From horse tendon to ECM scaffold. Horse superficial digital flexor tendon (SDFT) is a very interesting starting material to create a matrix as it already presents an organized structure and adequate biomechanical properties. The tendon has first to be cut to the desired dimension before being readily decellularized. Hematoxylin-eosin (HE) staining permits to control the effectiveness of decellularization while sirius red (SR) allows to evaluate the preservation of the structure mainly composed of collagen

are stained in purple and the extracellular matrix in pink. DAPI, with its specific binding to DNA, permits a second control with bright fluorescence where nuclear material is present, but it does not show the structure of tissue. To evaluate the efficiency of decellularization, it is also possible to analyze the specific dosage of DNA in the tissue. Concerning the impact on structure (mainly composed of aligned collagen), the evaluation is made easier by sirius red which interacts relatively specifically with collagen and allows observation of the fibers in colors varying from green to red [122, 123]. The maintenance of the architecture and orientation of the collagen fibers in the matrix are big advantages.

Despite structural assessment efficacy, histological sections would not permit to assess the final impact on mechanical properties of the treated tissues. Biomechanical properties are nevertheless important since the processed tissue would be destined to be grafted and would need to sustain and transmit forces while avoiding ruptures. Thus, an evaluation of the biomechanical properties of the matrix is of paramount importance.

Horse SDFT represents an attractive source to create ECM scaffolds as it has high availability, proper dimensions for sectioning, and good traceability. Detergent solutions have to be optimized as in general SDS treatments can be very efficient in cell removal but can lead to scaffold shrinkage and jellylike appearance (Fig. 27.9). Other detergents can be less efficient in cell removal but can allow a better conservation of the structure and therefore compromises in processing need to be done. Supplementary steps after detergent treatment could also improve the efficiency such as steps with 70% ethanol rinsing. The best treatment should be optimized before up-scaling the specimens to a graftable size and testing biomechanical parameters to have the best rigidity and resistance sufficient to fill gaps in tendinopathic rotator cuffs and to replace autografts in the hand for instance.

27.6 Future Aims and Conclusions

Translational research should provide solutions of off-the-shelf tendon engineered constructs and cellular therapies that could be available in different delivery formulations whenever and wherever needed (Fig. 27.9). The optimal result of new therapeutic strategies would be the achievement of a scar-free regenerated and repaired tissue for the patient.

In translational medicine, it is of utmost importance to first look at the potential clinical use and to always remember that the aim is to create a product which ultimately is beneficial for the patient. Security and efficacy are principal aspects for the final product. All processing steps must be in accordance with the evolving regulatory aspects. To be useful, the product must be available when the medical doctor requires it for the patient and it will be used readily if it can be handled in a simple manner. Moreover, the overall strategy should take into consideration the therapeutic benefit/economic investment ratio for acceptance into the clinic with insurance coverage.

Treatment of tendon injuries remains a challenge even today. Many unsatisfactory outcomes occur with high levels of adhesion and scar tissue found after repair. The proposed use of hFPTs was initiated with the knowledge that fetal tissue heals in a regenerative manner without scar formation and that the process of cell banking of large numbers of cells could be accomplished.

The evaluation of different hyaluronic acid hydrogels for cell delivery would allow injectable cell preparations that would be useful in case of tendinopathies or simple lacerations. Such formulations would permit to deliver the cells to the wounded site in the absence of surgery or around the repair site in the necessity of surgery.

Unfortunately, if the patient has an extended injury, injection of hydrogels would be insufficient and a scaffold or matrix with sufficient mechanical properties would be required to fill the gap. The use of horse SDFT could be a good starting material to create ECM scaffold.

Advancement of technology and interest in tendon regeneration and repair will promote more

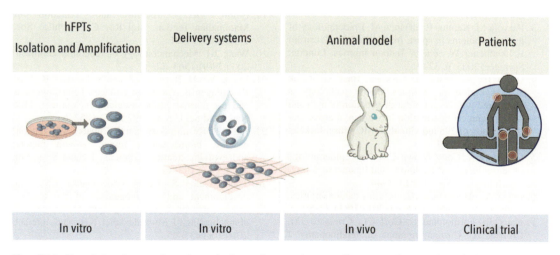

Fig. 27.9 Translational transplantation platform for hFPTs. The target of translational medicine is to develop new therapeutics for the patients. For cellular therapies, everything begins in culture flasks at the laboratory. Once the cells have been amplified, they need to be tested in vitro in order to characterize their capacities. If a cell source demonstrates interesting characteristics, a delivery system is searched. This one is also extremely important as it must allow a good ease of use in both preclinical animal experimentation and clinical use. The developed product combining the cells within the delivery system also has to be tested in vitro. If promising results are obtained, in vivo evaluation can be realized to assure efficacy and safety in animal model. Finally, only after all these steps have been realized with success it is possible to imagine a clinical application

complex cell-based therapies as alternative strategies or parallel strategies to surgical therapeutic techniques. Before reaching the patient, the different strategies developed in the laboratory will have to prove their efficacy and safety in vivo, on animal models. These techniques will then move forward to benefit the patient care and offer improved solutions in repairing, replacing, and restoring the function of damaged tissues and also play a role in the management of the pain associated.

Acknowledgements We would like to thank the Foundation S.A.N.T.E. and Foundation Family Sandoz for financing, in part, our Progenitor Transplantation Program. We also would like to thank the Orthopedic Hospital Foundation for the continued support of our laboratory and students.

References

1. Kannus P, Jozsa L. Histopathological changes preceding spontaneous rupture of a tendon. A controlled study of 891 patients. J Bone Joint Surg Am. 1991;73(10):1507–25.
2. Maffulli N, Barrass V, Ewen SW. Light microscopic histology of Achilles tendon ruptures. A comparison with unruptured tendons. Am J Sports Med. 2000;28(6):857–63.
3. Paavola M, Kannus P, Järvinen M. Epidemiology of tendon problems in sport. In: Maffulli N, Renström P, Leadbetter W, editors. Tendon Injuries. London: Springer; 2005. p. 32–9.
4. Gaida JE, Alfredson H, Kiss ZS, Bass SL, Cook JL. Asymptomatic Achilles tendon pathology is associated with a central fat distribution in men and a peripheral fat distribution in women: a cross sectional study of 298 individuals. BMC Musculoskelet Disord. 2010;11:41.
5. Clayton RA, Court-Brown CM. The epidemiology of musculoskeletal tendinous and ligamentous injuries. Injury. 2008;39(12):1338–44.
6. Egger AC, Berkowitz MJ. Achilles tendon injuries. Curr Rev Musculoskelet Med. 2017;10(1):72–80.
7. Kaux JF, Forthomme B, Goff CL, Crielaard JM, Croisier JL. Current opinions on tendinopathy. J Sports Sci Med. 2011;10(2):238–53.
8. de Jong JP, Nguyen JT, Sonnema AJ, Nguyen EC, Amadio PC, Moran SL. The incidence of acute traumatic tendon injuries in the hand and wrist: a 10-year population-based study. Clin Orthop Surg. 2014;6(2):196–202.
9. Kannus P. Structure of the tendon connective tissue. Scand J Med Sci Sports. 2000;10(6):312–20.

10. Sharma P, Maffulli N. Tendon injury and tendinopathy: healing and repair. J Bone Joint Surg Am. 2005;87(1):187–202.
11. Jozsa L, Kannus P, Balint JB, Reffy A. Three-dimensional ultrastructure of human tendons. Acta Anat (Basel). 1991;142(4):306–12.
12. O'Brien M. Structure and metabolism of tendons. Scand J Med Sci Sports. 1997;7(2):55–61.
13. Voleti PB, Buckley MR, Soslowsky LJ. Tendon healing: repair and regeneration. Annu Rev Biomed Eng. 2012;14:47–71.
14. Benjamin M, Toumi H, Ralphs JR, Bydder G, Best TM, Milz S. Where tendons and ligaments meet bone: attachment sites ('entheses') in relation to exercise and/or mechanical load. J Anat. 2006;208(4):471–90.
15. Apostolakos J, Durant TJ, Dwyer CR, Russell RP, Weinreb JH, Alaee F, Beitzel K, MB MC, Cote MP, Mazzocca AD. The enthesis: a review of the tendon-to-bone insertion. Muscles Ligaments Tendons J. 2014;4(3):333–42.
16. Thomopoulos S, Williams GR, Gimbel JA, Favata M, Soslowsky LJ. Variation of biomechanical, structural, and compositional properties along the tendon to bone insertion site. J Orthop Res. 2003;21(3):413–9.
17. Mienaltowski MJ, Birk DE. Structure, physiology, and biochemistry of collagens. Adv Exp Med Biol. 2014;802:5–29.
18. Nourissat G, Berenbaum F, Duprez D. Tendon injury: from biology to tendon repair. Nat Rev Rheumatol. 2015;11(4):223–33.
19. Elliott DH. Structure and Function of Mammalian Tendon. Biol Rev Camb Philos Soc. 1965;40:392–421.
20. Wang JH. Mechanobiology of tendon. J Biomech. 2006;39(9):1563–82.
21. Liu X, Wu H, Byrne M, Krane S, Jaenisch R. Type III collagen is crucial for collagen I fibrillogenesis and for normal cardiovascular development. Proc Natl Acad Sci U S A. 1997;94(5):1852–6.
22. James R, Kesturu G, Balian G, Chhabra AB. Tendon: biology, biomechanics, repair, growth factors, and evolving treatment options. J Hand Surg Am. 2008;33(1):102–12.
23. Thomopoulos S, Genin GM, Galatz LM. The development and morphogenesis of the tendon-to-bone insertion - what development can teach us about healing. J Musculoskelet Neuronal Interact. 2010;10(1):35–45.
24. Young BB, Gordon MK, Birk DE. Expression of type XIV collagen in developing chicken tendons: association with assembly and growth of collagen fibrils. Dev Dyn. 2000;217(4):430–9.
25. Izu Y, Ansorge HL, Zhang G, Soslowsky LJ, Bonaldo P, Chu ML, Birk DE. Dysfunctional tendon collagen fibrillogenesis in collagen VI null mice. Matrix Biol. 2011;30(1):53–61.

26. Butler DL, Grood ES, Noyes FR, Zernicke RF. Biomechanics of ligaments and tendons. Exerc Sport Sci Rev. 1978;6:125–81.

27. Yoon JH, Halper J. Tendon proteoglycans: biochemistry and function. J Musculoskelet Neuronal Interact. 2005;5(1):22–34.

28. Danielson KG, Baribault H, Holmes DF, Graham H, Kadler KE, Iozzo RV. Targeted disruption of decorin leads to abnormal collagen fibril morphology and skin fragility. J Cell Biol. 1997;136(3):729–43.

29. Ameye L, Young MF. Mice deficient in small leucine-rich proteoglycans: novel in vivo models for osteoporosis, osteoarthritis, Ehlers-Danlos syndrome, muscular dystrophy, and corneal diseases. Glycobiology. 2002;12(9):107R–16R.

30. Svensson L, Aszodi A, Reinholt FP, Fassler R, Heinegard D, Oldberg A. Fibromodulin-null mice have abnormal collagen fibrils, tissue organization, and altered lumican deposition in tendon. J Biol Chem. 1999;274(14):9636–47.

31. Chakravarti S, Magnuson T, Lass JH, Jepsen KJ, LaMantia C, Carroll H. Lumican regulates collagen fibril assembly: skin fragility and corneal opacity in the absence of lumican. J Cell Biol. 1998;141(5):1277–86.

32. Jelinsky SA, Archambault J, Li L, Seeherman H. Tendon-selective genes identified from rat and human musculoskeletal tissues. J Orthop Res. 2010;28(3):289–97.

33. Shukunami C, Takimoto A, Oro M, Hiraki Y. Scleraxis positively regulates the expression of tenomodulin, a differentiation marker of tenocytes. Dev Biol. 2006;298(1):234–47.

34. Murchison ND, Price BA, Conner DA, Keene DR, Olson EN, Tabin CJ, et al. Regulation of tendon differentiation by scleraxis distinguishes force-transmitting tendons from muscle-anchoring tendons. Development. 2007;134(14):2697–708.

35. Docheva D, Hunziker EB, Fassler R, Brandau O. Tenomodulin is necessary for tenocyte proliferation and tendon maturation. Mol Cell Biol. 2005;25(2):699–705.

36. Jozsa L, Lehto M, Kvist M, Balint JB, Reffy A. Alterations in dry mass content of collagen fibers in degenerative tendinopathy and tendon-rupture. Matrix. 1989;9(2):140–6.

37. Aydin A, Topalan M, Mezdegi A, Sezer I, Ozkan T, Erer M, Ozkan S. [Single-stage flexor tendoplasty in the treatment of flexor tendon injuries]. Acta Orthop Traumatol Turc. 2004;38(1):54–9.

38. Aurora A, McCarron J, Iannotti JP, Derwin K. Commercially available extracellular matrix materials for rotator cuff repairs: state of the art and future trends. J Shoulder Elbow Surg. 2007;16(5 Suppl):S171–8.

39. Verdan CE. Primary repair of flexor tendons. J Bone Joint Surg Am. 1960;42:647–57.

40. Killian ML, Cavinatto L, Galatz LM, Thomopoulos S. The role of mechanobiology in tendon healing. J Shoulder Elbow Surg. 2012;21(2):228–37.

41. Gimbel JA, Van Kleunen JP, Williams GR, Thomopoulos S, Soslowsky LJ. Long durations of immobilization in the rat result in enhanced mechanical properties of the healing supraspinatus tendon insertion site. J Biomech Eng. 2007;129(3):400–4.

42. Chen X, Song XH, Yin Z, Zou XH, Wang LL, Hu H, Cao T, Zheng M, Ouyang HW. Stepwise differentiation of human embryonic stem cells promotes tendon regeneration by secreting fetal tendon matrix and differentiation factors. Stem Cells. 2009;27(6):1276–87.

43. Watts AE, Yeager AE, Kopyov OV, Nixon AJ. Fetal derived embryonic-like stem cells improve healing in a large animal flexor tendonitis model. Stem Cell Res Ther. 2011;2(1):4.

44. Ferrara JL, Cooke KR, Teshima T. The pathophysiology of acute graft-versus-host disease. Int J Hematol. 2003;78(3):181–7.

45. Petersdorf EW, Shuler KB, Longton GM, Spies T, Hansen JA. Population study of allelic diversity in the human MHC class I-related MIC-A gene. Immunogenetics. 1999;49(7-8):605–12.

46. Mack GS. Osiris seals billion-dollar deal with Genzyme for cell therapy. Nat Biotechnol. 2009;27(2):106–7.

47. Miller School Physician/Scientists Receive $3M Grant to Treat Burn Wounds with Stem Cells. http://med.miami.edu/news/miller-school-physician-scientists-receive-3-million-defense-grant-to-treat. Accessed 9 Mar 2017

48. Ju YJ, Muneta T, Yoshimura H, Koga H, Sekiya I. Synovial mesenchymal stem cells accelerate early remodeling of tendon-bone healing. Cell Tissue Res. 2008;332(3):469–78.

49. Lacitignola L, Crovace A, Rossi G, Francioso E. Cell therapy for tendinitis, experimental and clinical report. Vet Res Commun. 2008;32(Suppl 1):S33–8.

50. Nixon AJ, Dahlgren LA, Haupt JL, Yeager AE, Ward DL. Effect of adipose-derived nucleated cell fractions on tendon repair in horses with collagenase-induced tendinitis. Am J Vet Res. 2008;69(7):928–37.

51. Smith RK. Mesenchymal stem cell therapy for equine tendinopathy. Disabil Rehabil. 2008;30(20-22):1752–8.

52. Longo UG, Lamberti A, Petrillo S, Maffulli N, Denaro V. Scaffolds in tendon tissue engineering. Stem Cells Int. 2012;2012:517165.

53. Beredjiklian PK, Favata M, Cartmell JS, Flanagan CL, Crombleholme TM, Soslowsky LJ. Regenerative versus reparative healing in tendon: a study of biomechanical and histological properties in fetal sheep. Ann Biomed Eng. 2003;31(10):1143–52.

54. Favata M, Beredjiklian PK, Zgonis MH, Beason DP, Crombleholme TM, Jawad AF, Soslowsky LJ. Regenerative properties of fetal sheep tendon are not adversely affected by transplantation into an adult environment. J Orthop Res. 2006;24(11):2124–32.

55. Young M. Stem cell applications in tendon disorders: a clinical perspective. Stem Cells Int. 2012;2012:637836.

56. Ellera Gomes JL, da Silva RC, Silla LM, Abreu MR, Pellanda R. Conventional rotator cuff repair complemented by the aid of mononuclear autologous stem cells. Knee Surg Sports Traumatol Arthrosc. 2012;20(2):373–7.

57. Awad HA, Boivin GP, Dressler MR, Smith FN, Young RG, Butler DL. Repair of patellar tendon injuries using a cell-collagen composite. J Orthop Res. 2003;21(3):420–31.

58. Bi Y, Ehirchiou D, Kilts TM, Inkson CA, Embree MC, Sonoyama W, Li L, Leet AI, Seo BM, Zhang L, Shi S, Young MF. Identification of tendon stem/progenitor cells and the role of the extracellular matrix in their niche. Nat Med. 2007;13(10):1219–27.

59. Xu W, Wang Y, Liu E, Sun Y, Luo Z, Xu Z, Liu W, Zhong L, Lv Y, Wang A, Tang Z, Li S, Yang L. Human iPSC-derived neural crest stem cells promote tendon repair in a rat patellar tendon window defect model. Tissue Eng Part A. 2013;19(21-22):2439–51.

60. Baksh N, Hannon CP, Murawski CD, Smyth NA, Kennedy JG. Platelet-rich plasma in tendon models: a systematic review of basic science literature. Art Ther. 2013;29(3):596–607.

61. Paoloni J, De Vos RJ, Hamilton B, Murrell GA, Orchard J. Platelet-rich plasma treatment for ligament and tendon injuries. Clin J Sport Med. 2011;21(1):37–45.

62. Taylor DW, Petrera M, Hendry M, Theodoropoulos JS. A systematic review of the use of platelet-rich plasma in sports medicine as a new treatment for tendon and ligament injuries. Clin J Sport Med. 2011;21(4):344–52.

63. Akhundov K, Pietramaggiori G, Waselle L, Darwiche S, Guerid S, Scaletta C, Hirt-Burri N, Applegate LA, Raffoul WV. Development of a cost-effective method for platelet-rich plasma (PRP) preparation for topical wound healing. Ann Burns Fire Disasters. 2012;25(4):207–13.

64. Kadner A, Hoerstrup SP, Tracy J, Breymann C, Maurus CF, Melnitchouk S, Kadner G, Zund G, Turina M. Human umbilical cord cells: a new cell source for cardiovascular tissue engineering. Ann Thorac Surg. 2002;74(4):S1422–8.

65. Kaviani A, Guleserian K, Perry TE, Jennings RW, Ziegler MM, Fauza DO. Fetal tissue engineering from amniotic fluid. J Am Coll Surg. 2003;196(4):592–7.

66. Kaviani A, Perry TE, Barnes CM, Oh JT, Ziegler MM, Fishman SJ, Fauza DO. The placenta as a cell source in fetal tissue engineering. J Pediatr Surg. 2002;37(7):995–9.

67. Applegate LA, Weber D, Simon J-P, Scaletta C, Hirt-Burri N, de Buys RA, Raffoul W, et al. Organ donation and whole-cell bioprocessing in the Swiss fetal progenitor cell transplantation platform. In: Saidi RF, editor. Organ Donation and Organ Donors: Issues, Challenges and Perspectives. Hauppauge NY: Nova Science Publishers, Inc.; 2013. p. 125–48.

68. Rosser AE, Bachoud-Levi AC. Clinical trials of neural transplantation in Huntington's disease. Prog Brain Res. 2012;200:345–71.

69. Schackel S, Pauly MC, Piroth T, Nikkhah G, Dobrossy MD. Donor age dependent graft development and recovery in a rat model of Huntington's disease: histological and behavioral analysis. Behav Brain Res. 2013;256:56–63.

70. Lindvall O. Developing dopaminergic cell therapy for Parkinson's disease--give up or move forward? Mov Disord. 2013;28(3):268–73.

71. Wirth ED 3rd, Reier PJ, Fessler RG, Thompson FJ, Uthman B, Behrman A, Beard J, Vierck CJ, Anderson DK. Feasibility and safety of neural tissue transplantation in patients with syringomyelia. J Neurotrauma. 2001;18(9):911–29.

72. Akesson E, Piao JH, Samuelsson EB, Holmberg L, Kjaeldgaard A, Falci S, Sundström E, Seiger A. Long-term culture and neuronal survival after intraspinal transplantation of human spinal cord-derived neurospheres. Physiol Behav. 2007;92(1-2):60–6.

73. Iwai H, Nori S, Nishimura S, Yasuda A, Takano M, Tsuji O, Fujiyoshi K, Toyama Y, Okano H, Nakamura M. Transplantation of neural stem/progenitor cells at different locations in mice with spinal cord injury. Cell Transplant. 2014;23(11):1451–64.

74. Mothe AJ, Tator CH. Review of transplantation of neural stem/progenitor cells for spinal cord injury. Int J Dev Neurosci. 2013;31(7):701–13.

75. Touraine JL, Raudrant D, Golfier F, Rebaud A, Sembeil R, Roncarolo MG, Bacchetta R, D'Oiron R, Lambert T, Gebuhrer L. Reappraisal of in utero stem cell transplantation based on long-term results. Fetal Diagn Ther. 2004;19(4):305–12.

76. Gridelli B, Vizzini G, Pietrosi G, Luca A, Spada M, Gruttadauria S, Cintorino D, Amico G, Chinnici C, Miki T, Schmelzer E, Conaldi PG, Triolo F, Gerlach JC. Efficient human fetal liver cell isolation protocol based on vascular perfusion for liver cell-based therapy and case report on cell transplantation. Liver Transpl. 2012;18(2):226–37.

77. Zaret KS, Grompe M. Generation and regeneration of cells of the liver and pancreas. Science. 2008;322(5907):1490–4.

78. Khan AA, Shaik MV, Parveen N, Rajendraprasad A, Aleem MA, Habeeb MA, Srinivas G, Raj TA, Tiwari SK, Kumaresan K, Venkateswarlu J, Pande G, Habibullah CM. Human fetal liver-derived stem cell transplantation as supportive modality in the management of end-stage decompensated liver cirrhosis. Cell Transplant. 2010;19(4):409–18.

79. Montanucci P, Pennoni I, Pescara T, Basta G, Calafiore R. Treatment of diabetes mellitus with microencapsulated fetal human liver (FH-B-TPN) engineered cells. Biomaterials. 2013;34(16):4002–12.

80. Hohlfeld J, de Buys RA, Hirt-Burri N, Chaubert P, Gerber S, Scaletta C, Hohlfeld P, Applegate LA. Tissue engineered fetal skin constructs for paediatric burns. Lancet. 2005;366(9488):840–2.

81. De Buys Roessingh AS, Hohlfeld J, Scaletta C, Hirt-Burri N, Gerber S, Hohlfeld P, Gebbers JO, Applegate LA. Development, characterization, and use of a fetal skin cell bank for tissue engineering in wound healing. Cell Transplant. 2006;15(8-9):823–34.

82. Ramelet AA, Hirt-Burri N, Raffoul W, Scaletta C, Pioletti DP, Offord E, Mansourian R, Applegate LA. Chronic wound healing by fetal cell therapy may be explained by differential gene profiling observed in fetal versus old skin cells. Exp Gerontol. 2009;44(3):208–18.

83. Pioletti DP, Montjovent MO, Zambelli PY, Applegate L. Bone tissue engineering using foetal cell therapy. Swiss Med Wkly. 2006;136(35-36):557–60.

84. Darwiche S, Scaletta C, Raffoul W, Pioletti DP, Applegate LA. Epiphyseal chondroprogenitors provide a stable cell source for cartilage cell therapy. Cell Med. 2012;4(1):23–32.

85. Tenorio DM, Scaletta C, Jaccoud S, Hirt-Burri N, Pioletti DP, Jaques B, Applegate LA. Human fetal bone cells in delivery systems for bone engineering. J Tissue Eng Regen Med. 2011;5(10):806–14.

86. Grognuz A, Scaletta C, Farron A, Pioletti DP, Raffoul W, Applegate LA. Stability enhancement using hyaluronic acid gels for delivery of human fetal progenitor tenocytes. Cell Med. 2016;8(3):87–97.

87. Grognuz A, Scaletta C, Farron A, Raffoul W, Applegate LA. Human fetal progenitor tenocytes for regenerative medicine. Cell Transplant. 2016;25(3):463–79.

88. Laurent-Applegate LA. Preparation of parental cell bank from foetal tissue. WO 2013008174 A1. 2013. www.google.com.na/patents/WO2013008174A1?cl=en. Accessed 9 Mar 2017

89. Prentice HG, Blacklock HA, Janossy G, Gilmore MJ, Price-Jones L, Tidman N, Trejdosiewicz LK, Skeggs DB, Panjwani D, Ball S, et al. Depletion of T lymphocytes in donor marrow prevents significant graft-versus-host disease in matched allogeneic leukaemic marrow transplant recipients. Lancet. 1984;1(8375):472–6.

90. Asai S, Otsuru S, Candela ME, Cantley L, Uchibe K, Hofmann TJ, Zhang K, Wapner KL, Soslowsky LJ, Horwitz EM, Enomoto-Iwamoto M. Tendon progenitor cells in injured tendons have strong chondrogenic potential: the CD105-negative subpopulation induces chondrogenic degeneration. Stem Cells. 2014;32(12):3266–77.

91. Le Blanc K, Tammik C, Rosendahl K, Zetterberg E, Ringden O. HLA expression and immunologic properties of differentiated and undifferentiated mesenchymal stem cells. Exp Hematol. 2003;31(10):890–6.

92. Devine SM, Bartholomew AM, Mahmud N, Nelson M, Patil S, Hardy W, Sturgeon C, Hewett T, Chung T, Stock W, Sher D, Weissman S, Ferrer K, Mosca J, Deans R, et al. Mesenchymal stem cells are capable of homing to the bone marrow of non-human primates following systemic infusion. Exp Hematol. 2001;29(2):244–55.

93. Bartholomew A, Sturgeon C, Siatskas M, Ferrer K, McIntosh K, Patil S, Hardy W, Devine S, Ucker D, Deans R, Moseley A, Hoffman R. Mesenchymal stem cells suppress lymphocyte proliferation in vitro and prolong skin graft survival in vivo. Exp Hematol. 2002;30(1):42–8.

94. Streit M, Braathen LR. Apligraf - a living human skin equivalent for the treatment of chronic wounds. Int J Artif Organs. 2000;23(12):831–3.

95. Ober C. HLA and pregnancy: the paradox of the fetal allograft. Am J Hum Genet. 1998;62(1):1–5.

96. Lomas AJ, Ryan CN, Sorushanova A, Shologu N, Sideri AI, Tsioli V, Fthenakis GC, Tzora A, Skoufos I, Quinlan LR, O'Laighin G, Mullen AM, Kelly JL, Kearns S, Biggs M, Pandit A, Zeugolis DI. The past, present and future in scaffold-based tendon treatments. Adv Drug Deliv Rev. 2015;84:257–77.

97. Schulze-Tanzil G, Al-Sadi O, Ertel W, Lohan A. Decellularized tendon extracellular matrix-a valuable approach for tendon reconstruction? Cell. 2012;1(4):1010–28.

98. Ingram JH, Korossis S, Howling G, Fisher J, Ingham E. The use of ultrasonication to aid recellularization of acellular natural tissue scaffolds for use in anterior cruciate ligament reconstruction. Tissue Eng. 2007;13(7):1561–72.

99. Badylak S, Toombs J, Shelbourne K, Hiles M, Lantz G, Van Sickle D. Small intestinal submucosa as an intra-articular ligamentous graft material: a pilot study in dogs. Vet Comp Orthop Traumatol. 1994;7(3):36–40.

100. Badylak SF, Tullius R, Kokini K, Shelbourne KD, Klootwyk T, Voytik SL, Kraine MR, Simmons C. The use of xenogeneic small intestinal submucosa as a biomaterial for Achilles tendon repair in a dog model. J Biomed Mater Res. 1995;29(8):977–85.

101. Derwin KA, Baker AR, Spragg RK, Leigh DR, Iannotti JP. Commercial extracellular matrix scaffolds for rotator cuff tendon repair. Biomechanical, biochemical, and cellular properties. J Bone Joint Surg Am. 2006;88(12):2665–72.

102. Longo UG, Lamberti A, Maffulli N, Denaro V. Tendon augmentation grafts: a systematic review. Br Med Bull. 2010;94:165–88.

103. Barber FA, Herbert MA, Coons DA. Tendon augmentation grafts: biomechanical failure loads and failure patterns. Art Ther. 2006;22(5):534–8.

104. Valentin JE, Badylak JS, McCabe GP, Badylak SF. Extracellular matrix bioscaffolds for orthopaedic applications. A comparative histologic study. J Bone Joint Surg Am. 2006;88(12):2673–86.

105. Deeken CR, White AK, Bachman SL, Ramshaw BJ, Cleveland DS, Loy TS, Grant SA. Method of preparing a decellularized porcine tendon using tributyl phosphate. J Biomed Mater Res B Appl Biomater. 2011;96(2):199–206.

106. Cartmell JS, Dunn MG. Effect of chemical treatments on tendon cellularity and mechanical properties. J Biomed Mater Res. 2000;49(1):134–40.

107. Zhang AY, Bates SJ, Morrow E, Pham H, Pham B, Chang J. Tissue-engineered intrasynovial tendons: optimization of acellularization and seeding. J Rehabil Res Dev. 2009;46(4):489–98.

108. Omae H, Zhao C, Sun YL, An KN, Amadio PC. Multilayer tendon slices seeded with bone marrow stromal cells: a novel composite for tendon engineering. J Orthop Res. 2009;27(7):937–42.

109. Kryger GS, Chong AK, Costa M, Pham H, Bates SJ, Chang J. A comparison of tenocytes and mesenchymal stem cells for use in flexor tendon tissue engineering. J Hand Surg Am. 2007;32(5):597–605.

110. Xing S, Liu C, Xu B, Chen J, Yin D, Zhang C. Effects of various decellularization methods on histological and biomechanical properties of rabbit tendons. Exp Ther Med. 2014;8(2):628–34.

111. Tischer T, Vogt S, Aryee S, Steinhauser E, Adamczyk C, Milz S, Martinek V, Imhoff AB. Tissue engineering of the anterior cruciate ligament: a new method using acellularized tendon allografts and autologous fibroblasts. Arch Orthop Trauma Surg. 2007;127(9):735–41.

112. Burk J, Erbe I, Berner D, Kacza J, Kasper C, Pfeiffer B, Winter K, Brehm W. Freeze-thaw cycles enhance decellularization of large tendons. Tissue Eng Part C Methods. 2014;20(4):276–84.

113. Youngstrom DW, Barrett JG, Jose RR, Kaplan DL. Functional characterization of detergent-decellularized equine tendon extracellular matrix for tissue engineering applications. PLoS One. 2013;8(5):e64151.

114. Huang Q, Ingham E, Rooney P, Kearney JN. Production of a sterilised decellularised tendon allograft for clinical use. Cell Tissue Bank. 2013;14(4):645–54.

115. Pridgen BC, Woon CY, Kim M, Thorfinn J, Lindsey D, Pham H, Chang J. Flexor tendon tissue engineering: acellularization of human flexor tendons with preservation of biomechanical properties and biocompatibility. Tissue Eng Part C Methods. 2011;17(8):819–28.

116. Raghavan SS, Woon CY, Kraus A, Megerle K, Choi MS, Pridgen BC, Pham H, Chang J. Human flexor tendon tissue engineering: decellularization of human flexor tendons reduces immunogenicity in vivo. Tissue Eng Part A. 2012;18(7-8):796–805.

117. Crapo PM, Gilbert TW, Badylak SF. An overview of tissue and whole organ decellularization processes. Biomaterials. 2011;32(12):3233–43.

118. Gilbert TW, Sellaro TL, Badylak SF. Decellularization of tissues and organs. Biomaterials. 2006;27(19):3675–83.

119. Chen J, Xu J, Wang A, Zheng M. Scaffolds for tendon and ligament repair: review of the efficacy of commercial products. Expert Rev Med Devices. 2009;6(1):61–73.

120. Junqueira LCU, Bignolas G, Brentani RR. Picrosirius staining plus polarization microscopy, a specific method for collagen detection in tissue sections. Histochem J. 1979;11(4):447–55.

121. Junqueira LCU, Montes GS, Sanchez EM. The influence of tissue section thickness on the study of collagen by the Picrosirius-polarization method. Histochemistry. 1982;74(1):153–6.

122. Adzick NS, Harrison MR, Glick PL, Beckstead JH, Villa RL, Scheuenstuhl H, Goodson WH 3rd. Comparison of fetal, newborn, and adult wound healing by histologic, enzyme-histochemical, and hydroxyproline determinations. J Pediatr Surg. 1985;20(4):315–9.

123. Calve S, Dennis RG, Kosnik PE 2nd, Baar K, Grosh K, Arruda EM. Engineering of functional tendon. Tissue Eng. 2004;10(5–6):755–61.

Part V

Nerve Regeneration

Current Trends and Future Perspectives for Peripheral Nerve Regeneration

28

Georgios N. Panagopoulos,
Panayiotis D. Megaloikonomos,
and Andreas F. Mavrogenis

28.1 Introduction

Peripheral nerve surgery is annually performed in 100,000 patients in the United States and Europe; the estimated costs for injuries to a median and an ulnar nerve are roughly $70,000 and $45,000, respectively, with 87% of these costs being the result of lost production [1]. Traction-related injury secondary to a motor vehicle accident and lacerations by sharp objects or long bone fractures are the most common mechanisms of injury in civilians [2–4], while blast injuries from explosives or gunshot wounds are the main causes of nerve injury seen in the austere environment and military personnel [5–7].

Advances in microsurgical techniques, clinical management protocols, better understanding of the pathophysiology of nerve injury, and deeper knowledge of the internal topography of peripheral nerves have contributed to a substantial progress in peripheral nerve surgery in the last few decades [3, 8]. Primary, end-to-end epineural microsurgical nerve repair with tension-free coaptation of nerve ends is still considered the gold standard microsurgical treatment. However, this is usually possible only in cases of sharp nerve division with minimal gap and neuronal tissue loss [9, 10]. When tension-free primary repair is not possible for irreducible nerve gaps, autogenous nerve grafting, biological or artificial nerve conduits, or nerve transfers can be considered (Table 28.1) [8, 10]. Binocular loupes or an operating microscope for magnification, fine skin hooks, plastic slings, light clips, malleable retractors, colored background material, as well as appropriate microsurgical instruments for work under the microscope (needle holders, jeweler's forceps, straight or curved microscissors, microsutures, and bipolar diathermy) are available.

Currently, experimental research using in vitro and in vivo animal models [11] has enabled the development of reproducible methods of peripheral nerve regeneration evaluation [12–14], the deeper understanding of the molecular basis of neuronal growth [15], and the acceleration of peripheral nerve regeneration with pharmacological agents or growth factors, stem cell-based therapies (stem cell-derived Schwann cells), bioengineered nerve conduits, and gene therapy. This chapter attempts a brief overview of the basic principles of nerve repair, current concepts, and future perspectives of peripheral nerve regeneration.

G. N. Panagopoulos · P. D. Megaloikonomos
A. F. Mavrogenis (✉)
First Department of Orthopaedics, National and Kapodistrian University of Athens, School of Medicine, Athens, Greece
e-mail: afm@otenet.gr

© Springer Nature Switzerland AG 2019
D. Duscher, M. A. Shiffman (eds.), *Regenerative Medicine and Plastic Surgery*,
https://doi.org/10.1007/978-3-030-19962-3_28

Table 28.1 Advantages and disadvantages of nerve grafts, transfers, and conduits

Type of repair	Advantages	Disadvantages
Nerve autografts	Bridging nerve gap, nonimmunogenic, variety of donor nerves available	Sensory loss, scarring, neuroma formation, donor-site morbidity
Nerve allografts	Availability, bridging nerve gap, no donor-site morbidity	Potential side effects of host immunosuppression
Nerve transfers	No donor-site morbidity, earlier reinnervation due to proximity of donor nerves to target motor endplates	Possible loss of function from donor nerve site, donor muscle no longer an acceptable donor for muscle transfer
Nerve conduits	Availability, no donor-site morbidity, bridging nerve gap, wrapping material (barrier), neurotrophic factor accumulation	Variable outcomes, use limited to short nerve gaps

28.2 Advances in Microsurgical Nerve Repair

Epineural nerve repair entails gross fascicular matching between the proximal and distal nerve ends; correct alignment can be achieved by taking into account the internal nerve topography and fascicles, and the surface epineural blood vessel patterns. Fascicular and grouped fascicular nerve repair can be performed for a more accurate approximation of regenerating axons in nerves with a more consistent motor and sensory topography; these techniques require intraneural dissection and direct matching of fascicular groups, on the expense of more dissection and potential soft-tissue disruption, with more surgical trauma and scarring [9, 16]. Surface vessels, electrical stimulation in the awake patient, and histological staining with acetylcholine esterase and carbonic anhydrase have been used to achieve better matching of nerve axons and fascicular groups [17]. Tissue adhesives, such as fibrin glue, have been employed to supplement or replace sutures. Nerve glue creates a gel-like clot, applied as an adhesive cylinder around the approximated nerve ends to minimize trauma to the nerve ends, and create a possible barrier to invading scar tissue. Material intervening between nerve ends does not seem to block nerve regeneration [18].

Concerns regarding the potential for neuroma formation with epineural and fascicular/group fascicular nerve repair have led to the development of the epineural sleeve techniques [19]. In the original technique, the free edge of the epineurium of the distal nerve stump is picked up gently and rolled back distally. Subsequently, a 2 mm nerve segment of the distal stump is trimmed off, thus creating a 2 mm free segment of epineural sleeve. The epineural sleeve is then anchored to the proximal stump epineurium 2 mm proximal to the coaptation site with two 10-0 sutures placed 180° apart. Finally, the epineural sleeve is pulled over the proximal nerve stump and sutures are knotted down to secure the repair. The same end-to-end repair technique can be employed during nerve grafting. Even though not widely popularized yet, it is believed that this technique creates less tension at the repair site, provides better alignment for the regenerating fascicles, and creates a biological chamber where axoplasmic fluid accumulates, providing a perfect milieu of growth factors for nerve regeneration [20]. Similarly, a novel epineural sheath jacket technique, involving coverage of the proximal nerve stump with a piece of epineurium before transposing it proximally, has been proposed for neuroma management, as an alternative to the traditional muscle burying technique [21].

28.3 Advances in Nerve Grafting

Severe neurotmetic lesions, such as gunshot wounds, or axonotmetic stretch injuries may create a significant gap between the proximal and distal stumps making primary end-to-end nerve repair impossible; in these cases, nerve

28.4 Nerve Autografts

Nerve autografts are considered the gold standard in nerve repair of critical (>3 cm) nerve gaps. They provide appropriate neurotrophic factors and viable Schwann cells for axonal regeneration [8]. They can be single, cable, trunk, interfascicular, or vascularized [23–27]. Motor and sensory fascicles should be properly realigned, based on the internal topography of the nerve stump and nerve autograft [8]. A grouped fascicular repair should be preferred because fashioning of multiple "fascicular fingers" provides a better match and alignment. A rule of thumb on how much tension at the repair site is acceptable suggests that if flexion of a joint is necessary to shorten the distance between the nerve ends, the resulting tension will be too great, and further dissection or a longer graft should be considered. As the graft may shrink slightly, it should be about 15% longer than the gap to bridge. Fibrin glue may be used to reduce the total number of sutures and the likelihood of suture-induced fibrosis.

Donor nerve autografts are usually harvested from expendable sensory nerve sites. The sural nerve is the most common donor nerve providing 30–40 cm of graft followed by the medial and lateral antebrachial cutaneous nerves, the dorsal cutaneous branch of the ulnar nerve, the superficial sensory branch of the radial nerve, the superficial and deep peroneal nerves, the intercostal nerves, and the posterior and lateral cutaneous nerves of the thigh (Table 28.2) [28, 29]. The harvested nerve graft undergoes Wallerian degeneration, but creates a supportive structure for the ingrowing axons, providing mechanical guidance [30]. Limitations of nerve autografting include sacrificing a functioning (usually sensory) nerve for a more important injured nerve (usually motor), sensory loss and scarring at the donor site, size and fascicle mismatch, scarring from suture-induced fibrosis, and potential for neuroma formation and pain at the repair site [31]. Use of a noncritical portion of a proximally injured nerve as an autograft, aiming to reduce donor-site morbidity, has been proposed [8, 32].

28.5 Nerve Allografts

Nerve allografts are readily accessible, without donor-site morbidity, however, at an increased cost. Cold preservation, irradiation, and lyophilization have been used to reduce allograft antigenicity [33–40]. Recently, nerve allografts have been decellularized by a process of chemical detergent, enzyme degradation, and irradiation,

Table 28.2 Common nerve autografts, length of harvesting, and sensory defect

Nerve autograft	Length (cm)	Sensory defect
Sural	30–40	Dorsal aspect of lower leg and lateral foot
Medial antebrachial cutaneous	10–12 (above elbow); 8–10 (below elbow)	Medial forearm
Lateral antebrachial cutaneous	10–12	Lateral forearm
Superficial sensory branch of the radial	25	Radial dorsal hand
Dorsal cutaneous branch of ulnar	4–6	Dorsal/ulnar hand
Posterior interosseous	6	No apparent deficit
Anterior interosseous	6	No apparent deficit
Lateral femoral cutaneous	10–20	Anterolateral thigh
Saphenous	40 cm	Medial lower leg and foot, prepatellar skin
Posterior cutaneous of forearm	2–5	Posterolateral forearm

resulting in an acellular nerve scaffold with no requirements for immunosuppression. Currently, there is only one commercially available decellularized nerve allograft (Avance® Nerve Graft, Axogen, Inc., Alachua, FL). Even though the literature currently supports their use for small diameters (1–2 mm) and short gap lengths (up to 3 cm), many believe that allografts may have the potential for successful repair of a longer nerve gap length [41].

28.6 Nerve Conduits

Biological (autogenous and nonautogenous) and synthetic (absorbable and nonabsorbable) nerve guidance channels (nerve conduits) have been used to overcome the limitations of nerve grafts [42]. A nerve conduit is a tubular structure designed to bridge the gap of a sectioned nerve that is not amenable to primary end-to-end nerve repair, to protect the nerve from the surrounding tissue and scar formation, and to guide the regenerating axons into the distal nerve stump [42]. Nerve conduits guide nerve regeneration by a combined mechanical (conduit lumen and wall) and chemical effect (accumulation of neurotropic and neurotrophic factors), thus favorably conditioning the nerve injury microenvironment [42].

Several types of nerve conduits are currently available as a bridge for small nerve defects (<3 cm), and as a wrapping material around nerve repair (Table 28.3) [43].

Biologic autogenous nerve conduits include arteries, veins, muscle, tendon, and epineural sheath [42, 44]. Biological nonautogenous nerve conduits include type I collagen (NeuraGen®, Integra LifeSciences Co.; NeuroFlex™, Collagen Matrix, Inc.; and others), gelatin (a protein deriving from collagen), silk fibroin, and polysaccharides, such as chitosan, alginate, and agarose hydrogel-based conduits [45–48]. Synthetic absorbable nerve conduits include aliphatic polyester- and copolyester-based conduits such as poly(glycolic acid) (PGA), poly(L-lactic acid) (PLLA), poly(ε-caprolactone-co-lactide) (PLCL), poly(caprolactone) (PCL), and polyvinyl alcohol (PVA) [3, 49]. Synthetic nonabsorbable nerve conduits include silicone and expanded polytetrafluoroethylene (ePTFE or Gore-Tex®)-based conduits [50, 51].

Introduction of luminal additives into conduits has been studied to enhance nerve conduit efficacy [52]; these include cellular components (Schwann cells, bone stromal cells, fibroblasts), structural components (fibrin, laminin, collagen), and neurotrophic factors (FGF, NGF, GGF, CNTF, VEGF, GDNF,

Table 28.3 Material and structure of commercially available nerve conduits

Nerve conduit	Material	Structure	Length (cm)
NeuraGen® (Integra LifeSciences Co)	Collagen type I	Semipermeable, fibrillar collagen structure	2–3
NeuroFlex™ (Collagen Matrix, Inc.)	Collagen type I	Flexible, semipermeable tubular collagen matrix	2.5
NeuroMatrix™ (Collagen Matrix Inc.)	Collagen type I	Semipermeable tubular collagen matrix	2.5
NeuraWrap™ (Integra LifeSciences Co)	Collagen type I	Longitudinal slit in a tubular wall structure	2–4
NeuroMend™ (Collagen Matrix Inc.)	Collagen type I	Semipermeable collagen wrap	2.5–5
Neurotube® (Synovis Micro Companies)	PGA	Absorbable woven PGA mesh tube	2–4
Neurolac™ (Polyganics, BV)	PLCL	PLCL tubular structure	3
Salutunnel™ (Salumedica LCC)	PVA	PVA tubular structure	6.35
AxoGuard® Nerve Connector (AxoGen Inc.)	Extracellular matrix	Porcine intestine submucosa	1–1.5
AxoGuard® Nerve Protector (Axogen Inc.)	Extracellular matrix	Porcine intestine submucosa	2–4

PGA polyglycolic acid, *PLCL* poly(epsilon-caprolactone-co-lactide), *PVA* polyvinyl alcohol

NT-3) [53–69]. More recently, a minimally processed human umbilical cord membrane has been introduced in clinical practice as a surgical solution in peripheral nerve injury (Avive™ Soft Tissue Membrane, Axogen, Inc., Alachua, FL). Its intent is to protect a nerve repair by preventing or minimizing postoperative scar and adhesion formation. Clinical results are still pending [70, 71].

28.7 Advances in Microsurgical Nerve Transfers

A nerve transfer is the surgical coaptation of a healthy nerve donor to an injured nerve [72]. Indications for nerve transfers include brachial plexus injuries, especially avulsion type with long distance from target motor endplates, delayed presentation, segmental loss of nerve function, and very broad zone of injury with dense scarring [73]. Popular goals for reinnervation via nerve transfer are shoulder abduction, scapular stabilization, elbow flexion and extension, and intrinsic hand function [72, 73]. Advantages of nerve transfers include (1) lack of autograft-associated donor-site morbidity, (2) earlier reinnervation from proximity of donor nerves to target motor endplates, (3) neurorrhaphy and dissection in uninjured and unscarred tissue beds, (4) unaltered muscle biomechanics, and (5) faster nerve recovery and motor reeducation. Disadvantages include (1) possible loss of function in the donor nerve site, and (2) the fact that donor muscle is no longer an acceptable donor for muscle transfer [8, 73, 74].

28.8 End-to-Side Nerve Transfer

End-to-side coaptation or neurorrhaphywas initially described in 1873 [8, 75] and revived in the early 1990s [76]. Instead of nerve repair or grafting, the distal stump of the injured nerve is transferred and coaptated to the side of an uninjured donor nerve. The rationale is that collateral axonal sprouting from a healthy nerve can invade the stump of an injured nerve, when the two are sutured together in an end-to-side fashion [77–79]. In 1984, Terzis et al. [80] described the reverse end-to-side neurorrhaphy named the "babysitter procedure." They dissected a normal motor fascicle from the hypoglossal nerve and transferred it to the side of an injured in-continuity facial nerve to gain some reinnervation of the facial musculature [80]. Similar encouraging results have been reported with this technique by other surgeons as well [81].

28.9 Advances in Molecular Biology

Advances in molecular biology have shown that targeting specific steps in molecular pathways such as the PI3K (phosphatidylinositol-3 kinase)/Akt signaling cascade, the Ras-ERK (extracellular signal-regulated kinase) pathway, the cyclic AMP (cAMP)/protein kinase A (PKA), and the Rho–ROK signaling may allow for purposeful pharmacological intervention, potentially leading to a better functional recovery [15]. Small molecules, peptides, hormones, neurotoxins, and growth factors such as erythropoietin (EPO) [82], tacrolimus (FK506) [83], acetyl-L-carnitine (ALCAR) [84], *N*-acetyl cysteine (NAC) [85], ibuprofen [86], melatonin [87], transthyretin [88], and others may improve nerve regeneration by reducing neuronal death and promoting axonal outgrowth [15, 89–93].

Extensive experimental studies have been conducted or are underway regarding the ideal clinical application, timing, and route of administration of FK506 in peripheral nerve regeneration. FK506 is a macrocyclic lactone produced by the bacterium *Streptomyces tsukubaensis*, isolated in 1984 from a soil sample in Tsukuba, Japan [94]. It is an FDA-approved immunosuppressant used to prevent allograft rejection after liver, kidney, and other solid organ transplantation. It has been found to have a 10–100 times stronger immunosuppressant effect than cyclosporine-A (CsA), with fewer side effects [95]. Interestingly, FK506 has been found to enhance nerve regeneration by increasing axonal outgrowth and inducing Schwann cell proliferation

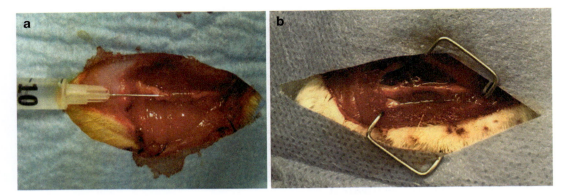

Fig. 28.1 Experimental surgery. (**a**) A 1 cm nerve gap at the left sciatic nerve of a rat treated with a collagen type I nerve conduit augmented with FK506. (**b**) Successful nerve regeneration at 3 months

[39]. The suggested mechanism of neuroregenerative effect is binding to FKBP-12 and inhibiting calcineurin, as well as increasing expression of GAP-43 and TGF β-1 [37]. Particularly promising is the potential use of FK506 in situations in which a sufficient amount of autologous nerve graft is difficult to obtain. This is frequently true in the case of an extensive nerve injury. Regardless of the type of graft used, the rate of recovery is slow. Thus, the availability of a drug to prevent rejection of the graft and speed regeneration would be a great therapeutic advantage. FK506, a drug that has both immunosuppressive and nerve regenerative properties meets both of these criteria. Furthermore, the potential development of a route of administration and a dosing scheme that would minimize its potentially serious systemic side effects would make the use of FK506 even more appealing (Fig. 28.1).

28.10 Stem Cell Therapies

Stem cells differentiated into a Schwann cell-like phenotype can aid in axonal guidance and remyelination, as well as enhance growth factor and extracellular matrix production [96]. Stem cells may be directly transplanted at the injury site (around nerve stumps, at a bridging nerve graft, into the lumen of a nerve conduit, or in a scaffold) at their undifferentiated state, or they can undergo a short period of in vitro differentiation into Schwann cell-like cells (Fig. 28.2) [97].

However, the necessity of stem cell differentiation is currently debated [98]; undifferentiated stem cells seem to demonstrate equally good results, and may also undergo in vivo differentiation in response to local stimuli [53].

Stem cells can be harvested from variable tissues including embryonic, fetal, neural, bone marrow, adipose tissue (Fig. 28.3), skin, hair follicle, and dental pulp [96, 99–116]. Embryonic stem cells (ESCs) were first isolated from human blastocysts in 1998 [117]. They can form derivatives of all three embryonic germ layers, and have great differentiation potential and long-term proliferation capacity [96]. However, their neural differentiation, possible immunogenicity and tumorigenicity, as well as potential ethical controversy are limitations hindering their clinical applications [96]. Nerve stem cells have been first isolated from adult murine brain in the early 1990s [118]. They naturally differentiate into neurons and glial cells, but this almost exclusively occurs during embryogenesis, or in very limited locations in the central nervous system after injury [119]. Bone marrow-derived stem cells (BMSCs) are readily accessible, have no potential ethical concerns, and are more clinically applicable than ESCs and neural stem cells (NSCs). However, harvest is invasive and painful, and their proliferation capacity and differentiation potential are inferior [96].

Particular attention has been given lately to adipose-derived stem cells (ADSCs). Adipose tissue contains a stromal population known as the

28 Current Trends and Future Perspectives for Peripheral Nerve Regeneration

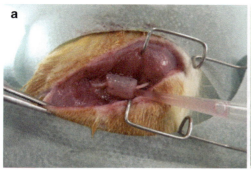 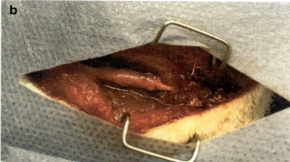

Fig. 28.2 Experimental surgery. (**a**) A 1 cm nerve gap at the left sciatic nerve of a rat treated with a fibrin glue custom-made nerve conduit augmented with Schwann cell-like phenotype ADSCs. (**b**) Successful nerve regeneration at 3 months

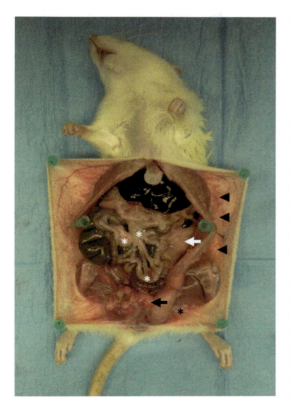

Fig. 28.3 Experimental surgery. Adipose tissue harvesting sites in a rat: subcutaneous (black arrowheads), inguinal (black asterisk), epididymal/testicular (black arrow), mesenteric (white asterisks), and perirenal (white arrow) adipose tissue

stromal vascular fraction (SVF), which can be isolated by centrifugation of collagenase-digested adipose tissue [120]. Cultured SVF can give rise to multipotent precursor cells [121]. In light of their easier harvest, superior stem cell fraction, differentiation potential, and proliferation capacity, ADSCs have now supplanted BMSCs, and are considered the preferred option for preclinical studies [96].

28.11 Gene Therapies

Gene therapy is the introduction of a foreign therapeutic gene into living cells in order to treat a disease [122]. This foreign gene is termed a transgene, and its expression is driven by a promoter. The most efficient way to insert a transgene into a cell is the use of a specially modified viral vector such as herpes simplex, adenovirus, lentivirus, and adeno-associated viral vectors (AAV) [122]. The main targets for gene therapy in peripheral nerve injury are Schwann cells, fibroblasts, and denervated muscle. The aim of gene therapy is to obtain a sort of transcriptional reprogramming so that more neurotrophic factors, cell adhesion or extracellular matrix molecules, and transcription factors are produced.

28.12 Clinical Applications of Research

Low-intensity electrical stimulation has been shown to improve nerve regeneration, probably due to an increased production of BDNF and NGF, and a subsequent enhancement of myelin production [123]. A recent randomized pilot study in human subjects showed that brief low-frequency electrical stimulation immediately following decompression surgery in patients suffering from severe carpal tunnel syndrome resulted in early and complete reinnervation of the thenar muscles [124].

Considerable interest exists in the potential therapeutic value of low-power laser irradiation (laser phototherapy) for peripheral nerve recovery and preservation of denervated muscle. Animal studies on peripheral nerve injury model showed that laser phototherapy decreases scar tissue formation at the injury site, decreases degeneration in corresponding motor neurons of the spinal cord, and significantly increases axonal growth and myelinization [125]. The restoration of the injured peripheral nerve prevents the progression of the muscle atrophy process, and makes functional restoration possible. However, restoring functions in the severe cases or long-term peripheral nerve injury is still difficult as progressive muscle atrophy starts shortly after nerve injury. For this reason, therapeutic solutions can lessen muscle degeneration during the period of nerve recovery, thereby increasing the chances of early recuperation of functional motor activity. It was found that in early stages of muscle atrophy, laser phototherapy might preserve the denervated muscle by maintaining creatine kinase activity and the amount of acetylcholine receptors [126, 127].

The use of olfactory ensheathing cells (OECs) as an adjunct to peripheral nerve regeneration is also promising [128]. OECs are specialized glial cells, which support axons that leave the olfactory epithelium and project through the olfactory nerve system into the olfactory bulb of the central nervous system. OECs are pluripotent, displaying Schwann cell and astrocyte-like properties. They possess the ability to phagocytose degenerating axons, create channels to guide new axon regeneration, and produce a variety of neurotrophic factors, including NGF, BDNF, PDGF, and neuropeptide Y, enhancing injured axon survival [129].

28.13 Rehabilitation After Nerve Repair

As with flexor tendon surgery, rehabilitation is essential to minimize complications and optimize function and quality of life after peripheral nerve injury. An ideal rehabilitation program should be tailored to each patient's needs, therapeutic goals, and stage of recovery. Physical modalities, such as heat, cold, ultrasound, transcutaneous electrical stimulation (TENS), and transdermal iontophoresis, have been shown to be effective in achieving pain control and reducing inflammation [130, 131]. Low-frequency electrical stimulation has been shown to partially prevent muscle atrophy and promote motor recovery [124]. A variety of splints and braces are used to maintain joint integrity. Sensory retraining and desensitization are used to facilitate somatosensory recovery [131]. Promotion of motor recovery and restoration of muscle strength require the development of appropriate exercise programs. Developmental disregard or learned nonuse should be avoided or at least minimized [132].

28.14 Reeducation After Nerve Repair

In the management of peripheral nerve injuries, emphasis is mostly given to the return of protective sensation and motor function, often at the expense of sensory rehabilitation. Sensory reeducation training is essential, allowing patients to learn all over again to interpret in a conscious level the altered neural impulses that reach the central nervous system through the injured hand [133]. In simple words, the brain must be reeducated to understand the new code that comes from the hand after injury repair. Without any training, the simplest touch feels strange. It has been shown that early implementation of strategies for sensory reeducation may maximize the preservation of the hand's cortical sensory map representation, thus improving the recovery of tactile discrimination [134, 135].

References

1. Kelsey JL. Upper extremity disorders: frequency, impact, and cost. New York: Churchill Livingstone; 1997.
2. Kouyoumdjian JA. Peripheral nerve injuries: a retrospective survey of 456 cases. Muscle Nerve. 2006;34(6):785–8.
3. Siemionow M, Brzezicki G. Current techniques and concepts in peripheral nerve repair. Int Rev Neurobiol. 2009;87:141–72.
4. Noble J, Munro CA, Prasad VS, Midha R. Analysis of upper and lower extremity peripheral nerve injuries in a population of patients with multiple injuries. J Trauma. 1998;45(1):116–22.
5. Maricevic A, Erceg M. War injuries to the extremities. Mil Med. 1997;162(12):808–11.
6. Razaq S, Yasmeen R, Butt AW, Akhtar N, Mansoor SN. The pattern of peripheral nerve injuries among Pakistani soldiers in the war against terror. J Coll Physicians Surg Pak. 2015;25(5):363–6.
7. Birch R, Misra P, Stewart MP, Eardley WG, Ramasamy A, Brown K, Shenoy R, Anand P, Clasper J, Dunn R, Etherington J. Nerve injuries sustained during warfare: Part I—Epidemiology. J Bone Joint Surg Br. 2012;94(4):523–8.
8. Ray WZ, Mackinnon SE. Management of nerve gaps: autografts, allografts, nerve transfers, and end-to-side neurorrhaphy. Exp Neurol. 2010;223(1):77–85.
9. Griffin JW, Hogan MV, Chhabra AB, Deal DN. Peripheral nerve repair and reconstruction. J Bone Joint Surg Am. 2013;95(23):2144–51.
10. Kline DG. Nerve surgery as it is now and as it may be. Neurosurgery. 2000;46(6):1285–93.
11. Tos P, Ronchi G, Papalia I, Sallen V, Legagneux J, Geuna S, Giacobini-Robecchi MG. Methods and protocols in peripheral nerve regeneration experimental research: Part I—Experimental models. Int Rev Neurobiol. 2009;87:47–79.
12. Costa LM, Simoes MJ, Mauricio AC, Varejao AS. Methods and protocols in peripheral nerve regeneration experimental research: Part IV—Kinematic gait analysis to quantify peripheral nerve regeneration in the rat. Int Rev Neurobiol. 2009;87:127–39.
13. Navarro X, Udina E. Methods and protocols in peripheral nerve regeneration experimental research: Part III—Electrophysiological evaluation. Int Rev Neurobiol. 2009;87:105–26.
14. Raimondo S, Fornaro M, Di Scipio F, Ronchi G, Giacobini-Robecchi MG, Geuna S. Methods and protocols in peripheral nerve regeneration experimental research: Part II—Morphological techniques. Int Rev Neurobiol. 2009;87:81–103.
15. Chan KM, Gordon T, Zochodne DW, Power HA. Improving peripheral nerve regeneration: from molecular mechanisms to potential therapeutic targets. Exp Neurol. 2014;261:826–35.
16. Grinsell D, Keating CP. Peripheral nerve reconstruction after injury: a review of clinical and experimental therapies. Biomed Res Int. 2014;2014:698256.
17. Isaacs J. Treatment of acute peripheral nerve injuries: current concepts. J Hand Surg Am. 2010;35(3):491–7.
18. Isaacs JE, McDaniel CO, Owen JR, Wayne JS. Comparative analysis of biomechanical performance of available "nerve glues". J Hand Surg Am. 2008;33(6):893–9.
19. Siemionow M, Tetik C, Ozer K, Ayhan S, Siemionow K, Browne E. Epineural sleeve neurorrhaphy: surgical technique and functional results—a preliminary report. Ann Plast Surg. 2002;48(3):281–5.
20. Siemionow MZ, Eisenmann-Klein M. Plastic and reconstructive surgery. Berlin: Springer; 2010.
21. Siemionow M, Bobkiewicz A, Cwykiel J, Uygur S, Francuzik W. Epineural sheath jacket as a new surgical technique for neuroma prevention in the rat sciatic nerve model. Ann Plast Surg. 2017;79(4):377–84.
22. Tsao B, Boulis N, Bethoux F, Murray B. Trauma of the nervous system. In: Daroff R, Fenichel G, Jankovic J, Mazziotta J, editors. Bradley's neurology in clinical practice. 6th ed. Philadelphia: Elsevier; 2012. p. 984–1002.
23. Colen KL, Choi M, Chiu DT. Nerve grafts and conduits. Plast Reconstr Surg. 2009;124(6 Suppl):e386–94.
24. Millesi H, Meissl G, Berger A. The interfascicular nerve-grafting of the median and ulnar nerves. J Bone Joint Surg Am. 1972;54(4):727–50.
25. Taylor GI, Ham FJ. The free vascularized nerve graft. A further experimental and clinical application of microvascular techniques. Plast Reconstr Surg. 1976;57(4):413–26.
26. Terzis JK, Kostopoulos VK. Vascularized nerve grafts and vascularized fascia for upper extremity nerve reconstruction. Hand (NY). 2010;5(1):19–30.
27. Farnebo S, Thorfinn J, Dahlin L. Peripheral nerve injuries of the upper extremity. In: Neligan P, editor. Plastic surgery, vol. 6. Philadelphia: Elsevier; 2013. p. 694–718.
28. Mackinnon SE, Dellon AL. Surgery of the peripheral nerve. New York: Thieme Medical Publishers; 1988.
29. Norkus T, Norkus M, Ramanauskas T. Donor, recipient and nerve grafts in brachial plexus reconstruction: anatomical and technical features for facilitating the exposure. Surg Radiol Anat. 2005;27(6):524–30.
30. Millesi H. Peripheral nerve surgery today: turning point or continuous development? J Hand Surg Br. 1990;15(3):281–7.
31. Moore AM, Ray WZ, Chenard KE, Tung T, Mackinnon SE. Nerve allotransplantation as it pertains to composite tissue transplantation. Hand (NY). 2009;4(3):239–44.
32. Ross D, Mackinnon SE, Chang YL. Intraneural anatomy of the median nerve provides "third web space" donor nerve graft. J Reconstr Microsurg. 1992;8(3):225–32.

33. Anderson PN, Turmaine M. Peripheral nerve regeneration through grafts of living and freeze-dried CNS tissue. Neuropathol Appl Neurobiol. 1986;12(4):389–99.

34. Evans PJ, Mackinnon SE, Best TJ, Wade JA, Awerbuck DC, Makino AP, Hunter DA, Midha R. Regeneration across preserved peripheral nerve grafts. Muscle Nerve. 1995;18(10):1128–38.

35. Lawson GM, Glasby MA. A comparison of immediate and delayed nerve repair using autologous freeze-thawed muscle grafts in a large animal model. The simple injury. J Hand Surg Br. 1995;20(5):663–700.

36. Doolabh VB, Mackinnon SE. FK506 accelerates functional recovery following nerve grafting in a rat model. Plast Reconstr Surg. 1999;103(7):1928–36.

37. Konofaos P, Terzis JK. FK506 and nerve regeneration: past, present, and future. J Reconstr Microsurg. 2013;29(3):141–8.

38. Gold BG. FK506 and the role of immunophilins in nerve regeneration. Mol Neurobiol. 1997;15(3):285–306.

39. Gold BG, Katoh K, Storm-Dickerson T. The immunosuppressant FK506 increases the rate of axonal regeneration in rat sciatic nerve. J Neurosci. 1995;15(11):7509–16.

40. Mackinnon SE, Doolabh VB, Novak CB, Trulock EP. Clinical outcome following nerve allograft transplantation. Plast Reconstr Surg. 2001;107(6):1419–29.

41. Sachanandani NF, Pothula A, Tung TH. Nerve gaps. Plast Reconstr Surg. 2014;133(2):313–9.

42. Konofaos P, Ver Halen JP. Nerve repair by means of tubulization: past, present, future. J Reconstr Microsurg. 2013;29(3):149–64.

43. Kehoe S, Zhang XF, Boyd D. FDA approved guidance conduits and wraps for peripheral nerve injury: a review of materials and efficacy. Injury. 2012;43(5):553–72.

44. Karacaoglu E, Yuksel F, Peker F, Guler MM. Nerve regeneration through an epineural sheath: its functional aspect compared with nerve and vein grafts. Microsurgery. 2001;21(5):196–201.

45. Chen YS, Chang JY, Cheng CY, Tsai FJ, Yao CH, Liu BS. An in vivo evaluation of a biodegradable genipin-cross-linked gelatin peripheral nerve guide conduit material. Biomaterials. 2005;26(18):3911–8.

46. Patel M, Mao L, Wu B, Vandevord PJ. GDNF-chitosan blended nerve guides: a functional study. J Tissue Eng Regen Med. 2007;1(5):360–7.

47. Uebersax L, Mattotti M, Papaloizos M, Merkle HP, Gander B, Meinel L. Silk fibroin matrices for the controlled release of nerve growth factor (NGF). Biomaterials. 2007;28(30):4449–60.

48. Jiang X, Lim SH, Mao HQ, Chew SY. Current applications and future perspectives of artificial nerve conduits. Exp Neurol. 2010;223(1):86–101.

49. Chiono V, Tonda-Turo C, Ciardelli G. Artificial scaffolds for peripheral nerve reconstruction. Int Rev Neurobiol. 2009;87:173–98.

50. Chen YS, Hsieh CL, Tsai CC, Chen TH, Cheng WC, Hu CL, Yao CH. Peripheral nerve regeneration using silicone rubber chambers filled with collagen, laminin and fibronectin. Biomaterials. 2000;21(15):1541–7.

51. Stanec S, Stanec Z. Reconstruction of upper-extremity peripheral-nerve injuries with ePTFE conduits. J Reconstr Microsurg. 1998;14(4):227–32.

52. Yan H, Zhang F, Chen MB, Lineaweaver WC. Conduit luminal additives for peripheral nerve repair. Int Rev Neurobiol. 2009;87:199–225.

53. Chen X, Wang XD, Chen G, Lin WW, Yao J, Gu XS. Study of in vivo differentiation of rat bone marrow stromal cells into Schwann cell-like cells. Microsurgery. 2006;26(2):111–5.

54. Raimondo S, Nicolino S, Tos P, Battiston B, Giacobini-Robecchi MG, Perroteau I, Geuna S. Schwann cell behavior after nerve repair by means of tissue-engineered muscle-vein combined guides. J Comp Neurol. 2005;489(2):249–59.

55. Nilsson A, Dahlin L, Lundborg G, Kanje M. Graft repair of a peripheral nerve without the sacrifice of a healthy donor nerve by the use of acutely dissociated autologous Schwann cells. Scand J Plast Reconstr Surg Hand Surg. 2005;39(1):1–6.

56. Evans GR, Brandt K, Katz S, Chauvin P, Otto L, Bogle M, Wang B, Meszlenyi RK, Lu L, Mikos AG, Patrick CW Jr. Bioactive poly(L-lactic acid) conduits seeded with Schwann cells for peripheral nerve regeneration. Biomaterials. 2002;23(3):841–8.

57. Mosahebi A, Wiberg M, Terenghi G. Addition of fibronectin to alginate matrix improves peripheral nerve regeneration in tissue-engineered conduits. Tissue Eng. 2003;9(2):209–18.

58. Keilhoff G, Goihl A, Stang F, Wolf G, Fansa H. Peripheral nerve tissue engineering: autologous Schwann cells vs. transdifferentiated mesenchymal stem cells. Tissue Eng. 2006;12(6):1451–165.

59. Nakayama K, Takakuda K, Koyama Y, Itoh S, Wang W, Mukai T, Shirahama N. Enhancement of peripheral nerve regeneration using bioabsorbable polymer tubes packed with fibrin gel. Artif Organs. 2007;31(7):500–8.

60. Matsumoto K, Ohnishi K, Kiyotani T, Sekine T, Ueda H, Nakamura T, Endo K, Shimizu Y. Peripheral nerve regeneration across an 80-mm gap bridged by a polyglycolic acid (PGA)-collagen tube filled with laminin-coated collagen fibers: a histological and electrophysiological evaluation of regenerated nerves. Brain Res. 2000;868(2):315–28.

61. Allmeling C, Jokuszies A, Reimers K, Kall S, Choi CY, Brandes G, Kasper C, Scheper T, Guggenheim M, Vogt PM. Spider silk fibres in artificial nerve constructs promote peripheral nerve regeneration. Cell Prolif. 2008;41(3):408–20.

62. Bunting S, Di Silvio L, Deb S, Hall S. Bioresorbable glass fibres facilitate peripheral nerve regeneration. J Hand Surg Br. 2005;30(3):242–7.

63. Xu X, Yee WC, Hwang PY, Yu H, Wan AC, Gao S, Boon KL, Mao HQ, Leong KW, Wang S. Peripheral

nerve regeneration with sustained release of poly(phosphoester) microencapsulated nerve growth factor within nerve guide conduits. Biomaterials. 2003;24(13):2405–12.

64. Wood MD, Moore AM, Hunter DA, Tuffaha S, Borschel GH, Mackinnon SE, Sakiyama-Elbert SE. Affinity-based release of glial-derived neurotrophic factor from fibrin matrices enhances sciatic nerve regeneration. Acta Biomater. 2009;5(4):959–68.

65. Ohta M, Suzuki Y, Chou H, Ishikawa N, Suzuki S, Tanihara M, Suzuki Y, Mizushima Y, Dezawa M, Ide C. Novel heparin/alginate gel combined with basic fibroblast growth factor promotes nerve regeneration in rat sciatic nerve. J Biomed Mater Res A. 2004;71(4):661–8.

66. Zhang J, Lineaweaver WC, Oswald T, Chen Z, Chen Z, Zhang F. Ciliary neurotrophic factor for acceleration of peripheral nerve regeneration: an experimental study. J Reconstr Microsurg. 2004;20(4):323–7.

67. Hobson MI, Green CJ, Terenghi G. VEGF enhances intraneural angiogenesis and improves nerve regeneration after axotomy. J Anat. 2000;197(Pt 4):591–605.

68. Kalbermatten DF, Kingham PJ, Mahay D, Mantovani C, Pettersson J, Raffoul W, Balcin H, Pierer G, Terenghi G. Fibrin matrix for suspension of regenerative cells in an artificial nerve conduit. J Plast Reconstr Aesthet Surg. 2008;61(6):669–75.

69. Fansa H, Schneider W, Wolf G, Keilhoff G. Influence of insulin-like growth factor-I (IGF-I) on nerve autografts and tissue-engineered nerve grafts. Muscle Nerve. 2002;26(1):87–93.

70. AxoGen, Inc. Announces commercial release and first clinical implant of Avive™ soft tissue membrane. 2016. *https://globenewswire.com/news-release/2016/11/21*. Accessed 13 Aug 2017.

71. Avive™ Soft Tissue Membrane. 2016. *http://www.axogeninc.com/products*. Accessed 13 Aug 2017.

72. Lee SK, Wolfe SW. Nerve transfers for the upper extremity: new horizons in nerve reconstruction. J Am Acad Orthop Surg. 2012;20(8):506–17.

73. Tung TH, Mackinnon SE. Nerve transfers: indications, techniques, and outcomes. J Hand Surg Am. 2010;35(2):332–41.

74. Wong AH, Pianta TJ, Mastella DJ. Nerve transfers. Hand Clin. 2012;28(4):571–7.

75. Tos P, Colzani G, Ciclamini D, Titolo P, Pugliese P, Artiaco S. Clinical applications of end-to-side neurorrhaphy: an update. Biomed Res Int. 2014;2014:646128.

76. Viterbo F, Trindade JC, Hoshino K, Mazzoni Neto A. End-to-side neurorrhaphy with removal of the epineural sheath: an experimental study in rats. Plast Reconstr Surg. 1994;94(7):1038–47.

77. Geuna S, Papalia I, Tos P. End-to-side (terminolateral) nerve regeneration: a challenge for neuroscien-

tists coming from an intriguing nerve repair concept. Brain Res Rev. 2006;52(2):381–8.

78. Tos P, Artiaco S, Papalia I, Marcoccio I, Geuna S, Battiston B. End-to-side nerve regeneration: from the laboratory bench to clinical applications. Int Rev Neurobiol. 2009;87:281–94.

79. Brenner MJ, Dvali L, Hunter DA, Myckatyn TM, Mackinnon SE. Motor neuron regeneration through end-to-side repairs is a function of donor nerve axotomy. Plast Reconstr Surg. 2007;120(1):215–23.

80. Terzis JK, Tzafetta K. "Babysitter" procedure with concomitant muscle transfer in facial paralysis. Plast Reconstr Surg. 2009;124(4):1142–56.

81. Davidge KM, Yee A, Moore AM, Mackinnon SE. The supercharge end-to-side anterior interosseous-to-ulnar motor nerve transfer for restoring intrinsic function: clinical experience. Plast Reconstr Surg. 2015;136(3):344e–52e.

82. Elfar JC, Jacobson JA, Puzas JE, Rosier RN, Zuscik MJ. Erythropoietin accelerates functional recovery after peripheral nerve injury. J Bone Joint Surg Am. 2008;90(8):1644–53.

83. Yan Y, Sun HH, Hunter DA, Mackinnon SE, Johnson PJ. Efficacy of short-term FK506 administration on accelerating nerve regeneration. Neurorehabil Neural Repair. 2012;26(6):570–80.

84. Kostopoulos VK, Davis CL, Terzis JK. Effects of acetylo-L-carnitine in end-to-side neurorrhaphy: a pilot study. Microsurgery. 2009;29(6):456–63.

85. Reid AJ, Shawcross SG, Hamilton AE, Wiberg M, Terenghi G. N-acetylcysteine alters apoptotic gene expression in axotomised primary sensory afferent subpopulations. Neurosci Res. 2009;65(2):148–55.

86. Mohammadi R, Hirsaee MA, Amini K. Improvement of functional recovery of transected peripheral nerve by means of artery grafts filled with diclofenac. Int J Surg. 2013;11(3):259–64.

87. Odaci E, Kaplan S. Melatonin and nerve regeneration. Int Rev Neurobiol. 2009;87:317–35.

88. Fleming CE, Saraiva MJ, Sousa MM. Transthyretin enhances nerve regeneration. J Neurochem. 2007;103(2):831–9.

89. Sun HH, Saheb-Al-Zamani M, Yan Y, Hunter DA, Mackinnon SE, Johnson PJ. Geldanamycin accelerated peripheral nerve regeneration in comparison to FK-506 in vivo. Neuroscience. 2012;223:114–23.

90. Udina E, Ladak A, Furey M, Brushart T, Tyreman N, Gordon T. Rolipram-induced elevation of cAMP or chondroitinase ABC breakdown of inhibitory proteoglycans in the extracellular matrix promotes peripheral nerve regeneration. Exp Neurol. 2010;223(1):143–52.

91. Sharma N, Coughlin L, Porter RG, Tanzer L, Wurster RD, Marzo SJ, Jones KJ, Foecking EM. Effects of electrical stimulation and gonadal steroids on rat facial nerve regenerative properties. Restor Neurol Neurosci. 2009;27(6):633–44.

92. Madura T, Kubo T, Tanag M, Matsuda K, Tomita K, Yano K, Hosokawa K. The Rho-associated kinase

inhibitor fasudil hydrochloride enhances neural regeneration after axotomy in the peripheral nervous system. Plast Reconstr Surg. 2007;119(2):526–35.

93. Zuo J, Neubauer D, Graham J, Krekoski CA, Ferguson TA, Muir D. Regeneration of axons after nerve transection repair is enhanced by degradation of chondroitin sulfate proteoglycan. Exp Neurol. 2002;176(1):221–8.

94. Kino T, Hatanaka H, Miyata S, Inamura N, Nishiyama M, Yajima T, Goto T, Okuhara M, Kohsaka M, Aoki H, et al. FK-506, a novel immunosuppressant isolated from a Streptomyces. II. Immunosuppressive effect of FK-506 in vitro. J Antibiot (Tokyo). 1987;40(9):1256–65.

95. Wang MS, Zeleny-Pooley M, Gold BG. Comparative dose-dependence study of FK506 and cyclosporin A on the rate of axonal regeneration in the rat sciatic nerve. J Pharmacol Exp Ther. 1997;282(2):1084–93.

96. Fairbairn NG, Meppelink AM, Ng-Glazier J, Randolph MA, Winograd JM. Augmenting peripheral nerve regeneration using stem cells: a review of current opinion. World J Stem Cells. 2015;7(1):11–26.

97. Dezawa M, Takahashi I, Esaki M, Takano M, Sawada H. Sciatic nerve regeneration in rats induced by transplantation of in vitro differentiated bone-marrow stromal cells. Eur J Neurosci. 2001;14(11):1771–6.

98. Hong SQ, Zhang HT, You J, Zhang MY, Cai YQ, Jiang XD, Xu RX. Comparison of transdifferentiated and untransdifferentiated human umbilical mesenchymal stem cells in rats after traumatic brain injury. Neurochem Res. 2011;36(12):2391–400.

99. Cui L, Jiang J, Wei L, Zhou X, Fraser JL, Snider BJ, Yu SP. Transplantation of embryonic stem cells improves nerve repair and functional recovery after severe sciatic nerve axotomy in rats. Stem Cells. 2008;26(5):1356–65.

100. Lee EJ, Xu L, Kim GH, Kang SK, Lee SW, Park SH, Kim S, Choi TH, Kim HS. Regeneration of peripheral nerves by transplanted sphere of human mesenchymal stem cells derived from embryonic stem cells. Biomaterials. 2012;33(29):7039–46.

101. Craff MN, Zeballos JL, Johnson TS, Ranka MP, Howard R, Motarjem P, Randolph MA, Winograd JM. Embryonic stem cell-derived motor neurons preserve muscle after peripheral nerve injury. Plast Reconstr Surg. 2007;119(1):235–45.

102. Pan HC, Chen CJ, Cheng FC, Ho SP, Liu MJ, Hwang SM, Chang MH, Wang YC. Combination of G-CSF administration and human amniotic fluid mesenchymal stem cell transplantation promotes peripheral nerve regeneration. Neurochem Res. 2009;34(3):518–27.

103. Cheng FC, Tai MH, Sheu ML, Chen CJ, Yang DY, Su HL, Ho SP, Lai SZ, Pan HC. Enhancement of regeneration with glia cell line-derived neurotrophic factor-transduced human amniotic fluid mesenchymal stem cells after sciatic nerve crush injury. J Neurosurg. 2010;112(4):868–79.

104. Gartner A, Pereira T, Alves MG, Armada-da-Silva PA, Amorim I, Gomes R, Ribeiro J, Franca ML, Lopes C, Carvalho RA, Socorro S, Oliveira PF, Porto B, Sousa R, Bombaci A, Ronchi G, Fregnan F, Varejao AS, Luis AL, Geuna S, Mauricio AC. Use of poly(DL-lactide-epsilon-caprolactone) membranes and mesenchymal stem cells from the Wharton's jelly of the umbilical cord for promoting nerve regeneration in axonotmesis: in vitro and in vivo analysis. Differentiation. 2012;84(5):355–65.

105. Guo BF, Dong MM. Application of neural stem cells in tissue-engineered artificial nerve. Otolaryngol Head Neck Surg. 2009;140(2):159–64.

106. Liard O, Segura S, Sagui E, Nau A, Pascual A, Cambon M, Darlix JL, Fusai T, Moyse E. Adult-brain-derived neural stem cells grafting into a vein bridge increases postlesional recovery and regeneration in a peripheral nerve of adult pig. Stem Cells Int. 2012;2012:128732.

107. Johnson TS, O'Neill AC, Motarjem PM, Nazzal J, Randolph M, Winograd JM. Tumor formation following murine neural precursor cell transplantation in a rat peripheral nerve injury model. J Reconstr Microsurg. 2008;24(8):545–50.

108. McKenzie IA, Biernaskie J, Toma JG, Midha R, Miller FD. Skin-derived precursors generate myelinating Schwann cells for the injured and dysmyelinated nervous system. J Neurosci. 2006;26(24):6651–60.

109. Marchesi C, Pluderi M, Colleoni F, Belicchi M, Meregalli M, Farini A, Parolini D, Draghi L, Fruguglietti ME, Gavina M, Porretti L, Cattaneo A, Battistelli M, Prelle A, Moggio M, Borsa S, Bello L, Spagnoli D, Gaini SM, Tanzi MC, Bresolin N, Grimoldi N, Torrente Y. Skin-derived stem cells transplanted into resorbable guides provide functional nerve regeneration after sciatic nerve resection. Glia. 2007;55(4):425–38.

110. Amoh Y, Aki R, Hamada Y, Niiyama S, Eshima K, Kawahara K, Sato Y, Tani Y, Hoffman RM, Katsuoka K. Nestin-positive hair follicle pluripotent stem cells can promote regeneration of impinged peripheral nerve injury. J Dermatol. 2012;39(1):33–8.

111. Martens W, Sanen K, Georgiou M, Struys T, Bronckaers A, Ameloot M, Phillips J. Lambrichts human dental pulp stem cells can differentiate into Schwann cells and promote and guide neurite outgrowth in an aligned tissue-engineered collagen construct in vitro. FASEB J. 2014;28(4):1634–43.

112. Salomone R, Bento RF, Costa HJ, Azzi-Nogueira D, Ovando PC, Da-Silva CF, Zanatta DB, Strauss BE, Haddad LA. Bone marrow stem cells in facial nerve regeneration from isolated stumps. Muscle Nerve. 2013;48(3):423–9.

113. Zhao Z, Wang Y, Peng J, Ren Z, Zhang L, Guo Q, Xu W, Lu S. Improvement in nerve regeneration through a decellularized nerve graft by supplementation with bone marrow stromal cells in fibrin. Cell Transplant. 2014;23(1):97–110.

114. di Summa PG, Kingham PJ, Raffoul W, Wiberg M, Terenghi G, Kalbermatten DF. Adipose-derived stem cells enhance peripheral nerve regeneration. J Plast Reconstr Aesthet Surg. 2010;63(9):1544–52.

115. Sun F, Zhou K, Mi WJ, Qiu JH. Repair of facial nerve defects with decellularized artery allografts containing autologous adipose-derived stem cells in a rat model. Neurosci Lett. 2011;499(2):104–8.

116. Ikeda M, Uemura T, Takamatsu K, Okada M, Kazuki K, Tabata Y, Ikada Y, Nakamura H. Acceleration of peripheral nerve regeneration using nerve conduits in combination with induced pluripotent stem cell technology and a basic fibroblast growth factor drug delivery system. J Biomed Mater Res A. 2014;102(5):1370–8.

117. Thomson JA, Itskovitz-Eldor J, Shapiro SS, Waknitz MA, Swiergiel JJ, Marshall VS, Jones JM. Embryonic stem cell lines derived from human blastocysts. Science. 1998;282(5391):1145–7.

118. Reynolds BA, Tetzlaff W, Weiss S. A multipotent EGF-responsive striatal embryonic progenitor cell produces neurons and astrocytes. J Neurosci. 1992;12(11):4565–74.

119. Paspala SA, Murthy TV, Mahaboob VS, Habeeb MA. Pluripotent stem cells—a review of the current status in neural regeneration. Neurol India. 2011;59(4):558–65.

120. Safford KM, Rice HE. Stem cell therapy for neurologic disorders: therapeutic potential of adipose-derived stem cells. Curr Drug Targets. 2005;6(1):57–62.

121. Guilak F, Lott KE, Awad HA, Cao Q, Hicok KC, Fermor B, Gimble JM. Clonal analysis of the differentiation potential of human adipose-derived adult stem cells. J Cell Physiol. 2006;206(1):229–37.

122. de Winter F, Hoyng S, Tannemaat M, Eggers R, Mason M, Malessy M, Verhaagen J. Gene therapy approaches to enhance regeneration of the injured peripheral nerve. Eur J Pharmacol. 2013;719(1–3):145–52.

123. Gordon T, Brushart TM, Chan KM. Augmenting nerve regeneration with electrical stimulation. Neurol Res. 2008;30(10):1012–22.

124. Gordon T, Amirjani N, Edwards DC, Chan KM. Brief post-surgical electrical stimulation accelerates axon regeneration and muscle reinner-

vation without affecting the functional measures in carpal tunnel syndrome patients. Exp Neurol. 2010;223(1):192–202.

125. Rochkind S. Phototherapy in peripheral nerve regeneration: from basic science to clinical study. Neurosurg Focus. 2009;26(2):E8.

126. Rochkind S, Geuna S, Shainberg A. Phototherapy and nerve injury: focus on muscle response. Int Rev Neurobiol. 2013;109:99–109.

127. Rochkind S, Shainberg A. Protective effect of laser phototherapy on acetylcholine receptors and creatine kinase activity in denervated muscle. Photomed Laser Surg. 2013;31(10):499–504.

128. Rochkind S, Geuna S, Shainberg A. Phototherapy in peripheral nerve injury: effects on muscle preservation and nerve regeneration. Int Rev Neurobiol. 2009;87:445–64.

129. Radtke C, Kocsis JD, Vogt PM. Transplantation of olfactory ensheathing cells for peripheral nerve regeneration. Int Rev Neurobiol. 2009;87:405–15.

130. Ramos LE, Zell JP. Rehabilitation program for children with brachial plexus and peripheral nerve injury. Semin Pediatr Neurol. 2000;7(1):52–7.

131. Scott KR, Ahmed A, Scott L, Kothari MJ. Rehabilitation of brachial plexus and peripheral nerve disorders. Handb Clin Neurol. 2013;110:499–514.

132. Smania N, Berto G, La Marchina E, Melotti C, Midiri A, Roncari L, Zenorini A, Ianes P, Picelli A, Waldner A, Faccioli S, Gandolfi M. Rehabilitation of brachial plexus injuries in adults and children. Eur J Phys Rehabil Med. 2012;48(3):483–506.

133. Mavrogenis AF, Spyridonos SG, Antonopoulos D, Soucacos PN, Papagelopoulos PJ. Effect of sensory re-education after low median nerve complete transection and repair. J Hand Surg Am. 2009;34(7):1210–5.

134. Rosen B, Lundborg G. Sensory re-education after nerve repair: aspects of timing. Handchir Mikrochir Plast Chir. 2004;36(1):8–12.

135. Paula MH, Barbosa RI, Marcolino AM, Elui VM, Rosen B, Fonseca MC. Early sensory re-education of the hand after peripheral nerve repair based on mirror therapy: a randomized controlled trial. Braz J Phys Ther. 2016;20(1):58–65.

The Regeneration of Peripheral Nerves Depends on Repair Schwann Cells

29

Kristján R. Jessen and Rhona Mirsky

29.1 Why Do Peripheral Nerves Regenerate?

It is a well-known dogma in neuroscience that peripheral nerve regenerates after injury, while comparable fiber tracts in the central nervous system CNS), for instance the spinal cord, do not. This difference in regenerative capacity is clearly demonstrated by comparing the outcome of a blunt injury (crush) of the sciatic nerve in rodents to a comparable injury of the spinal cord (contusion or crush). Crushing the sciatic nerve results in vigorous axon growth back towards the denervated targets and breakdown of redundant myelin, followed by re-formation of myelin after regeneration has taken place. The result is that after a relatively short time, 3–4 weeks, essentially normal nerve tissue is restored and sensory and motor functions of the limb are restored. The spinal cord injury on the other hand generates a lesion filled with extracellular matrix and fluid, retraction of injured axons, persistent presence of myelin debris, and no significant regeneration of axons [1–6].

Arguably, the cellular composition of nerve tracts in the peripheral nervous system (PNS) and CNS are quite comparable, both being composed of axons, glial cells, and myelin. Which factors, then, are responsible for this stark difference in regenerative potential? In particular, what goes on in the injured peripheral nerve to make repair potentially so effective? The brief answer is that peripheral nerves regenerate because of the flexible differentiation state of PNS neurons and Schwann cells, and the capacity of these cells to change their phenotype after injury to regenerative states, which are adapted to promote recovery. The neurons of the CNS, in contrast, show only limited capacity to switch to a regenerative mode. Similarly, oligodendrocytes, the cells that form CNS myelin, serve no helpful function for regeneration distal to injury, while the response of astrocytes to injury is hostile to axonal regrowth [2, 7, 8].

In the PNS, nerve damage, whether a nerve cut or crush, readily disturbs the normal differentiation state both of the injured neurons, and of the Schwann cells that ensheath the degenerating axons distal to the injury. In the neurons hundreds of genes, including a large number of transcription factors, change their expression resulting in a change in neuronal function from that of organizing cell-cell signaling to a cell devoted to rebuilding an axon. Classically this response has been referred to as the cell body reaction or the signaling to growth mode switch [8–10]. The myelin and non-myelin (Remak) Schwann cells distal to the injury site undergo a comparable, extensive change in gene expression, which

K. R. Jessen (✉) · R. Mirsky
Department of Cell and Developmental Biology,
University College London, London, UK
e-mail: k.jessen@ucl.ac.uk; r.mirsky@ucl.ac.uk

© Springer Nature Switzerland AG 2019
D. Duscher, M. A. Shiffman (eds.), *Regenerative Medicine and Plastic Surgery*,
https://doi.org/10.1007/978-3-030-19962-3_29

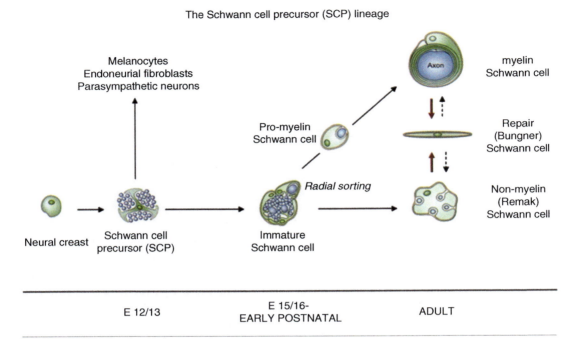

Fig. 29.1 The repair (Bungner) Schwann cell in a developmental context. The diagram shows the repair (Bungner) Schwann cell, and the key stages of Schwann cell development, in addition to other developmental options for the Schwann cell precursor [17]. Arrows indicate developmental and injury related transitions. Black uninterrupted arrows: normal development. Red arrows: the Schwann cell injury response. Stippled arrows: post-repair reformation of myelin and Remak cells. Embryonic dates (E) refer to mouse development (from Jessen et al. [16])

involves some thousands of genes. This reprograms myelin and Remak cells, changing their function from that of maintaining axonal ensheathment and myelin to that of supporting neuronal survival and promoting regeneration (Fig. 29.1) [11–14]. Because this injury-generated Schwann cell phenotype differs from other cells in the Schwann cell lineage, and because it is specialized for supporting nerve repair, we refer to these cells as repair Schwann cells (or Bungner cells since they form guidance tracks for regenerating axons called Bungner bands) [14–16].

29.2 The Schwann Cell Injury Response: Dedifferentiation, Activation, or Reprogramming?

The conversion of myelin and Remak cells to repair cells is referred to as the Schwann cell injury response. While it is clear that this conversion represents a comprehensive phenotypic change in Schwann cells, workers have disagreed about the nature of this change and, in particular, how it relates to the process of Schwann cell development [17]. On the one hand, the Schwann cell injury response has often been viewed as dedifferentiation, namely a reversal of the developmental process resulting in the generation of a cell comparable to that of earlier stages in the Schwann cell lineage [18, 19]. On the other hand, the injury response is frequently referred to as cell activation [20–23]. These terms suggest a contradiction, dedifferentiation indicating the loss of phenotypes, while activation suggests gain of phenotypes. It is now increasingly recognized that these terms, and their implied meaning, can be reconciled. This is due to the emerging realization that the Schwann cell injury response in fact involves both dedifferentiation and activation. After injury, myelin Schwann cells down-regulate their characteristic gene expression profile, representing dedifferentiation. But simultaneously, these cells acquire a number of novel, repair-supporting features; namely they activate a repair program as detailed in the following section, which is a response consistent with the concept of cell activation.

We have pointed out that the formation of repair Schwann cells, involving the combination of dedifferentiation and alternative differentiation, or activation, is not unique, but closely related to injury responses in other tissues, where cells also change phenotype to promote healing. We have termed this type of injury response adaptive cellular reprogramming [24]. The common feature of these cell type conversions is the combination of dedifferentiation and alternative differentiation, or activation, to generate new differentiation states, which compensate for injury or promote healing. For instance, in the newt the conversion of pigment epithelium to lens epithelium after eye injury includes the depigmentation and inhibition of melanogenesis (dedifferentiation) in combination with the gain (activation) of crystallin synthesis to form the transparent lens [25, 26]. In an example from mammals, the conversion of α-cells to β-cells in the pancreatic islets following destruction of β-cells is accomplished by downregulation of glucagon levels (dedifferentiation) and gain (activation) of insulin expression [27, 28]. Other mammalian examples include the conversion of supporting cells to hair cells after ear damage and conversion of fibroblasts to myofibroblasts in skin wounds [24].

Cell type conversions such as those listed above are typically referred to as transdifferentiation or direct reprogramming [29, 30]. The application of these concepts to the Schwann cell injury response reflects a more comprehensive understanding than using the terms dedifferentiation and activation, which refer only to a part of the cell type conversion taking place [24].

29.3 What Takes Place During Schwann Cell Reprogramming?

Most nerve injuries involve not only axonal transection but also the rupture of connective tissue and nerve sheaths resulting in a gap between the proximal and distal end of the nerve. If the gap is small, the proximal and distal stumps are reattached by suturing or gluing, while the repair of larger gaps may be attempted by insertion of artificial tissue bridges or inserted segments obtained from another nerve. Clinical issues associated with this nerve injury region include how to maximize the number of axons that succeed in crossing from the proximal to the distal stump, and how to prevent them from growing to the wrong targets (misrouting).

Having crossed the injury region, axons are faced with the task of regeneration along the distal stump to reach their target tissues, a process that can take many months depending on the proximal/distal location of the injury. Successful axonal growth through the distal stump takes place along the repair Schwann cells, which have now organized to form cellular columns, often referred to as regeneration tracks or Bungner bands, inside the basal lamina tubes, which formerly housed the myelinated or Remak fibers of uninjured nerves. We will now describe the generation of these repair-supportive cells from the Schwann cells of uninjured nerves in more detail. For simplicity this process will be discussed in terms of myelin Schwann cells only; similar principles are likely to hold for non-myelin (Remak) cells.

The Schwann cell injury response has two principal components. One of these is the reversal of myelin differentiation. This involves transcriptional downregulation of the central myelin transcription factor Egr2 (Krox20), enzymes of cholesterol synthesis, structural proteins such as myelin protein zero (MPZ), myelin basic protein (MBP), and membrane-associated proteins like myelin-associated glycoprotein (MAG) and periaxin [18, 19]. Conversely, the cells upregulate molecules that characterize pre-myelinating Schwann cells in developing nerves (immature Schwann cells) including L1, NCAM, p75NTR, and glial fibrillary acidic protein (GFAP) [18, 19].

The other, and functionally more important, part of the injury response is the activation of a group of repair-supportive phenotypes. This repair program includes a number of components: (1) Upregulation of neurotrophic factors and other proteins that support the survival of injured neurons and accelerate axonal regeneration, such as GDNF, artemin, BDNF, NT3, NGF, VEGF, erythropoietin, pleiotrophin, p75NTR, and N-cadherin [5, 18, 31–34]. (2) Activation of an innate immune response, including the upregulation of cytokines including TNFα, LIF Il-1α, Il-1β, LIF, and MCP-1 [35, 36]: This promotes interaction between repair cells and immune cells and recruitment of macrophages to the nerve. This immune response is likely to promote nerve regeneration through more than one mechanism. In addition to activating macrophages, cytokines such as Il-6 and LIF can also act on neurons to promote axonal regeneration [37–39]. In addition, invasion of macrophages into nerves and ganglia provides a secondary source of cytokines and stimulates the formation of blood vessels in the distal stump [40–42]. Macrophages also cooperate with Schwann cells to degrade myelin debris that potentially inhibits axon growth during myelin clearance (see further below) [36, 43]. (3) Reorganization of cell morphology, involving two- to threefold elongation of myelin cells and often elaboration of long parallel cellular processes (Fig. 29.2) [44]: This allows repair cells to align in a partly overlapping fashion to form compact cellular columns (Bungner bands). They act as obligatory regeneration tracks for regenerating axons, providing essential substrate and guidance cues, which allow regenerating axons to reach their targets [45]. (4) Activation of autophagy for myelin breakdown [46].

The dedifferentiation and repair programs described above reprogram Schwann cells of uninjured nerves to cells specialized to support regeneration. It is important or note that, just like the phenotype of myelin and Remak cells, the phenotype of these repair (Bungner) Schwann cells remains unstable and dependent on environmental signals. Thus, in regenerated nerves, when repair cells are redundant, they reassociate with axons and return to the myelin and Remak phenotypes seen before injury. The repair cells are therefore a transient population, produced on demand and present only as long as needed [14].

29.4 Repair Cells Activate Myelin Autophagy to Break Down Redundant Myelin

In contrast to CNS fiber tracts, where growth-inhibiting myelin remains for a long time, injured peripheral nerves quickly clear the redundant myelin sheaths [41, 43, 45, 47–49]. While Schwann cell and macrophage phagocytosis play a part, the initial step in this process is the activation of myelin autophagy. Therefore, during Schwann cell reprogramming, myelin cells not only switch off myelin maintenance, but also switch on an intracellular pathway for myelin destruction.

At the initiation of myelin clearance is the activation of an actin-dependent process, which divides myelin sheaths into oval-shaped intracellular segments, which break down into smaller fragments [50]. We find that these fragments are delivered to lysosomes for digestion by a selective, mTOR-independent autophagy, myelinophagy, which is strongly activated in Schwann cells of injured nerves (Fig. 29.3) [46]. This process is distinct from the classical, mTOR-dependent starvation autophagy mechanisms [46, 51]. The identification of autophagy as a major mechanism for destruction of myelin by Schwann cells

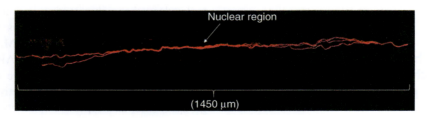

Fig. 29.2 The structure of repair Schwann cells. An example of a branched repair Schwann cell in the distal stump (4 week transected tibial nerve without re-innervation (adapted from Gomez-Sanchez et al. [44])

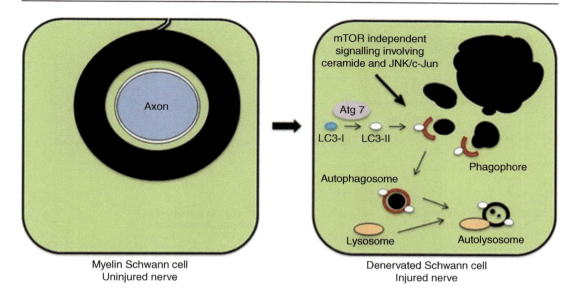

Fig. 29.3 Outline of myelinophagy. The left hand diagram depicts a transverse section through a myelin Schwann cell in an uninjured nerve. The myelin sheath is in direct continuity with the Schwann cell membrane and a component of the Schwann cell. The right hand diagram depicts a myelin Schwann cell after nerve injury and axonal degeneration. The myelin sheath has broken up into myelin ovoid and smaller fragments lying in the Schwann cell cytoplasm. The proposed role of autophagy in digesting these fragments is illustrated (from Gomez-Sanchez et al. [46])

is consistent with the widely held notion that autophagy is a central mechanism by which cells digest their own components, organelles, and large macromolecular complexes.

The existence of a pathway in Schwann cells which, if activated, will destroy myelin is obviously a potential hazard. Future work will show whether this pathway is liable to be inappropriately activated, and whether this plays a part in demyelinating disease. Recent observations indicate that this may be the case, since in a mouse model of CMT1A, the most common hereditary demyelinating neuropathy in man, uninjured nerves show evidence of enhanced autophagy activation [46]. Classical, mTOR-dependent starvation autophagy has also been shown in Schwann cells, where it is beneficial, helping to degrade protein aggregates that form in some myelin mutants [52].

It is intriguing that in the transected optic nerve, where myelin is not degraded after injury in line with that found in the rest of the CNS, there is little evidence of the activation of autophagy [46].

29.5 c-Jun Has a Major Role in Repair Schwann Cells

In uninjured nerves, the transcription factor c-Jun is expressed at low levels, although it is detectable in nuclei of both myelin, but especially Remak, cells [15, 53–55]. In the Schwann cells of injured nerves, c-Jun levels rise rapidly and can become 80–100-fold higher than the levels seen before injury (Fig. 29.4) [56, 57]. The functional significance of this became clear when nerve regeneration was shown to be severely affected in c-Jun cKO mice, in which c-Jun has been genetically inactivated in Schwann cells selectively [15]. Schwann cell development is not affected in these mice. But axonal regeneration and functional recovery after injury are strikingly compromised or absent. This regeneration failure reflects the key role of c-Jun in the Schwann cell injury response, where c-Jun activation is required for the normal execution of both the dedifferentiation and repair programs [15].

During the dedifferentiation of myelin cells, c-Jun is needed for the normal suppres-

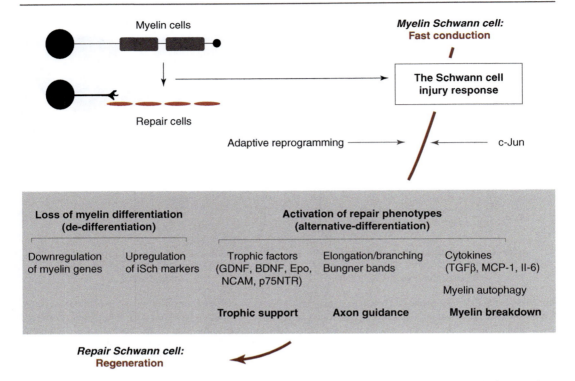

Fig. 29.4 c-Jun promotes the conversion of myelin cells to repair cells. Distal to nerve injury, the down-regulation of the myelin program and activation of the repair program combine to convert myelin cells of unperturbed nerves to the repair Schwann cells found in the distal stump of injured nerves. These cells are designed to keep injured neurons alive, organize myelin breakdown, and promote axon growth and to guide them back to their targets. c-Jun acts as a global amplifier of the repair Schwann cell phenotype

sion of myelin genes, including the *MPZ* and *MBP* genes and the gene encoding the pro-myelin transcription factor Egr2 (Krox20). The ability of c-Jun to downregulate Egr2 (Krox20) and other myelin factors, which was discovered before its importance for regeneration was realized, gave rise to the notion that c-Jun, and a number of other transcriptional regulators, such as Notch, Sox2, Pax3, and Id4, functioned as negative regulators of myelination [19, 53, 58].

c-Jun is also essential for the normal activation of the repair program [15, 31]. In c-Jun cKO mice, the normal upregulation of factors involved in regeneration and neuronal survival fails in the Schwann cells distal to injury. This includes GDNF, artemin and BDNF, p75NTR, and N-cadherin. GDNF and artemin have also been shown to be direct targets of c-Jun [31]. A central role for Schwann cell c-Jun activation in supporting the survival of injured neurons is shown by the observation that in c-Jun cKO mice substantial numbers of DRG sensory neurons and facial motoneurons die after sciatic and facial nerve injury, respectively. The nerves in c-Jun cKO mice also show long-term delay in myelin clearance, reflecting the role of c-Jun in stimulating myelin autophagy [46]. The regeneration tracks (Bungner bands) in c-Jun cKO mice are structurally disorganized, and in culture c-Jun is necessary for what has become known as the "typical" narrow, bi/tripolar Schwann cell morphology, since c-Jun-negative cells tend to be flattened and sheet-forming. Evidence is emerging that epigenetic mechanisms such as histone methylation state and miRNA also take part in the activation of the repair program, since demethylation of H3K27 and downregulation of key miRNAs have been implicated in the activation of important injury genes including Shh, Igfbp2, Olig1, and GDNF [59, 60].

A number of other signaling mechanisms have been implicated in controlling the Schwann cell injury response. This includes Notch [61] and the ERK1/2 MAPK signaling pathway, which may be particularly important for activation of cytokines and macrophage recruitment [62–67]. The other main MAPK signaling pathways, p38 and JNK, are also activated by nerve injury and promote dedifferentiation of myelin cells [53, 68, 69]. Inappropriate activation of Notch or Raf/ERK in Schwann cells is sufficient to cause demyelination, even in uninjured nerves [61, 66].

29.6 c-Jun and Demyelinating Neuropathy

In human nerves, as in rodents, c-Jun is found at low levels in normal Schwann cells. c-Jun is also upregulated in various human neuropathic conditions that do not involve nerve cut or crush [70]. In some of these cases, c-Jun may take part in promoting demyelination, since c-Jun has the potential to downregulate myelin genes as discussed earlier. This is supported by the finding that near-30-fold elevation of c-Jun, genetically engineered in uninjured nerves, results in dysmyelination [71]. c-Jun can therefore be considered a candidate gene for causing demyelinating pathology. In contrast, other studies show that modest c-Jun elevation in Schwann cells of distressed nerves can be beneficial and provide increased neurotrophic support. Thus c-Jun is elevated in uninjured nerves of a mouse model of the most common human genetic demyelinating neuropathy CMT1A [55]. If c-Jun is genetically inactivated in these CMT1A mice, they exhibit distal sensory axonopathy and deterioration in sensory-motor performance. In another mouse model human genetic demyelinating disease, CMT1X, Schwann cells of uninjured nerves also show c-Jun elevation, and a c-Jun target gene, GDNF, a factor that supports neuronal survival, is also elevated [54]. Further, in a mouse mutant involving inactivation of the LKB1 kinase in Schwann cells and showing axonal damage without overt demyelination, c-Jun protein is elevated

as well as c-Jun-associated injury genes including GDNF and Shh [72].

In both the CMT1A and CMTX mouse mutants, the c-Jun elevation is seen in the nuclei of Schwann cells that retain myelin differentiation. Further, we have found that a mouse engineered to show about sevenfold overexpression of c-Jun protein in Schwann cells nevertheless has relatively normal myelin sheaths and nerve architecture [71].

Taken together these observations raise several points. First, high c-Jun elevation can damage myelin sheaths, and a factor in demyelinating disease. Second, more modest but significant c-Jun elevation is compatible with myelin differentiation and does not cause demyelination. Third, in uninjured nerves of mice which are distressed by genetic disease, such as the CMT1A of CMTX mice, Schwann cells respond by relatively modest c-Jun activation, which is low enough to be compatible with the myelin differentiation, but high enough to activate repair-related genes, representing a graded neuroprotective Schwann cell response to nerve distress that does not involve the overt generation of the repair phenotype associated with nerve injury. Even in nerves engineered to show high c-Jun expression we have not seen evidence of tumor formation [71].

29.7 The Role of Repair Cells in Regeneration Failures

Although rodent nerves often regenerate impressively and human nerves also show clear capacity for regeneration, the clinical outcomes after nerve injury are frequently poor, and in humans nerve injuries remain an important problem. There are a number of reasons for these regeneration failures, including target atrophy, neuronal death, axonal misrouting, and difficulty of getting sufficient axon numbers across the injury site into the distal stump. Two further issues are directly relevant to the repair Schwann cells discussed here. One is the chronic denervation of repair cells. The other is aging.

In regenerating nerves of humans and other larger animals, the more distal Schwann cells often remain chronically denervated for many months while axons slowly elongate through the nerve. During this time, the axon-free distal nerve stump gradually loses its repair-supportive capacity as repair Schwann cells deteriorate, downregulate the expression of repair genes, and reduce in number. This decline is one of the key reasons for regeneration failure in humans [73, 74]. Age-dependent deterioration of regeneration capacity is also well documented both in humans and in other animals, and rodent studies indicate that this is for the most part due to reduced activation of the repair Schwann cell phenotype [75].

Since the fading of the repair cell phenotype during chronic denervation and subdued repair cell activation in older animals are major causes of deficient nerve repair, the mechanisms that promote and sustain the expression of the repair phenotype are attractive targets for interventions to improve regeneration. In this context, it is interesting that levels of c-Jun, a global amplifier of repair cell properties as explained earlier, gradually decline during chronic denervation. Similarly c-Jun activation in older animals is reduced compared with that seen in younger animals. Furthermore, prevention of impaired c-Jun levels after chronic denervation and in injured nerves of older animals is sufficient to correct the regeneration deficits seen in these conditions [76]. The STAT3 pathway is also involved in maintaining the repair phenotype during chronic denervation [77] and additional pathways that regulate repair cells will undoubtedly be identified in future studies. The targeting of these molecular mechanisms is likely to prove an effective way to develop therapeutics for improving the outcome of nerve injuries [78].

References

1. Beattie MS, Bresnahan JC, Komon J, Tovar CA, Van Meter M, Anderson DK, Faden AI, Hsu CY, Noble LJ, Salzman S, Young W. Endogenous repair after spinal cord contusion injuries in the rat. Exp Neurol. 1997;148:453–63.

2. Vargas ME, Barres BA. Why is Wallerian degeneration in the CNS so slow? Annu Rev Neurosci. 2007;30:153–79.
3. Plemel JR, Duncan G, Chen K-W, Shannon C, Park S, Sparling JS, Tetzlaff W. A graded forceps crush spinal cord injury model in mice. J Neurotrauma. 2008;25:350–70.
4. Glenn TD, Talbot WS. Signals regulating myelination in peripheral nerves and the Schwann cell response to injury. Curr Opin Neurobiol. 2013;23:1041–8.
5. Scheib J, Höke A. Advances in peripheral nerve regeneration. Nat Rev Neurol. 2013;9:668–76.
6. Brosius Lutz A, Barres BA. Contrasting the glial response to axon injury in the central and peripheral nervous systems. Dev Cell. 2014;28:7–17.
7. Bradke F, Fawcett JW, Spira ME. Assembly of a new growth cone after axotomy: the precursor to axon regeneration. Nat Rev Neurosci. 2012;13:183–93.
8. Doron-Mandel E, Fainzilber M, Terenzio M. Growth control mechanisms in neuronal regeneration. FEBS Lett. 2015;589:1669–77.
9. Blesch A, Lu P, Tsukada S, Alto LT, Roet K, Coppola G, Geschwind D, Tuszynski MH. Conditioning lesions before or after spinal cord injury recruit broad genetic mechanisms that sustain axonal regeneration: superiority to camp-mediated effects. Exp Neurol. 2012;235:162–73.
10. Fu SY, Gordon T. The cellular and molecular basis of peripheral nerve regeneration. Mol Neurobiol. 1997;14:67–116.
11. Nagarajan R, Le N, Mahoney H, Araki T, Milbrandt J. Deciphering peripheral nerve myelination by using Schwann cell expression profiling. Proc Natl Acad Sci U S A. 2002;99:8998–9003.
12. Bosse F, Hasenpusch-Theil K, Küry P, Müller HW. Gene expression profiling reveals that peripheral nerve regeneration is a consequence of both novel injury-dependent and reactivated developmental processes. J Neurochem. 2006;96:1441–57.
13. Barrette B, Calvo E, Vallières N, Lacroix S. Transcriptional profiling of the injured sciatic nerve of mice carrying the Wld(S) mutant gene: identification of genes involved in neuroprotection, neuroinflammation, and nerve regeneration. Brain Behav Immun. 2010;24:1254–67.
14. Jessen KR, Mirsky R. The repair Schwann cell and its function in regenerating nerves. J Physiol. 2016;594:3521–31.
15. Arthur-Farraj PJ, Latouche M, Wilton DK, Quintes S, Chabrol E, Banerjee A, Woodhoo A, Jenkins B, Rahman M, Turmaine M, Wicher GK, Mitter R, Greensmith L, Behrens A, Raivich G, Mirsky R, Jessen KR. c-Jun reprograms Schwann cells of injured nerves to generate a repair cell essential for regeneration. Neuron. 2012;75:633–47.
16. Jessen KR, Lloyd AC MR. Schwann cells: development and role in nerve repair. Cold Spring Harb Perspect Biol. 2015;7(7):a020487.

17. Jessen KR, Mirsky R. The origin and development of glial cells in peripheral nerves. Nat Rev Neurosci. 2005;6:671–82.
18. Chen ZL, Yu WM, Strickland S. Peripheral regeneration. Annu Rev Neurosci. 2007;30:209–33.
19. Jessen KR, Mirsky R. Negative regulation of myelination: relevance for development, injury, and demyelinating disease. Glia. 2008;56:1552–65.
20. Armstrong SJ, Wiberg M, Terenghi G, Kingham PJ. ECM molecules mediate both Schwann cell proliferation and activation to enhance neurite outgrowth. Tissue Eng. 2007;13:2863–70.
21. Campana WM. Schwann cells: activated peripheral glia and their role in neuropathic pain. Brain Behav Immun. 2007;21:522–7.
22. Webber C, Zochodne D. The nerve regenerative microenvironment: early behavior and partnership of axons and Schwann cells. Exp Neurol. 2010;223:51–9.
23. Allodi I, Udina E, Navarro X. Specificity of peripheral nerve regeneration: interactions at the axon level. Prog Neurobiol. 2012;98:16–37.
24. Jessen KR, Mirsky R, Arthur-Farraj P. The role of cell plasticity in tissue repair: adaptive cellular reprogramming. Dev Cell. 2015;34:613–20.
25. Shen C-N, Burke ZD, Tosh D. Transdifferentiation, metaplasia and tissue regeneration. Organogenesis. 2004;1:36–44.
26. Tsonis PA, Madhavan M, Tancous EE, Del Rio-Tsonis K. A newt's eye view of lens regeneration. Int J Dev Biol. 2004;48(8–9):975–80.
27. Thorel F, Népote V, Avril I, Kohno K, Desgraz R, Chera S, Herrera PL. Conversion of adult pancreatic α-cells to β-cells after extreme β-cell loss. Nature. 2010;464:1149–54.
28. Chera S, Baronnier D, Ghila L, Cigliola V, Jensen JN, Gu G, Furuyama K, Thorel F, Gribble FM, Reimann F, Herrera PL. Diabetes recovery by age-dependent conversion of pancreatic δ-cells into insulin producers. Nature. 2014;514:503–7.
29. Sisakhtnezhad S, Matin MM. Transdifferentiation: a cell and molecular reprogramming process. Cell Tissue Res. 2012;348:379–96.
30. Graf T, Enver T. Forcing cells to change lineages. Nature. 2009;426:587–94.
31. Fontana X, Hristova M, Da Costa C, Patodia S, Thei L, Makwana M, Spencer-Dene B, Latouche M, Mirsky R, Jessen KR, Klein R, Raivich G, Behrens A. c-Jun in Schwann cells promotes axonal regeneration and motoneuron survival via paracrine signaling. J Cell Biol. 2012;198:127–41.
32. Brushart TM, Aspalter M, Griffin JW, Redett R, Hameed H, Zhou C, Wright M, Vyas A, Höke A. Schwann cell phenotype is regulated by axon modality and central-peripheral location, and persists in vitro. Exp Neurol. 2013;247:272–81.
33. Boyd JG, Gordon T. Neurotrophic factors and their receptors in axonal regeneration and functional recovery after peripheral nerve injury. Mol Neurobiol. 2003;27:277–324.

34. Wood MD, Mackinnon SE. Pathways regulating modality-specific axonal regeneration in peripheral nerve. Exp Neurol. 2015;265:171–5.
35. Martini R, Fischer S, López-Vales R, David S. Interactions between Schwann cells and macrophages in injury and inherited demyelinating disease. Glia. 2008;56:1566–77.
36. Rotshenker S. Wallerian degeneration: the innate-immune response to traumatic nerve injury. J Neuroinflammation. 2011;8:109.
37. Hirota H, Kiyama H, Kishimoto T, Taga T. Accelerated Nerve Regeneration in Mice by upregulated expression of interleukin (IL) 6 and IL-6 receptor after trauma. J Exp Med. 1996;183:2627–34.
38. Cafferty WB, Gardiner NJ, Gavazzi I, Powell J, McMahon SB, Heath JK, Munson J, Cohen J, Thompson SW. Leukemia inhibitory factor determines the growth status of injured adult sensory neurons. J Neurosci. 2001;21:7161–70.
39. Bauer S, Kerr BJ, Patterson PH. The neuropoietic cytokine family in development, plasticity, disease and injury. Nat Rev Neurosci. 2007;8:221–32.
40. Barrette B, Hébert MA, Filali M, Lafortune K, Vallières N, Gowing G, Julien JP, Lacroix S. Requirement of myeloid cells for axon regeneration. J Neurosci. 2008;28:9363–76.
41. Niemi JP, DeFrancesco-Lisowitz A, Roldán-Hernández L, Lindborg JA, Mandell D, Zigmond RE. A critical role for macrophages near axotomized neuronal cell bodies in stimulating nerve regeneration. J Neurosci. 2013;33:16236–48.
42. Cattin AL, Burden JJ, Van Emmenis L, Mackenzie FE, Hoving JJ, Garcia Calavia N, Guo Y, McLaughlin M, Rosenberg LH, Quereda V, Jamecna D, Napoli I, Parrinello S, Enver T, Ruhrberg C, Lloyd AC. Macrophage-induced blood vessels guide Schwann cell-mediated regeneration of peripheral nerves. Cell. 2015;162:1127–39.
43. Hirata K, Kawabuchi M. Myelin phagocytosis by macrophages and nonmacrophages during Wallerian degeneration. Microsc Res Tech. 2002;57:541–7.
44. Gomez-Sanchez JA, Pilch KS, van der Lans M, Fazal SV, Benito C, Wagstaff LJ, Mirsky R, Jessen KR. After nerve injury, lineage tracing shows that myelin and Remak Schwann cells elongate extensively and branch to form repair Schwann cells, which shorten radically on re-myelination. J Neurosci. 2017;37:9086–99.
45. Stoll G, Müller HW. Nerve injury, axonal degeneration and neural regeneration: basic insights. Brain Pathol. 1999;9:313–25.
46. Gomez-Sanchez JA, Carty L, Iruarrizaga-Lejarreta M, Palomo-Irigoyen M, Varela-Rey M, Griffith M, Hantke J, Macias-Camara N, Azkargorta M, Aurrekoetxea I, De Juan VG, Jefferies HB, Aspichueta P, Elortza F, Aransay AM, Martínez-Chantar ML, Baas F, Mato JM, Mirsky R, Woodhoo A, Jessen KR. Schwann cell autophagy, myelinophagy, initiates myelin clearance from injured nerves. J Cell Biol. 2015;210:153–68.

47. Perry VH, Tsao JW, Fearn S, Brown MC. Radiation-induced reductions in macrophage recruitment have only slight effects on myelin degeneration in sectioned peripheral nerves of mice. Eur J Neurosci. 1995;7:271–80.

48. Ramaglia V, Wolterman R, de Kok M, Vigar MA, Wagenaar-Bos I, King RH, Morgan BP, Baas F. Soluble complement receptor 1 protects the peripheral nerve from early axon loss after injury. Am J Pathol. 2008;172:1043–52.

49. Vargas ME, Watanabe J, Singh SJ, Robinson WH, Barres BA. Endogenous antibodies promote rapid myelin clearance and effective axon regeneration after nerve injury. Proc Natl Acad Sci U S A. 2010;107:11993–8.

50. Jung J, Cai W, Lee HK, Pellegatta M, Shin YK, Jang SY, Suh DJ, Wrabetz L, Feltri ML, Park HT. Actin polymerization is essential for myelin sheath fragmentation during Wallerian degeneration. J Neurosci. 2011;31:2009–15.

51. Suzuki K, Lovera M, Schmachtenberg O, Couve E. Axonal degeneration in dental pulp precedes human primary teeth exfoliation. J Dent Res. 2015;94:1446–53.

52. Rangaraju S, Verrier JD, Madorsky I, Nicks J, Dunn WA Jr, Notterpek L. Rapamycin activates autophagy and improves myelination in explant cultures from neuropathic mice. J Neurosci. 2010;30:11388–97.

53. Parkinson DB, Bhaskaran A, Droggiti A, Dickinson S, D'Antonio M, Mirsky R, Jessen KR. Krox-20 inhibits Jun-NH2-terminal kinase/c-Jun to control Schwann cell proliferation and death. J Cell Biol. 2004;164:385–94.

54. Klein D, Groh J, Wettmarshausen J, Martini R. Nonuniform molecular features of myelinating Schwann cells in models for CMT1: distinct disease patterns are associated with NCAM and c-Jun upregulation. Glia. 2014;62:736–50.

55. Hantke J, Carty L, Wagstaff LJ, Turmaine M, Wilton DK, Quintes S, Koltzenburg M, Baas F, Mirsky R, Jessen KR. c-Jun activation in Schwann cells protects against loss of sensory axons in inherited neuropathy. Brain. 2014;137:2922–37.

56. De Felipe C, Hunt SP. The differential control of c-Jun expression in regenerating sensory neurons and their associated glial cells. J Neurosci. 1994;14:2911–23.

57. Shy ME, Shi Y, Wrabetz L, Kamholz J, Scherer SS. Axon-Schwann cell interactions regulate the expression of c-jun in Schwann cells. J Neurosci Res. 1996;43:511–25.

58. Parkinson DB, Bhaskaran A, Arthur-Farraj P, Noon LA, Woodhoo A, Lloyd AC, Feltri ML, Wrabetz L, Behrens A, Mirsky R, Jessen KR. c-Jun is a negative regulator of myelination. J Cell Biol. 2018;181:625–37.

59. Lin HP, Oksuz I, Hurley E, Wrabetz L, Awatramani R. Microprocessor complex subunit DiGeorge syndrome critical region gene 8 (Dgcr8) is required for Schwann cell myelination and myelin maintenance. J Biol Chem. 2015;290:24294–307.

60. Ma KH, Hung HA, Srinivasan R, Xie H, Orkin SH, Svaren J. Regulation of peripheral nerve myelin maintenance by gene repression through polycomb repressive complex 2. J Neurosci. 2015;35:8640–52.

61. Woodhoo A, Alonso MB, Droggiti A, Turmaine M, D'Antonio M, Parkinson DB, Wilton DK, Al-Shawi R, Simons P, Shen J, Guillemot F, Radtke F, Meijer D, Feltri ML, Wrabetz L, Mirsky R, Jessen KR. Notch controls embryonic Schwann cell differentiation, postnatal myelination and adult plasticity. Nat Neurosci. 2009;12:839–47.

62. Sheu JY, Kulhanek DJ, Eckenstein FP. Differential patterns of ERK and STAT3 phosphorylation after sciatic nerve transection in the rat. Exp Neurol. 2000;166:392–402.

63. Harrisingh MC, Perez-Nadales E, Parkinson DB, Malcolm DS, Mudge AW, Lloyd AC. The Ras/Raf/ERK signalling pathway drives Schwann cell dedifferentiation. EMBO J. 2004;23:3061–71.

64. Fischer S, Weishaupt A, Troppmair J, Martini R. Increase of MCP-1 (CCL2) in myelin mutant Schwann cells is mediated by MEK-ERK signaling pathway. Glia. 2008;56:836–43.

65. Groh J, Heinl K, Kohl B, Wessig C, Greeske J, Fischer S, Martini R. Attenuation of MCP-1/CCL2 expression ameliorates neuropathy in a mouse model for Charcot-Marie-Tooth 1X. Hum Mol Genet. 2010;19:3530–43.

66. Napoli I, Noon LA, Ribeiro S, Kerai AP, Parrinello S, Rosenberg LH, Collins MJ, Harrisingh MC, White IJ, Woodhoo A, Lloyd AC. A central role for the ERK-signaling pathway in controlling Schwann cell plasticity and peripheral nerve regeneration in vivo. Neuron. 2012;73:729–42.

67. Shin YK, Jang SY, Park JY, Park SY, Lee HJ, Suh DJ, Park HT. The Neuregulin-Rac-MKK7 pathway regulates antagonistic c-jun/Krox20 expression in Schwann cell dedifferentiation. Glia. 2013;61:892–904.

68. Myers RR, Sekiguchi Y, Kikuchi S, Scott B, Medicherla S, Protter A, Campana WM. Inhibition of p38 MAP kinase activity enhances axonal regeneration. Exp Neurol. 2003;184:606–14.

69. Yang DP, Kim J, Syed N, Tung YJ, Bhaskaran A, Mindos T, Mirsky R, Jessen KR, Maurel P, Parkinson DB, Kim HA. p38 MAPK activation promotes denervated Schwann cell phenotype and functions as a negative regulator of Schwann cell differentiation and myelination. J Neurosci. 2012;32:7158–68.

70. Hutton EJ, Carty L, Laurá M, Houlden H, Lunn MP, Brandner S, Mirsky R, Jessen K, Reilly MM. c-Jun expression in human neuropathies: a pilot study. J Peripher Nerv Syst. 2011;16:295–303.

71. Fazal SV, Gomez-Sanchez JA, Wagstaff LA, Musner N, Otto G, Janz M, Mirsky R, Jessen KR. Graded elevation of c-Jun in Schwann cells in vivo: gene dosage determines effects on development, re-myelination, tumorigenesis and hypomyelination. J Neurosci. 2017;37(50):12297–313.

72. Beirowski B, Babetto E, Golden JP, Chen YJ, Yang K, Gross RW, Patti GJ, Milbrandt J. Metabolic regula-

tor LKB1 is crucial for Schwann cell-mediated axon maintenance. Nat Neurosci. 2014;17:1351–61.

73. Sulaiman OA, Gordon T. Role of chronic Schwann cell denervation in poor functional recovery after nerve injuries and experimental strategies to combat it. Neurosurgery. 2009;65(4 Suppl):A105–14.

74. Höke A. Neuroprotection in the peripheral nervous system: rationale for more effective therapies. Arch Neurol. 2006;63:1681–5.

75. Painter MW, Brosius Lutz A, Cheng YC, Latremoliere A, Duong K, Miller CM, Posada S, Cobos EJ, Zhang AX, Wagers AJ, Havton LA, Barres B, Omura T, Woolf CJ. Diminished Schwann cell repair responses underlie age-associated impaired axonal regeneration. Neuron. 2014;83:331–43.

76. Wagstaff LJ, Gomez-Sanchez JA, Mirsky R, Jessen KR. The relationship between Schwann cell c-Jun and regeneration failures due to ageing and long-term injury. Glia. 2017;65:E532.

77. Benito C, Davis CM, Gomez-Sanchez JA, Turmaine M, Meijer D, Poli V, Mirsky R, Jessen KR. STAT3 controls the long-term survival and phenotype of repair Schwann cells during nerve regeneration. J Neurosci. 2017;37:4255–69.

78. Heinen A, Beyer F, Tzekova N, Hartung HP, Küry P. Fingolimod induces the transition to a nerve regeneration promoting Schwann cell phenotype. Exp Neurol. 2015;271:25–35.

Adipose-Derived Stem Cells (ASCs) for Peripheral Nerve Regeneration

30

Mathias Tremp and Daniel F. Kalbermatten

30.1 Introduction

Each year, numerous surgical procedures are performed for peripheral nerve repair, with a significant amount of lost working days with corresponding economic consequences [1]. Nerve lesions without defect or with a short gap are normally reconstructed by end-to-end coaptation. Traumatic injuries resulting in longer peripheral nerve lesions however often require a graft to bridge the gap. Nevertheless, full recovery might never be achieved, especially with extended lesions [2]. Schwann cells (SCs) play a prominent role in peripheral nerve regeneration [3] and are thus an attractive therapeutic target. Nerve injury disrupts the normal SC–axon interaction, resulting in dedifferentiation of the SCs and activation of a growth-promoting phenotype [4]. Proliferating SCs release neurotrophic factors [5] and form the bands of Büngner to direct regenerating axons across the lesion. When seeded in artificial nerve conduits, SCs have been shown to enhance nerve regeneration [6–8]. However, cultured SCs have limited clinical application. The requirement for

nerve donor material evokes additional morbidity and the time required to culture and expand the cells delays treatment. Instead, the ideal transplantable cell should be easily accessible, proliferate rapidly in culture, and successfully integrate into host tissue with immunological tolerance [9]. Mesenchymal stem cells (MSCs) are an attractive cell source for the regeneration of nerve tissue as they are able to self-renew with a high growth rate and possess multipotent differentiation. Moreover, adipose-derived stem cells (ASCs) represent a readily available source for isolation of potentially useful stem cells and thus are very attractive for tissue engineering purposes [10]. In culture, they showed to have an impressive developmental plasticity, including the ability to undergo multilineage differentiation and self-renewal [11]. When ASCs are compared with BM-MSCs, further similarities have been demonstrated in regard to their growth kinetics, cell senescence, gene transduction efficiency [12], as well as CD surface marker expression [13–15] and gene transcription [14]. However, compared to BM-MSCs, ASCs have significant advantages for tissue engineering application, because of the tissue accessibility, multipotency, and ease of isolation without donor-site injury or painful procedures. In this chapter, we discuss the potential use of ASCs in the field of peripheral nerve regeneration and discuss the tissue engineering approach.

M. Tremp · D. F. Kalbermatten (✉)
Department of Plastic, Reconstructive, Aesthetic and Hand Surgery, University Basel, Basel, Switzerland
e-mail: Mathias.Tremp@usb.ch;
daniel.kalbermatten@usb.ch

© Springer Nature Switzerland AG 2019
D. Duscher, M. A. Shiffman (eds.), *Regenerative Medicine and Plastic Surgery*,
https://doi.org/10.1007/978-3-030-19962-3_30

30.2 Technique

30.2.1 The Short-Term Regeneration Potential After Human and Autologous Stem Cell Transplantation in a Rat Sciatic Nerve Injury Model

The ability of human stem cells to improve peripheral nerve regeneration was recently investigated and compared with autologous stem cell transplantation in a rat sciatic nerve injury model. Evaluation was performed using MRI and immunocytochemistry [16]. In a first step, a sciatic nerve lesion creating a 10 mm gap was used in female Sprague Dawley rats. In group 1 (negative control group), a fibrin conduit was inserted with culture medium (CM) alone, containing DMEM/F-12, 1% penicillin/streptomycin, and 10% FBS. All study animals (groups 2–7, seven per group) were treated with either rat (r)ASCs, differentiated rASCs (SC-like cells), human (h-) ASCs from the superficial and deep layer, stromal vascular fraction (SVF), or SCs only (positive control group) (Fig. 30.1). After 2 weeks, nerve regeneration was analyzed by immunocytochemistry. Moreover, imaging analysis (MRI) was performed after 2 and 4 weeks to monitor nerve regeneration. The results showed that sciatic nerve axonal regeneration was different depending on the type of cells that were injected in the fibrin conduits. A significant longer regeneration distance in the SC fibrin conduit group was observed compared with all the other groups ($p < 0.0001$). Human ASCs from the superficial layer as well as SVF have inferior results when compared with autogenous ASCs. No significant difference was noticed between the hASC groups from the superficial and deep layers. Importantly, autogenous ASCs and hASCs from the deep abdominal layer similarly improved nerve regeneration. Differentiated rASCs showed longer nerve growth than rASCs by PGP 9.5, but not by S100. No significant nerve regeneration was achieved in the animals in which an empty fibrin conduit with CM only was used. A strong correlation was found between the length of the regenerating axon front measured by MRI and the length of the regenerating axon front measured by immunocytochemistry ($r = 0.74$, $p = 0.09$) [16].

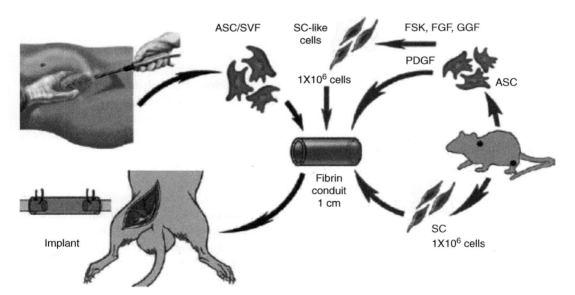

Fig. 30.1 Overall study design. *ASC/SVF* adipose-derived stem cell/stromal vascular fraction, *SC* Schwann cell, *FSK* forskolin, *FGF* fibroblast growth factor, *GGF* glial growth factor (neuregulin-1b1), *PDGF* platelet-derived growth factor [16]

30.2.2 Peripheral Nerve Repair: Multimodal Comparison of the Long-Term Regenerative Potential of ASCs in a Biodegradable Conduit

In a recent work by Kappos et al. [17], a broad overview over promising transplantable cells in peripheral nerve repair according to the current state of research was presented, including rASCs, differentiated rASCs (drASCs), rSCs, hASCs from the superficial (hASCsup) and deep (hASCdeep) abdominal layers, as well as hSVF, and, as a control, an autograft to bridge a 10 mm gap in the sciatic nerve of rats. The diverse cell types were examined long-term (12 weeks) in a multimodal way containing functional and morphometric as well as MRI analyses. The sciatic nerve injury model was used, creating a 10 mm gap in the left nerve of female Sprague Dawley rats that was bridged through a biodegradable fibrin conduit by the microsurgical suture technique. 1 Mio of each of the abovementioned cells was introduced into the conduits. The results showed that rats whose nerve had been repaired with drASCs had a higher mean sciatic functional index (SFI) than those with undifferentiated rASCs and an even higher mean SFI than rats with an autograft. Superficial abdominal layer human ASCs (hASCsup) had a higher mean SFI than deep (hASCdeep) abdominal layer human ASCs. While human abdominal layer ASCs (hASCsup/hASCdeep) and human SVF cells did not differ from rSCs in SFI, rASCs had a lower mean SFI and drASCs had a higher mean SFI than rSCs. In line with those results, rats whose left nerve had been repaired with drASCs had a higher relative muscle weight than those with rASCs. All other contrasts were nonsignificant. Rats with an autograft had the largest relative muscle weights.

30.2.3 Regeneration Patterns Influence Hind Limb Automutilation After Sciatic Nerve Repair Using Stem Cells in Rats

Despite common hind limb autotomy in operated animals, the rat sciatic nerve model is the most commonly used model by researchers working on peripheral nerve injury and regeneration [18]. The phenomenon of hind limb autotomy in operated animals, leading to amputation of part of one or several toes, was analyzed by Walland et al. in 1979 [19] and widely observed since. Hind limb autotomy is often underestimated in literature with little attention in peripheral nerve regeneration research. Autotomy usually starts within the first 3 weeks after nerve injury and peaks between weeks 2 and 4. Depending on the type of lesion and rat strain, incidence can be as high as 90% [20], and can lead to considerable research and ethical issues. In a recent study by Hasselbach et al. [21], they applied adult stem cells in a fibrin conduit (1 cm sciatic gap) in a total sciatic axotomy model to improve nerve regeneration, investigating whether a correlation could be detected between stem cell effects on regeneration and limb autophagy. Experimental groups included empty fibrin conduits, fibrin conduits seeded with primary SCs, and fibrin conduits seeded with SC-like dMSCs and dASCs. Controls were represented by autografts and by sham rats. At 16 weeks postimplantation, regeneration pattern was analyzed on histological sections and related to eventual autophagy. Hind limbs were evaluated and scored according to autophagy Wall's scale and X-ray radiological evaluation. In this study, all regenerative cell lines significantly improved myelination at the mid-conduit level, compared to the empty tubes. However, dMSC could not significantly improve myelination at the distal stump, showing a more disorganized regeneration compared to both other cell groups and controls. Moreover, autophagy was correlated to this regeneration patterns, with higher autophagy scores in the empty and dMSC group.

30.2.4 Neurotrophic Activity of hASCs Isolated from Deep and Superficial Layers of Abdominal Fat

Our group has shown that rASC can be differentiated into cells resembling SCs, which promote neurite outgrowth in vitro [22] and enhance peripheral nerve regeneration in vivo [23]. Although some transdifferentiation of transplanted MSC might occur in vivo, the exact mechanisms behind the neuroprotective and growth-promoting effects of adult stem cells remain poorly investigated. However, some of the benefits elicited by transplanted cells might occur as a direct result of their production of neurotrophic factors.

The properties of hASC isolated from the deep and superficial layers of abdominal fat tissue obtained during abdominoplasty procedures were analyzed in a recent study [24]. In this research, cells from the superficial layer proliferated significantly faster than those from the deep layer. In both the deep and superficial layers, ASCs express the pluripotent stem cell markers oct4 and nanog and also the stro-1 cell surface antigen. Superficial layer ASCs induce the significantly enhanced outgrowth of neurite-like processes from neuronal cell lines when compared with that of deep-layer cells. However, analysis by reverse transcription with the polymerase chain reaction and by enzyme-linked immunosorbent assay (ELISA) has revealed that ASCs isolated from both layers express similar levels of the following neurotrophic factors: nerve growth factor (NGF), brain-derived neurotrophic factor (BDNF), and glial-derived neurotrophic factor (GDNF) (Fig. 30.2). Thus, human ASCs especially isolated from the superficial layer show promising potential for the treatment of traumatic nerve injuries and further analysis of their neurotrophic molecules is needed.

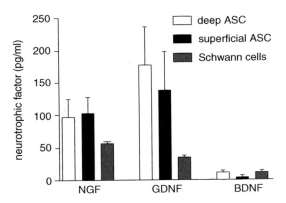

Fig. 30.2 Enzyme-linked immunosorbant assay (ELISA) was used to determine NGF, GDNF, and BDNF protein levels in cell culture supernatants from deep and superficial layer ASC and compared with expression levels in human SC cultures. Data are means ± SEM from five individual patients [24]

30.2.5 Regeneration Potential and Survival of Transplanted Undifferentiated ASCs in Peripheral Nerve Conduits

There is a lack of information about the regenerative properties of undifferentiated ASCs for the treatment of peripheral nerve injuries as well as their capacity for in vivo transdifferentiation. In a recent study by Erba et al. [25], these points were investigated using a nerve conduit filled with either a suspension of ASCs or a combination of cells and differentiating growth factors, in a rodent nerve injury model. It was also of interest to determine the survival of the transfected ASCs in vivo using a combination of markers. In their study, the results suggested that any regenerative effect of transplanted ASCs is more likely to be mediated by an initial boost of released growth factors and/or by an indirect effect on endogenous SCs activity [25, 26]. Santiago et al. [27] reported that while the transplantation of human ASCs in a rat sciatic nerve defect promoted nerve tissue regeneration and a decrease in muscle atrophy, the ASCs did not differentiate to SC-like cells at the site of injury. These results suggest that the regenerative effect of transplanted ASCs is likely due to an initial boost of released growth factors as well as an indirect effect on endogenous SC activity. Further evidence for peripheral nerve regeneration through the paracrine effects of ASCs is presented by Widgerow et al. [28] and Kingham et al. [29].

30.2.6 ASCs Enhance Peripheral Nerve Regeneration

In a study by di Summa et al. [23], fibrin nerve conduits were seeded with various cell types (primary SCs and adult stem cells differentiated to a SC-like phenotype) for repair of sciatic nerve injury. Two weeks after implantation, the conduits were removed and examined by immunohistochemistry for axonal regeneration (evaluated by PGP 9.5 expression) and SC presence (detected by S100 expression). The results showed a significant increase in axonal regeneration in the fibrin group seeded with SCs compared with the empty fibrin conduit. Differentiated ASCs also enhanced regeneration distance in a similar manner to dMSCs. These observations suggest that ASCs may provide an effective cell population, without the limitations of the donor-site morbidity associated with isolation of SCs, and could be a valuable route towards new methods to improve peripheral nerve repair [23].

30.2.7 Long-Term In Vivo Regeneration of Peripheral Nerves Through Bioengineered Nerve Grafts

In another study by di Summa et al. [30], long-term results were assessed where nerve fibrin conduits were seeded with various cell types: primary SCs, SC-like dMSC, and SC-like dASC. Two further control groups were fibrin conduits without cells and autografts. Conduits were used to bridge a 10 mm rat sciatic nerve gap in a long-term experiment (16 weeks). Functional and morphological properties of regenerated nerves were analyzed. A reduction in muscle atrophy was observed in the autograft and in all cell-seeded groups, when compared with the empty fibrin conduits. SCs showed significant

improvement in axon myelination and average fiber diameter of the regenerated nerves. dASCs were the most effective cell population in terms of improvement of axonal and fiber diameter, and evoked potentials at the level of the gastrocnemius muscle and regeneration of motoneurons, similar to the autografts. Given these results and other advantages of ASCs such as ease of harvest and relative abundance, dASCs could be a valuable translatable route towards novel methods to accelerate peripheral nerve repair [30].

30.3 Discussion

Peripheral nerve injury results in impaired sensory and motor functions and often leads to long-term disability [31]. It has been shown experimentally that even under the most optimal immediate nerve repair scenario only 50% of neurons regenerate their axons into the distal stump [31, 32]. One of the greatest clinical challenges today is to treat nerve defects in which there is a significant loss of tissue [31]. To date, there are few clinically relevant alternatives to nerve grafts and all of the currently marketed conduits failed to fully reproduce the biological properties of the nerve grafts [31].

Today, there are now a large number of studies describing the application of ASCs for peripheral nerve repair [31, 33]. ASCs are an attractive cell source for tissue regeneration because of their self-renewal ability, high growth rate, low immunogenicity, and multipotent differentiation properties, making them also ideal candidates for allogeneic transplantation [16, 34, 35]. ASCs might exert a positive influence on peripheral nerve regeneration and can be differentiated into SCs in vitro [31, 35]. ASCs have been shown to promote angiogenesis and neuronal survival in the spinal ganglion, neurotrophic factor provision, and protection of neurons [36, 37]. Interestingly, ASCs are suggested to be better at regenerating motor neurons specifically [27]. They can also be applied intravenously, spreading throughout the body, and get to their target organs by a homing response to the injured neurological tissue [38, 39].

Different strategies have been attempted to further improve the neurotrophic potential of ASCs such as electrical stimulation and highly conductive topographical materials [38]. Experiments using direct and indirect co-cultures of dASC and dorsal root ganglion (DRG) neurons showed that extracellular matrix (ECM) molecules influence cell viability, adhesion, and neurotrophic behavior of dASCs [40]. Furthermore, dASCs retain their regenerative capacity for longer than ASCs [41]. Moreover, treatment of dASCs with leukemia inhibitory factor increased the expression levels of the glial markers S100 and GFAP and of the myelin basic protein, suggesting that the myelinating potential can be ameliorated with the treatment with cytokines [42]. Furthermore, ASCs have been shown to produce a significant quantity of laminin which can activate SCs to enhance myelination in a mutant mouse model [43]. Interestingly, undifferentiated ASC-conditioned media were shown to promote neurite growth and prevent neuronal death, effects that have been associated with the release of neurotrophic factors such as NGF and BDNF [44, 45]. Ideally, for routine clinical application, fat is harvested from patients suffering from nerve injuries to derive and expand ASCs in vitro. Then, the cells could be directly transplanted into bioengineered nerve guides for the repair of the nerve damage, or transdifferentiated into dASC prior to reimplantation (Fig. 30.3).

Abbas et al. aimed to provide functional benefits for axons traversing a cross-face nerve autograft by seeding the local environment with ASCs in a model of facial paralysis. They showed that ASCs could be a clinically translatable route towards new methods to enhance recovery after cross-facial nerve grafting [46]. ASCs increased the number of axons that breach the coaptation site and enter the graft, optimized myelination to

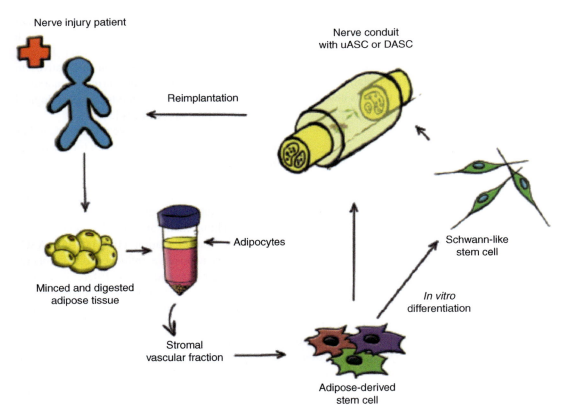

Fig. 30.3 Possible ideal clinical scenario: ASCs could be isolated from patients after peripheral nerve trauma and transplanted into nerve conduits with or without prior differentiation in SC-like cells [35]

achieve maximal functional and structural efficiency, and supported the functional integrity of denervated neuromuscular junctions [46].

In the recent years, many new therapeutic strategies for improving nerve repair are being developed. Almost all of these studies include the short gap injury model in rat sciatic nerve and there is clearly a need to translate these studies to large animal models [31]. The results showed that guidance conduit structures and living cells are essential for the repair of larger nerve gaps to provide trophic support and recreate the environment provided by the nerve autograft [47–49]. Recently, effective technology has been developed for the production of engineered neural tissue (EngNT), an aligned cellular biomaterial for nerve repair in the 15 mm rat sciatic nerve gap model [47, 50]. However, a key factor limiting the translation of the technology is the availability of SCs. Autologous human SCs are needed from invasive nerve biopsies and sufficient cell numbers for regeneration are only available after a time-consuming expansion time in vitro [47, 51]. ASCs are an accessible source of adult stem cells in large quantities with little donor-site morbidity or patient discomfort [47]. Moreover, rASCs differentiated into dASCs express a range of SC proteins; they can promote neurite outgrowth in vitro and enhance regeneration in vivo [11, 47, 52, 53]. Consequently, dASCs have been transplanted in fast-resorbing fibrin conduits, synthetic nerve tube conduits, or naturally occurring decellularized matrices [29, 41, 54]. The mechanism of the action of the stem cells has not been definitively elucidated yet, but studies suggest that they work by either releasing growth factors themselves or modulating the endogenous SCs [39, 55]. It is unclear to determine whether the dASCs themselves are responsible for myelinating the regeneration axons or if they enhance the myelination by the endogenous SCs. Nevertheless, in vitro studies showed that dASCs form myelin whereas undifferentiated ASCs do not form myelin [56].

In a recent study by Sowa et al. [57], ASCs and SCs were transplanted with gelatin hydrogel tubes at the artificially blunted sciatic nerve lesion in mice. The Cre-loxP-mediated fate tracking technology was used to visualize differentiation of the transplanted ASCs into the SC lineage in situ at the peripheral nerve injury lesion. In their study, the transplantation of ASCs promoted regeneration of axons, formation of myelin, and restoration of denervation muscle atrophy to levels comparable to those achieved by SC transplantation. Importantly, ASCs survived for at least 4 weeks after transplantation without differentiating into SCs. ASCs may produce more favorable specific matrix components or extracellular environment for peripheral nerve repair in vivo than SCs [55].

ASCs may offer more beneficial effects on surrounding structures than SCs by improving the environment of the injury site [57]. ASCs produce and release various angiogenic factors, such as vascular endothelial growth factor (VEGF) A, VEGF C, and hepatocyte growth factor, which are known to be responsible for neoangiogenesis and improvement of cellular niche for graft, at higher quantities than SCs [44, 58]. However, the potential disadvantages of undifferentiated ASCs are that they are heterogeneous and undifferentiated cell populations that could differentiate into undesired cell types, such as adipocytes, or develop a tumor (teratomas) after transplantation [27, 35, 57]. Thus, dASCs could prevent the risk of teratomas and in vivo differentiation towards undesired phenotype, and could potentially generate committed SC-like cells able to actively participate in the regeneration and re-myelination of the injured nerves [59].

30.4 Conclusions

ASCs may provide an effective cell population and could be a clinically translatable route towards new treatment strategies to enhance peripheral nerve repair. ASCs are easily accessible from the patients for use in an autologous cell therapy, eliminating the risk of rejection, and can be expanded in a controlled and reproducible manner. Regenerative effect of transplanted cells is more likely to be mediated by an initial boost of released growth factors and/or by an indirect effect on endogenous SC activity. However, there is a great debate whether undifferentiated or differentiated ASCs represent the most clinically viable option. Moreover, future studies in larger animals are needed to address long-term cell survival in tissue-engineered nerve conduits by means of functionalization of the coating with biological active substrates, combining a cell-based therapy and pharmacological intervention. It remains to be determined if culturing and expanding the cells in vitro are beneficial for transplantation strategies, or if a more immediate approach, using SVF obtained and transplanted on the day of nerve repair, would be a better option. That being said, it would be of interest to identify the specific subpopulation leading to the best outcome for nerve repair or generating better performing SC-like ASCs.

References

1. Chalfoun CT, Wirth GA, Evans GR. Tissue engineered nerve constructs: where do we stand? J Cell Mol Med. 2006;10:309–17.
2. Hu J, Zhu QT, Liu XL, Xu YB, Zhu JK. Repair of extended peripheral nerve lesions in rhesus monkeys using acellular allogenic nerve grafts implanted with autologous mesenchymal stem cells. Exp Neurol. 2007;204:658–66.
3. Ide C. Peripheral nerve regeneration. Neurosci Res. 1996;25:101–21.
4. Hall S. The response to injury in the peripheral nervous system. J Bone Joint Surg Br. 2005;87:1309–19.
5. Frostick SP, Yin Q, Kemp GJ. Schwann cells, neurotrophic factors, and peripheral nerve regeneration. Microsurgery. 1998;18:397–405.
6. Li Q, Ping P, Jiang H, Liu K. Nerve conduit filled with GDNF gene-modified Schwann cells enhances regeneration of the peripheral nerve. Microsurgery. 2006;26:116–21.
7. Mosahebi A, Fuller P, Wiberg M, Terenghi G. Effect of allogeneic Schwann cell transplantation on peripheral nerve regeneration. Exp Neurol. 2002;173:213–23.
8. Rutkowski GE, Miller CA, Jeftinija S, Mallapragada SK. Synergistic effects of micropatterned biodegradable conduits and Schwann cells on sciatic nerve regeneration. J Neural Eng. 2004;1:151–7.
9. Tohill M, Mantovani C, Wiberg M, Terenghi G. Rat bone marrow mesenchymal stem cells express glial markers and stimulate nerve regeneration. Neurosci Lett. 2004;362:200–3.
10. Sterodimas A, de Faria J, Nicaretta B, Pitanguy I. Tissue engineering with adipose-derived stem cells (ADSCs): current and future applications. J Plast Reconstr Aesthet Surg. 2010;63:1886–92.
11. Liu ZJ, Zhuge Y, Velazquez OC. Trafficking and differentiation of mesenchymal stem cells. J Cell Biochem. 2009;106:984–91.
12. De Ugarte DA, Morizono K, Elbarbary A, Alfonso Z, Zuk PA, Zhu M, Dragoo JL, Ashjian P, Thomas B, Benhaim P, Chen I, Fraser J, Hedrick MH. Comparison of multi-lineage cells from human adipose tissue and bone marrow. Cells Tissues Organs. 2003;174:101–9.
13. Gronthos S, Franklin DM, Leddy HA, Robey PG, Storms RW, Gimble JM. Surface protein characterization of human adipose tissue-derived stromal cells. J Cell Physiol. 2001;189:54–63.
14. Katz AJ, Tholpady A, Tholpady SS, Shang H, Ogle RC. Cell surface and transcriptional characterization of human adipose-derived adherent stromal (hADAS) cells. Stem Cells. 2005;23:412–23.
15. Zuk PA, Zhu M, Ashjian P, De Ugarte DA, Huang JI, Mizuno H, Alfonso ZC, Fraser JK, Benhaim P, Hedrick MH. Human adipose tissue is a source of multipotent stem cells. Mol Biol Cell. 2002;13:4279–95.
16. Tremp M, Meyer Zu Schwabedissen M, Kappos EA, Engels PE, Fischmann A, Scherberich A, Schaefer DJ, Kalbermatten DF. The regeneration potential after human and autologous stem cell transplantation in a rat sciatic nerve injury model can be monitored by MRI. Cell Transplant. 2015;24:203–11.
17. Kappos EA, Engels PE, Tremp M, Meyer zu Schwabedissen M, di Summa P, Fischmann A, von Felten S, Scherberich A, Schaefer DJ, Kalbermatten DF. Peripheral nerve repair: multimodal comparison of the long-term regenerative potential of adipose tissue-derived cells in a biodegradable conduit. Stem Cells Dev. 2015;24:2127–41.
18. Zeltser R, Beilin B, Zaslansky R, Seltzer Z. Comparison of autotomy behavior induced in rats by various clinically-used neurectomy methods. Pain. 2000;89:19–24.
19. Coderre TJ, Grimes RW, Melzack R. Deafferentation and chronic pain in animals: an evaluation of evidence suggesting autotomy is related to pain. Pain. 1986;26:61–84.
20. Weber RA, Proctor WH, Warner MR, Verheyden CN. Autotomy and the sciatic functional index. Microsurgery. 1993;14:323–7.
21. Haselbach D, Raffoul W, Larcher L, Tremp M, Kalbermatten DF, di Summa PG. Regeneration patterns influence hindlimb automutilation after sciatic nerve repair using stem cells in rats. Neurosci Lett. 2016;634:153–9.
22. Kingham PJ, Kalbermatten DF, Mahay D, Armstrong SJ, Wiberg M, Terenghi G. Adipose-derived stem cells differentiate into a Schwann cell phenotype and promote neurite outgrowth in vitro. Exp Neurol. 2007;207:267–74.
23. di Summa PG, Kingham PJ, Raffoul W, Wiberg M, Terenghi G, Kalbermatten DF. Adipose-derived stem cells enhance peripheral nerve regeneration. J Plast Reconstr Aesthet Surg. 2010;63:1544–52.
24. Kalbermatten DF, Schaakxs D, Kingham PJ, Wiberg M. Neurotrophic activity of human adipose stem cells

isolated from deep and superficial layers of abdominal fat. Cell Tissue Res. 2011;344:251–60.

25. Erba P, Mantovani C, Kalbermatten DF, Pierer G, Terenghi G, Kingham PJ. Regeneration potential and survival of transplanted undifferentiated adipose tissue-derived stem cells in peripheral nerve conduits. J Plast Reconstr Aesthet Surg. 2010;63:e811–7.

26. Bhangra KS, Busuttil F, Phillips JB, Rahim AA. Using stem cells to grow artificial tissue for peripheral nerve repair. Stem Cells Int. 2016;2016:7502178. 18 p.

27. Santiago LY, Clavijo-Alvarez J, Brayfield C, Rubin JP, Marra KG. Delivery of adipose-derived precursor cells for peripheral nerve repair. Cell Transplant. 2009;18:145–58.

28. Widgerow AD, Salibian AA, Lalezari S, Evans GR. Neuromodulatory nerve regeneration: adipose tissue-derived stem cells and neurotrophic mediation in peripheral nerve regeneration. J Neurosci Res. 2013;91:1517–24.

29. Kingham PJ, Kolar MK, Novikova LN, Novikov LN, Wiberg M. Stimulating the neurotrophic and angiogenic properties of human adipose-derived stem cells enhances nerve repair. Stem Cells Dev. 2014;23:741–54.

30. di Summa PG, Kalbermatten DF, Pralong E, Raffoul W, Kingham PJ, Terenghi G. Long-term in vivo regeneration of peripheral nerves through bioengineered nerve grafts. Neuroscience. 2011;181:278–91.

31. Kingham PJ, Reid AJ, Wiberg M. Adipose-derived stem cells for nerve repair: hype or reality? Cells Tissues Organs. 2014;200:23–30.

32. Welin D, Novikova LN, Wiberg M, Kellerth JO, Novikov LN. Survival and regeneration of cutaneous and muscular afferent neurons after peripheral nerve injury in adult rats. Exp Brain Res. 2008;186:315–23.

33. Hundepool CA, Nijhuis TH, Mohseny B, Selles RW, Hovius SE. The effect of stem cells in bridging peripheral nerve defects: a meta-analysis. J Neurosurg. 2014;121:195–209.

34. Gomillion CT, Burg KJ. Stem cells and adipose tissue engineering. Biomaterials. 2006;27:6052–63.

35. Faroni A, Terenghi G, Reid AJ. Adipose-derived stem cells and nerve regeneration: promises and pitfalls. Int Rev Neurobiol. 2013;108:121–36.

36. Dadon-Nachum M, Melamed E, Offen D. Stem cells treatment for sciatic nerve injury. Expert Opin Biol Ther. 2011;11:1591–7.

37. Masgutov RF, Masgutova GA, Zhuravleva MN, Salafutdinov II, Mukhametshina RT, Mukhamedshina YO, Lima LM, Reis HJ, Kiyasov AP, Palotas A, Rizvanov AA. Human adipose-derived stem cells stimulate neuroregeneration. Clin Exp Med. 2016;16:451–61.

38. Zack-Williams SD, Butler PE, Kalaskar DM. Current progress in use of adipose derived stem cells in peripheral nerve regeneration. World J Stem Cells. 2015;7:51–64.

39. Marconi S, Castiglione G, Turano E, Bissolotti G, Angiari S, Farinazzo A, Constantin G, Bedogni G, Bedogni A, Bonetti B. Human adipose-derived mes-

enchymal stem cells systemically injected promote peripheral nerve regeneration in the mouse model of sciatic crush. Tissue Eng Part A. 2012;18:1264–72.

40. di Summa PG, Kalbermatten DF, Raffoul W, Terenghi G, Kingham PJ. Extracellular matrix molecules enhance the neurotrophic effect of Schwann cell-like differentiated adipose-derived stem cells and increase cell survival under stress conditions. Tissue Eng Part A. 2013;19:368–79.

41. Scholz T, Sumarto A, Krichevsky A, Evans GR. Neuronal differentiation of human adipose tissue-derived stem cells for peripheral nerve regeneration in vivo. Arch Surg. 2011;146:666–74.

42. Razavi S, Mardani M, Kazemi M, Esfandiari E, Narimani M, Esmaeili A, Ahmadi N. Effect of leukemia inhibitory factor on the myelinogenic ability of Schwann-like cells induced from human adipose-derived stem cells. Cell Mol Neurobiol. 2013;33:283–9.

43. Carlson KB, Singh P, Feaster MM, Ramnarain A, Pavlides C, Chen ZL, Yu WM, Feltri ML, Strickland S. Mesenchymal stem cells facilitate axon sorting, myelination, and functional recovery in paralyzed mice deficient in Schwann cell-derived laminin. Glia. 2011;59:267–77.

44. Sowa Y, Imura T, Numajiri T, Nishino K, Fushiki S. Adipose-derived stem cells produce factors enhancing peripheral nerve regeneration: influence of age and anatomic site of origin. Stem Cells Dev. 2012;21:1852–62.

45. Zhao L, Wei X, Ma Z, Feng D, Tu P, Johnstone BH, March KL, Du Y. Adipose stromal cells-conditional medium protected glutamate-induced CGNs neuronal death by BDNF. Neurosci Lett. 2009;452:238–40.

46. Abbas OL, Borman H, Uysal CA, Gonen ZB, Aydin L, Helvacioglu F, Ilhan S, Yazici AC. Adipose-derived stem cells enhance axonal regeneration through cross-facial nerve grafting in a rat model of facial paralysis. Plast Reconstr Surg. 2016;138:387–96.

47. Georgiou M, Golding JP, Loughlin AJ, Kingham PJ, Phillips JB. Engineered neural tissue with aligned, differentiated adipose-derived stem cells promotes peripheral nerve regeneration across a critical sized defect in rat sciatic nerve. Biomaterials. 2015;37:242–51.

48. Gu X, Ding F, Yang Y, Liu J. Construction of tissue engineered nerve grafts and their application in peripheral nerve regeneration. Prog Neurobiol. 2011;93:204–30.

49. Wang X, Luo E, Li Y, Hu J. Schwann-like mesenchymal stem cells within vein graft facilitate facial nerve regeneration and remyelination. Brain Res. 2011;1383:71–80.

50. Georgiou M, Bunting SC, Davies HA, Loughlin AJ, Golding JP, Phillips JB. Engineered neural tissue for peripheral nerve repair. Biomaterials. 2013;34:7335–43.

51. Guest JD, Rao A, Olson L, Bunge MB, Bunge RP. The ability of human Schwann cell grafts to promote

regeneration in the transected nude rat spinal cord. Exp Neurol. 1997;148:502–22.

52. Jiang L, Zhu JK, Liu XL, Xiang P, Hu J, Yu WH. Differentiation of rat adipose tissue-derived stem cells into Schwann-like cells in vitro. Neuroreport. 2008;19:1015–9.

53. Gu JH, Ji YH, Dhong ES, Kim DH, Yoon ES. Transplantation of adipose derived stem cells for peripheral nerve regeneration in sciatic nerve defects of the rat. Curr Stem Cell Res Ther. 2012;7:347–55.

54. Zhang Y, Luo H, Zhang Z, Lu Y, Huang X, Yang L, Xu J, Yang W, Fan X, Du B, Gao P, Hu G, Jin Y. A nerve graft constructed with xenogeneic acellular nerve matrix and autologous adipose-derived mesenchymal stem cells. Biomaterials. 2010;31:5312–24.

55. Lopatina T, Kalinina N, Karagyaur M, Stambolsky D, Rubina K, Revischin A, Pavlova G, Parfyonova Y, Tkachuk V. Adipose-derived stem cells stimulate regeneration of peripheral nerves: BDNF secreted by these cells promotes nerve healing and axon growth de novo. PLoS One. 2011;6:e17899.

56. Wei Y, Gong K, Zheng Z, Liu L, Wang A, Zhang L, Ao Q, Gong Y, Zhang X. Schwann-like cell differentiation of rat adipose-derived stem cells by indirect co-culture with Schwann cells in vitro. Cell Prolif. 2010;43:606–16.

57. Sowa Y, Kishida T, Imura T, Numajiri T, Nishino K, Tabata Y, Mazda O. Adipose-derived stem cells promote peripheral nerve regeneration in vivo without differentiation into Schwann-like lineage. Plast Reconstr Surg. 2016;137:318e–30e.

58. Takeda K, Sowa Y, Nishino K, Itoh K, Fushiki S. Adipose-derived stem cells promote proliferation, migration, and tube formation of lymphatic endothelial cells in vitro by secreting lymphangiogenic factors. Ann Plast Surg. 2015;74:728–36.

59. Faroni A, Smith RJ, Reid AJ. Adipose derived stem cells and nerve regeneration. Neural Regen Res. 2014;9:1341–6.

Direct Reprogramming Somatic Cells into Functional Neurons: A New Approach to Engineering Neural Tissue In Vitro and In Vivo

31

Meghan Robinson, Oliver McKee-Reed, Keiran Letwin, and Stephanie Michelle Willerth

31.1 Introduction

The concept of reprogramming mature cells into other phenotypes has been around since the 1960s when Gurdon showed that the nuclei taken from the endothelium of tadpoles could be transplanted into tadpole embryos [1]. The properties of the embryo reprogrammed the nucleus, giving rise to a complete tadpole. This seminal work demonstrated the possibility of reprogramming and this process was called nuclear transfer. It also set the stage for more recent effects in this area of reprogramming cells. Dolly, the cloned sheep produced in 1996, also served as an important step in demonstrating the feasibility of reprogramming mammalian cells using the same nuclear transfer process [2]. While using embryos as a method of reprogramming was validated, researchers in 1981 confirmed that such repro-

gramming could be achieved using transcription factors, proteins that regulate which portions of DNA become converted to RNA, altering the protein expression patterns in a cell [3]. Transcription factors specify what types of mature cells will be produced during development [4]. In a 1987 study, the authors expressed the transcription factor MyoD (associated with the development of muscles) in fibroblasts, causing them to become myoblasts, the long tubelike cells that are found in muscle tissue. This work showed that successful cellular reprogramming could be achieved without the use of embryos.

The invention of induced pluripotent stem cells (iPSCs) also demonstrated the power of transcription factors as tools for cellular reprogramming. In 2006, Takahashi and Yamanaka showed that a combination of four transcription factors (Oct3/4, Sox2, Klf4, and c-Myc) converted mouse fibroblasts into iPSC lines, which possessed the property of pluripotency [5]. The next year, along with other research groups, they generated human iPSC lines from human fibroblasts, showing that such reprogramming could be accomplished in human cells [6, 7]. This reprogramming process works by virally expressing similar transcription factor patterns compared to those found in embryonic stem cells, which are also pluripotent [8]. These iPSC lines can differentiate into any cell type found in the organism from which the cells were derived. These iPSC lines also avoid the ethical issues associated with

M. Robinson · K. Letwin
Biomedical Engineering Program, University of Victoria, Victoria, BC, Canada
e-mail: meghanro@uvic.ca

O. McKee-Reed
Department of Biochemistry, University of Victoria, Victoria, BC, Canada

S. M. Willerth (✉)
Department of Mechanical Engineering and Division of Medical Sciences, University of Victoria, Victoria, BC, Canada
e-mail: willerth@uvic.ca

© Springer Nature Switzerland AG 2019
D. Duscher, M. A. Shiffman (eds.), *Regenerative Medicine and Plastic Surgery*,
https://doi.org/10.1007/978-3-030-19962-3_31

embryonic stem cells as they can be generated from somatic cells.

In 2010, the Wernig group demonstrated that it was possible to convert fibroblasts into cells they called induced neurons (iN) using a similar transcription factor-mediated reprogramming strategy [9]. The generation of iPSCs and iNs required viral overexpression of the necessary transcription factors to achieve reprogramming. Accordingly, the next section of this book chapter focuses on such viral reprogramming methods. More recent methods for altering transcription factor expression patterns include the use of microRNAs, gene editing, small molecules, and functionalized proteins, which will also be analyzed in this book chapter. Cellular reprogramming can help scientists to understand the different features and progression of neurodegenerative diseases like Parkinson's by reprogramming patient's cells into neurons for further characterization and study [10]. In addition to reprogramming somatic cells into neurons in vitro, reprogramming somatic cells into neurons in vivo using reprogramming also serves as an alternative tissue engineering strategy that could potentially restore function to the damaged nervous system [11, 12]. This book chapter covers the current state of reprogramming for applications in neural tissue using these different methods and closes by examining the barriers that must be addressed before such techniques can be translated for clinical applications.

31.2 Viral Mediated Reprogramming of Somatic Cells into Neural Phenotypes

31.2.1 Direct Reprogramming of Somatic Cells into Neurons Using Lentiviral Vectors

Lentiviral vectors, modified versions of wild-type retrovirus virus, retain only the cis-acting elements of the viral genes that allow RNA encapsidation, reverse transcription, and integration. This modification allows them to deliver a specific nucleic acid sequence that encodes the target gene to be expressed but removes their ability to replicate within the host cell [4]. Lentiviruses are unique among viruses in that they reverse transcribe their RNA into a triple-DNA strand which is then imported into the nucleus of the host cell without requiring cell division or disruption to the nuclear membrane. Once inside the nucleus, the triple-strand DNA integrates into the host genome through the action of the viral enzyme integrase, where it can then be transcribed into RNA and used to synthesize proteins [13]. Lentiviral expression of transcription factors was the first established method for direct cellular reprogramming. Once inserted into the host genome of a cell, the host cell transcribes the transcription factor DNA into proteins which go on to orchestrate the transcription of other genes, converting the cell from one phenotype to another [13, 14].

Wernig [15], who had previous experience in generating dopaminergic neurons from iPSCs, was the first to raise the question of whether it was possible to reprogram mature cells directly into divergent lineages. They isolated a set of three transcription factors able to successfully reprogram mouse fibroblasts into functioning neurons by screening a set of 19 genes expressed in neural tissues or involved in epigenetic reprogramming [9]. These three factors are Brn2, Ascl1, and Myt1l, often referred to as the BAM factors. Furthermore, they showed that Ascl1 alone was sufficient to activate the reprogramming process, while Brn2 and Myt1l assisted in maturation of functional properties. Conversion was seen to be rapid, taking only 20 days; however efficiency was low, ranging from 1.8 to 7.7%.

The Wernig group next applied the BAM factors to human fibroblasts, and found that although initial conversion into neurons was successful, maturation of functional properties required the help of a fourth transcription factor, NeuroD1 [16]. These induced neurons were mainly an excitatory neuronal subtype, along with a large percentage of peripheral neurons. They could be generated from fetal and adult fibroblasts with similar efficiencies of 2–4%, showing that even adult cells which possess less plasticity than fetal cells can be reprogrammed. Furthermore, viral

BAMN expression induced endogenous expression of BAMN, showing that once the reprogramming process is activated it is self-regulated and does not require further viral mediated expression. The Wernig group next demonstrated the possibility of reprogramming between cells derived from different germ layers. Adult mouse hepatocytes were successfully converted into neurons using the BAM factors with an efficiency of 3%. In this experiment, they silenced the hepatocyte transcriptome, confirming a complete lineage switch as opposed to a hybrid phenotype consisting of traits from both lineages [17]. Figure 31.1 shows that Ascl1 alone is sufficient to achieve neuronal reprogramming.

Following these experiments, several combinations of transcription factors have been identified for reprogramming specific neuronal subtypes for the treatment of neurodegenerative diseases and spinal cord injury. The rest of this section reviews some of the in vitro and in vivo reprogramming combinations discovered to date. Parkinson's disease results when the dopaminergic neurons of the midbrain degenerate, making these cells attractive cell therapy targets [18]. A screening process of 11 dopaminergic transcription factors plus the BAM factors identified Ascl1, Nurr1, and Lmx1a as the minimal gene set necessary to generate mature dopaminergic neurons from human fibroblasts [19]. Furthermore, these factors could reprogram fibroblasts taken from both healthy subjects and subjects suffering from Parkinson's disease, with respective efficiencies of $5 \pm 1\%$ and $3 \pm 1\%$.

Inhibitory interneuron transplants improve symptoms of epilepsy by correcting an imbalance between excitatory and inhibitory activity in cerebral neuronal networks [20]. Inhibitory interneurons also promote neuronal circuit plasticity and thus are promising therapeutic tools for treating a variety of other neurological disorders including Parkinson's disease and Alzheimer's disease. A screening process identified a set of five factors that can reprogram human fibroblasts into inhibitory interneurons: Ascl1, Dlx5, Lhx6, Foxg1, and Sox2. The first three are associated with inhibitory neuron development; however the last two, Foxg1 and Sox2, are not and so their necessity in direct reprogramming was unexpected. It is hypothesized that Foxg1 is needed to improve the ability of Ascl1 and Sox2 to bind to their targets, and thus it may be a "pioneer" factor for inhibitory interneurons, similar to the way that Ascl1 serves as a "pioneer" transcription factor for excitatory neurons.

In vivo studies began with a proof-of-concept experiment carried out by De La Rossaet al [21] who successfully converted postmitotic neocortical neurons into L5B neurons in mouse embryos using the transcription factor Fexf2. Studies have since identified various transcription factors to convert resident cells in damaged areas of the central nervous system into neurons. Spinal cord injury leads to degeneration of neurons and subsequent loss of motor function [22]. Astrocytes, a type of support cell found in abundance in the central nervous system, switch to a reactive state after spinal cord injury and form a scar in order to protect the integrity of the blood-brain barrier. This scar presents a physical barrier which prevents damaged neurons from re-establishing neuronal relays while secreting inhibitory proteins that suppress axonal regeneration around the injury. Reprogramming resident astrocytes into functional neurons near the injury site has been proposed as a strategy to re-establish neural connectivity and make the injury environment more favorable to axonal regeneration [23]. The transcription factor Sox2 was identified from a pool of 12 factors as the sole factor needed to reprogram resident mouse spinal cord astrocytes into neural precursor cells, with an efficiency of 6–8% as seen in Fig. 31.2 [24]. In this experiment, the cell source of the induced neural precursors was confirmed by genetic lineage tracing to be resident astrocytes. Subsequent treatment with valproic acid, a small molecule that inhibits histone deacetylase enzymes which regulate transcription, induced differentiation of these neural precursors into synapse-forming neurons. In a follow-up study, Sox2 was applied to mouse brains in vivo where it also succeeded in generating neural precursors from resident astrocytes. Promisingly, these cells persisted for months and could also be differentiated into mature neurons using valproic acid, or alternatively by treatment

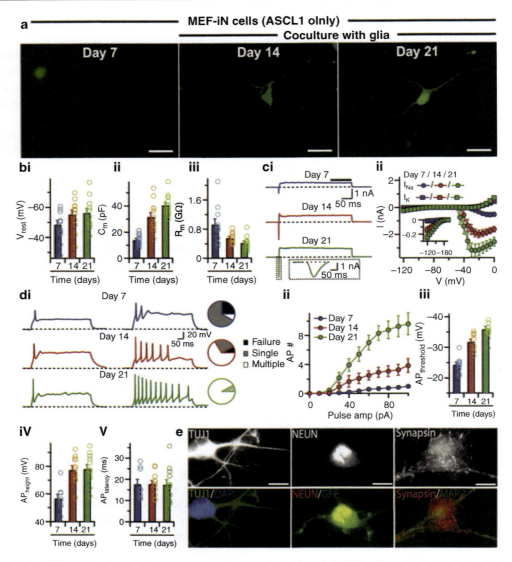

Fig. 31.1 ASCL1 alone is sufficient to generate functional induced neuronal cells from mouse embryonic fibroblasts referred to as MEF-iN cells. (**a**) Gradual development of the morphological complexity of ASCL1-induced single-factor MEF-iN cells at day 7 (left) and after co-culturing with glia until day 14 (middle) or day 21 (right). Scale bars, 10 μm. (**b**) Average values of resting membrane potential (V_{rest}, *i*), membrane capacitance (C_m, *ii*), and input resistance (R_m, *iii*) of ASCL1-induced single-factor MEF-iN cells from day 7 (blue), day 14 (red), and day 21 (green). Bar graphs represent mean values ± SEM ($n = 12$ for individual averages). Open circles of corresponding colors represent values measured from individual cells. (**c**) Example traces of Na⁺/K⁺ currents recorded at $V_{hold} = -70$ mV with a step voltage of 50 mV (*i*) and corresponding averages ± SEM ($n = 12$ for each point), (*ii*) for current–voltage (*I–V*) relationship (filled circles: Na⁺ current and filled squares: K⁺ current) recorded from single-factor MEF-iN cells at day 7 (blue), day 14 (red), and day 21 (green). The black line (upper panel, *i*) indicates time period used for calculating average K⁺ currents. The insets depict expanded views of Na⁺ current (bottom panel, *i*) and reversal of K⁺ current (*ii*). (**d**) Analysis of action potential (AP) firing properties from 1F-iN cells at day 7 (blue), day 14 (red), and day 21 (green). Example traces of single (left) or multiple (right) APs generated by a 90 pA step-current injection, with pie charts representing fraction of iN cells in each condition able to generate single AP (gray), multiple AP (white), or no AP (black) (*i*). Average values presented as means ± SEM ($n = 12$ for individual averages) for AP number with respect to current-pulse amplitude (*ii*), AP threshold (*iii*), AP height (*iv*), and AP latency (*v*). Open circles represent corresponding values measured from individual cells (*iii–v*). (**e**) Immunostaining analysis of 1F-iN cells at day 21 with indicated neuronal markers. Scale bars, 10 μm [63]

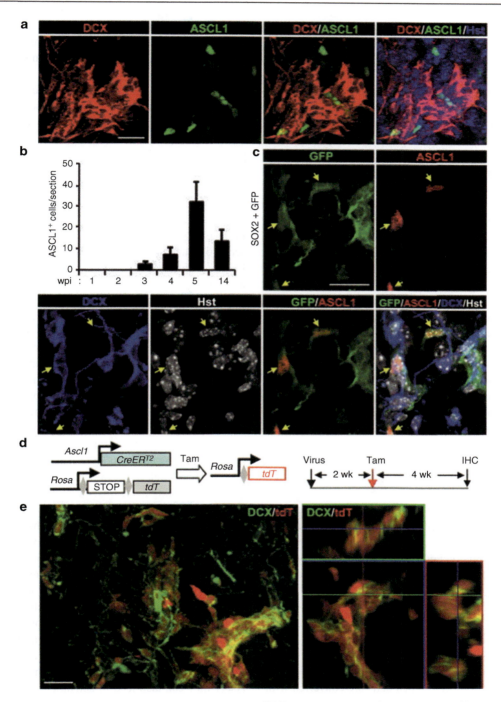

Fig. 31.2 SOX2 induces ASCL1+ neural progenitors from astrocytes in vivo. (**a**) The expression of ASCL1+ cells in striatal regions with DCX+iANBs at 5 wpi. The scale bar represents 20 μm. (**b**) A time course analysis of ASCL1+ cells in the reprogramming area (mean ± SD; $n = 3$ mice at each time point). (**c**) ASCL1 is detected in astrocytes transduced with SOX2 lentivirus. The co-expressed GFP marker is under the control of the human GFAP promoter. The scale bar represents 20 μm. (**d**) A genetic approach to trace derivatives of ASCL1+ progenitors. SOX2-driven reprogramming was induced in adult *Ascl1-CreER[T2];Rosa-tdTomato* (tdT) mice. (**e**) SOX2-induced DCX+ cells pass through an ASCL1+ progenitor stage. Confocal images show genetic labeling of iANBs. An orthogonal view is shown in the right panel. The scale bar represents 20 μm [25]

with growth factor brain-derived neurotrophic factor (BDNF) and the bone morphogenetic protein antagonist Nog [25]. The mechanism of this Sox2 program was seen to follow a distinct sequence involving Ascl1 and the neural precursor gene Doublecortin (Dcx) and give rise to mainly inhibitory interneurons [26].

Another potential application for in vivo reprogramming is the treatment of retinal degeneration [27]. Pollack et al. [28] discovered that Ascl1 can reprogram Muller glia, a type of retinal glial cell, into retinal neurons. Nonmammalian vertebrates regenerate injured retinal neurons by dedifferentiating Muller glia via activation of Ascl1. However, mammalian Muller glia cannot regenerate; instead they form a glial scar. Delivery of Ascl1 to mouse Muller glia cultures and intact retinal explants converted 5% of these cells into bipolar retinal neurons. Interestingly, Insm1 protein levels increased, a protein involved in non-mammalian vertebrate regeneration, suggesting that the conversion mechanism follows a similar process to that first noted in nonmammalian vertebrates. To this end, the study of nonmammalian vertebrates such as fish may hold promise for finding ways to improve the efficiency of retinal regeneration in humans. In another study, Hao et al. [29] tested the ability of the BAM factors to reprogram fibroblast-like cells from human retinas into neurons. While BAM factor expression was unable to reprogram these cells, further testing determined that the combination of Ascl1 with Pax6 was sufficient for reprogramming of these cells, generating a mix of inhibitory neurons and dopaminergic neurons, with an efficiency of 8.78%.

31.2.2 Limitations of Lentiviral Mediated Reprogramming

Directly reprogrammed cells are converted without an intermediate progenitor state or the need for cell division as would be necessary with directed differentiation from stem cells. Avoiding a pluripotent or mitotic state considerably reduces the risk of tumor formation, making direct reprogramming safer for clinical applications.

However, the underlying mechanisms of reprogramming remain unclear and therefore are difficult to predict and develop into a standardized medical practice. To this end, more work is needed to understand this powerful technology. One possible model put forth to describe the mechanism of reprogramming is that of "pioneer" transcription factors, factors which possess the ability to activate genes that are in a repressed chromatin state, also known as "silenced" genes [30]. In this model it is hypothesized that chromatin states at silenced loci fluctuate between repressed and active configurations, spending more time in a repressed state. "Pioneer" transcription factors bind weakly at these loci, stabilizing active configurations over time and allowing more accessible sites for further transcription factor binding. This process eventually leads to active transcription at these loci.

Another major limitation associated with lentiviral delivery is that they have a pattern of integrating within the core of transcribed genes in the host cell, which can lead to cancer-causing mutations. In other cases, it can lead to inappropriate expression of proteins, harming cell function and resulting in genotoxicity. To circumvent the dangers associated with insertional mutagenesis, research groups have modified lentiviral vectors to contain a defective integrase so that they remain as floating nuclear DNA circles within the nucleus; however, this method is less efficient. Additionally, lentiviral vectors harm the cell by eliciting an immune response if their viral capsids and their associated protein products are detected by the host's immune system [13]. Another major drawback is that viral mediated reprogramming remains difficult to control. After infection, protein expression levels from cell to cell can vary greatly. These inconsistent protein expression levels can cause incomplete activation of a transcriptional program, and in the case of Ascl1 they can also lead to activation of a myogenic program resulting in the generation of muscle cells (13, 14). To achieve greater control researchers are engineering switches to activate or deactivate transcription protein expression. Poulou et al. [31] have designed a novel OFF-ON activator of transcription factor expression using the yeast transcription factor Leu3p-αIPM as seen in Fig. 31.3. Acting as a repressor when bound to its UAS$_{Leu}$DNA element, it can become an activator of transcription in the presence of αIPM, a lipid-soluble and metabolically stable metabolite involved in leucine biosynthesis. Although it is only found in prokaryotes, fungi, and plants, it is fully functional and nontoxic in mammalian cells. The group tested the switch in mice using the transcription factor Sox2. After 2 days of Sox2 activation, mouse astrocytes converted into neural progenitor cells with a radial glial phenotype. Radial glial cell proteins nestin and RC2 were upregulated to 85 ± 13.89% and 93 ± 6.97%, respectively, and the astrocyte protein GFAP was downregulated to 25 ± 20%.

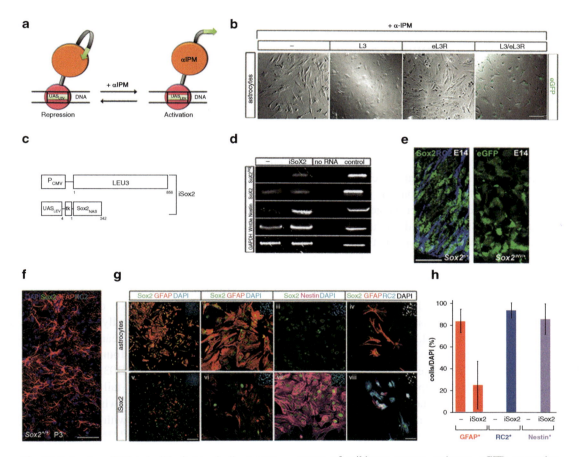

Fig. 31.3 Leu3p-α-IPM inducible fast-track direct reprogramming of astrocytes to neural stem cells. (**a**) Leu3p-α-IPM mode of action: Transcriptional repressor upon binding to the UAS$_{LEU}$ DNA element and transcriptional activator upon α-IPM ligand binding. (**b**) Superimposed bright-field and confocal images of transiently transfected primary P3 murine astrocytes (*i–iv*) with either Leu3p protein (L3) (*ii*) or UAS$_{LEU}$-eGFP reporter (eL3R) (*iii*) or both L3/eL3R (*iv*). eGFP is observed only in cell UAS$_{LEU}$ with both L3/eL3R (*iv*). Scale bar 75 μm. (**c**) Generation of iSox2 expression system under the control of Leu3p UAS$_{LEU}$ elements. (**d**) *iSox2* induces endogenous *Sox2*, *nestin*, and *wnt3a* 48 h post-transfection in P3 primary murine astrocytes. (**e**) Sox2$^+$ and RC2$^+$ neural progenitors in the proliferating zone of E14 cortex of wild-type mouse embryos. eGFP expression detected in the proliferating zone of E14 cortex after *Sox2* ablation in radial glia cells. Scale bar 75 μm. (**f**) Sox2 is not expressed in GFAP$^+$ cells in the proliferating zone of P3 mouse cortex. Scale bar 50 μm. (**g**) iSox2 (*v–viii*) reduces the astrocytic marker GFAP (*i, ii, iv–vi*) in P3 primary murine astrocytes and induces a Nestin$^+$ (*vii*) radial glia (RC2$^+$) (*viii*) NSC phenotype 48 h post-transfection. DAPI staining of nuclei is present in the upper right corner in all panels. Scale bars for panels *i* and *v* represent 250 μm and for panels *ii–iv* and *vi–viii* represent 75 μm. (**h**) Graphs depicting the GFAP$^+$ and either Nestin$^+$ or RC2$^+$ cells in untransfected astrocytes and in reprogrammed cultures. Reprinted with permission [31]

These cells also began to express endogenous Sox2. Overall, lentiviral mediated reprogramming provides a valuable tool for directly reprogramming cells into neural phenotypes. However, it has major limitations that must be addressed before such viral methods can be used for reprogramming in clinical settings with regard to engineering neural tissue from endogenous cells.

31.3 Small-Molecule-Mediated Reprogramming of Somatic Cells into Neural Phenotypes

31.3.1 Direct Reprogramming of Somatic Cells into Neurons Using Small-Molecule Cocktails

An increasingly successful approach for inducing neuronal transdifferentiation of somatic cells uses small molecules to alter the gene expression patterns [32]. Such cocktails induce epigenetic modifications, which are regulations of gene expression caused by alterations in DNA accessibility, changing it from one somatic type into neural lineages. As cells uptake the molecules directly, it avoids several of the complications associated with viral transfection mentioned earlier. For example, He et al. [33] produced an optimal medium known as 5C comprised of N2 supplement, bFGF, leukemia inhibitory factor, vitamin C, and β-mercaptoethanol. This media successfully converted mouse embryonic fibroblasts into Tuj1-positive neuronal-like cells that remained viable after implantation into a mouse brain, and rat astrocytes into electrophysically mature neuronal-like cells in vitro, which also facilitated recovery of brain injury when transplanted into mouse brain. Further, such 5C medium could induce neuronal characteristics when administered to human somatic cell types in vitro.

In another example, Pfisterer et al. [34] utilized human induced neurons to screen 307 compounds (kinase inhibitors, epigenetic modulators, Wnt pathway, nuclear receptors, and phosphatase inhibitors) and found a combination of 6 (Gsk3beta inhibitor kenpaullone, cAMP/PKA modulator prostaglandin E2 (PGE2), adenylyl cyclase activator forskolin, HDAC inhibitor BML210, SIRT1 activator amino resveratrol sulfate, and Src kinase inhibitor PP2) that successfully converted human fibroblasts to MAP2+ human induced neurons with neuronal morphology at concentrations very different from their toxic dose—a highly attractive feature for basal medium development in vitro. These identified conversion-optimizing compounds are not the only that are successful, as many others such as CHIR99021, valproic acid, and ROCK inhibitor Y-27632 work at a subthreshold level as defined by this research group, but show promise in small-molecule reprogramming nonetheless.

Researchers Han et al. [35] used small-molecule reprogramming to convert mouse embryonic fibroblasts into small-molecule induced neuronal stem cells (SMINS-MEF-7). Administrating seven small molecules (valproic acid, Bix01294, RG108, PD0325901, CHIR9901, vitamin C, A83-01) in a six-cycle protocol generated SMINS that stained positive for alkaline phosphatase (a marker for pluripotent embryonic stem, and related cells that exists in tissue specific isoforms), and neural stem cell (NSC) markers Sox2 and Nestin. These cells remained multipotent and morphologically indistinguishable from native NSCs for 2 years of expansion post-experiment. They also found a subset of three (Bix01294, RG108, PD0325901) factors sufficient for conversion of tail-tip fibroblasts (TTF) to SMINS-TTF-3 cells, which share common significant findings with their seven-molecule counterparts. The SMINS-TTF-3 cells express NSC marker genes Sox2, GFAP, Olig2, and Gli2 (which are not expressed by TTFs) and did not express pluripotent genes Oct4 and Nanog. Like most small-molecule-mediated reprogramming studies, the reprogramming mechanism remains unknown. Perhaps the most striking results of this study are the achieved multipotency of both SMINS-MEF-7/-TTF-3 cells with additional small-molecule reprogramming in vitro (both SMINS cell lines were differentiated to astrocytes (GFAP+), neurons (MAP2+), and oligodendrocytes (O4+)).

Li et al. [36] identified a small-molecule cocktail consisting of forskolin, cyclic AMP agonist ISX9, CHIR99021 a GSK3 inhibitor, and SB431542 out of 5000 candidate molecules for endogenous reprogramming of mouse fibroblasts into neurons. Next, Li screened 1500 more small molecules associated with neurite outgrowth and morphology and found that the BET family bromodomain inhibitor protein (I-BET151), which disrupts fibroblast-specific programs, improved reprogramming dramatically. The cocktail was then revised to contain forskolin, ISX9, CHIR99021, and I-BET151. The new combination had a conversion efficiency of >90% TUJ1+ cells with neurite outgrowth after 16 days of induction. Co-culturing the induced neuron-like cells with primary astrocytes for 2–3 weeks allowed further maturation marked by extended neurite outgrowth, a gene expression profile indicative of excitatory glutamatergic neurons, and membranes capable of forming functional synaptic connections with each other. Astonishingly, Li concluded that his chemically induced neurons bypassed an intermediary pluripotent state by witnessing a lack of 5-bromodeoxyuridine incorporation by the TUJ1+ cells throughout the reprogramming process—a finding that coincides with the speculative data of Han et al. [35]. The small molecules replace reported lineage reprogramming genes, activating desired cell-type-specific and silencing initial cell-type-specific gene expression, without the need for exogenous transgenes or cell-fate-specific factors (for example, microRNAs) [36]. This conclusion is therapeutically relevant and applicable to human somatic cell conversion, as demonstrated by Hu et al. [37] who identified a cocktail of seven small molecules (valproic acid, CHIR99021, Repsox, forskolin, SP600125, G06983, Y-27632) that convert human fibroblasts to neuronal cells.

Hu's method also demonstrated a progenitor bypass when the four-molecule cocktail was introduced at optimal concentrations and time course of administration. Hu also replaced the induction media with maturation media containing CHIR99021, forskolin, and dorsomorphin, and extra neurotrophic factors: BDNF, GDNF,

and NT3. The resulting induced neurons expressed the mature neuronal markers Tau, NeuN, and synapsin. The induction protocol was successfully applied to eight different human cell lines. All induced neuronal cells showed electrotonic potential, membrane current, functional glutamate and GABA receptors, and spontaneous calcium transients comparative to those of hiPSC-derived neurons, indicating physiological validity and therefore experimental reliability. Microarray analysis revealed that the induced neuronal cells expressed genes related to neuron differentiation, synapse, and synaptic transmission and downregulated expression of genes related to the extracellular regions, extracellular matrix, and motility. They concluded that CHIR99021 and SP600125 are critical to neuronal gene regulatory network upregulation since only fibroblast gene expression was obtained by removal of these compounds [37].

Zheng et al. [36] used a combination of four small molecules (A-83-01, thiazovivin, purmorphamine, and valproic acid) to reprogram mouse embryonic fibroblasts (MEF) to chemically induced neuron stem cells (iNSCs). TGFβ inhibition by A-83-01 improves efficiency of reprogramming of somatic cells to pluripotent stem cells and neural induction of said cells thereafter [38].

A recent study provides insight into how small-molecule reprogramming can bypass the need for a pluripotent state. While chemically reprogramming mouse fibroblasts to iPSCs, Deng et al. [39] defined an expandable extraembryonic endoderm (XEN)-like state, which could explain small-molecule-mediated reprogramming. Deng's group established a three-step chemical induction process to establish pluripotency from mouse somatic cells. They induced a XEN-like state by administering the following seven compound cocktails, VPA, TD114-2, 616452, tranylcypromine, forskolin, AM580, and EPZ004777, which activate the genes (Sall4, Sox17, Gata4, and Gata6). The XEN-like cells can be identified by the presence of hallmark downregulated fibroblast genes and captured in culture, a feature that makes this method attractive to researchers as they can exploit the plasticity of the newly defined intermediate XEN-like state during small-molecule reprogramming. Deng's group was the first to attempt this by recapitulating and applying chemical compounds that facilitate neural lineage differentiation and reprogramming for neural fate specification to the XEN-state cells with a >50% conversion efficiency after 12 days of induction.

31.3.2 Limitations of Using Small-Molecule-Mediated Reprogramming

While Sect. 3.1 discusses successful applications of small-molecule-mediated reprogramming, these methods still have limitations. Researchers are left with the challenge of finding the proper molecules, concentrations, combinations, methodologies (orchestrated timing of administration and validation protocols), and a basal medium that acts as an effective solvent and life support for the cells. Off-target effects caused by these small molecules also remain a concern.

31.4 MicroRNAs and CRISPR/Cas9 for Reprogramming Somatic Cells into Neural Phenotypes

Cellular reprogramming serves as a critical and promising new avenue of research as well as a key milestone for regenerative medicine. It has the potential to generate neural phenotypes for therapeutic uses as well as provide materials for systematic investigation [40–42]. In this section, we will discuss two methods of cellular reprogramming: miRNA (microRNA)-mediated reprogramming and CRISPR/Cas9-mediated genetic modification of protein expression. MicroRNAs are small, endogenous noncoding RNA molecules, typically 20–30 nucleotides in length, and are integral to the development and growth of plants and animals [41–43]. They regulate gene expression at a posttranscriptional level and affect many key cellular processes including metabolism, differentiation, and cell proliferation [2]. Their role in differentiation makes them an attractive method for altering transcription inside of cells (Fig. 31.4).

MicroRNAs function as a fine tuner of gene expression. They can silence the translation of mRNA by associating with the 3′ UTR (untranslated region) of mRNA (messenger RNA) and marking it for cleavage and/or translational repression by associating with the ribonucleoprotein RISC (RNA-induced silencing complex). Once the miRNA has associated with the RISC, it proceeds to find perfect, or near-perfect matches to the RNA guide (miRNA) and will cleave the mRNA on-site, regulating the overall expression in the cell [44]. Certain miRNAs enhance translation, such as miR-10a which interacts with the 5′ UTR of mRNAs encoding for ribosomal proteins [45].

An important aspect of cellular reprogramming is epigenetic modification, specifically DNA methylation. miRNA can target epigenetic regulators which in turn affect methylation patterns allowing transcriptional machinery to access these genes and promote reprogramming into neural phenotypes [41]. We can use this method of regulation to induce pluripotency of somatic cells and thereby influence and reprogram cell lines without affecting the nuclear DNA. MiRNAs promoting pluripotency generally occur in clustered regions of the DNA, specifically regions miRNA-302-367 [44, 46, 47]. Overexpressing this region in conjunction with optimal growing conditions can lead to the reprogramming of human fibroblasts into functional neurons (39). These regions are highly expressed in ESCs and their mode of action often involves transcriptional silencing/activation and/or chromatin remodeling to promote reprogramming [37].

Recent approaches include direct conversion or transdifferentiation of patient's somatic cells to neurons as well as iPSCs followed by differentiation into neurons [48]. The miRNAs associated with reprogramming fibroblasts to neurons are miRNA-9/9 and miRNA-124 in addition to the cluster mentioned previously (miRNA-302/367) [37, 39]. Reprogramming takes less time and is significantly more efficient than inducing pluripotency and then differentiating [39]. Using transcription factors, small molecules and fusion proteins can enhance the conversion

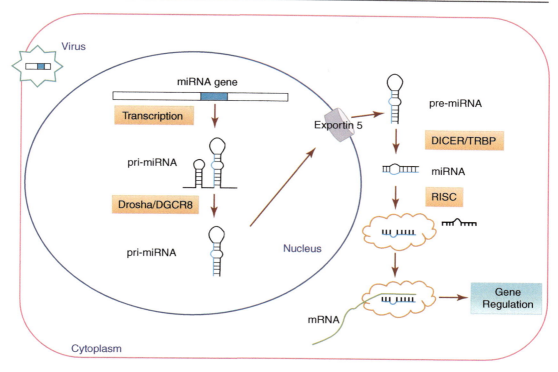

Fig. 31.4 The mechanism for microRNA-mediated gene expression regulation. Reprinted with permission

efficacy [2, 49]. Using transcription factors and miRNAs together, Richner et al. [49] have been able to directly convert human fibroblasts to striatal medium spiny neurons (MSNs).

Expressing the miR302/367 cluster increases the efficiency of reprogramming fibroblasts to iPSCs without the use of exogenous transcription factors. This method is more efficient by two orders of magnitude when compared to the standard Oct4/Sox2/Klf4/Myc-mediated methods.

Disadvantages of the miRNA method include specificity issues due to the multiple potential off targets in the transcriptome [44, 50]. Additionally, the uncontrollable expression after reprogramming induction can increase these off-targeting effects [44, 50]. The primary importance of miRNA is that they directly alter the adult transcriptome and proteome while leaving inheritable DNA unaltered. MiRNA leads to the increased efficiency of cellular reprogramming and decreases the necessity for the use of potentially harmful chemicals which can influence the cell's differentiation in a manner foreign to what it would experience in vivo.

CRISPR-Cas9 (clustered regularly interspaced short palindromic repeats (CRISPR)-associated protein 9) technology radically changed the field of genetic engineering. Until its discovery and implementation, humans relied on clumsy, blunt, and work-intensive methods to alter genomes using gene editing [51]. CRISPR-Cas9's programmable, RNA-guided, endonuclease activity allows precise and efficient genome editing in plants and animals. Its ease of use and customization have allowed many new possibilities for genetic studies in neurological diseases as seen in Fig. 31.5 [52]. CRISPR-Cas9 is a microbial adaptive immune system. The Cas enzyme has the endonuclease activity, and the CRISPR is the region of DNA which codes for the many RNA guides which are used to recognize infiltrating viral DNA/RNA [53]. There are three types of CRISPR systems, but we will focus on type II because of its ease of use and practical applications in the lab. The type II system requires a protospacer adjacent motif (PAM sequence), two RNAs, and a Cas9 enzyme. The two RNAs are a target-specific CRISPR RNA (crRNA) and a

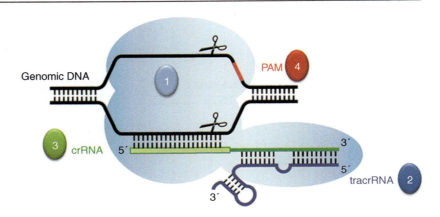

Fig. 31.5 CRISPR-Cas9. This illustrates the Cas9 protein (labeled 1) orchestrating a double-stranded break in a genomic strand of DNA. This double-stranded break can follow one of the two repair pathways, nonhomologous end joining (NHEJ), or homology-directed repair (HDR). Used with permission

trans-activating crRNA (tracrRNA); together these two make up the guide RNA (gRNA) which is named so because it guides the Cas9 enzyme to its specific genomic locus. The PAM sequence is essential to the enzymatic activity of the Cas9. The PAM sequence is a 2–6-nucleotide sequence immediately downstream of the 3′ target region [54]. Without the PAM sequence, the enzyme will not cut the DNA. Cas9 creates double-stranded breaks, cleaving the two phosphodiester bonds (the backbone) of the DNA [55]. Some CRISPR/Cas9 variants, which contain inactivating point mutations at the catalytic site, demonstrate the ability to function as a "nickase" [53]. When incubated with a native DNA plasmid, the mutant Cas9 enzyme performs single-stranded breaks yielding nicked open circular plasmids [53]. It is also possible to inactivate its enzymatic activity so it may be used as a gene location device if fused with a fluorescent protein, or can be used to recruit molecular machinery if fused with a transcription factor [56].

Genetic modification using CRISPR/Cas9 can use homology-directed repair (HDR) and nonhomologous end joining (NHEJ). HDR allows the insertion of DNA into the cut site, enabling the knock-in of genes, and auxotrophic/selection markers. NHEJ relies on the error-prone DNA repair mechanism of the cell. After the Cas9 performs a double-stranded break, the cell enacts its DNA repair which often causes a frameshift mutation resulting in an unreadable and discarded mRNA. NHEJ is a quick way to perform gene knockouts. The CRISPR/Cas9 system has been used to induce endogenous gene expression, allowing the direct conversion of fibroblasts to neuronal cells [57]. This effect is achieved through the overexpression of Brn2, Ascl1, and Myt1 which rapidly remodel the epigenetic state of specific regions of chromatin/DNA [16, 57]. This shows how transcriptional activation and epigenetic remodeling of native master transcription factors can convert between cell types [57]. Human neurons can be differentiated/reprogrammed in vitro from human fibroblasts, hPSCs, and hNPCs [52]. The two genes inactivated by Rubio et al. [52] using CRISPR-Cas9 were TSC2 and KCNQ2 with an 85% efficiency on gene targeting in differentiated cells. However, the new CRISPR/Cas9 system was employed by Rubio et al. [52] in combination with neurogenic factors to generate functional human neurons enriched for the gene modification of interest within 5 weeks [52]. This is a faster, more thorough method of reprogramming.

As we now know, CRISPR/Cas9 system uses DNA targets. However, scientists have recently discovered a single-component, programmable, RNA-guided, RNA-targeting CRISPR effector known as C2c2 [58]. It can be programmed to cleave ssRNA targets carrying the appropriate protospacers, yielding a programmable and easy-to-use system for effecting the expression of cells without altering genomic DNA. The identification of new Cas9 orthologs as well as the engineering of variant strains are leading to new specialized functions, flexibility, and targeting range due to size, PAM recognition, and catalytic variation [52, 59, 60].

31.4.1 Limitations of miRNA- and CRISPR-Cas9-Mediated Reprogramming of Somatic Cells into Neural Tissue

When working with eukaryotic organisms generally off targets increase with the complexity of the genome. When examining partially mismatched sites, four out of six CRISPR RGENs (RNA-guided endonucleases) displayed off-target alterations [61]. Yangfang et al. [61] demonstrated that RGENs, even with up to five base pair DNA-RNA mismatches with the genome, are still highly active in human cells. Doudna et al. [51] have evidence to refute this claim. Using immunoprecipitation assays and high-throughput sequencing, they showed that catalytically inactive RGENs will associate with many regions of the genome, but active RGENs rarely cleave at off-target sites. This implicates a decoupled binding and cleaving event. CRISPR-Cas9 has a limited ability to access regions of heterochromatin. Thus, it is important to note the location and the spatial arrangement of the DNA and how it will affect the binding efficiency.

31.5 Functionalized Transcription Factors as a Way to Directly Reprogram Somatic Cells into Neural Phenotypes

While transcription factors play important roles in cell maintenance and differentiation, they are typically unstable due to their transitory nature. Also, they are not efficiently taken up by cells if introduced exogenously. Modifying factors to allow efficient uptake and improve their stability could serve as a promising alternative to the use of viral based methods and small molecules for applications in reprogramming. iProgen Biotech has created a modification which allows for efficient uptake of transcription factors in human cells and has shown to effect rapid reprogramming using a novel intracellular protein delivery technology called IPTD shown in Fig. 31.6. This modification fuses a target protein to a secretion signal peptide, indicating that the protein should be retained by the cell. The secretion signal of a peptide is usually cleaved during maturation in the cell and so is protected by the addition of a cleavage inhibition sequence. While improving the stability of the

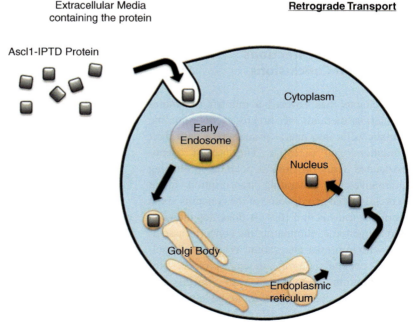

Fig. 31.6 Functionalizing a protein with the intracellular protein delivery technology enables it to be taken up from cell culture media into an early endosome. It is then transported back through the secretion pathway through the Golgi bodies and endoplasmic reticulum where it is then released into the cytoplasm. From the cytoplasm, Ascl1-IPTD can be transported into the nucleus where it regulates gene expression [64]

protein, this combination appears to allow for efficient entry into the cell via receptors on the cell surface, which is unique from conventional mechanisms that involve direct interaction with the phospholipid bilayer. The use of IPTD to deliver transcription factors means protein expression levels can be more precisely controlled and thus easier to translate to a clinical setting. A functionalized version of the transcription factor Ascl1, Ascl1-IPTD, was shown to effect rapid induction of mature neurons from human induced pluripotent stem cells in 12 days. These neurons were NeuN positive and exhibited morphologically mature features like neurite length and branching than those cultured for over 4 weeks using a standard neural differentiation protocol.

Ongoing work is being done to functionalize more transcription to enhance reprogramming efficiencies and generate specific neuronal subtypes. The use of antibodies to target surface cellular markers of specific cell populations is being investigated to allow reprogramming of a subset of a heterogeneous population of cell types for in vivo tissue engineering. Overall, the use of novel functionalized transcription factors serves as a promising, more clinically translatable approach for in vivo reprogramming for neural tissue engineering.

31.6 Future Directions and Conclusions

This chapter has covered a number of studies evaluating methods of directly reprogramming somatic cells into neural tissue along with their limitations and potential. The use of adeno-associated virus provides a potential promising alternative for generating transcription factor expression that avoids some of the concerns about lentiviral transfection [62]. It does not integrate into the genome, eliminating the concerns associated with improper integration. While small-molecule cocktails have shown promise for reprogramming applications, the mechanism behind these cocktails needs to be elucidated. Also, such combinations should be subjected to extensive preclinical testing to ensure that no harmful off-target effects are observed. Such dos-

ing of these cocktails may be hard to translate for clinical applications. Off-target effects also remain a concern with gene editing approaches as well, but this technology is evolving at a rapid pace. In terms of functionalized transcription factors, major concerns involve ensuring that the purity of the protein is sufficient for clinical applications along with dosing similar to the use of small molecules. In terms of future work, direct reprogramming using plasmid-based methods successfully converted adult human fibroblasts into induced neural stem cells. Such DNA-based methods of reprogramming will require more work to determine how to translate delivery and dosing to achieve in vivo reprogramming.

While these studies show the possibility of direct reprogramming, the question remains over whether or not reprogramming endogenous cells will result in enough functional cells to promote recovery after neurological disorders. Even if it is does not, these studies provide insight into how transcription factors alter cell behavior and such insights can be used to engineer exogenous cell therapies for transplantation. In summary, direct reprogramming of somatic cells into neural phenotypes serves as an exciting strategy for engineering tissues both in vitro and in vivo.

References

1. Gurdon JB. The developmental capacity of nuclei taken from intestinal epithelium cells of feeding tadpoles. Development. 1962;10(4):622–40.
2. Stahlhut M, Schambach A, Kustikova OS. Multimodal lentiviral vectors for pharmacologically controlled switching between constitutive single gene expression and tetracycline-regulated multiple gene collaboration. Hum Gene Ther Methods. 2017;28(4):191–204.
3. Gascon S, Paez-Gomez JA, Diaz-Guerra M, Scheiffele P, Scholl FG. Dual-promoter lentiviral vectors for constitutive and regulated gene expression in neurons. J Neurosci Methods. 2008;168(1):104–12.
4. Kafri T, van Praag H, Gage FH, Verma IM. Lentiviral vectors: regulated gene expression. Mol Ther. 2000;1(6):516–21.
5. Takahashi K, Yamanaka S. Induction of pluripotent stem cells from mouse embryonic and adult fibroblast cultures by defined factors. Cell. 2006;126(4):663–76.
6. Takahashi K, Tanabe K, Ohnuki M, Narita M, Ichisaka T, Tomoda K, et al. Induction of pluripotent stem cells from adult human fibroblasts by defined factors. Cell. 2007;131(5):861–72.

7. Yu J, Vodyanik MA, Smuga-Otto K, Antosiewicz-Bourget J, Frane JL, Tian S, Nie J, Jonsdottir GA, Ruotti V, Stewart R, Slukvin II, Thomson JA. Induced pluripotent stem cell lines derived from human somatic cells. Science. 2007;318(5858):1917–20.
8. Yamamizu K, Piao Y, Sharov AA, Zsiros V, Yu H, Nakazawa K, Schlessinger D, Ko MS. Identification of transcription factors for lineage-specific ESC differentiation. Stem Cell Reports. 2013;1(6):545–59.
9. Vierbuchen T, Ostermeier A, Pang ZP, Kokubu Y, Sudhof TC, Wernig M. Direct conversion of fibroblasts to functional neurons by defined factors. Nature. 2010;463(7284):1035–41.
10. Playne R, Connor B. Understanding Parkinson's disease through the use of cell reprogramming. Stem Cell Rev. 2017;13(2):151–69.
11. Srivastava D, DeWitt N. In vivo cellular reprogramming: the next generation. Cell. 2016;166(6):1386–96.
12. Chen G, Wernig M, Berninger B, Nakafuku M, Parmar M, Zhang CL. In vivo reprogramming for brain and spinal cord repair. eNeuro. 2015;2(5):ENEURO.0106-15.
13. Serguera C, Bemelmans AP. Gene therapy of the central nervous system: general considerations on viral vectors for gene transfer into the brain. Rev Neurol (Paris). 2014;170(12):727–38.
14. Treutlein B, Lee QY, Camp JG, Mall M, Koh W, Shariati SA, Sim S, Neff NF, Skotheim JM, Wernig M, Quake SR. Dissecting direct reprogramming from fibroblast to neuron using single-cell RNA-seq. Nature. 2016;534(7607):391–5.
15. Wernig M, Zhao JP, Pruszak J, Hedlund E, Fu D, Soldner F, Broccoli V, Constantine-Paton M, Isacson O, Jaenisch R. Neurons derived from reprogrammed fibroblasts functionally integrate into the fetal brain and improve symptoms of rats with Parkinson's disease. Proc Natl Acad Sci U S A. 2008;105(15):5856–61.
16. Pang ZP, Yang N, Vierbuchen T, Ostermeier A, Fuentes DR, Yang TQ, Citri A, Sebastiano V, Marro S, Südhof TC, Wernig M. Induction of human neuronal cells by defined transcription factors. Nature. 2011;476(7359):220–3.
17. Marro S, Pang ZP, Yang N, Tsai MC, Qu K, Chang HY, Südhof TC, Wernig M. Direct lineage conversion of terminally differentiated hepatocytes to functional neurons. Cell Stem Cell. 2011;9(4):374–82.
18. Jankovic J. Parkinson's disease: clinical features and diagnosis. J Neurol Neurosurg Psychiatry. 2008;79(4):368–76.
19. Caiazzo M, Dell'Anno MT, Dvoretskova E, Lazarevic D, Taverna S, Leo D, Sotnikova TD, Menegon A, Roncaglia P, Colciago G, Russo G, Carninci P, Pezzoli G, Gainetdinov RR, Gustincich S, Dityatev A, Broccoli V. Direct generation of functional dopaminergic neurons from mouse and human fibroblasts. Nature. 2011;476(7359):224–7.
20. Sebe JY, Baraban SC. The promise of an interneuron-based cell therapy for epilepsy. Dev Neurobiol. 2011;71(1):107–17.

21. De la Rossa A, Bellone C, Golding B, Vitali I, Moss J, Toni N, Lüscher C, Jabaudon D. In vivo reprogramming of circuit connectivity in postmitotic neocortical neurons. Nat Neurosci. 2013;16(2):193–200.
22. Fitch MT, Silver J. CNS injury, glial scars, and inflammation: inhibitory extracellular matrices and regeneration failure. Exp Neurol. 2008;209(2):294–301.
23. Sofroniew MV. Molecular dissection of reactive astrogliosis and glial scar formation. Trends Neurosci. 2009;32(12):638–47.
24. Su Z, Niu W, Liu ML, Zou Y, Zhang CL. In vivo conversion of astrocytes to neurons in the injured adult spinal cord. Nat Commun. 2014;5:3338.
25. Niu W, Zang T, Zou Y, Fang S, Smith DK, Bachoo R, Zhang CL. In vivo reprogramming of astrocytes to neuroblasts in the adult brain. Nat Cell Biol. 2013;15(10):1164–75.
26. Niu W, Zang T, Smith DK, Vue TY, Zou Y, Bachoo R, Johnson JE, Zhang CL. SOX2 reprograms resident astrocytes into neural progenitors in the adult brain. Stem Cell Reports. 2015;4(5):780–94.
27. Lamba DA, Karl MO, Reh TA. Strategies for retinal repair: cell replacement and regeneration. Prog Brain Res. 2009;175:23–31.
28. Pollak J, Wilken MS, Ueki Y, Cox KE, Sullivan JM, Taylor RJ, Levine EM, Reh TA. ASCL1 reprograms mouse Muller glia into neurogenic retinal progenitors. Development. 2013;140(12):2619–31.
29. Hao L, Xu Z, Sun H, Luo W, Yan Y, Wang J, Guo J, Liu Y, Chen S. Direct induction of functional neuronal cells from fibroblast-like cells derived from adult human retina. Stem Cell Res. 2017;23:61–72.
30. Vierbuchen T, Wernig M. Direct lineage conversions: unnatural but useful? Nat Biotechnol. 2011;29(10):892–907.
31. Poulou M, Mandalos NP, Karnavas T, Saridaki M, McKay RD, Remboutsika E. A "Hit and Run" approach to inducible direct reprogramming of astrocytes to neural stem cells. Front Physiol. 2016;7:127.
32. Xie X, Fu Y, Liu J. Chemical reprogramming and transdifferentiation. Curr Opin Genet Dev. 2017;46:104–13.
33. He S, Guo Y, Zhang Y, Li Y, Feng C, Li X, Lin L, Guo L, Wang H, Liu C, Zheng Y, Luo C, Liu Q, Wang F, Sun H, Liang L, Li L, Su H, Chen J, Pei D, Zheng H. Reprogramming somatic cells to cells with neuronal characteristics by defined medium both in vitro and in vivo. Cell Regen (Lond). 2015;4:12.
34. Pfisterer U, Ek F, Lang S, Soneji S, Olsson R, Parmar M. Small molecules increase direct neural conversion of human fibroblasts. Sci Rep. 2016;6:38290.
35. Han YC, Lim Y, Duffieldl MD, Li H, Liu J, Abdul Manaph NP, et al. Direct reprogramming of mouse fibroblasts to neural stem cells by small molecules. Stem Cells Int. 2016;2016:4304916.
36. Li X, Zuo X, Jing J, Ma Y, Wang J, Liu D, Zhu J, Du X, Xiong L, Du Y, Xu J, Xiao X, Wang J, Chai Z, Zhao Y, Deng H. Small-molecule-driven direct reprogramming of mouse fibroblasts into functional neurons. Cell Stem Cell. 2015;17(2):195–203.

37. Hu W, Qiu B, Guan W, Wang Q, Wang M, Li W, Gao L, Shen L, Huang Y, Xie G, Zhao H, Jin Y, Tang B, Yu Y, Zhao J, Pei G. Direct conversion of normal and Alzheimer's disease human fibroblasts into neuronal cells by small molecules. Cell Stem Cell. 2015;17(2):204–12.

38. Zheng J, Choi K-A, Kang PJ, Hyeon S, Kwon S, Moon JH, Hwang I, Kim YI, Kim YS, Yoon BS, Park G, Lee J, Hong S, You S. A combination of small molecules directly reprograms mouse fibroblasts into neural stem cells. Biochem Biophys Res Commun. 2016;476(1):42–8.

39. Li X, Liu D, Ma Y, Du X, Jing J, Wang L, Xie B, Sun D, Sun S, Jin X, Zhang X, Zhao T, Guan J, Yi Z, Lai W, Zheng P, Huang Z, Chang Y, Chai Z, Xu J, Deng H. Direct reprogramming of fibroblasts via a chemically induced XEN-like state. Cell Stem Cell. 2017;21(2):264. 273.e7

40. Yang H, Zhang L, An J, Zhang Q, Liu C, He B, Hao DJ. MicroRNA-mediated reprogramming of somatic cells into neural stem cells or neurons. Mol Neurobiol. 2017;54(2):1587–600.

41. Ha M, Kim VN. Regulation of microRNA biogenesis. Nat Rev Mol Cell Biol. 2014;15(8):509–24.

42. Ambros V. The functions of animal microRNAs. Nature. 2004;431(7006):350–5.

43. Bartel DP. MicroRNAs: target recognition and regulatory functions. Cell. 2009;136(2):215–33.

44. Judson RL, Babiarz JE, Venere M, Blelloch R. Embryonic stem cell-specific microRNAs promote induced pluripotency. Nat Biotechnol. 2009;27(5):459–61.

45. Orom UA, Nielsen FC, Lund AH. MicroRNA-10a binds the 5'UTR of ribosomal protein mRNAs and enhances their translation. Mol Cell. 2008;30(4):460–71.

46. Adlakha YK, Seth P. The expanding horizon of MicroRNAs in cellular reprogramming. Prog Neurobiol. 2017;148:21–39.

47. Zhou C, Gu H, Fan R, Wang B, Lou J. MicroRNA 302/367 cluster effectively facilitates direct reprogramming from human fibroblasts into functional neurons. Stem Cells Dev. 2015;24(23):2746–55.

48. Yoo AS, Sun AX, Li L, Shcheglovitov A, Portmann T, Li Y, Lee-Messer C, Dolmetsch RE, Tsien RW, Crabtree GR. MicroRNA-mediated conversion of human fibroblasts to neurons. Nature. 2011;476(7359):228–31.

49. Richner M, Victor MB, Liu Y, Abernathy D, Yoo AS. MicroRNA-based conversion of human fibroblasts into striatal medium spiny neurons. Nat Protoc. 2015;10(10):1543–55.

50. Anokye-Danso F, Trivedi CM, Juhr D, Gupta M, Cui Z, Tian Y, Zhang Y, Yang W, Gruber PJ, Epstein JA, Morrisey EE. Highly efficient miRNA-mediated reprogramming of mouse and human somatic cells to pluripotency. Cell Stem Cell. 2011;8(4):376–88.

51. Doudna JA, Charpentier E. Genome editing. The new frontier of genome engineering with CRISPR-Cas9. Science. 2014;346(6213):1258096.

52. Rubio A, Luoni M, Giannelli SG, Radice I, Iannielli A, Cancellieri C, Di Berardino C, Regalia G, Lazzari G, Menegon A, Taverna S, Broccoli V. Rapid and efficient CRISPR/Cas9 gene inactivation in human neurons during human pluripotent stem cell differentiation and direct reprogramming. Sci Rep. 2016;6:37540.

53. Jinek M, Chylinski K, Fonfara I, Hauer M, Doudna JA, Charpentier E. A programmable dual-RNA–guided DNA endonuclease in adaptive bacterial immunity. Science. 2012;337(6096):816–21.

54. Shah SA, Erdmann S, Mojica FJ, Garrett RA. Protospacer recognition motifs: mixed identities and functional diversity. RNA Biol. 2013;10(5):891–9.

55. Mali P, Esvelt KM, Church GM. Cas9 as a versatile tool for engineering biology. Nat Methods. 2013;10(10):957–63.

56. Esvelt KM, Mali P, Braff JL, Moosburner M, Yaung SJ, Church GM. Orthogonal Cas9 proteins for RNA-guided gene regulation and editing. Nat Methods. 2013;10(11):1116–21.

57. Black JB, Adler AF, Wang HG, D'Ippolito AM, Hutchinson HA, Reddy TE, Pitt GS, Leong KW, Gersbach CA. Targeted epigenetic remodeling of endogenous loci by CRISPR/Cas9-based transcriptional activators directly converts fibroblasts to neuronal cells. Cell Stem Cell. 2016;19(3):406–14.

58. Abudayyeh OO, Gootenberg JS, Konermann S, Joung J, Slaymaker IM, Cox DB, Shmakov S, Makarova KS, Semenova E, Minakhin L, Severinov K, Regev A, Lander ES, Koonin EV, Zhang F. C2c2 is a single-component programmable RNA-guided RNA-targeting CRISPR effector. Science. 2016;353(6299):aaf5573.

59. Kleinstiver BP, Prew MS, Tsai SQ, Topkar VV, Nguyen NT, Zheng Z, Gonzales AP, Li Z, Peterson RT, Yeh JR, Aryee MJ, Joung JK. Engineered CRISPR-Cas9 nucleases with altered PAM specificities. Nature. 2015;523(7561):481–5.

60. Zetsche B, Gootenberg JS, Abudayyeh OO, Slaymaker IM, Makarova KS, Essletzbichler P, Volz SE, Joung J, van der Oost J, Regev A, Koonin EV. Zhang F Cpf1 is a single RNA-guided endonuclease of a class 2 CRISPR-Cas system. Cell. 2015;163(3):759–71.

61. Fu Y, Foden JA, Khayter C, Maeder ML, Reyon D, Joung JK, Sander JD. High-frequency off-target mutagenesis induced by CRISPR-Cas nucleases in human cells. Nat Biotechnol. 2013;31(9):822–6.

62. Chan KY, Jang MJ, Yoo BB, Greenbaum A, Ravi N, Wu WL, Sánchez-Guardado L, Lois C, Mazmanian SK, Deverman BE, Gradinaru V. Engineered AAVs for efficient noninvasive gene delivery to the central and peripheral nervous systems. Nat Neurosci. 2017;20(8):1172–9.

63. Chanda S, Ang CE, Davila J, Pak C, Mall M, Lee QY, Ahlenius H, Jung SW, Südhof TC, Wernig M. Generation of induced neuronal cells by the single reprogramming factor ASCL1. Stem Cell Reports. 2014;3(2):282–96.

64. Willerth SM. Engineering personalized neural tissue using functionalized transcription factors. Neural Regen Res. 2016;11(10):1570–1.

Index

A

Abdominoplasty, 71, 83
Abnormal scar development, 20
Absorbable bone substitute materials, 219
Acellular dermal biological matrix, 130
Acellular human matrix, 142
Acellular scaffolds, 30
Acid-based (method B) and enzyme-based (method C)
    protocols, 74
Activated bone substitutes, 189
AdEasy technology, 264
Adhesion molecules, ECM, 8, 9
Adipogenesis, 71, 72
Adiponectin, 77
Adipose stem cells (ASCs), 390
Adipose tissue complex (ATC), 49, 68
Adipose tissue engineering, 72
Adipose tissue matrix preparation, 72, 73
Adipose tissue transfer, 39
Adipose-derived adult stem/stromal cells, 58–60
Adipose-derived cellular stromal vascular fraction
    (AD-cSVF), 54
Adipose-derived matrix (ADM), 72, 85
Adipose-derived stem cells (ADSCs), 40, 263, 288,
    373, 374
    in adipose tissue, 112, 113
    clinical challenges, 441
    dASCs, 442
    electrical stimulation, 442
    injured neurological tissue, 441
    microcarriers, 83
    rat sciatic nerve injury model
        abdominal fat, 440
        Cre-loxP-mediated fate tracking technology, 443
        decellularized matrices, 443
        hind limb autotomy, 439, 440
        in vivo transdifferentiation, 440, 441
        injury site, 443
        larger nerve gaps, 443
        long-term results, 441
        peripheral nerve regeneration, 438, 439
        regeneration, 441
        teratomas, 443
    requirement for, 437

Adult dermal repair system, 20
Adult mammalian skin damage, 19
Adult multipotent cells, 53
Adult stem cells and embryonic stem cells, 129
Adult tissue-derived MSCs, 129
Advanced BioHealing, 124
Advanced drug delivery systems (DDSs), 158
Adventitia, 47
Aesthetic-plastic surgical applications, 49
Aggrecan (ACAN) mRNA, 269
Aggregation and degranulation of platelets, 5
Aging
    autophagy, 318
    evidence, 317
    extrinsic changes, 316, 317
    ex-vivo pharmacological inhibition, 318
    intrinsic changes, 314, 316
    parabiosis experiments, 318
    pre-senescence state, 314, 315
    regeneration, 157–162
    Sarcopenia, 313
    satellite cells
        activation, 313
        cell cycle exit and return, 313
        heterogeneity, 312
        MRFs, 311
        Pax7, 311
        population of, 311
        quiescent state, 312, 313
    SCs, 317
    senescence/apoptosis, 317
Alcian blue-stained cultures, 265
Alizarin red stain procedure, 169
Alkaline phosphatase (ALP) assay, 266
Alkaline phosphatase stain procedure, 170
Alloantigenicity, 77
Alloderm®, 30, 142
Allogeneic mesenchymal stem cells,
    129, 130
Allograft adipose matrix (AAM), 72
Ambulatory surgical centers (ASC), 60
Amnion, 139
Angiogenesis, 196, 358
Angiogenesis-mediated effect, 186

© Springer Nature Switzerland AG 2019
D. Duscher, M. A. Shiffman (eds.), *Regenerative Medicine and Plastic Surgery*,
https://doi.org/10.1007/978-3-030-19962-3

Animal Experimentation Ethics Committee
of Bavaria, 212
Animal Study, surgical procedure, 212, 213
Applied regenerative efforts in wound healing and
orthopaedic applications, 64
Aquacel®, 140, 141
Articular cartilage
ACI, 251
ageing chondrocytes, 251
clinical settings, 255, 256
diarthrodial joints, 249
growth factors, 251, 253
integrins, 250
invasive procedures, 250
MSCs (*see* Mesenchymal stem cells)
osteoarthritis, 250
regeneration potential, 250
rehabilitation, 251
scaffolds, 251, 253–255
tidemark, 249
tissue homeostasis, 250
Artificial biological materials, 124
Asymmetrical cell division, 47, 53
ATP binding cassette (ABC) transporter family, 129
Autologous and non-autologous exosomes and
microvesicles, 64
Autologous bone marrow-derived mesenchymal stem
cells, 130
Autologous chondrocyte implantation (ACI), 251
Autologous dermo-epidermal (composite) skin
substitute, 126, 127
Autologous fat grafting, 71
Autotransplants, 197

B
Bacterial contamination, 27
Basal keratinocytes/stem cells, 93, 94
β-catenin pathway, 148, 151, 237, 239
Beta-tricalcium phosphate (β-TCP), 168, 169, 172
Bioactive compounds or drugs for skin regeneration,
local application, 159
Bioactive scaffolds, 30, 31
Biobrane®, 125, 139, 140
Biocellular applications, 57
Biocellular medicine, 48, 49
Biocellular regenerative matrix, 49, 58
Biocellular treatment, 61, 67, 68
Biochemical assays, 76, 77
Biocompatible dressing, 23
Biodegradation, 216
Bio-implants, 168
Biological and biomechanical processes, 5
Biological components, 49, 51–53
Biological therapies, 30
Biologic and cellular therapeutic concepts, 53–55, 57
Biomimetic wounds, 21
Biomimetics of fetal wound healing, 19
Bleomycin, 24
Body contouring procedures, 71, 83

Bone apposition rates after surgery, 215
Bone formation and implant resorption, 215, 216
Bone grafting, 168, 172, 181, 211
Bone grafting materials, 190
Bone implants, 171
Bone marrow-derived mesenchymal stem cells
(BM-MSCs), 111, 253
Bone marrow mesenchymal stem cells (BMSC), 77, 288,
289, 373, 374
Bone marrow stimulation (BMS), 374
Bone morphogenic protein 2 (BMP2), 77, 350
AdGFP, AdBMP2, and AdSox9
ALP assay, 266
immunocytochemical staining, 265, 266
matrix mineralization assay, 266
mouse fetal limb explant culture, 267
subcutaneous stem cell implantation, 267
Western blot analysis, 266
Alcian blue, 267
bone marrow stimulation techniques, 263
cartilaginous pathologies, 263
experimental protocols, 264
H&E, 267
HEK 293 and C3H10T1/2 cell lines, 264
histological evaluation, 267
Masson's trichrome, 267
micromass cell pellet, 265
MSCs, 263, 264
chondrogenic differentiation, 269, 270
chondrogenic differentiation, micromass culture,
264, 265
osteogenic differentiation, 270
osteogenenic differentiation, monolayer
culture, 265
treatment groups, 267
paraffin-embedded sections, 267
recombinant adenoviruses, 264
RNA isolation, 265
RT-PCR, 265
Safranin O-fast green, 267
semi-quantitative PCR, 265
Sox9, 264
chondrogenic differentiation, 269, 270
osteogenic differentiation, 270
treatment groups, 267
statistical analysis, 267
Bone morphogenetic proteins (BMP), 182, 184, 239,
240, 252
Bone repair and regeneration, 172, 212, 218, 219
Bone substitutes, 182, 189, 192, 193, 202, 216
Bone-related gene markers, 77
Bone tissue regeneration, 85, 86
Breast reconstruction, 84
Breast reduction procedures, 71
Burn wound Care, 137

C
Calcium sulfate (CS), 211, 212, 216, 217, 219
Canonical Wnt signaling pathway, 148, 149

Index 465

CCAAT-enhanced binding protein alpha (CEBPα), 83
CD45+antigen-presenting fibroblasts, 111
Cell adhesive proteins, 77
Cell-assisted lipotransfer (CAL), 71
Cell-based therapies, 5, 11
Cell delivery methods, 30
Cell enriched biocellular grafts, 67
Cell supplementation, 80
Cell therapies
  acute hand tendon injuries, 385
  acute tendon laceration, 390
  benefit/economic investment ratio, 403
  collagens, 388, 389
  elastin, 388
  GAGs, 389
  glycoproteins, 389
  GMP, 390
  healing process, 389
  inflammatory phase, 390
  innovative therapies, 392
  proteoglycans, 388, 389
  repair and regeneration
    Achilles heel tendon, 395
    adult cell sources, 393, 394
    characterization, 396–398
    ESCs, 392, 393
    fetal progenitor cells, 394, 395
    hydrogel solutions, 399, 400
    matrix solutions, 400–403
    patient safety, 392
    phenotypic stability, 399
    population homogeneity and stability, 397–399
    stability, 396, 397
    tissue regenerative properties, 395
    treatments, 396
  sport practice, 385
  structure and role, 386–388
  surgical intervention, 390
  tendon healing, 390, 391
  tendon injury, 389, 390
  translational research, 403
Cellular and molecular mechanisms of scarless healing, 5
Cellular-autologous products (CEA), 121, 123
  cost of production, 122
  IntegraTM combined use, 125, 126
  microskin autografting, 128
  Widely-Meshed Autografting, 127
Cellular based therapy, 57
Cellular/biocellular regenerative therapy, 61, 62
Cellular degradation, 218
Cellular stromal vascular fraction (cSVF), 55
Cellular therapy, 60, 63
Centrifugation, 57
Characterization assays, hDAM, 75
Chinese-originated microskin autografting, 127
Chondrocyte-derived progenitor cells (CDPC), 288
Chondroitin sulfate, 9
Chronic wound and musculoskeletal (MSK) applications, 58

Clinical trials, 67
Clinical-translatability and tissue-engineering applications, 79
Closed syringe microcannula lipoaspiration, 54
Clustered Regularly Interspaced Short Palindromic Repeats-CRISPR associated protein 9 (CRISPR-Cas9), 457–459
Coleman technique, 40
Collagen, 9
Collagen type II alpha 1 (Col2a1), 266
Colony forming efficiency assay, 122
Compact bio bone cells, 172
Condensed mesenchymal cell bodies (CMBs), 252–253
Contact dermatitis, 22
Cosmesis and function, 19
Cryosurgery, 24
Cultispher-S, 82
Cultured epidermal autografts (CEA), 118, 121
Cuono's method, 121, 127
Cytochemical staining of CFU-Fs, 170

D
dASCs, 442
Decellularization process, 72–75
Decellularization protocols, 83
Decellularized adipose tissue (DAT), 72
Decorin, 9
Deferoxamine (DFO), 29
Delipidization and decellularization of lipoasparate-derived ECM, 81
Dermal capillaries, 95
Dermal repair, 160
Dermal substitutes, 20, 24
  pre-clinical and clinical settings, 123
  using synthetic materials, 124
Dermalogen, 21
Dermis, 92, 94, 117
Dermoepidermal composite grafts, 21
Detergent-based adipose ECMs, 81
Detergent-based hDAM, 79, 81
Detractors, Cuono's method, 121, 122
Developmental signaling pathways in mammalian skin development and repair, 153
Dexamethasone, 252
Diabetic foot ulcers, 130
Direct gene transfer, 196
Dishevelled (Dvl), 237
DNA encoding, 197
DNA isolation/quantification, 76
Downregulating collagen synthesis, 24
Drug administration pathway, 159
Drug delivery routes for wound healing applications
  clinical practice, 29
  Deferoxamine, 29
  localized drug delivery, 29
  problems and preventive/solving strategies, 28
  systemic drug administration, 29
  systemic toxicity, 29
Drug delivery systems (DDS), 27, 157–162

E

Ear regeneration
history, 281
maturation process, 292, 293
microtia
congenital malformation, 281
grades, 281
incidence, 281
pathophysiology, 281
reconstruction of, 282
animal chondrocytes, 284, 286
BMSCs, 289
cell cultures, 284
cell expansion techniques, 284
costal cartilage, 282, 284
fixation, 284
growth factors, 292
human chondrocytes, 286, 287
Medpor, 284
microtia ear, 288
neonatal *vs.* adult chondrocytes, 289
scaffolds, 289–291
skin, 284
stem cell regeneration, 288
research, 293
results, 293
Electrospin silk fibroin mat (CSF), 376
Electrosprayed microcarriers, 82
Embryogenesis, 148
Embryonic skin development, 147, 148
Embryonic stem (ES) cells, 359
Embryonic stem cells (ESC), 372
Embryonic stem cells (ESCs), 392, 393
Embryonic-stage cells, 129
Encapsulation efficiency, 161
Endoplasmic reticulum (ER), 237
Endotenon, 386
Endothelial cells, 95
Endothelial cells in 3D organotypic co-cultures with
fibroblasts and keratinocytes, 103
Endothelial cells of dermal microvessels, 102
Engineered tendon matrix (ETM), 376
Enhancement of cellular and biologic therapies, 48
Enthesis, 386
Enzyme-linked immunosorbant assay (ELISA), 440
Epidermal and bulb stem cells, 148
Epidermal growth factor (EGF), 350
Epidermal homeostasis, 148
Epidermal regeneration, 91, 95
Epidermal stem cell depletion during graft
preparation, 129
Epidermal stem cells, 93, 129
Epidermis, 92, 93, 117
Epithelialization, 21
Erythropoietin (EPO), 415
Evidence-based wound care guidelines for biologic
therapies, 32
Evolution of Cell-Based Therapies, 46–48
Evolution of Regenerative Medicine in Plastic Surgery,
45, 46

Extracellular matrix (ECM), 8–10, 72
accumulation, 3
acellular biologic surgical meshes, 349, 350
adhesion molecules, 8
clinical outcome, 351, 352
musculotendinous injuries, 349
myogenesis, 350, 351
phenotypes, 8
proteins, 313
technical challenges, 349
Extrinsic aging, 157

F

False color computed tomography, 305
Fascicular fingers, 413
Fat grafting, 71
Feeder cell type with keratinocytes, 103
Fetal and adult wound healing, 5, 152
Fetal bovine serum (FBS), 362
Fetal cartilage-derived progenitor cells (FCPC), 288
Fetal dermis, 20
Fetal fibroblasts, 10
Fetal platelets, 5
Fetal scarless wound healing, 5, 19
Fibrillogenesis, 9
Fibroblast growth factor (FGF)-2, 293
Fibroblasts, 94, 95
Fibromodulin, 9
Fibronectin, 8
cellular deposition and remodeling, 79
unfolding, 79
Fibrosis, 10, 71
Flow-cytometry analysis, 169
Fluorescence colors of bone tissue specimens, 215
Fluorescence microscopy after polychrome sequence
labeling, 215
Fluorochrome labeling, 215
Flynn methods, 74
Flynn protocol, 74
Förster resonance energy transfer (FRET) analysis, 79
Fracture healing, 235
Fresh allograft skin, 138
Frizzled proteins, 148
Functional wound healing therapy, 19

G

Gene constructs, 194
Gene Ontology (GO), 78
Gene therapy, 193
Gene-activated bone substitutes, 193–198, 200
Gene-activated materials, 194, 196
Genes encoding growth factors/ hormones, 195
Genes encoding transcriptional factors, 196
GHK(glycyl-L-histidyl-L-lysine)-Cu peptide, 160
Glycerolized allograft, 138
Glycosaminoglycans (GAGs), 8, 370, 389
Good manufacturing processes (GMP), 390
Grading scars, 21

Index
467

Grafting, 42, 121
GraftJacket, 30
Granulocyte colony-stimulating factor (G-CSF), 378
Growth factor and progenitor based therapies, 30–32

H
Hair regeneration, 66
Harvested fat, processing before injection, 41, 42
Harvesting procedure, 71
Harvesting, transplantation tissue, 39–41
Healing phenotypes of adult and fetal skin, 4
Health care paradigm change, 68
Hedgehog (Hh) pathway, 240
Hedgehog (Hh), transforming growth factor-beta (TGF-beta), 152
Hematopoietic stem cells (HSC), 187
Hematoxylin and eosin (H&E) staining, 73, 77, 267
Hemostasis, 151
Heparin/fibrin-based delivery system (HBDS), 363
High density platelet concentrates (HD-PRP), 54
Histological staining, 76
Histomorphological analysis, 214, 215
Home-based Functional Electrical Stimulation (h-bFES), 304–306
Homeostasis, 53
Homology-directed repair (HDR), 458
Human allograft skin, 138
Human amniotic membrane, 139
Human bone mesenchymal stem cells (hBMSCs), 263
Human decellularized adipose extracellular matrix (hDAM), 72
    components, mass spectroscopy, 78
    isolation methods, 74
    mechanical properties, 79
    microcarriers, 82, 83
    porosity, 79
    supplementation, 80, 81
    3D architecture, 82, 83
    without cell supplementation, 84
Human epidermal and mesenchymal stem cells, 129
Human fetal progenitor tenocyte (hFPT), 397
Human induced pluripotent stem cells (hiPSCs), 129, 372
Human placenta-derived mesenchymal progenitor (stromal) cells (HPMSC), 374
Human skin structure, 98, 99
Human tendon-derived hydrogel (tHG), 362
Human umbilical cord-derived mesenchymal stem cell (HU-MSC), 110, 111
Human umbilical vein endothelial cells (HUVECs), 81
Hypertrophic/keloid scarring, 152
Hypodermal cells, 102
Hypodermis, 94

I
Immune cells, 102
Immune surveillance and synthesis of biological mediators, 91
Immunephenotyping by flowcytometry analysis, 170

Immunohistochemical staining (IHC), 77
Immunomodulatory effect of MSCs, 130
Immunostaining, 74
In murine fetal and embryonic skin wounds, 5
In vivo imaging system (IVIS) assays, 362
Induced neurons (iN), 448
Induced neuron stem cells (iNSCs), 455
Induced pluripotent stem cells (iPSCs), 112, 359, 393, 394, 447
Inflammatory cells, 5, 7
Inflammatory processes, 151
Inflammatory signaling molecules, 7
Inhibitory interneuron transplants, 449
Insulin-like growth factor-1 (IGF-1), 77, 252
Integra®, 142
Interferon gamma, 24
Intracellular cascade pathway of SDF-1 signal transduction, 188
Intracellular cascade pathway of VEGF signals, 186
Intracellular posttranscriptional mechanism, 194
Intracellular Wnt signaling, 148
Intrinsic skin aging, 157
iScript cDNA synthesis kit, 265

K
Keratinocyte growth and regeneration, 103
Keratinocyte proliferation and generation, 102, 103
Knockout techniques, 10

L
Laser-assisted liposuction (LAL), 40
Laser-assisted scar healing (LASH), 23
LEF/TCF DNA-binding transcription factors, 149
Lentiviral vectors
    iPSCs, 447
    neurodegenerative diseases, 448
    somatic cells
        CRISPR/Cas9, 456–459
        direct reprogramming, 448, 449, 452
        limitations, 452–454
        miRNA, 456, 457, 459
        plasmid-based methods, 460
        small molecule cocktails, 454–456
        small-molecule-mediated reprogramming, 456
        transcription factors, 459, 460
Ligament stem cells (LSC), 374
Lipid nanoparticles (LN), 161, 162
Lipoaspirate for soft tissue reconstruction, 39–42
Lipoaspirated tSVF via mechanical emulsification, 65
Lipofilling effect, 71
Liposomes, 160, 161
Liposuction, 71, 83
Loss-of-function mutations, 184

M
Macrophages, 5
Mammalian epidermal healing, 3

Masson's trichrome staining, 76
Mast cells, 7
Matriderm®, 142
MatrisomeDB, 78
Matrix metalloproteinases (MMPs), 9
Matrix porosity, 79
Matrix protein denaturation and degradation, 74
Mechanical based method, 74
Mechanical stress, 9
Mechanostress testing, 76, 79
Mechanotransduction, 9, 10
Meek method, 128
Mepilex Ag®, 140
Mesenchymal cells, human skin regeneration
    epidermal and dermal cell types, 91
    protective skin barrier, 91
Mesenchymal stem cells from gingiva (GMSC), 374
Mesenchymal stem/stromal cells (MSCs), 22, 23, 102,
        235, 263, 264, 373, 393
    adult tissues, 252
    BMPs, 239, 240
    chondrogenic potential, 252, 253
    clinical applications, 251
    CMBs, 253
    chondrogenic differentiation, 269, 270
    chondrogenic differentiation, micromass culture,
        264, 265
    gene variation, 239, 240
    growth factors, 238
    Hh pathway, 240
    immuno-modulation property, 253
    induction of, 238
    in vitro micromass culture method, 252
    lack of in-depth, 252
    liposomal vesicles, 238
    LRP5 receptor, 239
    miRNA, 240, 241
    mutations, 239
    non-canonical ligand, 239
    Notch pathway, 240
    osteoblast maturation, 238
    osteochondral defects, 251
    osteogenic differentiation, 239, 241, 270
    osteogenic differentiation, monolayer culture, 265
    progranulin, 240
    properties, 251
    skeletal healing, 238
    teratogenic and ethical issues, 251
    terminal differentiation, 251
    tissues types, 238
    TNFα, 240
    treatment groups, 267
    types, 251, 252
Meshing, 141, 142
Methylmetaloproteases (MMPs), 370
Micro-computed tomography (micro-CT), 214
Microcyn scar management hydrogel, 21
Micrograft transplantation, 128

Micronized & "NanofatTM" (emulsified AD-tSVF), 65, 66
MicroRNA (miRNA), 240, 241, 456, 457, 459
Microskin autografting, 127, 128
Microvascular capillary system, 47
Mineral matrix deposits and bone nodules of
    osteoblasts, 169
Mitogen-activated protein kinase (MAPK) pathway, 313
Modified Meek technique in combination with IntegraTM
    dermal template, 128
Monocytes, 5
Mouse embryonic fibroblasts (MEFs), 263
MSC-like identity of perivascular progenitor cells, 223
MSC-like pericyte population, 95
μDERM, 21
Multiphoton autofluorescence second harmonic
    generation (SHG) imaging, 79
Murine decellularized adipose tissue extracellular matrix
    (mDAM), 82
Muscle regeneration
    degeneration, 302
    h-bFES, 304–306
    implantable devices, 306
    long biphasic impulses, 306
    muscle plasticity, 302
    myogenesis
        denervated skeletal muscles, 302, 303
        long-term denervation, 303, 304
    myosins, 302
    in vivo protocols, 306
    pilot human studies, 306
    post-denervation atrophy, 302
    skeletal muscle results, 301
Muscle regulatory factors (MRFs), 311
MyoD, 447
MyoD1, 344
Myotendinous junction, 386

N
Nanoemulsions, 161
Nanoencapsulation techniques, 159
Necrosis, grafted adipocytes, 71
Negative and Positive Pressure Treatment, 23
Neovascularization, 71
Neovasculgen, 199
Neural crest stem cells (NCSC), 374
Neural stem cells (NSCs), 168
Neutrophil–endothelial cell interactions, 5
Nicotinamide adenine dinucleotide (NAD+), 316
Nondiabetic and diabetic hDAM on adipocyte metabolic
    function, 85
Nondiabetic hDAM, 85
Nonhomologous End Joining (NHEJ), 458
Nonhuman-derived acellular biological matrices, 31
Non-vascularized bone autograft, 201
Normothrophic bone callus, 201
Nuclear transfer, 447
Nucleated cells, 49

Index

O

Obesity-associated fibrosis, 77
Obesity-associated hDAM, 79
Olfactory ensheathing cells (OECs), 418
Oncological and inflammatory diseases, 181
Optical density (OD), 265
Oral stem cells, 374
Organotypic bone recovery, 188
Orthotopic models of bone formation, PPCs, 226
Osteoconduction, 172, 189, 194, 212, 219
Osteoconductive potential of CS, 218
Osteoconductive purposes, 216
Osteoconductive structure, 211
Osteogenic insufficiency, 188, 189, 198
Osteogenic protein 1 (OP-1), 241
Osteogenicity, 189
Osteoinduction, 183–184, 189, 194, 211
Osteopontin (OPN), 265
Osteoprotection, 189
Oxidative stress, 27

P

p38/MAPK signaling pathway, 113
Papillary dermal fibroblasts, 94
Paracrine keratinocyte-feeder cell cross-talk, 103
Paracrine secretory capabilities, 49
Parathyroid hormone-related peptide (pTHRP), 252
Partial-*vs.* full thickness burns, 143
Particle size separation, 76
Pericyte cell therapies for fibrous non-union, 226
Pericytes, 95, 101, 102, 224
    and adventitial cells, 225
    regional specification, 227
Periodontal ligament stem cells (PDLSC), 374
Periosteal-MSC, 253
Peripheral blood stem cells (PB-SCs), 167
Peripheral blood-MSCs (PBMSCs), 374
Peripheral nerve regeneration
    adipose tissue harvesting sites, 417
    advantages and disadvantages, 411, 412
    clinical applications, 417, 418
    end-to-side coaptation/neurorrhaphy, 415
    epineural nerve repair, 412
    experimental research, 411
    gene therapy, 417
    microsurgical techniques, 411
    molecular biology, 415, 416
    nerve allografts, 413, 414
    nerve autografts, 413
    nerve conduits, 414, 415
    nerve grafting, 412, 413
    nerve transfer, 415
    reeducation training, 418
    rehabilitation, 418
    stem cells, 416, 417
Peripheral nervous system (PNS), 425, 426
Perivascular cell therapy, 226

Perivascular-derived stem cells (PDSC), 374
Perivascular progenitor cells, 223, 224
    bone regeneration, 223–227
    in ectopic bone formation, 225, 226
Perivascular stem/stromal cell isolation (PSC), 225
Peroxisome proliferative activated receptor gamma
    (PPARγ), 77
Peroxisome proliferator-activated receptor delta
    (PPARD), 239
Pharmacokinetic properties, DDS, 158
Phosphate buffered saline (PBS), 265
Physical therapy, 348
PicoGreen assay, H&E staining, 74
Planar cell polarity (PCP) pathway, 148, 235
Platelet-derived growth factor (PDGF), 350
Platelet rich growth factors (PRGF), 255
Platelet-rich plasmas (PRPs), 378, 390, 394
Pluripotent and multipotent PB-SCs, 172
Polar solvent extraction, 74
Polychrome sequence marking, 217
Polycomb repressive complex 2 (PRC2), 314
Polyhydroxyalkanoates (PHA), 376
Poly-lactide-co-glycolic acid (PLGA), 376
Polymeric based treatments, 21
Porcine acellular dermal matrix (PADM), 333, 343, 344
Porcine small intestine submucosa (SIS), 376
Porcine xenograft, 139
Post-mastectomy breast reconstruction, 84
Postsurgical deformations, 181
Power-assisted liposuction (PAL), 40
Precursor replacement cell, 53
Pressure garments and bandages, 23
Primers, 171
Progenitor cell supplementation of hDAM, 83
Pro-inflammatory interleukins, 7
Proliferative phase of healing, 151, 152
Proteoglycan matrix modulators, 9
Proteomic capabilities, 74
Pyrvinium, 29, 33

Q

Quantitative muscle color–computed tomography
    (QMC-CT), 305

R

Radiation/solar damaged skin and small joint targeted
    applications in orthopaedic medicine, 66
Rat model for cavernous nerve injury, 84
Real time-polymerase chain reaction (RT-PCR)
    procedure, 170, 171, 265
Recombinant erythropoietin (rEPO), 29
Recombinant human epidermal growth factor
    (rhEGF), 162
Reconstructive surgery, 197, 211
Reduction mammoplasty, 83
Regeneration process, 147

Regenerative applications, hDAMs, 77, 78
Regenerative components, 71
Regenerative healing, 152
Regenerative medicine, 83
Regenerative properties, hDAM implantation, 77
Regranex, 31
Remodeling, wound healing, 152
Repairing process, 168
Reparative osteogenesis, 182–184, 189, 195–196, 202
Resorbable bone substitute materials, CS
          formulations, 212
Reticular dermal fibroblasts, 94
Retinoids, 23
Rib autograft, 200
RNA interference, 34
RNAi therapies, 33–35

S
Scaffolds, 162
Scar revision therapy, 24
Scar scales evaluation, 22
Scar treatment and prevention, 19
Scarless cutaneous wound healing, 4
Scarless fetal wounds, 12
Scars prevention, 22
Schwann cells
    activation, 427
    c-Jun, 429–431
    dedifferentiation, 426–428
    demyelinating neuropathy, 431
    myelin sheaths, 428, 429
    PNS, 425, 426
    repair cells, 431, 432
    repair programs, 427, 428
Sciatic functional index (SFI), 439
SF-decorin (SFD) scaffolds
    abdominal wall musculofascial, 342
    adjacent native tissue, 342
    amino acid sequence, 326
    biomaterials, 325
    collagen fibrillogenesis, 326
    fibrillar structure, 342
    gross evaluation, 328
    histological analysis, 335, 336, 338–341
    implanted materials, 343
    in vivo repair, 325
    mechanical properties, implants, 333–335
    mechanical testing, 328, 329
    MyoD1, 344
    omentum adhesions, 342
    PADM, 344
    PP mesh, 343
    preparation of, 326
    remodeling, 343
    results, 329, 330
    scanning electron microscope, 326, 327
    statistical analysis, 329

    surgical handling, 343
    vascularization, 329, 343, 344
    ventral hernia repair
        gross analysis, 330–333
        in vivo, 327, 328
        PADM, 327
        PP mesh, 327
Silenced genes, 452
Silicone treatment, 23
Skeletal muscle progenitor cells (SMPC), 374
Skin aging and radiation damage, 66
Skin allograft tissue banks, 130
Skin appendages, 117
Skin barrier, 158
Skin-based regenerative medicine, 131
Skin basement membrane, 94
Skin epithelial stem cell maintenance and/or
          determination, Wnt signaling, 153
Skin regeneration, 91, 157–162
Skin repair, 157–162
Skin substitutes, 137–143
Skin tissue engineering, severe burns
    acellular dermal matrix, 124
    AlloDerm®, 126
    Biobrane®, 124
    Dermagraft®, 124
    IntegraTM, 123
    MatriDerm®, 123
    Suprathel®, 125
Slow cycling basal keratinocyte stem cells, 93
Smad-mediated BMP action, 184, 185
Small-molecule therapies, 32, 33
Soft tissue repair, 45
Soft tissue replacement, 86
Soft-tissue filling, 71, 83
Sox9, 264
Split-thickness skin grafts (STSG), 141
Stem and stromal cells, adipose tissues, 66
Stem cells (SCs), *see* Mesenchymal stem cells (MSCs)
Stem cells and burn
    allogeneic tissue, 109
    autogenic tissue, 109
    biological mediators, 109
    clinical complications, 109
    co-morbidities, 109
    epidermal keratinocytes, 109
    induced pluripotent stem cells, 110
    pharmacological compounds, 109
    quality of life, 109
    synthetic substitutes, 109
    tumor formation, 109
Stem cell therapies, 11, 32, 35, 255
Stem/stromal cells and bioactive matrix, 63
Stem/stromal cells in ATC, 57
Stromal-derived factor-1 (SDF-1), 187
Stromal vascular fraction (SVF), 72, 255, 417
Suction-assisted liposuction (SAL), 40
Superficial digital flexor tendon (SDFT), 402

Index 471

Suprathel®, 140
Synthetic and biosynthetic materials, 139–141
Systemic circulation (SCs), 317

T
Targeted cell-based therapies, 46
Tendon-derived stem cells (TDSCs), 359, 374, 378
Tendon regeneration
    anatomy, 355, 356, 370, 371
    biomechanical evaluation, 357
    biophysical stimulation, 363
    blood supply, 357
    composition, 356
    culturing, 371
    differentiated cells, 371
    drug repositioning, 363
    growth factors, 360, 361
    horizons, 379
    repair and regeneration, 358, 371
    stem cells, 379
        ADSC, 373, 374
        allogeneic and autologous, 377
        BMSC, 373
        bone marrow extract/aspiration, 374
        characteristics, 372
        co-cultures, 377
        delivery methods, 378
        ESC, 372
        hiPSC, 372
        HPMSC, 374
        LSC, 374
        media, 379
        MSC, 373
        optimum number, 378
        oral stem cells, 374
        PBMSCs, 374
        PDSC, 374
        PRPs, 378
        scaffolds, 375, 376
        secretions, 375
        secretomes, 377, 378
        SMPCs, 374
        TDSCs, 374, 378
        tenocytes, 378
        trans-osseous drilling, 374, 375
        UC-MSC, 374
    tendon healing, 357, 358
    tenocytes, 371
    tissue engineering
        architectures and mechanical properties, 360
        biologic scaffolds, 360
        cell therapy, 362
        cells types, 358, 359
        electrospinning, 362
        growth factors, 362, 363
        matrix and cell production, 360
        synthetic scaffolds, 360, 361

Tenoblast, 370
Tenocytes, 370
3D organotypic models, 103
Tissue culture/expansion, 68
Tissue-engineered bone substitutes, 190–192
Tissue-engineered skin substitutes, 118–120
Tissue engineering and regenerative medicine, 81
    connexin 43, 12
    cytokines and growth factors, 11
    epidermal stem cells or mesenchymal stromal
        cells, 11
    growth factors and cytokines, 11
    mannose-6-phosphate, 11
    molecular antagonists, 11
    scaffolds, 12
    scarless wound healing, 11
    selective estrogen receptor modulator (SERM)
        tamoxifen, 11
    skin change with aging, 11
    of skin epidermis, 118, 121
    tension reduction, 12
Tissue-engineering applications, hDAM, 72
Tissue-engineering approaches, 83
Tissue homogenizer for mechanical disruption, 73
Tissue microenvironment, 171
Tissue regeneration and repair, Wnt signaling, 149, 150
Tissue regeneration applications, 5
Tissue regeneration in fetal epidermis, 4
Tissue stromal vascular fraction (tSVF), 55
Trabecular bone formation, 217
Transcutaneous electrical stimulation (TENS), 418
TransCyte, 140
Transepithelial water loss (TEWL), 23
Transfersomes, 161
Transforming growth factor-beta (TGF-β), 252
Transforming growth factor beta-1 (TGF-β1), 293
Transgene, 194
Translational medicine, 29
Transplantable graft tissue quality, 40
Tripalmitin, 216
TRIZOL reagent, 265
Tryptase, 7
Tumor necrosis factor (TNF)-α, 240
Tumorigenesis, 84

U
Ultrasound assisted liposuction (UAL), 40
Umbilical cord blood-MSC (UC-MSC), 374

V
Vascular anastomosis failure, 197
Vascular endothelial growth factor (VEGF), 7, 184–186,
    350, 358
Vascular network, 151
Vector control, 264
Viable allograft, 138

Visceral hDAM, 85
Vitamin C deficiency and scurvy, 27
Vitamin E, 22
Volumetric muscle loss (VML)
    ECM
        acellular biologic surgical meshes, 349, 350
        clinical outcome, 351, 352
        musculotendinous injuries, 349
        myogenesis, 350, 351
        technical challenges, 349
    stem cell-based approaches, 348
    therapeutic strategies, 347, 348
Von Kossa stain procedure, 170
von Willebrand factor, 336, 338, 340

W
Water jet-assisted liposuction (WAL), 40
Western blot analysis, 77
Wnt proteins, 33
Wnt/Ca^{2+} pathway, 148
Wnt-responsive interfollicular stem cells, 148
Wnt signaling pathway, 147–154
    aspects, 235
    β-catenin pathway, 235
    bone repair and regeneration, 232, 235
    Ca^{2+} pathway, 235
    canonical signaling, 236, 237
    downstream components, 235
    dysregulation, 235
    embryological development, 232–235
    MSCs (*see* Mesenchymal stem/stromal cells (MSCs))
    non-canonical pathway, 237
    N-terminal signal peptide, 235
    PCP pathway, 235
    pre-clinical models, 232
    skeletal physiology and regeneration, 242
    skeletal tissue regeneration and repair, 236
    therapeutic intervention, 241, 242
    xenopus ventralization, 236
Wound closure market, 20
Wound healing, 3, 6, 20, 27, 84, 85
    and epidermal regeneration, 91
    and scarring issues, 53
Wounds in theoral mucosal heal, 4

X
Xenograft, 139
Xenografted human bone marrow-derived mesenchymal stromal cells (BM-MSCs), 11

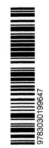